SOUTH WESTERN
Algebra 2
AN INTEGRATED APPROACH

SOLUTIONS MANUAL

GERVER, SGROI, CARTER, HANSEN
MOLINA & WESTEGAARD

JOIN US ON THE INTERNET
WWW: http://www.thomson.com
EMAIL: findit@kiosk.thomson.com

A service of I(T)P®

SOUTH
WESTERN
EDUCATIONAL
PUBLISHING

I(T)P® An International Thomson Publishing Company

Cincinnati • Albany • Bonn • Boston • Detroit • London • Madrid • Melbourne • Mexico City • New York
Philadelphia • Pacific Grove • Paris • San Francisco • Singapore • Tokyo • Toronto • Washington

Copyright © 1997
by South-Western Educational Publishing
Cincinnati, OH

The text of this publication, or any part thereof, may be reproduced for use in classes for which
South-Western Algebra 2: An Integrated Approach is the adopted text. It may not be reproduced in any
manner whatsoever for any other purpose without permission in writing from the publisher.

ISBN: 0-538-66520-3
1 2 3 4 5 6 7 8 MZ 03 02 01 00 99 98 97 96
Printed in the United States of America

I(T)P®
International Thomson Publishing

South-Western Educational Publishing is an ITP Company. The ITP logo is a registered trademark
used herein under license by South-Western Educational Publishing.

TABLE OF CONTENTS

JOIN US ON THE INTERNET

WWW: http://www.thomson.com

E-MAIL: findit@kiosk.thomson.com

Access South-Western Educational Publishing's complete catalog online through thomson.com. Internet users can search catalogs, examine subject-specific resource centers, and subscribe to electronic discussions lists. In addition, you'll find new product information and information on upcoming events in Mathematics.

For information on our products and services, point your web browser to:
http://www.swpco.com/swpco.html

For technical support, you may e-mail: hotline_education@kdc.com

South-Western Algebra 2: An Integrated Approach Internet Connection is a web site which accompanies the chapter projects. You can access the Internet Connection at:
www.swpco.com/swpco/algebra2.html

A service of I(T)P ®

Chapter 1 Modeling and Predicting

Data Activity, pages 2–3

1. $(6.70 + 2.20 + 0.90)(10,000) = \$980$

2. $2.40 - 1.60 = 0.80$ is the amount of increase in
cents/mile $\frac{0.80}{1.60}(100) = 50\%$ increase in maintenance
cost

3. $\frac{565 + 626 + 680 + 779 + 832 + 696}{6} \approx \$696$

4. 1993 total fixed cost
$= 724 + 179 + 2883 + 696 = \$4482$
$\frac{2883}{4482}(100) \approx 64.3\%$

5. Equations should be of the form $V = P - (0.15P)t$
where P is the original price and t is the number of
years. Graphs should be lines because the functions are
linear functions.

Lesson 1.1, pages 5–11

EXPLORE

1. t is the cost of 1 tire. So, $4t$ is the cost of 4 tires. The
equation models cost of tires + sales tax = total price

2. Subtract 14.94 from both sides of the equation
$4t + 14.94 = 274.74$ and obtain
$4t = 274.74 - 14.94 = \$259.80$

3. Divide both sides of the equation $4t = 259.80$ by 4
and obtain $t = \frac{259.80}{4} = \$64.95$

4. Answers will vary. One possible answer is the
following. Yes, first divide both sides by 4 and get
$t + 3.735 = 68.685$ and then subtract 3.735 from each
side and get $t = 64.95$.

TRY THESE

1. division property
2. addition property

3. subtraction property, multiplication property

4. addition property, division property

5. $m + 19 = 30$
$m = 11$ subtraction property
Check: $11 + 19 = 30$ ✓

6. $39 = 3p$
$13 = p$ division property
Check: $39 = 3(13)$ ✓

7. $-7 = -3 + y$
$-4 = y$ addition property
Check: $-7 = -3 + (-4)$ ✓

8. $15 = \frac{t}{3}$
$45 = t$ multiplication property

9. $2x - 7 = 11$
$2x = 18$ addition property
$x = 9$ division property
Check: $2(9) - 7 = 18 - 7 = 11$ ✓

10. $12 = 6 + 3h$
$6 = 3h$ subtraction property
$2 = h$ division property
Check: $6 + 3(2) = 6 + 6 = 12$ ✓

11. $5c = 18 + 8 - 6$
$5c = 20$ Simplify
$c = 4$ division property
Check: $5(4) = 20 = 18 + 8 - 6$ ✓

12. $\frac{m}{4} + 7 = 26$

$\frac{m}{4} = 19$ subtraction property
$m = 76$ multiplication property
Check: $\frac{76}{4} + 7 = 19 + 7 = 26$ ✓

13. $3x - 4 = 5$. Add 4 unit blocks to each side:
$3x - 4 + 4 = 5 + 4$. Remove zero pairs: $3x = 9$.
Group x-blocks with unit blocks: $x = 3$.

14. Answers will vary. The two sides of an equation are
equal. Two equations are equivalent if they have the
same solution.

15. $890 - 712 = 178$
markup is $178

16a. Use the Amount Formula, $P = \dfrac{M[(1 + r)^n - 1]}{r(1 + r)^n}$

$P = \dfrac{346.09[(1 + 0.00833)^{74} - 1]}{0.00833(1 + 0.00833)^{24}} \approx \$7500$

16b. $346.09(24) = 8306.16$
deferred payment price is $8306.16

16c. $8306.16 - 7500 = 806.16$
finance charge is $806.16

For Exercise 17. answers will vary due to method of
rounding.

17a. Use the Monthly Payment Formula,
$M = \dfrac{Pr(1 + r)^n}{(1 + r)^n - 1}$

$M = \dfrac{30000(0.00708)(1 + 0.00708)^{120}}{(1 + 0.00708)^{120} - 1} \approx \$371.89$

17b. $371.89(120) = 44,626.80$
deferred payment price $44,626.80

17c. $44,626.80 - 30,000 = 14,626.80$
finance charge $14,626.80

PRACTICE

1. $n - 8 = 15$
$n = 23$ addition property
Check: $23 - 8 = 15$ ✓

2. $-3 = 5 + d$
$-8 = d$ subtraction property
Check: $-3 = 5 + (-8)$ ✓

3. $8x = -20$
$x = -2.5$ division property
Check: $8(-2.5) = -20$ ✓

4. $4x + 2 = 26$
$4x = 24$ subtraction property
$x = 6$ division property
Check: $4(6) + 2 = 24 + 2 = 26$ ✓

5. $\frac{y}{5} = 8.4$
$y = 42$ multiplication property
Check: $\frac{42}{5} = 8.4$ ✓

6. $5 = 9 - 2e$
$-4 = -2e$ subtraction property
$2 = e$ division property
Check: $9 - 2(2) = 9 - 4 = 5$ ✓

7. $20k = 45 + 15$
$20k = 60$ simplify
$k = 3$ division property
Check: $20(3) = 60 = 45 + 15$ ✓

8. $\frac{y}{10} - 13 = 3$
$\frac{y}{10} = 16$ addition property
$y = 160$ multiplication property
Check: $\frac{160}{10} - 13 = 16 - 13 - 3$ ✓

9. $4(2 + x) = 28$
$2 + x = 7$ division property
$x = 5$ subtraction property
Check: $4(2 + 5) = 4(7) = 28$ ✓

10. $-3(9 - n) = -63$
$9 - n = 21$ division property
$-n = 12$ subtraction property
$n = -12$ multiplication property
Check: $-3[9 - (-12)] = -3(21) = -63$ ✓

11. $15 + k = 3(14 - 2)$
$15 + k = 36$ simplify
$k = 21$ subtraction property
Check: $15 + 21 = 36 = 3(14 - 2)$ ✓

12. $2a + 9a + 13 = 35$
$11a + 13 = 35$ simplify
$11a = 22$ subtraction property
$a = 2$ division property
Check: $2(2) + 9(2) + 13 = 4 + 18 + 13 = 35$ ✓

13. $P = R - C; R - C = P; R = P + C$

14. $F = ma; ma = F; a = \frac{F}{m}$

15. $A = \frac{1}{2}bh; \frac{1}{2}bh = A; bh = 2a; b = \frac{2A}{h}$

16. $F = \frac{9}{5}C + 32; \frac{9}{5}C + 32 = F; \frac{9}{5}C = F - 32;$
$C = \frac{5}{9}(F - 32)$

17. $P = 2l + 2w; 2l + 2w = P; 2w = P - 2l;$
$w = \frac{P - 2l}{2}$

18. $T = g - (w + p); g - (w + p) = T;$
$g - w - p = T; -w - p = T - g;$
$-w = T - g + p; w = -T + g - p$

19a. sticker price = dealer cost + markup $408 = 342 + m$

19b. $66 = m$; markup is $66

Answers to Exercises 20 and 21 will vary due to method of rounding.

20a. $P = \frac{M[(1 + r)^n - 1]}{r(1 + r)^n} =$
$\frac{97.35[(1 + 0.005)^{36} - 1]}{(0.005)(1 + 0.005)^{36}} \approx \$3200$

20b. deferred payment = $97.35(36) = \$3504.60$

20c. finance charge = $3504.60 - 3200 = \$304.60$

21a. $m = \frac{Pr(1 + r)^n}{(1 + r)^n - 1} =$
$\frac{8000(0.0065)(1 + 0.0065)^{60}}{(1 + 0.0065)^{60} - 1} \approx \$161.45$

21b. deferred payment price = $161.45(60) = \$9687.00$

21c. finance charge = $9687.00 - 8000 = \$1687$

EXTEND

22. Answers will vary. A possible equation is $\frac{1}{2}x - 5 = 1$.
An equation can be found by using the properties of equality and working backward.

23. a, c, d; They all have the same solution.
$5z - 3 = 17; 5z = 20; z = 4$
$5z - 5 = 15; 5z = 20; z = 4$
$5z + 7 = 27; 5z = 20; z = 4$
$z = 4$

24. $8x + 5 = 6x + 11; 2x + 5 = 11; 2x = 6; x = 3$
Check: $8(3) + 5 \overset{?}{=} 6(3) + 11$
$24 + 5 \overset{?}{=} 18 + 11$
$29 = 29$ ✓

25. $-5 + 13h = 12h - 15; 13h = 12h - 10; h = -10$
Check: $-5 + 13(-10) \overset{?}{=} 12(-10) - 15$
$-5 - 130 \overset{?}{=} -120 - 15$
$-135 = -135$ ✓

26. $4m + 9 = 7m + 24; 4m = 7m + 15; -3m = 15;$
$m = -5$
Check: $4(-5) + 9 \overset{?}{=} 7(-5) + 24$
$-20 + 9 \overset{?}{=} -35 + 24$
$-11 = -11$ ✓

27. $6(x - 4) = 3(x + 2); 6x - 24 = 3x + 6;$
$3x - 24 = 6; 3x = 30; x = 10$
Check: $6(10 - 4) \stackrel{?}{=} 3(10 + 2)$
$\qquad 6(6) \stackrel{?}{=} 3(12)$
$\qquad 36 = 36 \checkmark$

28. $8(5y + 3) = -7(3y + 14); 40y + 24 = -21y - 98;$
$61y + 24 = -98; 61y = -122; y = -2$
Check: $8[5(-2) + 3 \stackrel{?}{=} -7[3(-2) + 14]$
$\qquad 8(-10 + 3) \stackrel{?}{=} -7(-6 + 14)$
$\qquad -56 = -56 \checkmark$

29. $2c - (3 - c) = 5c - 4; 2c - 3 + c = 5c - 4;$
$3c - 3 = 5c - 4; -2c - 3 = -4; -2c = -1;$
$c = \dfrac{1}{2}$

Check: $2\left(\dfrac{1}{2}\right) - \left(3 - \dfrac{1}{2}\right) \stackrel{?}{=} 5\left(\dfrac{1}{2}\right) - 4$
$\qquad 1 - 2\dfrac{1}{2} \stackrel{?}{=} 2\dfrac{1}{2} - 4$
$\qquad -1\dfrac{1}{2} = -1\dfrac{1}{2} \checkmark$

30. $\dfrac{x}{2} + 7 = \dfrac{x}{3} - 2; 3x + 42 = 2x - 12;$
$x + 42 = -12; x - -54$
Check: $-\dfrac{54}{2} + 7 \stackrel{?}{=} \dfrac{-54}{3} - 2$
$\qquad -20 = -20 \checkmark$

31. $\dfrac{x + 5}{6} = \dfrac{x - 4}{3}; x + 5 = 2x - 8; -x + 5 = -8;$
$-x = -13; x = 13$
Check: $\dfrac{13 + 5}{6} \stackrel{?}{=} \dfrac{13 - 4}{3}$
$\qquad 3 - 3 \checkmark$

32. $8 + \dfrac{2x}{3} = \dfrac{3x}{4} + 7; 96 + 8x = 9x + 84;$
$96 = x + 84; 12 = x$
Check: $8 + \dfrac{2(12)}{3} \stackrel{?}{=} \dfrac{3(12)}{4} + 7$
$\qquad 8 + 8 \stackrel{?}{=} 9 + 7$
$\qquad 16 = 16 \checkmark$

33. markup $= 177 - 150 = 27$
percent markup $= \dfrac{27}{150}(100) = \dfrac{9}{50}(100) = 18\%$

34. monthly payment,
$M = \dfrac{8000(0.0067)(1 + 0.0067)^{36}}{(1 + 0.0067)^{36} - 1} \approx \$250.84$
deferred payment price $= 250.84(36) = \$9030.19$
finance charge $= 9030.19 - 8000 = \$1030.19$
% of loan $= \dfrac{1030.19}{8000}(100) \approx 12.8\%$

35. monthly payment $= \dfrac{4959.36}{4 \times 12} = 103.32$

$P = \dfrac{103.32[(1 + 0.007)^{48} - 1]}{0.007(1 + 0.007)^{48}} \approx \$4200$

36. For 4 years:
$M = \dfrac{12000(0.01042)(1 + 0.01042)^{48}}{(1 + 0.01042)^{48} - 1} \approx 318.98.$
deferred payment price $= 318.98(48) = 15,311.04$
For 5 years:
$M = \dfrac{12000(0.01042)(1 + 0.01042)^{60}}{(1 + 0.01042)^{60} - 1}$
$\approx 270(60) = 16,199.98$
saving $= 16,199.98 - 15,311.04 = \$888.44$

THINK CRITICALLY

37. The division property of equality excludes 0 as a divisor. In this case x can have any value, including 0. Thus a false or nonequivalent equation results.

38. $ax + b = 0; ax = -b; x = -\dfrac{b}{a}$
Answers will vary. If a one-variable equation can be transformed into an equivalent equation of the form $ax + b = 0$, then the equation has precisely one solution $x = -\dfrac{b}{a}$.

MIXED REVIEW

39. $2(2) - (-3) = 4 + 3 = 7$

40. $\dfrac{3}{4}(2) + (-3) - 5.5 = 1.5 - 3 - 5.5 = -7$

41. $5(2) + 3(-3) = 10 - 9 = 1$

42. $2^2 + [2 + (-3)]^3 = 4 + (-1)^3 = 4 - 1 = 3$

43. $150(0.27) = 40.5$ **44.** $\dfrac{60}{25}(100) = 240\%$

45. $0.75x = 87; x = \dfrac{87}{0.75} = 116$

46. $500(0.002) - 1$

47. $3n = 21; n = -7$
Check: $3(-7) = -21 \checkmark$

48. $5e + 2 = 27; 5e = 25; e = 5$
Check: $5(5) + 2 = 25 + 2 = 27$

49. $2(m - 4) - 20; m - 4 = 10; m = 14$
Check: $2(14 - 4) = 2(10) = 20 \checkmark$

50. $6y - 5 = 9y + 13; -3y - 5 = 13; -3y = -18$
$y = -6$
Check: $6(-6) - 5 \stackrel{?}{=} 9(-6) + 13$
$\qquad -36 - 5 \stackrel{?}{=} -54 + 13$
$\qquad -41 = -41 \checkmark$

ALGEBRAWORKS

1. $0.06x = 180; x = \dfrac{180}{0.06} = 3000$
monthly income about \$3000
$3000(12) = 36000$
yearly income about \$36,000

2. $\dfrac{250{,}000{,}000{,}000}{250{,}000{,}000} = 1000$

Americans owe per capita on cars about $1000.

3. Rico Martelli

$M = \dfrac{17800(0.0074)(1 + 0.0074)^{60}}{(1 + 0.0074)^{60} - 1} \approx \$368.42$

Chang Liu

$M = \dfrac{22{,}100(0.0074)(1 + 0.0074)^{60}}{(1 + 0.0074)^{60} - 1} \approx \$457.42$

Katrina Langsley

$M = \dfrac{15{,}200(0.0074)(1 + 0.0074)^{48}}{(1 + 0.0074)^{48} - 1} \approx \$377.35$

Ahmad Baghaii

$M = \dfrac{12{,}500(0.0074)(1 + 0.0074)^{48}}{(1 + 0.0074)^{48} - 1} \approx \$310.32$

Thomas Jones

$M = \dfrac{14{,}395(0.0074)(1 + 0.0074)^{60}}{(1 + 0.0074)^{60} - 1} \approx \$297.94$

4. Rico: $368.42(60) = \$22{,}105.20$
Chang: $457.42(60) = \$27{,}445.20$
Katrina: $377.35(48) = \$18{,}112.80$
Ahmad: $310.32(48) = \$14{,}895.36$
Thomas: $297.94(60) = \$17{,}876.40$

5. $22{,}105.20 + 27{,}445.20 + 18{,}112.80 +$
$14{,}895.36 + 17{,}876.40 = \$100{,}434.96$

6. $100{,}434.96 - 17{,}800 - 22{,}100 - 15{,}200 -$
$12{,}500 - 14{,}395 = \$18{,}439.96$

Lesson 1.2, pages 12–18

EXPLORE

1. $12{,}450(0.1) = \$1245$
$12{,}450(0.2) = \$2490$
$12{,}450(0.3) = \$3735$

2a. $12{,}450 - 1245 = \$11{,}205$
$12{,}450 - 2490 = \$9960$
$12{,}450 - 3735 = \$8715$

2b. $\dfrac{11{,}205(0.009)(1 + 0.009)^{48}}{(1 + 0.009)^{48} - 1} \approx \$288.51$

$\dfrac{9960(0.009)(1 + 0.009)^{48}}{(1 + 0.009)^{48} - 1} \approx \$256.46$

$\dfrac{8715(0.009)(1 + 0.009)^{48}}{(1 + 0.009)^{48} - 1} \approx \$224.40$

2c. $288.51(48) = \$13{,}848.48$
$256.46(48) = \$12{,}310.08$
$224.40(48) = \$10{,}771.20$

2d. $13{,}848.48 + 1245 = \$15{,}093.48$
$12{,}310.08 + 2490 = \$14{,}800.08$
$10{,}771.20 + 3735 = \$14{,}506.20$

3. $15{,}093.48 - 14{,}800.08 = \$293.40$
$15{,}093.48 - 14{,}506.20 = \$587.28$

TRY THESE

1. $3n - 5 = 3 * n - 5$

2. $k^4 = k \wedge 4$

3. $\dfrac{(a + h)}{2} = (a + h)/2$

4. $m - b^2 = m - b \wedge 2$

5. $2(8 - 3) = 2(5) = 10$
$2(8) - 3 - 16 - 3 - 13$

6. $0.6(15) = 9$
$9^2 = 81$

7. $\dfrac{5.1}{6.8} = \dfrac{51}{68} = \dfrac{3}{4}$
$24 - 3.1(5.1) = 8.19$

8. $\dfrac{12 - 2}{2} = 5$
$12 + 3(2)^3 = 12 + 3(8) = 36$

9. $= A2 * B2; \; 9(6) = 54$
$= 2 * A2 + 2 * B2; \; 2(9) + 2(6) = 30$

10. $= A3 * B; \; 12(5) = 60$
$= 2 * A3 + 2 * B3; \; 2(12) + 2(5) = 24 + 10 = 34$

11. $= A4 * B4; \; 7.2(3.9) = 28.08$
$= 2 * A4 + 2 * B4;$
$2(7.2) + 2(3.9) = 14.4 + 7.8 = 22.2$

12. $A3 + A3 * (B3 / 100); \; 3.20 + 3.20\left(\dfrac{4}{100}\right) = 3.328$

13. $A4 + A4 * (B4 / 100); \; 5.75 + 5.75\left(\dfrac{6}{100}\right) = 6.095$

14. $A5 + A5 * (B5 / 100); \; 8 + 8\left(\dfrac{5.25}{100}\right) = 8.42$

15.

	A	B	C
1	Sticker	Dealer's	Markup
2	Price	Cost	
3	276	240	= B3 − A3; 36
4	276	225	= B4 − A4; 51
5	276	208	= B5 − A5; 68

16. To calculate the payments on a single specific loan there is no advantage. The advantage arises when a spreadsheet is used to show the payments on many similar loans. This allows a potential borrower to compare the costs of the loans in order to determine which one is best for his or her purposes.

PRACTICE

1. $\dfrac{P}{x - y} = P/(x - y)$

2. $d^3 + v^2 = d \wedge 3 + v \wedge 2$

3. $4a - \dfrac{3}{c} = 4 * a - 3/c$

4. $\dfrac{2hk^{5p}}{3(2 - b)^{5p}} = 2 * h * k \wedge (5 * p) / (3 * (2 - b) \wedge (5 * p))$

5. C1: $4 + 2(9) = 22$ **6.** C2: $20^{\frac{6}{3}} = 20^2 = 400$

D1: $\dfrac{4(9)}{2}(4) = 72$ D2: $\dfrac{6}{\frac{20}{20}} = \dfrac{6}{1} = 6$

7. C3: $2(12.5) - 5(2.4) = 25 - 12 = 13$
D3: $(12.5 - 1.7)(2.4 + 3) = 58.32$

8. C4: $256(0.75)^2 = 144$
D4: $[(256)(0.75)]^2 = 36{,}864$

9. $= 12 * A3 * B3; 12(2)(189.64) = 4551.36$

10. $= 12 * A4 * B4; 12(3)(152.40) = 5486.40$

11. $= 12 * A5 * B5; 12(4)(115.16) = 5527.68$

12. $C3: = A3 * \left(\dfrac{B3}{100}\right); 12{,}340(0.1) = 1234$

$D3: = A3 - C3; 12{,}340 - 1234 = 11{,}106$

13. $C4: = A4 * \left(\dfrac{B4}{100}\right); 12{,}340(0.15) = 1851$

$D4: = A4 - C4; 12{,}340 - 1851 = 10{,}489$

14. $C5: = A5 * \left(\dfrac{B5}{100}\right); 12{,}340(0.2) = 2468$

$D5: = A5 - C; 12{,}340 - 2468 = 9872$

15. $C3: = 0.01 * B3 / 12; \dfrac{(0.01)(6)}{12} = 0.005$

$D3: = A3 * C3 * (1 + C3) \wedge 60 / ((1 + C3) \wedge 60 - 1);$

$\dfrac{6825(0.005)(1 + 0.005)^{60}}{(1 + 0.005)^{60} - 1} = 131.95$

16. $C4: = 0.01 * B4/12; \dfrac{0.01(9)}{12} = 0.0075$

$D4: = A4 * C4 * (1 + C4) \wedge 60 / ((1 + C4) \wedge 60 - 1);$

$\dfrac{6825(0.0075)(1 + 0.0075)^{60}}{(1 + 0.0075)^{60} - 1} = 141.68$

17. $C5: = 0.01 * B5 / 12; \dfrac{0.01(12)}{12} = 0.01$

$D5: = A5 * C5 * (1 + C5) \wedge 60 / ((1 + C5) \wedge 60 - 1);$

$\dfrac{6825(0.01)(1 + 0.01)^{60}}{(1 + 0.01)^{60} - 1} = 151.82$

18.

	A	B	C
1	Retail	Down	Amount to be
2	Price	Payment	Financed
3	27500	2500	$= A3 - B3; 25000$
4	27500	5000	$= A4 - B4; 22500$
5	27500	7500	$= A5 - B6; 20000$

19.

	A	B	C	D
1	Base 1	Base 2	Height	Area
2	4	6	10	$= 0.5 * C2 * (A2 + B2); 50$
3	4	6	15	$= 0.5 * C3 * (A3 + B3); 75$
4	4	6	20	$= 0.5 * C4 * (A4 + B4); 100$

EXTEND

20. Answers will vary.

21. The greater down payment and the shorter repayment period; 30% down, 3 years to pay

22.

	A	B	C	D	E	F	G
1	Retail	Down	Down	Amount	Repayment	Monthly	Total
2	Price	Payment, %	Payment	of Loan	Period, yrs	Payments	Financed
3							Price
4	19600	15	2940	16660	3	525.14	21845.17
5	19600	15	2940	16660	4	409.85	22613.02
6	19600	30	5880	13720	3	432.47	21448.96
7	19600	30	5880	13720	4	337.53	22081.31

23.

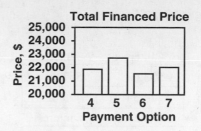

Total Financed Price

THINK CRITICALLY

24.

	A	B	C	D	E	F	G	H
1	Retail	Down	Down	Amount	Repayment	Annual	Monthly	Total
2	Price	Payment, %	Payment	of Loan	Period, years	Loan Rate	Payments	Financed
3								Price
4	39760	20	7952	31808	8	9.6	475.95	53643.20
5	39760	20	7952	31808	8	10.8	496.23	55590.08
6	39760	20	7952	31808	10	9.6	413.33	57551.60
7	39760	20	7952	31808	10	10.8	434.56	60099.20
8	39760	30	11928	27832	8	9.6	416.46	51908.16
9	39760	30	11928	27832	8	10.8	434.20	53611.20
10	39760	30	11928	27832	10	9.6	361.66	55327.20
11	39760	30	11928	27832	10	10.8	380.24	57556.80

MIXED REVIEW

25. $3(x - 5y) = 3x - 15y$

26. $4a(6a) = 24a^2$

27. $-2n(e - 3n) = -2ne + 6n^2$

28. $(a + b)(a + b) = a^2 + 2ab + b^2$

29. $C21: 2; D21: 2 - (2)(-3) = 2 + 6 = 8$

30. $C22: (12 - 10)^3 = 2^3 = 8; D22: \dfrac{12}{3 + 9} = 1$

31. B

ALGEBRAWORKS

1a. $\dfrac{ab}{a + b} = \dfrac{240(300)}{240 + 300} = 133\dfrac{1}{3}$ ohms.

1b. Yes. $a*b$ is calculated independent of the rest of the formula in either case.

2a. $\dfrac{5F}{4d^2} - 5 * F / (4 * d \wedge 2)$

2b. $\dfrac{RL}{d} - R = R * L / d - R$

2c. $2C + 1.57(D + d) + \dfrac{(D + d)}{4c} =$
$2 * C + 1.57 * (D + d) + (D + d) / (4 * c)$

3.

	A	B	C	D	E	F	G
1	Master	Earnings	Earnings	Earnings	Earnings	Earnings	Total
2	Wages, $	First 1000	Second 1000	Third 1000	Fourth 1000	Fifth 1000	Earnings,
3		50%	60%	70%	80%	90%	5000 hours
4	13.00	6500	7800	9100	10400	11700	45500
5	14.00	7000	8400	9800	11200	12600	49000
6	15.00	7500	9000	10500	12000	13500	52500
7	16.00	8000	9600	11200	12800	14400	56000

Lesson 1.3, pages 19–24

EXPLORE

1.

	6% interest	9% interest
10% down	533.63	557.80
20% down	474.34	495.82

2.

	6% interest	9% interest
10% down	19,210.68	20,080.80
20% down	17,076.24	17,849.52

Grid entries were obtained by multiplying each entry in the first grid by 36.

3.

	6% interest	9% interest
10% down	21,159.68	22,029.80
20% down	20,974.24	21.747.52

Grid entries were obtained by adding the down payment to each entry in the second grid.

TRY THESE

1. dimensions of B: 3×2

2. elements of A: 5, −5, 4, 7, 11, 0

3. $A + B = \begin{bmatrix} 5 + (-6) & -5 + (-10) \\ 4 + 4 & 7 + 8.5 \\ 11 + 20 & 0 + (-3) \end{bmatrix} = \begin{bmatrix} -1 & -15 \\ 8 & 15.5 \\ 31 & -3 \end{bmatrix}$

4. $C - B = \begin{bmatrix} 12.2 - (-6) & 15 - (-10) \\ 0 - 4 & 2 - 8.5 \\ 1 - 20 & -15.6 - (-3) \end{bmatrix} = \begin{bmatrix} 18.8 & 25 \\ -4 & -6.5 \\ -19 & -12.6 \end{bmatrix}$

5. $6C = \begin{bmatrix} 6(12.2) & 6(15) \\ 6(0) & 6(2) \\ 6(1) & 6(-15.6) \end{bmatrix} = \begin{bmatrix} 73.2 & 90 \\ 0 & 12 \\ 6 & -93.6 \end{bmatrix}$

6. $2(A - c) = 2\begin{bmatrix} 5 - 12.2 & -5 - 15 \\ 4 - 0 & 7 - 2 \\ 11 - 1 & 0 - (-15.6) \end{bmatrix} =$

$\begin{bmatrix} -14.4 & -40 \\ 8 & 10 \\ 20 & 31.2 \end{bmatrix}$

7. $M + N = \begin{bmatrix} 3 + 2 & 6 + 9 \\ 1 + 4 & 8 + 7 \end{bmatrix} = \begin{bmatrix} 5 & 15 \\ 5 & 15 \end{bmatrix} =$

$\begin{bmatrix} 2 + 3 & 9 + 6 \\ 4 + 1 & 7 + 8 \end{bmatrix} = N + M$

8. $A + C - B = \begin{bmatrix} 210 + 36 - 134 & 75 + 50 - 71 \\ 198 + 24 - 212 & 103 + 100 - 186 \\ 61 + 25 - 67 & 155 + 75 - 191 \end{bmatrix} =$

$\begin{bmatrix} 112 & 54 \\ 10 & 17 \\ 19 & 39 \end{bmatrix}$

9. Answers will vary.

PRACTICE

1. elements of C: 0, −5, −6, 20, 5, 2

2. dimensions of D: 1×5

3. $A + C = \begin{bmatrix} 7 + 0 & 7 + (-5) & 4 + (-6) \\ 13 + 20 & -2 + 5 & -8 + 2 \end{bmatrix} =$

$\begin{bmatrix} 7 & 2 & -2 \\ 33 & 3 & -6 \end{bmatrix}$

4. $8B = \begin{bmatrix} 8(15) & 8(0) & 8(-14) \\ 8(7) & 8(12) & 8(-3) \end{bmatrix} = \begin{bmatrix} 120 & 0 & -122 \\ 56 & 96 & -24 \end{bmatrix}$

5. $B - A = \begin{bmatrix} 15 - 7 & 0 - 7 & -14 - 4 \\ 7 - 13 & 12 - (-2) & -3 - (-8) \end{bmatrix} =$

$\begin{bmatrix} 8 & -7 & -18 \\ -6 & 14 & 5 \end{bmatrix}$

6. $0.75D = \begin{bmatrix} 0.75(28) & 0.75(42) & 0.75\left(\frac{4}{5}\right) & 0.75(-16) \end{bmatrix}$

$0.75(-8) \end{bmatrix} = \begin{bmatrix} 21 & 31.5 & \frac{3}{5} & -12 & -6 \end{bmatrix}$

7. $-C + B = \begin{bmatrix} 0 & 5 & 6 \\ -20 & -5 & -2 \end{bmatrix} + \begin{bmatrix} 15 & 0 & -14 \\ 7 & 12 & -3 \end{bmatrix} =$

$\begin{bmatrix} 15 & 5 & -8 \\ -13 & 7 & -5 \end{bmatrix}$

8. $2B + 3A = \begin{bmatrix} 30 & 0 & -28 \\ 14 & 24 & -6 \end{bmatrix} + \begin{bmatrix} 21 & 21 & 12 \\ 39 & -6 & -24 \end{bmatrix} =$

$\begin{bmatrix} 51 & 21 & -16 \\ 53 & 18 & -30 \end{bmatrix}$

9. $A + \begin{bmatrix} 0 & 0 & 0 \\ 0 & 0 & 0 \end{bmatrix} = A = \begin{bmatrix} 7 & 7 & 4 \\ 13 & -2 & -8 \end{bmatrix}$

10. Answers will vary. The matrix can have any number of rows but just one column. The dimensions will be $n \times 1$.

11. $s(J + K) = 3\left(\begin{bmatrix} 5 & 4 \\ -6 & 2 \end{bmatrix} + \begin{bmatrix} 2 & -3 \\ 5 & 4 \end{bmatrix}\right) =$

$3\begin{bmatrix} 7 & 1 \\ -1 & 6 \end{bmatrix} = \begin{bmatrix} 21 & 3 \\ -3 & 18 \end{bmatrix}$

$sJ + sK = 3\begin{bmatrix} 5 & 4 \\ -6 & 2 \end{bmatrix} + 3\begin{bmatrix} 2 & -3 \\ 5 & 4 \end{bmatrix} = \begin{bmatrix} 15 & 12 \\ -18 & 6 \end{bmatrix} +$

$\begin{bmatrix} 6 & -9 \\ 15 & 12 \end{bmatrix} = \begin{bmatrix} 21 & 3 \\ -3 & 18 \end{bmatrix}$

12. $36\begin{bmatrix} 332.15 & 195.40 & 215.87 & 280.60 \\ 95.98 & 244.55 & 356.81 & 303.76 \end{bmatrix} =$

$\begin{bmatrix} 11957.40 & 7034.40 & 7771.32 & 10101.6 \\ 3455.28 & 8803.80 & 12845.16 & 10935.36 \end{bmatrix}$

13. $F + G + H =$

$$\begin{bmatrix} 21 + 15 + 35 & 19 + 16 + 36 & 15 + 11 + 27 \\ 30 + 21 + 25 & 8 + 4 + 13 & 6 + 6 + 18 \end{bmatrix} =$$

$$\begin{bmatrix} 71 & 71 & 53 \\ 76 & 25 & 30 \end{bmatrix}$$

14. $F + G + H + N - S =$

$$\begin{bmatrix} 71 + 16 & 71 + 15 & 53 + 22 \\ 76 + 19 & 25 + 24 & 30 + 8 \end{bmatrix} - \begin{bmatrix} 20 & 24 & 13 \\ 33 & 21 & 19 \end{bmatrix} =$$

$$\begin{bmatrix} 87 & 86 & 75 \\ 95 & 49 & 38 \end{bmatrix} - \begin{bmatrix} 20 & 24 & 13 \\ 33 & 21 & 19 \end{bmatrix} = \begin{bmatrix} 67 & 62 & 62 \\ 62 & 28 & 19 \end{bmatrix}$$

15. If there are 468 branches then there are $\frac{468}{3} = 156$ 3-branch sets and if each of sets of 3 sold the same number as Fairview, Gateway, and High Point then the total sold is

$$156 \begin{bmatrix} 20 & 24 & 13 \\ 33 & 21 & 19 \end{bmatrix} = \begin{bmatrix} 3120 & 3744 & 2028 \\ 5148 & 3276 & 2964 \end{bmatrix}$$

16. Two matrices can be added if they have the same dimensions. To add, find the sum of each pair of corresponding elements.

EXTEND

17. $\begin{bmatrix} \frac{20}{4} & 6 - 3 \\ 2(-4) & 0 + 1 \end{bmatrix} = \begin{bmatrix} 5 & 3 \\ -8 & 1 \end{bmatrix}$; no

18. $\begin{bmatrix} 2 - 3 & \frac{12}{6} \\ 7 - 4 & -\frac{8}{2} \end{bmatrix} = \begin{bmatrix} -1 & 2 \\ 3 & -4 \end{bmatrix}$; yes

19. $5A = \begin{bmatrix} 14 & -8 & 20 \\ 55 & -28 & -60 \\ 2 & 0 & 100 \end{bmatrix}$

$A = \frac{1}{5}\begin{bmatrix} 14 & -8 & 20 \\ 55 & -28 & -60 \\ 2 & 0 & 100 \end{bmatrix} = \begin{bmatrix} 2.8 & -1.6 & 4 \\ 11 & -5.6 & -12 \\ 0.4 & 0 & 20 \end{bmatrix}$

20. $A = \begin{bmatrix} 9 & 2 & 7 & -1 \\ 4 & -11 & -12 & 5 \end{bmatrix} - \begin{bmatrix} 15 & 17 & -5 & 9 \\ 4 & 0 & 3 & -16 \end{bmatrix} =$

$\begin{bmatrix} -6 & -15 & 12 & -10 \\ 0 & -11 & -15 & 21 \end{bmatrix}$

21. $4A = \begin{bmatrix} 5 & 22 & -3 \\ 9 & -2 & -36 \end{bmatrix} - \begin{bmatrix} 5 & 6 & 1 \\ -3 & 2 & -8 \end{bmatrix} = \begin{bmatrix} 0 & 16 & -4 \\ 12 & -4 & -28 \end{bmatrix}$

$A = \frac{1}{4}\begin{bmatrix} 0 & 16 & -4 \\ 12 & -4 & -28 \end{bmatrix} = \begin{bmatrix} 0 & 4 & -1 \\ 3 & -1 & -7 \end{bmatrix}$

22. $9A = 2A + \begin{bmatrix} 19 & -18 \\ 4 & -43 \end{bmatrix} - \begin{bmatrix} 12 & 3 \\ -10 & -8 \end{bmatrix}$

$9A = 2A + \begin{bmatrix} 7 & -21 \\ 14 & -35 \end{bmatrix}$

$7A = \begin{bmatrix} 7 & -21 \\ 14 & -35 \end{bmatrix}$

$A = \frac{1}{7}\begin{bmatrix} 7 & -21 \\ 14 & -35 \end{bmatrix} = \begin{bmatrix} 1 & -3 \\ 2 & -5 \end{bmatrix}$

23. Answers will vary. No, because subtraction of real numbers is not commutative.

An example is $\begin{bmatrix} 1 & 2 \\ 3 & 4 \end{bmatrix} - \begin{bmatrix} 1 & 1 \\ 2 & 2 \end{bmatrix} = \begin{bmatrix} 0 & 1 \\ 1 & 2 \end{bmatrix}$ while

$\begin{bmatrix} 1 & 1 \\ 2 & 2 \end{bmatrix} - \begin{bmatrix} 1 & 2 \\ 3 & 4 \end{bmatrix} = \begin{bmatrix} 0 & -1 \\ -1 & -2 \end{bmatrix}$

24. directly above a_{34} is a_{24}

25. directly to the left of b_{76} is b_{75}

26. dimensions of A are $m \times n$
dimensions of B are also $m \times n$

27. The element in the ith row and jth column of $A + B$ is $a_{ij} + b_{ij} = (a + b)_{ij}$. Since this is true for all $1 \le i \le m$ and $1 \le j \le n$ the dimensions of $A + B$ are $m \times n$.

28. Answers will vary. Possible answer:

$\begin{bmatrix} 5 & 2 \\ 3 & -2 \\ 9 & 4 \end{bmatrix} + \begin{bmatrix} 4 & 3 \\ -3 & 3 \\ 6 & 1 \end{bmatrix} = \begin{bmatrix} 9 & 5 \\ 0 & 1 \\ 15 & 5 \end{bmatrix} = \begin{bmatrix} 4 & 3 \\ -3 & 3 \\ 6 & 1 \end{bmatrix} + \begin{bmatrix} 5 & 2 \\ 3 & -2 \\ 9 & 4 \end{bmatrix}$

29. Answers will vary. Possible answer:

$\left(\begin{bmatrix} 4 & 1 \\ 7 & 3 \end{bmatrix} + \begin{bmatrix} 5 & -2 \\ 1 & 4 \end{bmatrix} \right) + \begin{bmatrix} -6 & 3 \\ 4 & 2 \end{bmatrix} = \begin{bmatrix} 3 & 2 \\ 12 & 9 \end{bmatrix}$

$\begin{bmatrix} 4 & 1 \\ 7 & 3 \end{bmatrix} + \left(\begin{bmatrix} 5 & -2 \\ 1 & 4 \end{bmatrix} + \begin{bmatrix} -6 & 3 \\ 4 & 2 \end{bmatrix} \right) = \begin{bmatrix} 3 & 2 \\ 12 & 9 \end{bmatrix}$

30. Answers will vary. Possible answer:

$4 \left(\begin{bmatrix} 6 & 3 & -2 \\ 2 & 4 & 1 \end{bmatrix} + \begin{bmatrix} 3 & 1 & 4 \\ 6 & 5 & 9 \end{bmatrix} \right) = \begin{bmatrix} 36 & 16 & 8 \\ 32 & 36 & 40 \end{bmatrix}$

THINK CRITICALLY

31. Drawings will vary as the numbers $a, b, c, d, e,$ and f vary.

32. It is congruent to the original triangle but translated m units right or left and n units up or down.

33. It is similar to the original triangle but has sides twice as long.

34. Adding $\begin{bmatrix} m & m & m \\ n & n & n \end{bmatrix}$ translates the triangle m units left or right and n units up or down. Scalar multiplication by a factor of k multiplies the lengths of the sides of the triangle by k.

35. $3x - 4 = -x + 8$; $4x = 12$; $x = 3$

Check: $3(3) - 4 \overset{?}{=} -(3) + 8$

$9 - 4 \overset{?}{=} -3 + 8$

$5 = 5$ ✓

36. $4(c + 2) - 3(c - 3) = 11$; $4c + 8 - 3c + 9 = 11$;
$c + 17 = 11$; $c = -6$

Check: $4(-6 + 2) - 3(-6 - 3) = 4(-4) - 3(-9) =$
$-16 + 27 = 11$ ✓

37. $4m - \frac{1}{2} = 2m + 1$; $2m - \frac{1}{2} = 1$; $2m = \frac{3}{2}$; $m = \frac{3}{4}$

Check: $4\left(\frac{3}{4}\right) - \frac{1}{2} \overset{?}{=} 2\left(\frac{3}{4}\right) + 1$

$3 - \frac{1}{2} \overset{?}{=} \frac{3}{2} + 1$

$\frac{5}{2} = \frac{5}{2}$ ✓

38. $\frac{5x}{3} - 8 = 4x - 15$; $5x - 24 = 12x - 45$;
$-7x - 24 = -45$; $-7x = -21$; $x = 3$

Check: $\frac{5(3)}{3} - 8 \overset{?}{=} 4(3) - 15$

$5 - 8 \overset{?}{=} 12 - 15$

$-3 = -3$ ✓

39. $[4\ \ 2\ \ -5\ \ 8\ \ -3\ \ 1] + [11\ \ -4\ \ -6\ \ 0\ \ 5\ \ 2] =$
$[15\ \ -2\ \ -11\ \ 8\ \ 2\ \ 3]$

40. $-3[11\ \ -4\ \ -6\ \ 0\ \ 5\ \ 2] = [-33\ \ 12\ \ 18\ \ 0\ \ -15\ \ -6]$

41. $[11\ \ -4\ \ -6\ \ 0\ \ 5\ \ 2] - [4\ \ 2\ \ -5\ \ 8\ \ -3\ \ 1] =$
$[7\ \ -6\ \ -1\ \ -8\ \ 8\ \ 1]$

42. $2[4\ \ 2\ \ -5\ \ 8\ \ -3\ \ 1] + 3[11\ \ -4\ \ -6\ \ 0\ \ 5\ \ 2] =$
$[8\ \ 4\ \ -10\ \ 16\ \ -6\ \ 2] + [33\ \ -12\ \ -18\ \ 0\ \ 15\ \ 6] =$
$[41\ \ -8\ \ -28\ \ 16\ \ 9\ \ 8]$

43. C; $6(5)^2 - 29(5) - 5 = 6(25) - 145 - 5 =$
$150 - 145 - 5 = 0$

Lesson 1.4, pages 25–31

EXPLORE

1. Since more men were involved in accidents, men appear to be about twice as likely to be involved in an accident.

2. Men were involved in more accidents. Men in this group appear to be more likely to be involved in an accident.

3. More accidents occurred in the under-25 group. Drivers in this group are about $1\frac{1}{3}$ times as likely to have an accident.

4. Answers will vary. Since the younger driver is about four times as likely to be involved in an accident as the older driver, he or she should pay about four times as much for the same policy, or about $2000 a year.

1. 5, 6   **2.** 1, 2, 3, 4, 5, 6, 7, 8   **3.** $\frac{2}{8} = \frac{1}{4}$ or 0.25

4. There are 4 even numbers; 2, 4, 6, 8, So, the probability is $\frac{4}{8} = \frac{1}{2}$ or 0.5

5. $\frac{1}{8}$ or 0.125

6. There are 3 numbers less than 4; 1, 2, 3. So, the probability is $\frac{3}{8}$ or 0.375.

7. 1; you will definitely spin a number.

8. 0; it is not possible to spin 19.

9. $\frac{2}{6} = \frac{1}{3}$ or 1:3   **10.** $\frac{58}{200} = \frac{29}{100}$ or 0.29

11.

First roll   Second roll   First roll   Second roll

$1 \rightarrow$ 1, 2, 3, 4, 5, 6 $4 \rightarrow$ 1, 2, 3, 4, 5, 6

$2 \rightarrow$ 1, 2, 3, 4, 5, 6 $5 \rightarrow$ 1, 2, 3, 4, 5, 6

$3 \rightarrow$ 1, 2, 3, 4, 5, 6 $6 \rightarrow$ 1, 2, 3, 4, 5, 6

12. There are 36 out comes in the sample space. There are 3 odd numbers that can be rolled with 1, 3, and 5 which is a total of 9 outcomes in the event. So, the probability is $\frac{9}{36} = \frac{1}{4}$ or 0.25.

13. $\dfrac{\text{area of photo}}{\text{area of bulletin board}} = \dfrac{4(5)}{18(30)} = \dfrac{1}{27}$

14. Answers will vary. Theoretical probability is the mathematical likelihood that an event will occur. Experimental probability is what actually occurs in an experiment.

15. Answers will vary. Since there are 12 months the theoretical probability that someone was born in May is $\frac{1}{12}$. It can also be reasoned that since May has 31 days and there are 365 days in a year, the probability is $\frac{31}{365}$. and $\frac{1}{12} \approx 0.083$ and $\frac{31}{365} \approx 0.0851$.

16. The experimental probability that someone was born in May is $\frac{32}{256} = \frac{1}{8}$ or 0.125

PRACTICE

1–3. Examples will vary. 0: There will be 13 inches in a foot tomorrow morning; 0.5: a randomly chosen person will be female; 1: A tossed coin will come up heads or tails.

4.

First Throw	Second Throw	Sum		First Throw	Second Throw	Sum
1	1	2		4	1	5
	2	3			2	6
	3	4			3	7
	4	5			4	8
	5	6			5	9
2	1	3		5	1	6
	2	4			2	7
	3	5			3	8
	4	6			4	9
	5	7			5	10
3	1	4				
	2	5				
	3	6				
	4	7				
	5	8				

5. There are 25 outcomes in the sample space and 7 outcomes in the event. Therefore, the probability is $\frac{7}{25}$ or 0.28.

6. There are 3 multiples of 6; 6, 12, 18.

7. Outcomes in the sample space consist of the integers greater than or equal to 1 and less than or equal to 20.

8. $\frac{3}{20}$ or 0.15

9. There are 5 multiples of 4; 4, 8, 12, 16, 20. So, the probability is $\frac{5}{20} = \frac{1}{4}$ or 0.25

10. There are 7 numbers greater than 13. So, the probability is $\frac{7}{20}$ or 0.35.

11. There is no 30. So, the probability is 0.

12. There are 2 outcomes in the event. So, the probability is $\frac{2}{20} = \frac{1}{10}$ or 0.1.

13. You are certain to draw a number from 1 to 20. So, the probability is 1.

14. There are 3 multiples of 6 and 17 numbers not divisible by 6. So, the odds are $\frac{3}{17}$ or 3 to 17.

15. The experimental probability of drawing a multiple of 6 is $\frac{8}{50} = \frac{4}{25}$ or 0.16.

16. $\frac{\text{Area of center circle}}{\text{Area of board}} = \frac{\pi(4)^2}{\pi(12)^2} = \frac{16\pi}{144\pi} = \frac{1}{9}$

17. $\frac{23}{276} = \frac{1}{12}$ or approximately 0.083

18. The outcomes in the sample space is the total number of people, $28 + 19 + 35 + 24 + 16 + 20 = 142$ and the outcomes of the event consists of $35 + 16 + 24 + 20 = 95$. Therefore the probability is $\frac{95}{142}$ or approximately 0.669.

19. On 40% of the days in the past with conditions like those expected for the following day, it has rained.

EXTEND

20a. $1 - \frac{3}{5} = \frac{2}{5}$ **20b.** 3 to 2

21. 6 outcomes in the event
5 outcomes not in the event
11 outcomes in the sample space
Therefore, the probability is $\frac{6}{11}$.

22. Let A represent area of inner circle: $A = \pi(4)^2 = 16\pi$
Let B represent area of middle circle:
$B = \pi(8)^2 = 64\pi$
Let C represent area of outer circle:
$C = \pi(12)^2 = 144\pi$
Area of middle ring: $B - A = 64\pi - 16\pi = 48\pi$
Area of region not in middle ring: $144\pi - 48\pi = 96\pi$
Therefore the odds are $\frac{48\pi}{96\pi} = \frac{1}{2}$

THINK CRITICALLY

23. There are 12 outcomes that consist of 11 heads and one tail because the tail could come up on the first toss, second toss, third toss, and so on to the twelfth toss, with the other tosses being heads.

24. Let n = number of outcomes in event and let s = number of outcomes in sample space. Then P (will occur) $= \frac{n}{s}$, P (will not occur) $= \frac{s - n}{s}$, and sum $= \frac{n + (s - n)}{s} = \frac{s}{s} = 1$

MIXED REVIEW

25. $-3 = 5 - x; x - 3 = 5; x = 8$
Check: $5 - 8 = -3$ ✓

26. $2 + 6x = -1; 6x = -3; x = -\frac{1}{2}$
Check: $2 + 6\left(-\frac{1}{2}\right) = 2 - 3 = -1$ ✓

27. $-6 + n = 3; n = 9$
Check: $-6 + 9 = 3$ ✓

28. $\frac{2b - 3}{5} = 7; 2b - 3 = 35; 2b = 38; b = 19$
Check: $\frac{2(19) - 3}{5} = \frac{38 - 3}{5} = \frac{35}{5} = 7$

29. $\frac{1}{6}$ **30.** $\frac{3}{6} = \frac{1}{2}$ **31.** $\frac{2}{6} = \frac{1}{3}$ **32.** 0

33. D; $3\begin{bmatrix} 2 & 6 \\ -1 & 4 \end{bmatrix} = \begin{bmatrix} 6 & 18 \\ -3 & 12 \end{bmatrix} = \begin{bmatrix} 6 & \frac{36}{2} \\ -\frac{9}{3} & 12 \end{bmatrix}$

PROJECT CONNECTION

1. If n = number of groups, the probability of my group winning is $\frac{1}{n}$.

2. Answers will vary. If it appears some groups may not complete their models, the number n used in Exercise 1 will decrease and the probability of my group winning will increase.

ALGEBRAWORKS

1. Let x = the number involving men.
   $\frac{1}{2}x$ = the number involving women.
   $x + \frac{1}{2}x = 43,500; \frac{3}{2}x = 43,500; x = 29,000$

2. $0.48(43,500) = 20,880$

3. If 30% of the drivers are under 25 then 70% of the drivers are 25 or over. So, the odds that the driver is under 25 are $\frac{30}{70} = \frac{3}{7}$ or 3 to 7.

4. $100\% - 30\% = 70\%$ or $\frac{7}{10}$

5. Answers will vary. In general, policies cost less for females, drivers with experience, those who have taken driver's training, lower-priced cars, married people, drivers who drive less, drivers with no accidents or violations, and drivers who live in rural areas.

Lesson 1.5, page 32–37

EXPLORE

1. Bargain's claim is accurate. The average price of its new models is
   $$\frac{11,000 + 10,900 + 10,900 + 11,100 + 5,000 + 10,500}{6}$$
   $= \$9900$

2. Yes. Although the claim is accurate, it suggests that $10,000 prices are the rule, when in fact they are the exception.

5a. mean = $\dfrac{3.0 + 2.3 + 2.1 + 2.6 + 3.3 + 3.5 + 3.8 + 4.4 + 5.8 + 2.6 + 3.2 + 3.6}{12} = 3.35$

data in order: 2.1, 2.3, 2.6, 2.6, 3.0, 3.2, 3.3, 3.5, 3.6, 3.8, 4.4, 5.8

median = $\dfrac{3.2 + 3.3}{2} = 3.25$

mode = 2.6

range = $5.8 - 2.1 = 3.7$

5b. Answers will vary with various justifications.

6. 50% of the data is between 40 and 70

7. 70   8. 35   9. 60   10. $90 - 35 = 55$

11. 25% of the data is between 70 and 90.

12. from 35 to 40

3. A small number of extreme values can result in an average that is not close to any of the values. In this example the sticker price of $5000 is so low that it brings the average of all the prices down to a misleading price.

4. Answers will vary "always a bargain at Bargain. In an age of expensive cars, the price of a new car at Bargain remains around $11,000."

TRY THESE

1. mean = $\dfrac{46 + 28 + 49 + 53 + 29}{5} = 41$
   data in order 28, 29, 46, 49, 53
   median = 46
   mode: there is none
   range = $53 - 28 = 25$

2. mean =
   $\dfrac{8.5 + 11.2 + 9.3 + 14.6 + 15.0 + 11.2 + 9.3 + 9.4}{8}$
   $= 11.0625$
   data in order: 8.5, 9.3, 9.3, 9.4, 11.2, 11.2, 14.6, 15.0
   median = $\dfrac{9.4 + 11.2}{2} = \dfrac{20.6}{2} = 10.3$
   mode = 9.3 and 11.2
   range = $15.0 - 8.5 = 6.5$

3. mean = $\dfrac{488 + 466 + 517 + 581 + 404 + 496}{6} = 492$
   data in order: 404, 466, 488, 496, 517, 581
   median = $\dfrac{488 + 496}{2} = 492$
   mode: there is none
   range = $581 - 404 = 177$

4. mean =
   $\dfrac{-3.2 + 5.5 + (-7) + (-2) + 5.7 + 0 + 5.5 + 10 + (-1)}{9}$
   $= 1.5$
   data in order: $-7, -3.2, -2, -1, 0, 5.5, 5.5, 5.7, 10$
   median = 0
   mode = 5.5
   range = $10 - (-7) = 17$

13.

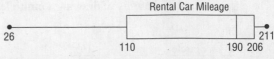

Rental Car Mileage

26   110   190 206   211

Half of the values are widely dispersed from the least value, 26, to the median, 190. The other half are tightly clustered between 190 and 211.

14. The mean represents a set of data well if all of the items are approximately equal. If some of the values in a set are greatly different from the rest the average does not represent the set well.
Examples of data will vary.

PRACTICE

1. $= \dfrac{20 + 28 + 9 + 14 + 21 + 8 + 33}{7} =$
$\dfrac{133}{7} = 19$
data in order: 8, 9, 14, 20, 21, 28, 33
median $= 20$
mode: there is none
range $= 33 - 8 = 25$

2. mean $= \dfrac{1.2 + 1.1 + 1.8 + 0.9 + 0.5 + 1.5 + 1.7 + 1.5}{8} =$
$\dfrac{10.2}{8} = 1.275$
data in order: 0.5, 0.9, 1.1, 1.2, 1.5, 1.5, 1.7, 1.8
median $= \dfrac{1.2 + 1.5}{2} = \dfrac{2.7}{2} = 1.35$
mode $= 1.5$
range $= 1.8 - 0.5 = 1.3$

3.
mean $= \dfrac{361 + 314 + 325 + 320 + 314 + 336 + 332 + 361}{8}$
$= \dfrac{2663}{8} = 332.875$
data in order: 314, 314, 320, 325, 332, 336, 361, 361
median $= \dfrac{325 + 332}{2} = \dfrac{657}{2} = 328.5$

mode $= 314$ and 361
range $= 361 - 314 = 47$

4.
mean $= \dfrac{4 + (-7) + (-1) + 5 + (-11) + (-2) + 1 + (-1)}{8}$
$= \dfrac{-12}{8} = -1.5$
data in order: $-11, -7, -2, -1, -1, 1, 4, 5$
median $= \dfrac{-1 + (-1)}{2} = -1$
mode $= -1$
range $= 5 - (-11) = 16$

5. mean $= \dfrac{6.68 + 6.83 + 6.65 + 6.82 + 6.77 + 6.81}{6}$
$= \dfrac{40.56}{6} = 6.76$
data in order: 6.65, 6.68, 6.77, 6.81, 6.82, 6.83
median $= \dfrac{6.77 + 6.81}{2} = 6.79$
mode: there is none
range: $6.83 - 6.65 = 0.18$

6.
mean $= \dfrac{9865 + 9009 + 9801 + 9871 + 9909 + 9686 + 9178}{7}$
$= \dfrac{67319}{7} = 9617$
data in order: 9009, 9178, 9686, 9801, 9865, 9871, 9909
median $= 9801$
mode: there is none
range $= 9909 - 9009 = 900$

7. mean $= \dfrac{53 + 57 + 68 + 51 + 58 + 61 + 79 + 49 + 53 + 62 + 71 + 55 + 57 + 46 + 60 + 48}{16}$
$= \dfrac{922}{6} = 57.625$
data in order: 46, 48, 49, 51, 51, 53, 53, 55, 57, 58, 60, 62, 68, 71, 79
median $= \dfrac{55 + 57}{2} = \dfrac{112}{2} = 56$
mode $= 51$ and 53
range $= 79 - 46 = 33$

8. Any answer is acceptable as long as you can justify your answer.

9.

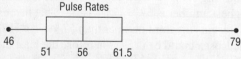
Pulse Rates
46   51   56   61.5   79

The data is more widely dispersed in the fourth quarter.

10. data in order: 13, 17, 18, 19, 21, 22, 22, 23, 23, 24, 25, 27, 29, 46, 55
first quartile: median of lower half $= 19$
median $= 23$
third quartile: median of upper half $= 27$

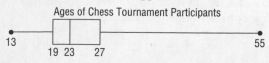
Ages of Chess Tournament Participants
13   19 23   27   55

Three-quarters of the values lie between 13 and 27. Values appear to be evenly dispersed in the first 75% of the data, with clustering in the interval 19–23.

11. Class 1 had the highest score, 98
Class 1 had the lowest score, 42

12. Class 1 had the higher median, 87

13. Class 2 had dispersion of grades in the middle 50% more even.

14. Class 2 achieved better results because 50% of its scores were 87 or above and less than 25% of Class 1's scores were 87 or above.

15. Examples will vary. Some examples of survey topics are hours spent watching TV or doing homework.

16. $IQR = 27 - 19 = 8$

17. $1.5 \cdot IQR = 1.5(8) = 12$
The outliers are 46 and 55.
$46 > 12 + 27 = 39$ and $55 > 12 + 27 = 39$

18. data in order: 39, 41, 41, 41, 42, 42, 42, 42, 45, 45, 45, 45
$$\text{mean} = \frac{39 + 3(41) + 4(42) + 4(45)}{1 + 3 + 4 + 4} = \frac{510}{12} = 42.5$$
$$\text{median} = \frac{42 + 42}{2} = 42$$
mode = 42 and 45

19. Any answer is acceptable as long as you can justify it.

20. Let x = the distance Janice must run to raise her mean.
$$\frac{x + 8(9)}{10} = 9$$
$x + 72 = 90$
$x = 18$ miles

21. Since the median is 13 and $\frac{5 + 12}{2} \neq 13$ or
$\frac{12 + 21}{2} \neq 13, \frac{n + 12}{2} = 13; n + 12 = 26; n = 14.$
Therefore, the mean $= \frac{5 + 12 + 14 + 21}{4} =$
$\frac{52}{4} = 13$

THINK CRITICALLY

22. 164; all of the numbers cannot be the mode.

23. 160 or 161; if $n = 160$ then 160 is the mode and if $n = 161$ then 161 is the mode.

24. any number other than 160, 161, or 164. For example if $n = 163$ then 160 and 161 are the two modes.

25. Answers will vary. For example {3, 3, 3, 4, 5, 5, 6, 7, 7, 7}. The first quartile is 3 and the third is 7 and 3 and 7 are the lowest and highest values. Therefore, the boxplot has no whiskers.

26. Since the mode is 2, two or more of the numbers must be 2. Since the median is 6, the middle number must be 6 and that means the middle number cannot be 2. This means there are exactly two 2's and one 6. Let the sum of the other two numbers be represented by x. Since the mean is 5.2, $\frac{x + 2(2) + 6}{5} = 5.2; x + 10 = 2;$
$x = 16$
The only two different single digit numbers that have a sum of 16 are 7 and 9.

27. To find mean speed, divide his total distance by the total time it took him to drive that distance.
Distance: $300 + 300 = 600$ mi
Time going: $\frac{300}{50} = 6$ h
Time returning: $\frac{300}{60} = 5$ h.
Total time: $6 + 5 = 11$ h
Average speed $= \frac{600}{11} = 54.5$ mi/h

A common mistake is to average the two rates and call this the mean speed, but this will not yield the correct answer.

MIXED REVIEW

28. $B - A = \begin{bmatrix} -5 - 2 & 1 - 4 & 8 - (-1) \\ -2 - (-5) & 4 - 3 & 3 - 7 \end{bmatrix}$
$= \begin{bmatrix} -7 & -3 & 9 \\ 3 & 1 & -4 \end{bmatrix}$

29. $B + \begin{bmatrix} 0 & 0 & 0 \\ 0 & 0 & 0 \end{bmatrix} = B = \begin{bmatrix} -5 & 1 & 8 \\ -2 & 4 & 3 \end{bmatrix}$

30. $-3A = \begin{bmatrix} -3(2) & -3(4) & -3(-1) \\ -3(-5) & -3(3) & -3(7) \end{bmatrix}$
$= \begin{bmatrix} -6 & -12 & 3 \\ 15 & -9 & -21 \end{bmatrix}$

31. $A - 2B = \begin{bmatrix} 2 & 4 & -1 \\ -5 & 3 & 7 \end{bmatrix} - \begin{bmatrix} -10 & 2 & 16 \\ -4 & 8 & 3 \end{bmatrix}$
$\begin{bmatrix} 2 - (-10) & 4 - 2 & -1 - 16 \\ -5 - (-4) & 3 - 8 & 7 - 3 \end{bmatrix}$

32. data in order: {1, 5, 6, 6, 8, 9, 11, 13, 14, 15}
$$\text{mean} = \frac{1 + 5 + 6 + 6 + 8 + 9 + 11 + 13 + 14 + 15}{10}$$
$= \frac{88}{10} = 8.8$
$\text{median} = \frac{8 + 9}{2} = \frac{17}{2} = 8.5$
mode = 6
first quartile = 6
third quartile = 13

33. data in order: {22, 22, 25, 26, 29, 30, 31, 31, 31, 31, 34, 36}
$\text{mean} = \frac{348}{12} = 29$
$\text{median} = \frac{30 + 31}{2} = \frac{61}{2} = 30.5$
mode = 31
first quartile $= \frac{25 + 26}{2} = \frac{51}{2} = 25.5$
third quartile $= \frac{31 + 31}{2} = \frac{62}{2} = 31$

34. E; If the range is 75 and the lowest number is 57 then the highest number is $57 + 75 = 132$. That means that if x represents the third number then $57 \leq x \leq 132$. So, the least x could be is 57. Therefore the least the mean could be is $\frac{57 + 57 + 132}{3} = 82 > 81$
So, I is impossible. II is true if $x = 57$. III is also true if $x = 57$.

PROJECT CONNECTION

1–4. Answers will vary. You may suggest that length-to-width, length-to-height, or length-to-weight ratios may be performance indicators.

EXPLORE

1. Answers will vary.

2. Answers will vary. Three possibilities are $(-3, -1)$, $(0, 1)$, and $(3, 3)$

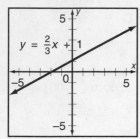

3. Answers will vary but the ratio $\frac{k}{h}$ will equal $\frac{2}{3}$.

4. The line intersects the y-axis at $y = 1$

5. Answers will vary. The equation should be written in the form $y = mx + b$

6. If the equation is written in the form $y = mx + b$, the value $\frac{k}{h}$ found in step 3 is equal to m, and the y-value found in step 4 is equal to b.

TRY THESE

1. Put equation into slope-intercept form.

   $2x + 5y = 10; 5y = -2x + 10; y = -\frac{2}{5}x + 2$

   x-intercept: Let $y = 0; 2x + 5(0) = 10; x = 5$

   y-intercept: $y = 2$

   slope: $m = -\frac{2}{5}$

2. $y = -3x + 12$

   x-intercept: Let $y = 0; 0 = -3x + 12; 3x = 12;$
   $x = 4$

   y-intercept: $y = 12$

   slope: $m = -3$

3. $0.3x - 6y - 15 = 0; 6y = 0.3x - 15;$
   $y = 0.05x - 2.5$

   x-intercept: Let $y = 0; 0 = 0.05x - 2.5; 0.05x = 2.5;$
   $x = 50$

   y-intercept: $y = -2.5$

   slope: $m = 0.05$

4. $-2x + 7y = 14$

   x-intercept: Let $y = 0; -2x + 7(0) = 14; x = -7$

   y-intercept: Let $x = 0; -2(0) + 7y = 14, y = 2$

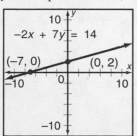

5. $x - 2y = 6;$
   $2y = x - 6;$
   $y = \frac{1}{2}x - 3$

   y-intercept: $y = -3;$

   slope: $m = \frac{1}{2}$

6. $6x + 4y = 480$, where x = number of adult tickets sold and y = number of children's tickets sold.

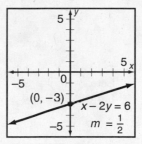
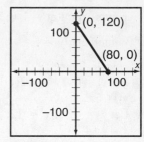

7. Move to the point on the graph where $y = 90$ and read the x-value on the x-axis, $x = 20$. So, 20 adult tickets were sold.

8.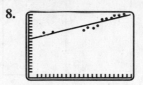

9. Equations will vary. $y = 0.4x + 11$, where x = Olympic number $(1904 = 1)$ and y = winning pole vault seconds, in feet

10. positive

11. Answers will vary
    $y = 0.4x + 11$; for $x = 23$;
    $y = 0.4(23) + 11 = 19.2$ ft

12. Answers will vary but most likely the prediction is close to the actual height.

13. Answers will vary. about 19.7 ft

14. no correlation; the number of doors a car has does not say anything about its price.

15. negative; the faster the speed, the lower the time

16. positive; the greater the price of a gallon, the greater the price of a tank.

17. Descriptions will vary. "Using intercepts" is better for equations in the form $Ax + By = C$ because each intercept can be found quickly. "Using the slope and y-intercept" is better for equations in the form $y = mx + b$ because the slope and y-intercept can be read from the equation.

PRACTICE

1. $y = \frac{1}{2}x - 2$

   x-intercept: Let $y = 0; 0 = \frac{1}{2}x - 2; \frac{1}{2}x = 2 \ x = 4$

   y-intercept: $y = -2$ slope: $m = \frac{1}{2}$

2. $4x + 5y = 10; 5y = -4x + 10; y = -\frac{4}{5}x + 2$

x-intercept: Let $y = 0; 4x = 10; x = \frac{5}{2}$

y-intercept: $y = 2$　　slope: $m = -\frac{4}{5}$

3. $2x - 4y + 3 = 0; 4y = 2x + 3; y = \frac{1}{2}x + \frac{3}{4}$

x-intercept: Let $y = 0; 2x + 3 = 0; 2x = -3; x = -\frac{3}{2}$

y-intercept: $y = \frac{3}{4}$　　slope: $m = \frac{1}{2}$

4. $-3x + 2y = 12$
Let $x = 0; 2y = 12;$
$y = 6$
Let $y = 0; -3x = 12;$
$x = -4$

5. $x + y = 9$
Let $x = 0; y = 9$
Let $y = 0; x = 9$

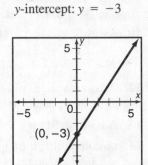

6. $y = -2x + 4$
Let $x = 0; y = 4$
Let $y = 0; 0 = -2x + 4;$
$2x = 4; x = 2$

7. $y = 2x - 3$
slope: $m = 2;$
y-intercept: $y = -3$

8. $y = -\frac{3}{4}x + 1$

slope: $m = -\frac{3}{4};$

y-intercept: $y = 1$

9. $2x - 5y + 15 = 0;$

$5y = 2x + 15;$

$y = \frac{2}{5}x + 3$

slope: $m = \frac{2}{5}$

y-intercept: $y = 3$

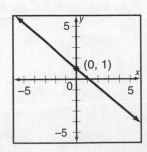 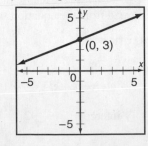

10.

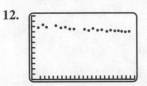

$x + 0.5y = 1500$, where
x = number of hardcovers
sold and y = number of
paperbacks sold.
Let $x = 0; 0.5y = 1500;$
$y = 3000$
Let $y = 0; x = 1500$

11. Find the point on the graph where $x = 750$ and read the number on the y-axis; $y = 1500$; So, 1500 paperbacks were sold.

12.

13. Answers will vary. Slope is negative and close to zero. y-intercept is close to 50.

14. The correlation is negative.

15. Answers will vary. Examples are 42.6 s; 43.2 s.

16. Answers will vary.

17. Answers will vary but probably will be about 43.0 s.

18. Answers will vary, and be about 8.4 s. This is not realistic because you cannot extrapolate too far away from the known data.

19. positive; an increase in diameter leads to an increase in the circumference

20. negative; if the area is fixed, an increase in width produces a decrease in length

21. negative; the greater your altitude the lower the air pressure

22. no correlation; a person's age has no effect on zip code

EXTEND

23. Answers will vary. See Exercises 19–22 or some examples.

24. y-intercept: $b = -7$
slope: $m = 5$
$y = 5x - 7$

25. $3x + 4y = 6; 4y = -3x + 6; y = -\frac{3}{4}x + \frac{3}{2}$

slope: $m = -\frac{3}{4}$

Since the line contains the point $(-8, 1)$, substituting into $y = mx + b$ yields $1 = -\frac{3}{4}(-8) + b;$

$1 = 6 + b; b = -5$

Therefore, the equation of the line is $y = -\frac{3}{4}x - 5$ or $3x + 4y = -20$

26. Since the line is perpendicular to $y = \frac{1}{2}x - 9$ which has slope $m = \frac{1}{2}$, its slope is -2. Substituting the coordinates $(-3, 1)$ into $y = mx + b$ yields $1 = -2(-3) + b; 1 = 6 + b, b = -5$. Therefore, the equation of the line is $y = -2x - 5$.

27a. Use the equation from Example 4 on page 41,
$y = -0.472x + 34$
Let $x = 22, y = -0.472(22) + 34 = -10.384 + 34 = 23.62$ mi/gal
Let $x = 37, y = -0.472(37) + 34 = -17.464 + 34 = 16.54$ mi/gal.

27b. $24.4 - 23.62 = 0.78$ is the difference in the predicted mileage and the actual mileage.
$\frac{0.78}{24.4}(100) = 3.2\%$
$16.54 - 12.3 = 4.24$ is the difference in the predicted mileage and the actual mileage.
$\frac{4.24}{12.3}(100) = 34.5\%$

27c. Answers will vary. You may conclude the percent error is higher, meaning the equation is not a very good prediction.

THINK CRITICALLY

28. slope: $m = \frac{y_2 - y_1}{x_2 - x_1} = \frac{7 - (-5)}{3 - (-3)} = \frac{12}{6} = 2$
Substitute into $y = mx + b$; using either $(3, 7)$ or $(-3, -5)$.
$7 = 2(2) + b; b + 4 = 7; b = 3$
So, the equation of the line is $y = 2x + 3$.

29. $m = \frac{y_2 - y_1}{x_2 - x_1} = \frac{6 - 4}{2 - 2} = \frac{2}{0}$
Division by 0 is undefined and the line is vertical. For this reason, vertical lines have undefined slope.

30. $600,000(0.05) = \$30,000$ per year
$\frac{30,000}{12} = \$2500$ per month

31. $y = 600,000 - 2500x$

32.
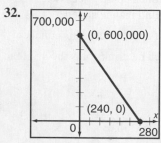

The restrictions are $x \geq 0$ and $y \geq 0$.

33. 5 years $= 5(12) = 60$ months
Let $x = 60; y = 600,000 - 2500(60) = \$450,000$

34. Let $y = 0$ and solve for x.
$0 = 600,000 - 2500x; 2500x = 600,000; x = 240$
months $= 20$ years

MIXED REVIEW

35. A; odds of 9:5 means there are 9 outcomes in the event and 5 outcomes not in the event. Therefore there are $5 + 9 = 14$ outcomes in the sample space. So, the probability is $\frac{9}{14}$.

36. data in order: $\{42, 44, 51, 55, 57, 57, 58, 60\}$
mean $=$
$\frac{42 + 44 + 57 + 55 + 57 + 57 + 58 + 60}{8} = 53$
median $= \frac{55 + 57}{2} = \frac{112}{5} = 56$
mode $= 57$
first quartile $= \frac{44 + 51}{2} = \frac{95}{2} = 47.5$
third quartile $= \frac{57 + 58}{2} = \frac{115}{2} = 57.5$

37. data in order $= \{0\ 1\ 1\ 1\ 2\ 3\ 3\ 3\ 4\ 4\ 4\ 5\ 5\ 6\}$
mean $=$
$\frac{0 + 3(1) + 2 + 3(3) + 3(4) + 2(5) + 6}{14} = 3$
median $= \frac{3 + 3}{2} = 3$
mode $= 1, 3,$ and 4
first quartile $= 1$
third quartile $= 4$

38. $5x - 6y = 30; 6y = 5x - 30; y = \frac{5}{6}x - 5$
x-intercept: Let $y = 0; 5x = 30; x = 6$
y-intercept: $y = -5$
slope: $m = \frac{5}{6}$

39. $y = 5x - 10$
x-intercept: Let $y = 0; 5x - 10 = 0; 5x = 10; x = 2$
y-intercept: $y = -10$
slope: $m = 5$

40. $2x + 2y + 8 = 0; 2y = -2x - 8; y = -x - 4$
x-intercept: Let $y = 0; 0 = -x - 4; x = -4$
y-intercept: $y = -4$
slope: $m = -1$

Lesson 1.7, pages 46–49

EXPLORE THE PROBLEM

1. You want to find the combined distance driven by all the vehicles during the year. The data you can use are total number of vehicles, mean number of miles driven per gallon, and mean number of gallons used. The irrelevant information is the total cost of gasoline used.

2. Find the product of the number of vehicles, the mean number of miles driver per gallon, and the mean number of gallons used. A calculator would simplify the work.

3. $185{,}000{,}000 = 1.85 \times 10^8$; $16.9 = 1.69 \times 10^1$; $683 = 6.83 \times 10^2$; Methods will vary.

$185{,}000{,}000 = \dfrac{185{,}000{,}000}{100{,}000{,}000} \times 100{,}000{,}000 = 1.85 \times 10^8$

$16.9 = \dfrac{16.9}{10} \times 10 = 1.69 \times 10^1$

$683 = \dfrac{683}{100} \times 100 = 6.83 \times 10^2$

Point out the following simple method for writing a number in scientific notation $a \times 10^n$. Move the decimal point to form the number a. The number of places moved is the absolute value of n. If you move the decimal point left, n is positive. If you move the decimal point right, n is negative.

4. $(1.85 \times 10^8)(1.69 \times 10^1)(6.83 \times 10^2) = 2.1354 \times 10^{12}$ miles

5. $2{,}135{,}400{,}000{,}000$ miles. $2.1354 \times 10^{12} = 2.1354 \times 1{,}000{,}000{,}000{,}000 = 2{,}135{,}400{,}000{,}000$. Point out that scientific notation can be converted to standard notation by reversing the decimal-moving process. Here, the decimal in 2.1354 is moved twelve places to the right to produce $2{,}135{,}400{,}000{,}000$.

6. Answers will vary. $16.9(683) \approx 15(700) \approx 10{,}000 = 10^4$, so the answer should be approximately $(1.85 \times 10^8) \times 10^4 = 1.85 \times 10^{12}$. This is reasonably close to the answer obtained, 2.1354×10^{12}.

INVESTIGATE FURTHER

7. Divide 462 by 100 and multiply 10^{13} by $100 = 10^2$. This is equivalent to multiplying by 1; 4.62×10^{15}.

8. $\dfrac{1}{185{,}000{,}000}$

9. 5.41×10^{-9}

10. $5.41 \times 10^{-9} = 0.00000000541$

11. 7.87×10^{-12} and 1.52×10^{11}

12. mean price $= (7.87 \times 10^{-12})(1.52 \times 10^{11}) = \$1.196$

13. Problems will vary. Find the mean cost of gasoline per vehicle in 1992.

$\dfrac{152{,}000{,}000{,}000}{185{,}000{,}000} = \dfrac{1.52 \times 10^{11}}{1.85 \times 10^8} \approx 0.822 \times 10^3 = 822$.

The mean cost was about $822 per vehicle.

APPLY THE STRATEGY

14. $\dfrac{1.3 \times 10^7}{3.2 \times 10^4} \approx 4.06 \times 10^2 = 406$ or about 400 times

15. $186{,}282 \times 60 \times 60 \times 24 \times 365 \approx 5.87 \times 10^{12}$ miles

16. From Exercise 4 above the total distance traveled by all motor vehicles during 1991 was 2.1354×10^{12} miles.

This is $\dfrac{2.1354 \times 10^{12}}{5.87 \times 10^{12}}$ of a light year.

$\dfrac{2.1354 \times 10^{12}}{5.87 \times 10^{12}} \times 100 \approx 36\%$ of a light year.

At this rate, they would travel 1 light-year in under 3 years.

17. $\dfrac{1.36 \times 10^{19}}{5.87 \times 10^{12}} \approx 2{,}300{,}000$ years

18. Light travels 186,282 miles per second. So, in 1 nanosecond it travels $\dfrac{186{,}282}{1{,}000{,}000{,}000}$

$= 1.86 \times 10^{-4}$ miles

Since 1 mile $= 5280$ ft, in 1 nanosecond light travels $(1.86 \times 10^{-4})(5280) = 0.98$ or approximately 1 ft.

19. $\dfrac{6.2 \times 10^9}{5.79 \times 10^7} \approx 1.07 \times 10^2$ or about 107 persons per square mile

20. $\dfrac{(5.79 \times 10^7)(5280)^2}{10^2} \approx 1.6142 \times 10^{13}$ or $16{,}142{,}000{,}000{,}000$

REVIEW PROBLEM SOLVING STRATEGIES

1.

tunnel: 1 mile long

train: 1 mi long

It takes a total of 2 hours. The front of the train takes 1 hour to reach the end of the tunnel, but then it takes another hour for the end of the train to emerge from the tunnel.

2a. two

2b. After 6000 km, take off two tires and put on the two new spares. After the next 6000 km, take off the two tires with 12,000 km on them and put back the two with 6000 km. At this point, all four tires will have 6000 km on them, so they can go the remaining 6000 km of the 18,000-km trip.

3a. Answers will vary; draw pictures, work with a specific number of ounces.

3b. Nick drank the same amount of each juice—1 glass. If you keep track of how much apple juice Nick poured into the glass you find that he added $\left(\dfrac{1}{6} + \dfrac{1}{3} + \dfrac{1}{2}\right)$ of a glass of apple juice. This equals 1 glass of apple juice which he drinks plus the original glass of cranberry juice.

Chapter Review, pages 50–51

1. c **2.** b **3.** d **4.** e **5.** a

6. $n - 3 = 5 + 5n$

$n = 8 + 5n$

$-4n = 8$

$n = -2$

Check: $-2 - 3 \overset{?}{=} 5 + 5(-2)$

$-5 = \overset{?}{=} 5 - 10$

$-5 = -5$ ✓

7. $15 = \dfrac{b}{5} + 19$

$\quad -4 = \dfrac{b}{5}$

$\quad -20 = b$

Check: $\dfrac{-20}{5} + 19 = -4 + 19 = 15$ ✓

8. $4x + 7 = 29$

$\quad 4x = 22$

$\quad x = 5\dfrac{1}{2}$

Check: $4\left(5\dfrac{1}{2}\right) + 7 = 22 + 7 = 29$ ✓

9. $3(y - 6) = -4(-23 + 2y)$

$\quad 3y - 18 = 92 - 8y$

$\quad 3y = 110 - 8y$

$\quad 11y = 110$

$\quad y = 10$

Check: $3(10 - 6) \overset{?}{=} -4[-23 + 2(10)]$

$\quad\quad 3(4) \overset{?}{=} -4(-3)$

$\quad\quad 12 \overset{?}{=} 12$ ✓

10. $M = \dfrac{10000(0.0079)(1 + 0.0079)^{48}}{(1 + 0.0079)^{48} - 1} \approx \$251.23$

11. deferred payment price $= (251.23)(48) = \$12,059.04$

12. C3: A3 * B3; $7(5.29) = 37.03$

C4: A4 * B4; $16(2.98) = 47.68$

D3: C3 * 1.05; $37.03(1.05) = 38.88$

D4: C4 * 1.05; $47.68(1.05) = 50.06$

13. $A + B = \begin{bmatrix} 2 + (-6) & -7 + (-8) & 5 + 1 \\ 13 + 0 & 21 + 15 & -11 + 9 \end{bmatrix} = \begin{bmatrix} -4 & -15 & 6 \\ 13 & 36 & -2 \end{bmatrix}$

14. $A - B = \begin{bmatrix} 2 - (-6) & -7 - (-8) & 5 - 1 \\ 13 - 0 & 21 - 15 & -11 - 9 \end{bmatrix} = \begin{bmatrix} 8 & 1 & 4 \\ 13 & 6 & -20 \end{bmatrix}$

15. $5B = \begin{bmatrix} 5(-6) & 5(-8) & 5(1) \\ 5(0) & 5(15) & 5(9) \end{bmatrix} = \begin{bmatrix} -30 & -40 & 5 \\ 0 & 75 & 45 \end{bmatrix}$

16. $B - 2A = \begin{bmatrix} -6 & -8 & 1 \\ 0 & 15 & 9 \end{bmatrix} - \begin{bmatrix} 4 & -14 & 10 \\ 26 & 42 & -22 \end{bmatrix} =$

$\begin{bmatrix} -6 - 4 & -8 - (-14) & 1 - 10 \\ 0 - 26 & 15 - 42 & 9 - (-22) \end{bmatrix} =$

$\begin{bmatrix} -10 & 6 & -9 \\ -26 & -27 & 31 \end{bmatrix}$

17. $1, 2, 3, 4, 5, 6$

18. $P(2 \text{ or } 3) = \dfrac{2}{6} = \dfrac{1}{3}$

19. $P(\text{even number}) = \dfrac{3}{6} = \dfrac{1}{2}$

20. data in order: 2, 2, 3, 4, 7, 11, 13

mean $= \dfrac{2 + 2 + 3 + 4 + 7 + 11 + 13}{7} = \dfrac{42}{7} = 6$

median $= 4$

mode $= 2$

range $= 13 - 2 = 11$

21. data in order: 151, 156, 177, 211

mean $= \dfrac{151 + 156 + 177 + 211}{4} = 173.75$

median $= \dfrac{156 + 177}{2} = 166.5$

mode: there is none

range: $211 - 151 = 60$

22. data in order: 3.2, 3.2, 3.5, 3.7, 3.7, 4.0

mean $= \dfrac{2(3.2) + 3.5 + 2(3.7) + 4.0}{6} = \dfrac{21.3}{6} = 3.55$

median $= \dfrac{3.5 + 3.7}{2} = \dfrac{7.2}{2} = 3.6$

mode $= 3.2$ and 3.7

range $= 4.0 - 3.2 = 0.8$

23. $7x - 5 = y$

x-intercept: Let $y = 0$; $7x - 5 = 0$; $7x = 5$; $x = \dfrac{5}{7}$

y-intercept: $y = -5$

slope: $m = 7$

24. $4x + 2y = 12$; $2y = -4x + 12$; $y = -2x + 6$

x-intercept: Let $y = 0$; $4x = 12$; $x = 3$

y-intercept: $y = 6$

slope: $m = -2$

25. $3(x - 2y) + 5 = 4y + 3$; $3x - 6y + 5 = 4y + 3$,

$3x - 6y + 2 - 4y$, $3x + 2 - 10y$; $y = \dfrac{3}{10}x + \dfrac{1}{5}$

x-intercept: Let $y = 0$; $3x + 5 = 3$; $3x = -2$;

$x = -\dfrac{2}{3}$

y-intercept: $y = \dfrac{1}{5}$ slope: $m = \dfrac{3}{10}$

26. positive; the greater the height, the greater the area

27. negative; the greater the velocity, the lower the time needed to travel 100 miles

28. 4.56×10^7

29. 3.07×10^{-4}

30. $9.1 = 9.1 \times 10^0$

31. 7.7777×10^{-1}

Chapter Assessment, pages 52–53

CHAPTER TEST

1. $6n + 11 = 3n + 23$; $3n + 11 = 23$; $3n = 12$;

$n = 4$

Check: $6(4) + 11 \overset{?}{=} 3(4) + 23$

$\quad\quad\quad 35 = 35$ ✓

2. $k + 20 = 5k + 44$; $-4k + 20 = 44$; $-4k = 24$;

$k = -6$

Check: $-6 + 20 \overset{?}{=} 5(-6) + 44$

$\quad\quad\quad 14 = 14$ ✓

3. $\frac{3x}{5} = 15; 3x = 75; x = 25$

   Check: $\frac{3(25)}{5} = 3(5) = 15$ ✓

4. $\frac{7c}{8} - (c - 2) = 12; 7c - 8c + 16 = 96; -c = 80;$
   $c = -80$

   Check: $\frac{7(-80)}{8} - (-80 - 2) = -70 + 82 = 12$ ✓

5. $4(2x - 5) = 3(2x + 8); 8x - 20 = 6x + 24;$
   $2x - 20 = 24; 2x = 44; x = 22$
   Check: $4[2(22) - 5] \stackrel{?}{=} 3[2(22) + 8]$
   $$4(39) \stackrel{?}{=} 3(52)$$
   $$156 = 156 ✓$$

6. $3(a - 7) - 3(2a - 4) = 4(a + 3)$
   $$3a - 21 - 6a + 12 = 4a + 12$$
   $$-3a - 9 = 4a + 12$$
   $$-7a - 9 = 12$$
   $$-7a = 21$$
   $$a = -3$$
   Check: $3[-3 - 7] - 3[2(-3) - 4] \stackrel{?}{=} 4[-3 + 3]$
   $$-30 + 30 \stackrel{?}{=} 4(0)$$
   $$0 = 0 ✓$$

7. $\frac{2x + 4}{3} - \frac{3x - 5}{5} = \frac{1}{2};$
   $10(2x + 4) - 6(3x - 5) = 15;$
   $20x + 40 - 18x + 30 = 15; 2x + 70 = 15;$
   $2x = -55 \; x = -\frac{55}{2}$

   Check: $\dfrac{2\left(-\frac{55}{2}\right) + 4}{3} - \dfrac{3\left(-\frac{55}{2}\right) - 5}{5}$

   $-17 + \frac{175}{10} = \frac{1}{2}$ ✓

8. $M = \frac{8200(0.0075)(1 + 0.0075)^{36}}{(1 + 0.0075)^{36} - 1} \approx \$260.76$

9. Total amount $= 260.76(36) = \$9387.36$

10. Cost of loan $= 9387.36 - 8200 = \$1187.36$

11. $4m + \frac{p^2}{4} = 4 * m + p \wedge 2 / 4$

12. C3: A3 * B3; 13(9) = 117
    C4: A4 * B4; 8.5(6) = 51
    D3: 2 * (A3 + B3); 2(13 + 9) = 44
    D4: 2 * (A4 + B4); 2(8.5 + 6) = 29

13. $A + B = \begin{bmatrix} 3 + (-5) & 4 + (-8) \\ 2 + 9 & 3 + 8 \\ -5 + (-4) & -6 + 9 \end{bmatrix} = \begin{bmatrix} -2 & -4 \\ 11 & 3 \\ -9 & 11 \end{bmatrix}$

14. $B - A = \begin{bmatrix} -5 - 3 & -8 - 4 \\ 9 - 2 & 9 - (-6) \\ -4 - (-5) & 8 - 3 \end{bmatrix} = \begin{bmatrix} -8 & -12 \\ 7 & 15 \\ 1 & 5 \end{bmatrix}$

15. $2A + 3B = \begin{bmatrix} 6 & 8 \\ 4 & -12 \\ -10 & 6 \end{bmatrix} + \begin{bmatrix} -15 & -24 \\ 27 & 27 \\ -12 & 24 \end{bmatrix} = \begin{bmatrix} -9 & -12 \\ 31 & 15 \\ -22 & 30 \end{bmatrix}$

16. Answers will vary. Data should have most terms close together plus one or two outliers. The mean gives equal weight to the extreme values and leans too heavily in their direction.
    For example {2, 2, 3, 3, 3, 3, 4, 47, 62}
    mean $= \dfrac{2(2) + 4(3) + 4 + 47 + 62}{9} = \dfrac{129}{9} \approx 14.3$
    median $= 3$ mode $= 3$

17. $P(3) = \frac{1}{8}$ 18. $P(\text{even number}) = \frac{4}{8} = \frac{1}{2}$

19. $P(\text{a number from 1 to 8}) = 1$

20. There are 3 numbers greater than 5 and 5 numbers less than or equal to 5. So, the odds of spinning a number greater than 5 is 3:5

21. Experimental probability is $\frac{9}{50}$.

22. data in order: 2, 3, 4, 4, 6, 8, 8, 9, 10
    mean $= \dfrac{2 + 3 + 4 + 4 + 6 + 8 + 8 + 9 + 10}{9} = \dfrac{54}{9} = 6$
    median $= 6$ mode $= 4$ and 8 range $= 10 - 2 = 8$

23. $P = \dfrac{\text{area of small circle}}{\text{area of large circle}} = \dfrac{\pi(2)^2}{\pi(5)^2} = \dfrac{4\pi}{25\pi} = \dfrac{4}{25}$ or 0.16

24. $3x - 5y + 10 = 0; 5y = 3x + 10; y = \frac{3}{5}x + 2$
    x-intercept: Let $y = 0; 3x + 10 = 0; 3x = -10;$
    $x = -\frac{10}{3}$ y-intercept $= 2$ slope: $m = \frac{3}{5}$

25. C; As the days since last Thanksgiving increases, the days until next Thanksgiving decreases. As the number of shirts increases the cost per shirt decreases.

26. 2.34×10^{11} 27. 1.2×10^{-5}

28. 45,500   29. 0.000000802

Cumulative Review, page 54

1. $2x + 3 = 11; 2x = 8; x = 4$
   Check: $2(4) + 3 = 8 + 3 = 11$ ✓

2. $\frac{z}{3} - 5 = 1; z - 15 = 3; z = 18$

   Check: $\frac{18}{3} - 5 = 6 - 5 = 1$ ✓

3. Joe let x represent the middle number and then $x - 1$ was the least and $x + 1$ the greatest. Gil let x represent the lowers number and then $x + 1$ was the middle and $x + 2$ the greatest
   $x - 1 + x + 1 = 26; 2x = 26; x = 13$
   Therefore, the numbers are 12, 13, and 14.

4.

	A	B	C
1	Retail	Down	Amount to be
2	Price	Payment	Financed
3	35,900	5,000	A3 − B3; 30,900
4	35,900	7,500	A4 − B4; 28,400
5	35,900	10,000	A5 − B5; 25,900

5. B; 3×4

6. $A + B = \begin{bmatrix} 3 + (-2) & 2 + (-5) & -1 + 0 \\ -4 + 1 & 0 + (-3) & 6 + (-4) \end{bmatrix} =$
$\begin{bmatrix} 1 & -3 & -1 \\ -3 & -3 & 2 \end{bmatrix}$

7. $-4B = \begin{bmatrix} -4(-2) & -4(-5) & -4(0) \\ -4(1) & -4(-3) & -4(-4) \end{bmatrix} = \begin{bmatrix} 8 & 20 & 0 \\ -4 & 12 & 16 \end{bmatrix}$

8. $2A - B = \begin{bmatrix} 6 & 4 & -2 \\ -8 & 0 & 12 \end{bmatrix} - \begin{bmatrix} -2 & -5 & 0 \\ 1 & -3 & -4 \end{bmatrix} =$
$\begin{bmatrix} 8 & 9 & -2 \\ -9 & 3 & 16 \end{bmatrix}$

9. $\frac{4}{8} = \frac{1}{2}$　　　**10.** $\frac{5}{8}$　　　**11.** $\frac{2}{8} = \frac{1}{4}$

12. There are 3 odd numbers and 3 even. So, the odds of getting an odd number are 3 to 3 = 1 to 1 or 1:1.

13. data in order: 26, 29, 48, 49, 73
mean $= \dfrac{26 + 29 + 48 + 49 + 73}{5} = \dfrac{225}{5} = 45$
median $= 48$　　　mode: there is none
range: $73 - 26 = 47$

14. range $= 25 - 17 = 8$　　**15.** median $= 19.5$

16. third quartile $= 22$

17. the percent of the data between 17 and 18 is 25%.

18. $y = 4x - 3$
x-intercept: Let $y = 0$; $4x - 3 = 0$; $4x = 3$; $x = \frac{3}{4}$
y-intercept $= -3$　　slope: $m = 4$

19. $2x + 3y = 7$; $3y = -2x + 7$; $y = -\frac{2}{3}x + \frac{7}{3}$
x-intercept: Let $y = 0$; $2x = 7$; $x = \frac{7}{2}$
y-intercept $= \frac{7}{3}$　　slope: $m = -\frac{2}{3}$

20.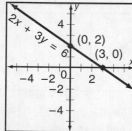

$2x + 3y = 6$;
x-intercept: Let $y = 0$;
$2x = 6$; $x = 3$;
y-intercept: Let $x = 0$;
$3y = 6$; $y = 2$

21.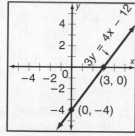

$3y = 4x - 12$;
x-intercept: Let $y = 0$;
$0 = 4x - 12$, $4x = 12$;
$x = 3$
y-intercept: Let $x = 0$;
$3y = -12$; $y = -4$

22.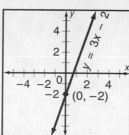

$y = 3x - 2$;
slope: $m = 3$;
y-intercept $= -2$

23.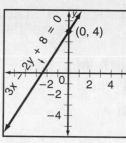

$3x - 2y + 8 = 0$;
$2y = 3x + 8$;
$y = \frac{3}{2}x + 4$;
slope: $m = \frac{3}{2}$;
y-intercept $= 4$

24. negative; If the area is fixed then as the base increases, the height decreases.

25. positive; As the sides increase the perimeter increases.

26. 1.00×10^{-24}

Standardized Test, page 55

1. B; $\frac{z}{3} = \frac{1}{3}$; $z = 1$
$3 \cdot \frac{z}{3} = \frac{1}{3} \cdot \frac{1}{3}$; $z = \frac{1}{9}$
The equations do not have the same solution. Therefore, they are not equivalent.

2. D; $35 - 20 = 15$　　　　**3.** A
$\frac{15}{20} = \frac{3}{4}$; $\frac{3}{4} \times 100 = 75\%$

4. B; $\frac{1}{1,000,000,000} = 1.0 \times 10^{-9}$

5. D; $B - 2A = \begin{bmatrix} 3 & -2 & -1 \\ -1 & 1 & -3 \end{bmatrix} - \begin{bmatrix} -4 & 2 & 8 \\ 6 & 0 & -6 \end{bmatrix}$
$= \begin{bmatrix} 7 & -4 & -9 \\ -7 & 1 & 3 \end{bmatrix}$

6. B　　　　　　　**7.** D; $\frac{5}{10} = \frac{1}{2}$

8. A; mean $= \dfrac{1 + 2 + 4 + 6 + 7}{5} = \dfrac{20}{5} = 4$
median $= 4$
mode: there is none

9. E; $2y = 3x - 5$; $y = \frac{3}{2}x - \frac{5}{2}$
slope: $m = \frac{3}{2}$; y-intercept $= -\frac{5}{2}$

10. C; $3y = 5x - 4$; $y = \frac{5}{3}x - \frac{4}{3}$; $m = \frac{5}{3}$

Chapter 2 Real Numbers, Equations, and Inequalities

Data Activity, pages 56–57

1. No. 6; 4.115 mm is approximately half of 8.252 mm.

2. $A = \pi r^2$; Area of No. 0 $= \pi \left(\dfrac{8.252}{2}\right)^2 \approx 53.48 \text{ mm}^2$;

   Area of No. 6 $= \pi \left(\dfrac{4.115}{2}\right)^2 \approx 13.30 \text{ mm}^2$ The area of No. 0 wire is approximately 4 times the area of No. 6 wire.

3. No. 8; 2.061 Ω/km is approximately 4 times 0.502 Ω/km

4. $4.132 \ \Omega/\text{km} = 4.132 \ \Omega/\text{km}\left(\dfrac{1 \text{ km}}{1000 \text{ m}}\right) =$

   $4.132 \times 10^{-3} \ \Omega/\text{m}$

5. Let x represent the number of meters of wire used.

   Use a proportion to find x. $\dfrac{1 \ \Omega}{x} = \dfrac{1.634 \ \Omega}{1 \text{ km}}$;

   $\dfrac{1 \ \Omega}{x} = \dfrac{1.634 \ \Omega}{1000 \text{ m}}$

   $1.634 \ x = 1000;\ x = \dfrac{1000}{1.634} \approx 612.0 \text{ m}$

6. Answers will vary. Students may make a scatterplot and determine the linear regression equation or they may find an average value for the change in diameter and use the point-slope or slope-intercept form of the equation. Have them compare with the actual value of 0.255 for number 30 wire.

Lesson 2.1, pages 59–61

THINK BACK/WORKING TOGETHER

For Exercises **1.–4.** lengths chosen will vary.

5. Explanations will vary. One possible explanation for why the triangles are congruent involves SAS and the fact that integer length sides are uniquely determined.

6. Drawing will vary.

EXPLORE

7. $\sqrt{2}$; Using the Pythagorean theorem, the length of the hypotenuse is $\sqrt{1^2 + 1^2} = \sqrt{2}$.

8. The length of $\overline{AD}$ is equal to the length of $\overline{AC}$. Point D is located at $\sqrt{2}$ on the number line. Identify D as $\sqrt{2}$.

9. $\sqrt{3}$; Using the Pythagorean theorem, the length of the hypotenuse is $\sqrt{1^2 + (\sqrt{2})^2} = \sqrt{1 + 2} = \sqrt{3}$

10. The length of AF is equal to the length of AE. Point F is located at $\sqrt{3}$ on the number line. Identify F as $\sqrt{3}$.

MAKE CONNECTIONS

11. $\sqrt{4} = 2$ 12. $\sqrt{5} \approx 2.24$ 13. $\sqrt{6} \approx 2.45$

14. $\sqrt{7} \approx 2.65$ 15. $\sqrt{8} \approx 2.83$ 16. $\sqrt{9} = 3$

17. $\sqrt{10} \approx 3.16$

18. Use the measure of $\sqrt{2}$ twice. $2\sqrt{2} \approx 2.83$.

19. From $\sqrt{2}$ measure $\sqrt{3}$ to the right. $\sqrt{3} + \sqrt{2} \approx 3.15$

20. From $\sqrt{10}$ measure 3 units to the left.
    $\sqrt{10} - 3 \approx 0.16$

21. From 3 measure $\sqrt{8}$ to the right. $3 + \sqrt{8} \approx 5.83$

SUMMARIZE

22. Consider the point of the compass as the center of a circle. The hypotenuse is the radius of the circle. The distance along any line segment from the center of the circle to a point on the circle is equal to the length of the radius. Therefore, the two line segments are equal lengths, the length of the radius.

23a. fourth triangle: 1, $\sqrt{4}$, $\sqrt{5}$
     fifth triangle: 1, $\sqrt{5}$, $\sqrt{6}$
     sixth triangle: 1, $\sqrt{6}$, $\sqrt{7}$

23b. eighth triangle: hypotenuse $= \sqrt{9} = 3$

23c. If the hypotenuse is $\sqrt{13}$ the triangle is the twelfth triangle. The nth triangle has a hypotenuse equal to $\sqrt{n + 1}$

24. The integers 4, 5, 6, 7, 8, and 9 have square roots from 2 to 3. There are an infinite number of square roots of real numbers from 2 to 3.

25. Answers will vary. Choose n such that $16 < n < 25$. For example $n = 17$. Since $16 < 17 < 25$, $\sqrt{16} < \sqrt{17} < \sqrt{25}$ or $4 < \sqrt{17} < 5$.

26. Answers will vary. If the number is a perfect square then its square root is a whole number. If the number is not a perfect square then its decimal equivalent does not repeat or terminate.

Lesson 2.2, pages 62–67

1. Drawing will vary. This is one possibility.

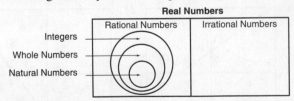

2. Answers will vary. Some examples are: -2 belongs to Integers, Rational numbers, and Real numbers; 6 belongs to Natural Numbers and Whole Numbers; and $\frac{1}{2}$ belongs to Rational Numbers and Real Numbers.

TRY THESE

1. rational, integer

2. rational, integer, whole number, natural number

3. rational   4. irrational   5. rational   6. rational

7. irrational 8. rational, integer

9. Irrational; 11 is not a perfect square; the number cannot be written as the ratio of two integers.

10. Rational; it can be written as the ratio $\frac{12}{13}$.

11. Irrational; 24 is not a perfect square; the number cannot be written as the ratio of two integers.

12. Rational; it can be written as $\frac{20}{1}$.

13. $\sqrt{14} \approx 3.74, \frac{15}{4} = 3.75$ so, $\sqrt{14} < \frac{15}{4}$.

14. $\sqrt{0.38} \approx 0.62$ so, $0.\overline{6} > \sqrt{0.38}$

15. $-\frac{41}{13} \approx -3.15, -\pi \approx -3.14$ so, $-\frac{41}{13} < -\pi$.

16. $\frac{7}{2} = 3.5$ so, $3.505005000\ldots > \frac{7}{2}$

For Exercises 17.–21. answers will vary. Some answers are given.

17. $\sqrt{87} \approx 9.3273$; so, any number between 9.3263 and 9.3273, such as 9.327

18. $\frac{19}{22} \approx 0.864$; so, any number between 0.864 and 0.87, such as 0.865

19. $\frac{26}{33} \approx 0.788$ and $\frac{35}{44} \approx 0.795$; so, any number between 0.788 and 0.795, such as 0.79

20. any number between 18.12121 and 18.12131, such as 18.12123

21. No. Integers include negative numbers.

22. Yes. Every integer can be written as the ratio of itself and 1.

23. $3\frac{1}{7} \approx 3.1428, \pi \approx 3.1415, 3\frac{11}{71} \approx 3.1408$ Therefore, $3\frac{11}{71} < \pi < 3\frac{1}{7}$.

24. Justin's; $\sqrt{148} \approx 12.1656$ is closer to $12\frac{1}{6} \approx 12.1667$ than 12.2

PRACTICE

1. rational   2. rational   3. irrational   4. rational

5. rational

6. rational, integer, whole number, natural number

7. rational, integer   8. rational

9. $-\sqrt{225}$ is rational. It can be written as the ratio of two integers, $\frac{-15}{1}$.

10. $\sqrt{\frac{94}{64}}$ is irrational because 94 is not a perfect square.

11. 0.75 is rational. It can be written as the ratio of two integers, $\frac{3}{4}$.

12. Let $x = -6.\overline{3}$
    $10x = -63.\overline{3}$ multiply both sides by 10
    $9x = -57$ subtract the first equation from the second
    $x = \frac{-57}{9} = \frac{-19}{3}$ divide both sides by 9
    Therefore, $-6.\overline{3}$ is rational. It can be written as the ratio of two integers.

13. $5\frac{1}{25}$ is rational. It can be written as the ratio of two integers, $\frac{126}{25}$.

14. Let $x = 0.5656\ldots$
    $100x = 56.5656\ldots$ multiply both sides by 100
    $99x = 56$ subtract the first equation from the second
    $x = \frac{56}{99}$
    Therefore, $0.5656\ldots$ is rational because it can be written as the ratio of two integers.

15. $-\sqrt{29}$ is irrational because 29 is not a perfect square.

16. $-\frac{79}{119}$ is rational because it can be written as the ratio of two integers, $-\frac{79}{119}$.

17. Let $x = 0.451451\ldots$
    $1000x = 451.451451\ldots$ multiply both sides by 1000
    $999x = 451$ subtract the first equation from the second
    $x = \frac{451}{999}$ divide both sides by 999
    Therefore, $0.451451\ldots x = \frac{451}{999}$

18. Since $\frac{122}{20} = 6.1, 6.090909 < \frac{122}{20}$.

19. Since $\sqrt{108} \approx 10.392, \sqrt{108} > 10.\overline{35}$

In Exercises 20.–22. answers will vary. Here are some possible answers.

20. $\sqrt{15} \approx 3.873$ so, 3.875 is between $3.\overline{8}$ and $\sqrt{15}$

21. $-\frac{17}{21} \approx -0.8095$ so, -0.8085 is between $-\frac{17}{21}$ and $-0.808080\ldots$

22. $\frac{539}{20} = 26.95$ so, 26.944 is between 26.94385073 and $\frac{539}{20}$.

23. Yes, it is about 6.7082, which is between $6.6666\ldots$ and 6.75.

24. $\frac{5}{6} = 8.333\ldots, \frac{7}{8} = 0.875$; $8.3333\ldots$ is less than 0.875 and 0.9, so by the trichotomy property, it is not between them.

25. $\frac{256}{81} \approx 3.1604$, $\sqrt{10} \approx 3.1623$ Therefore, $\pi \approx 3.14159$ is less than either $\frac{256}{81}$ or $\sqrt{10}$.

26. $\frac{7}{5} = 1.4$, $\frac{10}{8} = 1.25$ Therefore 5 in $\times$ 7 in is closer to 1.618.

EXTEND

27. $\sqrt{5} \approx 2.236$, $-\sqrt{10} \approx -3.162$, $-2\frac{2}{3} \approx -2.667$, $-\frac{215}{98} \approx 2.1939$ Therefore, the order is $-\sqrt{10}$, $-2\frac{2}{3}$, -2.6, $-\sqrt{5}$, -2.216215214, $-\frac{215}{98}$, $2.\overline{23}$, $\sqrt{5}$, 2.24

28. $\frac{9}{10} = 0.9$, $\sqrt{2} \approx 1.414$, $\sqrt{\frac{16}{25}} = \frac{4}{5} = 0.8$ Therefore, the order is -1.1, $0.6979989\ldots$, $\sqrt{\frac{16}{25}}$, $0.888\ldots$, $\frac{9}{10}$, $1.\overline{3}$, $\sqrt{2}$

29. $\sqrt{\frac{49}{100}} = \frac{7}{10} = 0.7$, $\frac{3}{4} = 0.75$, $\sqrt{11} \approx 3.317$.
Therefore, the order is -5.678, -0.5, -0.055, 0.5, $\sqrt{\frac{49}{100}}$, $\frac{3}{4}$, $\sqrt{11}$, $5.\overline{5}$.

30. Answers will vary; sample answer:

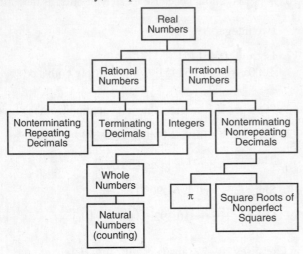

31. The set of rational numbers and irrational numbers have no numbers in common because either a number can be expressed as the ratio of two integers or it cannot.

32. Answers will vary. Heights are real numbers. For any two heights h_1 and h_2, exactly one of $h_1 < h_2$, $h_1 = h_2$, or $h_1 > h_2$ is true.

33. Wrought iron $<$ steel $<$ cast iron

34. $3\sqrt{16}\sqrt{81}\sqrt{49} = 3 \cdot 4 \cdot 9 \cdot 7 = 756 > 481$
So, $B > h$.

THINK CRITICALLY

35. No. Select any two consecutive whole numbers. No other whole numbers exist between them. Therefore, the set of whole numbers is not dense.

36. The horizontal lines from bottom to top are x-axis, y_1, y_2, y_7, y_3, y_8, y_5, y_6, and y_4.

37. Because $b = a$, $b - a = 0$, so dividing both sides of the equation by $b - a$ is division by 0. Division by 0 is undefined.

MIXED REVIEW

38. There are $9 \times 5 = 45$ keys, 10 of which are integers 0 through 9. So, the probability of pressing one of the integers is $\frac{10}{45}$. Therefore, the odds of pressing an integer is $\frac{10}{45 - 10}$ or 10:35.

39. You have 2 chances out of 45. So, the probability is $\frac{2}{45}$.

40. C; 2.8 appears on the list 2 times and the other numbers appear once each.

41. B; If the numbers are listed in order, 6.7 and 12.7 are the middle numbers in the list and their average is $\frac{6.7 + 12.7}{2} = 9.7$

42. $V = lwh$ Divide each side by wh to get $\frac{V}{wh} = l$.
So, $l = \frac{V}{wh}$

43. Use the result of Exercise 42. $l = \frac{V}{wh} = \frac{160.875}{(5.5)(4.5)} = 6.5$ m

44. $-0.4 = -\frac{4}{10} = -\frac{2}{5}$ Since -0.4 can be written as the ratio of two integers, it is rational.

45. $\sqrt{89}$ is irrational because 89 is not a perfect square.

46. Let $x = 6.77\ldots$
$10x = 67.77\ldots$
$9x = 67$
$x = \frac{67}{9}$

So, 6.77 is rational because it can be written as the ratio of two integers.

47. $0.118119120\ldots$ is non-repeating and non-terminating. Therefore, it is irrational.

Lesson 2.3, pages 68–73

EXPLORE

1. Yes; for any two elements p and q in set A, $p \$ q = q \$ p$.

2. Yes; for any two elements p and q in set A, $p \# q = q \# p$.

3. Yes; y is the identity element for $\$$ because y operated with any element yields that same element.

4. Whenever z is used with another element, the result is the other element. So, z is an identity element.

5. $x \# y = z$ **6.** $y \# x = z$ **7.** $z \# z = z$

8. $y \$ (x \# z) = y \$ x = x$ and
$(y \$ x) \# (y \$ z) = x \# z = x$ Therefore,
$y \$ (x \# z) = (y \$ x) \# (y \$ z)$

9. no; $y \# (x \$ z) = y \# y = x$ and
$(y \# x) \$ (y \# z) = z \$ y = z$

TRY THESE

1. $2.5 \cdot 6 - 108 \div 6^2 = 2.5 \cdot 6 - 108 \div 36 =$
$15 - 3 = 12$

2. $\dfrac{\sqrt{6 + 10}}{2^3} = \dfrac{\sqrt{16}}{2^3} = \dfrac{\sqrt{16}}{8} = \dfrac{4}{8} = \dfrac{1}{2}$

3. $\dfrac{13^2 + 20}{(5 + 2)4} = \dfrac{169 + 20}{7(4)} = \dfrac{189}{28} = 6.75$

4. $\dfrac{72 - 12}{15} + \left(\dfrac{1}{5} \cdot 5^2 \cdot 3\right) = \dfrac{60}{15} + \left(\dfrac{1}{5} \cdot 25 \cdot 3\right) =$
$\dfrac{60}{15} + 15 = 4 + 15 = 19$

5. $12 \div 3 + 3 \cdot 2 - 1 = 4 + 6 - 1 = 9$

6. $(7 - 0.5) - [(5) \cdot (4 - 9)] = 6.5 - (5 \cdot (-5)) =$
$6.5 - (-25) = 6.5 + 25 = 31.5$

7. identity property of multiplication

8. distributive property

9. associative property of multiplication

10. commutative property of multiplication

11. inverse property of multiplication

12. identity property of addition

13. No; the commutative property is not true for
subtraction. For example $6 - 2 = 4 \neq -4 = 2 - 6$.
No; the commutative property is not true for division.
For example $12 \div 3 = 4 \neq \dfrac{1}{4} = 3 \div 12$

14. $9 \times 8 + (7 \div 5)$ or $(9 \times 8) + 7 \div 5$ or
$(9 \times 8 + 7 \div 5) = 73.4$

15. If the right parentheses after 16 is removed,
the expression becomes
$162/2 * (-45 + 5) + 16 * (3 + 2.8)$

16. $\dfrac{\left(0.45 + \dfrac{1}{2}\right) \times 10^4}{1.7 \times 2} = \dfrac{0.95 \times 10^4}{3.4} = \dfrac{9500}{3.4} \approx 2794$ kg

PRACTICE

1. $\sqrt{49} + 3(4.5 + 2.75) = 7 + 3(7.25) =$
$7 + 21.75 = 28.75$

2. $4.8 \div (1.44 \div 1.2) \cdot 2^4 =$
$4.8 \div (1.2) \cdot 16 = 4 \cdot 16 = 64$

3. $(17 - 3)^2 + \dfrac{15}{16} = (14)^2 + \dfrac{15}{16} =$

$196 + \dfrac{15}{16} = 196\dfrac{15}{16}$

4. $[192 \div (7 + 5)]^2 - 200 = [192 \div 12]^2 - 200 =$
$16^2 - 200 = 256 - 200 = 56$

5. $45 \div 5 + 7 \cdot \sqrt{9} - 6^2 = 45 \div 5 + 7 \cdot 3 - 36 =$
$9 + 21 - 36 = -6$

6. $2^4 + 10^2 \cdot (46 \div 23) - 0.77 =$
$16 + 100 \cdot (2) - 0.77 =$
$16 + 200 - 0.77 = 215.23$

7. identity property of addition

8. identity property of multiplication

9. inverse property of multiplication

10. commutative property of addition

11. associative property of multiplication

12. distributive property

13. The associative property is not true for subtraction.
Here is a counterexample to illustrate why not.
$7 - (5 - 2) = 7 - 3 = 4$,
$(7 - 5) - 2 = 2 - 2 = 0$. Since $4 \neq 0$,
$7 - (5 - 2) \neq (7 - 5) - 2$ or in general,
$(a - b) - c \neq a - (b - c)$ The associative
property is not true for division. Here is a
counterexample. $(40 \div 20) \div 2 = 2 \div 2 = 1$,
$40 \div (20 \div 2) = 40 \div 10 = 4$. Since $1 \neq 4$,
$(40 \div 20) \div 2 \neq 40 \div (20 \div 2)$ or in general,
$(a \div b) \div c \neq a \div (b \div c)$

14. $A = \pi d^2 \div 4 = \pi(18)^2 \div 4 \approx 254.5$ ft^2

15. $A = \pi(36)^2 \div 4 \approx 1017.9$ ft^2

16. Answers will vary. Here is one possibility. Use the
formula $A = \pi d^2 \div 4$. Substitute 2 times the answer to
Exercise 14 for A and solve for d. $2(254) = \pi d^2 \div 4$;
$\pi d^2 = 4(2)(254)$; $d^2 = \dfrac{4(2)(254)}{\pi}$;

$d = \sqrt{\dfrac{4(2)(254)}{\pi}} \approx 25$ ft

EXTEND

17. $9x + 4(2x^2 - 7x) = 9x + 8x^2 - 28x = 8x^2 - 19x$

18. $2\left(\dfrac{7a + 11a}{3}\right) - 22 = 2\left(\dfrac{18a}{3}\right) - 22 =$
$2(6a) - 22 = 12a - 22$

19. $3^2 + [\sqrt{25} + (5r - 3r) + 1] =$
$9 + [5 + 2r + 1] = 9 + [6 + 2r] = 2r + 33$

20. $56 - \dfrac{1}{2}(6c + 8d) + 22 = 56 - 3c - 4d + 22 =$
$\qquad\qquad\qquad\qquad\qquad -3c - 4d + 78$

21. $\dfrac{a(9 + 5)}{7a} + 2a = \dfrac{14a}{7a} + 2a = 2a + 2$

22. $-32h + 4^2 \cdot h - \sqrt{36} = -32h + 16h - 6 =$
$-16h - 6$

23. Answers may vary. Here is one possibility:
$2(10 \times 36) + 2\left[\dfrac{1}{2}(10 \times 36)\right]$

24. $2(10 \times 36) + 2\left[\frac{1}{2}(10 \times 36)\right] = 720 + 360$

$= 1080$ in.^{2}

25. 1 ft$^2 = (12$ in.$) \times (12$ in.$) = 144$ in.^{2} Therefore,

1080 in$^2 = \frac{1080}{144}$ ft$^2 = 7.5$ ft^2

THINK CRITICALLY

26. $1 + 2^4 + 8 - 6\sqrt{2.2 + 1.8} \div 3 + 9 =$

$1 + 16 + 8 - 6\sqrt{4} \div 3 + 9 =$

$1 + 16 + 8 - 6(2) \div 3 + 9 =$

$1 + 16 + 8 - 12 \div 3 + 9 =$

$1 + 16 + 8 - 4 + 9 = 30$

27. $(1 + 2)^4 + 8 - 6\sqrt{2.2 + 1.8} \div (3 + 9)$

28. $1 + 2^4 + (8 - 6)(\sqrt{2.2 + 1.8} \div 3) + 9$

29. Doubling the diameter results in 4 times the area. This can be seen by substituting $2d$ for d in the formula $A = \pi d^2 \div 4$. This gives $A = \pi(2d)^2 \div 4 = \pi(4d^2) \div 4 = 4(\pi d^2) \div 4$.

30. When any number in the set is multiplied by any number in the set, even itself, the product is a number in the set. When any number in the set is added to any number in the set, including itself, the result is not always a number in the set. For example: $1 + 1 = 2$ and 2 is not in the set.

31. Answers may vary. One possible answer is

$(9 \cdot 3 \cdot 3) + (4 \cdot 4 \cdot 3) + \left(\frac{1}{2} \cdot 3 \cdot 3\right) +$

$\left(\frac{1}{2} \cdot 4 \cdot 3\right) - \left(\frac{2}{3} \cdot 3 \cdot 3\right) - \left(\frac{1}{3} \cdot 3 \cdot 3\right)$

$- \left(\frac{2}{3} \cdot 3 \cdot 3\right) - \left(\frac{1}{6} \cdot 3 \cdot 3\right)$

32. $81 + 48 + 4\frac{1}{2} + 6 - 6 - 3 - 6 - 1\frac{1}{2} = 123$ in.^{2}

33. If each dimension is doubled then the area is multiplied by 4, $4 \cdot 123$.

34. $4 \cdot 123 = 492$ in.^{2} or $(9 \cdot 6 \cdot 6) + (4 \cdot 8 \cdot 6) +$

$\left(\frac{1}{2} \cdot 6 \cdot 6\right) + \left(\frac{1}{2} \cdot 8 \cdot 6\right) - \left(\frac{2}{3} \cdot 6 \cdot 6\right) - \left(\frac{1}{3} \cdot 6 \cdot 6\right) -$

$\left(\frac{2}{3} \cdot 6 \cdot 6\right) - \left(\frac{1}{6} \cdot 6 \cdot 6\right) = 324 + 192 + 18 + 24 -$

$24 - 12 - 24 - 6 = 492$ in.^{2}

MIXED REVIEW

35. $\begin{bmatrix} -1 & 1 \\ 0 & 1 \end{bmatrix} - \begin{bmatrix} 4 & 4 \\ 0 & 3 \end{bmatrix} = \begin{bmatrix} -1 - 4 & 1 - 4 \\ 0 - 0 & 1 - 3 \end{bmatrix} = \begin{bmatrix} -5 & -3 \\ 0 & -2 \end{bmatrix}$

36. $\begin{bmatrix} -6 & 9 \\ -2 & 12 \end{bmatrix} + \begin{bmatrix} -6 & 5 \\ 10 & -17 \end{bmatrix} = \begin{bmatrix} -6 - 6 & 9 + 5 \\ -2 + 10 & 12 + (-17) \end{bmatrix}$

$= \begin{bmatrix} -12 & 14 \\ 8 & -5 \end{bmatrix}$

37. D. distributive property **38.** rational, integer

39. rational **40.** irrational

41. rational, integer, whole number, natural number

ALGEBRAWORKS

1. $T_f = \frac{F_w T_w + F_r T_r}{F_w + F_r} = \frac{0.35(26) + 0.88(21)}{0.35 + 0.88} = 22.4°$C. Since this is greater than 22°C, the answer is no.

2. Let $T_f = 22°$C (the maximum temperature that sustains trout) and solve for T_w. $22 = \frac{0.35 \cdot T_w + 0.88(21)}{0.35 + 0.88}$;

$22(0.35 + 0.88) = 0.35 T_w + 0.88(21)$;

$22(0.35 + 0.88) - 0.88(21) = 0.35 T_w$;

$T_w = \frac{22(0.35 + 0.88) - 0.88(21)}{0.35} \approx 24.5°$C

3. $\frac{F_w + 0.88(21)}{F_w + 0.88} = 22$

$F_w(26) + 0.88(21) = 22(F_w + 0.88)$

$26(F_w + 18.48 = 22 F_w + 19.36$

$4 F_w = 0.88$

$F_w = 0.22$

Therefore the flow of waste should be reduced to 0.22 m^3/s

4. $y = \frac{0.35x + (0.88)(21)}{0.35 + 0.88} = \frac{0.35x + 18.48}{1.23}$

5. The equation in Question 4. is a linear equation.

6. x represents the temperature of the waste that will result in a maximum sustainable value of 22°C. $x \approx 24.5$

Lesson 2.4, pages 74–81

EXPLORE

1. First find the sum of the digits for each number.

$4 + 7 = 11, 8 + 1 = 9, 3 + 4 = 7, 2 + 5 = 7,$

$7 + 0 = 7, 5 + 4 = 9, 3 + 8 = 11, 1 + 8 = 9.$

Now sort the set into subsets. $\{38, 47\}$; $\{18, 81, 36, 54\}$; $\{34, 70, 25\}$

2. Yes, any element in S has the same sum as itself.

3. Yes, if any number has the same sum as another, then the second has the same sum as the first.

4. Yes, if one number has the same sum as a second number, and the second has the same sum as a third number, then the first will have the same sum as the third.

1. $3\left(\frac{1}{2}x - 7\right) < -15$

$\quad \frac{1}{2}x - 7 < -5 \qquad$ division property

$\qquad \frac{1}{2}x < 2 \qquad$ addition property

$\qquad\quad x < 4 \qquad$ multiplication property

$-\frac{1}{3}(x + 7) \le -2$

$\qquad x + 7 \ge 6 \qquad$ multiplication property

$\qquad\quad x \ge -1 \qquad$ subtraction property

2. $6x + 8.2 = -12.6 + 10x; 6x + 20.8 = 10x;$
$20.8 = 4x; 5.2 = x; x = 5.2$
Check: $6(5.2) + 8.2 \stackrel{?}{=} -12.6 + 10(5.2)$
$\qquad 31.2 + 8.2 \stackrel{?}{=} -12.6 + 52$
$\qquad\qquad 39.4 = 39.4 ✓$

3. $2x - 4(3x - 5) = 5(-x + 2);$
$2x - 12x + 20 = -5x + 10;$
$-10x + 20 = -5x + 10; -5x + 20 = 10;$
$-5x = -10; x = 2$
Check: $2(2) - 4(3 \cdot 2 - 5) \stackrel{?}{=} 5(-2 + 2)$
$\qquad 4 - 4(6 - 5) \stackrel{?}{=} 5(0)$
$\qquad\qquad 4 - 4(1) \stackrel{?}{=} 0$
$\qquad\qquad\qquad 0 = 0 ✓$

4. $\frac{5}{8}x - 9 = 7(x + 3); \frac{5}{8}x - 9 = 7x + 21;$

$\frac{5}{8}x = 7x + 30; \left(-6\frac{3}{8}\right)x = 30;$

$x = 30\left(-\frac{8}{51}\right) = -\frac{240}{51} = -4\frac{36}{51} = -4\frac{12}{17}$

Check: $\frac{5}{8}\left(-4\frac{12}{17}\right) - 9 \stackrel{?}{=} 7\left(-4\frac{12}{17} + 3\right)$

$\qquad \frac{5}{8}\left(-\frac{80}{17}\right) - 9 \stackrel{?}{=} 7\left(-1\frac{12}{17}\right)$

$\qquad\quad -\frac{50}{17} - 9 \stackrel{?}{=} 7\left(-\frac{29}{17}\right)$

$\qquad\qquad\quad -\frac{203}{17} = \frac{203}{17} ✓$

5. $\frac{5t + 80}{3t} = 3; 5t + 80 = 9t; 80 = 4t; 20 = t; t = 20$

Check: $\frac{5(20) + 80}{3(20)} = \frac{100 + 80}{60} = \frac{180}{60} = 3 ✓$

6. $4\left(\frac{1}{2}x - 15\right) = 12x + 40; 2x - 60 = 12x + 40;$

$-10x = 100; x = -10$

Check: $4\left[\frac{1}{2}(-10) - 15\right] \stackrel{?}{=} 12(-10) + 40$

$\qquad\quad 4[-5 - 15] \stackrel{?}{=} -120 + 40$

$\qquad\qquad 4(-20) \stackrel{?}{=} -80$

$\qquad\qquad\quad -80 = -80 ✓$

7. $\frac{5(m + 9)}{4} = -3m; 5(m + 9) = -12m;$

$5m + 45 = -12m; 17m = -45; m = -\frac{45}{17}$

Check: $\dfrac{5\left(-\frac{45}{17} + 9\right)}{4} \stackrel{?}{=} -3\left(-\frac{45}{17}\right)$

$\qquad \dfrac{5\left(\frac{108}{17}\right)}{4} \stackrel{?}{=} \frac{135}{17}$

$\qquad\quad \dfrac{\frac{540}{17}}{4} \stackrel{?}{=} \frac{135}{17}$

$\qquad\qquad \frac{135}{17} = \frac{135}{17} ✓$

8. $9.3 - 7k < -6k + 11.2; 9.3 - k < 11.2; -k < 1.9;$
$k > -1.9$

Check a number greater than -1.9.

Try -1. $\quad 9.3 - 7(-1) \stackrel{?}{<} -6(-1) + 11.2$
$\qquad\qquad 9.3 + 7 \stackrel{?}{<} 6 + 11.2$
$\qquad\qquad\qquad 16.3 < 17.2 ✓$

9. $2x + 5 \le 5x - 4; -3x + 5 \le -4, -3x \le -9;$
$x \ge 3$

Check a number greater than or equal to 3. Try 4.
$2(4) + 5 \stackrel{?}{\le} 5(4) - 4$
$\quad 8 + 5 \stackrel{?}{\le} 20 - 4$
$\qquad 13 \le 16 ✓$

10. $9 + 5y > -1 \quad$ and $\quad 4y + 15 < 35$
$\qquad 5y \ge -10 \qquad\qquad\qquad 4y < 20$
$\qquad\quad y \ge -2 \quad$ and $\qquad\quad y < 5$
$-2 \le y < 5$

Check a number between -2 and 5. Try 0.
$9 + 5(0) \stackrel{?}{\ge} -1, 9 \ge -1$ and $4(0) + 15 \stackrel{?}{<} 35,$
$15 < 35 ✓$

11. $-c \ge \frac{c + 9}{4}; -4c \ge c + 9; -5c \ge 9; c \le -\frac{9}{5}$

Check a number less than $-\frac{9}{5}$. Try -2.

$-(-2) \stackrel{?}{\ge} \frac{-2 + 9}{4}; 2 \stackrel{?}{\ge} \frac{7}{4}; 2 \ge 1\frac{3}{4} ✓$

12. $6.2x - 0.1 \geq \dfrac{x + 5.5}{0.2}$; $1.24x - 0.02 \geq x + 5.5$;

$0.24x - 0.02 \geq 5.5$; $0.24x \geq 5.52$; $x \geq 23$

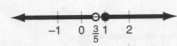

Check a number greater than 23. Try 25.

$6.2(25) - 0.1 \overset{?}{\geq} \dfrac{25 + 5.5}{0.2}$

$154.9 \geq 152.5$ ✓

13. $7n - 6 \geq 1$ or $\dfrac{5n}{3} < 1$

$\qquad 7n \geq 7 \qquad\quad 5n < 3$

$\qquad n \geq 1$ or $n < \dfrac{3}{5}$

Check a number less than $\dfrac{3}{5}$. Try 0. $7(0) - 6 \overset{?}{\geq} 1$ or

$\dfrac{5(0)}{3} \overset{?}{<} 1$; $-6 \geq 1$ or $0 < 1$. ✓

Check a number greater than 1. Try 2. $7(2) - 6 \geq 1$

or $\dfrac{5(2)}{3} < 1$; $8 \geq 1$ or $\dfrac{10}{3} < 1$ ✓

14. no; $-6(6) + 5 = -31 \neq 4(6) - 2 = 22$

15. yes; $\dfrac{5(-7)}{12} = -\dfrac{35}{12} = -2\dfrac{11}{12} \leq 3(-7 + 9) = $

$3(2) = 6$

16. yes; $\dfrac{2(0 + 4)}{3} = \dfrac{2 \cdot 4}{3} = \dfrac{8}{3} \leq 0 + 6 = 6$

17. $t \geq 90°$ **18.** $c > 200$ **19.** $40 \leq p \leq 60$

20. $\$90 \leq t \leq \$110$

21. Let b represent the cost of each binder.
$5b + 6.95 \leq 25.00$; $5b \leq 18.05$; $b \leq \$3.61$

PRACTICE

1. subtraction property of inequality

2. multiplication property of inequality

3. commutative property of addition

4. addition property of inequality

5. $4x + 9 = 2x - 1$; $2x + 9 = -1$; $2x = -10$;
$x = -5$
Check: $4(-5) + 9 \overset{?}{=} 2(-5) - 1$
$\qquad -20 + 9 \overset{?}{=} -10 - 1$
$\qquad\qquad -11 = -11$ ✓

6. $3z + 1 = 4z - 4$; $-z + 1 = -4$; $-z = -5$; $z = 5$
Check: $3(5) + 1 \overset{?}{=} 4(5) - 4$
$\qquad 15 + 1 \overset{?}{=} 20 - 4$
$\qquad\quad 16 = 16$ ✓

7. $\dfrac{2}{5}(3x + 9) = -74 + 2x$;

$\dfrac{5}{2}\left(\dfrac{2}{5}\right)(3x + 9) = \dfrac{5}{2}(-74 + 2x)$;

$3x + 9 = -185 + 5x$; $-2x + 9 = -185$;
$-2x = -194$; $x = 97$

Check: $\dfrac{2}{5}[3(97) + 9] \overset{?}{=} -74 + 2(97)$

$\dfrac{2}{5}[291 + 9] \overset{?}{=} -74 + 194$

$\dfrac{2}{5}(300) \overset{?}{=} 120$

$120 = 120$ ✓

8. $2(4m + 1) = 3(7m - 2)$; $8m + 2 = 21m - 6$;

$8m + 8 = 21m$; $8 = 13m$; $m = \dfrac{8}{13}$

Check: $2\left[4\left(\dfrac{8}{13}\right) + 1\right] \overset{?}{=} 3\left[7\left(\dfrac{8}{13}\right) - 2\right]$

$2\left(\dfrac{32}{13} + 1\right) \overset{?}{=} 3\left(\dfrac{56}{13} - 2\right)$

$2\left(\dfrac{45}{13}\right) \overset{?}{=} 3\left(\dfrac{30}{13}\right)$

$\dfrac{90}{13} = \dfrac{90}{13}$ ✓

9. $\dfrac{1}{4}y - 1 = 2 + \dfrac{1}{5}y$; $\dfrac{1}{4}y = 3 + \dfrac{1}{5}y$; $\dfrac{1}{20}y = 3$; $y = 60$

Check: $\dfrac{1}{4}(60) - 1 \overset{?}{=} 2 + \dfrac{1}{5}(60)$

$15 - 1 \overset{?}{=} 2 + 12$

$14 = 14$ ✓

10. $0.2(4 - 3b) + 0.3b = 2.6$;
$0.8 - 0.6b + 0.3b = 2.6$; $0.8 - 0.3b = 2.6$;
$-0.3b = 1.8$; $b = -6$
Check: $0.2[4 - 3(-6)] + 0.3(-6) = 0.2[4 + 18] -$
$1.8 = 0.2(22) - 1.8 = 4.4. - 1.8 = 2.6$ ✓

11. $0.7x - 0.1 < 2$; $0.7x < 2.1$; $x < 3$

Check a number less than 3.
Try 0. $0.7(0) - 0.1 \overset{?}{<} 2$
$\qquad\qquad -0.1 < 2$ ✓

12. $-1.75x \leq 1.25x - 15$; $-3x \leq -15$; $x \geq 5$

Check a number greater than 5.
Try 10. $-1.75(10) \overset{?}{\leq} 1.25(10) - 15$
$\qquad\qquad -17.5 \overset{?}{\leq} 12.5 - 15$
$\qquad\qquad -17.5 \leq -2.5$ ✓

13. $\frac{x+3}{2} < 1$ or $9x + 2 > 11$

$\quad x + 3 < 2 \qquad\qquad 9x > 9$

$\quad\quad x < -1$ or $\qquad x > 1$

Check a number less than -1.

Try -3. $\frac{-3+3}{2} \overset{?}{<} 1$

$\qquad\qquad 0 < 1$ ✓

Check a number greater than 1.

Try 2. $9(2) + 2 \overset{?}{>} 11$

$\qquad\qquad 20 > 11$ ✓

14. $-6(x+2) \le \frac{1}{3}(9x+15)$; $-6x - 12 \le 3x + 5$;

$-9x \le 17$; $x \ge -\frac{17}{9}$

Check a number greater than $-\frac{17}{9}$.

Try 0. $-6(0+2) \overset{?}{\le} \frac{1}{3}(9(0)+15)$

$\qquad\qquad -12 \le 5$ ✓

15. $\frac{1}{2}x - 2 > 6\left(-\frac{1}{4}x + 9\right)$; $\frac{1}{2}x - 2 > -\frac{3}{2}x + 54$;

$2x - 2 > 54$; $2x > 56$; $x > 28$.

Check a number greater than 28.

Try 32. $\frac{1}{2}(32) - 2 \overset{?}{>} 6\left[-\frac{1}{4}(32) + 9\right]$

$\qquad 16 - 2 \overset{?}{>} 6(-8 + 9)$

$\qquad\qquad 14 > 6$ ✓

16. $\frac{9x}{3} \ge 12$ and $\quad 5(x-3) < 2x$

$\qquad\qquad\qquad\qquad 5x - 15 < 2x$

$3x \ge 12 \qquad\qquad 3x - 15 < 0$

$x \ge 4 \qquad 4 \le x < 5 \qquad x < 5$

Check a number between 4 and 5.

Try $4\frac{1}{2}$. $\frac{9\left(4\frac{1}{2}\right)}{3} \overset{?}{\ge} 12$; $\frac{40\frac{1}{2}}{3} \overset{?}{\ge} 12$; $13\frac{1}{2} \ge 12$ and

$5\left(4\frac{1}{2} - 3\right) \overset{?}{<} 2\left(4\frac{1}{2}\right)$; $5\left(1\frac{1}{2}\right) \overset{?}{<} 9$; $7\frac{1}{2} < 9$ ✓

17. Answers will vary. One possible answer follows. The two types of compound inequalities are the conjunction consisting of two inequalities joined by **and** and the disjunction consisting of two inequalities joined by **or.** Both types are solved by solving each of their two parts separately. For the conjunction the solution is those

numbers that solve both parts simultaneously or the intersection of the two separate solutions. For the disjunction the solution is those numbers that solve either one of the two inequalities or the union of the two separate solutions.

18. b; $-16 - 4x \le 0$ and $\quad 11x - 5 \le 61$

$\qquad\quad -4x \le 16 \qquad\qquad 11x \le 66$

$\qquad\qquad x \ge -4$ and $\qquad\qquad x \le 6$

19. a; $\frac{22x - 6}{4} \le -29$ or $4(3x+7) \ge 64$

$\quad 22x - 6 \le -116 \qquad 12x + 28 \ge 64$

$\quad\quad 22x \le -110 \qquad\qquad 12x \ge 36$

$\qquad x \le -5$ or $\qquad\qquad x \ge 3$

20. c; $\frac{3}{2}x - 9 \le -3$ and $2x > -8$

$\qquad\quad \frac{3}{2}x \le 6$

$\qquad\qquad x \le 4$ and $\quad x > -4$

21. d; $\frac{x}{2} \le -0.5$ or $-2x < -10$

$\qquad x < -1.0$ or $\qquad x > 5$

22. $p \le 28$ **23.** $f \le 40$ **24.** $450 < p < 500$

25. Let x represent the amount Joanna can spend on snacks.
$2(6.25) + x \le 18$; $12.50 + x \le 18$; $x \le \$5.50$

26. Let x represent the least amount Stan must earn.
$\frac{x}{3} + 45 \ge 389$; $\frac{x}{3} \ge 344$; $x \ge \$1032$

EXTEND

27. R is reflexive because every employee's wage is the same as his or her wage. R is symmetric; if worker a makes the same as worker b, worker b makes the same as worker a. Finally, R is transitive; if r makes the same as s, and s makes the same as t, then r must make the same as t.

28. $\frac{3}{4}(2n+9) > \frac{3}{8}$ and $\quad \frac{15n}{6} + 1 \le -9$

$\quad 2n + 9 > \frac{1}{2} \qquad\qquad\quad \frac{15n}{6} \le -10$

$\quad\quad 2n > -8\frac{1}{2} \qquad\qquad\qquad n \le -4$

$\quad\quad\quad n > -4\frac{1}{4}$

$\quad -4\frac{1}{4} < n \le -4$

Check a number between $-4\frac{1}{4}$ and -4.

Try $n = -4\frac{1}{8} = -\frac{33}{8}$

$\frac{3}{4}\left[2\left(-\frac{33}{8}\right) + 9\right] = \frac{3}{4}\left[-\frac{33}{4} + 9\right] = \frac{3}{4}\left(\frac{3}{4}\right) = \frac{9}{16} > -4\frac{1}{4}$

and $\frac{15\left(-\frac{33}{8}\right)}{6} + 1 = -\frac{\frac{495}{8}}{6} + 1 = -\frac{495}{48} + 1 =$

$-\frac{447}{48} = -9\frac{5}{16} \le -4$ ✓

29. $15 < 7x + 9 < 26; 6 < 7x < 17; \frac{6}{7} < x < \frac{17}{7};$

$\frac{6}{7} < x < 2\frac{3}{7}$

Check a number between $\frac{6}{7}$ and $2\frac{3}{7}$. Try $x = 1$.

$7x + 9 = 7(1) + 9 = 16$ and $15 < 16 < 26.$ ✓

30. $14\frac{5}{8} \leq \text{IPP} \leq 16\frac{3}{4}$ **31.** $55 \leq \text{BNN} \leq 56\frac{1}{2}$

32. $28\frac{1}{8} \leq \text{HDD} \leq 35\frac{1}{8}$

33. Let x represent the least amount Mr. Stanford should invest in bonds. So, $5000 - x$ is the amount he will invest in a bank certificate.
$0.08x + 0.06(5000 - x) \geq 380;$

$0.08x + 300 - 0.06x \geq 380; 0.02x \geq 80; x \geq \$4000$

THINK CRITICALLY

34. false; it is the union

35. true   **36.** true   **37.** true

38. Examples will vary. There are no numbers that meet both conditions of the inequality. A possible example is $x < 2$ and $x > 3$.

39. Examples will vary. All numbers satisfy one inequality or the other. A possible example is $x > 2$ or $x < 3$.

40. $0; a + 0 = 0 + a = a$ for any real number a

41. $1; a \cdot 1 = 1 \cdot a = a$ for any real number a

42. $\frac{1}{a}; a\left(\frac{1}{a}\right) = \frac{1}{a}(a) = 1$ for any real number a

43. $-a; a + (-a) = -a + a = 0$ for any real number a.

44. D; $0.262262226\ldots$ is nonrepeating and nonterminating.

45. $21 - 8x = 6x + 49; 21 - 14x = 49; -14x = 28$
$x = 2$
Check: $21 - 8(-2) \overset{?}{=} 6(-2) + 49$
$21 + 16 \overset{?}{=} -12 + 49$
$37 = 37$ ✓

46. $\frac{x + 5}{8} = x - 9; x + 5 = 8x - 72; -7x = -77;$

$x = 11$

Check: $\frac{11 + 5}{8} \overset{?}{=} 11 - 9$

$\frac{16}{8} \overset{?}{=} 2$

$2 = 2$ ✓

47. $3(2x - 4) = -5x + 13; 6x - 12 = -5x + 13;$
$11x - 12 = 13; 11x = 25; x = \frac{25}{11}$

Check: $3\left[2\left(\frac{25}{11}\right) - 4\right] \overset{?}{=} -5\left(\frac{25}{11}\right) + 13$

$3\left(\frac{50}{11} - 4\right) \overset{?}{=} -\frac{125}{11} + 13$

$3\left(\frac{6}{11}\right) \overset{?}{=} \frac{18}{11}$

$\frac{18}{11} = \frac{18}{11}$ ✓

48. The range of the prices of hotel rooms is \$35 to \$150. The medium price is \$90 and $\frac{1}{4}$ of the rooms are \$50 or less per night and $\frac{1}{4}$ are \$105 per night or more.

ALGEBRAWORKS

1. $E = kL(T - t) = 0.000012(3000)(105 - 68) = 1.332$ ft

$\frac{1.332}{3000} \times 100 = 0.044\%$

2. at 97°F $E = 0.000012(3000)(97 - 68) = 1.044$ ft
and at 70°F
$E = 0.000012(3000)(70 - 68) = 0.072$ ft

3. $0.072 \leq E \leq 1.044$

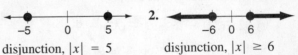

Lesson 2.5, pages 82–88

EXPLORE

1.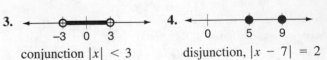
disjunction, $|x| = 5$

2.
disjunction, $|x| \geq 6$

3.
conjunction $|x| < 3$

4.
disjunction, $|x - 7| = 2$

TRY THESE

1. $-9 < x + 4$ and $x + 4 < 9$

2. $x - 3 < -6$ or $x - 3 > 6$

3. $2.5x - 1 \leq -5.2$ or $2.5x - 1 \geq 5.2$

4. $\frac{1}{2}x - 5 \leq 7$ and $\frac{1}{2}x - 5 \geq -7$

5. $0.5x - 9 = 11$ or $0.5x - 9 = -11$
     $0.5x = 20$ or $0.5x = -2$
       $x = 40$ or $x = -4$

Check: $|0.5(40) - 9| = |20 - 9| = |11| = 11$
        $|0.5(-4) - 9| = |-2 - 9| = |-11| = 11$

6. $3x + 1 = 2x + 3$ or $3x + 1 = -(2x + 3)$
  $x + 1 = 3$ or $3x + 1 = -2x - 3$
    $x = 2$ or $3x = -2x - 4$
            $5x = -4$
            $x = -\dfrac{4}{5}$

Check: $|3(2) + 1| \stackrel{?}{=} 2(2) + 3$
  $|6 + 1| \stackrel{?}{=} 4 + 3$
  $7 = 7$ ✓

$\left|3\left(-\dfrac{4}{5}\right) + 1\right| \stackrel{?}{=} 2\left(-\dfrac{4}{5}\right) + 3$

$\left|-\dfrac{12}{5} + 1\right| \stackrel{?}{=} -\dfrac{8}{5} + 3$

$\left|-\dfrac{7}{5}\right| \stackrel{?}{=} \dfrac{7}{5}$

$\dfrac{7}{5} = \dfrac{7}{5}$ ✓

7. $4x - 7 = 37$ or $4x - 7 = -37$
  $4x = 44$ or $4x = -30$

  $x = 11$ or $x = -\dfrac{15}{2} = -7.5$

Check: $|4(11) - 7| = |44 - 7| = |37| = 37$ ✓
$|4(-7.5) - 7| = |-30 - 7| = |-37| = 37$ ✓

8. $3x + 7 = 2x - 4$ or $3x + 7 = -(2x - 4)$
  $x + 7 = -4$ or $3x + 7 = -2x + 4$
    $x = -11$ or $3x = -2x - 3$
            $5x = -3$
            $x = -\dfrac{3}{5}$

Check: $|3(-11) + 7| \stackrel{?}{=} 2(-11) - 4$
  $|-33 + 7| \stackrel{?}{=} -22 - 4$
  $|-26| \stackrel{?}{=} -26$
  $26 \neq -26$

$\left|3\left(-\dfrac{3}{5}\right) + 7\right| \stackrel{?}{=} 2\left(-\dfrac{3}{5}\right) - 4$

$\left|-\dfrac{9}{5} + 7\right| \stackrel{?}{=} -\dfrac{6}{5} - 4$

$\left|\dfrac{26}{5}\right| \stackrel{?}{=} -\dfrac{26}{5}$

$\dfrac{26}{5} \neq -\dfrac{26}{5}$

Therefore, there are no solutions. These solutions are extraneous

9. $|7x - 6| = -12$ There are no solutions because the absolute value of a quantity can never be negative.

10. $2|x + 2| = 4x - 6$; $|x + 2| = 2x - 3$
  $x + 2 = 2x - 3$ or $x + 2 = -(2x - 3)$
    $x = 2x - 5$ or $x + 2 = -2x + 3$
    $-x = -5$ or $x = -2x + 1$
    $x = 5$ or $3x = 1$
            $x = \dfrac{1}{3}$

Check: $2|5 + 2| \stackrel{?}{=} 4(5) - 6$
  $2|7| \stackrel{?}{=} 20 - 6$
  $14 = 14$ ✓

$2\left|\dfrac{1}{3} + 2\right| \stackrel{?}{=} 4\left(\dfrac{1}{3}\right) - 6$

$2\left|\dfrac{7}{3}\right| \stackrel{?}{=} \dfrac{4}{3} - 6$

$\dfrac{14}{3} \neq -\dfrac{14}{3}$

So, the solution is $x = 5$

11. $4|3 - x| - 8 = 8$; $4|3 - x| = 16$; $|3 - x| = 4$
  $3 - x = 4$ or $3 - x = -4$
    $-x = 1$ or $-x = -7$
    $x = -1$ or $x = 7$
Check:
$4|3 - (-1)| - 8 = 4|4| - 8 = 16 - 8 = 8$ ✓
$4|3 - 7| - 8 = 4|-4| - 8 = 16 - 8 = 8$ ✓
So, the solutions are $x = -1$ or $x = 7$

12. $-|x - 2| = -4$; $|x + 2| = 4$
  $x + 2 = 4$ or $x + 2 = -4$
    $x = 2$ or $x = -6$
Check: $-|2 + 2| = -|4| = -4$
$-|-6 + 2| = -|-4| = -4$
So, the solutions are $x = 2$ or $x = -6$

13. Sample answer: To solve an absolute value equation such as $|x - 2| = 4$, set $x - 2$, the expression inside the absolute value symbols, equal to 2 and to -2 and solve both equations for x. To solve an absolute value inequality such as $|x - 2| < 4$, solve the inequality $-4 < x - 2 < 4$. To solve an inequality such as $|x - 2| > 4$, solve the inequalities $x - 2 < -4$ and $x - 2 > 4$.

14. $-7 < 2x - 4 < 7$
  $-3 < 2x \quad < 11$

  $-\dfrac{3}{2} < x \quad < \dfrac{11}{2}$

Check a number between $-\dfrac{3}{2}$ and $\dfrac{11}{2}$.

Try $x = 0.$ $|2 \cdot 0 - 4| = |-4| = 4 < 7$

So, the solution is $-\dfrac{3}{2} < x < \dfrac{11}{2}$

15. $5|3 - x| < 30$; $|3 - x| < 6$
  $-6 < 3 - x < 6$
  $-9 < \quad -x < 3$
  $-3 < \quad x < 9$

Check a number between -3 and 9. Try $x = 1$.
$5|3 - 1| = 5(2) = 10 < 30$ ✓

16. $3x \leq -21$ or $3x \geq 21$

$\quad\ x \leq -7$ or $\quad x \geq 7$

Check a number less than -7. Try $x = -9$.

$|3(-9)| = |-27| = 27 \geq 21.$ ✓

Check a number greater than 7. Try $x = 10$.

$|3(10)| = 30 \geq 21$ ✓

17. $0.5|3x + 5| < 5; |3x + 5| < 10$

$\quad -10 < 3x + 5 < 10$

$\quad -15 < 3x \qquad < 5$

$\quad\ -5 < x \qquad < \dfrac{5}{3}$

Check a number between -5 and $\dfrac{5}{3}$. Try $x = -1$.

$0.5|3(-1) + 5| = 0.5|2| = 1 < 5$ ✓

18. No solution; the absolute value of a quantity cannot be less than zero.

19. $\left(4 - \dfrac{1}{2}x\right) \leq -2$ or $\left(4 - \dfrac{1}{2}x\right) \geq 2$

$\quad -\dfrac{1}{2}x \leq -6$ or $\quad -\dfrac{1}{2}x \geq -2$

$\quad\qquad x \geq 12$ or $\quad\qquad x \leq 4$

Check a number less than 4. Try $x = 1$.

$\left|4 - \dfrac{1}{2}(1)\right| = \left|4 - \dfrac{1}{2}\right| = 3\dfrac{1}{2} \geq 2$ ✓

Check a number greater than 12. Try $x = 15$.

$\left|4 - \dfrac{1}{2}(15)\right| = \left|4 - 7\dfrac{1}{2}\right| = \left|-3\dfrac{1}{2}\right| = 3\dfrac{1}{2} \geq 2$ ✓

20. $\dfrac{2x + 1}{3} \leq -5$ or $\dfrac{2x + 1}{3} \geq 5$

$\quad 2x + 1 \leq -15$ or $\quad 2x + 1 \geq 15$

$\quad\ \ 2x \leq -16$ or $\quad\qquad 2x \geq 14$

$\quad\qquad x \leq -8$ or $\quad\qquad\ \ x \geq 7$

Check a number less than -8. Try -10.

$\left|\dfrac{2(-10) + 1}{3}\right| = \left|\dfrac{-20 + 1}{3}\right| = \dfrac{19}{3} = 6\dfrac{1}{3} \geq 5$ ✓

Check a number greater than 7. Try 10.

$\left|\dfrac{2(10) + 1}{3}\right| = \left|\dfrac{20 + 1}{3}\right| = 7 \geq 5$ ✓

21. $|2x + 1| + 9 \leq 14; |2x + 1| \leq 5$

$\quad -5 \leq 2x + 1 \leq 5$

$\quad -6 \leq 2x \qquad\ \leq 4$

$\quad -3 \leq \ x \qquad\ \leq 2$

Check a number between -3 and 2. Try $x = 0$.

$|2(0) + 1| + 9 = |1| + 9 = 10 \leq 14$ ✓

22. b; $x + 2 < -1$ or $x + 2 > 1$

$\qquad\quad x < -3$ or $\qquad x > -1$

23. c; $-1 < x + 2 < 1$

$\qquad\ -3 < x \qquad < -1$

24. a; $x + 2 = -1$ or $x + 2 = 1$

$\qquad\quad x = -3$ or $\qquad x = -1$

25. $|w - 420| \leq 2$ and $|l - 594| \leq 2$

26. $|x - 925| \leq 70$, where x is the amount of fertilizer used for each plant.

$\quad -70 \leq x - 925 \leq 70$

$\quad\ 855 \leq x \qquad\quad \leq 995$

PRACTICE

1. $-1\dfrac{1}{4} \leq z + \dfrac{1}{2}$ and $z + \dfrac{1}{2} \leq 1\dfrac{1}{4}$

2. $-12.3 < w + 9.3$ and $w + 9.3 < 12.3$

3. $4p - 7 \leq -22$ or $4p - 7 \geq 22$

4. $c + 14 < -19$ or $c + 14 > 19$

5. $2x - 4 = -6$ or $2x - 4 = 6$

$\quad\ 2x = -2$ or $\qquad 2x = 10$

$\quad\quad x = -1$ or $\qquad\ x = 5$

Check: $|2(-1) - 4| = |-2 - 4| = |-6| = 6$ ✓

$\qquad\ |2(5) - 4| = |10 - 4| = |6| = 6$ ✓

6. $\dfrac{1}{2}x - 8 = -12$ or $\dfrac{1}{2}x - 8 = 12$

$\quad\ \dfrac{1}{2}x = -4$ or $\qquad \dfrac{1}{2}x = 20$

$\quad\quad x = -8$ or $\qquad\quad x = 40$

Check: $\left|\dfrac{1}{2}(-8) - 8\right| = |-4 - 8| = |-12| = 12$ ✓

$\qquad\ \left|\dfrac{1}{2}(40) - 8\right| = |20 - 8| = |12| = 12$ ✓

7. $4x - 5 = -(x + 17)$ or $4x - 5 = x + 17$
$\quad\quad 4x - 5 = -x - 17$ or $3x - 5 = 17$
$\quad\quad 5x - 5 = -17$ $\quad\quad$ or $\quad\quad 3x = 22$
$\quad\quad\quad 5x = -12$ $\quad\quad$ or $\quad\quad x = \dfrac{22}{3}$
$\quad\quad\quad\quad x = -\dfrac{12}{5}$

Check: $\left|4\left(-\dfrac{12}{5}\right) - 5\right| \overset{?}{=} -\dfrac{12}{5} + 17$

$\quad\quad\quad \left|-14\dfrac{3}{5}\right| \overset{?}{=} -2\dfrac{2}{5} + 17$

$\quad\quad\quad\quad 14\dfrac{3}{5} = 14\dfrac{3}{5} \checkmark$

$\quad\quad \left|4\left(\dfrac{22}{3}\right) - 5\right| \overset{?}{=} \dfrac{22}{3} + 17$

$\quad\quad \left|29\dfrac{1}{3} - 5\right| \overset{?}{=} 7\dfrac{1}{3} + 17$

$\quad\quad\quad\quad 24\dfrac{1}{3} = 24\dfrac{1}{3} \checkmark$

8. $5x + 4 = 9x - 12$ or $5x + 4 = -(9x - 12)$
$\quad\quad 5x = 9x - 16$ or $5x + 4 = -9x + 12$
$\quad -4x = -16$ $\quad$ or $14x + 4 = 12$
$\quad\quad x - 4$ $\quad\quad$ or $\quad\quad 14x = 8$
$\quad\quad\quad\quad\quad\quad\quad\quad\quad x = \dfrac{8}{14} = \dfrac{4}{7}$

Check: $|5(4) + 4| \overset{?}{=} 9(4) - 12$
$\quad\quad\quad\quad 24 = 24 \checkmark$

$\quad\quad \left|5\left(\dfrac{4}{7}\right) + 4\right| \overset{?}{=} 9\left(\dfrac{4}{7}\right) - 12$

$\quad\quad\quad\quad 6\dfrac{6}{7} \neq -6\dfrac{6}{7}$

So, the solution is $x = 4$

9. $9x - 8 = -11x$ or $9x - 8 = 11x$
$\quad 20x - 8 = 0$ $\quad$ or $-2x - 8 = 0$
$\quad\quad 20x = 8$ $\quad$ or $\quad -2x = 8$
$\quad\quad\quad x = \dfrac{2}{5}$ or $\quad\quad x = -4$

Check: $\left|9\left(\dfrac{2}{5}\right) - 8\right| \overset{?}{=} -11\left(\dfrac{2}{5}\right)$

$\quad\quad\quad \left|\dfrac{18}{5} - 8\right| \overset{?}{=} -\dfrac{22}{5}$

$\quad\quad\quad\quad \dfrac{22}{5} \neq -\dfrac{22}{5}$

$\quad\quad |9(-4) - 8| \overset{?}{=} -11(-4)$

$\quad\quad\quad |-36 - 8| \overset{?}{=} 44$

$\quad\quad\quad\quad\quad 44 = 44 \checkmark$

So, the solution is $x = -4$

10. No solution; it is not possible for the absolute value of a quantity to be a negative number.

11. $15x = 0$
$\quad\quad x = 0$
Check: $|15(0)| = |0| = 0 \checkmark$
So, the solution is $x = 0$

12. $2|2x + 10| = 4x + 30$; $|2x + 10| = 2x + 15$
$\quad 2x + 10 = 2x + 15$ or $2x + 10 = -(2x + 15)$
$\quad\quad\quad 10 \neq 15$ $\quad\quad\quad\quad$ $2x + 10 = -2x - 15$
$\quad\quad\quad\quad\quad\quad\quad\quad\quad\quad 4x + 10 = -15$
$\quad\quad\quad\quad\quad\quad\quad\quad\quad\quad\quad 4x = -25$
$\quad\quad\quad\quad\quad\quad\quad\quad\quad\quad\quad x = -\dfrac{25}{4}$

Check: $2\left|2\left(-\dfrac{25}{4}\right) + 10\right| \overset{?}{=} 4\left(-\dfrac{25}{4}\right) + 30$

$\quad\quad 2\left|-\dfrac{25}{2} + 10\right| \overset{?}{=} -25 + 30$

$\quad\quad\quad 2\left|-\dfrac{5}{2}\right| \overset{?}{=} 5$

$\quad\quad\quad\quad\quad 5 = 5 \checkmark$

So, the solution is $x = -\dfrac{25}{4}$

13. $-15 < 5x - 10 < 15$
$\quad -5 < 5x \quad\quad < 25$
$\quad -1 < x \quad\quad\quad < 5$

Check a number between -1 and 5. Try $x = 2$.
$|5(2) - 10| = |0| = 0 < 15 \checkmark$

14. No solution; it is impossible for the absolute value of a quantity to be less than zero.

15. $5x < 0$ or $5x > 0$
$\quad x < 0$ or $\quad x > 0$

Check any number greater than 0. Try 3.
$|5(3)| = 15 > 0 \checkmark$

Check any number less than 0. Try -3.
$|5(-3)| = |-15| = 15 > 0 \checkmark$

16. $\quad -7 < -0.5k + 1 < 7$
$\quad\quad -8 < \quad -0.5k \quad < 6$
$\quad -16 < \quad\quad -k \quad\quad < 12$
$\quad -12 < \quad\quad\quad k \quad\quad < 16$

Check a number between -12 and 16. Try 2.
$|-0.5(2) + 1| = |-1 + 1| = 0 < 7 \checkmark$

17. $7|3x - 7| \geq 35$; $|3x - 7| \geq 5$;
$3x - 7 \leq -5$ or $3x - 7 \geq 5$
$3x \leq 2$ or $3x \geq 12$
$x \leq \frac{2}{3}$ or $x \geq 4$

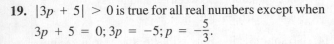

Check a number less than $\frac{2}{3}$. Try $x = 0$.
$7|3(0) - 7| = 7|-7| = 49 \geq 35$ ✓

Check a number greater than 4. Try $x = 5$.
$7|3(5) - 7| = 7|15 - 7| = 56 \geq 35$ ✓

18. $|5x - 4| - 8 > -3$; $|5x - 4| > 5$;
$5x - 4 < -5$ or $5x - 4 > 5$
$5x < -1$ or $5x > 9$
$x < -\frac{1}{5}$ or $x > \frac{9}{5}$

Check a number less than $-\frac{1}{5}$. Try $x = -1$.
$|5(-1) - 4| - 8 = |-9| - 8 = 1 > -3$ ✓
Check a number greater than $\frac{9}{5}$. Try $x = 2$.
$|5(2) - 4| - 8 = |10 - 4| - 8 =$
$6 - 8 = -2 > -3$ ✓

19. $|3p + 5| > 0$ is true for all real numbers except when
$3p + 5 = 0$; $3p = -5$; $p = -\frac{5}{3}$.

Check a number different from $p = -\frac{5}{3}$. Try $p = 3$.
$|3(3) + 5| \overset{?}{>} 0$
$14 > 0$ ✓

20. $-\frac{1}{2} + x \leq -3$ or $-\frac{1}{2} + x \geq 3$
$x \leq -2\frac{1}{2}$ or $x \geq 3\frac{1}{2}$

Check a number less than $-2\frac{1}{2}$. Try $x = -3$.
$\left|-\frac{1}{2} + (-3)\right| = \left|-3\frac{1}{2}\right| = 3\frac{1}{2} \geq 3$ ✓
Check a number greater than $3\frac{1}{2}$. Try $x = 4$.
$\left|-\frac{1}{2} + 4\right| = 3\frac{1}{2} \geq 3$ ✓

21. $13 + |5c + 7| < 13$; $|5c + 7| < 0$
No solution; it is impossible for the absolute value of a quantity to be less than zero.

22. $-12 + \left|\frac{3x + 2}{4}\right| < -9$; $\left|\frac{3x + 2}{4}\right| < 3$;
$-3 < \frac{3x + 2}{4} < 3$
$-12 < 3x + 2 < 12$
$-14 < 3x < 10$
$-4\frac{2}{3} < x < 3\frac{1}{3}$

Check a number between $-4\frac{2}{3}$ and $3\frac{1}{3}$. Try $x = 2$
$-12 + \left|\frac{3(2) + 2}{4}\right| = -12 + |2| = -10 < -9$ ✓

23. $-15 + 3|2n + 5| \leq 42$; $3|2n + 5| \leq 57$;
$|2n + 5| \leq 19$; $-19 \leq 2n + 5 \leq 19$
$-24 \leq 2n \leq 14$
$-12 \leq n \leq 7$

Check a number between -12 and 7. Try $n = 2$.
$-15 + 3|2(2) + 5| = -15 + 3(9) = 12 \leq 42$ ✓

24. $-4x < -20$ or $-4x > 20$
$x > 5$ or $x < -5$

Check a number less than -5. Try $x = -7$.
$|-4(-7)| = 28 > 20$ ✓

Check a number greater than 5. Try $x = 6$.
$|-4(6)| = |-24| = 24 > 20$ ✓

25. b; $-2 < x + 6 < 2$
$-8 < x < -4$

26. c; $2x - 4 < -2$ or $2x - 4 > 2$
$2x < 2$ or $2x > 6$
$x < 1$ or $x > 3$

27. a; $14 + |2x + 1| \leq 19$; $|2x + 1| \leq 5$;
$-5 \leq 2x + 1 \leq 5$
$-6 \leq 2x \leq 4$
$-3 \leq x \leq 2$

28. $|w - 594| \leq 2$ and $|l - 841| \leq 3$

29. $|p - 19.4| \leq 1.25$
$-1.25 \leq p - 19.4 \leq 1.25$
$18.15 \leq p \leq 20.65$

30. Answers may vary. Descriptions may include an interval with endpoints, such as $-1 \leq x \leq 1$; an interval without endpoints, such as $3 < x < 10$; the whole number line, such as $x < 2$ or $x > 0$; two rays, such as $x \leq -2$ or $x \geq 3$; and the empty set, such as $x < 1$ and $x > 1$.

31. $|x - 4| \geq 2$ **32.** $|2x| < 6$ **33.** $|5x - 6| > 5$

34. $|3(2x - 1)| < 3$ **35.** $|3x - 2| < 5$

36. $|5.6y - 9| \geq 17$

37.
$$x + 3 \geq -(x + 3) \quad \text{and} \quad x + 3 \leq x + 3$$
$$x + 3 \geq -x - 3 \qquad\qquad 0 \leq 0$$
$$2x + 3 \geq -3$$
$$2x \geq -6$$
$$x \geq -3$$

Check a number greater than -3. Try 2.
$$|2 + 3| \overset{?}{\leq} 2 + 3$$
$$5 \leq 5 \checkmark$$

Check a number less than -3.
$$\text{Try } x = -4; |-4 + 3| \overset{?}{\leq} -4 + 3$$
$$|-1| \overset{?}{\leq} -1$$
$$1 \nleq -1$$

So, the solution is $x \geq -3$.

38.
$$5 - x > 5 - x \quad \text{or} \quad 5 - x < -(5 - x)$$
$$0 \ngtr 0 \qquad\qquad 5 - x < -5 + x$$
$$-x < 10 + x$$
$$-2x < -10$$
$$x > 5$$

So, the solution is $x > 5$.

Check a number greater than 5.

$$\text{Try } x = 7. \; |5 - 7| \overset{?}{>} 5 - 7$$
$$|-2| \overset{?}{>} -2$$
$$2 > -2 \checkmark$$

39.
$$-(2x + 4) < 12 - 3x \quad \text{and} \quad 12 - 3x < 2x + 4$$
$$-2x - 4 < 12 - 3x \quad \text{and} \quad 12 \qquad < 5x + 4$$
$$x - 4 < 12 \qquad\quad \text{and} \qquad 8 \qquad < 5x$$
$$x < 16 \qquad\qquad \text{and} \qquad \tfrac{8}{5} < x$$

Check a number between $\frac{8}{5}$ and 16. Try $x = 3$.

$$|12 - 3(3)| \overset{?}{<} 2(3) + 4$$
$$|12 - 9| \overset{?}{<} 6 + 4$$
$$3 < 10 \checkmark$$

So, the solution is $\frac{8}{5} < x < 16$.

40.
$$4 - x \geq 3x \quad \text{or} \quad 4 - x \leq -3x$$
$$4 \geq 4x \quad \text{or} \quad 4 + 2x \leq 0$$
$$1 \geq x \quad \text{or} \qquad 2x \leq -4$$
$$x \leq 1 \quad \text{or} \qquad x \leq -2$$

Check a number less than 1. Try $x = 0$.
$$|4 - 0| \overset{?}{\geq} 3(0); 4 \geq 0 \checkmark$$

So, the solution is $x \leq 1$. This solution includes $x \leq -2$.

41.
$$x - 6 < 2x \qquad \text{and} \quad x - 6 > -2x$$
$$x < 2x + 6 \quad \text{and} \qquad x > -2x + 6$$
$$-x < 6 \qquad \text{and} \qquad 3x > 6$$
$$x > -6 \qquad \text{and} \qquad x > 2$$

Check a number greater than 2. Try $x = 5$.
$$|5 - 6| \overset{?}{<} 2(5); 1 < 10 \checkmark$$

So, the solution is $x > 2$.

42.
$$2x + 1 \geq x + 2 \quad \text{or} \quad 2x + 1 \leq -(x + 2)$$
$$x + 1 \geq 2 \qquad \text{or} \quad 2x + 1 \leq -x - 2$$
$$x \geq 1 \qquad \text{or} \quad 3x + 1 \leq -2$$
$$3x \leq -3$$
$$x \leq -1$$

Check a number greater than 1. Try $x = 3$.
$$|2(3) + 1| \overset{?}{\geq} 3 + 2; 7 \geq 5 \checkmark$$

Check a number less than -1. Try -2.
$$|2(-2) + 1| \overset{?}{\geq} (-2) + 2; 3 \geq 0 \checkmark$$

43. $-5 \leq x - 4 \leq 5; -1 \leq x \leq 9;$ midpoint is 4

44. $-14 \leq c - 22 \leq 14; 8 \leq c \leq 36;$ midpoint is 22

45. $-30 \leq m - 12 \leq 30; -18 \leq m \leq 42;$ midpoint is 12

46. $-3 \leq r - 9 \leq 3; 6 \leq r \leq 12;$ midpoint is 9

47. The solution is $(b - c) \leq x \leq (b + c)$ and the midpoint is b.

48. $|x| = 3; x = -3$ or $x = 3$
$|x - 2| = 3; x - 2 = 3$ or $x - 2 = -3; x = 5$ or $x = -1$. The solutions of $|x - 2| = 3$ are the solutions of $|x| = 3$ translated 2 units to the right.

49. No. Examples will vary. If $|x - a| = b$ then $x = a + b$ or $-b + a$. If $|x| - a = b$ then $x = a + b$ or $-(a + b); -b + a \neq -(a + b)$

50. $|t - 12.4| \leq 0.2$ **51.** $|t - 11.8| \leq 0.2$

52. $|s - 7.4| \leq 0.1$

53a. $-1 \leq \dfrac{h - 67.5}{2.6} \leq 1$

$$-2.6 \leq h - 67.5 \leq 2.6$$
$$64.9 \leq h \qquad\quad \leq 70.1$$

53b. Since the other 25% is not in the interval in part a., $h > 70.1$ or $h < 64.9$.

THINK CRITICALLY
For Exercises 54. through 57., answers will vary. Sample answers are given.

54.
$$x < -5 \quad \text{or} \qquad x > 3$$
$$x + 1 < -4 \quad \text{or} \quad x + 1 > 4$$
So, $|x + 1| \geq 4$

55. $-4 \leq x \leq 6; -5 \leq x - 1 \leq 5$
So, $|x - 1| \leq 5$

56. $-4 < x < 4$ so, $|x| < 4$.

57. $x < -1$ or $x > 5$; $x - 2 < -3$ or $x - 2 > 3$
So, $|x - 2| > 3$

58. $6 \le x \le 12$; $-3 \le x - 9 \le 3$; abs $(x - 9) \le 3$

59. $x \le -2$ or $x \ge 8$; $x - 3 \ge -5$ or $x - 3 \ge 5$;
abs $(x - 3) > 5$

MIXED REVIEW

60. 0; $-7 + 0 = 0 + (-7) = -7$

61. -62; $62 + (-62) = -62 + 62 = 0$

62. 1; $1\left(\dfrac{12}{13}\right) = \dfrac{12}{13}(1) = \dfrac{12}{13}$

63. $\dfrac{5}{17}$; $\dfrac{17}{5} \cdot \dfrac{5}{17} = \dfrac{5}{17} \cdot \dfrac{17}{5} = 1$

64. $S = \dfrac{1}{2}gt^2$

$2s = gt^2$

$\dfrac{2s}{t^2} = g$

$g = \dfrac{2s}{t^2}$

PROJECT CONNECTION

1. Answers will vary.

2. 50 cm

3. Solve the equations $a + b = d_1$ and $a - b = d_2$ simultaneously for a and b.

$\begin{cases} a + b = d_1 \\ a - b = d_2 \end{cases}$

$2a = d_1 + d_2$ add the second equation to the first

$a = \dfrac{d_1 + d_2}{2}$ divide both sides by 2

$2b = d_1 - d_2$ subtract the second equation from the first

$b = \dfrac{d_1 - d_2}{2}$ divide both sides by 2

4. For $|x - a| = b$, $a = \dfrac{d_1 + d_2}{2}$ represents the fulcrum marking on the stick (in this case, 50 cm) and $b = \dfrac{d_1 - d_2}{2}$ represents the distance each bag of coins is from the fulcrum.

5. Answers will vary.

6. $x - 50 = 18.5$ or $x - 50 = -18.5$
       $x = 68.5$ or $x = 31.5$
Therefore, the bags are at 31.5 cm and 68.5 cm.

ALGEBRAWORKS

1. Answers will vary. A possible answer is use a bar graph.

2. $-60 < d - 60 < 60$
     $0 < d$ < 120 or $d > 0$ and $d < 120$

3. The sound level at a rock concert is close to the upper limit of audibility, at the level of pain.

Lesson 2.6, pages 89–93

EXPLORE

1. Both graphs are bar graphs that show the same data.

2. One graph shows production by year and the other shows production by energy source.

3. Answers will vary. Some samples follow. How did the production of natural gas and petroleum compare in 1990? They were about the same. What is the trend in the production of petroleum? Production is steadily declining.

4. Answers will vary. Here are possible answers. She might select the graph on the left to compare production of petroleum to production of natural gas by the year. She might use the graph of the right to compare the number of BTUs of each energy source.

TRY THESE

1. Graphs will vary. One possibility is to use a double line graph, one line showing production and the other showing consumption.

2. Graphs will vary. One possibility is a double circle graph, one showing the percent of energy produced by resources now and the other showing the percent produced in 2010.

3. Answers will vary. It may also be helpful for client to know the availability of various energy sources in the area and also regional regulations and standards.

PRACTICE

1. Graphs will vary.

2. Graphs will vary. Here is one possibility.

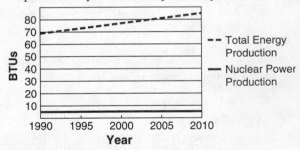

3. Answers will vary. They should expand on the information in the headline. Specific numbers may be included in the article.

4. Answers will vary. Possibly coal could be selected because it has the smallest difference between consumption and production in 2000. This can be shown on a bar graph.

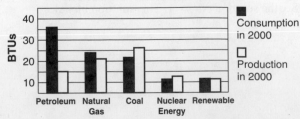

5. Answers will vary. A circle graph can be used when you want to illustrate a quantity as part of a whole. A bar graph may be more useful to compare two quantities with each other rather than with the whole.

6. The graph shows Total Production and Total Imports compared to Total Consumption

7. Total Production + Total Imports > Total Consumption.

8. Stacked bars illustrate data that are added.

9. Answers will vary. One possibility is to graph total energy consumption as the sum of each type of consumption.

10. Answers will vary. Some ideas that can be incorporated into the answers include using a bar graph to compare quantities at a certain period of time, such as production of items; using line graphs to illustrate the change in quantities over a period of time, such as the growth in production of an item over a certain number of years; and finally using a circle graph to show the proportion of the whole that certain quantities exhibit, such as the comparative costs that go into producing an item.

THINK CRITICALLY

11. Answers will vary. One possibility is: Sources Vary as Energy Production Increases.

12. Answers will vary. One possibility is a stacked bar graph that shows total energy production as the sum of its parts. Another possibility is a circle graph that shows what portion each form of energy is of the total.

13. The graph shows energy consumption for three forms of energy from 1990 to 2010.

14. Answers will vary. One possibility is 40 quadrillion BTUs.

15. Answers will vary. One possibility is 32 quadrillion BTUs.

16. Answers will vary. One possibility is 22 quadrillion BTUs.

17. Answers will vary. A different scale may be used or a different type of graph may be drawn.

MIXED REVIEW

18. B; $(100 - 8) \cdot (-3) + 1 = 92 \cdot (-3) + 1 = -276 + 1 = -275$

19. D; $121 - 15 + 1.49 = 107.49$

20. $V = \pi r^2 h$

    $\dfrac{V}{\pi r^2} = h \text{ or } h = \dfrac{V}{\pi r^2}$

21. $h = \dfrac{1386}{\pi (7)^2} \approx 9$

22. disjunction

23. conjunction

PROJECT CONNECTION

1.–2. Answers will vary.

3. $N_1 \cdot D_1 = 2 \cdot 20 = 40$
   Each product is approximately 40.

4. Let x represent the distance from the fulcrum.
   $5x = 6 \cdot 11$
   $x = \dfrac{66}{5}$
   $x \approx 13.2$ cm from the fulcrum

Lesson 2.7, pages 94–97

EXPLORE THE PROBLEM

1. $\dfrac{x}{2} + 1$ represents what Roger took. $x - \left(\dfrac{x}{2} + 1\right)$ represents what Marina had left.

2. $x - \left(\dfrac{x}{2} + 1\right) = 1$

3. $x - \dfrac{x}{2} - 1 = 1$

    $x - \dfrac{x}{2} = 2$

    $\dfrac{1}{2}x = 2$

    $x = 4$

   Check: Marina gave away $\dfrac{1}{2}$ of 4 which is 2, plus 1 makes 3, leaving her with 1.

4. The description of the quantity Phyllis took is similar to the description of the quantity Roger took.

5. The equation for the number Marina had before Phyllis took some is similar to the equation for the number Marina had before Roger took his. Let y represent the amount Marina had before Phyllis took some.

   $\dfrac{y}{2} + 1$ represents what Phyllis took.

   $y - \left(\dfrac{y}{2} + 1\right) = 4$

6. $y - \dfrac{y}{2} - 1 = 4$ 7. $w - \left(\dfrac{w}{2} + 1\right) = 10$

   $\dfrac{y}{2} = 5$ $w - \dfrac{w}{2} - 1 = 10$

   $y = 10$ $\dfrac{w}{2} = 11$

                             $w = 22$

   Check: Dan took half of 22 plus 1, which is 12. This leaves Marina with 10. Phyllis took half of 10 plus 1, which is 6. This leaves Marina with 4. Roger took half of 4 plus 1, which is 3. This leaves Marina with 1.

INVESTIGATE FURTHER

8. $x - 250 - 300 = 1210$ 9. $y + 3y = 1760$
   $x - 550 = 1210$ $4y = 1760$
   $x = 1760$ $y = 440$

10. Let z represent the number of radios before the shipment to McPherson's.
    $\dfrac{1}{2}z = 440; z = 880$

11. $r + 10r = 880$
$11r = 880$
$r = 80$

Check: The day began with 80 radios and there were 10 times that number produced. So, the new total number of radios is 880. Half that total or 440 were sent to McPherson's, leaving 440, and 3 times that amount is 1320 giving a total of 1760 radios on hand. At the end of the day two shipments of 250 and 300 were sent for a total of 550. This leaves 1760 minus 550 which is equal to 1210 radios.

12. Answers will vary; when the result or end number is known and the steps for solving are known and can be reversed.

APPLY THE STRATEGY

13. Let x represent the base price.
$x + 1170$ represents the price with options.
$x + 1170 + 0.07(x + 1170)$ represents the price with sales tax added.
$x + 1170 + 0.07(x + 1170) + 175 + 52$ represents the final price.
$x + 1170 + 0.07(x + 1170) + 175 + 52 = 13{,}067$;
$x + 1170 + 0.07x + 81.9 + 227 = 13{,}067$;
$1.07x + 1478.9 = 13{,}067$; $1.07x = 11{,}588.1$;
$x = \$10{,}830$

14. Let t represent the length of time the plane climbed at 1000 ft/min. $1000t$ represents the gain in altitude during this time.
$1000t - 1300 + \left(2\frac{1}{4}\right)(800)$ represents the total gain in altitude.
$1000t - 1300 + \left(2\frac{1}{4}\right)800 = 6000$;
$1000t - 1300 + 1800 = 6000$
$1000t + 500 = 6000$
$1000t = 5500$
$t = 5.5 \text{ min or } 5\frac{1}{2} \text{ min}$

15. Let x represent the number of students in Mr. Miller's class.
$\frac{1}{2}x$ represents the number that worked on science projects.
$\frac{1}{2}x$ represents the number of students who did not work on science projects.
$\frac{1}{3}\left(\frac{1}{2}x\right)$ represents the number of students who wrote short stories.
$\frac{2}{3}\left(\frac{1}{2}x\right)$ represents the number of students who did not write short stories.
$\frac{1}{2}\left[\frac{2}{3}\left(\frac{1}{2}x\right)\right]$ represents the number of students who acted in the school play.

$\frac{1}{2}\left[\frac{2}{3}\left(\frac{1}{2}x\right)\right]$ represents the number of students who did not act in the play. $\frac{1}{2}\left[\frac{2}{3}\left(\frac{1}{2}x\right)\right] = 5$; $\frac{1}{6}x = 5$; $x = 30$ students in Mr. Miller's class.

16. Let x represent Chan's age.
$[(7x + 7) \div 7] - 7 = 7$; $(7x + 7) \div 7 = 14$;
$7x + 7 = 98$; $7x = 91$; $x = 13$ years old

17. Let x represent the total value of the prize.
$.5x$ represents the value of the stove.
$.6(.5x)$ represents the value of the refrigerator.
$.4(.5x)$ represents the value of the remaining prizes, $\$300 + \$220 = \$520$.
Therefore $.4(.5x) = 520$; $.2x = 520$; $x = \$2600$.

18. Let x represent the number of shells Sylvia gave Lewis. $x + 2$ represents the number of shells Lewis ended up with. Nico also ended up with $x + 2$ shells, but since he doubled the number Sylvia gave him, he started with $\frac{1}{2}(x + 2)$. Using the same reasoning Mark started with $x + 2 + 2$ and Oscar started with $2(x + 2)$. Since Sylvia gave her brothers 54 shells,
$x + \frac{1}{2}(x + 2) + x + 2 + 2 + 2(x + 2) = 54$;
$x + \frac{1}{2}x + 1 + x + 2 + 2 + 2x + 4 = 54$;
$\left(4\frac{1}{2}\right)x + 9 = 54$; $\left(4\frac{1}{2}\right)x = 45$; $x = 10$.
Therefore, Lewis got $x = 10$, Mark got $x + 2 + 2 = 14$, Nico got $\frac{1}{2}(x + 2) = 6$ and Oscar got $2(x + 2) = 24$.

19. Let x represent the price of the tape player in 1985.
Price in 1990 $x - 30$
Price in 1993 $x - 30 - .2(x - 30)$
Price in 1994 $x - 30 - .2(x - 30) + 8$
Price in 1995 $\$27$ or $\frac{3}{4}$ the 1994 price. Therefore,
$\frac{3}{4}[x - 30 - .2(x - 30) + 8] = 27$
$[x - 30 - .2x + 6 + 8] = 36$
$0.8x - 16 = 36$
$0.8x = 52$
$x = \$65$

20. Errol, Fran, and Gloria each end up with 24 apples. Since Gloria gave Fran and Errol enough to double their numbers of apples, before that step Fran and Errol each had 12 apples, so Gloria had $24 + 24 = 48$ apples. Before this step Errol and Gloria had doubled their apples with the ones given to them by Fran. So, working backward again we conclude that Errol had 6 apples and Gloria had 24. So, Fran had to have 42 apples. On the first step Errol gave enough apples to double the number Fran and Gloria had. Therefore Fran started with 21, Gloria started with 12, and Errol started with $6 + 21 + 12 = 39$ apples.

21. Problems will vary. See the above problems for some examples.

REVIEW PROBLEM SOLVING STRATEGIES

1. Since Vernon and Laura ate the same type and Laura and Carl ate different types, then Carl and Vernon ate different types. Start a table to keep track of the information so far.

Type 1	Type 2
Vernon	Carl
Laura	
Janet	Katherine

Since Carl and Janet ate a different type, put Janet under Type 1. But since there are only two chicken sandwiches and three tuna, this means that Vernon, Laura, and Janet had tuna and Carl and Katherine had chicken.

2a. Start a list and look for a pattern. Starting with the lowest numbers the impossible scores are 1, 2, 3, 5, 6, 7, 10, 11, 14, 15, 19, 23. These are numbers that cannot be written in the form $4n + 9m$ where n and m are integers greater than or equal to 0 (not both 0)

2b. Using trial and error it can now be determined that 23 is the maximum impossible score because any number greater than 23 can be written in the form $4n + 9m$.

3a. 3 ways; each one a complete row.

3b. 6 ways; 3 in the bottom row and 3 in the top.

3c. 28 shaped like this

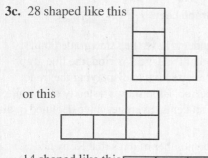

or this

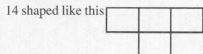

14 shaped like this

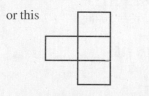

or this

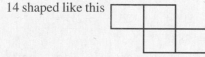

14 shaped like this

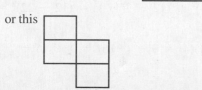

or this

3d. $3 + 6 + 28 + 14 + 14 = 65$ ways.

Chapter Review, pages 98–99

1. c **2.** e **3.** b **4.** a **5.** d

6. By the Pythagorean theorem,
$$AB = \sqrt{2^2 + 1^2} = \sqrt{4 - 1} = \sqrt{3}.$$

7. Mark AB off twice along the number line.

8. rational; $0.125 = \frac{125}{1000} = \frac{1}{8}$

9. rational; Let $x = 0.\overline{17}$
$$\frac{100x = 17.\overline{17}}{99x = 17}$$
$$x = \frac{17}{99}$$

10. rational; $\sqrt{625} = 25 = \frac{25}{1}$ **11.** irrational

12. $\frac{40 + 2}{10 - 3} + 15 \times 4 = \frac{42}{7} + 15 \times 4 = 6 + 60 = 66$

13. $3 \times 4^2 + (3 + 4)^2 = 3 \cdot 16 + (7)^2 =$
$48 + 49 = 97$

14. associative property of multiplication

15. commutative property of addition

16. $14a + 1.3 = 2.7 + 7a$
$7a + 1.3 = 2.7$
$7a = 1.4$
$a = 0.2$
Check: $14(0.2) + 1.3 \overset{?}{=} 2.7 + 7(0.2)$
$2.8 + 1.3 \overset{?}{=} 2.7 + 1.4$
$4.1 - 4.1$ ✓

17. $r - 2(4r - 5) = 1 - r$
$r - 8r + 10 = 1 - r$
$-7r + 10 = 1 - r$
$-6r + 10 = 1$
$-6r = -9$
$r = \frac{3}{2}$
Check: $\frac{3}{2} - 2\left|4\left(\frac{3}{2}\right) - 5\right| \overset{?}{=} 1 - \frac{3}{2}$
$\frac{3}{2} - 2[6 - 5] \overset{?}{=} -\frac{1}{2}$
$\frac{3}{2} - 2(1) \overset{?}{=} -\frac{1}{2}$
$-\frac{1}{2} = -\frac{1}{2}$ ✓

18. $\frac{4(z + 8)}{3} = -4z$
$4(z + 8) = -12z$
$4z + 32 = -12z$
$16z + 32 = 0$
$16z = -32$
$z = -2$
Check: $\frac{4(-2 + 8)}{3} \overset{?}{=} -4(-2)$
$\frac{4(6)}{3} \overset{?}{=} 8$
$8 = 8$ ✓

19. $4x + 3 \geq 6x + 9$

$\quad -2x + 3 \geq 9$

$\qquad -2x \geq 6$

$\qquad\quad x \leq -3$

Check a number less than -3.

Try $x = -4$. $\quad 4(-4) + 3 \overset{?}{\geq} 6(-4) + 9$

$\qquad\qquad\qquad -16 + 3 \overset{?}{\geq} -24 + 9$

$\qquad\qquad\qquad\qquad -13 \geq -15 \checkmark$

20. $\quad 8 < -3y + 2 \leq 14$

$\quad\ 6 < -3y \qquad \leq 12$

$-2 > \quad y \qquad \geq -4$

$-4 \leq \quad y \qquad < -2$

Check a number between -4 and -2. Try $x = -3$.

$8 \overset{?}{<} -3(-3) + 2 \overset{?}{\leq} 14$

$8 \overset{?}{<} \qquad 9 + 2 \overset{?}{\leq} 14$

$8 < 11 \qquad\qquad \leq 14 \checkmark$

21. $2t - 1 \leq -7 \quad$ or $\quad \dfrac{2t}{3} > 2$

$\qquad 2t \leq -6 \quad$ or $\quad 2t > 6$

$\qquad\ t \leq -3 \quad$ or $\quad t > 3$

Check a number less than -3. Try $t = -4$.

$2(-4) - 1 = -8 - 1 = -9 \leq -7 \checkmark$

Check a number greater than 3.

Try $t = 4$. $\quad \dfrac{2(4)}{3} = \dfrac{8}{3} = 2\dfrac{2}{3} > 2 \checkmark$

22. $3x - 2 = 5 \quad$ or $\quad 3x - 2 = -5$

$3x \qquad = 7 \quad$ or $\qquad 3x = -3$

$\quad x = \dfrac{7}{3} \quad$ or $\qquad\quad x = -1$

Check: $3\left(\dfrac{7}{3}\right) - 2 = 7 - 2 = 5 \checkmark$

$\qquad\ 3(-1) - 2 = -3 - 2 = -5 \checkmark$

23. $-7 \leq 2y - 1 \leq 7$

$\ -6 \leq 2y \qquad \leq 8$

$\ -3 \leq \ y \qquad \leq 4$

Check a number between -3 and 4. Try $y = 0$.

$|2(0) - 1| = |-1| = 1 \leq 7 \checkmark$

24. $2z - |z + 2| = 11$

$\quad -|z + 2| = 11 - 2z$

$\qquad |z + 2| = 2z - 11$

$z + 2 = 2z - 11 \quad$ or $\quad z + 2 = -(2z - 11)$

$-z + 2 = -11 \qquad$ or $\quad z + 2 = -2z + 11$

$\quad -z = -13 \qquad$ or $\quad 3z + 2 = 11$

$\qquad z = 13 \qquad$ or $\qquad 3z = 9$

$\qquad\qquad\qquad\qquad\qquad\ z = 3$

Check: $2(13) - |13 + 2| = 26 - |15| =$

$\qquad\qquad 26 - 15 = 11 \checkmark$

$\qquad 2(3) - |3 + 2| = 6 - 5 = 1$

$z = 3$ is extraneous.

So, the solution is $z = 13$

25.

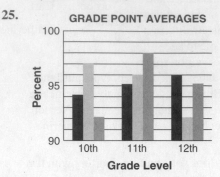

26.

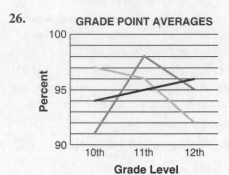

27. Although all 3 students have the same grade point average at the end of the 3-year period, the line graph appears to be more effective in displaying the trend over the 3-year period. Blue shows a steady rise, while both of the other students' averages have declined in the 12th grade.

28. Let x represent the number of days that Ruby had to work.

$4(5.25)x + 0.06(4)(5.25)x = 1113$

$\qquad\qquad 21x + 1.26x = 1113$

$\qquad\qquad\qquad\ 22.6x = 1113$

$\qquad\qquad\qquad\qquad\ x = 50 \text{ days}$

Chapter Assessment, page 100–101

CHAPTER TEST

1. c; $\sqrt{3^2 - 1^2} = \sqrt{9 - 1} = \sqrt{8}$

2. $3 \cdot 5^3 - (3 - 5)^3 = 3 \cdot 125 - (-2)^3 =$

$375 - (-8) = 383$

3. $(12 - 2\sqrt{121})^2 = (12 - 2(11))^2 = (12 - 22)^2 =$

$(-10)^2 = 100$

4. $(-3)^2 - 4(-1)(4) = 9 - (-16) = 9 + 16 = 25$

5. $\dfrac{-(-3) \pm \sqrt{(-3)^2 - 4(-1)(4)}}{2(-1)} = \dfrac{3 \pm \sqrt{9 + 16}}{-2} =$

$\dfrac{3 \pm \sqrt{25}}{-2} = \dfrac{3 \pm 5}{-2} = -4, 1$

6. $3a - 7.2 = 2.7 + 6a; -3a - 7.2 = 2.7; -3a = 9.9;$
$a = -3.3$
Check: $3(-3.3) - 7.2 \overset{?}{=} 2.7 + 6(-3.3)$
$-9.9 - 7.2 \overset{?}{=} 2.7 - 19.8$
$-17.1 = -17.1 ✓$

7. $\dfrac{3(y - 7)}{5} = -6y; 3(y - 7) = -30y; y - 7 = -10y;$

$-11y = -7; y = \dfrac{7}{11}$

Check: $\dfrac{3\left(\dfrac{7}{11} - 7\right)}{5} \overset{?}{=} -6\left(\dfrac{7}{11}\right)$

$\dfrac{3\left(-\dfrac{70}{11}\right)}{5} \overset{?}{=} -\dfrac{42}{11}$

$-\dfrac{210}{55} \overset{?}{=} -\dfrac{42}{11}$

$-\dfrac{42}{11} = -\dfrac{42}{11} ✓$

8. $4(2x + 1) - 7 = 26 - 3(-2x - 5)$
$8x + 4 - 7 = 26 + 6x + 15$
$8x - 3 = 6x + 41$
$x = 44$
$x = 22$
Check: $4[2(22) + 1] - 7 \overset{?}{=} 26 - 3[-2(22) - 5]$
$4(44 + 1) - 7 \overset{?}{=} 26 - 3(-44 - 5)$
$4(45) - 7 \overset{?}{=} 26 - 3(-49)$
$180 - 7 \overset{?}{=} 26 + 147$
$173 = 173 ✓$

9. $4m - 1 = 7$ or $4m - 1 = -7$
$4m = 8$ or $4m = -6$

$m = 2$ or $m = -\dfrac{3}{2}$

Check: $|4(2) - 1| = |8 - 1| = |7| = 7 ✓$

$\left|4\left(-\dfrac{3}{2}\right) - 1\right| = |-6 - 1| = |-7| = 7 ✓$

10. $3 - 2x = 11$ or $3 - 2x = -11$
$-2x = 8$ or $-2x = -14$
$x = -4$ or $x = 7$
Check: $|3 - 2(-4)| = |3 + 8| = 11 ✓$
$|3 - 2(7)| = |3 - 14| = |-11| = 11 ✓$

11. $3z + 2 = 4z + 5$ or $3z + 2 = -(4z + 5)$
$-z + 2 = 5$ or $3z + 2 = -4z - 5$
$-z = 3$ or $3z = -4z - 7$
$z = -3$ or $7z = -7$
$z = -1$

Check: $|3(-3) + 2| \overset{?}{=} 4(-3) + 5$
$|-9 + 2| \overset{?}{=} -12 + 5$
$|-7| \overset{?}{=} -7$
$7 \neq -7$
-3 is an extraneous root

$|3(-1) + 2| \overset{?}{=} 4(-1) + 5$
$|-3 + 2| \overset{?}{=} -4 + 5$
$|-1| \overset{?}{=} 1$
$1 = 1 ✓$
So, the solution is $z = -1$

12. $|3k + 8| + 4 = 0; |3k + 8| = -4$
No solution; the absolute value of a quantity cannot be negative.

13. To solve an absolute value equation, you use an equivalent compound equation. It is possible that values satisfying the equivalent equation do not satisfy the original equation and, thus, it is essential to check for extraneous solutions.

14. c; $-22 \leq 7x - 1 < 13$
$-21 \leq 7x \quad < 14$
$-3 \leq x \quad < 2$

Check a number between -3 and 2. Try $x = 1$.
$-22 \leq 7(1) - 1 < 13; -22 \leq 6 < 13 ; ✓$

15. $8t - 4 > 2t + 8$
$8t > 2t + 12$
$6t > 12$
$t > 2$

16. $2x > -8$ and $x - 3 \leq 0$
$x > -4$ and $x \leq 3$

17. $9 \leq -2z + 3 < 13$
$6 \leq -2z \quad < 10$
$-3 \geq z \quad > -5$
$-5 < \quad z \quad \leq -3$

18. $\dfrac{3y}{2} > 6$ or $y + 3 \leq 1$

$y > 4$ or $y \leq -2$

19. D; $-9 < 2x - 1 < 9$
$-8 < 2x \quad < 10$
$-4 < x \quad < 5$

20. $-7 < 2y - 3 < 7$
$-4 < 2y \quad < 10$
$-2 < y \quad < 5$

Check a number between -2 and 5. Try $y = 1$.
$|2(1) - 3| = |-1| = 1 < 7$

21. $5 + |2x + 1| > -4$

$\quad\quad |2x + 1| > -9$

The solution set is all real numbers because the absolute value of any quantity is positive or greater than 0. Check any number. Try $x = -3$.

$5 + |2(-3) + 1| = 5 + |-6 + 1| = 5 + 5 = 10 > -9$

22. $x > |2x + 1|; |2x + 1| < x;$

$-x < 2x + 1 \quad$ and $\quad 2x + 1 < x$

$-3x < 1$

$\quad x > -\dfrac{1}{3} \quad\quad$ and $\quad x + 1 < 0$

$\quad\quad\quad\quad\quad\quad\quad\quad\quad\quad x < -1$

no solution

23.

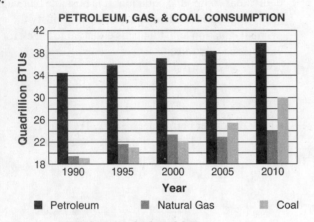

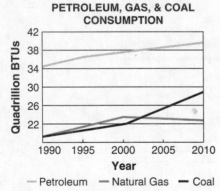

The bar graph illustrates the consumption of petroleum is by far the greatest over the entire period and has increased by an almost constant rate. Natural gas and coal were consumed in about equal amounts in 1990; gas was somewhat ahead of coal in 1995 and 2000; but after 2000, the consumption of coal is greater than that of natural gas. This relationship between the consumption of natural gas and coal is more apparent in the line graph.

24. Let x represent the distance Mia can drive the car.

$39 + .35x = 60$

$\quad\quad .35x = 21$

$\quad\quad\quad\quad x = 60 \text{ mi}$

25. Let x represent the cost for each ticket.

$300 + .3(200)x \le 540$

$\quad\quad .3(200)x \le 240$

$\quad\quad\quad\quad 60x \le 240$

$\quad\quad\quad\quad\quad x \le 4$

The most the committee can charge is \$4.00.

PERFORMANCE ASSESSMENT

USE OPERATION TABLES

a. The operation $\oplus$ is "normal" addition until you get a sum of 5 or more, when you subtract 5.

b.

$\otimes$	0	1	2	3	4
0	0	0	0	0	0
1	0	1	2	3	4
2	0	2	4	1	3
3	0	3	1	4	2
4	0	4	3	2	1

c. *Closure Property:* From the tables, the system is closed under $\oplus$ and $\otimes$.

Associative Property: Use the tables to test some cases.

Addition

$(1 \oplus 3) \oplus 4 \overset{?}{=} 1 \oplus (3 \oplus 4)$

$4 \oplus 4 \overset{?}{=} 1 \oplus 2$

$3 = 3$

The operation $\oplus$ appears to be associative.

Multiplication

$(2 \otimes 4) \otimes 3 \overset{?}{=} 2 \otimes (4 \otimes 3)$

$3 \otimes 3 \overset{?}{=} 2 \otimes 2$

$4 = 4$

The operation $\otimes$ appears to be associative.

Commutative Property: After using the tables to test some cases, both operations appear to be commutative.

Distributive Property: Use the tables to test some cases. The system appears to be distributive for $\otimes$ over $\oplus$.

Identity: 0 is the identity for $\oplus$.

$\quad\quad\quad\quad$ 1 is the identity for $\otimes$.

Inverses: Each element has an inverse under $\oplus$. The additive inverse of 0 is 0; the additive inverse of 1 is 4; the additive inverse of 2 is 3; the additive inverse of 4 is 1. It is not required that the additive identity, 0, have a multiplicative inverse. Each of the other elements has an inverse under $\otimes$. The multiplicative inverse of 1 is 1; the multiplicative inverse of 2 is 3; the multiplicative inverse of 3 is 2; the multiplicative inverse of 4 is 4. Thus, the system is a field.

FILL IN THE NUMBER LINE

The procedure does not work when one solution is a subset of the other solution.

Cumulative Review, page 102

1. D; $\sqrt{x} > 2$ and $\sqrt{x} \le 3$
   $x > 4$ and $\phantom{\sqrt{x}}x \le 9$

2. irrational

3. rational; $3.625 = 3\dfrac{625}{1000} = 3\dfrac{5}{8} = \dfrac{29}{8}$

4. $5 \cdot 3^2 - (2 - 5) = 5 \cdot 9 - (-3) = 45 + 3 = 48$

5. $3(9 - 2)^3 - \dfrac{1 - 4}{-3} = 3(7)^3 - \dfrac{-3}{-3} =$

   $3(343) - 1 = 1029 - 1 = 1028$

6. additive inverse

7. commutative property of addition

8. $\dfrac{4(y - 2)}{3} - -2y, 4y - 8 = 6y; 10y - 8 = 0;$

   $10y = 8; y = \dfrac{4}{5}$

   Check: $\dfrac{4\left(\frac{4}{5} - 2\right)}{3} \stackrel{?}{=} -2\left(\dfrac{4}{5}\right)$

   $\dfrac{4\left(-\frac{6}{5}\right)}{3} \stackrel{?}{=} -\dfrac{8}{5}$

   $-\dfrac{24}{15} \stackrel{?}{=} -\dfrac{8}{5}$

   $-\dfrac{8}{5} = -\dfrac{8}{5}\ \checkmark$

9. $7x - 3(2x - 1) = 4 - x$
   $7x - 6x + 3 = 4 - x$
   $x + 3 = 4 - x$
   $2x + 3 = 4$
   $2x = 1$
   $x = \dfrac{1}{2}$

Check: $7\left(\dfrac{1}{2}\right) - 3\left[2\left(\dfrac{1}{2}\right) - 1\right] \stackrel{?}{=} 4 - \dfrac{1}{2}$

$3\dfrac{1}{2} - 3[1 - 1] \stackrel{?}{=} 3\dfrac{1}{2}$

$3\dfrac{1}{2} - 0 \stackrel{?}{=} 3\dfrac{1}{2}$

$3\dfrac{1}{2} = 3\dfrac{1}{2}\ \checkmark$

10. $6 \le 3z - 12 < 15$
    $18 \le 3z < 27$
    $6 \le z < 9$

    $6 \le z < 9$

Check a number between 6 and 9. Try $z = 8$.
$6 \le 3(8) - 12 \stackrel{?}{\le} 15$
$6 \le 12 < 15\ \checkmark$

11. $4x + 1 > 9$ or $\dfrac{2x}{5} \le 0$

    $ 4x > 8$ or $2x \le 0$
    $ x > 2$ or $x \le 0$

Check a number less than 0. Try $x = -1$
$\dfrac{2(-1)}{5} = -\dfrac{2}{5} \le 0.\ \checkmark$

Check a number greater than 2. Try $x = 3$.
$4(3) + 1 = 13 > 9\ \checkmark$

12. $2|4z + 1| = 18; |4z + 1| = 9;$
    $4z + 1 = 9$ or $4z + 1 = -9$
    $4z = 8$ or $4z = -10$
    $z = 2$ or $z = -\dfrac{5}{2}$

    Check: $2|4(2) + 1| = 2|9| = 18\ \checkmark$

    $2\left|4\left(-\dfrac{5}{2}\right) + 1\right| = 2|-10 + 1| =$

    $2|-9| = 2 \cdot 9 = 18\ \checkmark$

13. $-7 \le 2x - 3 \le 7$
    $-4 \le 2x \le 10$
    $-2 \le x \le 5$

    Check a number between -2 and 5. Try $x = 1$.
    $|2(1) - 3| = |-1| = 1 \le 7\ \checkmark$

14. $3a - |a + 4| = 6; -|a + 4| = -3a + 6;$
$|a + 4| = 3a - 6;$

$a + 4 = 3a - 6$ or $a + 4 = -(3a - 6)$
$a = 3a - 10$ or $a + 4 = -3a + 6$
$-2a = -10$ or $4a + 4 = 6$
$a = 5$ or $4a = 2$
$a = \dfrac{1}{2}$

Check: $3(5) - |5 + 4| = 15 - 9 = 6$ ✓

$3\left(\dfrac{1}{2}\right) - \left|\dfrac{1}{2} + 4\right| = \dfrac{3}{2} - \dfrac{9}{2} = -3$

$\dfrac{1}{2}$ is an extraneous solution

So, there is one element in the solution set.

15. Answers will vary. *Bar graphs,* which are used to compare data at a given time. *Line graphs,* which are effective for showing trends and fluctuations in data. Multiple line graphs are useful for comparisons. *Circle graphs,* which are used to relate two or more values by representing them as part of a whole.

16. Use the monthly payment formula on page 6,

$M = \dfrac{Pr(1 + r)^n}{(1 + r)^n - 1}.$

$M = \dfrac{15,500(0.0067)(1 + 0.0067)^{36}}{(1 + 0.0067)^{36} - 1} \approx \$486$

Answers may vary due to rounding.

17. Total cost $= 486(3)(12) = \$17,496$

18. Total finance charge $=$ total cost $-$ amount financed $= 17,496 - 15,500 = \$1996$

19. $\begin{bmatrix} 12 + 2 & 6 + 0 & 10 + 3 \\ 0 + 5 & 15 + 15 & 2 + 12 \\ 9 + 8 & 3 + 12 & 8 + 9 \end{bmatrix} =$

$\begin{array}{c} \\ \text{red} \\ \text{white} \\ \text{blue} \end{array} \begin{array}{ccc} \text{S} & \text{M} & \text{L} \\ \end{array}$
$\begin{array}{c} \\ \text{red} \\ \text{white} \\ \text{blue} \end{array}\begin{bmatrix} 14 & 6 & 13 \\ 5 & 30 & 14 \\ 17 & 15 & 17 \end{bmatrix}$

20. Area of ring $= \pi(50)^2 = 2500$ ft^2
Area of equipment region $= 15 \cdot 10 = 150$ ft^2
Let P represent the probability that the towel lands in the equipment region.

$P = \dfrac{150}{2500\pi} \approx 0.0191$ or about 1.9%

21. mean $= \dfrac{5(7.7) + 7.8 + 8.0 + 8.9 + 9.0 + 9.4}{10} = 8.16$

The mean of 8.16 is not a good measure of central tendency since it is greater than 7 of the 10 values. The mode of 7.7 would be the best measure of central tendency since it represents 5 of the 10 values. The median of 7.75 could also be used.

1. Since the first triangle has hypotenuse equal to $\sqrt{2}$; the second, $\sqrt{3}$; and the third, $\sqrt{4}$, continuing this pattern means the 15th triangle has hypotenuse equal to $\sqrt{15 + 1} = \sqrt{16} = 4$

2. Let $x = 2.\overline{7}$
$\dfrac{10x = 27.\overline{7}}{9x = 25}$ multiply both sides by 10

 Subtract the first equation from the second

$x = \dfrac{25}{9}$ divide both sides by 9

3. 2; because $2 + 2 = 0$

4. $3 + x = 1 + 1; 3 + x = 2; x = 2 + (-3)$ Since the inverse of 3 is 1, $x = 2 + (-3) = 2 + 1 = 3$

5. $\dfrac{3x + 1}{4} = 10 - 6\left(\dfrac{1}{3}x - 1\right) + \dfrac{3}{4}$

$\dfrac{3x + 1}{4} = 10 - 2x + 6 + \dfrac{3}{4}$

$3x + 1 = 40 - 8x + 24 + 3$
$3x + 1 = 67 - 8x$
$3x = 66 - 8x$
$11x = 66$
$x = 6$

6. $2 < 3x - 1 \le 8$
$3 < 3x \le 9$
$1 < x \le 3$
The integer solutions are 2 and 3 $ 2 \cdot 3 = 6$

7. $3z - 1 = 8$ or $3z - 1 = -8$
$3z = 9$ or $3z = -7$
$z = 3$ or $z = -\dfrac{7}{3}$

Check: $|3(3) - 1| = |9 - 1| = 8$ ✓

$\left|3\left(-\dfrac{7}{3}\right) - 1\right| = |-7 - 1| = |-8| = 8$ ✓

$3 + \left(-\dfrac{7}{3}\right) = \dfrac{2}{3}$

8. $2x + 1 = 7$ or $2x + 1 = -7$
$2x = 6$ or $2x = -8$
$x = 3$ or $x = -4$

Check: $|2(3) + 1| = |6 + 1| = 7$ ✓
$|2(-4) + 1| = |-8 + 1| = |-7| = 7$ ✓
$3 + (-4) = -1$

9. $-3 < x - 1 < 3$
$-2 < x < 4$

The integer solutions are $-1, 0, 1, 2, 3$
$-1 + 0 + 1 + 2 + 3 = 5$

10. $-9 \leq 2x + 7 \leq 9$

$-16 \leq 2x \qquad \leq 2$

$-8 \leq x \qquad \leq 1$

The integer solutions are $-8, -7, -6, -5,$
$-4, -3, -2, -1, 0, 1$
$-8 + (-7) + (-6) + (-5) + (-4) + (-3) + (-2)$
$+ (-1) + 0 + 1 = -35$

11. $2x - |x + 1| = 7$

$-|x + 1| = 7 - 2x$

$|x + 1| = 2x - 7$

$x + 1 = 2x - 7 \quad$ or $\quad x + 1 = -(2x - 7)$

$x = 2x - 8 \quad$ or $\quad x + 1 = -2x + 7$

$-x = -8 \qquad$ or $\qquad 3x + 1 = 7$

$x = 8 \qquad\qquad$ or $\qquad 3x = 6$

$x = 2$

Check: $2(8) - |8 + 1| = 16 - 9 = 7$ ✓

$2(2) - |2 + 1| = 4 - 3 = 1 \neq 7$

2 is an extraneous solution.

So, the solution is $x = 8$

12. this year total sales $= 100 + 200 + 250 = 550$
last year total sales $= 250 + 100 + 300 = 650$
Difference $= 550 - 650 = -100$

13. By the monthly payment formula on page 6,

$$M = \frac{Pr(1 + r)^n}{(1 + r)^n - 1}.$$

For the 8.1% loan,

$$M = \frac{85000(0.00675)(1 + 0.00675)^{240}}{(1 + 0.00675)^{240} - 1} \approx \$716$$

For the 7.8% loan,

$$M = \frac{85000(0.0065)(1 + 0.0065)^{240}}{(1 + 0.0065)^{240} - 1} \approx \$700$$

Therefore, the difference is $\$716 - \$700 = \$16$.

14. $2A = \begin{bmatrix} 6 & -6 & 4 \\ -2 & 8 & 0 \end{bmatrix}$

$2A - B = \begin{bmatrix} 6 - 3 & -6 - 3 & 4 - 0 \\ -2 - (-1) & 8 - (-4) & 0 - 2 \end{bmatrix}$

$= \begin{bmatrix} 3 & -9 & 4 \\ -1 & 12 & -2 \end{bmatrix}$

Since $C = 2A - B$,

$\begin{bmatrix} 9 & -9 & x \\ -1 & 12 & y \end{bmatrix} = \begin{bmatrix} 3 & -9 & 4 \\ -1 & 12 & -2 \end{bmatrix}$

Therefore, $x = 4$ and $y = -2$.
So, $x + y = 4 + (-2) = 2$

To do Exercises 15. and 16. first the data must be placed in order.

80	72	64	60	54	47
77	71	63	58	50	45
75	70	62	57	49	43
73	66	60	56	48	40

15. mean $= \dfrac{\text{Sum of data entries}}{\text{Number of data entries}} = \dfrac{1440}{24} = 60$

median $=$ average of middle data entries

$= \dfrac{60 + 60}{2} = \dfrac{120}{2} = 60$

16. 1st quartile $=$ median of lower half of data
$= 50$
3rd quartile $=$ median of upper half of data
$= 70$
range $= 80 - 40 = 40$

Chapter 3 Functions and Graphs

Data Activity, pages 104–105

1. 1991-1992; 17,000 − 12,700 = $4400

2. $28,300; $219,700 (highest value) − $191,400 (lowest value) = $28,300

3. About 7,400 which is the median or middle number of the data set.

4. Answers will vary.

5.
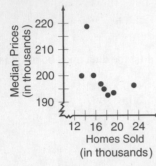

Answers will vary. Students may observe a negative correlation, there are more sales at lower prices.

6. Answers will vary. Students could show a vertical pictograph for homes sold and an annotated line graph for median sale price.

Lesson 3.1, pages 107–111

THINK BACK/WORKING TOGETHER

1. *F,* and *C, D* and *E*

2. Infinitely many

3. *A* and *E, B* and *D*

4. Infinitely many

5. $\overline{BE}$

6. $\overline{AD}$

7. 6

EXPLORE

8.
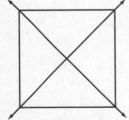

9.
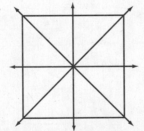

Other axes of symmetry are the vertical and horizontal lines drawn throughout the midpoint.

10.
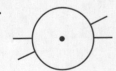

11.

12.

13. A, H, I, M, O, T, U, V, W, X, Y

14. B, C, D, E, H, I, K, O, X

15. H, I, N, O, S, X, Z

16. Clubs, hearts and spades, 2, 4, 10, jack, queen and king of diamonds; all cards except 7.

17. None

MAKE CONNECTIONS

18. b

19. c

20. a, c

21.
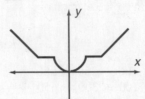

22.
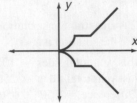

23.
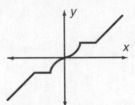

24a.
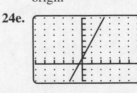

y-axis

24b.
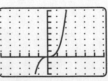

y-axis

24c.
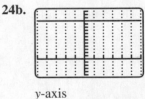

origin

24d.
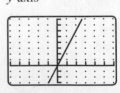

origin

24e.

24f.
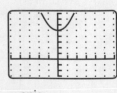

y-axis

24g.

24h.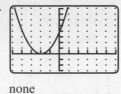

25.

26. Vertical line, $x = a$; horizontal line, $y = 0$, point $(a, 0)$; infinite number of lines of symmetry; one point of symmetry, $(a, 0)$

SUMMARIZE

27. Answers will vary. Possible answer; a figure folded along the x-axis is symmetric along the y-axis, a figure folded along the y-axis is symmetric along the y-axis, a figure rotated $180°$ along the point of origin that coincides with itself is symmetric along the point of origin.

28. Answers will vary.

29. Draw a line segment from each point perpendicular to the y-axis and extend an equal distance into Quadrant II and draw corresponding points.

30. Answers will vary. Possible solution, a circle centered at point of origin.

31. $(-3, 4); (-a, b)$ **32.** $(a, -b)$ **33.** $(-a, -b)$

34. $(0, 2)$ **35.** $x = -2$

PROJECT CONNECTION

1, 2, 4, and **5.** Answers will vary.

3. This function is the average of high and low prices and does not account for other prices.

ALGEBRAWORKS

1. $750.00 = \frac{3}{4}(1,000)$.

2. $6286; 900(0.50) deck + 600(3) 1st floor + 1500(2) 2nd floor + 400(1) Dormer + 252(1) garage + 512(0.75) pool

3. $1925; 750 land + 150 full and half bathroom + 125 fireplace + 200 CAC + 700 gas heat

4. $7928.24 = 1925 + 0.955018(6286)$

Lesson 3.2, pages 112–118

EXPLORE/WORKING TOGETHER

1. For each point (x, y) on the graph there is a corresponding point $(-x, y)$.

2. y-axis symmetry (examine graph).

3. $x^2 = (-x)^2$ so $2x^2 - 17 = 2(-x)^2 - 17$

TRY THESE

1.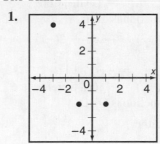

Domain: $\{-3, -1, 1\}$
Range: $\{-2, 4\}$
Quadrants: II, III, IV

2.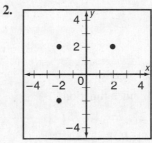

Domain: $\{-2, 2\}$
Range: $\{-2, 2\}$
Quadrants: I, II, III

3. not a function   **4.** function

5. Answers will vary. All functions are relations, not all relations are functions. A function is a relation in which each element in the domain is paired with one and only one element in the range.

6. $f(x) = x + 2$; D: $\{1, 2, 3\}$; R: $\{1, 4, 5\}$

7. $f(x) = x - 3$; D: $\{-2, 0, 5\}$; R: $\{-5, -3, 2\}$

8. $f(x) = -3x$; D: $\{-4, 1, 5\}$; R: $\{-15, -3, 12\}$

9. $f(x) = 2x$; D: $\{-2, 0, 4\}$; R: $\{-4, 0, 8\}$

10. not a function   **11.** function

12. $-47; (-3)^3 - 3(-3)^2 + 7 = -47$

13. $7; 0^3 - 3(0)^2 + 7 = 7$

14. $57; (5)^3 - 3(5)^2 + 7 = 57$

15.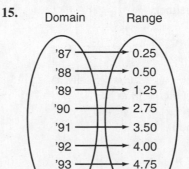

Relation is a function

16. $\frac{500}{3}\pi cm^3 = 53.05 cm^3$

It represents the volume of a sphere with radius 5 cm.

17.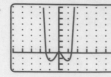

$f(-x) = (-x)^4 - 2(-x)^2$
$\quad\quad = x^4 - 2x^2$
$\quad\quad = f(x)$
$f(-x) = f(x)$ therefore function is even.

18.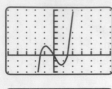

$g(-x) = (-x)^3 - 2(-x)$
$\quad\quad = -x^3 + 2x$
$\quad\quad = -g(x)$
$g(-x) = -g(x)$, therefore the function is ODD.

19.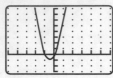

neither

PRACTICE

1. function

2. function

3. not a function

4. not a function

5. $f(x) = x + 3$; D: $\{-2, 1, 2\}$; R: $\{1, 4, 5\}$

6. $f(x) = x - 1$; D: $\{-2, 0, 4\}$; R: $\{-3, -1, 3\}$

7. $f(x) = \frac{1}{2}x$; D: $\left\{-2, -\frac{1}{2}, 1\right\}$; R: $\left\{-1, -\frac{1}{4}, \frac{1}{2}\right\}$

8. $f(x) = \frac{1}{4}x$; D: $\{-1, 2, 16\}$; R: $\left\{-\frac{1}{4}, \frac{1}{2}, 4\right\}$

9. $f(x) = x^2 - 1$; D: $\{-1, 3, 5\}$; R: $\{0, 8, 24\}$

10. $f(x) = x^3 - 3$; D: $\{-2, 2, 4\}$; R: $\{-11, 5, 61\}$

11. Answers will vary.

12. $T = 0.20x$; I

13. function

14. not a function

15. not a function

16. function

17. not a function

18. function

19. 1215

20. -10

21. -10.125

22. 7638

23.

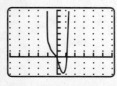

$f(x) = 3(-x)^4 - 2(-x)^2 - 2(-x)$
$f(-x) = 3x^4 - 2x^2 + 2x$
neither

24.

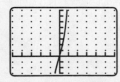

$g(-x) = 2(-x)^3 + 4(-x)$
$g(-x) = -2x^3 - 4x$
$g(-x) = -g(x)$
odd

25.

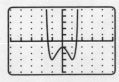

$h(-x) = (-x)^6 - 3(-x)^2 - 1$
$h(-x) = x^6 - 3x^2 - 1$
$h(-x) = h(x)$
even

26. Let x represent temperature on the thermometer in °C. The rule is $f(x) = 2x$ where x is all real numbers from -50 to 50. The function is odd since
$f(-x) = 2(-x) = -2x = -f(x)$

27. If Division is the domain the relation is a function. If Percentage is the domain, the relation is not a function since one element in the domain corresponds to two elements in the range.

28. $(2, 1)$

29. $(2, -1)$

30. $(2, -1)$

31.

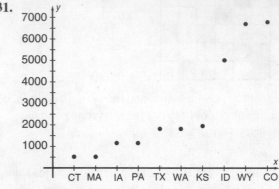

It is a function.

EXTEND

32. function, one-to-one

33. function, one-to-one

34. function, not one-to-one

35a. $L = 100 - x$

35b. $A(x) = x(100 - x)(x)$ *is a function*

35c. 475, 2176, 2500, 475

35d. $0 < x < 100$

35e. 2500 ft²

35f. no

36a. $108 - 4x$

36b. $v = x^2(108 - 4x)$; v is a function of x.

36c. 2200 in.³; 11,200 in.³

36d. $0 < x < 27$

36e. $v = 11,664$ in.³ = 18 in.

THINK CRITICALLY

37. If it is symmetric about the x−axis, then (x, y) and $(x, -y)$ are on the graph. Thus, a domain element will be paired with more than one range element. It cannot be a function.

38. yes

39. No, examples will vary. Let
$f(x) = 2x + 1$ so $f(2) = 5, f(7) = 15,$
$f(2 + 7) = f(9) = 19$ and $5 + 15 \neq 19$.

40. yes, $f(a) = Ca, f(b) = Cb$ so
$Ca + Cb = C(a + b) = f(a + b)$.

41. 3183 ft²

MIXED REVIEW

42. C

43. $f(x) = -3x$; D: $\{-5, -1, 3\}$; R: $\{-9, 3, 15\}$

44. $f(x) = \frac{1}{4}x$; D: $\{0, 12, 20\}$; R: $\{0, 3, 5\}$

Lesson 3.3, pages 119–125

EXPLORE/WORKING TOGETHER

1.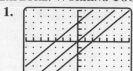

2. Up 3 units

3. Down 2 units

4.

	$x = 0$	$x = 1$	$x = 2$	$x = 3$	$x = 4$	$x = 5$
y_1	0	1	2	3	4	5
y_2	3	4	5	6	7	8
y_3	–2	–1	0	1	2	3

5. y_2 is 3 more than y_1.

6. y_3 is 2 less than y_1.

7. y_4 will be 10 units above y_1;
y_5 will be 9 units below y_1.

TRY THESE

1. Since the graph crosses the y-axis at $(0, 2)$ its y-intercept is 2. Select any two points on the graph and calculate the slope m.
$$m = \frac{y_1 - y_2}{x_1 - x_2} = \frac{0 - 2}{2 - 0} = \frac{-2}{2} = -1$$
Thus, an equation of the line is $y = -x + 2$.

2. The y-intercept is -1 and the slope m is -2 so an equation of the line is $y = -2x - 1$.

3. The y-intercept is 4 and the slope m is $\frac{1}{2}$ so an equation of the line is $y = \frac{1}{2}x + 4$.

4. For point-slope form, find the slope, $m = \frac{y_1 - y_2}{x_1 - x_2}$, then use the point-slope form, $y - y_1 = m(x - x_1)$,
$y - 4 = -2(x + 1)$
For slope-intercept find the y-intercept and the slope m,
$y = -2x + 2$.
For standard form express in terms of $Ax + By = C$;
$2x + y = 2$

5. point-slope:
$y - 6 = 5(x - 3)$ find slope, put in terms of:
$$y - y_1 = m(x - x_1)$$
slope-intercept:
$y = -5x - 9$ find slope and y-intercept
standard form:
$-5x + y = -9$ use form $Ax + By = C$

6. point-slope:
$y + 12 = -\frac{1}{4}(x - 4)$ find slope, put in terms of:
$$y - y_1 = m(x - x_1)$$
slope-intercept:
$y = \frac{1}{4}x - 11$ find slope and y-intercept
standard form:
$x + 4y = -44$ use form $Ax + By = C$

7. slope-intercept:
$y = -\frac{5}{8}x; m = \frac{5 - 0}{-8 - 0} = -\frac{5}{8}; y - 0 = -\frac{5}{8}(x - 0);$
$y = -\frac{5}{8}x$

8. slope-intercept:
$y = -\frac{1}{2}x + 7; m = \frac{8 - 4}{-2 - 6} = \frac{4}{-8} = -\frac{1}{2};$
$y - 8 = -\frac{1}{2}(x + 2); y = -\frac{1}{2}x + 7$

9. slope-intercept:
$y = -3x - 8; m = \frac{10 - -5}{-6 - -1} = \frac{15}{-5} = -3;$
$y - 10 = -3(x + 6); y = -3x - 8$

10. Use standard form, $Ax + By = C, 3x + 7y = 51;$
$m = \frac{9 - 6}{-4 - 3} = -\frac{3}{7}; y - 9 = -\frac{3}{7}(x + 4);$
$y = -\frac{3}{7}x + \frac{51}{7}; 7y = -3x + 51; 3x + 7y = 51$

11. Answers will vary.

12a. Use slope-intercept, point-slope or standard form to write the linear equation, $y = 15x + 200$.

12b. Use slope-intercept form. $y = mx + b$, to find $m = 15$. It represents the increase in cost for each additional guest.

13. Use the slope-intercept, $y = mx + b$ to determine an equation of the line $y - 3x - 2$. (Parallel lines have the same slope; $m = 3x; b = -2$.)

14. Use $y = mx + b, y = -\frac{1}{3}x + 6$. (Perpendicular lines are negative reciprocal slopes; $m = -\frac{1}{3}; b = 6$).

15. Set equations in slope-intercept form: $y = mx + b$, if slopes are the same, they are vertical translations of each other. Yes, it is a vertical translation and it is translated $1\frac{1}{2}$ units down.

16. Use slope-intercept form, yes it is a vertical translation up 3 units.

17. There is not enough information; need slope or another point on the line to find the slope.

18. Use graphing calculator to express the information as a linear equation, $y = -0.859x + 74.004$; the negative correlation is -0.998.

19. Use the equation from Exercise 18.

$f(x) = -0.859x + 74.004$

$f(35) = -0.859(35) + 74.004 = 43.9$

$f(45) = -0.859(45) + 74.004 = 35.3$

$f(55) = -0.859(55) + 74.004 = 26.8$

PRACTICE

1. Use $m = \dfrac{y_1 - y_2}{x_1 - x_2} = \dfrac{(3 - 6)}{(2 - 5)} = \dfrac{-3}{-3} = 1$

2. $\dfrac{(6 - 7)}{(4 - 3)} = \dfrac{-1}{1} = -1$

3. $\dfrac{(-5 - -6)}{7 - 2} = \dfrac{1}{5}$

4. $\dfrac{(-4 - 0)}{(2 - -6)} = \dfrac{-4}{8} = -\dfrac{1}{2}$

5. Use the point-slope form $y - y_1 = m(x - x_1)$; the slope-intercept, $y = mx + b$ and the standard form, $Ax + By = C$ to determine each of the equations.

$y - 3 = -3(x + 2)$

$y = -3x - 3$

$3x + y = -3$

6. $y - 7 = 2(x - 1)$

$y = 2x + 5$

$-2x + y = 5$

7. $y + 1 = -\dfrac{1}{2}(x - 6)$

$y = -\dfrac{1}{2}x + 2$

$x + 2y = 4$

Use the form $Ax + By = C$.

8. $8x - 3y = 42; m = \dfrac{10 - 2}{9 - 6} = \dfrac{8}{3};$

$y - 10 = \dfrac{8}{3}(x - 9); y = \dfrac{8}{3}x - 14; 3y = 8x - 42;$

$8x - 3y = 42$

9. $2x + 5y = 0; m = \dfrac{-4 - -2}{10 - 5} = -\dfrac{2}{5};$

$y + 4 = -\dfrac{2}{5}(x - 10); y = -\dfrac{2}{5}x; 5y = -2x;$

$2x + 5y = 0$

10. $x + 5y = 42; m = \dfrac{9 - 6}{-3 - 12} = \dfrac{3}{-15} = -\dfrac{1}{5};$

$y - 9 = -\dfrac{1}{5}(x + 3); y = -\dfrac{1}{5}x + \dfrac{42}{5};$

$5y = -x + 42; x + 5y = 42$

11. Use a linear equation to describe this function, $y = -10x + 700$; substituting 20 for x, the number of games that will sell at the price of $20 will be 500.

12. A parallel line has the same slope, so $y = \dfrac{1}{4}x$.

13. A perpendicular line has a negative inverse of the slope so, $y = -4x - 7$.

For numbers 14–17 put equations in form of slope-intercept, if slopes are the same they are vertical translations of each other.

14. yes, up 2 units

$6x + 3y = 15; 3y = -6x + 15; y = -2x + 5;$

$-7y = 14x - 49; y = -2x + 7; 7 - 5 = 2$

15. no

$y - 7 = 2x; y = 2x + 7; 4y = -2x + 12;$

$y = -\dfrac{1}{2}x + 3; 2x \neq -\dfrac{1}{2}x$

16. no

$y + 5 = 4x; y = 4x - 5; 4 + y = 5x; y = 5x - 4;$

$4x \neq 5x$

17. yes, down 1 unit

$6y = 9x + 6; y = \dfrac{3}{2}x + 1; 10y = 15x; y = \dfrac{3}{2}x;$

$1 - 0 = 1$

18. Answers will vary.

19. Use information in the chart to find a linear equation on your calculator, $y = 9.81x - 10.33$. The correlation is strong if it is close to 1; it is 0.87.

EXTEND

20. $m_{AB} = 2$ and $m_{AC} = -\dfrac{1}{2}$, and their product is -1. Thus triangle ABC is a right triangle.

21. $m_{AB} = -1$, $m_{AC} = \dfrac{4}{5}$, and $m_{AB} = 0$. No product is -1 and there is no right angle. Thus triangle ABC is not a right triangle.

22. $m_{AB} = -\dfrac{1}{2}, m_{BC} = 2, m_{CD} = -\dfrac{1}{2}, m_{AD} = 2$. The product of the slopes of AB and BC is $(-\dfrac{1}{2})(2) = -1$. Thus $ABCD$ is a rectangle because the opposite sides are parallel and it has a right angle.

23. $m_{AB} = 1, m_{BC} = -2, m_{CD} = 1$, and $m_{AD} = -2$, Since the opposite sides are parallel, $ABCD$ is a parallelogram.

24. No product of two slopes is -1.

THINK CRITICALLY

Use standard form, $Ax + By = C$ to answer the following.

$y = -\dfrac{A}{B}x + \dfrac{C}{B}; Ax + By = C; By = -Ax + C;$

$y = -\dfrac{A}{B}x + \dfrac{C}{B}$

25a. the slope: $-\dfrac{A}{B}$

25b. the y-intercept: $\dfrac{C}{B}$

25c. the x-intercept: $\dfrac{C}{A}$

$y = -\dfrac{A}{B}x + \dfrac{C}{B}$; x-intercept when $y = 0$

$0 = -\dfrac{A}{B}x + \dfrac{C}{B}; -\dfrac{A}{B}x = -\dfrac{C}{B}; x = \dfrac{C}{A}$

26. $m = \dfrac{8 - 2}{5 - 3} = \dfrac{6}{2} = 3; y - 2 = 3(x - 3);$

$y = 3x - 7; 3x - y = 7$

27. $Ax + By = C$

$By = -Ax + C$

$y = -\dfrac{A}{B}x + C$

The slope is $-\dfrac{A}{B}$. The slope of a line perpendicular to

this line is $\dfrac{B}{A}$. The line passes through the origin so the

y-intercept $C = 0$. The equation is $y = \dfrac{B}{A}x$ or

$Ay - Bx = 0$.

28. $3x + 2ky + 9 = 0$

$2ky = -3x - 9$

$y = -\dfrac{3}{2k} - x - \dfrac{9}{2k}$

$-\dfrac{9}{2k} = -6$

$-9 = -12k$

$\dfrac{3}{4} = k$

29. $n = 4 \cdot 12 = 48; r = 0.08 \div 12 \approx 0.00667;$

$P = 10{,}000; M = \dfrac{Pr(1 + r)^n}{(1 + r)^n - 1} =$

$\dfrac{10{,}000(0.00667)(1.00667)^{48}}{(1.00667)^{48} - 1} = \$244.13;$ the deferred

payment price is $48(244.13) = \$11{,}718.24$.

30. $n = 3 \cdot 12 = 36; r = 0.09 \div 12 = 0.0075;$

$P = 8950; M = \dfrac{Pr(1 + r)^n}{(1 + r)^n - 1} =$

$\dfrac{8950(0.0075)(1.0075)^{36}}{(1.0075)^{48} - 1} = \$284.61;$ the deferred

payment price is $36(284.61) = \$10245.96$

31. There are two posssible chances of drawing a red ten from the deck of 52 cards, $\dfrac{2}{52} = 0.038$ so C is right.

Lesson 3.4, pages 126–131

THINK BACK

1. Use the absolute value and the negative value or disjunction for a, b, c.

1a. $4x = 10$ and $4x = -10$

1b. $2x + 7 = 5$ and $2x + 7 = -5$

1c. $3x - 1 = 11$ and $3x - 1 = -11$

2a. $-5, 5$

$3x = 15; 3x = -15$

$x = 5; \quad x = -5$

2b. $1, 7$

$4 - x = 3; 4 - x = -3$

$-x = -1; -x = -7$

$x = 1; \quad x = 7$

2c. $-2, 5$

$2x - 3 = 7; 2x - 3 = -7$

$2x = 10; \quad 2x = -4$

$x = 5; \quad x = -2$

3. 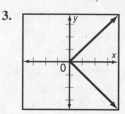 **4.**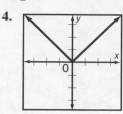

EXPLORE/WORKING TOGETHER

5.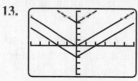

6. The graph of f is a straight line, the graph of g is V-shaped.

7. $(0, 0)$; point of origin **8.** upward

9. For the graph of f, the symmetries are the point of origin, point symmetry and for the graph of g, y-axis symmetry.

10. The absolute value function yields y-values ≥ 0. The y-values are positive only in Quadrants I and II.

11. By "flipping" over the x-axis.

12. The reflection is the negative of the function so, $-x$

13. 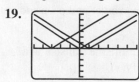 **14.** Examine the graph to find the vertex of each function. f: $(0,0)$; g: $(0, 4)$; h: $(0, -2)$.

15. They all have the y-axis as their line of symmetry and are all V-shaped.

16. Look at graph to find relationship to other graphs. g is 4 units above f.

17. 2 units below.

18. If C is positive, the graph of $j(x) = |x| + C$ is translated up C units from the graph of f. If C is negative, the graph is translated C units down.

19. 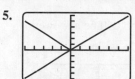 **20.** Examine the graph to determine this. f: $(0,0)$; g: $(-3, 0)$; h: $(1, 0)$.

21. Look at the graph to answer this. They are all V-shaped, their lines of symmetry are f: y − axis, g: $x = -3$, h: $x = 1$.

22. Examine the graph, 3 units left.

23. Examine the graph, 1 unit right.

24. If b is positive, the graph of $j(x) = |x - b|$ is translated b units right. If b is negative, the graph is translated b units left.

25. If is V-shaped, it is shifted 1 unit left and 4 units down; $(-1, -4)$; symmetric to $x = -1$.

26. Answers will vary.

27. 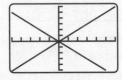 The graph of g is V-shaped, but the V is narrower or steeper than f. The graph of h is V-shaped, but the V is wider than for f.

28a. In the case of $a > 1$, the graph will be narrower

28b. If $0 < a < 1$ the graph will be wider.

29. g is a reflection because the graph of $f(x) = |x|$ is "flipped" over the x-axis to obtain $g(x) = -|x|$,

30. Examine the graph, the graph of f is a reflection of g over the x-axis.

31. Graph the equation $g(x) = 5|x - 3| + 2$ to find the vertex, $(3, 2)$; the x-value is 3.

32. By examining the graph or by substituting the x-coordinate of the vertex into the equation $y = a|x - b| + c$ and solve for y.

33. The equation of the line is $x =$ "the x-coordinate of the vertex."

34. Solve the equation, setting the absolute value to zero to determine the vertex or graph the equation and examine to determine vertex. The line of symmetry is x.

34a. $(0, 6)$; $x = 0$ **34b.** $(8, 0)$; $x = 8$

34c. $(1, 4)$; $x = 1$

MAKE CONNECTIONS

35. Determine how the graphs differ without graphing by utilizing the information that any whole positive number will produce a narrower graph and any positive fraction will produce a wider graph and a negative number will invert the graph.

35a. V is wider **35b.** V is narrower and opens down

35c. V is narrower

36. Examine the equations to determine the translation of each function from the graph of $f(x) = |x|$.

36a. 4 units left; 2 units up.

36b. 3 units right; 1 unit down.

36c. 3 units right.

37. From the graph $f(x) = |x|$ which is V-shaped with a vertex of $(0, 0)$ determine which of the translations match with each graph.

37a. I (3 units to the left)

37b. III (3 units up)

37c. II (2 units to the right and 1 unit up).

38. Use the information that a fraction times the $|x|$ will yield a wider graph than $|x|$ and a whole number multiplied by $|x|$ yields a narrower graph and a negative number times $|x|$ will produce a downward V.

38a. II **38b.** I **38c.** III

39. Using the information on positions of translations write functions that represents the graphs.

39a. $g(x) = |x + 5| - 2$ **39b.** $g(x) = |x - 7| + 3$

39c. $g(x) = |x - 6| - 1$ **39d.** $g(x) = |x + 8| + 4$

40. Determine the function of the form $f(x) = |x - b| + C$ by examining each graph.

40a. $f(x) = |x + 4| - 1$ **40b.** $f(x) = |x| + 4$

40c. $f(x) = -\frac{1}{4}|x| - 4$

41. Use your graphing utility on your calculator to graph the equations $y = |4 - x|$ and $y = 3$, find their points of intersection by examining the graph; $(1, 3)$ and $(7, 3)$.

SUMMARIZE

42. Answers will vary. **43.** Answers will vary.

44. No, the graphs of h and j are reflections of each other over the x-axis.

45. The graph of $f(x) = -x$ is a line, the graph of $f(x) = |-x|$ is a V. The portion of $f(x) = -x$ in Quadrant IV is reflected into Quadrant I for $f(x) = |-x|$. The line of reflections is the x-axis.

46. Answers will vary.

47a. **47b.**

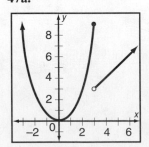

48. A piecewise function is a function that is defined differently over various parts of its domain.
$f(x) = |x|$ as a piecewise function is:
$$f(x) = \begin{cases} x, & x \geq 0 \\ -x, & x < 0 \end{cases}$$

1.

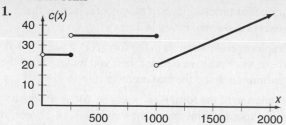

2. The commission for transactions over $2000 is 20%, so the equation of the function is, $C(x) = 0.02x$; Domain is D: $\{x > 1000\}$; Range is R: $\{y > 20\}$.

3. The flat rate for a $500 investment is $35 (see chart) to make the 2% commission rate equivalent to the flat rate the investor would have to invest $1750.00.
$C(35) = 0.02x$
$\dfrac{35}{0.02} = x$; $1750 = x$

4. The value of this commission is $64, the stockbroker wants the client to make money so they will give him repeat business.

Lesson 3.5, pages 132–136

THINK BACK

1. Even functions are symmetric about the y-axis.

2. Odd functions are symmetric about the point of origin.

3. Answers will vary.   **4.** Answers will vary.

5. Answers will vary.

EXPLORE / WORKING TOGETHER

6.

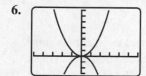

7. f is U shaped and opens upward; g is U shaped and opens downward; both have vertices at $(0, 0)$ and both are symmetric about the y-axis.

8. Reflect it about the x-axis.

9a.   **9b.**   **9c.**   **9d.**

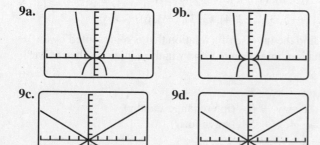

10. The graph of $-f(x)$ is a reflection of the graph of f over the x-axis.

11.

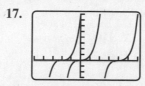

12. Examine the graph to find the vertices, f: $(0, 0)$; g: $(0, 3)$; h: $(0, -1)$.

13. Examine the graph, they are all U shaped and symmetric about the y-axis.

14. Examine the graph, g is 3 units up in relation to the graph of f.

15. h is 1 unit down in relation to the graph of f.

16. If c is positive, the graph of $j(x) = f(x) + c$ is up c units from the graph of f. If c is negative the graph is translated c units down.

17.

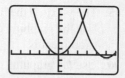

18. Examine the graph, f: $(0, 0)$, g: $(-2, 0)$; h: $(4, 0)$

19. Examine the graph, g is 2 units left in relation to f.

20. Examine the graph, h is 4 units right in relation to f.

21. If $b > 0$ the graph of $f(x - b)$ is b units to the right. If $b > 0$ the graph of $f(x - b)$ is b units to the left.

22. The graph is the same shape; shifted 5 units right and $\frac{1}{2}$ units down from the graph of f, $\left(5, -\frac{1}{2}\right)$; $x = 5$.

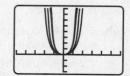

23. Answers will vary.   **24.** Answers will vary.

25. The graph of g is narrower than the graph of f; the graph of h is narrower than the graphs of f and g; the graph of j is wider than the graph of f.

26. The graph of $k(x) = af(x)$ is a dilation of the graph of f. If $a > 1$, the graph is narrower. If $0 < a < 1$, the graph is wider.

27. Answers will vary.

28. Examine your graphs from 27. U shape; symmetric about the y-axis.

29. Examine the graphs, lie in quadrants I and II.

30. Raising any number x to a positive even power yields a positive number.

31. In both cases the values are positive.

32. Answers will vary.   **33.** Answers will vary; origin.

34. I and III

35. Negative values when $x < 0$, and when $x > 0$ the values are positive.

36a. $g(x) = -x^5$ **36b.** $g(x) = |x|$

36c. $g(x) = -x + 1$ **36d.** $g(x) = x - 3$

36e. $g(x) = -\sqrt{x}$ **36f.** $g(x) = -\dfrac{1}{x}$

37a. right and up   **37b.** left and down   **37c.** left and up

38a. wider   **38b.** narrower, reflected and about x-axis

38c. narrower

39. Examine the graphs,

39a. III   **39b.** II   **39c.** I

40. Examine the graphs,

40a. II   **40b.** III   **40c.** I

41. Examine the graphs,

41a. I   **41b.** III   **41c.** II

42. Using the form $g(x) = (x - b)^{11} + c$

42a. $g(x) = (x + 8)^{11} + 1$

42b. $g(x) = (x - 5)^{11} - 4$

SUMMARIZE

43. Answers will vary.

44. No, $f(x)$ is a translation 1 unit down from $y(x)$.

45. The y-axis is the line of reflection for $f(-x)$ and $f(x)$. Use the graphing capabilities on the calculator to examine this.

46. Use the graphing capabilities on your calculator to graph the function $h(t) = 200t - 16t^2$ where $t = 0$, the rocket hits the ground when $h(t) = 0$.

46a. $t = 12.5$ sec.

46b. Graph the equation $j(t) = 200(t - 3) - 16(t - 3)^2$ on the calculator and use the trace feature to determine the height of Amy's rocket when John's hits the ground, 456 feet.

47a. 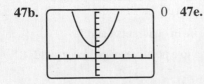 1   **47d.** 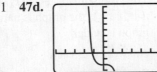 1

47b. 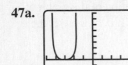 0   **47e.** 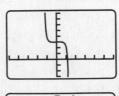 1

47c. 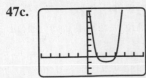 2   **47f.** 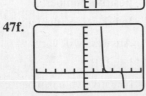 1

48. Examine graphs from question 47, the minimum number of times a graph of this form can cross the x-axis is 0, the maximum is 2.

49. Graph several functions in the form $f(x) = a(x - b)^n + c$ where n is a positive odd number, you will find that the minimum is 1 and the maximum is 1.

50. A function of the form $f(x) = a(x - b)^n + c$ with an even exponent n intersects the x-axis 0, 1, or 2 times. If n is odd the function will intersect the x-axis only once.

Lesson 3.6, pages 137–142

EXPLORE/WORKING TOGETHER

1. Answers will vary, some possible answers include
$f: \{(-5, -2), (-3, 2), (0, 8), (1, 10)\}$
$g: \{(-2, -5), (2, -3), (8, 0), (10, 1)\}$

2. Answers will vary but the coordinates of the points that satisfy one equation satisfy the other equation if they are reversed.

3. $f(x)$ multiplies the input by 2 and then adds 8.; $g(x)$ subtracts 8 from the input and divides by 2.

4. f and g coincide, see graph:

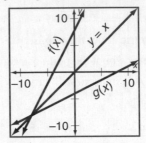

TRY THESE

The inverse of a relation can be found by interchanging the first and second coordinates of each ordered pair in the relation.

Example: (2, 1) inverse is (1, 2).

1. $\{(2, 1), (4, 3), (6, 5), (8, 7)\}$

2. $\{(4, 2), (6, 3), (8, 4), (10, 5)\}$

3. $\{(5, 2), (7, 3), (5, -4), (9, 2)\}$

4. $\{(-1, 3), (-2, 4), (3, 3), (-3, 6)\}$

To find the inverse of a relation that is represented by an equation, interchange x and y in the equation. Then solve for y.

5. $y = x - 5$
    $x = y - 5$ (interchange x and y)
    $y = x + 5$ (solve for y)

6. $y = x - 4$

7. $y = -2x + 4$

8. $y = -4x + 16$

9. Answers will vary.

10. f^{-1} is the inverse function of f. Find f^{-1} and graph it along with f and $y = x$.

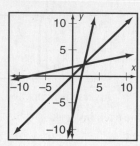

$f^{-1}(x) = \frac{1}{5}x + \frac{8}{5}$, see graph

$$f(x) = 5x - 8$$
$$x = 5y - 8$$
$$5y = x + 8$$
$$y = \frac{1}{5}x + \frac{8}{5}$$
$$f^{-1}(x) = \frac{1}{5}x + \frac{8}{5}$$

11.

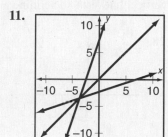

$f^{-1}(x) = \frac{1}{3}x - \frac{7}{3}$

$$f(x) = 3x + 7$$
$$x = 3y + 7$$
$$3y = x - 7$$
$$y = \frac{1}{3}x - \frac{7}{3};$$
$$f^{-1}(x) = \frac{1}{3}x - \frac{7}{3}$$

12.

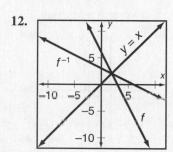

$f^{-1}(x) = -\frac{1}{2}x + 3$

$$f(x) = -2x + 6$$
$$x = -2y + 6$$
$$-2y = x - 6$$
$$y = -\frac{1}{2}x + 3;$$
$$f^{-1}(x) = -\frac{1}{2}x + 3$$

13.

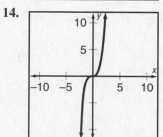

$f^{-1}(x) = -\frac{1}{3}x + 1$

$$f(x) = -3x + 3$$
$$x = -3y + 3$$
$$-3y = x - 3$$
$$y = -\frac{1}{3}x + 1;$$
$$f^{-1}(x) = -\frac{1}{3}x + 1$$

14.

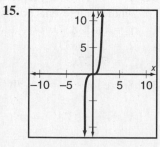

The inverse is a function. Use the horizontal line test.

15.

The inverse is a function. Use the horizontal line test.

16.

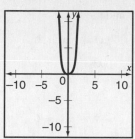

17.

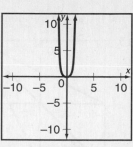

The inverse is not a function. Use the horizontal line test.

The inverse is not a function. Use the horizontal line test.

Examine the graphs to determine each inverse.

18. d   **19.** c   **20.** b   **21.** a

22. Find the inverse of the function:
$$f(x) = 4x - 9$$
$$x = 4y - 9 \quad \text{Substitute } x \text{ for } y$$
$$y = \frac{1}{4}x + \frac{9}{4} \quad \text{Solve for } y$$
$$f^{-1}(x) = \frac{1}{4}x + \frac{9}{4}$$

Graph to find that the inverse is a function.

PRACTICE

For 1–4., invert the x and y values to find the inverse of each relation or function.

1. $\{(1, 4), (0, 5), (6, 6), (3, 7)\}$

2. $\{(3, 4), (5, 6), (7, 8), (9, 10)\}$

3. $\{(4, -1), (2, -3), (1, -1), (2, 2)\}$

4. $\{(-4, 4), (-1, 5), (3, 4), (-3, 3)\}$

Invert x and y and solve for y.

5. $y = x + 2$ **6.** $y = x - 7$ **7.** $y = -x + 2$
  $x = y + 2$ $x = y - 7$ $x = -y + 2$
  $y = x - 2$ $y = x + 7$ $y = -x + 2$

8. $y = -x - 8$ **9.** $y = 8x - 4$ **10.** $y = 6x - 3$
  $x = -y - 8$ $x = 8y - 4$ $x = 6y - 3$
  $y = -x + 8$ $y = \frac{1}{8}x + \frac{1}{2}$ $y = \frac{1}{6}x + \frac{1}{2}$

11. $y = 7x + 15$ **12.** $y = 3x - 10$
  $x = 7y + 15$ $x = 3y - 10$
  $y = \frac{1}{7}x - \frac{15}{7}$ $y = \frac{1}{3}x + \frac{10}{3}$

13. Answers will vary.

14. Inverse is $y = \frac{9}{5}x + 32$, use the horizontal line test to find that yes it is a function.

$$y = \frac{5}{9}(x - 32); \ x = \frac{5}{9}(y - 32); \ x = \frac{5}{9}y - \frac{160}{9};$$
$$\frac{5}{9}y = x + \frac{160}{9}; \ y = \frac{9}{5}x + 32$$

Find f^{-1} and graph along with f and the line $y = x$.

15.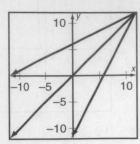

$f^{-1}(x) = \frac{1}{2}x + 6$

$f(x) = 2x - 12$
$x = 2y - 12$
$2y = x + 12$
$y = \frac{1}{2}x + 6$

$f^{-1}(x) = \frac{1}{2}x + 6$

16.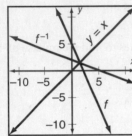

$f^{-1}(x) = \frac{1}{4}x - 2$

$f(x) = 4x + 8$
$x = 4y + 8$
$4y = x - 8$
$y = \frac{1}{4}x - 2$

$f^{-1}(x) = \frac{1}{4}x - 2$

17.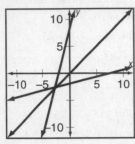

$f^{-1}(x) = -\frac{1}{3}x + 2$

$f(x) = -3x + 6$
$x = -3y + 6$
$-3y = x - 6$
$y = -\frac{1}{3}x + 2$

$f^{-1}(x) = -\frac{1}{3}x + 2$

18.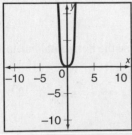

$f^{-1}(x) = -\frac{1}{2}x + 5$

$f(x) = -2x + 10$
$x = -2y + 10$
$-2y = x - 10$
$y = -\frac{1}{2}x + 5$

$f^{-1}(x) = -\frac{1}{2}x + 5$

Graph each function f. Use the horizontal line test to determine whether the inverse is also a function.

19.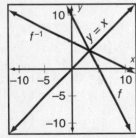

Inverse is not a function.

20.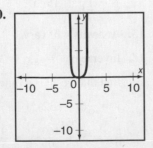

Inverse is not a function.

21.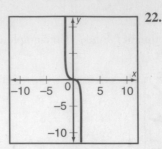

Inverse is a function.

22.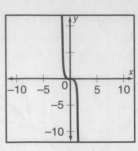

Inverse is a function.

Examine the graphs to determine each inverse.

23. b   **24.** a   **25.** d   **26.** c

Graph the inverse and use the vertical line test to determine if the inverse is a function.

27.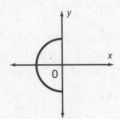

Inverse is not a function.

28.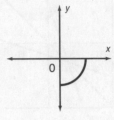

Inverse is a function.

29.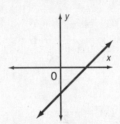

Inverse is a function.

30.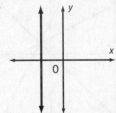

Inverse is not a function.

31. Find the inverse of the function $E = MC^2$ by substituting E for M and then solving for M:
$E = MC^2$
$M = EC^2$
$M = \frac{E}{C^2}$

Graph and determine if it is a function using the horizontal line test; the inverse is a function.

32. $y = 0.013837x$, x represents inches; $y = 72.27x$, where x represents points; the inverse is a function.
$y = 0.013837x$
$x = 0.013837y$
$y = 72.27x$

33. Solve for r, $r = \sqrt[3]{\frac{3V}{4\pi}}$; It is a function.

$V = \frac{4\pi r^3}{3}$; $3V = 4\pi r^3$; $\frac{3V}{4\pi} = r^3$; $r = \sqrt[3]{\frac{3V}{4\pi}}$

EXTEND

It is possible to restrict the domain to ensure that it has an inverse function.

34. $\{x: x \geq 0\}$ or
$\{x: x \leq 0\}\{x: x \geq 5\}$

35. $\{x: x \geq 0\}$ or
$\{x: x \leq 0\}\{x: x \geq 0\}$

36. $\{x: x \geq -6\}$
$\{x: x \geq 0\}$

37. $\{x: x \geq -3\}$
$\{x: x \geq 0\}$

38. Solve for r, $r = \dfrac{C}{2\pi}$, yes this is a function (use horizontal line test).

$C = 2\pi r; r = \dfrac{C}{2\pi}$

THINK CRITICALLY

39. Examine the graphs to discover they are perpendicular to $y = x$.

40. If $m \neq 0$, the graph of $y = mx + b$ is a non-horizontal line that passes the horizontal line test. If $m = 0$, the graph $y = b$ is a horizontal line that would not pass the horizontal line test.

41. $f(x) = \dfrac{x + 2}{x}$

$x = \dfrac{y + 2}{y}$

$y = \dfrac{2}{x - 1} \ (x \neq 1)$

$f^{-1}(x) = \dfrac{2}{x - 1} \ (x \neq 1)$

42. $f(x) = \dfrac{x - 3}{x}$

$x = \dfrac{y - 3}{y}$

$y = \dfrac{-3}{x - 1} \ (x \neq 1)$

$f^{-1}(x) = \dfrac{-3}{x - 1} \ (x \neq 1)$

43. No, if $f(x) = mx + b$ then $f^{-1}(x) = \dfrac{1}{m}x - \dfrac{b}{m}$, perpendicular lines have slopes whose product is -1. Since $m \cdot \dfrac{1}{m} \neq -1$, the function with slope $\neq 0$ is not perpendicular to its inverse.

44. If the inverse is not a function, then when a message is decoded, one input could have two inputs, it would be hard to determine which was the correct and which was the incorrect message.

MIXED REVIEW

45. A parallel line to the line $y = 5x + 2$ has the same slope: $y = 5x - 3$

46. $y = -\dfrac{1}{5}x + 4$ The perpendicular line has the negative inverse of the slope m.

47. Median: 9   3, 5, 7, 8, ⑨, 10, 11, 12, 25

48. Median: 0 $-4, -3, -2, -1,$ ⓪$, 1, 4, 7, 9$

49. $M = 250$
$r = 0.07 \div 12 \approx 0.00583333$
$n = 4 \cdot 12 = 48$
$P = \dfrac{M[(1 + r)^n - 1]}{r(1 + r)^n} =$
$\dfrac{250[(1.00583333)^{48} - 1]}{0.00583333(1.00583333)^{48}} = \$10,440.05; \text{B}$

EXPLORE

1. 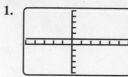 When x increases $h(x)$ remains at 1.

2. 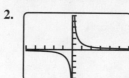 As x increases in $f(x) = \dfrac{1}{x}$, the values decrease toward O.

3. 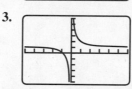 As x increases in, the function $g(x) = \dfrac{1}{x} + 1$, the values decrease toward 1; g is a vertical translation of f by 1 unit up.

4. 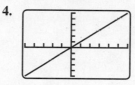 As x increases in $j(x) = x$, the values increase.

5. 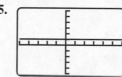 As x increases, the values remain at 1.

6. 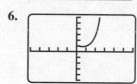 As x increases, the values increase.

7. Answers will vary. The number 1 raised to any exponent is 1, and $1 + \dfrac{1}{x}$ approaches 1 as x increases. As the base decreases toward 1, the exponent increases so it is difficult to determine what will have more influence.

8. 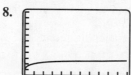 The values increase as x increases and then remain at around 2.718.

TRY THESE

1. $(f + g)(x) = f(x) + g(x)$
$= 4x + x + 6$
$= 5x + 6$
$(f - g)(x) = f(x) - g(x)$
$= 4x - (x + 6)$
$= 3x - 6$
$(f \cdot g)(x) = f(x) \cdot g(x)$
$= 4x(x + 6)$
$= 4x^2 + 24x$

2. $(f + g)(x) = f(x) + g(x)$
$= 5x + x - 8$
$= 6x - 8$
$(f - g)(x) = f(x) - g(x)$
$= 5x - (x - 8)$
$= 4x + 8$
$(f \cdot g)(x) = f(x) \cdot g(x)$
$= 5x(x - 8)$
$= 5x^2 - 40x$

3. $(f + g)(x) = f(x) + g(x)$
$= 2x - 1 + 3x^2$
$= 3x^2 + 2x - 1$
$(f - g)(x) = f(x) - g(x)$
$= 2x - 1 - (3x^2)$
$= -3x^2 + 2x - 1$
$(f \cdot g)(x) = f(x) \cdot g(x)$
$= 2x - 1(3x^2)$
$= 6x^3 - 3x^2$

4. $(f + g)(x) = f(x) + g(x)$
$= 3x + 2 + 2x^2$
$= 2x^2 + 3x + 2$
$(f - g)(x) = f(x) - g(x)$
$= 3x + 2 - (2x^2)$
$= -2x^2 + 3x + 2$
$(f \cdot g)(x) = f(x) \cdot g(x)$
$= 3x + 2(2x^2)$
$= 6x^3 + 4x^2$

5. $D_f = x \geq 0; D_g =$ all real numbers
$\left(\dfrac{f}{g}\right)(x) = \dfrac{\sqrt{x}}{x + 3}; x \geq 0$ and $x \neq -3$

6. $\left(\dfrac{g}{f}\right)(x) = \dfrac{x + 3}{\sqrt{x}}; x > 0$

7.
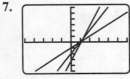
$(f + g)(x) = 3x - 3$

8.
$(f + g)(x) = x^2 + x - 2$

9. $(f \circ g)(3) = f(g(3))$
$= f(3 \cdot 3)$
$= (9)^2$
$= 81$
$(g \circ f)(3) = g(f(3))$
$= g(3^2)$
$= 3 \cdot (3^2)$
$= 3 \cdot 9$
$= 27$

10. $(f \circ g)(3) = f(g(3))$
$= f(-2 \cdot 3)$
$= 2(-6)^2$
$= 72$
$(g \circ f)(3) = g(f(3))$
$= g(2 \cdot (3)^2)$
$= -2 \cdot 18$
$= -36$

11. $(f \circ g)(x) = f(g(x))$
$= f(-x)$
$= 4 \cdot (-x^2)$
$= 4x^2$
$(g \circ f)(x) = g(f(x))$
$= g(4x^2)$
$= -4x^2$

12. $(f \circ g)(x) = f(g(x))$
$= f(-3x)$
$= 5 \cdot (-3x)^2$
$= 45x^2$
$(g \circ f)(x) = g(f(x))$
$= g(5x^2)$
$= -3(5x^2)$
$= -15x^2$

13. $(f \circ g)(x) = f(g(x))$
$= f(-x^2)$
$= -x^2 + 1$
$(g \circ f)(x) = g(f(x))$
$= g(x + 1)$
$= -(x + 1)^2$
$= -x^2 - 2x - 1$

14. $(f \circ g)(x) = f(g(x))$
$= f(-2x^2)$
$= -2x^2 - 2$
$(g \circ f)(x) = g(f(x))$
$= g(x - 2)$
$= -2(x - 2)^2$
$= -2x^2 + 8x - 8$

15. Answers will vary.

16. Use the graphics on p.147 to answer these.
16a. 1
16b. $g(x) = x^2$
16c. $f(x) = x - 3$
16d. $(f \circ g)(x) = x^2 - 3$
Any two functions f and g are inverse functions if and only if $(f \circ g)(x) = x$ for all x in the domain of g and $(g \circ f)(x) = x$ for all x in the domain of f.

17. yes

$$f(x) = 2x - 4; f(g(x)) = f\left(\frac{1}{2}x + 2\right) =$$
$$2\left(\frac{1}{2}x + 2\right) - 4 = x + 4 - 4 = x$$
$$g(x) = \frac{1}{2}x + 2; g(f(x)) = g(2x - 4) =$$
$$\frac{1}{2}(2x - 4) + 2 = x - 2 + 2 = x$$

18. no

$$f(x) = \frac{1}{4}x - 6; f(g(x)) = f(4x + 6) =$$
$$\frac{1}{4}(4x + 6) - 6 = x + \frac{3}{2} - 6 = x - \frac{9}{2}$$
$$g(x) = 4x + 6; g(f(x)) = g\left(\frac{1}{4}x - 6\right) =$$
$$4\left(\frac{1}{4}x - 6\right) + 6 = x - \frac{3}{2} + 6 = x + \frac{9}{2}$$

19. yes

$$f(x) = -x + 25; f(g(x)) = f(-x + 25) =$$
$$-(-x + 25) + 25 = x - 25 + 25 = x$$
$$g(x) = -x + 25; g(f(x)) = g(-x + 25) =$$
$$-(-x + 25) + 25 = x - 25 + 25 = x$$

20a. $f(x) = x - 10$ (price of suit before sale)

20b. $g(x) = 0.70x$ (price of suit to nonemployee during sale) $1 - 0.3 = 0.7$

20c. $h(x) = f(g(x)) = 0.70x - 10$ or
$h(x) = g(f(x)) = 0.70x - 7$

20d. $h(x) = 0.70x - 10$, Greg saves \$3 if the store takes 30% and then gives him the employee discount.

PRACTICE

1. $(f + g)(x) = f(x) + g(x)$
$\qquad = 3x + x - 4$
$\qquad = 4x - 4$
$(f - g)(x) = f(x) - g(x)$
$\qquad = 3x - (x - 4)$
$\qquad = 2x + 4$
$(f \cdot g)(x) = f(x) \cdot f(g)$
$\qquad = 3x(x - 4)$
$\qquad = 3x^2 - 12x$

2. $(f + g)(x) = f(x) + g(x)$
$\qquad = 6x + x + 10$
$\qquad = 7x + 10$
$(f - g)(x) = f(x) - g(x)$
$\qquad = 6x - (x + 10)$
$\qquad = 5x - 10$
$(f \cdot g)(x) = f(x) \cdot g(x)$
$\qquad = 6x(x + 10)$
$\qquad = 6x^2 + 60x$

3. $(f + g)(x) = f(x) + g(x)$
$\qquad = 3x - 2 + 4x^2$
$\qquad = 4x^2 + 5x - 2$
$(f - g)(x) = f(x) - g(x)$
$\qquad = 5x - 2 - 4x^2$
$\qquad = -4x^2 + 5x - 2$
$(f \cdot g)(x) = f(x) \cdot g(x)$
$\qquad = 5x - 2(4x^2)$
$\qquad = 20x^3 - 8x^2$

4. $(f + g)(x) = f(x) + g(x)$
$\qquad = 4x - 4 + 5x^2$
$\qquad = 5x^2 + 4x - 4$
$(f - g)(x) = f(x) - g(x)$
$\qquad = 4x - 4 - 5x^2$
$\qquad -5x^2 + 4x - 4$
$(f \cdot g)(x) = f(x) \cdot g(x)$
$\qquad = 4x - 4(5x^2)$
$\qquad = 20x^3 - 20x^2$

5. $\dfrac{f}{g}(x) = \dfrac{3x + 1}{2x + 1}; \left\{x: x \neq -\dfrac{1}{2}\right\};$

$\dfrac{g}{f}(x) = \dfrac{2x + 1}{3x + 1}; \left\{x: x \neq -\dfrac{1}{3}\right\}$

6. $\dfrac{f}{g}(x) = \dfrac{4x - 3}{x + 2}; \{x: x \neq -2\}$

$\dfrac{g}{f}(x) = \dfrac{x + 2}{4x - 3}; \left\{x: x \neq -\dfrac{3}{4}\right\}$

7. $\dfrac{f}{g}(x) = \dfrac{\sqrt{x - 3}}{x^2}; \{x: x \geq 3\}$

$\dfrac{g}{f}(x) = \dfrac{x^2}{\sqrt{x - 3}}; \{x: x > 3\}$

8. $\dfrac{f}{g}(x) = \dfrac{\sqrt{x + 2}}{2x^2}; \{x: x \geq -2 \text{ and } x \neq 0\}$

$\dfrac{g}{f}(x) = \dfrac{2x^2}{\sqrt{x + 2}}; \{x: x > -2\}$

9.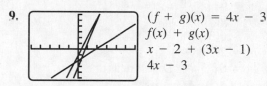
$(f + g)(x) = 4x - 3$
$f(x) + g(x)$
$x - 2 + (3x - 1)$
$4x - 3$

10.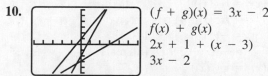
$(f + g)(x) = 3x - 2$
$f(x) + g(x)$
$2x + 1 + (x - 3)$
$3x - 2$

11.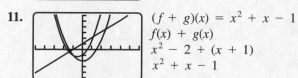
$(f + g)(x) = x^2 + x - 1$
$f(x) + g(x)$
$x^2 - 2 + (x + 1)$
$x^2 + x - 1$

12.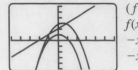

$(f + g)(x) = -x^2 + x + 3$
$f(x) + g(x)$
$-x^2 + 1 + (x + 2)$
$-x^2 + x + 3$

13.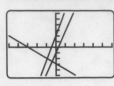

$(f - g)(x) = 5x + 2$
$f(x) - g(x)$
$4x - 1 - (-x - 3)$
$4x - 1 + x + 3$
$5x + 2$

14.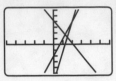

$(f - g)(x) = 5x - 6$
$f(x) - g(x)$
$3x - 2 - (-2x + 4)$
$3x - 2 + 2x - 4$
$5x - 6$

15. $(f \circ g)(4) = f(g(4))$
$= f(4(4) + 1)$
$= -2(17)^2$
$= -578$
$(g \circ f)(4) = g(f(4))$
$= g(-2(4)^2)$
$= 4(-32) + 1$
$= -127$

16. $(f \circ g)(4) = f(g(4))$
$= f(2(4) - 6)$
$= -4(2)^2$
$= -16$
$(g \circ f)(4) = g(f(4))$
$= g(-4(4)^2)$
$= 2(-64) - 6$
$= -134$

17. $(f \circ g)(x) = f(g(x))$
$= f(2x - 4)$
$= 3(2x - 4)$
$= 6x - 12; D: \{\text{all reals}\}$
$(g \circ f)(x) = g(f(x))$
$= g(3x)$
$= 2(3x) - 4$
$= 6x - 4 \quad D: \{\text{all reals}\}$

18. $(f \circ g)(x) = f(g(x))$
$= f(5x + 4)$
$= -(5x + 4) + 1$
$= -5x - 3 \quad D: \{\text{all reals}\}$
$(g \circ f)(x) = g(f(x))$
$= g(-x + 1)$
$= 5(-x + 1) + 4$
$= -5x + 9 \quad D: \{\text{all reals}\}$

19. $(f \circ g)(x) = f(g(x))$
$= f\left(\frac{1}{2}x^3\right)$
$= 6\left(\frac{1}{2}x^3\right)^2$
$= \left(\frac{3}{2}x^6\right) \quad D: \{\text{all reals}\}$
$(g \circ f)(x) = g(f(x))$
$= g(6x^2)$
$= \frac{1}{2}(6x^2)^3$
$= (108x^6) \quad D: \{\text{all reals}\}$

20. $(f \circ g)(x) = f(g(x))$
$= f(x^2 - 3)$
$= \frac{2}{x^2 - 3} \quad D: \{x: x \neq \pm\sqrt{3}\}$
$(g \circ f)(x) = g(f(x))$
$= g\left(\frac{2}{x}\right)$
$= \left(\frac{2}{x}\right)^2 - 3$
$= \frac{4}{x^2} - 3; D: \{x: x \neq 0\}$

21. $(f \circ g)(x) = f(g(x))$
$= f(x + 3)$
$= 5(x + 3)^2$
$= 5x^2 + 30x + 45 \quad D: \{\text{all reals}\}$
$(g \circ f)(x) = g(f(x))$
$= g(5x^2)$
$= 5x^2 + 3; D: \{\text{all reals}\}$

22. $(f \circ g)(x) = f(g(x))$
$= f(5x)$
$= \sqrt{5x + 4}; D: \left\{x: x \geq -\frac{4}{5}\right\}$
$(g \circ f)(x) = g(f(x))$
$= g\left(\sqrt{x + 4}\right)$
$= 5\left(\sqrt{x + 4}\right)$
$= 5\sqrt{x + 4}; D: \{x: x \geq -4\}$

One can determine if any two functions f and g are inverse functions if and only if $(f \circ g)(x) = x$ for all x in the domain of g and $(g \circ f)(x) = x$ for all the x in the domain of f.

23. yes

$f(x) = 3x + 9; f(g(x)) = f\left(\frac{1}{3}x - 3\right) =$

$3\left(\frac{1}{3}x - 3\right) + 9 = x - 9 + 9 = x$

$g(x) = \frac{1}{3}x - 3; g(f(x)) = g(3x + 9) =$

$\frac{1}{3}(3x + 9) - 3 = x + 3 - 3 = x$

24. yes

$f(x) = 5x + 20; f(g(x)) = f\left(\frac{1}{5}x - 4\right) =$

$5\left(\frac{1}{5}x - 4\right) + 20 = x - 20 + 20 = x$

$g(x) = \frac{1}{5}x - 4; g(f(x)) = g(5x + 20) =$

$\frac{1}{5}(5x + 20) - 4 = x + 4 - 4 = x$

25. yes

$f(x) = -x + 16; f(g(x)) = f(-x + 16) =$
$-(-x + 16) + 16 = x - 16 + 16 = x$
$g(x) = -x + 16; g(f(x)) = g(-x + 16) =$
$-(-x + 16) + 16 = x - 16 + 16 = x$

26. no

$f(x) = \frac{4}{x}; f(g(x)) = f\left(-\frac{4}{x}\right) = \frac{4}{-\frac{4}{x}} =$

$\frac{4x}{-4} = -x$

$g(x) = -\frac{4}{x}; g(f(x)) = g\left(\frac{4}{x}\right) = \frac{-4}{\frac{4}{x}} = \frac{-4x}{4} = -x$

27. no

$f(x) = x^2 - 4; f(g(x)) = f\left(\sqrt{x - 2}\right) =$

$\left(\sqrt{x - 2}\right)^2 - 4 = x - 2 - 4 = x - 6$

$g(x) = \sqrt{x - 2}; g(f(x)) = g(x^2 - 4) =$

$\sqrt{(x^2 - 4) - 2} = \sqrt{x^2 - 6}$

28. yes

$f(x) = x^2 - 5; f(g(x)) = f\left(\sqrt{x + 5}\right) =$

$\left(\sqrt{x + 5}\right)^2 - 5 = x + 5 - 5 = x$

$g(x) = \sqrt{x + 5}; g(f(x)) = g(x^2 - 5) =$

$\sqrt{(x^2 - 5) + 5} = \sqrt{x^2} = x$

29a. Monthly net earnings are $v(x) = 0.05x$

29b. Broker's fee function is $b(x) = 0.08x$

29c. $t(x) = b(v(x))$ represents the broker's fee as a function of Elisa's monthly earnings $T(x) = 0.004x$
$b(v(x)) = b(0.05x) = 0.08(0.05x) = 0.004x$

30. $v(x) = 0.05(4800) = \$240$
$t(x) = b(v(x))$
$\quad = b(.05x)$
$\quad = 0.08(0.05x)$
$\quad = 0.004x$
$\quad = 19.20$
$240.00 + 19.20 = \$259.20$

31a. $r(x) = x - 200$ (cost after rebate function)

31b. $C(x) = 0.85x$ (discounted cost function)

31c. $r(C(x)) = 0.85x - 200$ (composite function for final cost) $r(C(x)) = r(.85x) = .85x - 200$

31d. $C(r(x)) = 0.85x - 170$ (final cost if you take the rebate first) $C(r(x)) = C(x - 200) = .85(x - 200) = .85x - 170$

31e. $C(r(x)) = C(x - 200)$
$\quad = 0.85(x - 200)$
$\quad = 0.854 - 170$
$\quad = 0.85(5000) - 170$
$\quad = \$4080$ (rebate)

31e. $r(C(x)) = r(0.85x)$
$\quad = 0.85x - 200$
$\quad = 0.85 (5000) - 200$
$\quad = \$4050$ (discount first)
Take the discount first.

32. Answers will vary.

EXTEND

33. No, if $g(x) = 4\sqrt{x}$ and $f(x) = x + 2$, then $(f \circ g)(x) = 4\sqrt{x} + 2$ also.

34. "Decompose" the composite function for 34–37.
$f(x) = 3x - 5; g(x) = x^3; f(g(x)) = f(x^3) = 3x^3 - 5 = h(x)$

35. $f(x) = x - 6; g(x) = \sqrt{x}; f(g(x)) = f(\sqrt{x}) = \sqrt{x} - 6 = h(x)$

36. $f(x) = x + 9; g(x) = \sqrt{x}; f(g(x)) = f(\sqrt{x}) = \sqrt{x} + 9 = h(x)$

37. $f(x) = x^2 + x + 1; g(x) = x^2; f(g(x)) = f(x^2) = x^4 + x^2 + 1 = h(x)$

38. $y(p) = 137.1676p$, where p represents the number of British pounds, y represents U.S. dollars; $137.1676 = 8717 \div 0.6355$

39. $p(f) = 8.0407f$, where f represents the number of French francs, p represents British pounds and D represents U.S. dollars; $f = 0.1957D; \frac{f}{0.1957} = D$;

$D = 0.6355p; \left(\frac{f}{0.1957}\right)\left(\frac{1}{0.6355}\right) = p$

40. $R(x) = 250x + 0.625$

41. $\$3.00; R(x) = 250(0.0095) + 0.625$

42. $S(t) = 9\pi t^4$

$S(r(t)) = 4\pi\left(\frac{3}{2}t^2\right)^2$

$\quad = 4\pi\frac{9}{4}t^4$

$\quad = 9\pi t^4$

43. $S(10) = 9\pi(10)^4 = 282{,}743.3 mm^2$

THINK CRITICALLY

44. Let $f(x) = ax + b$ and $g(x) = cx + d$, where $a, b, c,$ and d are constants, then $f(g(x)) = a(cx + d) + b = acx + ad + b$, which is a linear function with slope ac and y-intercept $ad + b$.

45. Let $f(x) = h$ and $g(x) = k$, where h and k are constants, then $f(g(x)) = f(k) = h$.

46. No, for example, if $f(x) = x$ and $g(x) = x$, then $(f - g)(x) = 0$, which is not one-to-one.
$f(x) - g(x)$
$x - x = 0$

47. $h(x)$ will equal 0 if $f(x) = 0$ and/or $g(x) = 0$.

48. $(f \circ g)(x^2) = f(g(x^2))$
$= f(3x^2 + 5)$
$= 9x^4 + 30x^2 + 25$

49. $(g \circ f)(x^2) = g(f(x^2))$
$= g(x^2)^2$
$= 3(x^4) + 5$
$= 3x^4 + 5$

50. $(f \circ f)(x) = f(f(x))$
$= f(x^2)$
$= (x^2)^2$
$= x^4$

51. $(g \circ g)(x) = g(g(x))$
$= g(3x + 5)$
$= 3(3x + 5) + 5$
$= 9x + 20$

MIXED REVIEW

52. $-32 < x < 44$
$-38 < x - 6 < 38$
$-32 < x < 44$

53. $x < -58$ or $x > 42$
$x + 8 > 50$ or $x + 8 < -50$
$x > 42$ or $x < -58$

54. $\$250,000(0.30) = \$75,000$ downpayment
$m = \dfrac{Pr(1 + r)^n}{(1 + r)^n - 1}, p = \$175,000; r = 0.10 \div 12;$
$n = 360$
$m = 1535.750248$
$1535.750248 \times 360 = 552,870.0893$
$552,870.0893 + 75,000 = \$627,870.09$

55. $\dfrac{6}{52} = 0.115$

56. A; $f(g(-2)) = f(2(-2) + 8) = -3(4)^2 = -48$
$g(f(-2)) = g(-3(-2)^2) = 2(-12) + 8 = -16$

57. a: positive (points upward)
b: positive (right)
c: negative (below y-axis)
n: odd (observe shape)

58. a: positive (upward)
b: negative (left of x-axis)
c: negative (below y-axis)
d: odd (observe shape)

59. a: negative (downward)
b: negative (left of x-axis)
c: positive (above y-axis)
d: even (observe U shape)

ALGEBRAWORKS

1. $300,000 \times 0.05 = \$15,000$

2. $f(x) = 0.03x$, where x is the selling price of the house.

3. $100,000(0.03) = 3,000$
$3,000 \times 2 = 6,000$
$6,000 \times 12 = \$72,000/\text{yr}$

4a. $110,000(0.06) = 6,600$
$102,000(0.06) = 6,120$
$6,120 \div 2 = 3,060$
$6,600 - 3,060 = \$3,540$ difference

4b. $110,000 - 6,600 = \$103,400$ left

4c. $102,000 - 6,120 = \$95,880$ left

4d. $102,000 - 3,060 = \$98,940$ left

5. $W = \dfrac{0.05(12\,MD)}{52}$

6. Disadvantages: irregular pay, no base salary, dependence on economy. Advantages: flexible hours, if you're good, you can make a lot of money.

Lesson 3.8, pages 152–155

EXPLORE THE PROBLEM

1. $S = kP$

2. Substitute known values for S and P into $S = kP$ and solve for k.

3. $S = kP, 1.75 = 25k$
$k = \dfrac{1.75}{25}, k = 0.07; S = 0.07P; k$ represents a sales tax rate of 7%.

4. $\$6.02; \dfrac{6.02}{86} = 0.07$ and $\dfrac{1.75}{25} - 0.07$; the ratios are the same and equal the constant of variation.

5. $\dfrac{1.75}{25} = \dfrac{S}{135}; 236.25 = 25S; \$9.45 = S$; check using the equation method.

6. $\dfrac{1.75}{25} = \dfrac{74.13}{P}; 1.75p = 1853.25; p = \$1059.$

7. A linear function.

8. A direct variation is a case of $y = mx + b$ where $b = 0$ and $m = k$, so k is the slope of the graph. A direct variation will always contain the point $(0, 0)$.

INVESTIGATE FURTHER

9. Using $A = \frac{1}{2}bh$; $A = \frac{1}{2} \cdot 5 \cdot 5$; $A = 12.5$ units^2.

10. Use $A = \frac{1}{2}bh$ to find Area for chart, A over x for second column:

Area	$\dfrac{\text{Area}}{X}$
0.5	0.5
2	1
45	1.5
8	2
18	3

11. No, if $x = 2$ the area is doubled, but not for any other values of x. If x is tripled, the area is not tripled.

$x = 1; A = .5$ $x = 1; A = .5$
$x = 2; A = 2$ $x = 3; A = 45$
   not double   not triple

12. No, the ratio $\dfrac{\text{area}}{x}$ is not constant.

13. Use $A = bh$ to complete chart:

Area	$\dfrac{\text{Area}}{X}$
4	4
8	4
12	4
16	4
20	4
24	4

APPLY THE STRATEGY

14. yes, $x = 2$, $A = 8$ is doubled, $x = 4$, $A = 16$.
yes, $x = 2$, $A = 8$ is tripled, $x = 6$, $A = 24$.

15. yes; the ratio $\dfrac{\text{Area}}{x}$ is constant.

16. Answers will vary.

17. When $x = 9$, $y = 30$; so $30 = 9k$; $k = \dfrac{10}{3}$
$y = \dfrac{10}{3}(24)$; $y = 80$

18. When $x = 14$, $y = 103.6$; so $103.6 = 14k$;
$k = 7.4$
$y = 7.4(10)$; $y = 74$

19. When $x = 20$, $y = 6.8$; so $6.8 = 20k$;
$k = 0.34$
$17 = 0.34x$;
$x = 50$

20a. $I = 0.08L$

20b. 128; $I = 0.08(1600) = 128$

21a. $k = 2.54$ or $k = 0.3937$; $\dfrac{139.7}{55} = 2.54$

21b. 304.8 cm; $k = 2.54$; $2.54(120) = 304.8$

22. 3.25 in.; $1.5(8) = 12$; $3.25(8) = 26$; $k = 8$.

23. 5013 bracelets; $4488.75 \div 225 = 19.95$; $k = 19.95$; $100{,}000 \div 19.95 = 5012.53$, so 5013 bracelets are needed.

24a. $A = \dfrac{1}{2}x^2$

24b. 98 sq. units; $A = \dfrac{1}{2}(14)^2 = 98$.

25a. yes, $b = ka$ so $k = \dfrac{b}{a}$, Then $k(2a) = \dfrac{b}{a}(2a) = 2b$, so $(2b, 2a)$ is on the graph.

25b. $k = \dfrac{b}{a}$, then $k(2a + 1) = \dfrac{b}{a}(2a + 1)$. By the distributive property, $\dfrac{b}{a}(2a) + \dfrac{b}{a}$ which is equal to $2a + 1$ if and only if $\dfrac{b}{a} = 1$. So, $(a + 2, b + 2)$ is on the graph if $k = 1$; otherwise it is not.

REVIEW PROBLEM SOLVING STRATEGIES

1. \$17.50; The strategy is work backward. Karen had no money left after paying \$20 for sneakers, so she must have had \$10 and borrowed \$10 before buying the sneakers. Since the jeans cost \$20, she had \$30 before buying the jeans. Since half of that was borrowed, she had \$15 after buying the jacket. Before buying the jacket, she had \$35, half of which she had borrowed. So she had \$17.50 when she first arrived at the flea market.

2. 1:2; Let t = sum of teachers ages
      r = sum of rock singers ages
      m = number of teachers
      n = number of rock singers

$\dfrac{t}{m} = 50$; $t = 50m$; $\dfrac{r}{n} = 35$; $r = 35n$;

$\dfrac{t + r}{n + m} = 40$;

$\dfrac{50m + 35n}{m + n} = 40$

$50m + 35n = 40m + 40n$
$\quad\quad 10m = 5n$
$\quad\quad \dfrac{m}{n} = \dfrac{1}{2}$

3. A possible arrangement is

A1	B2	C3	D4	E5
C4	D5	E1	A2	B3
E2	A3	B4	C5	D1
B5	C1	D2	E3	A4
D3	E4	A5	B1	C2

Chapter Review, pages 156–157

1. d   **2.** c   **3.** b   **4.** a
5. $(2, -3)$

6. $(-2, -3)$

7. $(-2, 3)$

8. not a function

$1 \begin{smallmatrix} \nearrow 2 \\ \searrow 6 \end{smallmatrix}$ A number in the domain cannot correspond to two numbers in the range.

9. function

10. $f(0) = 3(0)^3 - 2(0)^2 - 0 + 4 = 4$

11. $f(-1) = 3(-1)^3 - 2(-1)^2 - (-1) + 4 =$
       $3(1) - 2(1) + 1 + 4 = 0$

12. $f(2) = 3(2)^3 - 2(2)^2 - 2 + 4 =$
       $3(8) - 2(4) - 2 + 4 = 18$

13. $f(-3) = 3(-3)^3 - 2(-3)^2 - (-3) + 4 =$
       $3(-27) - 18 + 3 + 4 = -92$

14. $y = -\dfrac{3}{2}x, \dfrac{(y_1 - y_2)}{(x_1 - x_2)} = \dfrac{(6 - 0)}{(-4 - 0)} = -\dfrac{3}{2}$

    $y - 0 = -\dfrac{3}{2}(x - 0)$

           $y = -\dfrac{3}{2}x$

Standard form is $Ax + By = C$; $3x + 2y = 0$.

15. $y = -x, \dfrac{(-1 - -7)}{(1 - 7)} = \dfrac{-6}{6} = -1$

    $y + 1 = -1(x - 1)$

        $y = -x$

Standard form: $x + y = 0$

16. $y = -\dfrac{1}{4}x + \dfrac{13}{4}, \left(\dfrac{(2 - 3)}{(5 - 1)} = -\dfrac{1}{4}\right)$

    $y - 3 = -\dfrac{1}{4}(x - 1)$

        $y = -\dfrac{1}{4}x + \dfrac{13}{4}$

Standard form: $x + 4y = 13$

17. $y = -\dfrac{1}{6}x - 7$, a perpendicular line has a negative

reciprocal of slope, m.

    $m = -\dfrac{1}{6}; b = -7$

18. $f(x) = |x|; g(x) = |x - 7| - 4$

19. $f(x) = |x|; g(x) = -|x|$

20. $g(x) = (x - 2)^4 + 9$; right 2 units and up 9 units.

21. $h(x) = (x + 5)^4 + 7$; left 5 units and up 7 units.

22. $j(x) = (x - 3)^2 - 6$; right 3 units and down 6 units.

23. $y = x - 5$ **24.** $y = 2x - 4$ **25.** $y = -4x + 8$
    $x = y - 5$ $x = 2y - 4$ $x = -4y + 8$
    $y = x + 5$ $y = \dfrac{1}{2}x + 2$ $y = -\dfrac{1}{4}x + 2$

26. $y = -x + 4$
    $x = -y + 4$
    $y = -x + 4$

27. $\{(3, 9), (5, 1), (6, -2), (-3, -3)\}$; yes it is a function.

28. $f^{-1}(x) = -\dfrac{x}{3} + \dfrac{5}{3}$ $f(x) = -3x + 5$

                              $x = -3y + 5$

                              $y = -\dfrac{x}{3} + \dfrac{5}{3}$

yes it is a function.

29. $(f + g)(-1) = f(-1) + g(-1)$
                 $= 3(-1)^2 + (-1 - 5)$
                 $= 3 + -6$
                 $= -3$

30. $(f \circ g)(-1) = f(-1) \cdot g(-1)$
                 $= 3(-1)^2 \cdot (-1 - 5)$
                 $= 3 \cdot -6$
                 $= -18$

31. $(f \circ g)(-1) = f(g(-1))$
                 $= f(-1 - 5)$
                 $= 3(-6)^2$
                 $= 108$

32. $(g \circ f)(-1) = g(f(-1))$
                 $= g(3(-1)^2)$
                 $= 3 - 5$
                 $= -2$

33. $S = kp$; $58.50 = .05p$; $p = \$1,170$

34. $y = 70$ when $x = 20$, when $x = 35, y = 122.5$;
    $k = 3.5$
    $y = kx$

    $70 = k(20)$ $y = \dfrac{7}{2}x$

     $k = \dfrac{7}{2}$ $\dfrac{7}{2}(3.5) = 122.5$

Chapter Assessment, pages 158–159

CHAPTER TEST

1. B $(a, -b)$ has x-axis symmetry with (a, b).

2. origin   Graph nos. 2-5 and examine to

3. y-axis   determine what type of

4. none   symmetry each graph has.

5. origin

6. Answers will vary. $\nearrow 4$ A number in the domain

7. Not a function $3 \rightarrow 5$ may not correspond to
                              $\searrow 6$ more than one number

8. function   in the range.

9. $f(-5) = 2(-5)^3 - 3(-5)^2 - (-5) + 7$
           $= 2(-125) - 75 + 12$
           $= -313$

10. $f(0) = 2(0)^3 - 3(0)^2 - 0 + 7$
       $= 7$

11. $f(-2) = 2(-2)^3 - 3(-2)^2 - (-2) + 7$
$= 2(-8) - 12 + 9$
$= -19$

12. $f(1) = 2(1)^3 - 3(1)^2 - 1 + 7$
$= 2 - 3 - 1 + 7$
$= 5$

13. $y = -3x + 19; m = \dfrac{10 - 4}{3 - 5} = -\dfrac{6}{2} = -3;$
$y - 10 = -3(x - 3); y = -3x + 19$

14. $y = -2x - 4$ (parallel line has same slope).
$m = -2, b = -4$

15. right 7 units and down 6 units

16. Graph the function $f(x) = |x + 3|$. The vertex is $(-3, 0)$ and the line of symmetry is $x = -3$.

17. $g(x) = |x| + 2$; vertex is $(0, 2)$; line of symmetry is y-axis or $x = 0$.

18. $g(x) = |x + 9| - 3$

19. $g(x) = -x^7$

20. $g(x) = |2x|$

21. narrower (whole number multiplied by x^4 makes it narrower).

22. wider (positive fraction multiplied by x^4 makes graph wider).

23. Answers will vary.

24. $\{(5, 5), (1, 2), (2, 4), (1, 6)\}$; no

25. D $(f(g(x))$ and $g(f(x))$ must be equal to x to be inverse functions.
$f(x) = 3x\ g(x) = x + 2$

26. $(f + g)(x) = 4x + 2$;{all real numbers}
$f(x) + g(x) = 3x + (x + 2) = 4x + 2$

27. $(f - g)(x) = 2x - 2$;{all real numbers}
$f(x) - g(x) = 3x - (x + 2) = 3x - x - 2 = 2x - 2$

28. $(f \cdot g)(x) = 3x^2 + 6x$;{all real numbers}
$f(x) \cdot g(x) = 3x(x + 2) = 3x^2 + 6x$

29. $(f/g)(x) = \dfrac{3x}{x + 2}$; $\{x: x \neq -2\}$
$\dfrac{f(x)}{g(x)} = \dfrac{3x}{x + 2}$

30. $(f \circ g)(x) = 3x + 6$;{all real numbers}
$f(g(x)) = f(x + 2) = 3(x + 2) = 3x + 6$

31. $(g \circ f)(x) = 3x + 2$;{all real numbers}
$g(f(x)) = g(3x) = 3x + 2$

32. $k = 4.5$, so $y = 67.5$
$y = kx; 36 = k(8); k = 4.5; 15(4.5) = 67.5$

33. $k = 2.5$, so $x = 36$
$y = kx; 45 = k(18); k = 2.5; 90 = 2.5x; x = 36$

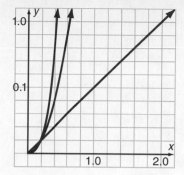

a. 0, 1

b. $0 < x < 1$

c. $x > 1$

Cumulative Review, page 160

1.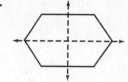

2.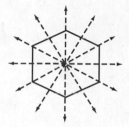

3. D; Each point in the Domain corresponds with only one value in the range.

4. If there is no vertical line that intersects the graph in more than one place, then the graph represents a function. Exactly one point of intersection between the graph and a vertical line means that for the particular value of x, there is exactly one value of y. If this is true for every value of x, then the definition of function is satisfied.

5. $f(-1) = |(-1)^3 - 1| = |-1 - 1| = |-2| = 2$

6. $m = \dfrac{7 - 3}{0 + 2} = \dfrac{4}{2} = 2; y - 3 = 2(x + 2)$ or
$y - 7 = 2x$

7. These two linear equations produce parallel lines since they have the same slope, $-\dfrac{3}{2}$.

The graph of the second equation is a vertical shift of the first equation by 6 units downward (the y-intercept of the first is 4 and the y-intercept of the second is -2).

$2y + 3x = 8; \quad 6y = -9x - 12 \qquad$ Since $m = m$

$2y = -3x + 8 \quad y = -\dfrac{3}{2}x - 2 \qquad\qquad -\dfrac{3}{2} = -\dfrac{3}{2}$

$y = -\dfrac{3}{2}x + 4 \qquad\qquad\qquad$ then $4 - -2 = 6$

8. $g(x) = |x + 6| - 3$

9. $(-4, 9), (5, 9)$

10. C (Examine the graphs).

11. $f^{-1}(x) = 3x - 1$ $\qquad y = \dfrac{x + 1}{3}$

$$x = \dfrac{y + 1}{3}$$
$$y = 3x - 1$$

12. $f(g(-3)) = f(3(-3) + 1)$
$$= 2(-8)^2$$
$$= 128$$

13. 66 kilograms ($k = 6; 6(11) = 66$)
$y = kx; 48 = 8k; k = 6; 11(6) = 66$

14. Make a function, $f(x) = 39 + 0.35x$;
$60 = 39 + 0.35x; x = 60$ miles

15. $4; 3(4) - |4 + 1| = 7; 12 - 5 = 7$;
$$3x - |x + 1| = 7$$
$$3x - x - 1 = 7$$
$$2x = 8$$
$$x = 4$$

16. $\dfrac{1}{2}; \dfrac{26}{52} = \dfrac{1}{2}$

17. $\begin{bmatrix} 1 & 3 & 4 \\ 2 & -2 & -3 \end{bmatrix} + \begin{bmatrix} 0 & 9 & -5 \\ 3 & -4 & 6 \end{bmatrix} =$

$\begin{bmatrix} 1 + 0 & 3 + 9 & 4 + (-5) \\ 2 + 3 & -2 + (-4) & -3 + 6 \end{bmatrix} = \begin{bmatrix} 1 & 12 & -1 \\ 5 & -6 & 3 \end{bmatrix}$

18. $\begin{bmatrix} 1 & 3 & 4 \\ 2 & -2 & -3 \end{bmatrix} - \begin{bmatrix} 0 & 9 & -5 \\ 3 & -4 & 6 \end{bmatrix} =$

$\begin{bmatrix} 1 - 0 & 3 - 9 & 4 - (-5) \\ 2 - 3 & -2 - (-4) & -3 - 6 \end{bmatrix} = \begin{bmatrix} 1 & -6 & 1 \\ -1 & 2 & -9 \end{bmatrix}$

Standardized Test, page 161

1. B

2. C

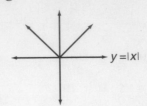

3. B

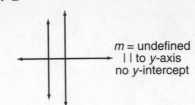

m = undefined
| | to y-axis
no y-intercept

4. D $(-4, 5)$

5. B (reflection is negative)

6. C $y - 2 = 3x; x - 2 = 3y; y = \dfrac{x - 2}{3}$

7. B $2 < 5.5$

8. E

9. C $\pi r^2 = A; \pi(2r)^2 = 4\pi r^2 = A; \dfrac{4\pi r^2}{\pi r^2} = 4$

10. B

11. C (Examine the graph).

Chapter 4 Systems of Linear Equations

Data Activity, pages 162–163

1. $\dfrac{708.1 + 699.3 + 752.8}{3} \approx 720.1$; 720.1 billion

2. Answers will vary. Rounding to the nearest hundred billion: $500 + 100 + 200 + 600 + 200 + 100 = 1700$ billion; 5 Students may also round to other places or use clustering.

3a. $\dfrac{63.2 - 63.5}{63.5} \approx -0.0047$; -0.5%

3b. $\dfrac{450.3 - 436.2}{436.2} \approx 0.0323$; 3.2%

3c. $\dfrac{158.0 - 172.6}{172.6} \approx -0.0845$; -8.5%

4. $0.144 \times 209.0 = 30.096$; $30.1 billion

5. Answers will vary.

PROJECT

1.–3. No solutions, these are activities.

Lesson 4.1, pages 165–167

THINK BACK

1–5. Answers will vary. Possible answers given.

1. The line should go up from left to right.

2. The line should go down from left to right.

3. The line should be parallel to the y-axis.

4. The line should be parallel to the x-axis.

5. The graph of an absolute value equation is an example.

EXPLORE

6.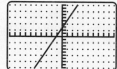

7. Answers will vary.

8. Answers will vary. If the lines did not intersect, the partner's chosen line may have been parallel to the original line. Another possibility is that lines may intersect outside the perimeter of the calculator's viewing screen.

9. Answers will vary. Substituting the coordinates in each equation will show whether a true equation results.

10. Answers will vary.

11. Answers will vary. Use equations of lines that have different slopes.

12. The system has exactly one solution. The coordinates of the point represent the solution.

MAKE CONNECTIONS

13.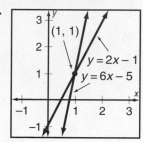

one point of intersection; $(1, 1)$

14. yes; $(1, 1)$

15.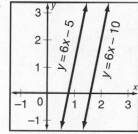

no points of intersection

16. no; The lines do not intersect because they have the same slope and different y-intercepts.

17.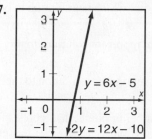

The lines coincide, so they intersect at an infinite number of points.

18. no; There are an infinite number of ordered pairs that satisfy both the equations because the links coincide.

19a. The graphs intersect at a single point.

19b. The graphs coincide so they are the same line.

19c. The graphs do not intersect so they are parallel.

SUMMARIZE

20. Graph both lines on the same coordinate plane. The point of intersection of the lines will satisfy both equations.

21.

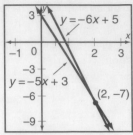

$6x + y = 5, y = -6x + 5;$
$5x + y = 3, y = -5x + 3;$
The solution is $(2, -7)$.

22.

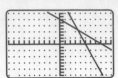

23a. Center City: $y = 100x + 950$;
Vista View: $y = 150x + 700$

23b.

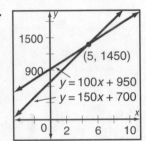

The solution is $(5, 1450)$.

23c. The population of both cities in 5 years will be 1450.

24. If the two lines have different slopes, the lines will intersect at exactly one point which is a solution to both equations.

25. b; There are no points that represent solutions to the pair of equations.

26. c; There is one point that represents a solution to the pair of equations.

27. a; There is an infinite number of points that represent solutions to the pair of equations.

28. Answers will vary. Answers should be lines that coincide. One way to find such a system of equations is to multiply your first equation by a constant. Use your original equation and your new equation for the system of equations.

29. no; If the lines have the same y-intercept, then they will intersect or coincide. The y-intercept is a solution that satisfies both equations.

30. Answers will vary. Answers should include the idea that solutions can sometimes be difficult to pinpoint exactly on a graph.

31.

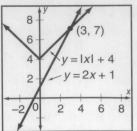

$y - 4 = |x|, y = |x| + 4;$
$y = 2x + 1;$
The solution is $(3, 7)$.

32.

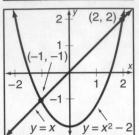

$y = x^2 - 2;$
$y - x = 0, y = x;$
The solutions are $(-1, -1)$ and $(2, 2)$.

Lesson 4.2, pages 168–175

EXPLORE

1.

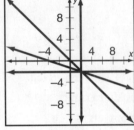

System 1 in red.
System 2 in black.

2. Both system intersect at the same point. The two systems are equivalent.

3. $x + y = 0, y = -x; x + 3y = -4, x + 3(-x) = -4, x - 3x = -4, -2x = -4, x = 2$

4. $x + y = 0, 2 + y = 0, y = -2$

5. $x = 2; y = -2;$ The systems are equivalent. The ordered pair $(2, -2)$ is the solution for both systems.

TRY THESE

1. $x - 2y = 0$ $4x + y = 9$
       $x = 2y$ $4(2) + 1 \overset{?}{=} 9$
                     $9 = 9$

     $4x + y = 9$
   $4(2y) + y = 9$ $x - 2y = 0$
     $8y + y = 9$ $2 - 2(1) \overset{?}{=} 0$
        $9y = 9$ $0 = 0$
         $y = 1$

    $x - 2y = 0$ The solution is
  $x - 2(1) = 0$
     $x - 2 = 0$ $(2, 1)$.
        $x = 2$

2. $x - 9y = -34$
$x = 9y - 34$

$7x + 5y = 68$
$7(9y - 34) + 5y = 68$
$63y - 238 + 5y = 68$
$68y = 306$
$y = 4.5$

$x - 9y = -34$
$x - 9(4.5) = -34$
$x = 6.5$

$7x + 5y = 68$
$7(6.5) + 5(4.5) \overset{?}{=} 68$
$68 = 68$

$x - 9x = -34$
$6.5 - 9(4.5) \overset{?}{=} -34$
$-34 = -34$

The solution is (6.5, 4.5).

3. $\frac{x}{4} - \frac{y}{5} = 9$

$\frac{x}{4} - \frac{5x}{5} = 9$

$5x - 20x = 180$
$-15x = 180$
$x = -12$

$y = 5x$
$y = 5(-12)$
$y = -60$

$\frac{x}{4} - \frac{y}{5} = 9$

$\frac{-12}{4} - \frac{-60}{5} \overset{?}{=} 9$

$-3 + 12 \overset{?}{=} 9$
$9 = 9$

$y = 5x$
$-60 \overset{?}{=} 5(-12)$
$-60 = -60$

The solution is $(-12, -60)$

4. $2x + 2y = 8$
$\underline{-2x + 3y = -3}$
$5y = 5$
$y = 1$

$2x + 2y = 8$
$2x + 2(1) = 8$
$2x = 6$
$x = 3$

$2x + 2y = 8$
$2(3) + 2(1) \overset{?}{=} 8$
$8 = 8$

$2x - 3y = 3$
$2(3) - 3(1) \overset{?}{=} 3$
$3 = 3$

The solution is (3, 1).

5. $-6x - 2y = -30$
$\underline{2x + 2y = 16}$
$-4x = -14$
$x = 3.5$

$3x + y = 15$
$3(3.5) + y = 15$
$y = 4.5$

$3x + y = 15$
$3(3.5) + 4.5 \overset{?}{=} 15$
$15 = 15$

$2y + 2x = 16$
$2(3.5) + 2(4.5) \overset{?}{=} 16$
$16 = 16$

The solution is (3.5, 4.5).

6. $0.9x + 0.6y = 2.4$
$\underline{0.4x - 0.6y = 2.8}$
$1.3x = 5.2$
$x = 4$

$0.3x + 0.2y = 0.8$
$0.3(4) + 0.2y = 0.8$
$0.2y = -0.4$
$y = -2$

$0.3x + 0.2y = 0.8$
$0.3(4) + 0.2(-2) \overset{?}{=} 0.8$
$0.8 = 0.8$

$0.2x - 0.3y = 1.4$
$0.2(4) - 0.3(-2) \overset{?}{=} 1.4$
$1.4 = 1.4$

The solution is $(4, -2)$

7a. $1600x + 2400y = 67{,}200$
$2000x + 2400y = 73{,}200$

7b. $1600x + 2400y = 67{,}200$
$\underline{-2000x - 2400y = -73{,}200}$
$-400x = -6000$
$x = 15$

$1600x + 2400y = 67{,}200$
$1600(15) + 2400y = 67{,}200$
$2400y = 43{,}200$
$y = 18$

$1600x + 2400y = 67{,}200$
$1600(15) + 2400(18) \overset{?}{=} 67{,}200$
$67{,}200 = 67{,}200$

$2000x + 2400y = 73{,}200$
$2000(15) + 2400(18) \overset{?}{=} 73{,}200$
$73{,}200 = 73{,}200$

The lawn seats cost $15.00 and the reserved seats cost $18.00.

8. $x + 2y = 7$
$x = -2y + 7$

$2x - 3y = -7$
$2(-2y + 7) - 3y = -7$
$-4y + 14 - 3y = -7$
$-7y = -21$
$y = 3$

$x + 2y = 7$
$x + 2(3) = 7$
$x + 6 = 7$
$x = 1$
(1, 3); independent

9. $-4x + 6y = 14$
$\underline{4x - 6y = -14}$
$0 = 0$

true equation; infinite solutions; dependent

10. $-4x + 6y = 14$
$\underline{4x - 6y = 8}$
$0 = 22$

false equation; no solution; inconsistent

11.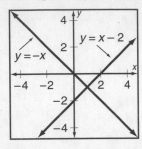

intersecting lines
one solution
consistent
independent
$\begin{cases} y = -x \\ y = x - 2 \end{cases}$

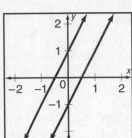

parallel lines
no solution
inconsistent
$\begin{cases} y = 2x + 1 \\ y = 2x - 1 \end{cases}$

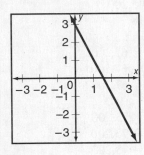

identical lines
infinite solutions
consistent
dependent
$\begin{cases} 3y = 9 - 6x \\ -6y = 12x - 18 \end{cases}$

PRACTICE

1. $x + y = 10$ $\qquad$ $x + y = 10$
$ x = 10 - y$ $\qquad$ $8 + 2 \overset{?}{=} 10$
$ 10 = 10$

$x - y = 6$ $\qquad$ $x - y = 6$
$10 - y - y = 6$ $\qquad$ $8 - 2 \overset{?}{=} 6$
$-2y = -4$ $\qquad$ $ 6 = 6$
$y = 2$

 The solution is

$x + y = 10$
$x + 2 = 10$ $\qquad$ $(8, 2).$
$x = 8$

2. $x + y = 0.5$
$y = 0.5 - x$

$3x - y = 5$
$3x - (0.5 - x) = 5$
$3x - 0.5 + x = 5$
$4x = 5.5$
$x = 1.375$

$x + y = 0.5$
$1.375 + y = 0.5$
$y = -0.875$

$3x - y = 5$
$3(1.375) - (-0.875) \overset{?}{=} 5$
$5 = 5$

$x + y = 0.5$
$1.375 + (-0.875) \overset{?}{=} 0.5$
$0.5 = 0.5$

The solution is $(1.375, -0.875)$.

3. $x + y = 12$
$x = 12 - y$

$0.2x + 0.5y = 3.6$
$0.2(12 - y) + 0.5y = 3.6$
$2.4 - 0.2y + 0.5y = 3.6$
$0.3y = 1.2$
$y = 4$

$x + y = 12$
$x + 4 = 12$
$x = 8$

$x + y = 12$
$8 + 4 \overset{?}{=} 12$
$12 = 12$

$0.2x + 0.5y = 3.6$
$0.2(8) + 0.5(4) \overset{?}{=} 3.6$
$3.6 = 3.6$

The solution is $(8, 4)$.

4. Let x represent the amount of $3.00 per pound coffee beans. Let y represent the amount of $4.25 per pound coffee beans.

$$\begin{cases} x + y = 25 \\ 3x + 4.25y = 25(3.50) \end{cases}$$

$$x + y = 25$$
$$x = 25 - y$$

$$3x + 4.25y = 25(3.50)$$
$$3(25 - y) + 4.25y = 87.50$$
$$75 - 3y + 4.25y = 87.50$$
$$1.25y = 12.50$$
$$y = 10$$

$$x + y = 25$$
$$x + 10 = 25$$
$$x = 15$$

$$x + y = 25$$
$$15 + 10 \overset{?}{=} 25$$
$$25 = 25$$

$$3x + 4.25y = 25(3.50)$$
$$3(15) + 4.25(10) \overset{?}{=} 25(3.50)$$
$$87.50 = 87.50$$

To produce the new blend, mix 15 lb of $3.00 per pound coffee beans and 10 lb of $4.25 per pound coffee beans.

5.
$$\begin{aligned} x - 2y &= 6 \\ -x + 3y &= -4 \\ \hline y &= 2 \end{aligned}$$

$$x - 2y = 6$$
$$10 - 2(2) \overset{?}{=} 6$$
$$6 = 6$$

$$x - 2y = 6$$
$$x - 2(2) = 6$$
$$x = 10$$

$$-x + 3y = -4$$
$$-10 + 3(2) \overset{?}{=} -4$$
$$-4 = -4$$

The solution is $(10, 2)$.

6.
$$\begin{aligned} 9x + 3y &= -3 \\ 2x - 3y &= -8 \\ \hline 11x &= -11 \\ x &= -1 \end{aligned}$$

$$9x + 3y = -3$$
$$9(-1) + 3(2) \overset{?}{=} -3$$
$$-3 = -3$$

$$9x + 3y = -3$$
$$9(-1) + 3y = -3$$
$$3y = 6$$
$$y = 2$$

$$2x - 3y = -8$$
$$2(-1) - 3(2) \overset{?}{=} -8$$
$$-2 - 6 = -8$$
$$-8 = -8$$

The solution is $(-1, 2)$.

7.
$$\begin{aligned} 3x + 4y &= 4 \\ 2x - 4y &= 4 \\ \hline 5x &= 8 \\ x &= \frac{8}{5} \end{aligned}$$

$$x - 2y = 2$$
$$\frac{8}{5} - 2y = 2$$
$$-2y = \frac{2}{5}$$
$$y = -\frac{1}{5}$$

$$3x + 4y = 4$$
$$3\left(\frac{8}{5}\right) + 4\left(-\frac{1}{5}\right) \overset{?}{=} 4$$
$$\frac{24}{5} - \frac{4}{5} \overset{?}{=} 4$$
$$4 = 4$$

$$x - 2y = 2$$
$$\frac{8}{5} - 2\left(-\frac{1}{5}\right) \overset{?}{=} 2$$
$$\frac{8}{5} + \frac{2}{5} \overset{?}{=} 2$$
$$2 = 2$$

The solution is $\left(\frac{8}{5}, -\frac{1}{5}\right)$.

8.
$$\begin{aligned} 0.6x - 0.4y &= 8 \\ -0.6x + 0.6y &= 12 \\ \hline 0.2y &= 20 \\ y &= 100 \end{aligned}$$

$$0.3x - 0.2y = 4$$
$$0.3x - 0.2(100) = 4$$
$$0.3x = 24$$
$$x = 80$$

$$0.3x - 0.2y = 4$$
$$0.3(80) - 0.2(100) \overset{?}{=} 4$$
$$4 = 4$$

$$-0.6x + 0.6y = 12$$
$$-0.6(80) + 0.6(100) \overset{?}{=} 12$$
$$12 = 12$$

The solution is $(80, 100)$.

9. Answers will vary. Multiply equation 1 by 5 and equation 2 by -2 or multiply equation 1 by 7 and equation 2 by -3.

$$\begin{aligned} 10x + 15y &= 85 \\ -10x - 14y &= -58 \\ \hline y &= 27 \end{aligned}$$

$$2x + 3y = 17$$
$$2(-32) + 3(27) \overset{?}{=} 17$$
$$17 = 17$$

$$2x + 3y = 17$$
$$2x + 3(27) = 17$$
$$2x = -64$$
$$x = -32$$

$$5x + 7y = 29$$
$$5(-32) + 7(27) \overset{?}{=} 29$$
$$29 = 29$$

The solution is $(-32, 27)$.

10. Answers will vary. Multiply equation 1 by 3 and equation 2 by 5 or multiply equation 1 by 7 and equation 2 by 9,

$$
\begin{array}{rl}
15x - 27y = & 21 \\
-15x + 35y = & -25 \\
\hline
8y = & -4 \\
y = & -\dfrac{1}{2}
\end{array}
$$

$$5x - 9y = 7$$

$$5\left(\dfrac{1}{2}\right) - 9\left(-\dfrac{1}{2}\right) \overset{?}{=} 7$$

$$7 = 7$$

$$7y - 3x = -5$$

$$
\begin{array}{l}
5x - 9y = 7 \\
5x - 9\left(-\dfrac{1}{2}\right) = 7 \\
5x + \dfrac{9}{2} = 7 \\
5x = \dfrac{5}{2} \\
x = \dfrac{1}{2}
\end{array}
$$

$$7\left(-\dfrac{1}{2}\right) - 3\left(\dfrac{1}{2}\right) \overset{?}{=} -5$$

$$-5 = -5$$

The solution is $\left(\dfrac{1}{2}, -\dfrac{1}{2}\right)$.

11. The solution to the system will be $x = 8$ and $y = 5$. Explanations will vary. They should include the idea that multiplying both sides of an equation by the same real number produces an equation equivalent to the original equation.

12. Let x represent the number of grams of 12-carat alloy. Let y represent the number of grams of 18-carat alloy.
$$\begin{cases} 12x + 18y = 14(15) \\ x + y = 15 \end{cases}$$

$$
\begin{array}{l}
x + y = 15 \\
\quad y = 15 - x
\end{array}
$$

$$
\begin{array}{l}
12x + 18y = 14(15) \\
12x + 18(15 - x) = 210 \\
12x + -270 - 18x = 210 \\
\qquad\qquad -6x = -60 \\
\qquad\qquad\quad x = 10
\end{array}
$$

$$
\begin{array}{l}
x + y = 15 \\
10 + y = 15 \\
\quad\; y = 5
\end{array}
$$

To obtain 15 g of 14-carat gold, mix 10 g of 12-carat gold and 5 g of 18-carat gold.

13.
$$
\begin{array}{rl}
-2x + 6y = & 6 \\
-2x + 6y = & 12 \\
\hline
0 = & 18
\end{array}
$$

false equation; no solution; inconsistent

14.
$$
\begin{array}{l}
1 + y = x \\
\quad\;\; y = x - 1
\end{array}
$$

$$
\begin{array}{l}
y - 2x = 5 \\
x - 1 - 2x = 5 \\
\qquad -x = 6 \\
\qquad\; x = -6
\end{array}
$$

$$
\begin{array}{l}
1 + y = x \\
1 + y = -6 \\
\quad\; y = -7
\end{array}
$$

$(-6, -7)$; independent

15.
$$
\begin{array}{rl}
-12x + 6y = & 57 \\
-12x - 6y = & 3 \\
\hline
-24x = & 60
\end{array}
$$

$$x = -\dfrac{60}{24}$$

$$x = -2.5$$

$$
\begin{array}{l}
4x - 2y = -19 \\
4(-2.5) - 2y = -19 \\
\qquad\qquad -2y = -9 \\
\qquad\qquad\quad y = \dfrac{-9}{-2} \\
\qquad\qquad\quad y = 4.5
\end{array}
$$

$(-2.5, 4.5)$; independent

16.
$$
\begin{array}{rl}
-4x + 6y = & -12 \\
-4x + 6y = & -12 \\
\hline
0 = & 0
\end{array}
$$

true equation; infinite solutions; dependent

17. Let x represent one of the angles. Let y represent the other angle.

$$\begin{cases} x + y = 90 \\ x - y = 16 \end{cases}$$

$$
\begin{array}{rl}
x + y = & 90 \\
x - y = & 16 \\
\hline
2x = & 106 \\
x = & 53
\end{array}
$$

$$
\begin{array}{l}
x + y = 90 \\
53 + y = 90 \\
\quad\;\; y = 37
\end{array}
$$

The measures of the angles are 53° and 37°.

18. Let x represent one angle. Let y represent the other angle.
$$\begin{cases} y = 5x - 6 \\ x + y = 180 \end{cases}$$

$$
\begin{array}{l}
x + y = 180 \\
x + 5x - 6 = 180 \\
\qquad\quad 6x = 186 \\
\qquad\quad\; x = 31
\end{array}
$$

$$
\begin{array}{l}
y = 5x - 6 \\
y = 5(31) - 6 \\
y = 149
\end{array}
$$

The measures of the angles are 31° and 149°.

19. Let x represent the plane's speed. Let y represent the wind's speed.
$$\begin{cases} 5x + 5y = 3000 \\ 6x - 6y = 3000 \end{cases}$$

$$
\begin{array}{rl}
30x + 30y = & 18{,}000 \\
30x - 30y = & 15{,}000 \\
\hline
60x = & 33{,}000 \\
x = & 550
\end{array}
$$

$$
\begin{array}{l}
5x + 5y = 3000 \\
5(550) + 5y = 3000 \\
\qquad\quad 5y = 250 \\
\qquad\quad\; y = 50
\end{array}
$$

The plane's speed is 550 mi/h and the wind's speed is 50 mi/h.

20.
$$2(x - y) - 4y = 2 \qquad -6x - 8y = 2(-17 + y)$$
$$2x - 2y - 4y = 2 \qquad -6x - 8y = -34 + 2y$$
$$2x - 6y = 2 \qquad -6x - 10y = -34$$

$$6x - 18y = 6 \qquad\qquad 2x - 6y = 2$$
$$\underline{-6x - 10y = -34} \qquad 2x - 6(1) = 2$$
$$-28y = -28 \qquad\qquad 2x = 8$$
$$y = 1 \qquad\qquad x = 4$$

$$2(x - y) - 4y = 2$$
$$2(4 - 1) - 4(1) \overset{?}{=} 2$$
$$2 = 2$$

$$-6x - 8y = 2(-17 + y)$$
$$-6(4) - 8(1) \overset{?}{=} 2(-17 + 1)$$
$$-32 = -32$$

The solution is $(4, 1)$.

21. $\frac{1}{5}(x + y) = 6 - \frac{3}{2}y$

$$2(x + y) = 60 - 15y$$
$$2x + 2y = 60 - 15y$$
$$2x + 17y = 60$$

$$\frac{3}{5}(x + 2) = \frac{1}{2}y + \frac{16}{5}$$
$$6(x + 2) = 5y + 32$$
$$6x + 12 = 5y + 32$$
$$6x - 5y = 20$$

$$-6x - 51y = -180$$
$$\underline{6x \quad 5y = 20}$$
$$-56y = -160$$
$$y = \frac{-160}{-56}$$
$$y = \frac{20}{7}$$

$$6x - 5\left(\frac{20}{7}\right) = 20$$
$$6x - \frac{100}{7} = 20$$
$$42x - 100 = 140$$
$$x = \frac{240}{42}$$
$$x = \frac{40}{7}$$

$$\frac{1}{5}(x + y) = 6 - \frac{3}{2}y$$
$$\frac{1}{5}\left(\frac{40}{7} + \frac{20}{7}\right) \overset{?}{=} 6 - \frac{3}{2}\left(\frac{20}{7}\right)$$
$$\frac{1}{5}\left(\frac{60}{7}\right) \overset{?}{=} 6 - \frac{30}{7}$$
$$\frac{12}{7} = \frac{12}{7}$$

$$\frac{3}{5}(x + 2) = \frac{1}{2}y + \frac{16}{5}$$
$$\frac{3}{5}\left(\frac{40}{7} + 2\right) \overset{?}{=} \frac{1}{2}\left(\frac{20}{7}\right) + \frac{16}{5}$$
$$\frac{3}{5}\left(\frac{54}{7}\right) = \frac{10}{7} + \frac{16}{5}$$
$$\frac{162}{7} = \frac{162}{7}$$

The solution is $\left(\frac{40}{7}, \frac{20}{7}\right)$.

22. Let x represent the number of checks written. Let y represent the total monthly charge.
$$\begin{cases} y = 13.00 + 10x \\ y = 14.25 + 10x \end{cases}$$
$$-y = -13.00 - 10x$$
$$\underline{\quad y = \quad 14.25 + 10x}$$
$$0 = \quad 1.25$$

false equation; The system is inconsistent. The checking plan at Guarantee Trust will never be cheaper than the plan at Standard Bank.

THINK CRITICALLY

23. no; The solution to a system is the intersection of two lines. Two lines intersect at either one point or infinite points.

24.
$$ax + by = 2 \qquad\qquad ax + by = 2$$
$$\underline{-ax + 2by = 1} \qquad ax + b\left(\frac{1}{b}\right) = 2$$
$$3by = 3 \qquad\qquad ax + 1 = 2$$
$$y - \frac{3}{3b} \qquad\qquad ax = 1$$
$$y = \frac{1}{b} \qquad\qquad x = \frac{1}{a}$$
$$\left(\frac{1}{a}, \frac{1}{b}\right)$$

25.
$$x + y = 2a \qquad\qquad x + y = 2a$$
$$\underline{x - y = \quad 2b} \qquad a + b + y = 2a$$
$$2x = 2a + 2b \qquad\qquad y = a - b$$
$$x = a + b$$
$$(a + b, a - b)$$

26.
$$9x + 3y = 3a \qquad\qquad 3x + y = a$$
$$\underline{x - 3y = \quad b} \qquad \underline{-3x + 9y = \quad -3b}$$
$$10x = 3a + b \qquad\qquad 10y = a - 3b$$
$$x = \frac{3a + b}{10} \qquad\qquad y = \frac{a - 3b}{10}$$
$$\left(\frac{3a + b}{10}, \frac{a - 3b}{10}\right)$$

27.
$$12x - 18y = 3 \qquad\qquad 3 - 3c = 0$$
$$\underline{12x - 18y = 3c} \qquad\qquad -3c = -3$$
$$0 = 0 \qquad\qquad c = 1$$

28.
$$\frac{3}{2}cx - 21y = -9 \qquad \frac{3}{2}cx - 15x = 0$$
$$\underline{-15x + 21y = 9} \qquad 3cx = 30x$$
$$0 = 0 \qquad\qquad c = 10$$

29.
$$2x + 6y = 8 \qquad 6y - cy = 0$$
$$\underline{-2x - cy = 3} \qquad -cy = -6y$$
$$0 = 11 \qquad\qquad c = 6$$

30.
$$0.5x - y = 0.5c \qquad 0.5c + 2 \neq 0$$
$$\underline{-0.5x + y = -2} \qquad 0.5c \neq -2$$
$$0 \neq 0 \qquad\qquad c \neq -4$$

MIXED REVIEW

31. $y = 2x + 3; m = 2; b = 3$

32. $2y = -6x + 10, y = -3x + 5; m = -3; b = 5$

33. $3x - 4y = 12, -4y = -3x + 12, y = \frac{3}{4}x - 3;$
$m = \frac{3}{4}; b = -3$

34. $3y + 10 = 0, 3y = -10, y = -\frac{10}{3}; m = 2; b = -\frac{10}{3}$

35. **A.** $\quad 3x + y = 2.5$ $\qquad$ **B.** $\quad x + y = 2.5$
$\qquad\quad 3(1) + 0.5 = 2.5 \qquad\qquad 3(1) + 2 = 2.5$
$\qquad\qquad\quad 3.5 \neq 2.5 \qquad\qquad\qquad\quad 5 \neq 2.5$

$\quad$ **C.** $3x + y = 2.5$ $\qquad$ **D.** $\qquad x + y = 2.5$
$\qquad\quad 3(0.5) + 1 = 2.5 \qquad\qquad 3(-0.5) + 1 = 2.5$
$\qquad\qquad 2.5 = 2.5 \qquad\qquad\qquad\quad -0.5 \neq 2.5$

$\qquad 2x - y = 0$
$\qquad 2(0.5) + 1 = 0$
$\qquad\qquad 0 = 0$

The correct answer is C.

PROJECT CONNECTION

1–3. No solutions, these are activities.

4. $U\%$ of C represents cost of posterboard used for boxes. $W\%$ of C represents the cost of waste.

5. Answers will vary.

ALGEBRAWORKS

1. $\frac{1}{2}$

2. $\frac{1}{f}$ would represent the portion of the job that Frank could complete in 1 h.

3. $\frac{1}{d} + \frac{1}{f} = \frac{1}{2}$

4. $(1)\frac{1}{d} + (4)\frac{1}{f} = 1$

5. $x + y = \frac{1}{2}; x + 4y = 1$

6. $x + y = \frac{1}{2} \qquad\qquad x = \frac{1}{d}$
$\qquad y = \frac{1}{2} - x \qquad\qquad \frac{1}{3} = \frac{1}{d}$
$\qquad\qquad\qquad\qquad\qquad\quad d = 3$

$x + 4y = 1$

$x + 4\left(\frac{1}{2} - x\right) = 1$
$\qquad x + 2 - 4x = 1 \qquad\quad y = \frac{1}{f}$
$\qquad\qquad -3x = -1 \qquad\quad \frac{1}{6} = \frac{1}{f}$
$\qquad\qquad\quad x = \frac{1}{3} \qquad\qquad f = 6$

$x + y = \frac{1}{2}$ $\qquad$ Daria would need 3h to paint
$\frac{1}{3} + y = \frac{1}{2}$ $\qquad$ the dollhouse. Frank would
$\qquad\quad y = \frac{1}{6}$ $\qquad$ need 6h.

Lesson 4.3, pages 176–182

EXPLORE

1. 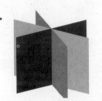 $\qquad$ **2.**

3.  $\qquad$ no $\quad$ **4.**  $\qquad$ no

5. 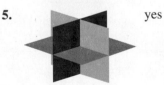 $\qquad$ yes

TRY THESE

1.
$$4x - y + z = 6$$
$$4(2) - 1 + (-1) \overset{?}{=} 6$$
$$6 = 6$$

$$2x + y + 2z = 3$$
$$2(2) + 1 + 2(-1) \overset{?}{=} 3$$
$$3 = 3$$

$$3x - 2y + z = 3$$
$$3(2) - 2(1) + (-1) \overset{?}{=} 3$$
$$3 = 3$$

Yes, $(2, 1, -1)$ is a solution.

2.
$$4x + 2y + 5z = 6$$
$$4(1) + 2(2) + 5(3) \overset{?}{=} 6$$
$$23 \neq 6$$
No, $(1, 2, 3)$ is not a solution.

3.
$$4x + 2y + 5z = 6$$
$$4(2) + 2(-1) + 5(0) \overset{?}{=} 6$$
$$6 = 6$$

$$2x - y + z = 5$$
$$2(2) - (-1) + 0 \overset{?}{=} 5$$
$$5 = 5$$

$$x + 2y - z = 2$$
$$2 + 2(-1) - 0 \overset{?}{=} 2$$
$$0 \neq 2$$

No, $(2, -1, 0)$ is not a solution.

4.
$$x - y + 2z = -3 \rightarrow 2x - 2y + 4z = -6$$
$$\underline{x + 2y + 3z = 4 \rightarrow x + 2y + 3z = 4}$$
$$3x + 7z = -2$$

$$x + 2y + 3z = 4 \rightarrow -x - 2y - 3z = -4$$
$$\underline{2x + 2y + z = -3 \rightarrow 4x + 2y + 2z = -6}$$
$$3x - z = -10$$

$$\begin{cases} 3x + 7z = -2 \rightarrow 3x - 7z = -2 \\ 3x - z = -10 \rightarrow \underline{-3x + z = 10} \end{cases}$$
$$8z = 8$$
$$z = 1$$

$$3x - z = -10 \qquad\qquad x - y + 2z = -3$$
$$3x - 1 = -10 \qquad\qquad -3 - y + 2(1) = -3$$
$$3x = -9 \qquad\qquad -3 - y + 2 = -3$$
$$x = -3 \qquad\qquad\qquad y = 2$$
$$\qquad\qquad\qquad\qquad\qquad\qquad y = 2$$

$$(-3, 2, 1)$$

5.
$$2x + 3y + 12z = 4 \rightarrow 4x + 6y + 24z = 8$$
$$\underline{4x - 6y + 6z = 1 \rightarrow 4x - 6y + 6z = 1}$$
$$8x + 30z = 0$$

$$2x + 3y + 12z = 4 \rightarrow 2x + 3y + 12z = 4$$
$$\underline{x + y + z = 1 \rightarrow -3x - 3y - 3z = -3}$$
$$-x + 9z = 1$$

$$\begin{cases} 8x + 30z = 9 \rightarrow 8x + 30z = 9 \\ -x + 9z = 1 \rightarrow \underline{-8x + 72z = 8} \end{cases}$$
$$102z = 17$$

$$z = \frac{17}{102}$$
$$z = \frac{1}{6}$$

$$8x + 30z = 9$$
$$8x + 30\left(\frac{1}{6}\right) = 9$$
$$8x + 5 = 9$$
$$x = \frac{4}{8}$$
$$x = \frac{1}{2}$$

$$x + y + z = 1$$
$$\frac{1}{2} + y + \frac{1}{6} = 1$$
$$y = 1 - \frac{6}{12} - \frac{2}{12}$$
$$y = \frac{4}{12} = \frac{1}{3}$$

$$\left(\frac{1}{2}, \frac{1}{3}, \frac{1}{6}\right)$$

6.
$$5x - 3y + 2z = 13 \rightarrow 20x - 12y + 8z = 52$$
$$\underline{2x + 4y - 3z = -9 \rightarrow 6x + 12y - 9z = -27}$$
$$26x - z = 25$$

$$2x + 4y - 3z = -9 \rightarrow 2x + 4y - 3z = -9$$
$$\underline{4x - 2y + 5z = 13 \rightarrow 8x - 4y + 10z = 26}$$
$$10x + 7z = 17$$

$$26x - z = 25 \rightarrow 182x - 7z = 175$$
$$\underline{10x + 7z = 17 \rightarrow 10x + 7z = 17}$$
$$192x = 192$$
$$x = 1$$

$$26x - z = 25$$
$$26(1) - z = 25$$
$$-z = -1$$
$$z = 1$$

$$5x - 3y + 2z = 13$$
$$5(1) - 3y + 2(1) = 13$$
$$5 - 3y + 2 = 13$$
$$-3y = 6$$
$$y = -2$$

$$(1, -2, 1)$$

7a.
$$\begin{cases} a + b + c = 5700 \\ a + b = 3400 \\ b + c = 4200 \end{cases}$$

7b.
$$a + b + c = 5700 \rightarrow -a - b - c = -5700$$
$$\underline{a + b = 3400 \rightarrow a + b = 3400}$$
$$-c = -2300$$
$$c = 2300$$

$$b + c = 4200$$
$$b + 2300 = 4200$$
$$b = 1900$$

$$a + b = 3400$$
$$a + 1900 = 3400$$
$$a = 1500$$

Mill A can produce 1500 board-feet. Mill B can produce 1900 board feet. Mill C can produce 2300 board-feet.

8. Answers will vary. Answers should include the idea that to solve a system of two equations you need the same two variables in each equation.

9.
$$-x + 4y - 3z = 2 \rightarrow -2x + 8y - 6z = 4$$
$$2x - 8y + 6z = 1 \rightarrow \underline{2x - 8y + 6z = 1}$$
$$0 = 5$$

False equation; no solution; inconsistent

10.
$$2x - 3y + 3z = -15 \rightarrow 4x - 6y + 6z = -30$$
$$3x + 2y - 5z = 19 \rightarrow \underline{9x + 6y - 15z = 57}$$
$$13x \qquad - 9z = 27$$

$$3x + 2y - 5z = 19 \rightarrow 6x + 4y - 10z = 38$$
$$5x - 4y - 2z = -2 \rightarrow \underline{5x - 4y - 2z = -2}$$
$$11x \qquad - 12z = 36$$

$$\begin{cases} 13x - 9z = 27 \rightarrow 143x - 99z = 297 \\ 11x - 12z = 36 \rightarrow \underline{-143x - 156z = 468} \end{cases}$$
$$- 255z = 765$$
$$z = -3$$

$$13x - 9x = 27 \qquad\qquad 2x - 3y + 3z = -15$$
$$13x - 9(-3) = 27 \qquad 2(0) - 3y + 3(-3) = -15$$
$$13x = 0 \qquad\qquad -3y - 9 = -15$$
$$x = 0 \qquad\qquad -3y = -6$$
$$y = 2$$

$$2x - 3y + 3z = -15$$
$$2(0) - 3(2) + 3(-3) \overset{?}{=} -15$$
$$-15 = -15$$

$$3x + 2y - 5z = 19$$
$$3(0) + 2(2) - 5(-3) \overset{?}{=} 19$$
$$19 = 19$$

$$5x - 4y - 2z = -2$$
$$5(0) - 4(2) - 2(-3) \overset{?}{=} -2$$
$$-2 = -2$$

The solution is $(0, 2, -3)$; independent

11.
$$y + z = 7 \rightarrow y + z = 7$$
$$x - z = 2 \rightarrow \underline{x \qquad - z = 2}$$
$$x + y \qquad = 9$$

$$\begin{cases} x + y = 9 \\ x + y = 9 \end{cases}; \text{true equation;}$$

infinite solutions; dependent

PRACTICE

1. Answers will vary.

2.
$$x + y + z = 6 \rightarrow x + y + z = 6$$
$$2x - y + 3z = 9 \rightarrow \underline{2x - y + 3z = 9}$$
$$3x \qquad + 4z = 15$$

$$2x - y + 3z = 6 \rightarrow 4x - 2y + 6z = 18$$
$$x + 2y + 2z = 11 \rightarrow \underline{x + 2y + 2z = 11}$$
$$5x \qquad + 8z = 29$$

$$3x + 4z = 15 \rightarrow -6x - 8z = -30$$
$$5x + 8z = 29 \rightarrow \underline{5x + 8z = 29}$$
$$- x \qquad = -1$$
$$x = 1$$

$$3x + 4z = 15 \qquad\qquad x + y + z = 6$$
$$3(1) + 4z = 15 \qquad\qquad 1 + y + 3 = 6$$
$$4z = 12 \qquad\qquad\qquad y = 2$$
$$z = 3$$

$$x + y + z = 6 \qquad\qquad 2x - y + 3z = 9$$
$$1 + 2 + 3 \overset{?}{=} 6 \qquad 2(1) - 2 + 3(3) \overset{?}{=} 9$$
$$6 = 6 \qquad\qquad\qquad 9 = 9$$

$$x + 2y + 2z = 11$$
$$1 + 2(2) + 2(3) \overset{?}{=} 11$$
$$11 = 11$$

The solution is $(1, 2, 3)$.

3.
$$0.2x + 0.3y - 0.1z = 0.1 \rightarrow$$
$$0.1x + 0.2y + 0.2z = 0.5 \rightarrow$$

$$0.2x + 0.3y - 0.1z = 0.1$$
$$\underline{-0.2x - 0.4y - 0.4z = -1}$$
$$- 0.1y - 0.5z = -0.9$$

$$0.1x + 0.2y + 0.2z = 0.5 \rightarrow$$
$$0.1x - 0.1y + 0.1z = 0.6 \rightarrow$$

$$0.1x + 0.2y + 0.2z = 0.5$$
$$\underline{- 0.1x + 0.1y - 0.1z = -0.6}$$
$$0.3y + 0.1z = -0.1$$

$$\begin{cases} -0.1y + 0.5z = -0.9 \rightarrow -0.3y - 1.5z = -2.7 \\ 0.3y + 0.1z = -0.1 \rightarrow \underline{0.3y + 0.1z = -0.1} \end{cases}$$
$$-1.4z = -2.8$$
$$z = 2$$

$$0.3y + 0.1z = -0.1$$
$$0.3y + 0.1(2) = -0.1$$
$$0.3y = -0.3$$
$$y = -1$$

$$0.2x + 0.3y - 0.1z = 0.1$$
$$0.2x + 0.3(-1) - 0.1(2) = 0.1$$
$$0.2x = 0.6$$
$$x = 3$$

$$0.2x + 0.3y - 0.1z = 0.1$$
$$0.2(3) + 0.3(-1) - 0.1(2) \overset{?}{=} 0.1$$
$$0.1 = 0.1$$

$$0.1x + 0.2y + 0.2z = 0.5$$
$$0.1(3) + 0.2(-1) + 0.2(2) \overset{?}{=} 0.5$$
$$0.5 = 0.5$$

$$0.1x - 0.1y + 0.1z = 0.6$$
$$0.1(3) - 0.1(-1) + 0.1(2) \overset{?}{=} 0.6$$
$$0.6 = 0.6$$

The solution is $(3, -1, 2)$.

4. $2x - 3y + z = 5$
$\underline{x + 3y + 8z = 22}$
$3x \qquad + 9z = 27$

$x + 3y + 8z = 22 \rightarrow x + 3y + 8z = 22$
$2x - y + 2z = 9 \rightarrow \underline{6x - 3y + 6z = 27}$
$\qquad\qquad\qquad\qquad 7x \qquad + 14z = 49$

$\begin{cases} 3x + 9z = 27 \rightarrow -21x - 63z = -189 \\ 7x + 14z = 49 \rightarrow \underline{21x + 42z = 147} \end{cases}$
$\qquad\qquad\qquad\qquad\qquad -21z = -42$
$\qquad\qquad\qquad\qquad\qquad\qquad z = 2$

$7x + 14z = 49 \qquad 2x - 3y + z = 5$
$7x + 14(2) = 49 \qquad 2(3) - 3y + 2 = 5$
$\qquad 7x = 21 \qquad\qquad\qquad -3y = -3$
$\qquad\ \ x = 3 \qquad\qquad\qquad\quad y = 1$

$\qquad 2x - 3y + z = 5$
$2(3) - 3(1) + 2 \overset{?}{=} 5$
$\qquad\qquad\qquad 5 = 5$

$\qquad\ x + 3y + 8z = 22$
$3 + 3(1) + 8(2) \overset{?}{=} 22$
$\qquad\qquad\qquad 22 = 22$

$\qquad 2x - y + 2z = 9$
$2(3) - 1 + 2(2) \overset{?}{=} 9$
$\qquad\qquad\qquad 9 = 9$

The solution is $(3, 1, 2)$.

5. $2x - y + 3z = 4 \rightarrow 4x - 2y + 6z = 8$
$\ x + 2y - z = -3 \rightarrow \underline{x + 2y - z = -3}$
$\qquad\qquad\qquad\qquad\quad 5x \qquad + 5z = 5$

$2x - y + 3z = 4 \rightarrow 6x - 3y + 9z = 12$
$4x + 3y + 2z = -5 \rightarrow \underline{4x + 3y + 2z = -5}$
$\qquad\qquad\qquad\qquad 10x \qquad + 11z = 7$

$\begin{cases} 5x + 5z = 5 \rightarrow -10x - 10z = -10 \\ 10x + 11z = 7 \rightarrow \underline{10x + 11z = 7} \end{cases}$
$\qquad\qquad\qquad\qquad\qquad\qquad z = -3$

$5x + 5z = 5 \qquad\qquad x + 2y - z = -3$
$5x + 5(-3) = 5 \qquad 4 + 2y - (-3) = -3$
$\qquad 5x = 20 \qquad\qquad\qquad 2y = -10$
$\qquad\ \ x = 4 \qquad\qquad\qquad\quad y = -5$

$\qquad\ \ 2x - y + 3z = 4$
$2(4) - (-5) + 3(-3) \overset{?}{=} 4$
$\qquad\qquad\qquad\qquad 4 = 4$

$\qquad\quad x + 2y - z = -3$
$4 + 2(-5) - (-3) \overset{?}{=} -3$
$\qquad\qquad\qquad\quad -3 = -3$

$\qquad\ \ 4x + 3y + 2z = -5$
$4(4) + 3(-5) + 2(-3) \overset{?}{=} -5$
$\qquad\qquad\qquad\qquad 5 = -5$

The solution is $(4, -5, -3)$.

6. $2x + 5y + 2z = 9 \rightarrow 6x + 15y + 6z = 27$
$4x - 7y - 3z = 7 \rightarrow \underline{8x - 14y - 6z = 14}$
$\qquad\qquad\qquad\qquad\quad 14x + y \qquad = 41$

$4x - 7y - 3z = 7 \rightarrow -8x + 14y + 6z = -14$
$3x - 8y - 2z = 9 \rightarrow \underline{9x - 24y - 6z = 27}$
$\qquad\qquad\qquad\qquad\quad x - 10y \qquad = 13$

$\begin{cases} 14x + y = 41 \rightarrow 140x + 10y = 410 \\ \ x - 10y = 13 \rightarrow \underline{x - 10y = 13} \end{cases}$
$\qquad\qquad\qquad\qquad\qquad 141x \qquad = 423$
$\qquad\qquad\qquad\qquad\qquad\quad x = 3$

$x - 10y = 13 \qquad\qquad 2x + 5y + 2z = 9$
$3 - 10y = 13 \qquad 2(3) + 5(-1) + 2z = 9$
$\quad -10y = 10 \qquad\qquad\qquad\quad 2z = 8$
$\qquad\ y = -1 \qquad\qquad\qquad\qquad z = 4$

$\qquad\quad 2x + 5y + 2z = 9$
$2(3) + 5(-1) + 2(4) \overset{?}{=} 9$
$\qquad\qquad\qquad\qquad 9 = 9$

$\qquad\quad 4x - 7y - 3z = 7$
$4(3) - 7(-1) - 3(4) \overset{?}{=} 7$
$\qquad\qquad\qquad\qquad 7 = 7$

$\qquad\quad 3x - 8y - 2z = 9$
$3(3) - 8(-1) - 2(4) \overset{?}{=} 9$
$\qquad\qquad\qquad\qquad 9 = 9$

The solution is $(3, -1, 4)$.

7. $\frac{1}{2}x + \frac{1}{2}y + z = \frac{1}{2} \ \rightarrow x + y + 2z = 1$
$\frac{1}{2}x - \frac{1}{4}y - \frac{1}{4}z = 0 \rightarrow \underline{2x - y - z = 0}$
$\qquad\qquad\qquad\qquad\quad 3x \qquad + z = 1$

$\frac{1}{2}x - \frac{1}{4}y - \frac{1}{4}z = 0 \ \rightarrow 2x - y - z = 0$
$\frac{1}{4}x + \frac{1}{12}y + \frac{1}{6}z = \frac{1}{6} \rightarrow \underline{3x + y + 2z = 2}$
$\qquad\qquad\qquad\qquad\quad 5x \qquad + z = 2$

$3x + z = 1 \rightarrow -3x - z = -1$
$5x + z = 2 \qquad \underline{5x + z = 2}$
$\qquad\qquad\qquad\quad 2x \qquad = 1$

$\qquad\qquad\qquad\quad x = \frac{1}{2}$

$\qquad 3x + z = 1$
$3\left(\frac{1}{2}\right) + z = 1$
$\qquad\quad z = -\frac{1}{2}$

$$\frac{1}{2}x + \frac{1}{2}y + z = \frac{1}{2}$$

$$\frac{1}{2}\left(\frac{1}{2}\right) + \frac{1}{2}y + \left(-\frac{1}{2}\right) = \frac{1}{2}$$

$$\frac{1}{2}y = \frac{3}{4}$$

$$y = \frac{6}{4} = \frac{3}{2}$$

$$\frac{1}{2}x + \frac{1}{2}y + z = \frac{1}{2}$$

$$\frac{1}{2}\left(\frac{1}{2}\right) + \frac{1}{2}\left(\frac{3}{2}\right) + \left(-\frac{1}{2}\right) \overset{?}{=} \frac{1}{2}$$

$$\frac{1}{2} = \frac{1}{2}$$

$$\frac{1}{2}x - \frac{1}{4}y - \frac{1}{4}z = 0$$

$$\frac{1}{2}\left(\frac{1}{2}\right) - \frac{1}{4}\left(\frac{3}{2}\right) - \frac{1}{4}\left(-\frac{1}{2}\right) \overset{?}{=} 0$$

$$0 = 0$$

$$\frac{1}{4}x + \frac{1}{12}y + \frac{1}{6}z = \frac{1}{6}$$

$$\frac{1}{4}\left(\frac{1}{2}\right) + \frac{1}{12}\left(\frac{3}{2}\right) + \frac{1}{6}\left(-\frac{1}{2}\right) \overset{?}{=} \frac{1}{6}$$

$$\frac{1}{6} \overset{?}{=} \frac{1}{6}$$

The solution is $\left(\frac{1}{2}, \frac{3}{2}, -\frac{1}{2}\right)$.

8.
$$\begin{array}{r} 4x + 2y - 3z = 6 \rightarrow 8x + 4y - 6z = 12 \\ x - 4y + z = -4 \rightarrow \underline{x - 4y + z = -4} \\ 9x \qquad - 5z = 8 \end{array}$$

$$\begin{cases} 9x - 5z = 8 \rightarrow \\ -x + 2z = 2 \rightarrow \end{cases} \begin{cases} 9x - 5z = 8 \\ \underline{-9x + 18z = 18} \\ 13z = 26 \\ z = 2 \end{cases}$$

$$\begin{array}{ll} -x + 2z = 2 & x - 4y + z = -4 \\ -x + 2(2) = 2 & 2 - 4y + 2 = -4 \\ -x = -2 & -4y = -8 \\ x = 2 & y = 2 \end{array}$$

$$4x + 2y - 3z = 6$$
$$4(2) + 2(2) - 3(2) \overset{?}{=} 6$$
$$6 = 6$$

$$x - 4y + z = -4$$
$$2 - 4(2) + 2 \overset{?}{=} -4$$
$$-4 = -4$$

$$-x + 2z = 2$$
$$-2 + 2(2) \overset{?}{=} 2$$
$$2 = 2$$

The solution is $(2, 2, 2)$.

9.
$$\begin{array}{r} 4y - z = -13 \rightarrow 8y - 2z = -26 \\ 3y + 2z = 4 \rightarrow \underline{3y + 2z = 4} \\ 11y \qquad = -22 \\ y = -2 \end{array}$$

$$3y + 2z = 4$$
$$3(-2) + 2z = 4$$
$$2z = 10$$
$$z = 5$$

$$6x - 5y - 2z = 0$$
$$6x - 5(-2) - 2(5) = 0$$
$$6x = 0$$
$$x = 0$$

$$4y - z = -13$$
$$4(-2) - 5 \overset{?}{=} -13$$
$$-13 = -13$$

$$3y + 2z = 4$$
$$3(-2) + 2(5) \overset{?}{=} 4$$
$$4 = 4$$

$$6x - 5y - 2z = 0$$
$$6(0) - 5(-2) - 2(5) \overset{?}{=} 0$$
$$0 = 0$$

The solution is $(0, -2, 5)$.

10.
$$\begin{array}{r} 3y + 2z = 9 \rightarrow 3y \qquad + 2z = 9 \\ 4x + z = 5 \rightarrow \underline{-8x - 2z = -10} \\ 3y - 8x \qquad = -1 \end{array}$$

$$\begin{cases} 3y - 8x = -1 \rightarrow \\ x + y = 7 \rightarrow \end{cases} \begin{array}{r} 3y - 8x = -1 \\ \underline{-3y - 3x = -21} \\ -11x = -22 \\ x = 2 \end{array}$$

$$\begin{array}{ll} x + y = 7 & 4x + z = 5 \\ 2 + y = 7 & 4(2) + z = 5 \\ y = 5 & z = -3 \end{array}$$

$$\begin{array}{ll} x + y = 7 & 3y + 2z = 9 \\ 2 + 5 \overset{?}{=} 7 & 3(5) + 2(-3) \overset{?}{=} 9 \\ 7 = 7 & 9 = 9 \end{array}$$

$$4x + z = 5$$
$$4(2) + (-3) \overset{?}{=} 5$$
$$5 = 5$$

The solution is $(2, 5, -3)$.

11. Let a represent the amount of Mix A.
Let b represent the amount of Mix B.
Let c represent the amount of Mix C.

$$\begin{cases} 0.2a + 0.1b + 0.15c = 23 \\ 0.02a + 0.06b + 0.05c = 6.2 \\ 0.15a + 0.10b + 0.05c = 16 \end{cases}$$

$$0.2a + 0.1b + 0.15c = 23 \rightarrow$$
$$0.02a + 0.06b + 0.05c = 6.2 \rightarrow$$

$$0.2a + 0.1b + 0.15c = 23$$
$$\underline{-0.06a - 0.18b - 0.15c = -18.6}$$
$$0.14a - 0.08b = 4.4$$

$$0.02a + 0.06b + 0.05c = 6.2 \rightarrow$$
$$0.15a + 0.10b + 0.05c = 16 \rightarrow$$

$$-0.02a - 0.06b - 0.05c = -6.2$$
$$\underline{0.15a + 0.10b + 0.05c = 16}$$
$$0.13a + 0.04b = 9.8$$

$$0.14a - 0.08b = 4.4 \rightarrow 0.14a - 0.08b = 4.4$$
$$0.13a + 0.04b = 9.8 \rightarrow \underline{0.26a + 0.08b = 19.6}$$
$$0.40a = 24.0$$
$$a = 60$$

$$0.13a + 0.04b = 9.8$$
$$0.13(60) + 0.04b = 9.8$$
$$0.04b = 2$$
$$b = 50$$

$$0.2a + 0.1b + 0.15c = 23$$
$$0.2(60) + 0.1(50) + 0.15c = 23$$
$$0.15c = 6$$
$$c = 40$$

The veterinarian should mix 60 g of Mix A, 50 g of Mix B, and 40 g of Mix C.

12. $x + 2y + z = 1 \rightarrow x + 2y + z = 1$
$x - y + z = 1 \rightarrow \underline{2x - 2y + 2z = 2}$
$ 3x + 3z = 3$

$$x - y + z = 1$$
$$\underline{2x + y + 2z = 2}$$
$$3x + 3z = 3$$

$3x + 3z = 3 \rightarrow 3x + 3z = 3$
$3x + 3z = 3 \rightarrow \underline{-3x - 3z = -3}$
$ 0 = 0$

true equation; infinite solutions; dependent

13. $x + z = 0 \rightarrow x + z = 0$
$y + z = 2 \rightarrow \underline{-y - z = -2}$
$ x - y = -2$

$x + y + 2z = 3 \rightarrow x + y + 2z = 3$
$y + z = 2 \rightarrow \underline{-2y - 2z = -4}$
$ x - y = -1$

$x - y = -2 \rightarrow x - y = -2$
$x - y = -1 \rightarrow \underline{-x + y = 1}$
$ 0 = -1$

false equation; no solution; inconsistent

14. $8x + 6z = -1 \rightarrow 8x + 6z = -1$
$6y - 6z = -1 \rightarrow \underline{+ 6y - 6z = -1}$
$ 8x + 6y = -2$

$\begin{cases} 8x + 6y = -2 \rightarrow & 8x + 6y = -2 \\ 4x + 9y = 8 \rightarrow & \underline{-8x - 18y = -16} \end{cases}$
$ -12y = -18$

$$y = \frac{18}{12} = \frac{3}{2}$$

$8x + 6y = -2 \qquad\qquad 8x + 6z = -1$
$8x + 6\left(\frac{3}{2}\right) = -2 \qquad 8\left(-\frac{11}{8}\right) + 6z = -1$
$8x = -11 \qquad\qquad 6z = 10$
$x = -\frac{11}{8} \qquad\qquad z = \frac{10}{6} = \frac{5}{3}$

$\left(-\frac{11}{8}, \frac{3}{2}, \frac{5}{3}\right)$; independent

EXTEND

15. Let x represent the number of quarts picked on Thursday. Let y represent the number of quarts picked on Friday. Let z represent the number of quarts picked on Saturday.

$$\begin{cases} x + y + z = 87 \\ y = x + 15 \\ z = y - 3 \end{cases}$$

$y = x + 15 \qquad \rightarrow -x + y = 15$
$z = y - 3 \qquad \rightarrow \underline{- y + z = -3}$
$ -x + z = 12$

$x + y + z = 87 \qquad \rightarrow x + y + z = 87$
$z = y - 3 \qquad\quad \rightarrow \underline{- y + z = -3}$
$ x + 2z = 84$

$-x + z = 12 \qquad \rightarrow -x + z = 12$
$x + 2z = 84 \qquad \rightarrow \underline{x + 2z = 84}$
$ 3z = 96$
$ z = 32$

$x + 2z = 84 \qquad\qquad y = x + 15$
$x + 2(32) = 84 \qquad y = 20 + 15$
$x = 20 \qquad\qquad\quad y = 35$

She picked 20 qt on Thursday, 35 qt on Friday, and 32 qt on Saturday.

16. Let l represent the number of model L produced. Let m represent the number of model M produced. Let n represent the number of model N produced.

$$\begin{cases} 2l + m + 3n = 100 \\ 2l + 3m + 2n = 100 \\ l + m + 2n = 65 \end{cases}$$

$$2l + m + 3n = 100 \quad \rightarrow \quad -2l - m - 3n = -100$$
$$2l + 3m + 2n = 100 \quad \rightarrow \quad \underline{\;\;2l + 3m + 2n = \;\;100\;}$$
$$2m - n = 0$$

$$2l + m + 3n = 100 \quad \rightarrow \quad 2l + m + 3n = \;\;100$$
$$l + m + 2n = 65 \quad \rightarrow \quad \underline{-2l - 2m - 4n = -130}$$
$$-m - n = -30$$

$$2m - n = 0 \quad \rightarrow \quad 2m - \;\;n = \;\;0$$
$$-m - n = -30 \quad \rightarrow \quad \underline{-2m - 2n = -60}$$
$$-3n = -60$$
$$n = 20$$

$$2m - n = 0 \qquad\qquad l + m + 2n = 65$$
$$2m - 20 = 0 \qquad\qquad l + 10 + 2(20) = 65$$
$$2m = 20 \qquad\qquad\qquad\qquad l = 15$$
$$m = 10$$

Each week there are 15 model Ls, 10 model Ms and 20 model Ns produced.

17. This system is two parallel planes intersected by a third plane.

18. Answers will vary. Since equation 2 is the sum of equation 1 and equation 3, the system is dependent and will have infinitely many solutions.

THINK CRITICALLY

19. Let a represent the amount of Mix A. Let b represent the amount of Mix B. Let c represent the amount of Mix C.

$$\begin{cases} 0.20a + 0.10b + 0.15c = 18.5 \\ 0.02a + 0.06b + 0.05c = 4.9 \\ 0.15a + 0.10b + 0.05c = 13 \end{cases}$$

$$0.20a + 0.10b + 0.15c = 18.5 \rightarrow$$
$$0.02a + 0.06b + 0.05c = 4.9 \rightarrow$$

$$0.20a + 0.10b + 0.15c = \;\;18.5$$
$$\underline{-\,0.06a - 0.18b - 0.15c = -14.7}$$
$$0.14a - 0.08b \qquad\qquad = 3.8$$

$$0.02a + 0.06b + 0.05c = \;\;4.9 \rightarrow$$
$$0.15a + 0.10b + 0.05c = 13 \rightarrow$$

$$-0.02a - 0.06b - 0.05c = -\,4.9$$
$$\underline{\;\;0.15a + 0.10b + 0.05c = \;\;13\;}$$
$$0.13a + 0.04b \qquad\qquad = 8.1$$

$$0.14a - 0.08b = 3.8 \rightarrow 0.14a - 0.08b = 3.8$$
$$0.13a + 0.04b = 8.1 \rightarrow \underline{0.26a + 0.08b = 16.2}$$
$$0.40a \qquad\qquad = 20.0$$
$$a = 50$$

$$0.13a + 0.04b = 8.1$$
$$0.13(50) + 0.04b = 8.1$$
$$0.04b = 1.6$$
$$b = 40$$

$$0.20a + 0.10b + 0.15c = 18.5$$
$$0.20(50) + 0.10(40) + 0.15c = 18.5$$
$$0.15c = 4.5$$
$$c = 30$$

To get the desired mix, she should mix 50 g of Mix A, 40 g of Mix B, and 30 g of Mix C.

20. dependent; Answers will vary. Answers should include the idea that the graph would show two planes intersecting in a line.

21. $\begin{cases} 2 = C - A - B \\ -6 = C - 3A - 2B \\ -4 = C - 6A - 0B \end{cases}$

$$2 = C - A - B \rightarrow -2 = -C + A + B$$
$$-6 = C - 3A - 2B \rightarrow \underline{-6 = \;\;C - 3A - 2B}$$
$$-8 = \qquad\; -2A - B$$

$$-6 = C - 3A - 2B \rightarrow 6 = -C + 3A + 2B$$
$$-4 = C - 6A - 0B \rightarrow \underline{-4 = \;\;C - 6A - 0B}$$
$$2 = \qquad -3A + 2B$$

$$-8 = -2A - B \rightarrow -16 = -4A - 2B$$
$$2 = -3A + 2B \rightarrow \underline{\;\;2 = -3A + 2B\;}$$
$$-14 = -7A$$
$$2 = A$$

$$2 = -3A + 2B \qquad\quad 2 = C - A - B$$
$$2 = -3(2) + 2B \qquad 2 + C - 2 - 4$$
$$8 = 2B \qquad\qquad\quad 2 = C - 6$$
$$B = 4 \qquad\qquad\qquad C = 8$$
$$A = 2; B = 4; C = 8$$

22. Answers will vary. Possible answers: 3 planes intersecting in a line and 3 planes coinciding, 2 coinciding planes so all 3 planes intersect in a line.

23. Answers will vary. Possible answers: 3 parallel planes, 2 coinciding planes with the third plane parallel to them, 2 parallel planes with a third plane intersecting them in 2 parallel lines, 3 planes intersecting in 3 parallel lines.

24. $\begin{cases} 2l - m - 3n = -1 \\ 2l - m + n = -9 \\ l + 2m - 4n = 17 \end{cases}$

$$2l - m - 3n = -1 \rightarrow -2l + m + 3n = \;\;1$$
$$2l - m + n = -9 \rightarrow \underline{\;\;2l - m + \;\;n = -9\;}$$
$$4n = -8$$
$$n = -2$$

$$2l - m + n = -9 \rightarrow 2l - m + n = -9$$
$$l + 2m - 4n = 17 \rightarrow \underline{-2l - 4m + 8n = -34}$$
$$-5m + 9n = -43$$

$$-5m + 9n = -43$$
$$-5m + 9(-2) = -43$$
$$-5m = -25$$
$$m = 5$$

$$l + 2m - 4n = 17$$
$$l + 2(5) - 4(-2) = 17$$
$$l = -1$$

$-1 = \dfrac{1}{x}$	$5 = \dfrac{1}{y}$	$-2 = \dfrac{1}{z}$
$-x = 1$	$5y = 1$	$-2z = 1$
$x = -1$	$y = \dfrac{1}{5}$	$z = -\dfrac{1}{2}$

$$\left(-1, \frac{1}{5}, -\frac{1}{2}\right)$$

25a.

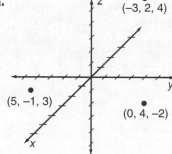

$(-3, 2, 4)$

$(5, -1, 3)$

$(0, 4, -2)$

25b.
$$2x + 4y + 6z = 12$$
$$2x + 4(0) + 6(0) = 12$$
$$2x = 12$$
$$x = 6$$

$$2x + 4y + 6z = 12$$
$$2(0) + 4y + 6(0) = 12$$
$$4y = 12$$
$$y = 3$$

$$2x + 4y + 6z = 12$$
$$2(0) + 4(0) + 6z = 12$$
$$6z = 12$$
$$z = 2$$

$(6, 0, 0); (0, 3, 0); (0, 0, 2)$

25c.

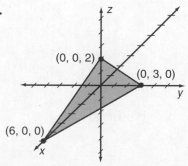

$(0, 0, 2)$

$(0, 3, 0)$

$(6, 0, 0)$

26. $2x - 2y = 6, -2y = -2x + 6, y = x - 3$

27. $2x - 5y = 7x - 40, -5y = 5x - 40, y = 8 - x$

28. $(y - 2) = 7x + 25, y = 7x + 27$

29. $10(x + y) = 2(5x - 5), 10x + 10y = 10x - 10,$
$10y = -10, y = -1$

30. 1×3 **31.** 3×1 **32.** 2×2 **33.** 2×3

34.
$$x - 2y + 4z = -10 \rightarrow \quad 3x - 6y + 12z = -30$$
$$-3x + 6y - 12z = 20 \rightarrow \underline{-3x + 6y - 12z = \quad 20}$$
$$0 = -10$$

false equation; no solution; inconsistent

35.
$$2x - 8y + 2z = -10 \rightarrow \quad 2x - 8y + 2z = -10$$
$$-x + 4y - z = 5 \rightarrow \underline{-2x + 8y - 2z = \quad 10}$$
$$0 = 0$$

$$-x + 4y - z = 5 \rightarrow \quad -3x + 12y - 3z = \quad 15$$
$$3x - 12y + 3z = -15 \rightarrow \underline{3x - 12y + 3z = -15}$$
$$0 = 0$$

true equations; infinite solutions; dependent

36. Let x represent the smallest angle, then $2x$ represents the largest angle, and $2x - 20$ represents the third angle.

$$x + 2x + 2x - 20 = 180$$
$$5x - 20 = 180$$
$$5x = 200$$
$$x = 40$$

Since x represents the smallest angle, the correct answer is B.

37. $(f \circ g)(x) = f[g(x)]$
$$g(x) = 4x^2$$
$$f[g(x)] = 2(4x^2) + 3$$
$$= 8x^2 + 3$$

38. $(g \circ f)(x) = g[f(x)]$
$$f(x) = 2x + 3$$
$$g[f(x)] = 4(2x + 3)^2$$
$$= 4(4x^2 + 12x + 9)$$
$$= 16x^2 + 48x + 36$$

39. $y = 2x + 3; x = 2y + 3, 2y = x - 3, y = \dfrac{x - 3}{2}$

40. $2x + 3 + 4x^2, 4x^2 + 2x + 3$

41. 2 units right, 6 units up

42. same vertex at $(0, 0)$, narrower

43. 8 units left, 1 unit down

EXPLORE

1. $2, 1, -5; 1, -3, 1$

2. The elements of the matrix correspond to the coefficients of the variables and constants of the system of equations.

3. $2x + y = -5 \rightarrow \quad 2x + y = -5$
   $x - 3y = 1 \rightarrow \underline{-2x + 6y = -2}$
   $\qquad\qquad\qquad\qquad 7y = -7$
   $\qquad\qquad\qquad\qquad\quad y = -1$

   $\qquad\qquad 2x + y = -5$
   $\qquad 2x + (-1) = -5$
   $\qquad\qquad\qquad 2x = -4$
   $\qquad\qquad\qquad\quad x = -2$

   $(-2, -1)$

4. $\begin{bmatrix} 1 & -3 & 1 \\ 2 & 1 & -5 \end{bmatrix}$ 5. $\begin{bmatrix} 1 & -3 & 1 \\ 0 & 7 & -7 \end{bmatrix}$

6. $\begin{bmatrix} 1 & -3 & 1 \\ 0 & 1 & -1 \end{bmatrix}$

7. Use the elements as the coefficients and constants.

   $\begin{cases} x - 3y = 1 \\ \quad y = -1 \end{cases}$

   $x - 3y = 1, x - 3(-1) = 1, x = -2; (-2, -1)$

8. The matrices represent equivalent systems. The first matrix can be transformed into the second matrix.

TRY THESE

1. $\begin{bmatrix} 1 & 2 & 7 \\ 1 & -1 & -2 \end{bmatrix}$ 2. $\begin{bmatrix} 2 & 4 & -3 & -18 \\ 3 & 1 & -1 & -5 \\ 1 & -2 & 4 & 14 \end{bmatrix}$

3. $\begin{bmatrix} 2 & -3 & 0 & 12 \\ 0 & 3 & 1 & -12 \\ 5 & 0 & -3 & 3 \end{bmatrix}$ 4. $\begin{cases} x - 2y = 7 \\ \quad y = -2 \end{cases}$

   $\qquad\qquad\qquad x - 2y = 7$
   $\qquad\qquad x - 2(-2) = 7$
   $\qquad\qquad\qquad x + 4 = 7$
   $\qquad\qquad\qquad\qquad x = 3$
   $\qquad$ The solution is $(3, -2)$.

5. $\begin{cases} x = 0.65 \\ y = 1.2 \\ z = 0.43 \end{cases}$

   The solution is $(0.65, 1.2, 0.43)$.

6. $\begin{cases} x - 4z = 5 \\ y - 12z = 13 \\ \quad z = -\dfrac{1}{2} \end{cases}$

   $\quad y - 12z = 13 \qquad\qquad x - 4z = 5$
   $\quad y - 12\left(-\dfrac{1}{2}\right) = 13 \qquad x - 4\left(-\dfrac{1}{2}\right) = 5$
   $\qquad y + 6 = 13 \qquad\qquad x + 2 = 5$
   $\qquad\qquad y = 7 \qquad\qquad\qquad x = 3$

   The solution is $\left(3, 7, -\dfrac{1}{2}\right)$.

7. $\begin{bmatrix} 2 & 4 & 22 \\ 6 & -3 & -9 \end{bmatrix}$ 8. $\begin{bmatrix} 6 & -3 & -9 \\ 1 & 2 & 11 \end{bmatrix}$

9. $\begin{bmatrix} 6 & -3 & -9 \\ 0 & 5 & 25 \end{bmatrix}$

10. Answers will vary. Students may explain that the row operations are analogous to those used with the elimination method.

11. $\begin{bmatrix} 2 & 4 & -10 & -2 \\ 3 & 9 & -21 & 0 \\ 1 & 5 & -12 & 1 \end{bmatrix} \rightarrow \begin{bmatrix} 1 & 2 & -5 & -1 \\ 0 & 3 & -6 & 3 \\ 0 & 3 & -7 & 2 \end{bmatrix} \rightarrow$

    $\begin{bmatrix} 1 & 0 & -1 & -3 \\ 0 & 1 & -2 & 1 \\ 0 & 0 & -1 & -1 \end{bmatrix} \rightarrow \begin{bmatrix} 1 & 0 & 0 & -2 \\ 0 & 1 & 0 & 3 \\ 0 & 0 & 1 & 1 \end{bmatrix};$

    $(-2, 3, 1);$ independent; consistent

12. $\begin{bmatrix} 1 & -2 & 1 & 4 \\ 3 & -6 & 3 & 12 \\ -2 & 4 & -2 & -8 \end{bmatrix} \rightarrow \begin{bmatrix} 1 & -2 & 1 & 4 \\ 0 & 0 & 0 & 0 \\ 0 & 0 & 0 & 0 \end{bmatrix};$

    infinite solutions; dependent; consistent

13. $\begin{bmatrix} 1 & 5 & -3 & 4 \\ 4 & -3 & 2 & 4 \\ 8 & -6 & 4 & 14 \end{bmatrix} \rightarrow \begin{bmatrix} 1 & 5 & -3 & 4 \\ 0 & -23 & 14 & -12 \\ 0 & -46 & 28 & -18 \end{bmatrix} \rightarrow$

    $\begin{bmatrix} 1 & 0 & \dfrac{1}{23} & \dfrac{32}{23} \\ 0 & 1 & -\dfrac{14}{23} & \dfrac{12}{23} \\ 0 & 0 & 0 & -42 \end{bmatrix};$ no solution; inconsistent

14. $\begin{bmatrix} 2 & 7 & 15 & -12 \\ 4 & 7 & 13 & -10 \\ 3 & 6 & 12 & -9 \end{bmatrix} \rightarrow \begin{bmatrix} 1 & \dfrac{7}{2} & \dfrac{15}{2} & -6 \\ 0 & -7 & -17 & 14 \\ 0 & -\dfrac{9}{2} & -\dfrac{21}{2} & 9 \end{bmatrix} \rightarrow$

    $\begin{bmatrix} 1 & 0 & -1 & 1 \\ 0 & 1 & \dfrac{17}{7} & -2 \\ 0 & 0 & 3 & 0 \end{bmatrix} \rightarrow \begin{bmatrix} 1 & 0 & 0 & 1 \\ 0 & 1 & 0 & -2 \\ 0 & 0 & 1 & 0 \end{bmatrix};$

    $(1, -2, 0);$ independent; consistent

15a. Let x represent the amount invested at 7%. Let y represent the amount invested at 8%. Let z represent the amount invested at 9%.

$$\begin{cases} 0.07x + 0.08y + 0.09z = 212 \\ x + y + z = 2500 \\ z - y = 1100 \end{cases}$$

15b. $\begin{bmatrix} 0.07 & 0.08 & 0.09 & | & 212 \\ 1 & 1 & 1 & | & 2500 \\ 0 & -1 & 1 & | & 1100 \end{bmatrix}$

15c. $\begin{bmatrix} 0.07 & 0.08 & 0.09 & | & 212 \\ 1 & 1 & 1 & | & 2500 \\ 0 & -1 & 1 & | & 1100 \end{bmatrix} \rightarrow$

$\begin{bmatrix} 1 & 1 & 1 & | & 2500 \\ 0 & 0.01 & 0.02 & | & 37 \\ 0 & -1 & 1 & | & 1100 \end{bmatrix} \rightarrow$

$\begin{bmatrix} 1 & 0 & -1 & | & -1200 \\ 0 & 1 & 2 & | & 3700 \\ 0 & 0 & 3 & | & 4800 \end{bmatrix} \rightarrow \begin{bmatrix} 1 & 0 & 0 & | & 400 \\ 0 & 1 & 0 & | & 500 \\ 0 & 0 & 1 & | & 1600 \end{bmatrix};$

Jason invested \$400 at 7%, \$500 at 8% and \$1600 at 9%.

PRACTICE

1. $\begin{bmatrix} 3 & 4 & | & 7 \\ -5 & 2 & | & 10 \end{bmatrix}$

2. $\begin{bmatrix} \frac{1}{3} & \frac{1}{2} & 1 & | & -1 \\ 1 & -1 & \frac{1}{5} & | & 1 \\ 1 & 1 & 1 & | & 5 \end{bmatrix}$

3. $\begin{bmatrix} 1 & -3 & 0 & | & 6 \\ 0 & 1 & 2 & | & 2 \\ 7 & -3 & -5 & | & 14 \end{bmatrix}$

4. $\begin{cases} x = \frac{5}{7} \\ y = \frac{8}{7} \end{cases}, \left(\frac{5}{7}, \frac{8}{7}\right)$

5. $\begin{cases} x + 2y + 3z = 4 \\ y + 2z = 4 \\ z = 2 \end{cases}$

$\begin{aligned} y + 2z &= 4 \\ y + 2(2) &= 4 \\ y &= 0 \end{aligned}$

$\begin{aligned} x + 2y + 3z &= 4 \\ x + 2(0) + 3(2) &= 4 \\ x &= -2 \end{aligned}$

$(-2, 0, 2)$

6. $\begin{cases} x - 3y + 2z = 0 \\ y + 0.7z = 0.7 \\ z = 1 \end{cases}$

$\begin{aligned} y + 0.7z &= 0.7 \\ y + 0.7(1) &= 0.7 \\ y &= 0 \end{aligned}$

$\begin{aligned} x - 3y + 2z &= 0 \\ x - 3(0) + 2(1) &= 0 \\ x &= -2 \end{aligned}$

$(-2, 0, 1)$

7. If the augmented matrix has a row containing all zeros except for the last element, the system is inconsistent. If the matrix has a row containing all zeros, then the matrix is dependent.

8. $\begin{bmatrix} 1 & 4 & | & 8 \\ 3 & 5 & | & 3 \end{bmatrix} \rightarrow \begin{bmatrix} 1 & 4 & | & 8 \\ 0 & -7 & | & -21 \end{bmatrix} \rightarrow \begin{bmatrix} 1 & 0 & | & -4 \\ 0 & 1 & | & 3 \end{bmatrix};$

$(-4, 3)$; independent; consistent

9. $\begin{bmatrix} -4 & 12 & | & 36 \\ 1 & -3 & | & 9 \end{bmatrix} \rightarrow \begin{bmatrix} 1 & -3 & | & -9 \\ 0 & 0 & | & 18 \end{bmatrix};$

no solution; inconsistent

10. $\begin{bmatrix} \frac{1}{3} & \frac{1}{5} & | & 2 \\ \frac{1}{3} & -\frac{1}{2} & | & -\frac{1}{3} \end{bmatrix} \rightarrow \begin{bmatrix} 1 & \frac{3}{5} & | & 6 \\ 0 & -\frac{7}{10} & | & -\frac{7}{3} \end{bmatrix} \rightarrow \begin{bmatrix} 1 & 0 & | & 4 \\ 0 & 1 & | & \frac{10}{3} \end{bmatrix};$

$\left(4, \frac{10}{3}\right)$; independent; consistent

11. $\begin{bmatrix} 1 & 1 & 1 & | & 1 \\ 1 & 2 & 3 & | & 4 \\ 4 & 5 & 6 & | & 7 \end{bmatrix} \rightarrow \begin{bmatrix} 1 & 1 & 1 & | & 1 \\ 0 & 1 & 2 & | & 3 \\ 0 & 1 & 2 & | & 3 \end{bmatrix} \rightarrow \begin{bmatrix} 1 & 0 & -1 & | & -2 \\ 0 & 1 & 2 & | & 3 \\ 0 & 0 & 0 & | & 0 \end{bmatrix};$

infinite solutions; dependent; consistent

12. $\begin{bmatrix} 4 & -1 & -3 & | & 1 \\ 8 & 1 & -1 & | & 5 \\ 2 & 1 & 2 & | & 5 \end{bmatrix} \rightarrow \begin{bmatrix} 1 & -\frac{1}{4} & -\frac{3}{4} & | & \frac{1}{4} \\ 0 & 3 & 5 & | & 3 \\ 0 & \frac{3}{2} & \frac{7}{2} & | & \frac{9}{2} \end{bmatrix} \rightarrow$

$\begin{bmatrix} 1 & 0 & -\frac{1}{3} & | & \frac{1}{2} \\ 0 & 1 & \frac{5}{3} & | & 1 \\ 0 & 0 & 1 & | & 3 \end{bmatrix} \rightarrow \begin{bmatrix} 1 & 0 & 0 & | & \frac{3}{2} \\ 0 & 1 & 0 & | & -4 \\ 0 & 0 & 1 & | & 3 \end{bmatrix};$

$(1.5, -4, 3)$; independent; consistent

13. $\begin{bmatrix} 1 & -1 & 5 & | & -6 \\ 0 & 6 & -16 & | & 28 \\ 1 & 3 & 2 & | & 5 \end{bmatrix} \rightarrow \begin{bmatrix} 1 & -1 & 5 & | & -6 \\ 0 & 6 & -16 & | & 28 \\ 0 & 4 & -3 & | & 11 \end{bmatrix} \rightarrow$

$\begin{bmatrix} 1 & 0 & \frac{14}{6} & | & \frac{8}{6} \\ 0 & 1 & -\frac{16}{6} & | & \frac{28}{6} \\ 0 & 0 & \frac{46}{6} & | & -\frac{46}{6} \end{bmatrix} \rightarrow \begin{bmatrix} 1 & 0 & 0 & | & 1 \\ 0 & 1 & 0 & | & 2 \\ 0 & 0 & 1 & | & -1 \end{bmatrix};$

$(1, 2, -1)$; independent; consistent

14. Let x represent the amount of 5% hydrochloric acid. Let y represent the amount of 20% hydrochloric acid.

$\begin{cases} x + y = 10 \\ 0.05x + 0.20y = (0.125)10 \end{cases}$

$\begin{bmatrix} 1 & 1 & | & 10 \\ 0.05 & 0.20 & | & 1.25 \end{bmatrix} \rightarrow \begin{bmatrix} 1 & 1 & | & 10 \\ 0 & 0.15 & | & 0.75 \end{bmatrix}$

$\begin{bmatrix} 1 & 0 & | & 5 \\ 0 & 1 & | & 5 \end{bmatrix};$ To obtain the correct mixture, add 5 oz of

5% hydrochloric acid and 5 oz of 20% hydrochloric acid.

15. Let x represent the number of pounds of apples. Let y represent the number of pounds of cheese. Let z represent the number of pounds of tomatoes.

$\begin{cases} x - 2y = 0 \\ -y + z = 1 \\ 0.70x + 1.50y + 0.80z = 8.20 \end{cases}$

$\begin{bmatrix} 1 & -2 & 0 & | & 0 \\ 0 & -1 & 1 & | & 1 \\ 0.70 & 1.50 & 0.80 & | & 8.20 \end{bmatrix} \rightarrow$

$\begin{bmatrix} 1 & -2 & 0 & | & 0 \\ 0 & -1 & 1 & | & 1 \\ 0 & 2.9 & 0.80 & | & 8.20 \end{bmatrix} \rightarrow$

$$\begin{bmatrix} 1 & 0 & -2 & | & -2 \\ 0 & 1 & -1 & | & -1 \\ 0 & 0 & 3.7 & | & 11.1 \end{bmatrix} \rightarrow \begin{bmatrix} 1 & 0 & 0 & | & 4 \\ 0 & 1 & 0 & | & 2 \\ 0 & 0 & 1 & | & 3 \end{bmatrix};$$

Rosa bought 4 lb of apples, 2 lb of cheese and 3 lb of tomatoes.

16. $\begin{cases} 3x - 2y = 0 \\ x + y = 5 \end{cases} \rightarrow \begin{bmatrix} 3 & -2 & | & 0 \\ 1 & 1 & | & 5 \end{bmatrix} \rightarrow \begin{bmatrix} 1 & 1 & | & 5 \\ 0 & -5 & | & -15 \end{bmatrix} \rightarrow$

$\begin{bmatrix} 1 & 0 & | & 2 \\ 0 & 1 & | & 3 \end{bmatrix}; (2, 3)$

17. $\begin{cases} x - 3y + 4z = 0 \\ 2x + 2y + z = 4 \\ 3x - 4y + 2z = 25 \end{cases} \rightarrow \begin{bmatrix} 1 & -3 & 4 & | & 0 \\ 2 & 2 & 1 & | & 4 \\ 3 & -4 & 2 & | & 25 \end{bmatrix} \rightarrow$

$\begin{bmatrix} 1 & -3 & 4 & | & 0 \\ 0 & 8 & -7 & | & 4 \\ 0 & 5 & -10 & | & 25 \end{bmatrix} \rightarrow \begin{bmatrix} 1 & 0 & -2 & | & 15 \\ 0 & 1 & -2 & | & 5 \\ 0 & 0 & 9 & | & -36 \end{bmatrix} \rightarrow$

$\begin{bmatrix} 1 & 0 & 0 & | & 7 \\ 0 & 1 & 0 & | & -3 \\ 0 & 0 & 1 & | & -4 \end{bmatrix}; (7, -3, -4)$

EXTEND

18. $\begin{bmatrix} -2 & 2 & 2 & 2 & | & -10 \\ 1 & 1 & 1 & 1 & | & -5 \\ 3 & 1 & -1 & 4 & | & -2 \\ 1 & 3 & -2 & 2 & | & -6 \end{bmatrix}$

19. substitution:
$4x - y - 3z = 1$
$-y = 1 + 3z - 4x$
$y = -1 - 3z + 4x$

$8x + y - z = 5$
$8x + (-1 - 3z + 4x) - z = 5$
$12x - 4z = 6$

$2x + y + 2z = 5$
$2x + (-1 - 3z + 4x) + 2z = 5$
$6x - z = 6$

$12x - 4z = 6$
$6x - z = 6$

$12x - 4z = 6x - z$
$6x = 3z$
$2x = z$

$6x - z = 6 \qquad 6x - z = 6$
$6x - 2x = 6 \qquad 6\left(\dfrac{3}{2}\right) - z = 6$
$4x = 6 \qquad 9 - z = 6$
$x = \dfrac{6}{4} = \dfrac{3}{2} \qquad z = 3$

$4x - y - 3z = 1$
$4\left(\dfrac{3}{2}\right) - y - 3(3) = 1$
$-y = 4$
$y = -4$

$\left(\dfrac{3}{2}, -4, 3\right)$

elimination:

$\begin{array}{ll} 4x - y - 3z = 1 & 4x - y - 3z = 1 \\ 8x + y - z = 5 & 2x + y + 2z = 5 \\ \hline 12x \quad - 4z = 6 & 6x \quad - z = 6 \end{array}$

$\begin{cases} 12x - 4z = 6 \\ 6x - z = 6 \end{cases} \rightarrow \begin{array}{l} 12x - 4z = 6 \\ -12x + 2z = -12 \\ \hline -2z = -6 \\ z = 3 \end{array}$

$\begin{array}{ll} 6x - z = 6 & 4x - y - 3z = 1 \\ 6x - 3 = 6 & 4\left(\dfrac{3}{2}\right) - y - 3(3) = 1 \\ 6x = 9 & -y = 4 \\ x = \dfrac{9}{6} = \dfrac{3}{2} & y = -4 \end{array}$

$\left(\dfrac{3}{2}, -4, 3\right)$

augmented matrices: see Exercise 12
Answers will vary. Possible answers: the more variables the more likely augmented matrices will be easiest.

20a. $\begin{cases} I_1 - I_2 + I_3 = 0 \\ 3I_1 + 3I_2 = 6 \\ 3I_2 + 3I_3 = 12 \end{cases} \rightarrow \begin{bmatrix} 1 & -1 & 1 & | & 0 \\ 3 & 3 & 0 & | & 6 \\ 0 & 3 & 3 & | & 12 \end{bmatrix} \rightarrow$

$\begin{bmatrix} 1 & -1 & 1 & | & 0 \\ 0 & 6 & -3 & | & 6 \\ 0 & 3 & 3 & | & 12 \end{bmatrix} \rightarrow \begin{bmatrix} 1 & 0 & \frac{1}{2} & | & 1 \\ 0 & 1 & -\frac{1}{2} & | & 1 \\ 0 & 0 & \frac{9}{2} & | & 9 \end{bmatrix} \rightarrow$

$\begin{bmatrix} 1 & 0 & 0 & | & 0 \\ 0 & 1 & 0 & | & 2 \\ 0 & 0 & 1 & | & 2 \end{bmatrix}; I_1 = 0$ amperes, $I_2 = 2$ amperes,

$I_3 = 2$ amperes

20b. $\begin{cases} I_1 - I_2 + I_3 = 0 \\ 4I_1 + I_2 = 6 \\ I_2 + 4I_3 = 12 \end{cases} \rightarrow \begin{bmatrix} 1 & -1 & 1 & | & 0 \\ 4 & 1 & 0 & | & 6 \\ 0 & 1 & 4 & | & 12 \end{bmatrix} \rightarrow$

$\begin{bmatrix} 1 & -1 & 1 & | & 0 \\ 0 & 5 & -4 & | & 6 \\ 0 & 1 & 4 & | & 12 \end{bmatrix} \rightarrow \begin{bmatrix} 1 & 0 & 5 & | & 12 \\ 0 & 1 & 4 & | & 12 \\ 0 & 0 & -24 & | & -54 \end{bmatrix} \rightarrow$

$\begin{bmatrix} 1 & 0 & 0 & | & \frac{3}{4} \\ 0 & 1 & 0 & | & 37 \\ 0 & 0 & 1 & | & \frac{9}{4} \end{bmatrix}; I_1 = \dfrac{3}{4}$ amperes, $I_2 = 3$ amperes,

$I_3 = \dfrac{9}{4}$ amperes

21. Let x represent the shortest side. Let y represent the longest side. Let z represent the third side.

$\begin{array}{ll} x + y + z = 56 & x + y + z = 56 \\ x + z - 12 = y \rightarrow & x - y + z = 12 \\ 3x - 26 = y & 3x - y \quad = 26 \end{array}$

$$\begin{bmatrix} 1 & 1 & 1 & | & 56 \\ 1 & -1 & 1 & | & 12 \\ 3 & -1 & 0 & | & 26 \end{bmatrix} \rightarrow \begin{bmatrix} 1 & 1 & 1 & | & 56 \\ 0 & -2 & 0 & | & -44 \\ 0 & -4 & -3 & | & 142 \end{bmatrix} \rightarrow$$

$$\begin{bmatrix} 1 & 0 & 1 & | & 34 \\ 0 & 1 & 0 & | & 22 \\ 0 & 0 & -3 & | & -54 \end{bmatrix} \rightarrow \begin{bmatrix} 1 & 0 & 0 & | & 16 \\ 0 & 1 & 0 & | & 22 \\ 0 & 0 & 1 & | & 18 \end{bmatrix};$$

The lengths of 3 sides are 16 cm, 18 cm, and 22 cm.

22. $\begin{cases} w + x + y + 2z = 6 \\ w + 2x + z = -2 \\ w + x + 3y - 2z = 12 \\ w + x - 4y + 5z = -16 \end{cases} \rightarrow$

$$\begin{bmatrix} 1 & 1 & 1 & 2 & | & 6 \\ 1 & 2 & 0 & 1 & | & -2 \\ 1 & 1 & 3 & -2 & | & 12 \\ 1 & 1 & -4 & 5 & | & -16 \end{bmatrix} \rightarrow \begin{bmatrix} 1 & 1 & 1 & 2 & | & 6 \\ 0 & 1 & -1 & -1 & | & -8 \\ 0 & 0 & 2 & -4 & | & 6 \\ 0 & 0 & -5 & 3 & | & -22 \end{bmatrix} \rightarrow$$

$$\begin{bmatrix} 1 & 0 & 2 & 3 & | & 14 \\ 0 & 1 & -1 & -1 & | & -8 \\ 0 & 0 & 2 & -4 & | & 6 \\ 0 & 0 & -5 & 3 & | & -22 \end{bmatrix} \rightarrow \begin{bmatrix} 1 & 0 & 0 & 7 & | & 8 \\ 0 & 1 & 0 & -3 & | & -5 \\ 0 & 0 & 1 & -2 & | & 3 \\ 0 & 0 & 0 & -7 & | & -7 \end{bmatrix} \rightarrow$$

$$\begin{bmatrix} 1 & 0 & 0 & 0 & | & 1 \\ 0 & 1 & 0 & 0 & | & -2 \\ 0 & 0 & 1 & 0 & | & 5 \\ 0 & 0 & 0 & 1 & | & 1 \end{bmatrix}; (1, -2, 5, 1)$$

23. yes; The first matrix can be transformed to the second using the following row operations: interchange R1 and R2; $-2(R1) + -(R2); -1(R1) + R3; -\frac{1}{3}(R2);$ $-4(R2) + R1; 7(R2) + R3; -3(R3);$ $-\frac{1}{3}(R3) + R1; -\frac{2}{3}(R3) + R2.$

24. $\sqrt{\dfrac{3240}{2560}} = 1 + i$

Answers may vary. With c, the new row 2 will be $[0 \quad -2 \quad -3 \,|\, -14]$, so with the next step he can get $[0 \quad 1 \quad \frac{3}{2} \,|\, 7]$ by multiplying by $-\frac{1}{2}$.

MIXED REVIEW

25. $3240 = 2560(1 + i)^2$

$\sqrt{\dfrac{3240}{2560}} = 1 + i$

$1.125 = 1 + i$

$0.125 = i$

The interest rate is 12.5%.

26. D; $m = -\dfrac{7}{3}$

Lesson 4.5, pages 189–195

EXPLORE

1. $\begin{bmatrix} 62 & 74 & 123 \\ 57 & 76 & 135 \end{bmatrix}; \begin{bmatrix} 19 \\ 25 \\ 19 \end{bmatrix}; 2 \times 3; 3 \times 1$

2. Week 1: Strategy: $62 \times \$19 = \$1178$
Simulation: $74 \times \$25 = \$1850$
Arcade: $123 \times \$19 = \$2337$

Week 2: Strategy: $57 \times \$19 = \$1083$
Simulation: $76 \times \$25 = \$1900$
Arcade: $135 \times \$19 = \$2565$
Multiply the number of games sold by the price of the game.

3. week 1: $\$1178 + \$1850 + \$2337 = \$5365$
week 2: $\$1083 + \$1900 + \$2565 = \$5548$

Add the totals spent on each type of game for that week.

4. $\begin{bmatrix} 5365 \\ 5548 \end{bmatrix}; 2 \times 1$

5. The number of rows in the totals matrix equals the number of rows in the games sold matrix. The number of columns in the totals matrix equals the number of columns in the price matrix.

TRY THESE

1. $AB_{2 \times 2}$

2. $AB_{5 \times 2}$

3. not defined

4. $AB_{2 \times 5}$

5. $AB = [-3(1) + 7(4) + 2(-5)] = [15];$

$$BA = \begin{bmatrix} 1(-3) & 1(7) & 1(2) \\ 4(-3) & 4(7) & 4(2) \\ -5(-3) & 5(7) & -5(2) \end{bmatrix} = \begin{bmatrix} -3 & 7 & 2 \\ -12 & 28 & 8 \\ 15 & -35 & -10 \end{bmatrix}$$

6. $AB = \begin{bmatrix} -8(1) + 3(2) & -8(-4) + 3(0) \\ -4(1) + 4(2) & -4(-4) + 4(0) \end{bmatrix} =$

$\begin{bmatrix} -2 & 32 \\ 4 & 16 \end{bmatrix}; BA = \begin{bmatrix} 1(-8) + (-4)(-4) & 1(3) + (-4)(4) \\ 2(-8) + 0(-4) & 2(3) + 0(4) \end{bmatrix} =$

$\begin{bmatrix} 8 & -13 \\ -16 & 6 \end{bmatrix}$

7. $AB = \begin{bmatrix} 4(2) + (-3)(0) + 1(-4) & 4(1) + (-3)(1) + 1(7) \\ -5(2) + 2(0) + 2(-4) & -5(1) + 2(1) + 2(7) \end{bmatrix}$

$= \begin{bmatrix} 4 & 8 \\ -18 & 11 \end{bmatrix}$; $BA =$

$\begin{bmatrix} 2(4) + 1(-5) & 2(-3) + 1(2) & 2(1) + 1(2) \\ 0(4) + 1(-5) & 0(-3) + 1(2) & 0(1) + 1(2) \\ -4(4) + 7(-5) & -4(-3) + 7(2) & -4(1) + 7(2) \end{bmatrix} =$

$\begin{bmatrix} 3 & -4 & 4 \\ -5 & 2 & 2 \\ -51 & 26 & 10 \end{bmatrix}$

8. $AB =$

$\begin{bmatrix} 1(2) + 0(0) + 2(-1) & 1(2) + 0(0) + 2(-1) \\ 1(2) + 0(0) + 2(-1) & 1(2) + 0(0) + 2(-1) \\ 1(2) + 0(0) + 2(-1) & 1(2) + 0(0) + 2(-1) \end{bmatrix}$

$\begin{bmatrix} 1(2) + 0(0) + 2(-1) \\ 1(2) + 0(0) + 2(-1) \\ 1(2) + 0(0) + 2(-1) \end{bmatrix} =$

$\begin{bmatrix} 0 & 0 & 0 \\ 0 & 0 & 0 \\ 0 & 0 & 0 \end{bmatrix}$; $BA =$

$\begin{bmatrix} 2(1) + 2(1) + 2(1) & 2(0) + 2(0) + 2(0) \\ 0(1) + 0(1) + 0(1) & 0(0) + 0(0) + 0(0) \\ -1(1) + (-1)(1) + (-1)(1) & -1(0) + -1(0) + -1(0) \end{bmatrix}$

$\begin{bmatrix} 2(2) + 2(2) + 2(2) \\ 0(2) + 0(2) + 0(2) \\ -1(2) + (-1)(2) + (-1)(2) \end{bmatrix} =$

$\begin{bmatrix} 6 & 0 & 12 \\ 0 & 0 & 0 \\ -3 & 0 & -6 \end{bmatrix}$

9. For both AB and BA to be defined, the number of columns in A must equal to number of rows in B and the number of rows in A must equal the number of columns in B. Examples will vary. See exercises 5–8.

10. $AB =$

$\begin{bmatrix} -1(2) + 2(4) + 1(1) & -1(-2) + 2(3) + 1(-3) \\ 2(2) + 4(4) + (-3)(1) & 2(-2) + 4(3) + (-3)(-3) \end{bmatrix}$

$\begin{bmatrix} -1(2) + 2(0) + 1(4) \\ 2(2) + 4(0) + (-3)(4) \end{bmatrix} = \begin{bmatrix} 7 & 5 & 2 \\ 17 & 17 & -8 \end{bmatrix}$

11. BA is not defined.

12. $BC =$

$\begin{bmatrix} 2(2) + (-2)(-1) + 2(1) & 2(3) + (-2)(2) + 2(3) \\ 4(2) + 3(-1) + 0(1) & 4(3) + 3(2) + 0(3) \\ 1(2) + (-3)(-1) + 4(1) & 1(3) + (-3)(2) + 4(3) \end{bmatrix}$

$\begin{bmatrix} 2(1) + (-2)(-1) + 2(2) \\ 4(1) + 3(-1) + 0(2) \\ 1(1) + (-3)(-1) + 4(2) \end{bmatrix} = \begin{bmatrix} 8 & 8 & 8 \\ 5 & 18 & 1 \\ 9 & 9 & 12 \end{bmatrix}$

13. $CB = \begin{bmatrix} 2(2) + 3(4) + 1(1) \\ -1(2) + 2(4) + (-1)(1) \\ 1(2) + 3(4) + 2(1) \end{bmatrix}$

$\begin{bmatrix} 2(-2) + 3(3) + 1(-3) & 2(2) + 3(0) + 1(4) \\ -1(-2) + 2(3) + (-1)(-3) & -1(2) + 2(0) + (-1)(4) \\ 1(-2) + 3(3) + 2(-3) & 1(2) + 3(0) + 2(4) \end{bmatrix} =$

$\begin{bmatrix} 17 & 2 & 8 \\ 5 & 11 & -6 \\ 16 & 1 & 10 \end{bmatrix}$

14. $AC = \begin{bmatrix} -1(2) + 2(-1) + 1(1) & -1(3) + 2(2) + 1(3) \\ 2(2) + 4(-1) + (-3)(1) & 2(3) + 4(2) + (-3)(3) \end{bmatrix}$

$\begin{bmatrix} -1(1) + 2(-1) + 1(2) \\ 2(1) + 4(-1) + (-3)(2) \end{bmatrix} = \begin{bmatrix} -3 & 4 & -1 \\ -3 & 5 & -8 \end{bmatrix}$

15. CA is not defined.

16. $B + C = \begin{bmatrix} 2 & -2 & 2 \\ 4 & 3 & 0 \\ 1 & -3 & 4 \end{bmatrix} + \begin{bmatrix} 2 & 3 & 1 \\ -1 & 2 & -1 \\ 1 & 3 & 2 \end{bmatrix} = \begin{bmatrix} 4 & 1 & 3 \\ 3 & 5 & -1 \\ 2 & 0 & 6 \end{bmatrix}$;

$A(B + C) =$

$\begin{bmatrix} -1(4) + 2(3) + 1(2) & -1(1) + 2(5) + 1(0) \\ 2(4) + 4(3) + (-3)(2) & 2(1) + 4(5) + (-3)(0) \end{bmatrix}$

$\begin{bmatrix} -1(3) + 2(-1) + 1(6) \\ 2(3) + 4(-1) + (-3)(6) \end{bmatrix} = \begin{bmatrix} 4 & 9 & 1 \\ 14 & 22 & -16 \end{bmatrix}$

17. from Exercise 10, $AB = \begin{bmatrix} 7 & 5 & 2 \\ 17 & 17 & -8 \end{bmatrix}$; from Exercise 14,

$AC = \begin{bmatrix} -3 & 4 & -1 \\ -3 & 5 & -8 \end{bmatrix}$; $AB + AC = \begin{bmatrix} 7 & 5 & 2 \\ 17 & 17 & -8 \end{bmatrix} +$

$\begin{bmatrix} -3 & 4 & -1 \\ -3 & 5 & -8 \end{bmatrix} = \begin{bmatrix} 4 & 9 & 1 \\ 14 & 22 & -16 \end{bmatrix}$

18.
$$[6 \quad 3 \quad 1 \quad 2 \quad 2] \times \begin{bmatrix} 3 & 1 & 2 & 3 \\ 2 & 2 & 3 & 1 \\ 3 & 0 & 2 & 2 \\ 0 & 1 & 0 & 1 \\ 0 & 0 & 1 & 0 \end{bmatrix} =$$

$$\begin{aligned} &[6(3) + 3(2) + 1(3) + 2(0) + 2(0) \\ &6(1) + 3(2) + 1(0) + 2(1) + 2(0) \\ &6(2) + 3(3) + 1(2) + 2(0) + 2(1) \\ &6(3) + 3(1) + 1(2) + 2(1) + 2(0)] = \end{aligned}$$

$[27 \quad 14 \quad 25 \quad 25]$; Bears scored 27 points, Lions scored 14 points,

Tigers scored 25 points and Falcons scored 25 points.

PRACTICE

1. AB is not defined.

2. $AB_{2 \times 3}$

3. $AB_{3 \times 5}$

4. AB is not defined.

5. In matrix multiplication, the number of columns in the first matrix must be equal to the number of rows in the second matrix. In matrix addition and subtraction, both matrix dimensions must be equal.

6. $[4(2) + (-2)(3) + (3)(-5)] = [-13]$

7. $[5(2) + 2(-5)] = [0]$

8. Matrix is not defined.

9. $\begin{bmatrix} 4(3) + (-2)(4) \\ 0(3) + 3(4) \\ -7(3) + 5(4) \end{bmatrix} = \begin{bmatrix} 4 \\ 12 \\ -1 \end{bmatrix}$

10. Matrix is not defined.

11.
$$\begin{bmatrix} 1(3) + (-5)(0) + 0(5) & 1(0) + (-5)4 + 0(-3) \\ 4(3) + 1(0) + (-2)(5) & 4(0) + 1(4) + (-2)(-3) \\ 0(3) + (-1)(0) + 3(5) & 0(0) + (-1)(4) + 3(-3) \end{bmatrix}$$

$$\begin{aligned} &1(-1) + (-5)(2) + 0(1) \\ &4(-1) + 1(2) + (-2)(1) \\ &0(-1) + (-1)(2) + 3(1) \end{aligned} = \begin{bmatrix} 3 & -20 & -11 \\ 2 & 10 & -4 \\ 15 & -13 & 1 \end{bmatrix}$$

12. $B + C = \begin{bmatrix} 2 & -1 \\ 3 & 1 \end{bmatrix} + \begin{bmatrix} 3 & 1 \\ -2 & 0 \end{bmatrix} = \begin{bmatrix} 5 & 0 \\ 1 & 1 \end{bmatrix}$;

$$A(B + C) = \begin{bmatrix} 1(5) + 2(1) & 1(0) + 2(1) \\ 0(5) + (-3)(1) & 0(0) + (-3)(1) \end{bmatrix} =$$

$$\begin{bmatrix} 7 & 2 \\ -3 & -3 \end{bmatrix}; \begin{bmatrix} 1(2) + 2(3) & 1(-1) + 2(1) \\ 0(2) + (-2)(3) & 0(-1) + (-3)(1) \end{bmatrix} =$$

$$\begin{bmatrix} 8 & 1 \\ -9 & -3 \end{bmatrix}; AC = \begin{bmatrix} 1(3) + (2)(-2) & 1(1) + 2(0) \\ 0(3) + (-3)(-2) & 0(1) + (-3)(0) \end{bmatrix} =$$

$AB + AC = \begin{bmatrix} 7 & 2 \\ -3 & -3 \end{bmatrix}$; yes, $A(B + C) = AB + AC$

13. from exercise 12, $B + C = \begin{bmatrix} 5 & 0 \\ 1 & 1 \end{bmatrix}$; $(B + C)A =$

$$\begin{bmatrix} 5(1) + 0(0) & 5(2) + (0)(-3) \\ 1(1) + 1(0) & 1(2) + 1(-3) \end{bmatrix} = \begin{bmatrix} 5 & 10 \\ 1 & -1 \end{bmatrix};$$

$$BA = \begin{bmatrix} 2(1) + (-1)(0) & 2(2) + (-1)(-3) \\ 3(1) + 1(0) & 3(2) + 1(-3) \end{bmatrix} = \begin{bmatrix} 2 & 7 \\ 3 & 3 \end{bmatrix};$$

$$CA = \begin{bmatrix} 3(1) + 1(0) & 3(2) + 1(-3) \\ -2(1) + 0(0) & -2(2) + 0(-3) \end{bmatrix} = \begin{bmatrix} 3 & 3 \\ -2 & -4 \end{bmatrix};$$

$$BA + CA = \begin{bmatrix} 2 & 7 \\ 3 & 3 \end{bmatrix} + \begin{bmatrix} 3 & 3 \\ -2 & -4 \end{bmatrix} = \begin{bmatrix} 5 & 10 \\ 1 & -1 \end{bmatrix};$$ yes,

$(B + C)A = BA + CA$

14. $[19.95(10) + 5.50(32) + 11.25(20)$
$19.95(23) + 5.50(31) + 11.25(47)$
$19.95(16) + 5.50(55) + 11.25(51)] =$
$[600.50 \quad 1158.10 \quad 1195.45]$; The product matrix gives the total daily sales for each store.

EXTEND

15. Answers will vary. Possible answers: $AB_{2 \times 2}$:
$A_{2 \times 2} \times B_{2 \times 2}$ or $A_{2 \times 3} \times B_{3 \times 2}$; $AB_{3 \times 1}$: $A_{3 \times 4} \times B_{4 \times 1}$ or
$A_{3 \times 1} \times B_{1 \times 1}$; $AB_{2 \times 4}$: $A_{2 \times 2} \times B_{2 \times 4}$ or $A_{1 \times 2} \times B_{2 \times 4}$

16. $[1(1) + 2(0) + 0(3) \quad 1(4) + 2(1) + 0(2)$

$1(3) + 2(2) + 0(1)] = [1 \quad 6 \quad 7]; [1 \quad 6 \quad 7] \begin{bmatrix} 1 \\ 2 \\ -1 \end{bmatrix} =$

$[1(1) + 6(2) + 7(-1)] = [6]$

17. $A + B = \begin{bmatrix} -1 & 0 \\ 2 & 1 \end{bmatrix} + \begin{bmatrix} 1 & -1 \\ 0 & 2 \end{bmatrix} = \begin{bmatrix} 0 & -1 \\ 2 & 3 \end{bmatrix}$;

$A - B = \begin{bmatrix} -1 & 0 \\ 2 & 1 \end{bmatrix} - \begin{bmatrix} 1 & -1 \\ 0 & 2 \end{bmatrix} = \begin{bmatrix} -2 & 1 \\ 2 & -1 \end{bmatrix}$;

$(A + B)(A - B) = \begin{bmatrix} 0(-2) + (-1)(2) \\ 2(-2) + 3(2) \end{bmatrix}$

$\begin{aligned} 0(1) + (-1)(-1) \\ 2(1) + 3(-1) \end{aligned} = \begin{bmatrix} -2 & 1 \\ 2 & -1 \end{bmatrix}$;

$A^2 = \begin{bmatrix} 1(1) + 0(2) & -1(0) + 0(1) \\ 2(-1) + 1(2) & 2(0) + 1(1) \end{bmatrix} = \begin{bmatrix} 1 & 0 \\ 0 & 1 \end{bmatrix}$;

$B^2 = \begin{bmatrix} 1(1) + (-1)(0) & 1(-1) + (-1)(2) \\ 0(1) + 2(0) & 0(-1) + 2(2) \end{bmatrix} = \begin{bmatrix} 1 & -3 \\ 0 & 4 \end{bmatrix}$;

$A^2 - B^2 = \begin{bmatrix} 1 & 0 \\ 0 & 1 \end{bmatrix} - \begin{bmatrix} 1 & -3 \\ 0 & 4 \end{bmatrix} = \begin{bmatrix} 0 & 3 \\ 0 & -3 \end{bmatrix}$; no,

$(A + B)(A - B) \neq A^2 - B^2$

18. from Exercise 17, $A + B = \begin{bmatrix} 0 & -1 \\ 2 & 3 \end{bmatrix}$; from Exercise 17,

$A - B = \begin{bmatrix} -2 & 1 \\ 2 & -1 \end{bmatrix}$; $(A + B)(A - B) =$

$\begin{bmatrix} 0(-2) + (-1)(2) & 0(1) + (-1)(-1) \\ 2(-2) + 3(2) & 2(1) + 3(-1) \end{bmatrix} =$

$\begin{bmatrix} -2 & 1 \\ 2 & -1 \end{bmatrix}$; from Exercise 17, $A^2 = \begin{bmatrix} 1 & 0 \\ 0 & 1 \end{bmatrix}$;

$BA = \begin{bmatrix} 1(-1) + (-1)(2) & 1(0) + (-1)(1) \\ 0(-1) + 2(2) & 0(0) + 2(1) \end{bmatrix} =$

$\begin{bmatrix} -3 & -1 \\ 4 & 2 \end{bmatrix}$; $AB = \begin{bmatrix} -1(1) + 0(0) & -1(-1) + 0(2) \\ 2(1) + 1(0) & 2(-1) + 1(2) \end{bmatrix} =$

$\begin{bmatrix} -1 & 1 \\ 2 & 0 \end{bmatrix}$; from Exercise 17, $B^2 = \begin{bmatrix} 1 & -3 \\ 0 & 4 \end{bmatrix}$;

$A^2 + BA - AB - B^2 = \begin{bmatrix} 1 & 0 \\ 0 & 1 \end{bmatrix} + \begin{bmatrix} -3 & -1 \\ 4 & 2 \end{bmatrix} -$

$\begin{bmatrix} -1 & 1 \\ 2 & 0 \end{bmatrix} - \begin{bmatrix} 1 & -3 \\ 0 & 4 \end{bmatrix} = \begin{bmatrix} -2 & 1 \\ 2 & -1 \end{bmatrix}$; yes,

$(A + B)(A - B) = A^2 + BA - AB - B^2$

19a. $\begin{bmatrix} 0 & -1 \\ 1 & 0 \end{bmatrix}\begin{bmatrix} -1 & 0 & 4 \\ 1 & -3 & 2 \end{bmatrix} =$

$\begin{bmatrix} (0)(-1) + (-1)(1) & 0(0) + (-1)(-3) \\ 1(-1) + 0(1) & 1(0) + 0(3) \end{bmatrix}$

$\begin{matrix} 0(4) + (-1)(2) \\ 1(4) + 0(2) \end{matrix} = \begin{bmatrix} -1 & -3 & -2 \\ -1 & 0 & 4 \end{bmatrix}$

19b. $(-1, -1), (3, 0), (-2, 4)$

19c.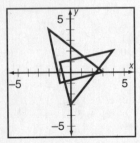

Original triangle is $(-1, 1)$ $(0, -3)$ $(4, 2)$. Rotated triangle is $(-1, -1)$ $(3, 0)$ $(-2, 4)$.

20. $\begin{bmatrix} 2000 & 1000 \\ 1800 & 800 \end{bmatrix}\begin{bmatrix} 8000 & 1600 & 2400 \\ 9500 & 2500 & 2400 \end{bmatrix} =$

$\begin{bmatrix} 2000(8000) + 1000(9500) & 2000(1600) + 1000(2500) \\ 1800(8000) + 800(9500) & 1800(1600) + 800(2500) \end{bmatrix}$

$\begin{matrix} 2000(2400) + 1000(2400) \\ 1800(2400) + 800(2400) \end{matrix}$

$= \begin{bmatrix} 25,500,000 & 5,700,000 & 7,200,000 \\ 22,000,000 & 4,880,000 & 6,240,000 \end{bmatrix}$;

The top row of the product matrix represents the total amounts of money paid by undergraduates on tuition, room and board. The lower row represents the total amounts of money paid by graduates on tuition, room and board.

THINK CRITICALLY

21. $5(6) + (-2)(-4) = x, x = 38$;
$1(2) + 7(3) = y, y = 23$

22. $1(x) + 2(-1) = 3, x - 2 = 3, x = 5$;
$3(-3) + 4(2) = y, y = -1$

23. $\begin{cases} -4x(1) + y(2) = 4 \\ -4x(-4) + y(0) = 16 \end{cases} \rightarrow \begin{cases} -4x + 2y = 4 \\ 16x = 16 \end{cases}$

$16x = 16, x = 1; -4x + 2y = 4, -4(1) + 2y = 4,$
$2y = 8, y = 4$

24. $4(2) + 2y(-4) = 0, 8 - 8y = 0, -8y = -8, y = 1$;
$2x(2) + 2(-4) = 0, 4x - 8 = 0, 4x = 8, x = 2$

25. B is a 2×4 matrix. To multiply A and B, the number of rows in B must equal the number of columns in A. Therefore, the number of rows in B is 2. To multiply the product AB and C, the number of columns in AB must equal the number of rows in C. Therefore, the number of columns in AB, which is also equal to the number of columns in B, is 4.

26a. $\begin{bmatrix} 0.042 & 0.09 & 0.147 \\ 0.086 & 0.106 & 0.019 \\ 0.034 & 0.043 & 0.021 \\ 0.113 & 0.15 & 0.074 \end{bmatrix}\begin{bmatrix} 8791.8 \\ 3455.6 \\ 203.8 \end{bmatrix} =$

$\begin{bmatrix} 0.042(8791.8) + 0.09(3455.6) + 0.147(203.8) \\ 0.086(8791.8) + 0.106(3455.6) + 0.019(203.8) \\ 0.034(8791.8) + 0.043(3455.6) + 0.021(203.8) \\ 0.113(8791.8) + 0.15(3455.6) + 0.074(203.8) \end{bmatrix} =$

$\begin{bmatrix} 710.2 \\ 1126.3 \\ 451.8 \\ 1526.9 \end{bmatrix}$; China produced 710.2t, Canada produced 1126.3t, Peru produced 451.8t, and Russia produced 1526.9t.

26b. $\begin{bmatrix} 0.042 & 0.09 & 0.147 \\ 0.086 & 0.106 & 0.019 \\ 0.034 & 0.043 & 0.021 \\ 0.113 & 0.15 & 0.074 \end{bmatrix}\begin{bmatrix} 9231.1 \\ 4001.0 \\ 183.5 \end{bmatrix} =$

$\begin{bmatrix} 0.042(9231.1) + 0.09(4001.0) + 0.147(183.5) \\ 0.086(9231.1) + 0.106(4001.0) + 0.019(183.5) \\ 0.034(9231.1) + 0.043(4001.0) + 0.021(183.5) \\ 0.113(9231.1) + 0.15(4001.0) + 0.074(183.5) \end{bmatrix}$;

$\begin{bmatrix} 774.8 \\ 1221.5 \\ 489.8 \\ 1656.8 \end{bmatrix}; \begin{bmatrix} 774.8 \\ 1221.5 \\ 489.8 \\ 1656.8 \end{bmatrix} - \begin{bmatrix} 710.2 \\ 1126.3 \\ 451.8 \\ 1526.8 \end{bmatrix} = \begin{bmatrix} 64.6 \\ 95.2 \\ 38.0 \\ 129.9 \end{bmatrix}$;

China produced 64.6*t* more, Canada produced 95.2*t* more, Peru produced 38*t* more, and Russia produced 129.9*t* more.

27a. $A^2 = \begin{bmatrix} 1 & 2 \\ 0 & 0 \end{bmatrix} \times \begin{bmatrix} 1 & 2 \\ 0 & 0 \end{bmatrix} = \begin{bmatrix} 1 & 2 \\ 0 & 0 \end{bmatrix} = \begin{bmatrix} 1 & 2 \\ 0 & 0 \end{bmatrix}; A^2 = A$

27b. $A^5 = A \cdot A \cdot A \cdot A \cdot A = A$;
$A^n = A_1 \cdot A_2 \cdots A_n = A$; Working from the right, the product of each pair of matrices is A.

27c. Answers will vary. $\begin{bmatrix} \frac{1}{2} & \frac{1}{2} \\ \frac{1}{2} & \frac{1}{2} \end{bmatrix}$

MIXED REVIEW

28. additive identity   **29.** multiplicative inverse

30. additive inverse   **31.** multiplicative identity

32.

$x = -3 \text{ or } x = 3$

33.

$-3 < x < 3$

34.

$x \leq -3 \text{ or } x \geq 3$

35.

$-5 < x < 1$

PROJECT CONNECTION

1–2. No solutions, these are activities.

3–4. Answers will vary.

Lesson 4.6, pages 196–202

EXPLORE

1. $AI = \begin{bmatrix} 11(1) + 3(0) & 11(0) + 3(1) \\ 7(1) + 2(0) & 7(0) + 2(1) \end{bmatrix} = \begin{bmatrix} 11 & 3 \\ 7 & 2 \end{bmatrix};$

$IA = \begin{bmatrix} 1(11) + 0(7) & 1(3) + 0(2) \\ 0(11) + 1(7) & 0(3) + 1(2) \end{bmatrix} = \begin{bmatrix} 11 & 3 \\ 7 & 2 \end{bmatrix};$

The two products are the same and are equal to matrix A.

2. $BI = \begin{bmatrix} 2(1) + -3(0) & 2(0) + (-3)(1) \\ -7(1) + 11(0) & -7(0) + 11(1) \end{bmatrix} = \begin{bmatrix} 2 & 3 \\ -7 & 11 \end{bmatrix};$

$IB = \begin{bmatrix} 1(2) + 0(-7) & 1(-3) + 0(11) \\ 0(2) + 1(-7) & 0(-3) + 1(11) \end{bmatrix} = \begin{bmatrix} 2 & -3 \\ -7 & 11 \end{bmatrix};$

The two products are the same and are equal to matrix B.

3. When you multiply a number, a, by 1, the product is a. When you multiply a matrix, A, by 1, the product is A.

4. $AB = \begin{bmatrix} 11(2) + 3(-7) & 11(-3) + 3(11) \\ 7(2) + 2(-7) & 7(-3) + 2(11) \end{bmatrix} = \begin{bmatrix} 1 & 0 \\ 0 & 1 \end{bmatrix};$

$BA = \begin{bmatrix} 2(11) + (-3)(7) & 2(3) + (-3)(2) \\ -7(11) + 11(7) & -7(3) + 11(2) \end{bmatrix} = \begin{bmatrix} 1 & 0 \\ 0 & 1 \end{bmatrix};$

The two products are the same and are equal to the matrix I.

TRY THESE

1. $\begin{bmatrix} 5(2) + 3(-3) & 5(-3) + 3(5) \\ 3(2) + 2(-3) & 3(-3) + 2(5) \end{bmatrix} = \begin{bmatrix} 1 & 0 \\ 0 & 1 \end{bmatrix};$ true

2. This matrix multiplication is not defined.; false

3. $\begin{bmatrix} 2\left(\frac{7}{3}\right) + (-1)\left(-\frac{10}{3}\right) + 1(-3) & 2(-1) + (-1)(0) + 1(2) \\ 1\left(\frac{7}{3}\right) + (-2)\left(-\frac{10}{3}\right) + 3(-3) & 1(-1) + (-2)(0) + 3(2) \\ 4\left(\frac{7}{3}\right) + 1\left(-\frac{10}{3}\right) + 2(-3) & 4(-1) + 1(0) + 2(2) \end{bmatrix}$

$\begin{bmatrix} 2\left(\frac{1}{5}\right) + (-1)\left(\frac{5}{3}\right) + 1(1) \\ 1\left(\frac{1}{5}\right) + (-2)\left(\frac{5}{3}\right) + 3(1) \\ 4\left(\frac{1}{5}\right) + 1\left(\frac{5}{3}\right) + 2(1) \end{bmatrix} = \begin{bmatrix} 5 & 0 & \frac{2}{3} \\ 0 & 5 & \frac{2}{3} \\ 0 & 0 & 11 \end{bmatrix};$ false

4. $\begin{bmatrix} \frac{1}{2}(1) + \left(-\frac{2}{3}\right)(0) & \frac{1}{2}(0) + \left(-\frac{2}{3}\right)(1) \\ \frac{5}{6}(1) + \frac{1}{3}(0) & \frac{5}{6}(0) + \frac{1}{3}(1) \end{bmatrix} =$

$\begin{bmatrix} \frac{1}{2} & -\frac{2}{3} \\ \frac{5}{6} & \frac{1}{3} \end{bmatrix} \overset{?}{=} \begin{bmatrix} 1\left(\frac{1}{2}\right) + 0\left(\frac{5}{6}\right) & 1\left(-\frac{2}{3}\right) + 0\left(\frac{1}{3}\right) \\ 0\left(\frac{1}{2}\right) + 1\left(\frac{5}{6}\right) & 0\left(-\frac{2}{3}\right) + 1\left(\frac{1}{3}\right) \end{bmatrix} =$

$\begin{bmatrix} \frac{1}{2} & -\frac{2}{3} \\ \frac{5}{6} & \frac{1}{3} \end{bmatrix};$ true

5. $\begin{bmatrix} 8 & 5 & | & 1 & 0 \\ 5 & 3 & | & 0 & 1 \end{bmatrix} \rightarrow \begin{bmatrix} 1 & \frac{5}{8} & | & \frac{1}{8} & 0 \\ 0 & -\frac{1}{8} & | & -\frac{5}{8} & 1 \end{bmatrix} \rightarrow \begin{bmatrix} 1 & 0 & | & -3 & 5 \\ 0 & 1 & | & 5 & -8 \end{bmatrix};$

$A^{-1} = \begin{bmatrix} -3 & 5 \\ 5 & -8 \end{bmatrix}$

6. $\begin{bmatrix} 2 & 4 & | & 1 & 0 \\ 4 & 8 & | & 0 & 1 \end{bmatrix} \rightarrow \begin{bmatrix} 1 & 2 & | & \frac{1}{2} & 0 \\ 0 & 0 & | & -2 & 1 \end{bmatrix};$ does not exist

7. $\begin{bmatrix} 2 & -4 & | & 1 & 0 \\ 1 & 3 & | & 0 & 1 \end{bmatrix} \rightarrow \begin{bmatrix} 1 & -2 & | & \frac{1}{2} & 0 \\ 0 & 5 & | & -\frac{1}{2} & 1 \end{bmatrix} \rightarrow \begin{bmatrix} 1 & 0 & | & \frac{3}{10} & \frac{2}{5} \\ 0 & 1 & | & -\frac{1}{10} & \frac{1}{5} \end{bmatrix};$

$A^{-1} = \begin{bmatrix} \frac{3}{10} & \frac{2}{5} \\ -\frac{1}{10} & \frac{1}{5} \end{bmatrix}$

8. $\begin{bmatrix} 0 & 0 & 3 & | & 1 & 0 & 0 \\ 0 & -2 & 0 & | & 0 & 1 & 0 \\ 4 & 0 & 0 & | & 0 & 0 & 1 \end{bmatrix} \rightarrow \begin{bmatrix} 4 & 0 & 0 & | & 0 & 0 & 1 \\ 0 & -2 & 0 & | & 0 & 1 & 0 \\ 0 & 0 & 3 & | & 1 & 0 & 0 \end{bmatrix} \rightarrow$

$\begin{bmatrix} 1 & 0 & 0 & | & 0 & 0 & \frac{1}{4} \\ 0 & 1 & 0 & | & 0 & -\frac{1}{2} & 0 \\ 0 & 0 & 1 & | & \frac{1}{4} & 0 & 0 \end{bmatrix};$ $A^{-1} = \begin{bmatrix} 0 & 0 & \frac{1}{4} \\ 0 & -\frac{1}{2} & 0 \\ \frac{1}{3} & 0 & 0 \end{bmatrix}$

9. $\begin{bmatrix} 2 & -1 & 0 & | & 1 & 0 & 0 \\ 3 & 0 & 1 & | & 0 & 1 & 0 \\ -2 & 4 & 0 & | & 0 & 0 & 1 \end{bmatrix} \rightarrow \begin{bmatrix} 1 & -\frac{1}{2} & 0 & | & \frac{1}{2} & 0 & 0 \\ 0 & \frac{3}{2} & 1 & | & -\frac{3}{2} & 1 & 0 \\ 0 & 3 & 0 & | & 1 & 0 & 1 \end{bmatrix} \rightarrow$

$\begin{bmatrix} 1 & 0 & \frac{1}{3} & | & 0 & \frac{1}{3} & 0 \\ 0 & 1 & \frac{2}{3} & | & -1 & \frac{2}{3} & 0 \\ 0 & 0 & -2 & | & 4 & -2 & 1 \end{bmatrix} \rightarrow \begin{bmatrix} 1 & 0 & 0 & | & \frac{2}{3} & 0 & \frac{1}{6} \\ 0 & 1 & 0 & | & \frac{1}{3} & 0 & \frac{1}{3} \\ 0 & 0 & 1 & | & -2 & 1 & -\frac{1}{2} \end{bmatrix};$

$A^{-1} = \begin{bmatrix} \frac{2}{3} & 0 & \frac{1}{6} \\ \frac{1}{3} & 0 & \frac{1}{3} \\ -2 & 1 & -\frac{1}{2} \end{bmatrix}$

10. $\begin{bmatrix} 1 & -4 & 8 & | & 1 & 0 & 0 \\ 1 & -3 & 2 & | & 0 & 1 & 0 \\ 2 & -7 & 10 & | & 0 & 0 & 1 \end{bmatrix} \rightarrow \begin{bmatrix} 1 & -4 & 8 & | & 1 & 0 & 0 \\ 0 & 1 & -6 & | & -1 & 1 & 0 \\ 0 & 1 & -6 & | & 2 & -2 & 1 \end{bmatrix} \rightarrow$

$\begin{bmatrix} 1 & 0 & -16 & | & -3 & 4 & 0 \\ 0 & 1 & -6 & | & -1 & 1 & 0 \\ 0 & 0 & 0 & | & 3 & -3 & 1 \end{bmatrix};$ does not exist

11. $A^{-1} = \begin{bmatrix} \frac{5}{22} & \frac{1}{11} \\ -\frac{1}{22} & \frac{2}{11} \end{bmatrix};$ $X = \begin{bmatrix} \frac{5}{22} & \frac{1}{11} \\ -\frac{1}{22} & \frac{2}{11} \end{bmatrix}\begin{bmatrix} -1 \\ 1 \end{bmatrix},$

$X = \begin{bmatrix} \frac{5}{22}(-1) + \frac{1}{11}(1) \\ -\frac{1}{22}(-1) + \frac{2}{11}(1) \end{bmatrix}, X = \begin{bmatrix} -\frac{3}{22} \\ \frac{5}{22} \end{bmatrix}$

12. $A^{-1} \begin{bmatrix} -\frac{1}{2} & \frac{1}{2} & \frac{1}{2} \\ 1 & 0 & -1 \\ \frac{3}{2} & -\frac{1}{2} & -\frac{1}{2} \end{bmatrix}; X = \begin{bmatrix} -\frac{1}{2} & \frac{1}{2} & \frac{1}{2} \\ 1 & 0 & -1 \\ \frac{3}{2} & -\frac{1}{2} & -\frac{1}{2} \end{bmatrix}\begin{bmatrix} 1 \\ 3 \\ 4 \end{bmatrix},$

$X = \begin{bmatrix} -\left(\frac{1}{2}\right)(1) + \frac{1}{2}(3) + \frac{1}{2}(4) \\ 1(1) + 0(3) + (-1)(4) \\ \frac{3}{2}(1) + \left(-\frac{1}{2}\right)(3) + \left(-\frac{1}{2}\right)(4) \end{bmatrix}, X = \begin{bmatrix} 3 \\ -3 \\ -2 \end{bmatrix}$

PRACTICE

1. The product is equal to A.

2. The product is the identity matrix.

3. $AB = \begin{bmatrix} 0(0) + (-1)(-1) & 0(1) + (-1)(0) \\ 1(0) + 0(-1) & 1(1) + 0(0) \end{bmatrix} =$

$\begin{bmatrix} 1 & 0 \\ 0 & 1 \end{bmatrix};$ yes

4. $AB = \begin{bmatrix} 2(-0.3) + (-4)(0.1) & 2(-0.4) + (-4)(0.2) \\ 1(-0.3) + 3(0.1) & 1(-0.4) + 3(0.2) \end{bmatrix} =$

$\begin{bmatrix} -1 & -1.6 \\ 0 & 0.2 \end{bmatrix};$ no

5. $AB =$

$\begin{bmatrix} 3\left(\frac{3}{8}\right) + 1\left(-\frac{1}{8}\right) + 0\left(-\frac{1}{4}\right) & 3\left(\frac{1}{8}\right) + 1\left(-\frac{3}{8}\right) + 0\left(\frac{1}{4}\right) \\ -1\left(\frac{3}{8}\right) + (-3)\left(-\frac{1}{8}\right) + 6\left(-\frac{1}{4}\right) & -1\left(\frac{1}{8}\right) + (-3)\left(-\frac{3}{8}\right) + 6\left(\frac{1}{4}\right) \\ -2\left(\frac{3}{8}\right) + 2\left(-\frac{1}{8}\right) + 4\left(-\frac{1}{4}\right) & -2\left(\frac{1}{8}\right) + 2\left(-\frac{3}{8}\right) + 4\left(\frac{1}{4}\right) \end{bmatrix}$

$\begin{matrix} 3\left(-\frac{1}{4}\right) + 1\left(\frac{3}{4}\right) + 0\left(\frac{1}{2}\right) \\ -1\left(-\frac{1}{4}\right) + (-3)\left(\frac{3}{4}\right) + 6\left(\frac{1}{2}\right) \\ -2\left(-\frac{1}{4}\right) + 2\left(\frac{3}{4}\right) + 4\left(\frac{1}{2}\right) \end{matrix} = \begin{bmatrix} \frac{7}{8} & 0 & 0 \\ -\frac{3}{2} & \frac{5}{2} & 1 \\ -2 & 0 & 4 \end{bmatrix};$ no

6. $AB = \begin{bmatrix} -1(-7) + 18(-4) + 6(11) \\ -7(-7) + 136(-4) + 45(11) \\ 1(-7) + (-21)(-4) + (-7)(11) \end{bmatrix}$

$\begin{matrix} -1(0) + 18(1) + 6(-3) \\ -7(0) + 136(1) + 45(-3) \\ 1(0) + (-21)(1) + (-7)(-3) \end{matrix}$

$\begin{matrix} -1(-6) + 18(3) + 6(-10) \\ -7(-6) + 136(3) + 45(-10) \\ 1(-6) + (-21)(3) + (-7)(-10) \end{matrix} = \begin{bmatrix} 1 & 0 & 0 \\ 0 & 1 & 0 \\ 0 & 0 & 1 \end{bmatrix};$ yes

7. $\begin{bmatrix} -2 & 6 & | & 1 & 0 \\ -1 & 3 & | & 0 & 1 \end{bmatrix} \rightarrow \begin{bmatrix} 1 & -3 & | & -\frac{1}{2} & 0 \\ 0 & 0 & | & -\frac{1}{2} & 1 \end{bmatrix};$ does not exist

8. $\begin{bmatrix} 1 & 1 & | & 1 & 0 \\ 3 & 4 & | & 0 & 1 \end{bmatrix} \rightarrow \begin{bmatrix} 1 & 1 & | & 1 & 0 \\ 0 & 1 & | & -3 & 1 \end{bmatrix} \rightarrow \begin{bmatrix} 1 & 0 & | & 4 & -1 \\ 0 & 1 & | & -3 & 1 \end{bmatrix};$

$A^{-1} = \begin{bmatrix} 4 & -1 \\ -3 & 1 \end{bmatrix}$

9. $\begin{bmatrix} 14 & 8 & | & 1 & 0 \\ 6 & 4 & | & 0 & 1 \end{bmatrix} \rightarrow \begin{bmatrix} 1 & \frac{8}{14} & | & \frac{1}{14} & 0 \\ 0 & \frac{8}{14} & | & -\frac{6}{14} & 1 \end{bmatrix} \rightarrow \begin{bmatrix} 1 & 0 & | & \frac{1}{2} & -1 \\ 0 & 1 & | & -\frac{3}{4} & \frac{7}{4} \end{bmatrix};$

$A^{-1} = \begin{bmatrix} \frac{1}{2} & -1 \\ -\frac{3}{4} & \frac{7}{4} \end{bmatrix}$

10. $\begin{bmatrix} 1 & 0 & 0 & | & 1 & 0 & 0 \\ 0 & 4 & 7 & | & 0 & 1 & 0 \\ 0 & 1 & 2 & | & 0 & 0 & 1 \end{bmatrix} \rightarrow \begin{bmatrix} 1 & 0 & 0 & | & 1 & 0 & 0 \\ 0 & 1 & 2 & | & 0 & 0 & 1 \\ 0 & 4 & 7 & | & 0 & 1 & 0 \end{bmatrix} \rightarrow$

$\begin{bmatrix} 1 & 0 & 0 & | & 1 & 0 & 0 \\ 0 & 1 & 2 & | & 0 & 0 & 1 \\ 0 & 0 & -1 & | & 0 & 1 & -4 \end{bmatrix} \rightarrow \begin{bmatrix} 1 & 0 & 0 & | & 1 & 0 & 0 \\ 0 & 1 & 0 & | & 0 & 2 & -7 \\ 0 & 0 & 1 & | & 0 & -1 & 4 \end{bmatrix};$

$A^{-1} = \begin{bmatrix} 1 & 0 & 0 \\ 0 & 2 & -7 \\ 0 & -1 & 4 \end{bmatrix}$

11. $\begin{bmatrix} -2 & 5 & 3 & | & 1 & 0 & 0 \\ 4 & -1 & 3 & | & 0 & 1 & 0 \\ 4 & -10 & -6 & | & 0 & 0 & 1 \end{bmatrix} \rightarrow \begin{bmatrix} 1 & -\frac{5}{2} & -\frac{3}{2} & | & -\frac{1}{2} & 0 & 0 \\ 0 & 9 & 9 & | & 2 & 1 & 0 \\ 0 & 0 & 0 & | & 2 & 0 & 1 \end{bmatrix};$

does not exist

12. $\begin{bmatrix} -2 & 2 & 3 & | & 1 & 0 & 0 \\ 1 & -1 & 0 & | & 0 & 1 & 0 \\ 0 & 1 & 4 & | & 0 & 0 & 1 \end{bmatrix} \rightarrow$

$\begin{bmatrix} 1 & -1 & -\frac{3}{2} & | & -\frac{1}{2} & 0 & 0 \\ 0 & 0 & \frac{3}{2} & | & \frac{1}{2} & 1 & 0 \\ 0 & 1 & 4 & | & 0 & 0 & 1 \end{bmatrix} \rightarrow \begin{bmatrix} 1 & 0 & \frac{5}{2} & | & -\frac{1}{2} & 0 & 1 \\ 0 & 1 & 4 & | & 0 & 0 & 1 \\ 0 & 0 & \frac{3}{2} & | & \frac{1}{2} & 1 & 0 \end{bmatrix} \rightarrow$

$\begin{bmatrix} 1 & 0 & 0 & | & -\frac{4}{3} & -\frac{5}{3} & 1 \\ 0 & 1 & 0 & | & -\frac{4}{3} & -\frac{8}{3} & 1 \\ 0 & 0 & 1 & | & \frac{1}{3} & \frac{2}{3} & 0 \end{bmatrix}; A^{-1} = \begin{bmatrix} -\frac{4}{3} & -\frac{5}{3} & 1 \\ -\frac{4}{3} & -\frac{8}{3} & 1 \\ \frac{1}{3} & \frac{2}{3} & 0 \end{bmatrix}$

13. Only square matrices can have inverses.

14. Answers will vary. Explanations should include the following steps: 1. Write an augmented matrix that has A on the left and I on the right. 2. Use row operations to reduce the matrix until I appears on the left. A^{-1} will be the matrix on the right. 3. Find the product of $A^{-1}B$. The product is equal to X.

15. $A^{-1} = \begin{bmatrix} \frac{1}{11} & -\frac{2}{11} \\ \frac{3}{22} & \frac{5}{22} \end{bmatrix}; x = \begin{bmatrix} \frac{1}{11} & -\frac{2}{11} \\ \frac{3}{22} & \frac{5}{22} \end{bmatrix}\begin{bmatrix} 10 \\ -16 \end{bmatrix},$

$x = \begin{bmatrix} \frac{1}{11}(10) + \left(-\frac{2}{11}\right)(-16) \\ \frac{3}{22}(10) + \frac{5}{22}(-16) \end{bmatrix}, x = \begin{bmatrix} \frac{42}{11} \\ -\frac{25}{11} \end{bmatrix}$

16. $A^{-1} = \begin{bmatrix} \frac{8}{11} & \frac{1}{11} & \frac{2}{11} \\ \frac{5}{11} & \frac{2}{11} & -\frac{4}{11} \\ -\frac{14}{11} & \frac{1}{11} & \frac{9}{11} \end{bmatrix}; x = \begin{bmatrix} \frac{8}{11} & \frac{1}{11} & -\frac{2}{11} \\ \frac{5}{11} & \frac{2}{11} & -\frac{4}{11} \\ -\frac{14}{11} & \frac{1}{11} & \frac{9}{11} \end{bmatrix}\begin{bmatrix} -5 \\ 15 \\ -7 \end{bmatrix},$

$x = \begin{bmatrix} \frac{8}{11}(-5) + \frac{1}{11}(15) + \left(-\frac{2}{11}\right)(-7) \\ \frac{5}{11}(-5) + \frac{2}{11}(15) + \left(-\frac{4}{11}\right)(-7) \\ -\frac{14}{11}(-5) + \frac{1}{11}(15) + \frac{9}{11}(-7) \end{bmatrix}, x = \begin{bmatrix} -1 \\ 3 \\ 2 \end{bmatrix}$

EXTEND

17. $A^{-1} = \frac{1}{(4)(2) - (7)(1)}\begin{bmatrix} 2 & -7 \\ -1 & 4 \end{bmatrix} = 1\begin{bmatrix} 2 & -7 \\ -1 & 4 \end{bmatrix} =$

$\begin{bmatrix} 2 & -7 \\ -1 & 4 \end{bmatrix}$

18. A^{-1} does not exist since
$ad - bc = 6(4) - 12(2) = 0.$

19. $A^{-1} = \frac{1}{5(11) - 9(6)}\begin{bmatrix} 11 & -9 \\ -6 & 5 \end{bmatrix} = 1\begin{bmatrix} 11 & -9 \\ -6 & 5 \end{bmatrix} =$

$\begin{bmatrix} 11 & -9 \\ -6 & 5 \end{bmatrix}$

20. $A^{-1} = \begin{bmatrix} -3 & 2 \\ 5 & -3 \end{bmatrix}; X = \begin{bmatrix} -3 & 2 \\ 5 & -3 \end{bmatrix}\begin{bmatrix} 25 \\ 40 \end{bmatrix},$

$X = \begin{bmatrix} -3(25) + 2(40) \\ 5(25) + (-3)(40) \end{bmatrix}, X = \begin{bmatrix} 5 \\ 5 \end{bmatrix}; x = 5; y = 5;$

Both employees are paid $5.00 per hour.

THINK CRITICALLY

21. In regular multiplication, a real number multiplied by its multiplicative inverse equals the multiplicative identity, or 1. In matrix multiplication, a matrix multiplied by its inverse matrix equals the identity matrix, I.

22. $[a]x = [1], x = \left[\frac{1}{a}\right]; A^{-1} = \left[\frac{1}{a}\right]$

23. $A^{-1} = \frac{1}{ab - 0}\begin{bmatrix} b & 0 \\ 0 & a \end{bmatrix} = \frac{1}{ab}\begin{bmatrix} b & 0 \\ 0 & a \end{bmatrix} = \begin{bmatrix} \frac{1}{a} & 0 \\ 0 & \frac{1}{b} \end{bmatrix}$

24. $\begin{bmatrix} 0 & 0 & a & | & 1 & 0 & 0 \\ 0 & b & 0 & | & 0 & 1 & 0 \\ c & 0 & 0 & | & 0 & 0 & 1 \end{bmatrix} \rightarrow \begin{bmatrix} c & 0 & 0 & | & 0 & 0 & 1 \\ 0 & b & 0 & | & 0 & 1 & 0 \\ 0 & 0 & a & | & 1 & 0 & 0 \end{bmatrix} \rightarrow$

$\begin{bmatrix} 1 & 0 & 0 & | & 0 & 0 & \frac{1}{c} \\ 0 & 1 & 0 & | & 0 & \frac{1}{b} & 0 \\ 0 & 0 & 1 & | & \frac{1}{a} & 0 & 0 \end{bmatrix}; A^{-1} = \begin{bmatrix} 0 & 0 & \frac{1}{c} \\ 0 & \frac{1}{b} & 0 \\ \frac{1}{a} & 0 & 0 \end{bmatrix}$

25. $AX = B$

$A^{-1}(AX) = A^{-1}B$ multiplication property
$(A^{-1}A)X = A^{-1}B$ associative property
$1X = A^{-1}B$ definition of multiplicative inverse
$X = A^{-1}B$ definition of multiplicative identity

26. $\begin{bmatrix} 3 & 2 \\ 5 & 4 \end{bmatrix}X + \begin{bmatrix} 1 \\ 4 \end{bmatrix} = \begin{bmatrix} 2 \\ 1 \end{bmatrix}, \begin{bmatrix} 3 & 2 \\ 5 & 4 \end{bmatrix}X = \begin{bmatrix} -3 \\ -3 \end{bmatrix};$

$A^{-1} = \begin{bmatrix} 2 & 1 \\ -\frac{5}{2} & \frac{3}{2} \end{bmatrix}; X = \begin{bmatrix} 2 & -1 \\ -\frac{5}{2} & \frac{3}{2} \end{bmatrix}\begin{bmatrix} -3 \\ -3 \end{bmatrix},$

$X = \begin{bmatrix} 2(-3) + (-1)(-3) \\ -\frac{5}{2}(-3) + \frac{3}{2}(-3) \end{bmatrix}, X = \begin{bmatrix} -3 \\ 3 \end{bmatrix}$

MIXED REVIEW

27. rational   **28.** irrational   **29.** rational   **30.** irrational

31. B; $y = 2x + 5;$
$$x = 2y + 5, 2y = x - 5, y = \frac{1}{2}x - \frac{5}{2},$$
$$y = \frac{x - 5}{2}$$

32. Let x represent the number of subcompacts sold. Let y represent the number of compacts sold. Let z represent the number of full-size cars sold.

$$\begin{cases} x + y + z = 20 \\ x = y + 4 \\ z = y + 1 \end{cases} \rightarrow \begin{cases} x + y + z = 20 \\ x - y \quad\;\; = 4 \\ -y + z = 1 \end{cases}$$

$$\begin{bmatrix} 1 & 1 & 1 & 20 \\ 1 & -1 & 0 & 4 \\ 0 & -1 & 1 & 1 \end{bmatrix} \rightarrow \begin{bmatrix} 1 & 1 & 1 & 20 \\ 0 & -2 & -1 & -16 \\ 0 & -1 & 1 & 1 \end{bmatrix} \rightarrow \begin{bmatrix} 1 & 0 & \frac{1}{2} & 12 \\ 0 & 1 & \frac{1}{2} & 8 \\ 0 & 0 & \frac{3}{2} & 9 \end{bmatrix} \rightarrow$$

$$\begin{bmatrix} 1 & 0 & 0 & 9 \\ 0 & 1 & 0 & 5 \\ 0 & 0 & 1 & 6 \end{bmatrix};$$ Jorge sold 9 subcompacts, 5 compacts, and 6 full-size cars.

ALGEBRAWORKS

1. 1-A   8-H   15-O   22-V
   2-B   9-I   16-P   23-W
   3-C   10-J   17-Q   24-X
   4-D   11-K   18-R   25-Y
   5-E   12-L   19-S   26-Z
   6-F   13-M   20-T
   7-G   14-N   21-U

   [8 9] [0 13] [15 13]

2. $[8 \;\; 9]\begin{bmatrix} 1 & 2 \\ 2 & 5 \end{bmatrix} = [8(1) + 9(2) \quad 8(2) + 9(5)] = [26 \;\; 61];$

   $[0 \;\; 13]\begin{bmatrix} 1 & 2 \\ 2 & 5 \end{bmatrix} = [0(1) + 13(2) \;\; 0(2) + 13(5)] =$

   $[26 \;\; 65]; [15 \;\; 13]\begin{bmatrix} 1 & 2 \\ 2 & 5 \end{bmatrix} =$

   $[15(1) + 13(2) \quad 15(2) + 13(5)] = [41 \;\; 95]$

3. Find the inverse of A, and multiply each row matrix by A^{-1}. Then translate the resulting numbers into the corresponding letters.

4. $A^{-1} = \begin{bmatrix} 5 & -2 \\ -2 & 1 \end{bmatrix};$ $[39 \;\; 86]\begin{bmatrix} 5 & -2 \\ -2 & 1 \end{bmatrix} =$

   $[39(5) + 86(-2) \;\; 39(-2) + 86(1)] = [23 \;\; 8];$

   $[15 \;\; 30]\begin{bmatrix} 5 & -2 \\ -2 & 1 \end{bmatrix} = [15(5) + 30(-2) \;\; 15(-2) + 30(1)] =$

   $[15 \;\; 0]; [47 \;\; 113]\begin{bmatrix} 5 & -2 \\ -2 & 1 \end{bmatrix} =$

   $[47(5) + 113(-2) \;\; 47(-2) + 113(1)] = [9 \;\; 19];$

   $[43 \;\; 101]\begin{bmatrix} 5 & -2 \\ -2 & 1 \end{bmatrix} =$

   $[43(5) + 101(-2) \;\; 43(-2) + 101(1)] = [13 \;\; 15];$

$[13 \;\; 26]\begin{bmatrix} 5 & -2 \\ -2 & 1 \end{bmatrix} =$

$[13(5) + 26(-2) \quad 13(-2) + 26(1)] = [13 \;\; 0];$

$[23 \;\; 8] \;\; [15 \;\; 0] \;\; [9 \;\; 19] \;\; [13 \;\; 15] \;\; [13 \;\; 0];$
$\downarrow \;\; \downarrow \;\; \downarrow \quad\; \downarrow \;\; \downarrow \;\; \downarrow \;\; \downarrow \;\; \downarrow$
W H O   I S M O M

5. Answers will vary.

Lesson 4.7, pages 203–209

EXPLORE

1. $\begin{bmatrix} 11 & 3 \\ 7 & 2 \end{bmatrix}$ **2.** $\begin{bmatrix} x \\ y \end{bmatrix}$

3. $AX = \begin{bmatrix} 11 & 3 \\ 7 & 2 \end{bmatrix}\begin{bmatrix} x \\ y \end{bmatrix} = \begin{bmatrix} 11x + 3y \\ 7x + 2y \end{bmatrix};$ The product matrix is equivalent to the left side of the equations in the system.

4. $\begin{bmatrix} -4 \\ 5 \end{bmatrix}$ **5.** $\begin{bmatrix} 11 & 3 \\ 7 & 2 \end{bmatrix}\begin{bmatrix} x \\ y \end{bmatrix} = \begin{bmatrix} -4 \\ 5 \end{bmatrix}$

6. Multiply both sides of the equation by A^{-1}. The result is $X = A^{-1}B$

   $A^{-1} = \begin{bmatrix} 2 & -3 \\ -7 & 11 \end{bmatrix}$

   $A^{-1}B = \begin{bmatrix} 2 & -3 \\ -7 & 11 \end{bmatrix}\begin{bmatrix} -4 \\ 5 \end{bmatrix} = \begin{bmatrix} -23 \\ 83 \end{bmatrix}$

   $x = -23, y = 83$

TRY THESE

1. yes   **2.** no   **3.** yes   **4.** yes

5. $\begin{bmatrix} 3 & -7 \\ 7 & 3 \end{bmatrix}\begin{bmatrix} x \\ y \end{bmatrix} = \begin{bmatrix} 7 \\ 3 \end{bmatrix}$

6. $\begin{bmatrix} 1 & 5 & -10 \\ 2 & -1 & 3 \\ -4 & 6 & 12 \end{bmatrix}\begin{bmatrix} x \\ y \\ z \end{bmatrix} = \begin{bmatrix} 13 \\ 18 \\ 7 \end{bmatrix}$

7. $\begin{bmatrix} -1 & 0 & 1 \\ 0 & 4 & 3 \\ 1 & -1 & 0 \end{bmatrix}\begin{bmatrix} x \\ y \\ z \end{bmatrix} = \begin{bmatrix} 6 \\ -1 \\ 0 \end{bmatrix}$

8. To solve a linear system using an inverse matrix, find the product of the inverse matrix and the constants matrix.

9. $\begin{bmatrix} 1 & 3 \\ 2 & 5 \end{bmatrix}\begin{bmatrix} x \\ y \end{bmatrix} = \begin{bmatrix} 10 \\ 2 \end{bmatrix}, \begin{bmatrix} x \\ y \end{bmatrix} = \begin{bmatrix} -5 & 3 \\ 2 & -1 \end{bmatrix}\begin{bmatrix} 10 \\ 2 \end{bmatrix}, \begin{bmatrix} x \\ y \end{bmatrix} = \begin{bmatrix} -44 \\ 18 \end{bmatrix};$
$(-44, 18)$

10. $\begin{bmatrix} 2 & -3 \\ -1 & 2 \end{bmatrix}\begin{bmatrix} x \\ y \end{bmatrix} = \begin{bmatrix} 6 \\ -4 \end{bmatrix}, \begin{bmatrix} x \\ y \end{bmatrix} = \begin{bmatrix} 2 & 3 \\ 1 & 2 \end{bmatrix}\begin{bmatrix} 6 \\ -4 \end{bmatrix}, \begin{bmatrix} x \\ y \end{bmatrix} = \begin{bmatrix} 0 \\ 2 \end{bmatrix};$
$(0, 2)$

11. $\begin{bmatrix} 1 & 2 \\ 2 & -5 \end{bmatrix}\begin{bmatrix} x \\ y \end{bmatrix} = \begin{bmatrix} 5 \\ -8 \end{bmatrix}, \begin{bmatrix} x \\ y \end{bmatrix} = \begin{bmatrix} \frac{5}{9} & \frac{2}{9} \\ \frac{2}{9} & -\frac{1}{9} \end{bmatrix}\begin{bmatrix} 5 \\ -8 \end{bmatrix}, \begin{bmatrix} x \\ y \end{bmatrix} = \begin{bmatrix} 1 \\ 2 \end{bmatrix};$
$(1, 2)$

12. $\begin{bmatrix} 1 & 3 & 3 \\ 1 & 3 & 4 \\ 1 & 4 & 3 \end{bmatrix}\begin{bmatrix} x \\ y \\ z \end{bmatrix} = \begin{bmatrix} 14 \\ 17 \\ 15 \end{bmatrix}, \begin{bmatrix} x \\ y \\ z \end{bmatrix} = \begin{bmatrix} 7 & -3 & -3 \\ -1 & 0 & 1 \\ -1 & 1 & 0 \end{bmatrix}\begin{bmatrix} 14 \\ 17 \\ 15 \end{bmatrix}, \begin{bmatrix} x \\ y \\ z \end{bmatrix} =$
$\begin{bmatrix} 2 \\ 1 \\ 3 \end{bmatrix}; (2, 1, 3)$

13. $\begin{bmatrix} -1 & 3 & 1 \\ 2 & 5 & 0 \\ 3 & 1 & -2 \end{bmatrix}\begin{bmatrix} x \\ y \\ z \end{bmatrix} = \begin{bmatrix} 1 \\ 3 \\ -2 \end{bmatrix}, \begin{bmatrix} x \\ y \\ z \end{bmatrix} = \begin{bmatrix} -\frac{10}{9} & \frac{7}{9} & -\frac{5}{9} \\ \frac{4}{9} & -\frac{1}{9} & \frac{2}{9} \\ -\frac{13}{9} & \frac{1}{9} & -\frac{11}{9} \end{bmatrix}\begin{bmatrix} 1 \\ 3 \\ -2 \end{bmatrix},$
$\begin{bmatrix} x \\ y \\ z \end{bmatrix} = \begin{bmatrix} \frac{7}{3} \\ -\frac{1}{3} \\ \frac{13}{3} \end{bmatrix}; \left(\frac{7}{3}, -\frac{1}{3}, \frac{13}{3}\right)$

14. $\begin{bmatrix} 0 & 1 & -1 \\ 4 & 1 & 0 \\ 3 & -1 & 3 \end{bmatrix}\begin{bmatrix} x \\ y \\ z \end{bmatrix} = \begin{bmatrix} -4 \\ -3 \\ 1 \end{bmatrix}, \begin{bmatrix} x \\ y \\ z \end{bmatrix} = \begin{bmatrix} -\frac{3}{5} & \frac{2}{5} & -\frac{1}{5} \\ \frac{12}{5} & -\frac{3}{5} & \frac{4}{5} \\ \frac{7}{5} & -\frac{3}{5} & \frac{4}{5} \end{bmatrix}\begin{bmatrix} -4 \\ -3 \\ 1 \end{bmatrix},$
$\begin{bmatrix} x \\ y \\ z \end{bmatrix} = \begin{bmatrix} 1 \\ -7 \\ -3 \end{bmatrix}; (1, -7, -3)$

15a. $x + y + z = 200$
$75x + 100y + 50z = 15,000$
$x - 2y = 0$

$\begin{bmatrix} 1 & 1 & 1 \\ 75 & 100 & 50 \\ 1 & -2 & 0 \end{bmatrix}\begin{bmatrix} x \\ y \\ z \end{bmatrix} = \begin{bmatrix} 200 \\ 15,000 \\ 0 \end{bmatrix}, \begin{bmatrix} x \\ y \\ z \end{bmatrix} =$

$\begin{bmatrix} -1 & \frac{1}{50} & \frac{1}{2} \\ -\frac{1}{2} & \frac{1}{100} & -\frac{1}{4} \\ \frac{5}{2} & -\frac{3}{100} & -\frac{1}{4} \end{bmatrix}\begin{bmatrix} 200 \\ 15,000 \\ 0 \end{bmatrix}, \begin{bmatrix} x \\ y \\ z \end{bmatrix} =$

$\begin{bmatrix} 100 \\ 50 \\ 50 \end{bmatrix}; (100, 50, 50);$

The farmer should plant 100 acres of gladiolas, 50 acres of irises, and 50 acres of tulips.

15b. $\begin{bmatrix} 1 & 1 & 1 \\ 75 & 100 & 50 \\ 1 & -2 & 0 \end{bmatrix}\begin{bmatrix} x \\ y \\ z \end{bmatrix} = \begin{bmatrix} 250 \\ 15,000 \\ 0 \end{bmatrix}, \begin{bmatrix} x \\ y \\ z \end{bmatrix} = \begin{bmatrix} -1 & \frac{1}{50} & \frac{1}{2} \\ -\frac{1}{2} & \frac{1}{100} & -\frac{1}{4} \\ \frac{5}{2} & -\frac{3}{100} & -\frac{1}{4} \end{bmatrix}$

$\begin{bmatrix} 250 \\ 15,000 \\ 0 \end{bmatrix}, \begin{bmatrix} x \\ y \\ z \end{bmatrix} = \begin{bmatrix} 50 \\ 25 \\ 175 \end{bmatrix};$

$(50, 25, 175)$; The farmer should plant 50 acres of gladiolas, 25 acres of irises, and 175 acres of tulips.

PRACTICE

1. Multiply both equations by 1000 to clear decimals. Write a coefficient matrix A with the coefficients of x and y. Write a variable matrix X. Write a constants matrix B. Write the equation in the form $AX = B$.

$\begin{bmatrix} 135 & 405 \\ 40 & 877 \end{bmatrix}\begin{bmatrix} x \\ y \end{bmatrix} 5 \begin{bmatrix} 1000 \\ 3150 \end{bmatrix}$

2. $\begin{cases} -x - 5y = 10 \\ 3x - 3y = 3 \end{cases}$

3. $\begin{cases} x + 5z = 12 \\ -8x + 4y = 0 \\ 2z = 4 \end{cases}$

4. $\begin{cases} 0.4x + 0.8y - 0.2z = 7 \\ 1.2x - 0.4y + 0.4z = 9 \\ 0.6x + 0.4y - 0.4z = 4 \end{cases}$

5. $\begin{bmatrix} 8 & 5 \\ 5 & 3 \end{bmatrix}\begin{bmatrix} x \\ y \end{bmatrix} = \begin{bmatrix} -6 \\ 2 \end{bmatrix}, \begin{bmatrix} -3 & 5 \\ 5 & -8 \end{bmatrix}\begin{bmatrix} -6 \\ 2 \end{bmatrix}, \begin{bmatrix} x \\ y \end{bmatrix} = \begin{bmatrix} 28 \\ 46 \end{bmatrix};$

$(28, -46)$

6. $\begin{bmatrix} 1 & 2 & 3 \\ 2 & -3 & 4 \\ -3 & 5 & -6 \end{bmatrix}\begin{bmatrix} x \\ y \\ z \end{bmatrix} = \begin{bmatrix} -1 \\ 2 \\ 4 \end{bmatrix}, \begin{bmatrix} x \\ y \\ z \end{bmatrix} = \begin{bmatrix} -2 & 27 & 17 \\ 0 & 3 & 2 \\ 1 & -11 & -7 \end{bmatrix}\begin{bmatrix} -1 \\ 2 \\ 4 \end{bmatrix}, \begin{bmatrix} x \\ y \\ z \end{bmatrix}$

$= \begin{bmatrix} 124 \\ 14 \\ -51 \end{bmatrix}; (124, 14, -51)$

7. $\begin{bmatrix} 1 & 2 & -4 \\ 0 & 1 & -1 \\ 1 & -1 & 0 \end{bmatrix}\begin{bmatrix} x \\ y \\ z \end{bmatrix} = \begin{bmatrix} 16 \\ 4 \\ 1 \end{bmatrix}, \begin{bmatrix} x \\ y \\ z \end{bmatrix} = \begin{bmatrix} -1 & 4 & 2 \\ -1 & 4 & 1 \\ -1 & 3 & 1 \end{bmatrix}\begin{bmatrix} 16 \\ 4 \\ 1 \end{bmatrix}, \begin{bmatrix} x \\ y \\ z \end{bmatrix}$

$= \begin{bmatrix} 2 \\ 1 \\ -3 \end{bmatrix}; (2, 1, -3)$

8. Let x represent the amount of 15% acid solution. Let y represent the amount of 75% acid solution.

$$\begin{cases} x + y = 20 \\ 15x + 75y = 39(20) \end{cases} \rightarrow \begin{cases} x + y = 20 \\ 15x + 75y = 780 \end{cases}$$

$$\begin{bmatrix} 1 & 1 \\ 15 & 75 \end{bmatrix}\begin{bmatrix} x \\ y \end{bmatrix} = \begin{bmatrix} 20 \\ 780 \end{bmatrix}, \begin{bmatrix} x \\ y \end{bmatrix} = \begin{bmatrix} \frac{5}{4} & -\frac{1}{60} \\ -\frac{1}{4} & \frac{1}{60} \end{bmatrix}\begin{bmatrix} 20 \\ 780 \end{bmatrix},$$

$$\begin{bmatrix} x \\ y \end{bmatrix} = \begin{bmatrix} 12 \\ 8 \end{bmatrix}; (12, 8);$$ The chemist should mix 12 gal of 15% solution with 8 gal of 75% solution.

9. $\begin{bmatrix} 3 & 5 \\ 2 & 4 \end{bmatrix}\begin{bmatrix} x \\ y \end{bmatrix} = \begin{bmatrix} -4 \\ -2 \end{bmatrix}, \begin{bmatrix} x \\ y \end{bmatrix} = \begin{bmatrix} 2 & -2.5 \\ -1 & 1.5 \end{bmatrix}\begin{bmatrix} -4 \\ -2 \end{bmatrix}, \begin{bmatrix} x \\ y \end{bmatrix} =$

$\begin{bmatrix} -3 \\ 1 \end{bmatrix}; (-3, 1)$

10. $\begin{bmatrix} -1 & -3 & 0 \\ 0 & 5 & -2 \\ 2 & 0 & 2 \end{bmatrix}\begin{bmatrix} x \\ y \\ z \end{bmatrix} = \begin{bmatrix} 14 \\ 2 \\ -2 \end{bmatrix}, \begin{bmatrix} x \\ y \\ z \end{bmatrix} = \begin{bmatrix} 5 & 3 & 3 \\ -2 & -1 & -1 \\ -5 & -3 & -2.5 \end{bmatrix}\begin{bmatrix} 14 \\ 2 \\ -2 \end{bmatrix},$

$\begin{bmatrix} x \\ y \\ z \end{bmatrix} = \begin{bmatrix} 70 \\ -28 \\ -71 \end{bmatrix}; (70, -28, -71)$

11. $\begin{bmatrix} 0.1 & 0.3 & 0.1 \\ 0.1 & 0.5 & 0.2 \\ 0.2 & 0.6 & 0.3 \end{bmatrix}\begin{bmatrix} x \\ y \\ z \end{bmatrix} = \begin{bmatrix} 1.4 \\ 1.8 \\ 0.8 \end{bmatrix}, \begin{bmatrix} x \\ y \\ z \end{bmatrix} = \begin{bmatrix} 15 & -15 & 5 \\ 5 & 5 & -5 \\ -20 & 0 & 10 \end{bmatrix}\begin{bmatrix} 1.4 \\ 1.8 \\ 0.8 \end{bmatrix},$

$\begin{bmatrix} x \\ y \\ z \end{bmatrix} = \begin{bmatrix} -2 \\ 12 \\ -20 \end{bmatrix}; (-2, 12, -20)$

12a. Let x represent the amount invested at 7%. Let y represent the amount invested at 12%. Let z represent the amount invested at 6%.

$$\begin{cases} x + y + z = 80{,}000 \\ 0.07x + 0.12y + 0.06z = 6800 \\ y = z \end{cases}$$

$$\begin{cases} x + y + z = 80{,}000 \\ 7x + 12y + 6z = 680{,}000 \\ y - z = 0 \end{cases}$$

$$\begin{bmatrix} 1 & 1 & 1 \\ 7 & 12 & 6 \\ 0 & 1 & -1 \end{bmatrix}\begin{bmatrix} x \\ y \\ z \end{bmatrix} = \begin{bmatrix} 80{,}000 \\ 680{,}000 \\ 0 \end{bmatrix}, \begin{bmatrix} x \\ y \\ z \end{bmatrix} =$$

$$\begin{bmatrix} 4.5 & -0.5 & 1.5 \\ -1.75 & 0.25 & -0.25 \\ -1.75 & 0.25 & -1.25 \end{bmatrix}\begin{bmatrix} 80{,}000 \\ 680{,}000 \\ 0 \end{bmatrix}, \begin{bmatrix} x \\ y \\ z \end{bmatrix} = \begin{bmatrix} 20{,}000 \\ 30{,}000 \\ 30{,}000 \end{bmatrix};$$

(20,000, 30,000, 30,000); Shelly invested $20,000 at 7%, $30,000 at 12%, and $30,000 at 6%.

12b. $\begin{bmatrix} 1 & 1 & 1 \\ 7 & 12 & 6 \\ 0 & 1 & -1 \end{bmatrix}\begin{bmatrix} x \\ y \\ z \end{bmatrix} = \begin{bmatrix} 60{,}000 \\ 500{,}000 \\ 0 \end{bmatrix}, \begin{bmatrix} x \\ y \\ z \end{bmatrix} =$

$$\begin{bmatrix} 4.5 & -0.5 & 1.5 \\ -1.75 & 0.25 & -0.25 \\ -1.75 & 0.25 & -1.25 \end{bmatrix}\begin{bmatrix} 60{,}000 \\ 500{,}000 \\ 0 \end{bmatrix},$$

$\begin{bmatrix} x \\ y \\ z \end{bmatrix} = \begin{bmatrix} 20{,}000 \\ 20{,}000 \\ 20{,}000 \end{bmatrix}; (20{,}000, 20{,}000, 20{,}000);$

Shelly invested $20,000 at 7%, $20,000 at 12%, and $20,000 at 6%.

EXTEND

13. $\begin{bmatrix} x \\ y \end{bmatrix} = \frac{1}{10}\begin{bmatrix} 3 & 4 \\ -1 & 2 \end{bmatrix}\begin{bmatrix} 10 \\ 20 \end{bmatrix}, \begin{bmatrix} x \\ y \end{bmatrix} = \begin{bmatrix} 0.3 & 0.4 \\ -0.1 & 0.2 \end{bmatrix}\begin{bmatrix} 10 \\ 20 \end{bmatrix}, \begin{bmatrix} x \\ y \end{bmatrix}$

$= \begin{bmatrix} 11 \\ 3 \end{bmatrix}; (11, 3)$

14. $\begin{bmatrix} x \\ y \\ z \end{bmatrix} = \frac{1}{9}\begin{bmatrix} 3 & 1 & -2 \\ 0 & -3 & 6 \\ -3 & 2 & 5 \end{bmatrix}\begin{bmatrix} 2 \\ -5 \\ 5 \end{bmatrix}, \begin{bmatrix} x \\ y \\ z \end{bmatrix} = \begin{bmatrix} \frac{1}{3} & \frac{1}{9} & -\frac{2}{9} \\ 0 & -\frac{1}{3} & \frac{2}{3} \\ -\frac{1}{3} & \frac{2}{9} & \frac{5}{9} \end{bmatrix}$

$\begin{bmatrix} 2 \\ -5 \\ 5 \end{bmatrix}, \begin{bmatrix} x \\ y \\ z \end{bmatrix} = \begin{bmatrix} -1 \\ 5 \\ 1 \end{bmatrix}; (-1, 5, 1)$

15. $\begin{cases} 2w + 4x - 5y + 12z = 2 \\ 4w - x + 12y - z = 5 \\ -w + 4x + 2z = 13 \\ 2w + 10x + y = 5 \end{cases}$

16. $\begin{bmatrix} -8 & 3 \\ -4 & 4 \end{bmatrix}\begin{bmatrix} x \\ y \end{bmatrix} = \begin{bmatrix} -16 \\ -8 \end{bmatrix}, \begin{bmatrix} x \\ y \end{bmatrix} = \begin{bmatrix} -0.2 & 0.15 \\ -0.2 & 0.4 \end{bmatrix}\begin{bmatrix} -16 \\ -8 \end{bmatrix},$

$\begin{bmatrix} x \\ y \end{bmatrix} = \begin{bmatrix} 2 \\ 0 \end{bmatrix}; (2, 0)$

17. $\begin{bmatrix} -8 & 3 \\ -4 & 4 \end{bmatrix}\begin{bmatrix} x \\ y \end{bmatrix} = \begin{bmatrix} 2 \\ 4 \end{bmatrix}, \begin{bmatrix} x \\ y \end{bmatrix} = \begin{bmatrix} -0.2 & 0.15 \\ -0.2 & 0.4 \end{bmatrix}\begin{bmatrix} 2 \\ 4 \end{bmatrix},$

$\begin{bmatrix} x \\ y \end{bmatrix} = \begin{bmatrix} 0.2 \\ 1.2 \end{bmatrix}; \left(\frac{1}{5}, \frac{6}{5}\right)$

18. $\begin{bmatrix} -8 & 3 \\ -4 & 4 \end{bmatrix}\begin{bmatrix} x \\ y \end{bmatrix} = \begin{bmatrix} \frac{1}{6} \\ \frac{1}{2} \end{bmatrix}, \begin{bmatrix} x \\ y \end{bmatrix} = \begin{bmatrix} -0.2 & 0.15 \\ -0.2 & 0.4 \end{bmatrix}\begin{bmatrix} \frac{1}{6} \\ \frac{1}{2} \end{bmatrix},$

$\begin{bmatrix} x \\ y \end{bmatrix} = \begin{bmatrix} \frac{1}{24} \\ \frac{1}{6} \end{bmatrix}; \left(\frac{1}{24}, \frac{1}{6}\right)$

19. Answers will vary.

20. Let x represent the smallest angle. Let y represent the middle size angle. Let z represent the largest angle.

$$\begin{cases} x + y + z = 180 \\ x = z - 70 \\ 2y = z \end{cases} \rightarrow \begin{cases} x + y + z = 180 \\ x - z = -70 \\ 2y - z = 0 \end{cases}$$

$$\begin{bmatrix} 1 & 1 & 1 \\ 1 & 0 & -1 \\ 0 & 2 & -1 \end{bmatrix}\begin{bmatrix} x \\ y \\ z \end{bmatrix} = \begin{bmatrix} 180 \\ -70 \\ 0 \end{bmatrix}, \begin{bmatrix} x \\ y \\ z \end{bmatrix} = \begin{bmatrix} 0.4 & 0.6 & -0.2 \\ 0.2 & -0.2 & 0.4 \\ 0.4 & -0.4 & -0.2 \end{bmatrix}$$

$$\begin{bmatrix} 180 \\ -70 \\ 0 \end{bmatrix}, \begin{bmatrix} x \\ y \\ z \end{bmatrix} = \begin{bmatrix} 30 \\ 50 \\ 100 \end{bmatrix};$$ The measures of the angles are 30°, 50°, and 100°.

THINK CRITICALLY

21. Answers will vary. Answers should include the idea that the system has either an infinite number of solutions or no solutions.

22. $\begin{bmatrix} 1 & a \\ 2 & a \end{bmatrix}\begin{bmatrix} x \\ y \end{bmatrix} = \begin{bmatrix} 1 \\ 0 \end{bmatrix}, \begin{bmatrix} x \\ y \end{bmatrix} = \dfrac{1}{1(a) - a(2)}\begin{bmatrix} a & -a \\ -2 & 1 \end{bmatrix}\begin{bmatrix} 1 \\ 0 \end{bmatrix},$

$\begin{bmatrix} x \\ y \end{bmatrix} = -\dfrac{1}{a}\begin{bmatrix} a & -a \\ -2 & 1 \end{bmatrix}\begin{bmatrix} 1 \\ 0 \end{bmatrix}, \begin{bmatrix} x \\ y \end{bmatrix} = \begin{bmatrix} -1 & 1 \\ \frac{2}{a} & -\frac{1}{a} \end{bmatrix}\begin{bmatrix} 1 \\ 0 \end{bmatrix}, \begin{bmatrix} x \\ y \end{bmatrix} =$

$\begin{bmatrix} -1 \\ \frac{2}{a} \end{bmatrix}; \left(-1, \dfrac{2}{a}\right)$

23. $\begin{bmatrix} b & u \\ b & -a \end{bmatrix}\begin{bmatrix} x \\ y \end{bmatrix} = \begin{bmatrix} 2 \\ 10 \end{bmatrix}, \begin{bmatrix} x \\ y \end{bmatrix} = \dfrac{1}{b(-a) - a(b)}\begin{bmatrix} -a & -a \\ -b & b \end{bmatrix}\begin{bmatrix} 2 \\ 10 \end{bmatrix},$

$\begin{bmatrix} x \\ y \end{bmatrix} = -\dfrac{1}{2ab}\begin{bmatrix} -a & -a \\ -b & b \end{bmatrix}\begin{bmatrix} 2 \\ 10 \end{bmatrix}, \begin{bmatrix} x \\ y \end{bmatrix} = \begin{bmatrix} \frac{1}{2b} & \frac{1}{2b} \\ \frac{1}{2a} & -\frac{1}{2a} \end{bmatrix}\begin{bmatrix} 2 \\ 10 \end{bmatrix}, \begin{bmatrix} x \\ y \end{bmatrix} =$

$\begin{bmatrix} \frac{6}{b} \\ -\frac{4}{a} \end{bmatrix}; \left(\dfrac{6}{b}, -\dfrac{4}{a}\right)$

24. $\begin{cases} bx + 4y = 6 \\ bx + 3y = 4 \end{cases}; \begin{bmatrix} b & 4 \\ b & 3 \end{bmatrix}\begin{bmatrix} x \\ y \end{bmatrix} = \begin{bmatrix} 6 \\ 4 \end{bmatrix}, \begin{bmatrix} x \\ y \end{bmatrix} = \dfrac{1}{3b - 4b}$

$\begin{bmatrix} 3 & -4 \\ -b & b \end{bmatrix}\begin{bmatrix} 6 \\ 4 \end{bmatrix}, \begin{bmatrix} x \\ y \end{bmatrix} = -\dfrac{1}{b}\begin{bmatrix} 3 & -4 \\ -b & b \end{bmatrix}\begin{bmatrix} 6 \\ 4 \end{bmatrix}, \begin{bmatrix} x \\ y \end{bmatrix} = \begin{bmatrix} -\frac{3}{b} & \frac{4}{b} \\ 1 & -1 \end{bmatrix}$

$\begin{bmatrix} 6 \\ 4 \end{bmatrix}, \begin{bmatrix} x \\ y \end{bmatrix} = \begin{bmatrix} -\frac{2}{b} \\ 2 \end{bmatrix}; \left(-\dfrac{2}{b}, 2\right)$

25. $\begin{bmatrix} w \\ x \\ y \\ z \end{bmatrix} = \begin{bmatrix} 1 & -2 & 1 & 0 \\ 1 & -2 & 2 & -3 \\ 0 & 1 & -1 & 1 \\ -2 & 3 & -2 & 3 \end{bmatrix}\begin{bmatrix} 18 \\ 24 \\ 31 \\ 10 \end{bmatrix}, \begin{bmatrix} w \\ x \\ y \\ z \end{bmatrix} = \begin{bmatrix} 1 \\ 2 \\ 3 \\ 4 \end{bmatrix}; (1, 2, 3, 4)$

26a. Let x represent company one. Let y represent company 2. Let z represent company 3.

$$\begin{cases} x + 2y + 2z = 60 \\ 3x + y + 3z = 78 \\ 2x + y + 2z = 56 \end{cases}$$

$$\begin{bmatrix} 1 & 2 & 2 \\ 3 & 1 & 3 \\ 2 & 1 & 2 \end{bmatrix}\begin{bmatrix} x \\ y \\ z \end{bmatrix} = \begin{bmatrix} 60 \\ 78 \\ 56 \end{bmatrix}, \begin{bmatrix} x \\ y \\ z \end{bmatrix} = \begin{bmatrix} -1 & -2 & 4 \\ 0 & -2 & 3 \\ 1 & 3 & -5 \end{bmatrix}\begin{bmatrix} 60 \\ 78 \\ 56 \end{bmatrix},$$

$$\begin{bmatrix} x \\ y \\ z \end{bmatrix} = \begin{bmatrix} 8 \\ 12 \\ 14 \end{bmatrix};$$ To fill the order, Company 1 must work

8h, Company 2 must work 12h and Company 3 must work 14h.

26b. $\begin{bmatrix} x \\ y \\ z \end{bmatrix} = \begin{bmatrix} -1 & -2 & 4 \\ 0 & -2 & 3 \\ 1 & 3 & -5 \end{bmatrix}\begin{bmatrix} 42 \\ 52 \\ 39 \end{bmatrix}, \begin{bmatrix} x \\ y \\ z \end{bmatrix} = \begin{bmatrix} 10 \\ 13 \\ 3 \end{bmatrix};$ To fill the

order, Company 1 must work 10h, Company 2 must work 13h, and Company 3 must work 3h.

MIXED REVIEW

27. $2x - 5y = 7x - 40, -5y = 5x - 40, y = 8 - x$

28. $5(y - 2) = 7x + 25, 5y - 10 = 7x + 25,$

$5y = 7x + 35, y = \dfrac{7}{5}x + 7$

29. $10(x + y) = 2(5x - 5), 10x + 10y = 10x - 10;$

$10y = -10, y = -1$

30. $x - 3y = 2$

$\quad x = 3y + 2$

$\quad 6x + 5y = -34$

$6(3y + 2) + 5y = -34$

$18y + 12 + 5y = -34$

$\qquad 23y = -46$

$\qquad\quad y = -2$

$\quad x - 3y = 2$

$x - 3(-2) = 2$

$\qquad x = -4$

$(-4, -2)$

31. $\begin{cases} 0.3x + 0.2y = -0.9 \rightarrow \quad 6x + 4y = -18 \\ 0.2x - 0.3y = -0.6 \rightarrow \underline{-6x + 9y = \quad 18} \end{cases}$

$\qquad\qquad\qquad\qquad\qquad\quad 13y = 0$

$\qquad\qquad\qquad\qquad\qquad\qquad y = 0$

$0.3x + 0.2y = -0.9$

$0.3x + 0.2(0) = -0.9$

$\qquad 0.3x = -0.9$

$\qquad\quad x = -3$

$(-3, 0)$

32.

$$\frac{1}{5}x + \frac{1}{2}y = 6 \qquad\qquad \frac{1}{5}x + \frac{1}{2}y = 6$$

$$\frac{3}{5}x - \frac{1}{2}y = 2 \qquad\qquad \frac{1}{5}(10) + \frac{1}{2}y = 6$$

$$\overline{\frac{4}{5}x \qquad\quad = 8} \qquad\qquad \frac{1}{2}y = 4$$

$$x = 10 \qquad\qquad\qquad y = 8$$

$(10, 8)$

PROJECT CONNECTION

1–4. Answers will vary.

5. The first entry represents the total cost of materials used for the boxes and bags. The second entry represents the total cost of wasted materials.

ALGEBRAWORKS

1. $\begin{bmatrix} 0.15 & 0.10 & 0.05 \\ 0.25 & 0.10 & 0.25 \\ 0.60 & 0.80 & 0.70 \end{bmatrix}$

2. $\begin{bmatrix} 7100 \\ 14{,}300 \\ 52{,}600 \end{bmatrix}$

3. $\begin{bmatrix} 0.15 & 0.10 & 0.05 \\ 0.25 & 0.10 & 0.25 \\ 0.60 & 0.80 & 0.70 \end{bmatrix}\begin{bmatrix} x \\ y \\ z \end{bmatrix} = \begin{bmatrix} 7100 \\ 14{,}300 \\ 52{,}600 \end{bmatrix}$

4. $\begin{bmatrix} x \\ y \\ z \end{bmatrix} = \begin{bmatrix} \frac{26}{3} & 2 & -\frac{4}{3} \\ \frac{5}{3} & -5 & \frac{5}{3} \\ -\frac{28}{3} & 4 & \frac{2}{3} \end{bmatrix}\begin{bmatrix} 7100 \\ 14{,}300 \\ 52{,}600 \end{bmatrix}, \begin{bmatrix} x \\ y \\ z \end{bmatrix} = \begin{bmatrix} 20{,}000 \\ 28{,}000 \\ 26{,}000 \end{bmatrix};$

$(20{,}000, 28{,}000, 26{,}000)$; The safety budget spent $20,000 at the Western plant, $28,000 at the Southern plant, and $26,000 at the Eastern plant.

5. $\begin{bmatrix} x \\ y \\ z \end{bmatrix} = \begin{bmatrix} \frac{26}{3} & 2 & -\frac{4}{3} \\ \frac{5}{3} & -5 & \frac{5}{3} \\ -\frac{28}{3} & 4 & \frac{2}{3} \end{bmatrix}\begin{bmatrix} 6200 \\ 13{,}100 \\ 54{,}700 \end{bmatrix}, \begin{bmatrix} x \\ y \\ z \end{bmatrix} = \begin{bmatrix} 7000 \\ 36{,}000 \\ 31{,}000 \end{bmatrix};$

The new budgeted amounts would be $7000 for the Western plant, $36,000 for the Southern plant, and $31,000 for the Eastern plant.

Lesson 4.8, pages 210–215

EXPLORE

1. $a_1b_2x + b_1b_2y = c_2b_2; \; -a_2b_1x - b_2b_1y = -c_2b_1$

2. $a_1b_2x + b_1b_2y - a_2b_1x - b_2b_1y = c_1b_2 - c_2b_1$

$a_1b_2x - a_2b_1x = c_1b_2 - c_2b_1$

$(a_1b_2 - a_1b_1)x = c_1b_2 - c_2b_1$

$x = \dfrac{c_1b_2 - c_2b_1}{a_1b_2 - a_2b_1}$

3. $-a_1a_2x - a_2b_1y = -a_2c_1; \; a_1a_2x + a_1b_2y = a_1a_2$

4. $-a_1a_2x - a_2b_1y + a_1a_2x + a_1b_2y = -a_2c_1 + a_1c_2$

$-a_1b_1y + a_1b_2y = -a_2c_1 + a_1c_2$

$(-a_2b_1 + a_1b_2)y = -a_2c_1 + a_1c_2$

$y = \dfrac{a_1c_2 - a_2c_1}{a_1b_2 - a_2b_1}$

5. $\begin{bmatrix} a_1 & b_1 \\ a_2 & b_2 \end{bmatrix};$ They contain the same elements and the same number of rows and columns.

TRY THESE

1. $2(-1) - 3(4) = -2 - 12 = -14$

2. $(-3)(-2) - 4(-1) = 6 + 4 = 10$

3. $(-3)(-4) - 6(2) = 12 - 12 = 0$

4. $\begin{vmatrix} 1 & 2 & 0 \\ 0 & 2 & 1 \\ 1 & 1 & 1 \end{vmatrix}\begin{matrix} 1 & 2 \\ 0 & 2 \\ 1 & 1 \end{matrix} = 2 + 2 + 0 - 0 - 1 - 0 = 3$

5. $\begin{vmatrix} -1 & -2 & -3 \\ 3 & 4 & 2 \\ 0 & 1 & 2 \end{vmatrix}\begin{matrix} -1 & -2 \\ 3 & 4 \\ 0 & 1 \end{matrix} =$

$-8 + 0 + (-9) - 0 - (-2) - (-12) = -3$

6. $\begin{vmatrix} 1 & 1 & 2 \\ 5 & 5 & 7 \\ 3 & 3 & 1 \end{vmatrix}\begin{matrix} 1 & 1 \\ 5 & 5 \\ 3 & 3 \end{matrix}$

$= 5 + 21 + 30 - 30 - 21 - 5 = 0$

7. The minor of an element is the 2×2 determinant that remains when the row and column containing the element is removed from a 3×3 determinant.

$a_1 = \begin{vmatrix} 0 & -1 \\ 3 & 14 \end{vmatrix}; b_1 = \begin{vmatrix} 1 & -1 \\ 5 & 14 \end{vmatrix}; c_1 = \begin{vmatrix} 1 & 0 \\ 5 & 3 \end{vmatrix}$

$a_2 = \begin{vmatrix} -3 & 5 \\ 3 & 14 \end{vmatrix}; b_2 = \begin{vmatrix} 2 & 5 \\ 5 & 14 \end{vmatrix}; c_2 = \begin{vmatrix} 2 & -3 \\ 5 & 3 \end{vmatrix};$

$a_3 = \begin{vmatrix} -3 & 5 \\ 0 & -1 \end{vmatrix}; b_3 = \begin{vmatrix} 2 & 5 \\ 1 & -1 \end{vmatrix}; c_3 = \begin{vmatrix} 2 & -3 \\ 1 & 0 \end{vmatrix}$

8. $\begin{vmatrix} 4 & -4 & 6 \\ 2 & 8 & -3 \\ 0 & -5 & 0 \end{vmatrix} = 4\begin{vmatrix} 8 & -3 \\ -5 & 0 \end{vmatrix} - 2\begin{vmatrix} -4 & 6 \\ -5 & 0 \end{vmatrix} + 0$

$\begin{vmatrix} -4 & 6 \\ 8 & -3 \end{vmatrix} = -120$

9. $\begin{vmatrix} 3 & 2 & -2 \\ -2 & 1 & 4 \\ -4 & -3 & 3 \end{vmatrix} = 3\begin{vmatrix} 1 & 4 \\ -3 & 3 \end{vmatrix} - (-2)\begin{vmatrix} 2 & -2 \\ -3 & 3 \end{vmatrix} + (-4)$

$\begin{vmatrix} 2 & -2 \\ 1 & 4 \end{vmatrix} = 5$

10. $\begin{vmatrix} 2 & \frac{1}{2} & 2 \\ -1 & 4 & -3 \\ \frac{1}{2} & 1 & 1 \end{vmatrix} = 2\begin{vmatrix} 4 & -3 \\ 1 & 1 \end{vmatrix} - (-1)\begin{vmatrix} \frac{1}{2} & 2 \\ 1 & 1 \end{vmatrix} + \frac{1}{2}\begin{vmatrix} \frac{1}{2} & 2 \\ 4 & -3 \end{vmatrix} = 7.75$

11. Using the definition of the determinant, $a_1 b_2 - b_1 a_2$ gives $y(1) - m(x) = b, y - mx = b, y = mx + b$.

12. $D = \begin{vmatrix} 2 & 1 \\ 5 & 3 \end{vmatrix} = 2(3) - 5(1) = 1;$

$D_x = \begin{vmatrix} 1 & 1 \\ 2 & 3 \end{vmatrix} = 1(3) - 2(1) = 1;$

$D_y = \begin{vmatrix} 2 & 1 \\ 5 & 2 \end{vmatrix} = 2(2) - 5(1) = -1;$

$x = \frac{1}{1} = 1; y = \frac{-1}{1} = -1; (1, -1)$

13. $D = \begin{vmatrix} 1 & 1 \\ 1 & -1 \end{vmatrix} = 1(-1) - 1(1) = -2;$

$D_x = \begin{vmatrix} 0 & 1 \\ 0 & -1 \end{vmatrix} = 0(-1) - 0(1) = 0;$

$D_y = \begin{vmatrix} 1 & 0 \\ 1 & 0 \end{vmatrix} = 1(0) \quad 1(0) = 0;$

$x = \frac{0}{-2} = 0; y = \frac{0}{-2} = 0; (0, 0)$

14. $D = \begin{vmatrix} 3 & -2 \\ 3 & 2 \end{vmatrix} = 3(2) - 3(-2) = 12;$

$D_x = \begin{vmatrix} 7 & -2 \\ 9 & 2 \end{vmatrix} = 7(2) - 9(-2) = 32;$

$D_y = \begin{vmatrix} 3 & 7 \\ 3 & 9 \end{vmatrix} = 3(9) - 3(7) = 6;$

$x = \frac{32}{12} = \frac{8}{3}; y = \frac{6}{12} = \frac{1}{2}; \left(\frac{8}{3}, \frac{1}{2}\right)$

15. $D = \begin{vmatrix} 2 & 4 & 3 \\ 1 & -3 & 2 \\ -1 & 2 & -1 \end{vmatrix} = 2\begin{vmatrix} -3 & 2 \\ 2 & -1 \end{vmatrix} - 1\begin{vmatrix} 4 & 3 \\ 2 & -1 \end{vmatrix} + (-1)$

$\begin{vmatrix} 4 & 3 \\ -3 & 2 \end{vmatrix} = -9;$

$D_x = \begin{vmatrix} 6 & 4 & 3 \\ -7 & -3 & 2 \\ 5 & 2 & -1 \end{vmatrix} = 6\begin{vmatrix} -3 & 2 \\ 2 & -1 \end{vmatrix} - (-7)\begin{vmatrix} 4 & 3 \\ 2 & -1 \end{vmatrix} + 5$

$\begin{vmatrix} 4 & 3 \\ -3 & 2 \end{vmatrix} = 9;$

$D_y = \begin{vmatrix} 2 & 6 & 3 \\ 1 & -7 & 2 \\ -1 & 5 & -1 \end{vmatrix} = 2\begin{vmatrix} -7 & 2 \\ 5 & -1 \end{vmatrix} - 1\begin{vmatrix} 6 & 3 \\ 5 & -1 \end{vmatrix} + (-1)$

$\begin{vmatrix} 6 & 3 \\ -7 & 2 \end{vmatrix} = -18;$

$D_z = \begin{vmatrix} 2 & 4 & 6 \\ 1 & -3 & -7 \\ -1 & 2 & 5 \end{vmatrix} = 2\begin{vmatrix} -3 & -7 \\ 2 & 5 \end{vmatrix} - 1\begin{vmatrix} 4 & 6 \\ 2 & 5 \end{vmatrix} + (-1)$

$\begin{vmatrix} 4 & 6 \\ -3 & -7 \end{vmatrix} = 0;$

$x = \frac{9}{-9} = -1; y = \frac{-18}{-9} = 2; z = \frac{0}{-9} = 0; (-1, 2, 0)$

16. $D = \begin{vmatrix} 1 & 1 & -1 \\ -1 & 1 & 1 \\ 1 & 1 & 1 \end{vmatrix} = 1\begin{vmatrix} 1 & 1 \\ 1 & 1 \end{vmatrix} - (-1)\begin{vmatrix} 1 & -1 \\ 1 & 1 \end{vmatrix} + 1$

$\begin{vmatrix} 1 & -1 \\ 1 & 1 \end{vmatrix} = 4;$

$D_x = \begin{vmatrix} 2 & 1 & -1 \\ 3 & 1 & 1 \\ 4 & 1 & 1 \end{vmatrix} = 2\begin{vmatrix} 1 & 1 \\ 1 & 1 \end{vmatrix} - 3\begin{vmatrix} 1 & -1 \\ 1 & 1 \end{vmatrix} + 4$

$\begin{vmatrix} 1 & -1 \\ 1 & 1 \end{vmatrix} = 2;$

$D_y = \begin{vmatrix} 1 & 2 & -1 \\ -1 & 3 & 1 \\ 1 & 4 & 1 \end{vmatrix} = 1\begin{vmatrix} 3 & 1 \\ 4 & 1 \end{vmatrix} - (-1)\begin{vmatrix} 2 & -1 \\ 4 & 1 \end{vmatrix}$

$+ 1\begin{vmatrix} 2 & -1 \\ 3 & 1 \end{vmatrix} = 10;$

$D_z = \begin{vmatrix} 1 & 1 & 2 \\ -1 & 1 & 3 \\ 1 & 1 & 4 \end{vmatrix} = 1\begin{vmatrix} 1 & 3 \\ 1 & 4 \end{vmatrix} - (-1)\begin{vmatrix} 1 & 2 \\ 1 & 4 \end{vmatrix}$

$+ 1\begin{vmatrix} 1 & 2 \\ 1 & 3 \end{vmatrix} = 4;$

$x = \frac{2}{4} = \frac{1}{2}; y = \frac{10}{4} = \frac{5}{2}; z = \frac{4}{4} = 1; \left(\frac{1}{2}, \frac{5}{2}, 1\right)$

17. $D = \begin{vmatrix} 3 & 0 & 5 \\ 2 & 3 & 0 \\ 0 & 1 & -2 \end{vmatrix} = 3\begin{vmatrix} 3 & 0 \\ 1 & -2 \end{vmatrix} - 2\begin{vmatrix} 0 & 5 \\ 1 & -2 \end{vmatrix} + 0$

$\begin{vmatrix} 0 & 5 \\ 3 & 0 \end{vmatrix} = -8;$

$D_x = \begin{vmatrix} 0 & 0 & 5 \\ 1 & 3 & 0 \\ 11 & 1 & -2 \end{vmatrix} = 0\begin{vmatrix} 3 & 0 \\ 1 & -2 \end{vmatrix} - 1\begin{vmatrix} 0 & 5 \\ 1 & -2 \end{vmatrix} + 11$

$\begin{vmatrix} 0 & 5 \\ 3 & 0 \end{vmatrix} = -160;$

$D_y = \begin{vmatrix} 3 & 0 & 5 \\ 2 & 1 & 0 \\ 0 & 11 & -2 \end{vmatrix} = 3\begin{vmatrix} 1 & 0 \\ 11 & -2 \end{vmatrix} - 2\begin{vmatrix} 0 & 5 \\ 11 & -2 \end{vmatrix} + 0$

$\begin{vmatrix} 0 & 5 \\ 1 & 0 \end{vmatrix} = 104;$

$D_z = \begin{vmatrix} 3 & 0 & 0 \\ 2 & 3 & 1 \\ 0 & 1 & 11 \end{vmatrix} = 3\begin{vmatrix} 3 & 1 \\ 1 & 11 \end{vmatrix} - 2\begin{vmatrix} 0 & 0 \\ 1 & 11 \end{vmatrix} + 0$

$\begin{vmatrix} 0 & 0 \\ 3 & 1 \end{vmatrix} = 96;$

$x = \dfrac{-160}{-8} = 20; y = \dfrac{104}{-8} = -13; z = \dfrac{96}{-8} = -12;$

$(20, -13, -12)$

18a. $\begin{cases} x = 6.5y \\ y = 3z \\ x + y + z = 564 \end{cases} \rightarrow \begin{cases} x - 6.5y = 0 \\ y - 3z = 0 \\ x + y + z = 564 \end{cases}$

$x = \dfrac{D_x}{D} = \dfrac{\begin{vmatrix} 0 & -6.5 & 0 \\ 0 & 1 & -3 \\ 564 & 1 & 1 \end{vmatrix}}{\begin{vmatrix} 1 & -6.5 & 0 \\ 0 & 1 & -3 \\ 1 & 1 & 1 \end{vmatrix}};$

$y = \dfrac{D_y}{D} = \dfrac{\begin{vmatrix} 1 & 0 & 0 \\ 0 & 0 & -3 \\ 1 & 564 & 1 \end{vmatrix}}{\begin{vmatrix} 1 & -6.5 & 0 \\ 0 & 1 & -3 \\ 1 & 1 & 1 \end{vmatrix}}; z = \dfrac{D_z}{D} = \dfrac{\begin{vmatrix} 1 & -6.5 & 0 \\ 0 & 1 & 0 \\ 1 & 1 & 564 \end{vmatrix}}{\begin{vmatrix} 1 & -6.5 & 0 \\ 0 & 1 & -3 \\ 1 & 1 & 1 \end{vmatrix}};$

18b. $D = \begin{vmatrix} 1 & -6.5 & 0 \\ 0 & 1 & -3 \\ 1 & 1 & 1 \end{vmatrix} = 1 \begin{vmatrix} 1 & -3 \\ 1 & 1 \end{vmatrix} - 0 \begin{vmatrix} -6.5 & 0 \\ 1 & 1 \end{vmatrix}$

$\begin{vmatrix} -6.5 & 0 \\ 1 & -3 \end{vmatrix} = 23.5;$

$D_x = \begin{vmatrix} 0 & -6.5 & 0 \\ 0 & 1 & -3 \\ 564 & 1 & 1 \end{vmatrix} = 0 \begin{vmatrix} 1 & -3 \\ 1 & 1 \end{vmatrix} - 0 \begin{vmatrix} -6.5 & 0 \\ 1 & 1 \end{vmatrix}$

$+ 564 \begin{vmatrix} -6.5 & 0 \\ 1 & -3 \end{vmatrix} = 10{,}998;$

$D_y = \begin{vmatrix} 1 & 0 & 0 \\ 0 & 0 & -3 \\ 1 & 564 & 1 \end{vmatrix} = 1 \begin{vmatrix} 0 & -3 \\ 564 & 1 \end{vmatrix} - 0 \begin{vmatrix} 0 & 0 \\ 564 & 1 \end{vmatrix}$

$+ 1 \begin{vmatrix} 0 & 0 \\ 0 & -3 \end{vmatrix} = 1692;$

$D_z = \begin{vmatrix} 1 & -6.5 & 0 \\ 0 & 1 & 0 \\ 1 & 1 & 564 \end{vmatrix} = 1 \begin{vmatrix} 1 & 0 \\ 1 & 564 \end{vmatrix} - 0 \begin{vmatrix} -6.5 & 0 \\ 1 & 564 \end{vmatrix}$

$+ 1 \begin{vmatrix} -6.5 & 0 \\ 1 & 0 \end{vmatrix} = 564;$

$x = \dfrac{10{,}998}{23.5} = 468; y = \dfrac{1692}{23.5} = 72;$

$z = \dfrac{564}{23.5} = 24;$ (468,72,24); The average yearly rainfall is 468 in. for Mawsynram, 72 in. for Beijing, and 24 in for Paris.

PRACTICE

1. $6(4) - (-3)(0) = 24$

2. $(-3)\left(\dfrac{1}{2}\right) - (-5)(-1) = -\dfrac{13}{2}$

3. $1(1) - 0(0) = 1$

4. $\begin{vmatrix} 1 & 0 & 0 \\ -2 & 4 & 3 \\ 5 & -2 & 1 \end{vmatrix}\begin{matrix} 1 & 0 \\ -2 & 4 \\ 5 & -2 \end{matrix}$

$= 4 + 0 + 0 - 0 - (-6) - 0 = 10$

5. $\begin{vmatrix} 1 & 3 & 7 \\ -2 & 6 & 4 \\ 3 & 7 & -1 \end{vmatrix}\begin{matrix} 1 & 3 \\ -2 & 6 \\ 3 & 7 \end{matrix}$

$= -6 + 36 + (-98) - 126 - 28 - 6 = -228$

6. $\begin{vmatrix} 1 & 4 & 3 \\ 2 & 1 & 6 \\ 3 & -2 & 9 \end{vmatrix}\begin{matrix} 1 & 4 \\ 2 & 1 \\ 3 & -2 \end{matrix}$

$= 9 + 72 + (-12) - 9 - (-12) - 72 = 0$

7. Answers will vary. Possible answer: Begin by evaluating determinants D, D_x, D_y, and D_z.

$D = \begin{vmatrix} 2 & -3 & 5 \\ 1 & 2 & -1 \\ 5 & -1 & 4 \end{vmatrix} = 2 \begin{vmatrix} 2 & -1 \\ -1 & 4 \end{vmatrix} - 1 \begin{vmatrix} -3 & 5 \\ -1 & 4 \end{vmatrix} + 5$

$\begin{vmatrix} -3 & 5 \\ 2 & -1 \end{vmatrix} = -14;$

$D_x = \begin{vmatrix} 27 & -3 & 5 \\ -4 & 2 & -1 \\ 27 & -1 & 4 \end{vmatrix} = 27 \begin{vmatrix} 2 & -1 \\ -1 & 4 \end{vmatrix} - (-4)$

$\begin{vmatrix} -3 & 5 \\ -1 & 4 \end{vmatrix} + 27 \begin{vmatrix} -3 & 5 \\ 2 & -1 \end{vmatrix} = -28;$

$D_y = \begin{vmatrix} 2 & 27 & 5 \\ 1 & -4 & -1 \\ 5 & 27 & 4 \end{vmatrix} = 2 \begin{vmatrix} -4 & -1 \\ 27 & 4 \end{vmatrix} - 1 \begin{vmatrix} 27 & 5 \\ 27 & 4 \end{vmatrix} + 5$

$\begin{vmatrix} 27 & 5 \\ -4 & -1 \end{vmatrix} = 14;$

$D_z = \begin{vmatrix} 2 & -3 & 27 \\ 1 & 2 & -4 \\ 5 & -1 & 27 \end{vmatrix} = 2 \begin{vmatrix} 2 & -4 \\ -1 & 27 \end{vmatrix} - 1 \begin{vmatrix} -3 & 27 \\ -1 & 27 \end{vmatrix} + 5$

$\begin{vmatrix} -3 & 27 \\ 2 & -4 \end{vmatrix} = -56;$

Use Cramer's rule to find the solution.

$D_x = \dfrac{-28}{-14} = 2; D_y = \dfrac{14}{-14} = -1; D_z = \dfrac{-56}{-14} = 4;$
$(2, -1, 4)$

8. Let e represent the electrician's hours. Let c represent the carpenter's hours. Let p represent the plumber's hours.

$\begin{cases} e + c + p = 21.5 \\ 32.5e + 40.75c + 62p = 1059.75 \rightarrow \\ c + 2 = p \end{cases}$

$$\begin{cases} e + c + p = 21.5 \\ 32.5e + 40.75c + 62p = 1059.75 \\ c - p = -2 \end{cases}$$

$$D = \begin{vmatrix} 1 & 1 & 1 \\ 32.5 & 40.75 & 62 \\ 0 & 1 & -1 \end{vmatrix} = 1\begin{vmatrix} 40.75 & 62 \\ 1 & -1 \end{vmatrix}$$

$$- 32.5\begin{vmatrix} 1 & 1 \\ 1 & -1 \end{vmatrix} + 0\begin{vmatrix} 1 & 1 \\ 40.75 & 62 \end{vmatrix} = -37.75;$$

$$D_e = \begin{vmatrix} 21.5 & 1 & 1 \\ 1059.75 & 40.75 & 62 \\ -2 & 1 & -1 \end{vmatrix} =$$

$$21.5\begin{vmatrix} 40.75 & 62 \\ 1 & -1 \end{vmatrix} - 1059.75\begin{vmatrix} 1 & 1 \\ 1 & -1 \end{vmatrix} + (-2)$$

$$\begin{vmatrix} 1 & 1 \\ 40.75 & 62 \end{vmatrix} = -132.125; e = \frac{-132.125}{37.75} = 3.5;$$

The electrician worked 3.5h.

9. $D = \begin{vmatrix} 4 & 3 \\ 3 & -4 \end{vmatrix} = -25; D_x = \begin{vmatrix} 0 & 3 \\ 25 & -4 \end{vmatrix} - 75;$

$D_y = \begin{vmatrix} 4 & 0 \\ 3 & 25 \end{vmatrix} = 100; x = \frac{-75}{-25} = 3;$

$y = \frac{100}{-25} = -4; (3, -4)$

10. $D = \begin{vmatrix} -2 & 4 \\ 3 & -7 \end{vmatrix} = 2; D_x = \begin{vmatrix} 3 & 4 \\ 1 & -7 \end{vmatrix} = -25; D_y =$

$\begin{vmatrix} -2 & 3 \\ 3 & 1 \end{vmatrix} = -11; x = \frac{-25}{2}; y = \frac{-11}{2}; \left(-\frac{25}{2}, \frac{11}{2} \right)$

11. $D = 0$; not possible

12. $D = \begin{vmatrix} 0 & 1 & 4 \\ 3 & 0 & 1 \\ 0 & 5 & -1 \end{vmatrix} = 0\begin{vmatrix} 0 & 1 \\ 5 & -1 \end{vmatrix} - 3\begin{vmatrix} 1 & 4 \\ 5 & -1 \end{vmatrix} + 0$

$\begin{vmatrix} 1 & 4 \\ 0 & 1 \end{vmatrix} = 63;$

$D_x = \begin{vmatrix} 6 & 1 & 4 \\ 7 & 0 & 1 \\ 9 & 5 & -1 \end{vmatrix} = 6\begin{vmatrix} 0 & 1 \\ 5 & -1 \end{vmatrix} - 7\begin{vmatrix} 1 & 4 \\ 5 & -1 \end{vmatrix} + 9$

$\begin{vmatrix} 1 & 4 \\ 0 & 1 \end{vmatrix} = 126;$

$D_y = \begin{vmatrix} 0 & 6 & 4 \\ 3 & 7 & 1 \\ 0 & 9 & -1 \end{vmatrix} = 0\begin{vmatrix} 7 & 1 \\ 9 & -1 \end{vmatrix} - 3\begin{vmatrix} 6 & 4 \\ 9 & -1 \end{vmatrix} + 0$

$\begin{vmatrix} 6 & 4 \\ 7 & 1 \end{vmatrix} = 126;$

$$D_z = \begin{vmatrix} 0 & 1 & 6 \\ 3 & 0 & 7 \\ 0 & 5 & 9 \end{vmatrix} = 0\begin{vmatrix} 0 & 7 \\ 5 & 9 \end{vmatrix} - 3\begin{vmatrix} 1 & 6 \\ 5 & 9 \end{vmatrix} + 0$$

$$\begin{vmatrix} 1 & 6 \\ 0 & 7 \end{vmatrix} = 63;$$

$$x = \frac{126}{63} = 2; y = \frac{126}{63} = 2; z = \frac{63}{63} = 1;$$

13. $D_z = \begin{vmatrix} 2 & 2 & 1 \\ 1 & 3 & 2 \\ 1 & -1 & -1 \end{vmatrix} = 2\begin{vmatrix} 3 & 2 \\ -1 & -1 \end{vmatrix} - 1\begin{vmatrix} 2 & 1 \\ -1 & -1 \end{vmatrix} = 0;$

not possible

14. $D = \begin{vmatrix} 3 & 2 & -1 \\ 3 & -2 & 1 \\ 4 & -5 & -1 \end{vmatrix} = 3\begin{vmatrix} -2 & 1 \\ -5 & -1 \end{vmatrix} - 3\begin{vmatrix} 2 & -1 \\ -5 & -1 \end{vmatrix} + 4$

$\begin{vmatrix} 2 & -1 \\ -2 & 1 \end{vmatrix} = 42;$

$D_x = \begin{vmatrix} 4 & 2 & -1 \\ 5 & -2 & 1 \\ -1 & -5 & -1 \end{vmatrix} = 4\begin{vmatrix} -2 & 1 \\ -5 & -1 \end{vmatrix} - 5\begin{vmatrix} 2 & -1 \\ -5 & -1 \end{vmatrix}$

$+ (-1)\begin{vmatrix} 2 & -1 \\ -2 & 1 \end{vmatrix} = 63;$

$D_y = \begin{vmatrix} 3 & 4 & -1 \\ 3 & 5 & 1 \\ 4 & -1 & -1 \end{vmatrix} = 3\begin{vmatrix} 5 & 1 \\ -1 & -1 \end{vmatrix} - 3\begin{vmatrix} 4 & -1 \\ -1 & -1 \end{vmatrix}$

$+ 4\begin{vmatrix} 4 & -1 \\ 5 & 1 \end{vmatrix} = 39;$

$D_z = \begin{vmatrix} 3 & 2 & 4 \\ 3 & -2 & 5 \\ 4 & -5 & -1 \end{vmatrix} = 3\begin{vmatrix} -2 & 5 \\ -5 & -1 \end{vmatrix} - 3\begin{vmatrix} 2 & 4 \\ -5 & -1 \end{vmatrix} + 4$

$\begin{vmatrix} 2 & 4 \\ -5 & -1 \end{vmatrix} = 99;$

$x = \frac{63}{42} = \frac{3}{2}; y = \frac{39}{42} = \frac{13}{14} z = \frac{99}{42} = \frac{33}{14}; \left(\frac{3}{2}, \frac{13}{14}, \frac{33}{14} \right)$

EXTEND
15. yes; Examples will vary. Possible examples:

$\begin{vmatrix} 3 & 2 & -2 \\ -2 & 1 & 4 \\ 6 & 4 & -4 \end{vmatrix}$ and $\begin{vmatrix} 1 & 2 & 0 \\ 0 & 2 & 1 \\ 1 & 2 & 0 \end{vmatrix}$; Both are 0.

$\begin{vmatrix} 1 & 0 & 0 \\ -2 & 4 & 3 \\ 5 & -2 & 1 \end{vmatrix}$ and $\begin{vmatrix} 1 & 2 & 4 \\ 2 & 1 & 6 \\ 2 & 3 & 4 \end{vmatrix}$; Both are 10.

16. $x(3) - 5(-2) = 1, 3x + 10 = 1, 3x = -9, x = -3$

17. $(x + 3)(5) - (x - 3)(4) = -7,$
$5x + 15 - 4x + 12 = -7$
$x + 27 = -7, x = -34$

18. $2\begin{vmatrix} 3 & 2 \\ 1 & 1 \end{vmatrix} - (-1)\begin{vmatrix} x & -1 \\ 1 & 1 \end{vmatrix} + (-2)\begin{vmatrix} x & -1 \\ 3 & 2 \end{vmatrix} = -12,$
$2(3 \cdot 1 - 1 \cdot 2) + 1(x \cdot 1 - 1 \cdot (-1)) -$
$2(x \cdot 2 - 3 \cdot (-1)) = -12,$
$2(1) + 1(x + 1) - 2(2x + 3) =$
$-12, 2 + x + 1 - 4x - 6 = -12;$
$-3x - 3 = -12, -3x = -9, x = 3$

19. Answers will vary. A 3 × 3 matrix with a row or column of zeros will have a determinant that is equal to zero. The matrix represents a dependent or inconsistent system.

20. Answers will vary. A 3 × 3 matrix with two identical rows or columns will have a determinant that is equal to zero. The matrix represents a dependent or inconsistent system.

21. $\dfrac{1}{2}\begin{vmatrix} -1 & 4 & 1 \\ 4 & 8 & 1 \\ 1 & 1 & 1 \end{vmatrix}; D = \begin{vmatrix} -1 & 4 & 1 \\ 4 & 8 & 1 \\ 1 & 1 & 1 \end{vmatrix} = (-1)\begin{vmatrix} 8 & 1 \\ 1 & 1 \end{vmatrix} - 4$

$\begin{vmatrix} 4 & 1 \\ 1 & 1 \end{vmatrix} + 1\begin{vmatrix} 4 & 1 \\ 8 & 1 \end{vmatrix} = -23; |-23| = 23; \dfrac{1}{2}(23) = \dfrac{23}{2}$

22. Let l represent the price of a large balloon. Let m represent the price of a medium balloon. Let s represent the price of a small balloon.

$\begin{cases} l + 2m + 4s = 17 \\ 2l + m + 6s = 22 \\ 2l + 3m + 4s = 25 \end{cases}$

$D = \begin{vmatrix} 1 & 2 & 4 \\ 2 & 1 & 6 \\ 2 & 3 & 4 \end{vmatrix} = 1\begin{vmatrix} 1 & 6 \\ 3 & 4 \end{vmatrix} - 2\begin{vmatrix} 2 & 4 \\ 3 & 4 \end{vmatrix} + 2$

$\begin{vmatrix} 2 & 4 \\ 1 & 6 \end{vmatrix} = 10;$

$D_l = \begin{vmatrix} 17 & 2 & 4 \\ 22 & 1 & 6 \\ 25 & 3 & 4 \end{vmatrix} = 17\begin{vmatrix} 1 & 6 \\ 3 & 4 \end{vmatrix} - 22\begin{vmatrix} 2 & 4 \\ 3 & 4 \end{vmatrix} + 25$

$\begin{vmatrix} 2 & 4 \\ 1 & 6 \end{vmatrix} = 50;$

$D_m = \begin{vmatrix} 1 & 17 & 4 \\ 2 & 22 & 6 \\ 2 & 25 & 4 \end{vmatrix} = 1\begin{vmatrix} 22 & 6 \\ 25 & 4 \end{vmatrix} - 2\begin{vmatrix} 17 & 4 \\ 25 & 4 \end{vmatrix} + 2$

$\begin{vmatrix} 17 & 4 \\ 22 & 6 \end{vmatrix} = 30;$

$D_s = \begin{vmatrix} 1 & 2 & 17 \\ 2 & 1 & 22 \\ 2 & 3 & 25 \end{vmatrix} = 1\begin{vmatrix} 1 & 22 \\ 3 & 25 \end{vmatrix} - 2\begin{vmatrix} 2 & 17 \\ 3 & 25 \end{vmatrix} + 2$

$\begin{vmatrix} 2 & 17 \\ 1 & 22 \end{vmatrix} = 15;$

$l = \dfrac{50}{10} = 5; m = \dfrac{30}{10} = 3; s = \dfrac{15}{10} = 1.5; (5, 3, 1.5);$
The large balloons cost $5.00, medium balloons cost $3.00, and small cost $1.50.
$3l + 3m + 3s = 3(5) + 3(3) + 3(1.5) = 28.5.$ Ken paid $28.50 for his balloons.

THINK CRITICALLY

23. O; row 3 $= -\dfrac{1}{2}$ (row 4), 2 identical rows

24. An equation of the line through the two points is
$y - y_1 = \dfrac{y_2 - y_1}{x_2 - x_1}(x - x_1)$ which is equivalent to
$yx_2 - yx_1 - y_1x_2 - y_2x + y_2x_1 + y_1x = 0.$
The left hand side of the equation is the formula for the determinant of the matrix, so the determinant is equal to 0.

25. $(x_1 y_1) \rightarrow (6, 2); (x_2 y_2) \rightarrow (3, -2);$
$yx_2 - yx_1 - y_1x_2 - y_2x + y_2x_1 + y_1x = 0$
$y(3) - y(6) - (2)(3) - (-2)x + (-2)(6) + 2x = 0,$
$3y - 3y - 6y - 6 + 2x - 12 + 2x = 0$
$4x - 3y = 18$

26. $\begin{cases} x + 2y = 1 \\ 3x + 4y = 0 \end{cases}$

27. $D = \begin{vmatrix} a & b \\ b & a \end{vmatrix} = a^2 - b^2;$

$D_x = \begin{vmatrix} -1 & b \\ 1 & a \end{vmatrix} = -a - b;$

$D_y = \begin{vmatrix} a & -1 \\ b & 1 \end{vmatrix} = a + b;$

$x = \dfrac{-a - b}{a^2 - b^2} = \dfrac{-a - b}{(a - b)(a + b)} =$

$-\dfrac{a + b}{(a - b)(a + b)} = -\dfrac{1}{a - b}; y = \dfrac{a + b}{a^2 - b^2} =$

$\dfrac{a + b}{(a - b)(a + b)} = \dfrac{1}{a - b}; \left(-\dfrac{1}{a - b}, \dfrac{1}{a - b}\right)$

28. $D = \begin{vmatrix} a & b \\ 1 & 1 \end{vmatrix} = a - b;$

$D_x = \begin{vmatrix} \frac{b}{a} & b \\ \frac{1}{b} & 1 \end{vmatrix} = \dfrac{b}{a} - 1 = \dfrac{b - a}{a};$

$D_y = \begin{vmatrix} a & \frac{b}{a} \\ 1 & \frac{1}{b} \end{vmatrix} = \dfrac{a}{b} - \dfrac{b}{a} = \dfrac{a^2 - b^2}{ab};$

$x = \dfrac{\frac{b - a}{a}}{a - b} = \dfrac{b - a}{a} \cdot \dfrac{-1}{b - a} = -\dfrac{1}{a};$

$$y = \frac{\frac{a^2 - b^2}{ab}}{a - b} = \frac{a^2 - b^2}{ab} \cdot \frac{1}{a - b} = \frac{a + b}{ab};$$

$$\left(-\frac{1}{a}, \frac{a + b}{ab}\right)$$

MIXED REVIEW

29. $3 - 4x < -2$ or $3 - 4x > 2$

$\quad -4x < -5 \qquad\qquad -4x > -1$

$\qquad x > \dfrac{5}{4} \qquad\qquad\quad x < \dfrac{1}{4}$

The correct answer is A.

30. $f(6) = 6 + 32 = 38; f(10) = 10 + 32 = 42;$
$f(14) = 14 + 32 = 46; f(18) = 18 + 32 = 50;$ Yes it has an inverse; $x = y + 32, y = x - 32$

Lesson 4.9, pages 216–219

EXPLORE THE PROBLEM

1. An intersection of a row and column represent the number of ways in which supplies can be transferred directly from one plant to another.

2.

	F	L	B	P	W
F	0	0	1	0	1
L	1	0	1	0	0
B	1	1	0	1	1
P	1	0	1	0	1
W	1	0	1	1	0

3. $0 + 0 + 1 + 0 + 1 = 2$; The sum of the elements in the first row represent the number of plants that receive supplies directly from Fitchburg.

4. 2nd row: $1 + 0 + 1 + 0 + 0 = 2$; This sum represents the number of plants that receive supplies directly from Lowell.
3rd row: $1 + 1 + 0 + 1 + 1 = 4$; This sum represents the number of plants that receive supplies directly from Boston.
4th row: $1 + 0 + 1 + 0 + 1 = 3$; This sum represents the number of plants that receive supplies directly from Providence.
5th row: $1 + 0 + 0 + 1 + 0 = 2$; This sum represents the number of plants that receive supplies directly from Worcester.

5. $2 + 2 + 4 + 3 + 2 = 13$; This total represents the total number of ways in which supplies can be transferred directly from one plant to another without being routed through a third plant.

6. 3 ways; Find the sum of the elements in the third column.

INVESTIGATE FURTHER

7. There are 2 ways to move supplies from Fitchburg to Providence with one stop in-between.

8.

	F	L	B	P	W
F	2	1	0	2	1
L	1	1	1	1	2
B	3	0	3	1	2
P	2	1	1	2	2
W	1	0	2	0	2

The sum of the elements of the matrix yields 34, the total number of ways in which supplies can be transferred between two plants if there is one stop in-between.

9.

	F	L	B	P	W
F	2	1	0	2	1
L	1	1	1	1	2
B	3	0	3	1	2
P	2	1	1	2	2
W	1	0	2	0	2

The two matrices are the same. You could square the matrix instead of counting paths.

APPLY THE STRATEGY

10. System 1

	A	B	C	D
A	0	1	1	1
B	1	0	0	1
C	1	0	0	1
D	0	1	0	0

System 2

	L	M	N	O
L	0	1	1	0
M	1	0	1	1
N	0	1	0	0
O	0	1	1	0

No, both subways have eight direct connections.

11. System 1

	A	B	C	D
A	2	1	0	2
B	0	2	1	1
C	0	2	1	1
D	1	0	0	1

System 2

	L	M	N	O
L	1	1	1	1
M	0	3	2	0
N	1	0	1	1
O	1	1	1	1

Each matrix is the square of the corresponding original matrix; System 2 has more one-stop connections than System 1.
System 1: 16 one-stop connections;
System 2: 15 one-stop connections

12. System 1

	A	B	C	D
A	1	4	2	3
B	3	1	0	3
C	3	1	0	3
D	0	2	1	1

System 2

	L	M	N	O
L	3	5	5	3
M	2	9	8	2
N	3	2	3	3
O	3	5	5	3

Each matrix would represent the number of ways to travel between two stations making two stops enroute.

13.

	T	F	R	B	C
T	0	1	0	1	1
F	1	0	1	0	1
R	0	1	0	1	1
B	0	0	1	0	0
C	0	0	1	0	0

14. $1 + 0 + 1 + 0 + 1 = 3; 1 + 0 + 1 + 0 + 0 = 2$

15.
$$\begin{array}{c} \\ T \\ F \\ R \\ B \\ C \end{array}\begin{array}{ccccc} T & F & R & B & C \\ \begin{bmatrix} 1 & 0 & 3 & 0 & 1 \\ 0 & 2 & 1 & 2 & 2 \\ 1 & 0 & 3 & 0 & 1 \\ 0 & 1 & 1 & 1 & 1 \\ 0 & 1 & 1 & 1 & 1 \end{bmatrix} \end{array}$$

The sum of the elements of the matrix yields, 23 different ways in which Kevin and Angela could travel between two locations with one stop on the way.

16. Change the elements along the main diagonal to 0's then find the total of the sum of the rows.

17.

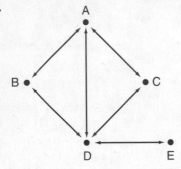

Answers will vary. All connections should be as shown.

REVIEW PROBLEM SOLVING STRATEGIES

1a.

7	8	9	10
8	9	10	11
9	10	11	12
10	11	12	13

$160 = 16 \cdot 10$. The sum equals the number of entries times the number that runs along the diagonal.

b. $(50)(50) \cdot 99 = 247{,}500$

c. $(100)(100) \cdot x = 3{,}000{,}000$
$$x = 300$$
Since x is the last element of a row of 100:
$300 - 99 = 201$; 201 must be the beginning number.

2a. six cuts

b. Every cube except one (the center cube) will have at least one face that was part of a face of the original block. Every one of these six faces of the center cube must be formed by cuts, which shows that the job cannot be done with fewer than six cuts.

3. No; upon returning the first item, she is due only half the cost of that item. Rhonda is considering the value of the item, and hence the item, to be hers even after it is returned.

1. b   **2.** a   **3.** c

4.

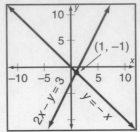

$(1, -1)$

$$\begin{array}{ll} 2x - y = 3 & y = -x \\ 2(1) - (-1) \overset{?}{=} 3 & 1 \overset{?}{=} -(-1) \\ 3 = 3 & 1 = 1 \end{array}$$

5.

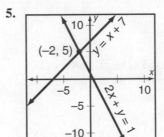

$(-2, 5)$

$$\begin{array}{ll} y = x + 7 & 2x + y = 1 \\ 5 = -2 + 7 & 2(-2) + 5 = 1 \\ 5 = 5 & 1 = 1 \end{array}$$

6.

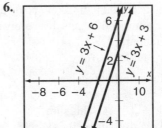

no solution

7.
$$\begin{array}{ll} 7x + 2y = 37 & y = 3x - 1 \\ 7x + 2(3x - 1) = 37 & y = 3(3) - 1 \\ 7x + 6x - 2 = 37 & y = 8 \\ 13x = 39 & \\ x = 3 & \end{array}$$

$$\begin{array}{ll} 7x + 2y = 37 & y = 3x - 1 \\ 7(3) + 2(8) = 37 & 8 = 3(3) - 1 \\ 37 = 37 & 8 = 8 \end{array}$$

The solution is $(3, 8)$.

8.
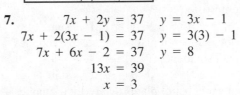
$$\begin{array}{llr} 5x - 2y = 20 & \rightarrow & 15x - 6y = 60 \\ 2x + 3y = 27 & \rightarrow & \underline{4x + 6y = 54} \\ & & 19x = 114 \\ & & x = 6 \end{array}$$

$2x + 3y = 27$ $5x - 2y = 20$
$2(6) + 3y = 27$ $5(6) - 2(5) \overset{?}{=} 20$
$3y = 15$ $20 = 20$
$y = 5$

$2x + 3y = 27$
$2(6) + 3(5) \overset{?}{=} 27$
$27 = 27$

The solution is $(6, 5)$.

9.
$$\begin{aligned} 3x + y + 2z &= 6 \\ x + y + 4z &= 3 \end{aligned} \quad \rightarrow \quad \begin{aligned} 3x + y + 2z &= 6 \\ -x - y - 4z &= -3 \\ \hline 2x \quad\quad - 2z &= 3 \end{aligned}$$

$$\begin{aligned} x + y + 4z = 3 &\rightarrow -3x - 3y - 12z = -9 \\ 2x + 3y + 2z = 2 &\rightarrow \underline{\;\;2x + 3y + 2z = 2\;\;} \\ & \quad\quad -x \quad\quad\quad -10z = -7 \end{aligned}$$

$$\begin{aligned} 2x - 2z = 3 &\rightarrow -10x + 10z = -15 \\ -x - 10z = -7 &\rightarrow \underline{\;\;-x - 10z = -7\;\;} \\ & \quad\quad -11x \quad\quad = -22 \\ & \quad\quad\quad\quad x = 2 \end{aligned}$$

$2x - 2z = 3$ $x + y + 4z = 3$
$2(2) - 2z = 3$ $2 + y + 4\left(\frac{1}{2}\right) = 3$
$-2z = -1$ $2 + y + 2 = 3$
$z = \frac{1}{2}$ $y = -1$

$3x + y + 2z = 6$ $x + y + 4z = 3$
$3(2) + (-1) + 2\left(\frac{1}{2}\right) \overset{?}{=} 6$ $2 + (-1) + 4\left(\frac{1}{2}\right) \overset{?}{=} 3$
        $6 - 6$ $3 - 3$

$2x + 3y + 2z = 2$
$2(2) + 3(-1) + 2\left(\frac{1}{2}\right) \overset{?}{=} 2$
              $2 = 2$

The solution is $\left(2, -1, \frac{1}{2}\right)$.

10. $\begin{bmatrix} 3 & 1 & | & 4 \\ 2 & 1 & | & 2 \end{bmatrix} \rightarrow \begin{bmatrix} 1 & \frac{1}{3} & | & \frac{4}{3} \\ 0 & \frac{1}{3} & | & -\frac{2}{3} \end{bmatrix} \rightarrow \begin{bmatrix} 1 & 0 & | & 2 \\ 0 & 1 & | & -2 \end{bmatrix}; \; (2, -2)$

11. $\begin{cases} 4x + y = 1 \\ 3x + 2y = 2 \end{cases}; \begin{bmatrix} 4 & 1 & | & 1 \\ 3 & 2 & | & 2 \end{bmatrix} \rightarrow \begin{bmatrix} 1 & \frac{1}{4} & | & \frac{1}{4} \\ 0 & \frac{5}{4} & | & \frac{5}{4} \end{bmatrix} \rightarrow$

$\begin{bmatrix} 1 & 0 & | & 0 \\ 0 & 1 & | & 1 \end{bmatrix}; \; (0, 1)$

12. $\begin{bmatrix} 1 & 0 & -3 & | & -2 \\ 3 & 1 & -2 & | & 5 \\ 2 & 2 & 1 & | & 4 \end{bmatrix} \rightarrow \begin{bmatrix} 1 & 0 & -3 & | & -2 \\ 0 & 1 & 7 & | & 1 \\ 0 & 2 & 7 & | & 8 \end{bmatrix} \rightarrow$

$\begin{bmatrix} 1 & 0 & -3 & | & 2 \\ 0 & 1 & 7 & | & 11 \\ 0 & 0 & -7 & | & -14 \end{bmatrix} \rightarrow \begin{bmatrix} 1 & 0 & 0 & | & 4 \\ 0 & 1 & 0 & | & -3 \\ 0 & 0 & 1 & | & 2 \end{bmatrix}; \; (4, -3, 2)$

13. $AB = [1(4) + 2(5) + 3(6)] = [32]$

14. $BA = \begin{bmatrix} 4(1) & 4(2) & 4(3) \\ 5(1) & 5(2) & 5(3) \\ 6(1) & 6(2) & 6(3) \end{bmatrix} = \begin{bmatrix} 4 & 8 & 12 \\ 5 & 10 & 15 \\ 6 & 12 & 18 \end{bmatrix}$

15. $AC = [1(1) + 2(2) + 3(3) \quad 1(4) + 2(5) + 3(6)] = [14 \quad 32]$

16. BC is not defined.

17. $\begin{bmatrix} 4 & -4 & | & 1 & 0 \\ 3 & 2 & | & 0 & 1 \end{bmatrix} \rightarrow \begin{bmatrix} 1 & -1 & | & \frac{1}{4} & 0 \\ 0 & 5 & | & -\frac{3}{4} & 1 \end{bmatrix} \rightarrow$

$\begin{bmatrix} 1 & 0 & | & \frac{1}{10} & \frac{1}{5} \\ 0 & 1 & | & -\frac{3}{20} & \frac{1}{5} \end{bmatrix}; A^{-1} = \begin{bmatrix} 0.1 & 0.2 \\ -0.15 & 0.2 \end{bmatrix}; A \cdot A^{-1} =$

$\begin{bmatrix} 4(0.1) + (-4)(-0.15) & 4(0.2) + (-4)(0.2) \\ 3(0.1) + 2(-0.15) & 3(0.2) + 2(0.2) \end{bmatrix} = \begin{bmatrix} 1 & 0 \\ 0 & 1 \end{bmatrix}$

18. $\begin{bmatrix} 6 & 3 & | & 1 & 0 \\ 8 & 4 & | & 0 & 1 \end{bmatrix} \rightarrow \begin{bmatrix} 1 & \frac{1}{2} & | & \frac{1}{6} & 0 \\ 0 & 0 & | & -\frac{4}{3} & 1 \end{bmatrix}$; no inverse

19. $\begin{bmatrix} -1 & 1 & 0 & | & 1 & 0 & 0 \\ -1 & 0 & 1 & | & 0 & 1 & 0 \\ 6 & 2 & -3 & | & 0 & 0 & 1 \end{bmatrix} \rightarrow$

$\begin{bmatrix} 1 & -1 & 0 & | & -1 & 0 & 0 \\ 0 & -1 & 1 & | & -1 & 1 & 0 \\ 0 & 8 & -3 & | & 6 & 0 & 1 \end{bmatrix} \rightarrow$

$\begin{bmatrix} 1 & 0 & -1 & | & 0 & -1 & 0 \\ 0 & 1 & -1 & | & 1 & -1 & 0 \\ 0 & 0 & 5 & | & -2 & 8 & 1 \end{bmatrix} \rightarrow$

$\begin{bmatrix} 1 & 0 & 0 & | & -\frac{2}{5} & \frac{3}{5} & \frac{1}{5} \\ 0 & 1 & 0 & | & \frac{3}{5} & \frac{3}{5} & \frac{1}{5} \\ 0 & 0 & 1 & | & -\frac{2}{5} & \frac{8}{5} & \frac{1}{5} \end{bmatrix};$

$A^{-1} \begin{bmatrix} -0.4 & 0.6 & 0.2 \\ 0.6 & 0.6 & 0.2 \\ -0.4 & 1.6 & 0.2 \end{bmatrix}; A \cdot A^{-1} =$

$\begin{bmatrix} -1(-0.4) + 1(0.6) + 0(-0.4) \\ -1(-0.4) + 0(0.6) + 1(-0.4) \\ 6(-0.4) + 2(0.6) + (-3)(-0.4) \end{bmatrix}$

$\begin{matrix} -1(0.6) + 1(0.6) + 0(1.6) & -1(0.2) + 1(0.2) + 0(0.2) \\ -1(0.6) + 0(0.6) + 1(1.6) & -1(0.2) + 0(0.2) + 1(0.2) \\ 6(0.6) + 2(0.6) + (-3)(1.6) & 6(0.2) + 2(0.2) + (-3)(0.2) \end{matrix} \Big]$

$= \begin{bmatrix} 1 & 0 & 0 \\ 0 & 1 & 0 \\ 0 & 0 & 1 \end{bmatrix}$

20. $\begin{bmatrix} 2 & -5 \\ 3 & -7 \end{bmatrix} X = \begin{bmatrix} 2 \\ 1 \end{bmatrix}; A^{-1} = \begin{bmatrix} -7 & 2 \\ -3 & 2 \end{bmatrix}; \begin{bmatrix} -7 & 5 \\ -3 & 2 \end{bmatrix}\begin{bmatrix} 2 & -5 \\ 3 & -7 \end{bmatrix}$

$X = \begin{bmatrix} -7 & 5 \\ -3 & 2 \end{bmatrix}\begin{bmatrix} 2 & -5 \\ 3 & -7 \end{bmatrix}, X = \begin{bmatrix} -7(2) + 5(1) \\ -3(2) + 2(1) \end{bmatrix},$

$X = \begin{bmatrix} -9 \\ -4 \end{bmatrix}; (-9, -4)$

21. $\begin{cases} 4x - 3y = 11 \\ 5x - 6y = 9 \end{cases}; \begin{bmatrix} 4 & -3 \\ 5 & -6 \end{bmatrix} X = \begin{bmatrix} 11 \\ 9 \end{bmatrix};$

$A^{-1} = \begin{bmatrix} \frac{2}{3} & -\frac{1}{3} \\ \frac{5}{9} & -\frac{4}{9} \end{bmatrix}; \begin{bmatrix} \frac{2}{3} & -\frac{1}{3} \\ \frac{5}{9} & -\frac{4}{9} \end{bmatrix}\begin{bmatrix} 4 & -3 \\ 5 & -6 \end{bmatrix} X = \begin{bmatrix} \frac{2}{3} & -\frac{1}{3} \\ \frac{5}{9} & -\frac{4}{9} \end{bmatrix}\begin{bmatrix} 11 \\ 9 \end{bmatrix},$

$X = \begin{bmatrix} \frac{2}{3}(11) + \left(-\frac{1}{3}\right)(9) \\ \frac{5}{9}(11) + \left(-\frac{4}{9}\right)(9) \end{bmatrix}, X = \begin{bmatrix} \frac{13}{3} \\ \frac{19}{9} \end{bmatrix}; \left(\frac{13}{3}, \frac{19}{9}\right)$

22. $\begin{bmatrix} 6 & -2 & 1 \\ 2 & -1 & 5 \\ 2 & 0 & -3 \end{bmatrix} X = \begin{bmatrix} 16 \\ -2 \\ 8 \end{bmatrix}; A^{-1} = \begin{bmatrix} -\frac{1}{4} & \frac{1}{2} & \frac{3}{4} \\ -\frac{4}{3} & \frac{5}{3} & \frac{7}{3} \\ -\frac{1}{6} & \frac{1}{3} & \frac{1}{6} \end{bmatrix};$

$\begin{bmatrix} -\frac{1}{4} & \frac{1}{2} & \frac{3}{4} \\ -\frac{4}{3} & \frac{5}{3} & \frac{7}{3} \\ -\frac{1}{6} & \frac{1}{3} & \frac{1}{6} \end{bmatrix}\begin{bmatrix} 6 & -2 & 1 \\ 2 & -1 & 5 \\ 2 & 0 & -3 \end{bmatrix} X = \begin{bmatrix} -\frac{1}{4} & \frac{1}{2} & \frac{3}{4} \\ -\frac{4}{3} & \frac{5}{3} & \frac{7}{3} \\ -\frac{1}{6} & \frac{1}{3} & \frac{1}{6} \end{bmatrix}\begin{bmatrix} 1 \\ -6 \\ -2 \end{bmatrix},$

$X = \begin{bmatrix} -\frac{1}{4}(16) + \frac{1}{2}(-2) + \frac{3}{4}(8) \\ -\frac{4}{3}(16) + \frac{5}{3}(-2) + \frac{7}{3}(8) \\ -\frac{1}{6}(16) + \frac{1}{3}(-2) + \frac{1}{6}(8) \end{bmatrix} = \begin{bmatrix} 1 \\ -6 \\ -2 \end{bmatrix};$

$(1, -6, -2)$

23. $3(2) - 4(7) = -22$

24. $3\begin{vmatrix} 1 & 6 \\ 9 & 5 \end{vmatrix} - 7\begin{vmatrix} 4 & 8 \\ 9 & 5 \end{vmatrix} + 2\begin{vmatrix} 4 & 8 \\ 1 & 6 \end{vmatrix} = 249$

25. $D = \begin{vmatrix} 1 & 1 \\ 5 & 2 \end{vmatrix} = 1(2) - 5(1) = -3; D_x = \begin{vmatrix} 4 & 1 \\ 11 & 2 \end{vmatrix} =$

$4(2) - 11(1) = -3; D_y = \begin{vmatrix} 1 & 4 \\ 5 & 11 \end{vmatrix} = 1(11) -$

$5(4) = -9; X = \frac{D_x}{D} = \frac{-3}{-3} = 1; y = \frac{D_y}{D} = \frac{-9}{-3} = 3;$

$(1, 3)$

26. $D = \begin{vmatrix} -5 & 16 \\ 3 & -9 \end{vmatrix} = (-5)(-9) - 3(16) = -3;$

$D_x = \begin{vmatrix} -7 & 16 \\ 4 & -9 \end{vmatrix} = (-7)(-9) - 4(16) = -1;$

$D_y = \begin{vmatrix} -5 & -7 \\ 3 & 4 \end{vmatrix} = (-5)(4) - 3(-7) = 1;$

$X = \frac{D_x}{D} = \frac{-1}{-3} = \frac{1}{3}; y = \frac{D_y}{D} = \frac{1}{-3} = -\frac{1}{3}; \left(\frac{1}{3}, -\frac{1}{3}\right)$

27.

$$\begin{array}{c} \quad\quad \text{A B C D E F G H} \\ \begin{array}{c} \text{A} \\ \text{B} \\ \text{C} \\ \text{D} \\ \text{E} \\ \text{F} \\ \text{G} \\ \text{H} \end{array} \begin{bmatrix} 0 & 1 & 1 & 1 & 0 & 0 & 0 & 0 \\ 1 & 0 & 1 & 1 & 0 & 0 & 0 & 0 \\ 1 & 1 & 0 & 1 & 0 & 0 & 0 & 0 \\ 1 & 1 & 1 & 0 & 1 & 1 & 1 & 1 \\ 0 & 0 & 0 & 1 & 0 & 1 & 1 & 1 \\ 0 & 0 & 0 & 1 & 1 & 0 & 1 & 0 \\ 0 & 0 & 0 & 1 & 1 & 1 & 0 & 1 \\ 0 & 0 & 0 & 1 & 1 & 0 & 1 & 0 \end{bmatrix} \end{array}$$; Adding the elements yields 30 nonstop routes.

Chapter Assessment, pages 222–223

CHAPTER TEST

1. Answers will vary. Answers should include the idea that there is no identity element for a non-square matrix.

2. $x + y = 7, x + 2 = 7, x = 5; (5, 2); B$

3.
$\quad y - x = 1 \quad\quad\quad x + 2y = -4$
$\quad -1 - (-2) = 1 \quad\quad -2 + 2(-1) = -4$
$\quad\quad\quad 1 = 1 \quad\quad\quad\quad\quad -4 = -4$

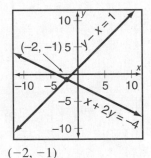

$(-2, -1)$

4.
$\quad 3x + 2y = 6 \quad\quad\quad\quad x = 4$
$\quad 3(4) + 2(-3) = 6 \quad\quad 4 = 4$
$\quad\quad\quad 6 = 6 \quad\quad\quad\quad\quad 4 = 4$

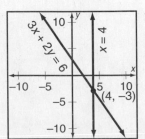

5. $2x + 2y = 8$ $3x + 3y = 12$
    $x + y = 4$ $3(4 - y) + 3y = 12$
    $x = 4 - y$ $12 = 12$
true equation; infinite solutions; consistent; dependent

6. $x + y = 8$ $2x + 2y = 3$
    $x = 8 - y$ $2(8 - y) + 2y = 3$
            $16 = 3$

false equation; no solution; inconsistent

7. $x + 3y = 10$ $x + 3y = 10$
    $\underline{4x - 3y = 5}$ $3 + 3y = 10$
    $5x \quad = 15$ $3y = 7$

    $x = 3$ $y = \dfrac{7}{3}$

    $\left(3, \dfrac{7}{3}\right)$

8. $\begin{cases} 2x + y = 6 \\ x - 3y = 10 \end{cases} \begin{matrix} \rightarrow \\ \rightarrow \end{matrix} \begin{matrix} 6x + 3y = 18 \\ \underline{x - 3y = 10} \\ 7x \quad = 28 \\ x = 4 \end{matrix}$

    $x - 3y = 10$
    $4 - 3y = 10$
    $-3y = 6$
    $y = -2$
    $(4, -2)$

9.

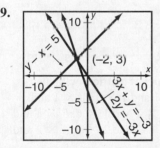

    $(-2, 3)$

10. $3x - 2y - 3z = -1 \rightarrow 3x - 2y - 3z = -1$
    $6x + y + 2z = 7 \rightarrow \underline{12x + 2y + 4z = 14}$
            $15x \quad + z = 13$

    $6x + y + 2z = 7 \rightarrow -18x - 3y - 6z = -21$
    $9x + 3y + 4z = 9 \rightarrow \underline{9x + 3y + 4z = 9}$
            $-9x \quad - 2z = -12$

    $15x + z = 13 \rightarrow 30x + 2z = 26$
    $-9x - 2z = -12 \rightarrow \underline{-9x - 2z = -12}$
            $21x \quad = 14$

            $x = \dfrac{14}{21} = \dfrac{2}{3}$

$-9\left(\dfrac{2}{3}\right) - 2z = -12 \qquad 3x - 2y - 3z = -1$

    $-2z = -6 \qquad 3\left(\dfrac{2}{3}\right) - 2y - 3(3) = -1$

    $z = 3 \qquad\qquad -2y = 6$
                $y = -3$

$\left(\dfrac{2}{3}, -33\right)$

11. $\begin{bmatrix} 1 & 2 & | & 4 \\ 3 & -1 & | & 5 \end{bmatrix} \rightarrow \begin{bmatrix} 1 & 2 & | & 4 \\ 0 & -7 & | & -7 \end{bmatrix} \rightarrow \begin{bmatrix} 1 & 0 & | & 2 \\ 0 & 1 & | & 1 \end{bmatrix}; (2, 1)$

12. $AB = \begin{bmatrix} 1(2) & 1(3) & 1(6) \\ 4(2) & 4(3) & 4(6) \\ 5(5) & 5(3) & 5(6) \end{bmatrix} = \begin{bmatrix} 2 & 3 & 6 \\ 8 & 12 & 24 \\ 10 & 15 & 30 \end{bmatrix}$

13. $BA = [2(1) + 3(4) + 6(5)] = [44]$

14. BC is not defined.

15. $2A = \begin{bmatrix} 2 \\ 8 \\ 10 \end{bmatrix}; 2A \cdot B = \begin{bmatrix} 2(2) & 2(3) & 2(6) \\ 8(2) & 8(3) & 8(6) \\ 10(2) & 10(3) & 10(6) \end{bmatrix} =$

$\begin{bmatrix} 4 & 6 & 12 \\ 16 & 24 & 48 \\ 20 & 30 & 60 \end{bmatrix}$

16. $\begin{bmatrix} 2 & 8 & | & 1 & 0 \\ 1 & 4 & | & 0 & 1 \end{bmatrix} \rightarrow \begin{bmatrix} 1 & 4 & | & 1 & 0 \\ 0 & 0 & | & -1 & 1 \end{bmatrix}$; no inverse

17. $\begin{bmatrix} 3 & 2 & 1 & | & 1 & 0 & 0 \\ 1 & 1 & -1 & | & 0 & 1 & 0 \\ 4 & 3 & 1 & | & 0 & 0 & 1 \end{bmatrix} \rightarrow \begin{bmatrix} 1 & 1 & -1 & | & 0 & 1 & 0 \\ 0 & -1 & 4 & | & 1 & -3 & 0 \\ 0 & -1 & 5 & | & 0 & -4 & 1 \end{bmatrix} \rightarrow$

$\begin{bmatrix} 1 & 0 & 3 & | & 1 & -2 & 0 \\ 0 & 1 & -4 & | & -1 & 3 & 0 \\ 0 & 0 & 1 & | & -1 & -1 & 1 \end{bmatrix} \rightarrow$

$\begin{bmatrix} 1 & 0 & 0 & | & 4 & 1 & -3 \\ 0 & 1 & 0 & | & -5 & -1 & 4 \\ 0 & 0 & 1 & | & -1 & -1 & 1 \end{bmatrix}; \begin{bmatrix} 4 & 1 & -3 \\ -5 & -1 & 4 \\ -1 & -1 & 1 \end{bmatrix}$

18. $\begin{bmatrix} 1 & -2 & 3 \\ 2 & 1 & 5 \\ 3 & -1 & -3 \end{bmatrix} X = \begin{bmatrix} 3 \\ 8 \\ -22 \end{bmatrix};$

$A^{-1} = \begin{bmatrix} -\dfrac{2}{55} & \dfrac{9}{55} & \dfrac{13}{55} \\ -\dfrac{21}{55} & \dfrac{12}{55} & -\dfrac{1}{55} \\ \dfrac{1}{11} & \dfrac{1}{11} & -\dfrac{1}{11} \end{bmatrix}; X = \begin{bmatrix} -\dfrac{2}{55} & \dfrac{9}{55} & \dfrac{13}{55} \\ -\dfrac{21}{55} & \dfrac{12}{55} & -\dfrac{1}{55} \\ \dfrac{1}{11} & \dfrac{1}{11} & -\dfrac{1}{11} \end{bmatrix} \begin{bmatrix} 3 \\ 8 \\ -22 \end{bmatrix},$

$X = \begin{bmatrix} -\dfrac{2}{55}(3) + \dfrac{9}{55}(8) + \dfrac{13}{55}(-22) \\ -\dfrac{21}{55}(3) + \dfrac{12}{55}(8) + \left(-\dfrac{1}{55}\right)(-22) \\ \dfrac{1}{11}(3) + \dfrac{1}{11}(8) + \left(-\dfrac{1}{11}\right)(-22) \end{bmatrix}, X = \begin{bmatrix} -4 \\ 1 \\ 3 \end{bmatrix};$

$(-4, 1, 3)$

19. $1\begin{vmatrix} 4 & -1 \\ 6 & 3 \end{vmatrix} - 1\begin{vmatrix} -2 & 5 \\ 6 & 3 \end{vmatrix} + 2\begin{vmatrix} -2 & 5 \\ 4 & -1 \end{vmatrix} = 18$

20. $0\begin{vmatrix} 3 & -2 \\ -1 & -3 \end{vmatrix} - 2\begin{vmatrix} 4 & 1 \\ -1 & -3 \end{vmatrix} + 4\begin{vmatrix} 4 & 1 \\ 3 & -2 \end{vmatrix} = -22$

21. $\begin{cases} x - 2y = 1 \\ x + 2y = 4 \end{cases}; D = \begin{vmatrix} 1 & -2 \\ 1 & 2 \end{vmatrix} = 1(2) - 1(-2) = 4;$

$D_x = \begin{vmatrix} 1 & -2 \\ 4 & 2 \end{vmatrix} = 1(2) - 4(-2) = 10;$

$D_y = \begin{vmatrix} 1 & 1 \\ 1 & 4 \end{vmatrix} = 1(4) - 1(1) = 3;$

$x = \dfrac{D_x}{D} = \dfrac{10}{4} = \dfrac{5}{2}; y = \dfrac{D_y}{D} = \dfrac{3}{4}; \left(\dfrac{5}{2}, \dfrac{3}{4}\right)$

22. $D = \begin{vmatrix} 1 & 1 & 1 \\ 2 & -1 & 2 \\ 5 & 2 & 2 \end{vmatrix} = 1\begin{vmatrix} -1 & 2 \\ 2 & 2 \end{vmatrix} - 2\begin{vmatrix} 1 & 1 \\ 2 & 2 \end{vmatrix}$

$+ 5\begin{vmatrix} 1 & 1 \\ -1 & 2 \end{vmatrix} = 9;$

$D_x = \begin{vmatrix} 3 & 1 & 1 \\ 6 & -1 & 2 \\ 0 & 2 & 2 \end{vmatrix} = 3\begin{vmatrix} -1 & 2 \\ 2 & 2 \end{vmatrix} - 6\begin{vmatrix} 1 & 1 \\ 2 & 2 \end{vmatrix}$

$+ 0\begin{vmatrix} 1 & 1 \\ -2 & 2 \end{vmatrix} = -18;$

$D_y = \begin{vmatrix} 1 & 3 & 1 \\ 2 & 6 & 2 \\ 5 & 0 & 2 \end{vmatrix} = 1\begin{vmatrix} 6 & 2 \\ 0 & 2 \end{vmatrix} - 2\begin{vmatrix} 3 & 1 \\ 0 & 2 \end{vmatrix}$

$+ 5\begin{vmatrix} 3 & 1 \\ 6 & 2 \end{vmatrix} = 0;$

$D_z = \begin{vmatrix} 1 & 1 & 3 \\ 2 & -1 & 6 \\ 5 & 2 & 0 \end{vmatrix} = 1\begin{vmatrix} -1 & 6 \\ 2 & 0 \end{vmatrix} - 2\begin{vmatrix} 1 & 3 \\ 2 & 0 \end{vmatrix}$

$+ 5\begin{vmatrix} 1 & 3 \\ -1 & 6 \end{vmatrix} = 45;$

$x = \dfrac{D_x}{D} - \dfrac{-18}{9} = -2; y = \dfrac{D_y}{D} = \dfrac{0}{9} = 0;$

$z = \dfrac{D_z}{D} = \dfrac{45}{9} = 5; (-2, 0, 5)$

23. $\begin{cases} 0.02A + 0.02B + 0.03C = 65 \\ 0.03A + 0.03B + 0.02C = 85 \\ 0.03A + 0.02B + 0.01C = 65 \end{cases}$
$0.02A + 0.02B + 0.03C = 65 \rightarrow$
$\qquad\qquad 0.06A + 0.06B + 0.09C = 195$
$0.03A + 0.03B + 0.02C = 85 \rightarrow$
$\qquad\underline{-0.06A - 0.06B - 0.04C = -170}$
$\qquad\qquad\qquad\qquad 0.05C = 25$
$\qquad\qquad\qquad\qquad\quad C = 500$

$0.03A + 0.03B + 0.02C = 85 \rightarrow$
$\qquad\qquad 0.03A + 0.03B + 0.02C = 85$
$0.03A + 0.02B + 0.01C = 65 \rightarrow$
$\qquad\underline{-0.03A - 0.02B - 0.01C = -65}$
$\qquad\qquad\qquad 0.01B + 0.01C = 20$

$0.01B + 0.01C = 20$
$0.01B + 0.01(500) = 20$
$0.01B = 15$
$B = 1500$

$0.02A + 0.02B + 0.03C = 65$
$0.02A + 0.02(1500) + 0.03(500) = 65$
$0.02A = 20$
$A = 1000$

Ms. Li bought 1000 shares of A, 1500 shares of B, and 500 shares of C.

24.
$$\begin{array}{c} \\ A \\ B \\ C \\ D \\ E \\ F \\ G \end{array} \begin{array}{c} A\;B\;C\;D\;E\;F\;G \\ \begin{bmatrix} 0 & 1 & 1 & 0 & 0 & 0 & 0 \\ 0 & 0 & 1 & 1 & 1 & 0 & 0 \\ 0 & 0 & 0 & 0 & 0 & 1 & 1 \\ 0 & 1 & 0 & 0 & 1 & 0 & 0 \\ 0 & 1 & 0 & 1 & 0 & 0 & 0 \\ 0 & 0 & 0 & 0 & 0 & 0 & 0 \\ 0 & 0 & 0 & 0 & 0 & 0 & 0 \end{bmatrix} \end{array}$$
; Adding the elements yields 11 direct calls that can be made.

PERFORMANCE ASSESSMENT

BUILD POLYGONS
Answers will vary.

CREATE MATRICES
Students should construct pairs of matrices with the following dimensions: 7×1 and 1×1, 1×1 and 1×7, 6×1 and 1×2, 2×1 and 1×6, 5×1 and 1×3, 3×1 and 1×5, 2×2 and 2×2, 4×1 and 1×4, 1×4 and 4×1.

TIC-TAC-TOE
Answers will vary.

DESIGN A NETWORK
Answers will vary.

PROJECT CONNECTION
1–2. Answers will vary.

3. Answers will vary with data collected from activities 1 and 2. Students must solve the system

$$\begin{cases} Ax + By = 720 \\ Cx + Dy = 720 \end{cases}$$

1. The graph of the system is 2 parallel lines. Therefore, there is no solution and the system is inconsistent.

2. $\begin{cases} y = 2x + 2 \\ 4x - y = 6 \end{cases} \rightarrow \begin{cases} -2x + y = 2 \\ 4x - y = 6 \end{cases} \rightarrow$

$$\begin{array}{r} -2x + y = 2 \\ 4x - y = 6 \\ \hline 2x = 8 \\ x = 4 \end{array}$$

$$4x - y = 6 \qquad \frac{y}{2} = x + 1 \qquad 4x - y = 6$$

$$4(4) - y = 6 \qquad \frac{10}{2} \overset{?}{=} 4 + 1 \qquad 4(4) - 10 \overset{?}{=} 6$$

$$-y = -10 \qquad 5 = 5 \qquad 6 = 6$$

$$y = 10$$

The solution is $(4, 10)$.

3. $\begin{array}{l} 3x - 4y = 12 \\ 4x - y = -10 \end{array} \rightarrow \begin{array}{r} 3x - 4y = 12 \\ -16x + 4y = 40 \\ \hline -13x = 52 \\ x = -4 \end{array}$

$$3x - 4y = 12 \qquad\qquad 3x - 4y = 12$$

$$3(-4) - 4y = 12 \qquad 3(-4) - 4(-6) \overset{?}{=} 12$$

$$-4y = 24 \qquad\qquad 12 = 12$$

$$y = -6$$

$$4x - y = -10$$

$$4(-4) - (-6) = -10$$

$$10 = 10$$

The solution is $(-4, -6)$.

4. Let h represent the hundreds digit. Let t represent the tens digit. Let u represent the units digit.

$$\begin{cases} h + t + u = 14 \\ u = h + t \\ 100u + 10t + h = 100h + 10t + u + 297 \end{cases}$$

$$\begin{cases} h + t + u = 14 \\ -h - t + u = 0 \\ -99h + 99u = 297 \end{cases}$$

$$\begin{array}{r} h + t + u = 14 \\ -h - t + u = 0 \\ \hline 2u = 14 \\ u = 7 \end{array}$$

$$\begin{array}{ll} -99h + 99u = 297 & h + t + u = 14 \\ -99h + 99(7) = 297 & 4 + t + 7 = 14 \\ -99h = -396 & t = 3 \\ h = 4 & \end{array}$$

The original number is 437.

5. C

6. $AB = \begin{bmatrix} 4(5) + 2(0) & 4(6) + 2(-7) \\ 3(5) + 1(0) & 3(6) + 1(-7) \end{bmatrix} = \begin{bmatrix} 20 & 10 \\ 15 & 10 \end{bmatrix}$

7. $\begin{bmatrix} 4 & 2 & | & 1 & 0 \\ 3 & 1 & | & 0 & 1 \end{bmatrix} \rightarrow \begin{bmatrix} 1 & \frac{1}{2} & | & \frac{1}{4} & 0 \\ 0 & -\frac{1}{2} & | & -\frac{3}{4} & 1 \end{bmatrix} \rightarrow \begin{bmatrix} 1 & 0 & | & -\frac{1}{2} & 1 \\ 0 & 1 & | & \frac{3}{2} & -2 \end{bmatrix};$

$$A^{-1} = \begin{bmatrix} -0.5 & 1 \\ 1.5 & -2 \end{bmatrix}$$

8. $\begin{bmatrix} 1 & -3 & 0 \\ 3 & -2 & 1 \\ 2 & 1 & 2 \end{bmatrix} X = \begin{bmatrix} -2 \\ 5 \\ 4 \end{bmatrix}; A^{-1} = \begin{bmatrix} -\frac{5}{7} & \frac{6}{7} & -\frac{3}{7} \\ -\frac{4}{7} & \frac{2}{7} & -\frac{1}{7} \\ 1 & -1 & 1 \end{bmatrix};$

$$X = \begin{bmatrix} -\frac{5}{7} & \frac{6}{7} & -\frac{3}{7} \\ -\frac{4}{7} & \frac{2}{7} & -\frac{1}{7} \\ 1 & -1 & 1 \end{bmatrix} \begin{bmatrix} -2 \\ 5 \\ 4 \end{bmatrix}, X =$$

$$\begin{bmatrix} -\frac{5}{7}(-2) + \frac{6}{7}(5) + \left(-\frac{3}{7}\right)(4) \\ -\frac{4}{7}(-2) + \frac{2}{7}(5) + \left(-\frac{1}{7}\right)(4) \\ 1(-2) + (-1)(5) + 1(4) \end{bmatrix}, X = \begin{bmatrix} 4 \\ 2 \\ -3 \end{bmatrix};$$

$$(4, 2, -3)$$

9. $7(-2) - (-3)(-4) = -26$

10. $1\begin{vmatrix} -1 & 6 \\ 4 & 7 \end{vmatrix} - 3\begin{vmatrix} 2 & 8 \\ 4 & 7 \end{vmatrix} + 5\begin{vmatrix} 2 & 8 \\ -1 & 6 \end{vmatrix} = 123$

11. $2y = 3x - 1$ $\qquad\qquad y - y_1 = m(x - x_1)$

$$y = \frac{2}{3}x - \frac{1}{2} \qquad\qquad y - 5 = \frac{2}{3}(x - (-4))$$

$$y - 5 = \frac{2}{3}x + \frac{8}{3}$$

$$3y - 15 = 2x + 8$$

$$3y - 2x = 22$$

12.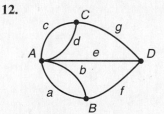

There are 9 ways to get from B to C crossing any bridge once.

$$\begin{array}{lll} a\,c & b\,c & f\,g \\ a\,d & b\,d & f\,e\,c \\ a\,e\,g & b\,e\,g & f\,e\,d \end{array}$$

13. The correct answers is D.

14. $1.1\overline{7} = 1\frac{17}{99} = \frac{116}{99}$

15. The system ($\{$integers$\}$, $+$, $\times$) is not a field since there are no multiplicative inverses.

16. $\frac{4}{6} = \frac{2}{3}$

17. $186{,}000 = 1.86 \times 10^5$;
$1.86 \times 60 \times 60 \times 24 \times 365 = 58{,}656{,}960$;
$58{,}656{,}960 \times 10^5 \approx 5.866 \times 10^{12}$

18. $f(x) = 9x + 11$
$y = 9x + 11$
$x = 9y + 11$
$-9y = 11 - x$
$y = \dfrac{x}{9} - \dfrac{11}{9}$
$f^{-1}(x) = \dfrac{x}{9} - \dfrac{11}{9}$

19. $g(x) = -x$
$y = -x$
$x = -y$
$y = -x$
$g^{-1}(x) = -x$

Standardized Test, page 225

1. C;
$x = y$
$x = 2x - 1$
$-x = -1$
$x = 1$

2. A;

$3n + 4p = 9 \qquad \rightarrow \quad 3n + 4p = 9$
$3m - p = 0 \qquad \qquad \underline{12m - 4p = 0}$
$\qquad\qquad\qquad\qquad\qquad 12m + 3n = 9$

$12m + 3n = 9 \quad \rightarrow \quad 12m + 3n = 9$
$2m + n = -1 \quad \rightarrow \quad \underline{-12m - 6n = 6}$
$\qquad\qquad\qquad\qquad\qquad\quad -3n = 15$
$\qquad\qquad\qquad\qquad\qquad\qquad n = -5$

$3m + n = -1 \qquad\qquad 3m - p = 0$
$3m + (-5) = -1 \qquad\quad 3\left(\dfrac{4}{3}\right) - p = 0$
$\qquad 3m = 4 \qquad\qquad\qquad\quad -p = -4$
$\qquad\quad m = \dfrac{4}{3} \qquad\qquad\qquad\quad p = 4$

$p \div m = 4 \div \dfrac{4}{3} = 4 \cdot \dfrac{3}{4} = 3$;

$p + n = 4 + -5 = -1$

3. D; $AB = [1(1) + 2(2) + 3(3) \quad 1(4) + 2(5) + 3(6)]$
$= [14 \quad 32]; A + B$ is not defined.

4. B; $I_{2 \times 2} = \begin{bmatrix} 1 & 0 \\ 0 & 1 \end{bmatrix}$; element row 2, column 1 $= 0$;

$I_{3 \times 3} = \begin{bmatrix} 1 & 0 & 0 \\ 0 & 1 & 0 \\ 0 & 0 & 1 \end{bmatrix}$; element row 2, column 2 $= 1$

5. B; $A^{-1} = \begin{bmatrix} \frac{1}{3} & \frac{1}{12} \\ -\frac{1}{3} & \frac{1}{6} \end{bmatrix}$; element row 2, column 1 $= -\dfrac{1}{3}$;

element row 2, column 2 $= \dfrac{1}{6}$

6. C; $\begin{vmatrix} -4 & 0.5 \\ 16 & 2 \end{vmatrix} = (-4)(2) - (16)(0.5) = -16$;

$\begin{vmatrix} 2 & -1 & 3 \\ 1 & 2 & 3 \\ 3 & -2 & 1 \end{vmatrix} = 2\begin{vmatrix} 2 & 3 \\ -2 & 1 \end{vmatrix} - 1\begin{vmatrix} -1 & 3 \\ -2 & 1 \end{vmatrix} +$

$3\begin{vmatrix} -1 & 3 \\ 2 & 3 \end{vmatrix} = -16$

7. A

8. B; parallel to z, $m = -1$, perpendicular to Z, $m = 1$

9. A; $f(x) = 3|2x - 3| - 4$; $v\left(\dfrac{3}{2}, -4\right)$

10. B; $(f \circ g)(3)$: $g(3) = 3 - 1 = 2$; $f[g(3)] = 2^2 = 4$;
$(g \circ f)(3)$: $f(3) = 3^2 = 9$; $g[f(3)] = 9 - 1 = 8$;

11. B; $f(x) = 7x^3$; $f(-3) = 7(-3)^3 = -189$;
$7x^3 = 189$, $x = 3$

12. A; $\sqrt{5} + \sqrt{11} \approx 5.6$; $\sqrt{16} = 4$

13. D

14. B; $P(7) = 0$; $P(\text{even}) = 1$

Chapter 5 Polynomials and Factoring

Data Activity, pages 226–227

1. $3883 + 807 + 2655 + 378 + 161 = 7884$

2. $\frac{807}{7884} = .1023$ or 10.2%

3. $2655(.797) \approx 2116$

4. Food since $3883(.293) \approx 1138$

5. Students should include survey questions that answer questions asked in the problem.

Lesson 5.1, pages 229–233

THINK BACK

1. $x^3 - x^2 + x - 3$

2. Remove 1 x^3 zero pair and 1 unit zero pair

EXPLORE

3a. **3b.** **3c.**

3d. $3x^3 - x^2 - x - 3$

4. $-x^3 - x^2 - 5x + 8$ **5.** $-2x^3 + 2x^2 - x + 4$

6a. **6b.**

6c. subtraction.

6d. $x^3 + x^2 + 4x - 2$.

7. $2x^3 - x + 5 - x^3 + 2x - 1 = x^3 + x + 4$

8. $-x^3 + 3x^2 - 2x + x^3 + 3x^2 - 2x - 1 = 6x^2 - 4x - 1$

9a. $2x(x - 1)$

9b.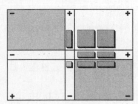

9c. $2x(x - 1) = 2x^2 - 2x$

10a. 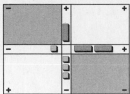 **10b.**

10c. the product of $(2x - 1)$ and $(x - 3)$

10d. $2x^2 - 7x + 3$

11. 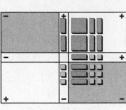 **12.**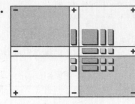

$2x^2 + x - 6$ 　　　　 $x^2 - 4$

13.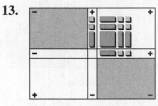

$x^2 + 4x + 4$

14. No; when the factors have the same first term and opposite last term.

15a.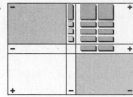

$2x^2$ 　　　　 $6x$

15b. $2x$

15c. 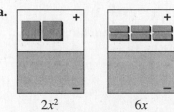 **15d.**

15e. $2x^2 + 6x = 2x(x + 3)$

16. $2x^2 + x = x(2x + 1)$

17. $-4x^2 + 2x = 2x(-2x + 1)$

18a. 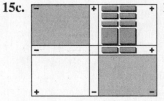 **18b.**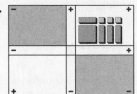

18c. $x^2 + 4x + 3 = (x + 3)(x + 1)$

19a.

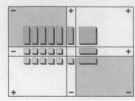

$x^2 - 6x + 5$

19b. $x^2 - 6x + 5 = (x - 1)(x - 5)$

19c. Variations in Algeblocks arrangements in Quadrants will only affect the order of the terms in their factored form.

20a.

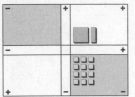

$x^2 + x - 12$

The blocks cannot be arranged into a square.

20b. $3x$ and $-3x$

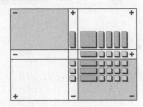

20c. $x^2 + x - 12 = (x - 3)(x + 4)$

21. Check student mats.
$x^2 + 2x - 8 = (x + 4)(x - 2)$

22. Check student mats
$2x^2 - 5x - 3 = (2x + 1)(x - 3)$

23. Answers will vary. Possible answer:
$(x^2 + 2x + 4) - (4x + 3) = x^2 - 2x + 1$
$x^2 + 2x + 4 - 4x - 3 = x^2 - 2x + 1$

24. Answers will vary. Possible answer:
$(x - 2)(x + 1) = x^2 - x - 2$
$\dfrac{x^2 - x - 2}{x - 2} = x + 1$ or $\dfrac{x^2 - x - 2}{x + 1} = x - 2$

25. Answers will vary. Possible answer:
$(x - 3)(x + 3) = x^2 - 9$
$\dfrac{x^2 - 9}{x - 3} = x + 3$ or $\dfrac{x^2 - 9}{x + 3} = x - 3$

26. Answers will vary. Possible answer:
$(x + 2)(x + 2) = x^2 + 4x + 4$
$\dfrac{x^2 + 4x + 4}{x + 2} = x + 2$

27. They have the same first terms and opposite second terms.

28. It is the difference of the first term squared and the second term squared.

29. Both have the same first terms, the same last terms, and the same signs.

30. Trinomial whose first and last terms are squared and whose middle term is the product of two and the terms of the binomial factor.

31. A trinomial whose first and last terms are squared and whose middle term the product of two and the terms of the binomial.

32. $x^2 - 25 = (x + 5)(x - 5)$;   C

33. $x^2 + 10x + 25 = (x + 5)(x + 5)$;   A

34. $x^2 - 10x + 25 = (x - 5)(x - 5)$;   B

SUMMARIZE

35. Answers will vary. Possible answer: Model $4x^2$ and $-12x$ in appropriate quadrants. Make rectangles so that they have the greatest common dimensions.
$\dfrac{4x^2 - 12x}{4x} = x - 3$
So $4x^2 - 12x = 4x(x - 3)$

36.

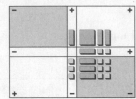

$x^2 - x - 6 = (x + 2)(x - 3)$

37. $2x^2 + 3x - 2 = (2x - 1)(x + 2)$

38. perimeter $= 6x + 4 + 3 + 2 + 3 + (4 - x)$
$\qquad\qquad\quad + (x - 2)$
$\qquad\qquad = 6x + 14$
Area $= 2x^2 + 4x + 6$

Lesson 5.2, pages 234–241

EXPLORE

1a. $a^6 \cdot a^4 = (a \cdot a \cdot a \cdot a \cdot a \cdot a) \cdot (a \cdot a \cdot a \cdot a)$

1b. a^{10} **1c.** $b^3 \cdot b^2 = (b \cdot b \cdot b) \cdot (b \cdot b)$ **1d.** b^5

1e. To Multiply powers with like bases, add exponents.
$a^n \cdot a^m = a^{n+m}$

2a. $(a^6)^4$ means a^6 is being multiplied by itself four times.

2b. $(a^6)^4 = a^6 \cdot a^6 \cdot a^6 \cdot a^6$ **2c.** a^{24}

2d. To raise a power to a power, multiply the exponents.

3a. $(ab)^5 = ab \cdot ab \cdot ab \cdot ab \cdot ab$

3b. $a^5 b^5$; commutative and associative.

3c. To raise several bases to an exponent, raise each base to that exponent. $(ab)^n = a^n b^n$

4a. $\left(\dfrac{a}{b}\right)^4 = \dfrac{a}{b} \cdot \dfrac{a}{b} \cdot \dfrac{a}{b} \cdot \dfrac{a}{b}$ **4b.** $\dfrac{a \cdot a \cdot a \cdot a}{b \cdot b \cdot b \cdot b} = \dfrac{a^4}{b^4}$

4c. To raise a fraction to an exponents, raise both the numerator and denominator to the exponent.
$\left(\dfrac{a}{b}\right)^n = \dfrac{a^n}{b^n}$

5a. $\dfrac{a^6}{a^4} = \dfrac{a \cdot a \cdot a \cdot a \cdot a \cdot a}{a \cdot a \cdot a \cdot a}$

5b. $\dfrac{d \cdot d \cdot d \cdot d \cdot a \cdot a}{d \cdot d \cdot d \cdot d} = a^2$

5c. To divide terms with like bases, subtract exponents.

$$\dfrac{a^n}{a^m} = a^{n-m}$$

6a. $\dfrac{a^4}{a^4} = \dfrac{d \cdot d \cdot d \cdot d}{d \cdot d \cdot d \cdot d} = 1$ **6b.** $\dfrac{a^4}{a^4} = a^{4-4} = a^0$

6c. $a^0 = 1$ **7a.** $\dfrac{a^4}{a^6} = \dfrac{d \cdot d \cdot d \cdot d}{d \cdot d \cdot d \cdot d \cdot a \cdot a} = \dfrac{1}{a^2}$

7b. $\dfrac{a^4}{a^6} = a^{4-6} = a^{-2}$ **7c.** $\dfrac{1}{a^2} = a^{-2}$

TRY THESE

1. $3x^2 \cdot 9x^6 = 3(9)x^{2+6} = 27x^8$

2. $(3x^6)^2 = (3)^2(x^6)^2 = 9x^{12}$

3. $9x^6 \div 3x^2 = \dfrac{9}{3}(x^{6-2}) = 3x^4$ **4.** $\left(\dfrac{1}{6}\right)^2 = \dfrac{1^2}{6^2} = \dfrac{1}{36}$

5. $y^{-4} = \dfrac{1}{y^4}$ **6.** $6x^{-2} = \left(\dfrac{6}{1}\right)\left(\dfrac{1}{x^2}\right) = \dfrac{6}{x^2}$

7. $\dfrac{3}{x^{-3}} = \left(\dfrac{3}{1}\right)\left(\dfrac{1}{\frac{1}{x^3}}\right) = \dfrac{3}{1} \cdot \dfrac{x^3}{1} = 3x^3$

8. $\dfrac{m^{-2}}{n^{-3}} = \dfrac{\frac{1}{m^2}}{\frac{1}{n^3}} = \dfrac{1}{m^2} \cdot \dfrac{n^3}{1} = \dfrac{n^3}{m^2}$

9. $2x^0 = 2 \cdot 1 = 2$ **10.** $(2x)^0 = 2^0 x^0 = 1 \cdot 1 = 1$

11. $2^0 x = 1x = x$ **12.** $(2 + x)^0 = 1$

13.

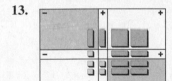

$(2x - 1)(x - 2) = 2x^2 - 5x + 2$

14. The length of the unmowed lawn is $(x - 10)$ ft.
$$40x - 30(x - 10) = 900$$
$$40x - 30x - 300 = 900$$
$$40x + (-30x + 300) = 900$$
$$10x + 300 = 900$$
$$10x = 600$$
$$x = 60 \text{ ft}$$

15. $(3x^2 + 5x - 1) + (-2x + 3) =$
$(3x^2) + (5x - 2x) + (-1 + 3) =$
$3x^2 + (5 - 2)x + (-1 + 3) = 3x^2 + 3x + 2$

16. $(-2x^2 + 4x^3 - 3) + (x^3 - 2) =$
$(4x^3 + x^3) + (-2x^2) + (-3 - 2) =$
$(4 + 1)x^3 + (-2x^2) + (-3 - 2) = 5x^3 - 2x^2 - 5$

17. $(7y^2 - 4y + 6) - (3y^2 - 2y - 7) =$
$(7y^2 - 4y + 6) + (-3y^2 + 2y + 7) =$
$(7 - 3)y^2 + (-4 + 2)y + (6 + 7) =$
$4y^2 - 2y + 13$

18. $3z^2(2z^2 - 4z + 5) =$
$3z^2(2z^2) + 3z^2(-4z) + 3z^2(5) =$
$6z^4 - 12z^3 + 15z^2$

19. $(3t + 2)(2t - 3) = 3t(2t - 3) + 2(2t - 3) =$
$6t^2 - 9t + 4t - 6 = 6t^2 - 5t - 6$

20. $(2w^2 - 3w + 4)(w - 3) =$
$2w^2(w - 3) - 3w(w - 3) + 4(w - 3) =$
$2w^3 - 6w^2 - 3w^2 + 9w + 4w - 12 =$
$2w^3 - 9w^2 + 13w - 12$

21. Subtraction is adding the opposite.

22. $\dfrac{(x - 2)(2x + 6)}{2} = \dfrac{x(2x + 6) - 2(2x + 6)}{2}$

$= \dfrac{2x^2 + 6x - 4x - 12}{2} = \dfrac{2x^2 + 2x - 12}{2}$

$= x^2 + x - 6$

23. $\dfrac{1}{2}(2x)(x + 5 + x + 8)$

$= x(2x + 13)$
$= 2x^2 + 13x$

PRACTICE

1. $-3^2(3)^2 = -9(9) = -81$

2. $(a + b)^2(a + b)^3 = (a + b)^5$

3. $(4m^2n^3)^2 = 16m^4n^6$ **4.** $\dfrac{16x^8}{4x^4} = 4x^4$

5. $2^{-3} = \dfrac{1}{2^3}$ **6.** $\dfrac{1}{2}y^{-2} = \dfrac{1}{2y^2}$ **7.** $\dfrac{-4}{x^{-4}} = -4x^4$

8. $\dfrac{b^{-3}}{\frac{1}{a^{-2}}} = \dfrac{1}{b^3} \div a^2 = \dfrac{1}{b^3} \cdot \dfrac{1}{a^2} = \dfrac{1}{a^2 b^3}$

9. $-3x^0 = -3(1) = -3$ **10.** $(-3x)^0 = 1$

11. $(-3)^0 x = 1 \cdot x = x$ **12.** $(x - 3)^0 = 1$

13. $2s(s - 4) = 2s^2 - 8s$

14.

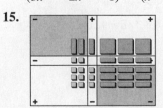

$(3x^3 - 2x^2 + 1) - (x^3 - 2x^2 - 2) = 2x^3 + 3$

15.

$(3x - 2)(x - 3) = 3x^2 - 11x + 6$

16. $(4y^2 - 2y + 5) + (-2y^2 + y - 3)$
$= 4y^2 - 2y^2 - 2y + y + 5 - 3 = 2y^2 - y + 2$

17. $(-2 + z^2) + (4z^3 - 3z^2 + 6)$
$= 4z^3 + z^2 - 3z^2 + 6 - 2 = 4z^3 - 2z^2 + 4$

18. $(2k^2 + 3k + 1) - (3k^2 - 2k - 5)$
$= 2k^2 - 3k^2 + 3k + 2k + 1 + 5$
$= -k^2 + 5k + 6$

19. $(-8m^2 - 3m + 3) - (-2m^2 - 5)$
$= -8m^2 + 2m^2 - 3m + 3 + 5$
$= -6m^2 - 3m + 8$

20. $-4h^2(h^3 - 2h^2 + h - 3)$
$= -4h^5 + 8h^4 - 4h^3 + 12h^2$

21. $3ab(2a^2 + 4ab - b^2)$
$= 6a^3b + 12a^2b^2 - 3ab^3$

22. $(3y - 4)(4y + 3)$
$= 3y(4y + 3) - 4(4y + 3)$
$= 12y^2 + 9y - 16y - 12$
$= 12y^2 - 7y - 12$

23. $(4m + 3)(2m - 1)$
$= 4m(2m - 1) + 3(2m - 1)$
$= 8m^2 - 4m + 6m - 3$
$= 8m^2 + 2m - 3$

24. $(2 - h)(3 + 2h)$
$= 2(3 + 2h) - h(3 + 2h)$
$= 6 + 4h - 3h - 2h^2$
$= 6 + h - 2h^2$

25. $(1 + r)(2 - 4r)$
$= 1(2 - 4r) + r(2 - 4r)$
$= 2 - 4r + 2r - 4r^2$
$= 2 - 2r - 4r^2$

26. $(5k + 2)(5k - 2)$
$= 25k^2 - 4$

27. $(3z - 4)(3z + 4)$
$= 9z^2 - 16$

28. $(a + b)(a + b)$
$= a^2 + 2ab + b^2$

29. $(2m + 3)(2m + 3)$
$= 4m^2 + 12m + 9$

30. $(x^2 + 3x - 4)(2x - 3)$
$= 2x^3 - 3x^2 + 6x^2 - 9x - 8x + 12$
$= 2x^3 + 3x^2 - 17x + 12$

31. $(r^2 + rt - t^2)(r - 2t)$
$= r^3 - 2r^2t + r^2t - 2rt^2 - rt^2 + 2t^3$
$= r^3 - r^2t - 3rt^2 + 2t^3$

32. No. Possible examples:
$(x + a)(x - a) = x^2 - a^2$;
$(x + y)(a + b) = ax - bx + ay + by$

EXTEND

33. $(4x + 5)(3x - 2) - x(x + 3)$
$= 12x^2 - 8x + 15x - 10 - x^2 - 3x$
$= 11x^2 + 4x - 10$

34. outer $= [4(5) + 5][3(5) - 2] = 325$
inner $= 5(5 + 3) = 40$
shaded $= 325 - 40 = 285$

35.

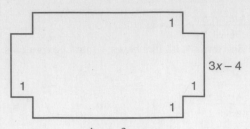

$4x + 3$

Volume $= (4x + 3)(3x - 4)(1)$
$\qquad = 12x^2 - 16x + 9x - 12$
$\qquad = 12x^2 - 7x - 12$

36. Volume $= 12(5)^2 - 7(5) - 12 = 253$ cubic units

37. $x^7; y^5$ **38.** $x^{-1}; y^8$ **39.** $x^2; y^{-13}$

40. Yes; any order of FOIL can be used as long as each term in one binomial is multiplied by each term of the other binomial.

THINK CRITICALLY

41. $(3 + 2)^3 \neq 3^3 + 2^3$
$(5)^3 \neq 27 + 8$
$125 \neq 35$
so $(a + b)^3 \neq a^3 + b^3$

42. area of *MATH* in shaded overlap is larger.
area in *HELP* not in overlap $= (5 - t)t$
area in *MATH* not in overlap $= (5 - t)5$
Since you can see $5 > t$, then $(5 - t)5 > (5 - t)t$.

MIXED REVIEW

43. 3

44. $2y = 8x + 9$
$y = 4x + \dfrac{9}{2}$
slope $= 4$

45. $y + 2x = 7$
$y = -2x + 7$
slope $= -2$

46. $f^{-1}(x) = -\dfrac{3}{2}x$

47. $y = \dfrac{x + 1}{2}$
$2y = x + 1$
$2y - 1 = x$
$f^{-1}(x) = 2x - 1$

48. $y = 3x + 2$
$y - 2 = 3x$
$\dfrac{y - 2}{3} = x$
$y = \dfrac{x - 2}{3}$

49.

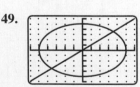

2 pts. of intersection

50.

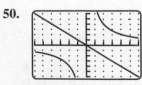

0 pts. of intersection

51.

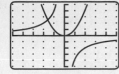

1 pt. of intersection

52. B; $-3^{-2} = -\dfrac{1}{3^2} = -\dfrac{1}{9}$

AlgebraWorks

1. Income $= 5x$

2. Profit $=$ Income $-$ Expenses
$$= 5x - 5300$$

3. Food Profit $= 0.50(0.30x) = 0.15x$

4. Profit $= 5x + 0.15x - 5300$
$$= 5.15x - 5300$$

5.

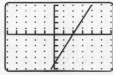

x scl: 500   y scl: 1000

5a. The y-intercept is the value for total cost, where their is no income. In this case, $5300.

5b. The x-intercept represents the break-even point. After 1,029 tickets are sold, the cinema will begin to make money.

Lesson 5.3, pages 242–248

Explore

1. $8x^2$ **2.** $12x^5$

3. GCF $= 4x^2$ **4.** GCF $= 5x^2$

5a. $(x + 2)(x - 2) = x^2 - 4$

5b. $(2x + 3)(2x - 3) = 4x^2 - 9$

6. Each product is the difference of two perfect squares.

7. The first terms are the same and the last terms are opposites.

8a. $(x + 2)(x + 2) = x^2 + 4x + 4$

8b. $(2x - 3)(2x - 3) = 4x^2 - 12x + 9$

9. The first and last terms are perfect squares; the middle term is twice the product of the terms of the factors.

10. The factors are identical.

Try These

1. $3y^2 - 6y = 3y(y - 2)$
Check: $3y(y - 2)$
$$= 3y(y) - 3y(2)$$
$$= 3y^2 - 6y$$

2. $21a^3b^2 - 14a^2b$
$$= 7a^2b(3ab - 2)$$
Check: $7a^2b(3ab - 2)$
$$= 7a^2b(3ab) - 7a^2b(2)$$
$$= 21a^3b^2 - 14a^2b$$

3. $3c(r - t) + 2d(r - t)$
$$= (3c + 2d)(r - t)$$
Check: $3c(r - t) + 2d(r - t)$

4. $35 \cdot 49 + 35 \cdot 51$
$$= 35(49 + 51)$$
$$= 35(100)$$
$$= 3500$$

5. $\dfrac{22}{7} \cdot 1600 - \dfrac{22}{7} \cdot 900$
$$= \dfrac{22}{7}(1600 - 900)$$
$$= \dfrac{22}{7}(700)$$
$$= 2200$$

6. $4h^2 - 25 = (2h - 5)(2h + 5)$

7. $0.16m^2 - 0.25$
$$= 0.01(16m^2 - 25)$$
$$= 0.01(4m - 5)(4m + 5)$$

8. $(a + b)^2 - c^2$
$$= [(a + b) - c][(a + b) + c]$$
$$= (a + b - c)(a + b + c)$$

9. $25 \cdot 15$
$$= (20 + 5)(20 - 5)$$
$$= 20^2 - 5^2$$
$$= 400 - 25$$
$$= 375$$

10.

$x^2 + 6x + 9 = (x + 3)(x + 3)$

11. $x^2 + 8x + 16$
$$= (x + 4)(x + 4)$$
$$= (x + 4)^2$$
Check: $(x + 4)(x + 4)$
$$= x(x + 4) + 4(x + 4)$$
$$= x^2 + 4x + 4x + 16$$
$$= x^2 + 8x + 16$$

12. $y^2 + 10y + 25$
$$= (y + 5)(y + 5)$$
$$= (y + 5)^2$$
Check: $(y + 5)(y + 5)$
$$= y(y + 5) + 5(y + 5)$$
$$= y^2 + 5y + 5y + 25$$
$$= y^2 + 10y + 25$$

13. $z^2 - 4z + 4$
$= (z - 2)(z - 2)$
$= (z - 2)^2$
Check: $(z - 2)(z - 2)$
$= z(z - 2) - 2(z - 2)$
$= z^2 - 2z - 2z + 4$
$= z^2 - 4z + 4$

14.

$x^2 - x - 6$
$= (x - 3)(x + 2)$

15. $x^2 + 4x + 3$
$= (x + 3)(x + 1)$
Check: $(x + 3)(x + 1)$
$= x(x + 1) + 3(x + 1)$
$= x^2 + x + 3x + 3$
$= x^2 + 4x + 3$

16. $a^2 - 2a - 8$
$= (a - 4)(a + 2)$
Check: $(a - 4)(a + 2)$
$= a(a + 2) - 4(a + 2)$
$= a^2 + 2a - 4a - 8$
$= a^2 - 2a - 8$

17. $m^2 + m - 12$
$= (m + 4)(m - 3)$
Check: $m(m - 3) + 4(m - 3)$
$= m^2 - 3m + 4m - 12$
$= m^2 + m - 12$

18.

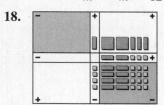

$2x^2 - 5x - 12 = (2x + 3)(x - 4)$

19. $x^2 - 36$
$= (x - 6)(x + 6)$

20. $3x^2 + 7x + 2$
$= (3x + 1)(x + 2)$
Check:
$(3x + 1)(x + 2)$
$= 3x(x + 2) + 1(x + 2)$
$= 3x^2 + 6x + x + 2$
$= 3x^2 + 7x + 2$

21. $2z^2 + 17z + 8$
$= (2z + 1)(z + 8)$
Check: $(2z + 1)(z + 8)$
$= 2z(z + 8) + 1(z + 8)$
$= 2z^2 + 16z + z + 8$
$= 2z^2 + 17z + 8$

22. $4m^2 + 4m - 15$
$= (2m + 5)(2m - 3)$
Check: $(2m + 5)(2m - 3)$
$= 2m(2m - 3) + 5(2m - 3)$
$= 4m^2 - 6m + 10m - 15$
$= 4m^2 + 4m - 15$

PRACTICE

1. $p + prt$
$= p(1 + rt)$

2. $28m^4n^3 - 70m^2n^4$
$= 14m^2n^3(2m^2 - 5n)$

3. $4r(y + z) + 7s(y + z)$
$= (4r + 7s)(y + z)$

4. $\pi r^2 + 2\pi rh$
$= \pi r(r + 2h)$

5. $9ab^2 - 6ab + 3a$
$= 3a(3b^2 - 2b + 1)$

6. $a(p - q) - b(p - q)$
$= (a - b)(p - q)$

7. $\frac{1}{2} \cdot 153 + \frac{1}{2} \cdot 47$
$= \frac{1}{2}(153 + 47)$
$= 100$

8. $\frac{1}{2} \cdot 7 \cdot 6.3 + \frac{1}{2} \cdot 7 \cdot 1.7$
$= \frac{1}{2} \cdot 7(6.3 + 1.7)$
$= \frac{1}{2} \cdot 7(8)$
$= \frac{1}{2} \cdot 56$
$= 28$

9. $p^2 - 100$
$= (p - 10)(p + 10)$

10. $\frac{1}{9}t^2 - \frac{1}{81}$
$= \frac{1}{9}\left(t^2 - \frac{1}{9}\right)$
$= \frac{1}{9}\left(t - \frac{1}{3}\right)\left(t + \frac{1}{3}\right)$

11. $(u + v)^2 - 169$
$= [(u + v) - 13][(u + v) + 13]$
$= (u + v - 13)(u + v + 13)$

12. $225 - y^2$
$= (15 - y)(15 + y)$

13. $1.44r^2 - 0.36s^4$
$= 0.36(4r^2 - s^4)$
$= 0.36(2r^2 - s^2)(2r^2 + s^2)$

14. $25 - (m - n)^2$
$= [5 - (m - n)][5 + (m - n)]$
$= (5 - m + n)(5 + m - n)$

15. 23^2
$= (20 + 3)^2$
$= 20^2 + 2(20)(3) + 3^2$
$= 400 + 120 + 9$
$= 529$

16.

$x^2 - 8x + 16$
$= (x - 4)(x - 4)$
$= (x - 4)^2$

17. $t^2 + 2t + 1$
$= (t + 1)(t + 1)$
$= (t + 1)^2$
Check: $t(t + 1) + 1(t + 1)$
$\qquad = t^2 + t + t + 1$
$\qquad = t^2 + 2t + 1$

18. $x^2 + 16x + 64$
$= (x + 8)(x + 8)$
$= (x + 8)^2$
Check: $x(x + 8) + 8(x + 8)$
$\qquad = x^2 + 8x + 8x + 64$
$\qquad = x^2 + 16x + 64$

19. $w^2 - 10w + 25$
$= (w - 5)(w - 5)$
$= (w - 5)^2$
Check: $w(w - 5) - 5(w - 5)$
$\qquad = w^2 - 5w - 5w + 25$
$\qquad = w^2 - 10w + 25$

20. $x^2 - 4x + 4$
$= (x - 2)(x - 2)$
$= (x - 2)^2$
Check: $x(x - 2) - 2(x - 2)$
$\qquad = x^2 - 2x - 2x + 4$
$\qquad = x^2 - 4x + 4$

21. $16 + 8m + m^2$
$= (4 + m)(4 + m)$
$= (4 + m)^2$
Check: $4(4 + m) + m(4 + m)$
$\qquad = 16 + 4m + 4m + m^2$
$\qquad = 16 + 8m + m^2$

22. $49 - 14s + s^2$
$= (7 - s)(7 - s)$
$= (7 - s)^2$
Check: $7(7 - s) - s(7 - s)$
$\qquad = 49 - 7s - 7s + s^2$
$\qquad = 49 - 14s + s^2$

23. $4x^2 + 4x + 1$
$= (2x + 1)(2x + 1)$
$= (2x + 1)^2$
Check: $2x(2x + 1) + 1(2x + 1)$
$\qquad = 4x^2 + 2x + 2x + 1$
$\qquad = 4x^2 + 4x + 1$

24. $9a^2 - 12a + 4$
$= (3a - 2)(3a - 2)$
$= (3a - 2)^2$
Check: $3a(3a - 2) - 2(3a - 2)$
$\qquad 9a^2 - 6a - 6a + 4$
$\qquad = 9a^2 - 12a + 4$

25. $1 + 6z + 9z^2$
$= (1 + 3z)(1 + 3z)$
$= (1 + 3z)^2$
Check: $1(1 + 3z) + 3z(1 + 3z)$
$\qquad = 1 + 3z + 3z + 9z^2$
$\qquad = 1 + 6z + 9z^2$

26.

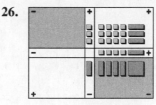

$12 - x - x^2$
$= (4 + x)(3 - x)$

27. $z^2 + 7z + 12$
$= (z + 4)(z + 3)$
Check: $z(z + 3) + 4(z + 3)$
$\qquad = z^2 + 3z + 4z + 12$
$\qquad = z^2 + 7z + 12$

28. $t^2 - 4t - 5$
$= (t - 5)(t + 1)$
Check: $t(t + 1) - 5(t + 1)$
$\qquad = t^2 + t - 5t - 5$
$\qquad = t^2 - 4t - 5$

29. $n^2 - 5n - 6$
$= (n - 6)(n + 1)$
Check: $n(n + 1) - 6(n + 1)$
$\qquad = n^2 + n - 6n - 6$
$\qquad = n^2 - 5n - 6$

30. $2 + 3m + m^2$
$= (1 + m)(2 + m)$
Check: $1(2 + m) + m(2 + m)$
$\qquad = 2 + m + 2m + m^2$
$\qquad = 2 + 3m + m^2$

31. $15 - 8n + n^2$
$= (5 - n)(3 - n)$
Check: $5(3 - n) - n(3 - n)$
$\qquad = 15 - 5n - 3n + n^2$
$\qquad = 15 - 8n + n^2$

32. $12 - r - r^2$
$= (4 + r)(3 - r)$
Check: $4(3 - r) + r(3 - r)$
$\quad\quad = 12 - 4r + 3r - r^2$
$\quad\quad = 12 - r - r^2$

33.

$5 + 8x - 4x^2$
$= (5 - 2x)(1 + 2x)$

34. $2x^2 + 5x + 2$
$= (2x + 1)(x + 2)$
Check: $2x(x + 2) + 1(x + 2)$
$\quad\quad = 2x^2 + 4x + x + 2$
$\quad\quad = 2x^2 + 5x + 2$

35. $2a^2 + 7a + 6$
$= (2a + 3)(a + 2)$
Check: $2a(a + 2) + 3(a + 2)$
$\quad\quad = 2a^2 + 4a + 3a + 6$
$\quad\quad = 2a^2 + 7a + 6$

36. $4x^2 - 4x - 15$
$= (2x + 3)(2x - 5)$
Check: $2x(2x - 5) + 3(2x - 5)$
$\quad\quad = 4x^2 - 10x + 6x - 15$
$\quad\quad = 4x^2 - 4x - 15$

37. $9 + 9x + 2x^2$
$= (3 + 2x)(3 + x)$
Check: $3(3 + x) + 2x(3 + x)$
$\quad\quad = 9 + 3x + 6x + 2x^2$
$\quad\quad = 9 + 9x + 2x^2$

38. $8 + 14r + 3r^2$
$= (2 + 3r)(4 + r)$
Check: $2(4 + r) + 3r(4 + r)$
$\quad\quad = 8 + 2r + 12r + 3r^2$
$\quad\quad = 8 + 14r + 3r^2$

39. $12 - 13a - 4a^2$
$= (4 + a)(3 - 4a)$
Check: $4(3 - 4a) + a(3 - 4a)$
$\quad\quad = 12 - 16a + 3a - 4a^2$
$\quad\quad = 12 - 13a - 4a^2$

40. $2lw + 2wh + 2lh$
$= 2(lw + wh + lh)$

41. $2\pi r^2 + 2\pi rh$
$= 2\pi r(r + h)$

EXTEND

42. $9x^2y^2 + 24xy + 16$
$= (3xy + 4)(3xy + 4)$
$= (3xy + 4)^2$

43. $4m^2n^2 - 20mn + 25$
$= (2mn - 5)(2mn - 5)$
$= (2mn - 5)^2$

44. $25a^4b^2 + 10a^2b + 1$
$= (5a^2b + 1)(5a^2b + 1)$
$= (5a^2b + 1)^2$

45. $7x^4 - 6x^2y - y^2$
$= (7x^2 + y)(x^2 - y)$

46. $10x^2y^2 + 11xyz + 3z^2$
$= (5xy + 3z)(2xy + z)$

47. $15c^2d^2 + 13cdf + 2f^2$
$= (3cd + 2f)(5cd + f)$

48a. $p^2 = \left(\dfrac{3}{4}\right)^2 = \dfrac{9}{16}; 9\!:\!16$

48b. $2pq = 2 \cdot \dfrac{3}{4} \cdot \dfrac{1}{4} = \dfrac{3}{8}; 3\!:\!8$

48c. $q^2 = \left(\dfrac{1}{4}\right)^2 = \dfrac{1}{16}; 1\!:\!16$

49. $\dfrac{15}{16} = \dfrac{x}{32{,}000}$
$16x = 480{,}000$
$\quad x = 30{,}000$

THINK CRITICALLY

50. $x^2 + 16x + 64; 64$ **51.** $y^2 - 10y + 25; 25$

52. $4z^2 + 12z + 9; 9$

53. $3k^{2a} - 10k^a + 3$
$= (3k^a - 1)(k^a - 3)$
Check: $\quad 3k^a(k^a - 3) - 1(k^a - 3)$
$\quad\quad = 3k^{2a} - 9k^a - k^a + 3$
$\quad\quad = 3k^{2a} - 10k^a + 3$

54. $(2x + 1)^2 - 3(2x + 1) - 10$
$= ((2x + 1) - 5)((2x + 1) + 2)$
$= (2x - 4)(2x + 3)$
$= 2(x - 2)(2x + 3)$
Check: $\quad 2(x - 2)(2x + 3)$
$\quad\quad \overset{?}{=} (2x + 1)^2 - 3(2x + 1) - 10$
$\quad\quad\quad 2(2x^2 - 4x + 3x - 6)$
$\quad\quad \overset{?}{=} (4x^2 + 4x + 1) - (6x + 3) - 10$
$\quad\quad\quad 4x^2 - 8x + 6x - 12$
$\quad\quad \overset{?}{=} 4x^2 + 4x - 6x + 1 - 3 - 10$
$\quad\quad\quad 4x^2 - 2x - 12 = 4x^2 - 2x - 12 \checkmark$

55. $(x^2 + x - 2)^2 - x^4$
$= (x^2 + x - 2 - x^2)(x^2 + x - 2 + x^2)$
$= (x - 2)(2x^2 + x - 2)$
Check: $\quad (x - 2)(2x^2 + x - 2)$
$\quad\quad \overset{?}{=} (x^2 + x - 2)^2 - x^4$
$\quad\quad\quad 2x^3 + x^2 - 2x - 4x^2 - 2x + 4$
$\quad\quad \overset{?}{=} (x^4 + 2x^3 - 3x^2 - 4x + 4) - x^4$
$\quad\quad\quad 2x^3 - 3x^2 - 4x + 4$
$\quad\quad = 2x^3 - 3x^2 - 4x + 4 \checkmark$

MIXED REVIEW

56.
$$ax + b = d - cx$$
$$ax + cx = d - b$$
$$x(a + c) = d - b$$
$$x = \frac{d - b}{a + c}$$

57. $S = \pi a(b + c)$

$$\frac{S}{\pi a} = b + c$$

$$\frac{S}{\pi a} - b = c \quad \text{or} \quad \frac{S - \pi ab}{\pi a} = c$$

58. $|2x + 1| = -3$

D; no real solution because absolute value solutions are always positive.

59. $f(a + 1) = \frac{a + 1 + 3}{a + 1 - 1}$

$$= \frac{a + 4}{a}; C$$

Lesson 5.4 pages 249–253

EXPLORE

1a. $(x - 1)(x^2 + x + 1)$
$= x(x^2 + x + 1) - 1(x^2 + x + 1)$
$= x^3 + x^2 + x - x^2 - x - 1$
$= x^3 - 1$

1b. $(x - 2)(x^2 + 2x + 4)$
$= x(x^2 + 2x + 4) - 2(x^2 + 2x + 4)$
$= x^3 + 2x^2 + 4x - 2x^2 - 4x - 8$
$= x^3 - 8$

2. Each product is a difference of the cubes of the items of the binomial.

3a. $(x - 3)(x^2 + 3x + 9)$ **3b.** $(x - 4)(x^2 + 4x + 16)$
   $= x^3 - 27$ $= x^3 - 64$

4a. $(x + 1)(x^2 - x + 1)$
$= x(x^2 - x + 1) + 1(x^2 - x + 1)$
$= x^3 - x^2 + x + x^2 - x + 1$
$= x^3 + 1$

4b. $(x + 2)(x^2 - 2x + 4)$
$= x(x^2 - 2x + 4) + 2(x^2 - 2x + 4)$
$= x^3 - 2x^2 + 4x + 2x^2 - 4x + 8$
$= x^3 + 8$

5. Each product is a sum of the cubes of the terms of the binomial.

6a. $(x + 3)(x^2 - 3x + 9)$
   $= x^3 + 27$

6b. $(x + 4)(x^2 - 4x + 16)$
   $= x^3 + 64$

TRY THESE

1. $3x^2 - 27$ Check: $3(x - 3)(x + 3)$
$= 3(x^2 - 9)$ $= (3x - 9)(x + 3)$
$= 3(x^2 - 3^2)$ $= 3x^2 + 9x - 9x - 27$
$= 3(x - 3)(x + 3)$ $= 3x^2 - 27$

2. $2y^2 - 4y - 30$ Check: $2(y - 5)(y + 3)$
$= 2(y^2 - 2y - 15)$ $= (2y - 10)(y + 3)$
$= 2(y - 5)(y + 3)$ $= 2y^2 + 6y - 10y - 30$
                $= 2y^2 - 4y - 30$

3. $6z^2 - 9z - 6$ Check: $3(2z + 1)(z - 2)$
$= 3(2z^2 - 3z - 2)$ $= (6z + 3)(z - 2)$
$= 3(2z + 1)(z - 2)$ $= 6z^2 - 12z + 3z - 6$
                $= 6z^2 - 9z - 6$

4a. No; $\pi(R - r)^2$ represents the area of a circle with radius $R - r$. If $R = 10$ and $r = 4$, then $\pi(R - r)^2 = 36\pi$. But the area of the shaded region is $100\pi - 16\pi = 84\pi$.

4b. area of shaded region $= \pi R^2 - \pi r^2$
$= \pi(R^2 - r^2)$
$= \pi(R - r)(R + r)$

5. $b^4 - 625$
$= (b^4 - 25^2)$
$= (b^2 - 25)(b^2 + 25)$
$= (b - 5)(b + 5)(b^2 + 25)$

6. $y^8 - 256$
$= (y^8 - 16^2)$
$= (y^4 - 16)(y^4 + 16)$
$= (y^2 - 4)(y^2 + 4)(y^4 + 16)$
$= (y - 2)(y + 2)(y^2 + 4)(y^4 + 16)$

7.

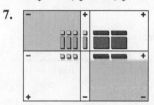

$2xy + 2y - 3x - 3 = (x + 1)(2y - 3)$

8. $4xy + 8xz + 6y + 12z$
$= (4xy + 6y) + (8xz + 12z)$
$= 2y(2x + 3) = 4z(2x + 3)$
$= (2y + 4z)(2x + 3)$
$= 2(y + 2z)(2x + 3)$
Check: $(2y + 4z)(2x + 3)$
$= 4xy + 6y + 8xz + 12z$
$= 4xy + 8xz + 6y + 12z$

9. $x^2 - 5x + 15y - 3xy$
$= x(x - 5) + 3y(5 - x)$
$= x(x - 5) - 3y(x - 5)$
$= (x - 3y)(x - 5)$
Check: $(x - 3y)(x - 5)$
$= x^2 - 5x - 3xy + 15y$
$= x^2 - 5x + 15y - 3xy$

10. $y^3 + 4y^2 - 9y - 36$

$= y^2(y + 4) - 9(y + 4)$

$= (y^2 - 9)(y + 4)$

$= (y - 3)(y + 3)(y + 4)$

Check: $(y - 3)(y + 3)(y + 4)$

$\qquad = (y^2 - 9)(y + 4)$

$\qquad = y^2 + 4y^2 - 9y - 36$

11. $z^2 + 8z + 16 - x^2$

$= (z + 4)(z + 4) - x^2$

$= (z + 4)^2 - x^2$

$= (z + 4 - x)(z + 4 + x)$

12. $(3xy + 6xz) + (4y + 8z)$

$= 3x(y + 2z) + 4(y + 2z)$

$= (3x + 4)(y + 2z)$

or

$(3xy + 4y) + (6xz + 8z)$

$= y(3x + 4) + 2z(3x + 4)$

$= (y + 2z)(3x + 4)$

Two methods of grouping result in the same prime factors.

13. r^3s^3

$= (r - s)(r^2 + rs + s^2)$

Check: $(r - s))(r^2 + rs + s^2)$

$\qquad = r^3 + r^2s + rs^2 - r^2s - rs^2 - s^2$

$\qquad = r^3 - s^3$

14. $m^3 + n^3$

$= (m + n)(m^2 - mn + n^2)$

Check: $(m + n)(n^2 - mn + n^2)$

$\qquad = m^3 - m^2n + mn^2 + m^2n - mn^2 + n^3$

$\qquad = m^3 + n^3$

15. $64x^3 - 1$

$= (4x - 1)(16x^2 + 4x + 1)$

Check: $(4x - 1)(16x^2 + 4x + 1)$

$\qquad = 64x^3 + 16x^2 + 4x - 16x^2 - 4x - 1$

$\qquad = 64x^3 - 1$

16. $y^6 + 0.001$

$= (y^2 + 0.1)(y^4 - 0.1y^2 + 0.01)$

$= 0.1(10y^2 + 1) \cdot 0.01(100y^4 - 10y^2 + 1)$

$= 0.001(10y^2 + 1)(100y^4 - 10y^2 + 1)$

Check: $0.001(1000y^6 - 100y^4 + 10y^2 + 100y^4 -$

$\qquad 10y^2 + 1$

$\qquad = 0.001(1000y^6 + 1)$

$\qquad = y^6 + 0.001$

PRACTICE

1. $5y^2 - 80$

$= 5(y^2 - 16)$

$= 5(y - 4)(y + 4)$

Check: $5(y - 4)(y + 4)$

$\qquad = (5y - 20)(y + 4)$

$\qquad = 5y^2 + 20y - 20y - 80$

$\qquad = 5y^2 - 80$

2. $175x^2 - 343$

$= 7(25x^2 - 49)$

$= 7(5x - 7)(5x + 7)$

Check: $7(5x - 7)(5x + 7)$

$\qquad = (35x - 49)(5x + 7)$

$\qquad = 175x^2 + 245x - 245x - 343$

$\qquad = 175x^2 - 343$

3. $3x^2 - 15x + 18$

$= 3(x^2 - 5x + 6)$

$= 3(x - 3)(x - 2)$

Check: $3(x - 3)(x - 2)$

$\qquad = (3x - 9)(x - 2)$

$\qquad = 3x^2 - 6x - 9x + 18$

$\qquad = 3x^2 - 15x + 18$

4. $5y^2 - 25y - 30$

$= 5(y^2 - 5y - 6)$

$= 5(y - 6)(y + 1)$

Check: $5(y - 6)(y + 1)$

$\qquad = (5y - 30)(y + 1)$

$\qquad = 5y^2 + 5y - 30y - 30$

$\qquad = 5y^2 - 25y - 30$

5. $4x^2 + 16x + 16$

$= 4(x^2 + 4x + 4)$

$= 4(x + 2)(x + 2)$

$= 4(x + 2)^2$

Check: $4(x + 2)(x + 2)$

$\qquad = (4x + 8)(x + 2)$

$\qquad = 4x^2 + 8x + 8x + 16$

$\qquad = 4x^2 + 16x + 16$

6. $3z^2 - 24z + 48$

$= 3(z^2 - 8z + 16)$

$= 3(z - 4)(z - 4)$

$= 3(z - 4)^2$

Check: $3(z - 4)(z - 4)$

$\qquad = (3z - 12)(z - 4)$

$\qquad = 3z^2 - 12z - 12z + 48$

$\qquad = 3z^2 - 24z + 48$

7. When the GCF is not removed, you get the factorization.

$4x^2 - 36 = (2x - 6)(2x - 6)$

$\qquad\qquad = 2(x - 3) \cdot 2(x - 3)$

$\qquad\qquad = 4(x - 3)(x - 3)$

If you had first removed the GCF, you would get the same factorization.

$4x^2 - 36 = 4(x^2 - 9)$

$\qquad\qquad = 4(x - 3)(x - 3)$

8a. $187 - 56x + 4x^2$

$= (11 - 2x)(17 - 2x)$

width $(11 - 2x)$ in., length $(17 - 2x)$ in.

8b. The width is 11 in., the length is 17 in., and the width of the border is x in.

8c. $187 - 56x + 4x^2 = 91$
$4x^2 - 56x + 96 = 0$
$(4x - 48)(x - 2) = 0$
$x = 12$ or $x = 2$
width $11 - 14 = 7$ in., length $17 - 4 = 13$ in.,
border $1(2) = 2$ in.

9. $t^4 - 81$
$= (t^2 - 9)(t^2 + 9)$
$= (t - 3)(t + 3)(t^2 + 9)$
Check: $(t - 3)(t + 3)(t^2 + 9)$
$= (t^2 - 9)(t^2 + 9)$
$= t^4 - 81$

10. $1296 - x^4$
$= (36 - x^2)(36 + x^2)$
$= (6 - x)(6 + x)(36 + x^2)$
Check: $(36 - x^2)(36 + x^2)$
$= 1296 - x^4$

11. $a^8 - 16$
$= (a^4 - 4)(a^4 + 4)$
$= (a^2 - 2)(a^2 + 2)(a^4 + 4)$
Check: $(a^4 - 4)(a^4 + 4)$
$= a^8 - 16$

12. $\frac{1}{27} - b^{12}$

$= \left(\frac{1}{3}\right)^3 - (b^4)^3$

$= \left(\frac{1}{3} - b^4\right)\left(\frac{1}{9} + \frac{1}{3}b^4 + b^8\right)$

$= \frac{1}{3}(1 - 3b^4) \cdot \frac{1}{9}(1 + 3b^4 + 9b^8)$

$= \frac{1}{27}(1 - 3b^4)(1 + 3b^4 + 9b^8)$

13.

$3y + 3xy - 2x - 2 = (x + 1)(3y - 2)$

14. $5ab + 10ac - 3b - 6c$
$= 5a(b + 2c) - 3(b + 2c)$
$= (5a - 3)(b + 2c)$
Check: $(5a - 3)(b + 2c)$
$= 5ab + 10c - 3b - 6c$

15. $2ax + 3 + x + 6a$
$= (2ax + x) + (3 + 6a)$
$= x(2a + 1) + 3(1 + 2a)$
$= (x + 3)(2a + 1)$
Check: $(x + 3)(2a + 1)$
$= 2ax + x + 6a + 3$
$= 2ax + 3 + x + 6a$

16. $y^2 - 3y + 6x - 2xy$
$= y(y - 3) + 2x(3 - y)$
$= y(y - 3) - 2x(y - 3)$
$= (y - 2x)(y - 3)$
Check: $(y - 2x)(y - 3)$
$= y^2 - 3y - 2xy + 6x$
$= y^2 - 3y + 6x - 2xy$

17. $10y - 5xy - 6x + 3x^2$
$= 5y(2 - x) - 3x(2 - x)$
$= (5y - 3x)(2 - x)$
Check: $(5y - 3x)(2 - x)$
$= 10y - 5xy - 6x + 3x^2$

18. $z^2 + 8z + 16 - q^2$
$= (z + 4)^2 - q^2$
$= (z - 3 - q)(z - 4 + q)$
Check: $(z - 4 - q)(z - 4 + q)$
$= z^2 - 4z + qz - 4z + 16$
$\quad - 4q - qz + 4q - q^2$
$= z^2 - 8z + 16 - q^2$

19. $t^2 - 10t + 25 - u^2$
$= (t - 5)^2 - u^2$
$= (t - 5 - u)(t - 5 + u)$
Check: $t^2 - 5t + tu - 5t + 25 - 5u - ut +$
$\quad 5u - u^2$
$= t^2 - 10t + 25 - u^2$

20. $36 - a^2 - 2a - 1$
$= 35 - a^2 - 2a$
$\quad - 35 - 2a - a^2$
$= (7 + a)(5 - a)$
Check: $35 - 7a + 5a - a^2$
$= 35 - 2a - a^2$
$= 36 - 2a - a^2 - 1$

21. $25 - r^2 - 4r - 4$
$= 21 - 4r - r^2$
$= (7 + r)(3 - r)$
Check: $21 - 7r + 3r - r^2$
$= 25 - r^2 - 4r - 4$

22. A polynomial of two terms may be factored by removing a common monomial factor, as the difference of two squares, the difference of two cubes, or the sum of two cubes.

23. $(2ax - bx - cx + 2ay - by + cy) \div (x + y)$
$= [x(2a - b + c) + y(2a - b + c)] \div (x + y)$
$= [(x + y)(2a - b + c)] \div (x + y)$
$= 2a - b + c =$ length.

EXTEND

24. $a^7 + b^7 = (a + b)(a^6 - a^5b + a^4b^2 - a^3b^3 + a^2b^4 - ab^5 + b^6)$

25. The first factor is $(a + b)$. The second factor has n terms. These terms are in descending powers of a beginning with a^{n-1}. The sum of the powers of a and b in any term is $n - 1$. The coefficients alternate between 1 and -1, beginning with 1.

26. $x^4 + 2x^2y^2 + y^4 - x^2y^2$
$= (x^2 + y^2)^2 - x^2y^2$
$= (x^2 + y^2 - xy)(x^2 + y^2 + xy)$

27. $9x^4 + 2x^2y^2 + y^4$
$= 9x^4 + 6x^2y^2 + y^4 - 4x^2y^2$
$= (3x^2 + y^2)^2 - 4x^2y^2$
$= (3x^2 + y^2 - 2xy)(3x^2 + y^2 + 2xy)$
Check: $9x^4 + 3x^2y^2 + 6x^3y + 3x^2y^2 + y^4 +$
    $2xy^3 - 6x^3y - 2xy^3 - 4x^2y^2$
$= 9x^4 + 2x^2y^2 + y^4$

28. $x^4 + 64$
$= x^4 + 16x^2 + 64 - 16x^2$
$= (x^2 + 8)^2 - 16x^2$
$= (x^2 + 8 - 4x)(x^2 + 8 + 4x)$
Check: $x^4 + 8x^2 + 4x^3 + 8x^2 + 64 + 32x -$
    $4x^3 - 32x - 16x^2$
$= x^4 + 64$

THINK CRITICALLY

29. $(x - 2)^3 + 64$
$= (x - 2 + 4)[(x - 2)^2 - 4(x - 2) + 16]$
$= (x + 2)(x^2 - 4x + 4 - 4x + 8 + 16)$
$= (x + 2)(x^2 - 8x + 28)$
Check: $x^3 - 8x^2 + 28x + 2x^2 - 16x + 56$
    $= x^3 - 6x^2 + 12x + 56$
        and
    $(x - 2)^3 + 64$
    $= x^3 - 6x^2 + 12x - 8 + 64$
    $= x^3 - 6x^2 + 12x + 56$

30. $x^4 + x^2 - 20$
$= (x^2 - 4)(x^2 + 5)$
$= (x - 2)(x + 2)(x^2 + 5)$
Check: $(x^2 - 4)(x^2 + 5)$
    $= x^4 + 5x^2 - 4x^2 - 20$
    $= x^4 + x^2 - 20$

31. $(3x - 1)^2 - 3(3x - 1) - 10$
$= ((3x - 1) + 2)((3x - 1) - 5)$
$= (3x - 1 + 2)(3x - 1 - 5)$
$= (3x + 1)(3x - 6)$
$= 9x^2 - 9x - 6$

32. $(y^2 + y - 2)^2 - (y^2 - y - 6)^2$
$= (y^2 + y - 2 + y^2 - y - 6) -$
    $(y^2 + y - 2 - y^2 + y + 6)$
$= (2y^2 - 8)(2y + 4)$
$= 2(y^2 - 4) \cdot 2(y + 2)$
$= 4(y^2 - 4)(y + 2)$
$= 4(y - 2)(y + 2)(y + 2)$
$= 4(y - 2)(y + 2)^2$

33. $(x - y)^3 - 27$
$= (x - y - 3)[(x - y)^2 + 3(x - y) + 9]$
$= (x - y - 3)(x^2 - 2xy + y^2 + 3x - 3y + 9)$

34. range $= 3212 - 1600 = 1612$

35. median $= 2007$

36. first quartile $= 1758$

37. D; $\dfrac{y - -2}{x - 4} = \dfrac{2}{3}$ $y = 0$
                        $x = 7$ $(7, 0)$

38. $2y = 5x + 3$ $5 = \dfrac{5}{2}(1) + b$
$y = \dfrac{5}{2}x + \dfrac{3}{2}$ $\dfrac{5}{1} - \dfrac{5}{2} = b$ $y = \dfrac{5}{2}x + \dfrac{5}{2}$
$m = \dfrac{5}{2}$ $b = \dfrac{5}{2}$

39. $2y = 5x + 3$
$y = \dfrac{5}{2}x + \dfrac{3}{2}$
$m = -\dfrac{2}{5}$ $3 = \left(-\dfrac{2}{1}\right)\left(-\dfrac{2}{5}\right) + b$
$\dfrac{3}{1} - \dfrac{4}{5} = b$
$b = \dfrac{11}{5}$ $y = -\dfrac{2}{5}x + \dfrac{11}{5}$

40. B

Lesson 5.5 pages 254–261

EXPLORE

1a.

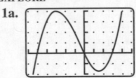

1b. 3 times; $x = -5, x = 0, x = 3$

1c. $x^3 + 2x^2 - 15x$
$= x(x^2 + 2x - 15)$
$= x(x + 5)(x - 3)$

1d. If you set each factor equal to 0 and solve, you get the x-intercepts.

2a.

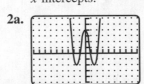

The graph crosses the x-axis at $x = \pm 1$ and $x = \pm 2$
$x^4 - 5x^2 + 4$
$= (x^2 - 4)(x^2 - 1)$
$= (x - 2)(x + 2)(x - 1)(x + 1)$

2b. Yes.

3a. $x = -4, 0, 2$

3b. $x(x + 4)(x - 2)$

3c. $(x^2 + 4x)(x - 2)$
$= x^3 - 2x^2 + 4x^2 - 8x$
$= x^3 + 2x^2 - 8x$

TRY THESE

1. $x^2 - 7x = 0$
$x(x - 7) = 0$
$x = 0, 7$
Check: $0^2 - 7(0) = 0$ ✓
$7^2 - 7(7) = 0$
$49 - 49 = 0$ ✓

2. $m^2 + 5m = 0$
$m(m + 5) = 0$
$m = -5, 0$
Check: $0^2 + 5(0) = 0$
$0 + 0 = 0$ ✓
$(-5)^2 + 5(-5) = 0$
$25 - 25 = 0$ ✓

3. $k^2 - 4 = 0$
$(k - 2)(k + 2) = 0$
$k = -2, 2$
Check: $(2)^2 - 4 = 0$
$4 - 4 = 0$ ✓
$(-2)^2 - 4 = 0$
$4 - 4 = 0$ ✓

4. $x^2 - 7x + 10 = 0$
$(x - 5)(x - 2) = 0$
$x = 2, 5$
Check: $(5)^2 - 7(5) + 10 = 0$
$25 - 35 + 10 = 0$
$-10 + 10 = 0$ ✓
$(2)^2 - 7(2) + 10 = 0$
$4 - 14 + 10 = 0$
$-10 + 10 = 0$ ✓

5. $a^2 + 7a + 12 = 0$
$(a + 4)(a + 3) = 0$
$a = -4, -3$
Check: $(-4)^2 + 7(-4) + 12 = 0$
$16 - 28 + 12 = 0$
$-12 + 12 = 0$ ✓
$(-3)^2 + 7(-3) + 12 = 0$
$9 - 21 + 12 = 0$
$-12 + 12 = 0$ ✓

6. $-c^2 + 2c + 15 = 0$
$-1(c^2 - 2c - 15) = 0$
$-1(c - 5)(c + 3) = 0$
$c = -3, 5$
Check: $-(5)^2 + 2(5) + 15 = 0$
$-25 + 10 + 15 = 0$
$-15 + 15 = 0$ ✓
$-(-3)^2 + 2(-3) + 15 = 0$
$-9 - 6 + 15 = 0$
$-15 + 15 = 0$ ✓

7. $t^2 = 8t$
$t^2 - 8t = 0$
$t(t - 8) = 0$
$t = 0, 8$
Check: $0^2 = 8(0)$
$0 = 0$ ✓
$8^2 = 8(8)$
$64 = 64$ ✓

8. $35 = z^2 + 2z$
$0 = z^2 + 2z - 35$
$0 = (z + 7)(z - 5)$
$z = -7, 5$
Check: $35 = (-7)^2 + 2(-7) = 49 - 14 = 35$ ✓
$35 = 5^2 + 2(5) = 25 + 10 = 35$ ✓

9. $r^2 - 5r - 20 = 4$
$r^2 - 5r - 24 = 0$
$(r - 8)(r + 3) = 0$
$r = -3, 8$
Check: $8^2 - 5(8) - 20 = 4$
$64 - 40 - 20 = 4$
$24 - 20 = 4$
$4 = 4$ ✓
$(-3)^2 - 5(-3) - 20 = 4$
$9 + 15 - 20 = 4$
$24 - 20 = 4$
$4 = 4$ ✓

10. Cynthia; Factors can only be set equal to zero. Subtract 12 from both sides of the equation to obtain $x^2 - 4x - 12 = 0$. Then factor; $(x - 6)(x + 2) = 0$. Apply the zero product property; $x = 6$ or $x = -2$.

11. $x = $ width of side strip
$(2x + 3)(4x + 5) = 45$
$8x^2 + 10x + 12x + 15 - 45 = 0$
$8x^2 + 22x - 30 = 0$
$2(4x^2 + 11x - 15) = 0$
$2(4x + 15)(x - 1) = 0$
$x = -\dfrac{15}{4}$ or $x = 1$

Since x is a length, it must be positive. So $x = 1$ in.

12. $x^3 + x^2 - 16x - 16 = 0$
$x^2(x + 1) - 16(x + 1) = 0$
$(x^2 - 16)(x + 1) = 0$
$(x - 4)(x + 4)(x + 1) = 0$
$x = -4, -1, 4$
Check: $(4)^3 + (4)^2 - 16(4) - 16 = 0$
$64 + 16 - 64 - 16 = 0$
$0 = 0$ ✓
$(-4)^3 + (-4)^2 - 16(-4) - 16 = 0$
$-64 + 16 + 64 - 16 = 0$
$0 = 0$ ✓
$(-1)^3 + (-1)^2 - 16(-1) - 16 = 0$
$-1 + 1 + 16 - 16 = 0$
$0 = 0$ ✓

13. $x^3 - x^2 - 25x + 25 = 0$
$x^2(x - 1) - 25(x - 1) = 0$
$(x^2 - 25)(x - 1) = 0$
$(x - 5)(x + 5)(x - 1) = 0$
$x = -5, 1, 5$
Check: $\quad (5)^3 - (5)^2 - 25(5) + 25 = 0$
$125 - 25 - 125 + 25 = 0$
$0 = 0 \checkmark$
$(-5)^3 - (-5)^2 - 25(-5) + 25 = 0$
$-125 - 25 + 125 + 25 = 0$
$0 = 0 \checkmark$
$(1)^3 - (1)^2 - 25(1) + 25 = 0$
$1 - 1 - 25 + 25 = 0$
$0 = 0 \checkmark$

14. $3x^3 + 30x^2 + 75x = 0$
$3x(x^2 + 10x + 25) = 0$
$3x(x + 5)(x + 5) = 0$
$x = -5, -5, 0$
Check: $\quad 3(0)^3 + 30(0)^2 + 75(0) = 0$
$0 = 0 \checkmark$
$3(-5)^3 + 30(-5)^2 + 75(-5) = 0$
$3(-125) + 30(25) - 375 = 0$
$-375 + 750 - 375 = 0$
$0 = 0 \checkmark$

15. $-2x^3 + 2x^2 + 12x = 0$
$-2x(x^2 - x - 6) = 0$
$-2x(x - 3)(x + 2) = 0$
$x = -2, 0, 3$
Check: $\quad -2(0)^3 + 2(0)^2 + 12(0) = 0$
$0 = 0 \checkmark$
$-2(3)^3 + 2(3)^2 + 12(3) = 0$
$-2(27) + 2(9) + 36 = 0$
$-54 + 18 + 36 = 0$
$0 = 0 \checkmark$
$-2(-2)^3 + 2(-2)^2 + 12(-2) = 0$
$-2(-8) + 2(4) - 24 = 0$
$16 + 8 - 24 = 0$
$0 = 0 \checkmark$

16. $(x - 0)(x - 1) = 0$
$x(x - 1) = 0$
$x^2 - x = 0$

17. $(x - 2)(x - 3) = 0$
$x^2 - 3x - 2x + 6 = 0$
$x^2 - 5x + 6 = 0$

18. $(x - 4)(x + 7) = 0$
$x^2 + 7x - 4x - 28 = 0$
$x^2 + 3x - 28 = 0$

19. $(x - 0)(x - 3)(x - 3) = 0$
$x(x^2 - 6x + 9) = 0$
$x^3 - 6x^2 + 9x = 0$

20. $(x - 0)(x - 1)(x + 2) = 0$
$x(x^2 + 2x - x - 2) = 0$
$x^3 + 2x^2 - x^2 - 2x = 0$
$x^3 + x^2 - 2x = 0$

21. $(x - 1)(x + 2)(x + 4) = 0$
$(x^2 + 2x - x - 2)(x + 4) = 0$
$(x^2 + x - 2)(x + 4) = 0$
$(x^2 + x - 2)(x + 4) = 0$
$x^3 + 4x^2 + x^2 + 4x - 2x - 8 = 0$
$x^3 + 5x^2 + 2x - 8 = 0$

PRACTICE

1. $-k^2 + 8k = 0$ $\qquad$ Check: $-0^2 + 8(0) = 0$
$-k(k - 8) = 0$ $\qquad\qquad\qquad 0 = 0 \checkmark$
$k = 0, 8$ $\qquad\qquad\qquad -8^2 + 8(8) = 0$
$\qquad\qquad\qquad\qquad\qquad -64 + 64 = 0$
$\qquad\qquad\qquad\qquad\qquad 0 = 0 \checkmark$

2. $2m^2 - 12m = 0$ $\qquad$ Check: $2(0)^2 - 12(0) = 0$
$2m(m - 6) = 0$ $\qquad\qquad\qquad 0 = 0 \checkmark$
$m = 0, 6$ $\qquad\qquad\qquad 2(6)^2 - 12(6) = 0$
$\qquad\qquad\qquad\qquad\qquad 2(36) - 72 = 0$
$\qquad\qquad\qquad\qquad\qquad 72 - 72 = 0 \checkmark$

3. $6z^2 + 18z = 0$ $\qquad$ Check: $6(0)^2 + 18(0) = 0$
$6z(z + 3) = 0$ $\qquad\qquad\qquad 0 = 0 \checkmark$
$z = -3, 0$ $\qquad\qquad\qquad 6(-3)^2 + 18(-3) = 0$
$\qquad\qquad\qquad\qquad\qquad 6(9) - 54 = 0$
$\qquad\qquad\qquad\qquad\qquad 54 - 54 = 0 \checkmark$

4. $y^2 - 121 = 0$ $\qquad$ Check: $(11)^2 - 121 = 0$
$(y - 11)(y + 11) = 0$ $\qquad\qquad 121 - 121 = 0$
$y = -11, 11$ $\qquad\qquad\qquad 0 = 0 \checkmark$
$\qquad\qquad\qquad\qquad\qquad (-11)^2 - 121 = 0$
$\qquad\qquad\qquad\qquad\qquad 121 - 121 = 0$
$\qquad\qquad\qquad\qquad\qquad 0 = 0 \checkmark$

5. $4p^2 - 36 = 0$ $\qquad$ Check: $4(3)^2 - 36 = 0$
$4(p^2 - 9) = 0$ $\qquad\qquad\qquad 4(9) - 36 = 0$
$4(p - 3)(p + 3) = 0$ $\qquad\qquad 36 - 36 = 0$
$p = -3, 3$ $\qquad\qquad\qquad 0 = 0 \checkmark$
$\qquad\qquad\qquad\qquad\qquad 4(-3)^2 - 36 = 0$
$\qquad\qquad\qquad\qquad\qquad 4(9) - 36 = 0$
$\qquad\qquad\qquad\qquad\qquad 36 - 36 = 0$
$\qquad\qquad\qquad\qquad\qquad 0 = 0 \checkmark$

6. $6g^2 - 96 = 0$ $\qquad$ Check: $6(4)^2 - 96 = 0$
$6(g^2 - 16) = 0$ $\qquad\qquad\qquad 6(16) - 96 = 0$
$6(g - 4)(g + 4) = 0$ $\qquad\qquad 96 - 96 = 0$
$g = -4, 4$ $\qquad\qquad\qquad 0 = 0 \checkmark$
$\qquad\qquad\qquad\qquad\qquad 6(-4)^2 - 96 = 0$
$\qquad\qquad\qquad\qquad\qquad 6(16) - 96 = 0$
$\qquad\qquad\qquad\qquad\qquad 96 - 96 = 0$
$\qquad\qquad\qquad\qquad\qquad 0 = 0 \checkmark$

7. $x^2 - 7x + 12 = 0$ Check: $(3)^2 - 7(3) + 12 = 0$
$(x - 3)(x - 4) = 0$ $\qquad$ $9 - 21 + 12 = 0$
$x = 3, 4$ $\qquad\qquad$ $0 = 0$ ✓
$\qquad\qquad\qquad$ $4^2 - 7(4) + 12 = 0$
$\qquad\qquad\qquad$ $16 - 28 + 12 = 0$
$\qquad\qquad\qquad$ $28 - 28 = 0$
$\qquad\qquad\qquad$ $0 = 0$ ✓

8. $y^2 - 10y + 21 = 0$ Check: $7^2 - 10(7) + 21 = 0$
$(y - 7)(y - 3) = 0$ $\qquad$ $49 - 70 + 21 = 0$
$y = 3, 7$ $\qquad\qquad$ $-21 + 21 = 0$
$\qquad\qquad\qquad$ $0 = 0$ ✓
$\qquad\qquad\qquad$ $3^2 - 10(3) + 21 = 0$
$\qquad\qquad\qquad$ $9 - 30 + 21 = 0$
$\qquad\qquad\qquad$ $0 = 0$ ✓

9. $-z^2 + 6z + 16 = 0$
$\qquad\qquad$ Check: $-8^2 + 6 \cdot 8 + 16 = 0$
$z^2 - 6z - 16 = 0$ $\qquad$ $-64 + 48 + 16 = 0$
$(z - 8)(z + 2) = 0$ $\qquad$ $-16 + 16 = 0$
$z = -2, 8$ $\qquad\qquad$ $0 = 0$ ✓
$\qquad\qquad\qquad$ $-(-2)^2 + 6(-2) + 16 = 0$
$\qquad\qquad\qquad$ $-4 - 12 + 16 = 0$
$\qquad\qquad\qquad$ $-16 + 16 = 0$
$\qquad\qquad\qquad$ $0 = 0$ ✓

10. $a^2 + 4a + 3 = 0$
$\qquad\qquad$ Check: $(-3)^2 + 4 \cdot -3 + 3 = 0$
$(a + 3)(a + 1) = 0$ $\qquad$ $9 - 12 + 3 = 0$
$a = -3, -1$ $\qquad\qquad$ $0 = 0$ ✓
$\qquad\qquad\qquad$ $(-1)^2 + 4(-1) + 3 = 0$
$\qquad\qquad\qquad$ $1 - 4 + 3 = 0$
$\qquad\qquad\qquad$ $0 = 0$ ✓

11. $2b^2 - b - 3 = 0$ $\quad$ Check: $2\left(\frac{3}{2}\right)^2 - \frac{3}{2} - 3 = 0$
$(2b - 3)(b + 1) = 0$
$b = -1, \frac{3}{2}$ $\qquad\qquad$ $\frac{2}{1}\left(\frac{9}{4}\right) - \frac{3}{2} - \frac{3}{1} = 0$
$\qquad\qquad\qquad$ $\frac{9}{2} - \frac{3}{2} - \frac{3}{1} = 0$
$\qquad\qquad\qquad$ $0 = 0$ ✓
$\qquad\qquad\qquad$ $2(-1)^2 - (-1) - 3 = 0$
$\qquad\qquad\qquad$ $2 + 1 - 3 = 0$
$\qquad\qquad\qquad$ $0 = 0$ ✓

12. $-6c^2 + c + 1 = 0$ Check: $-6\left(\frac{1}{2}\right)^2 + \frac{1}{2} + 1 = 0$
$6c^2 - c - 1 = 0$
$(2c - 1)(3c + 1) = 0$ $\qquad$ $-6\left(\frac{1}{4}\right) + \frac{1}{2} + 1 = 0$
$c = -\frac{1}{3}, \frac{1}{2}$ $\qquad\qquad$ $-\frac{3}{2} + \frac{1}{2} + 1 = 0$
$\qquad\qquad\qquad$ $0 = 0$ ✓
$\qquad\qquad\qquad$ $-6\left(-\frac{1}{3}\right)^2 - \frac{1}{3} + 1 = 0$
$\qquad\qquad\qquad$ $-6\left(\frac{1}{9}\right) - \frac{1}{3} + 1 = 0$
$\qquad\qquad\qquad$ $-\frac{2}{3} - \frac{1}{3} + 1 = 0$
$\qquad\qquad\qquad$ $0 = 0$ ✓

13. $6x^2 + 5x - 6 = 0$ Check: $6\left(\frac{2}{3}\right)^2 + 5\left(\frac{2}{3}\right) - 6 = 0$
$(3x - 2)(2x + 3) = 0$ $\qquad$ $6\left(\frac{4}{9}\right) + \frac{10}{3} - 6 = 0$
$x = -\frac{3}{2}, \frac{2}{3}$ $\qquad\qquad$ $\frac{24}{9} + \frac{30}{9} - 6 = 0$
$\qquad\qquad\qquad$ $0 = 0$ ✓
$\qquad\qquad$ $6\left(\frac{-3}{2}\right)^2 + 5\left(\frac{-3}{2}\right) - 6 = 0$
$\qquad\qquad\qquad$ $6\left(\frac{9}{4}\right) - \frac{15}{2} - 6 = 0$
$\qquad\qquad\qquad$ $\frac{27}{2} - \frac{15}{2} - 6 = 0$
$\qquad\qquad\qquad$ $0 = 0$ ✓

14. $m^2 = 9m$ $\qquad$ Check: $(0)^2 = 9(0)$ $\qquad$ $(9)^2 = 9(9)$
$m^2 - 9m = 0$ $\qquad\qquad$ $0 = 0$ ✓ $\qquad$ $81 = 81$ ✓
$m(m - 9) = 0$
$m = 0, 9$

15. $10 = k^2 + 3k$
$0 = k^2 + 3k - 10$
$0 = (k + 5)(k - 2)$
$k = -5, 2$
$\qquad$ Check: $10 = (-5)^2 + 3(-5) = 25 - 15 = 10$ ✓
$\qquad\qquad$ $10 = (2)^2 + 3(2) = 4 + 6 = 10.$ ✓

16. $15 = r^2 + 2r$ $\qquad$ Check: $15 = (-5)^2 + 2(-5)$
$0 = r^2 + 2r - 15$ $\qquad\qquad$ $15 = 25 - 10 = 15$ ✓
$0 = (r + 5)(r - 3)$ $\qquad\qquad$ $15 = (3)^2 + 2(3)$
$r = -5, 3$ $\qquad\qquad$ $15 = 9 + 6 = 15$ ✓

17. $x^2 - 3x - 25 = 3$ $\quad$ Check: $7^2 - 3(7) - 25 = 3$
$x^2 - 3x - 28 = 0$ $\qquad$ $49 - 21 - 25 = 3$
$(x - 7)(x + 4) = 0$ $\qquad\qquad\qquad$ $3 = 3$ ✓
$x = -4, 7$ $\qquad\qquad$ $(-4)^2 - 3(-4) - 25 = 3$
$\qquad\qquad\qquad$ $16 + 12 - 25 = 3$
$\qquad\qquad\qquad$ $3 = 3$ ✓

18. $z^2 - 11z + 28 = 10$ Check: $9^2 - 11(9) + 28 = 10$
$z^2 - 11z + 18 = 0$ $\qquad$ $81 - 99 + 28 = 10$
$(z - 9)(z - 2) = 0$ $\qquad\qquad$ $10 = 10$ ✓
$z = 2, 9$ $\qquad\qquad$ $2^2 - 11(2) + 28 = 10$
$\qquad\qquad\qquad$ $4 - 22 + 28 = 10$
$\qquad\qquad\qquad$ $10 = 10$ ✓

19. The equation $x^2 + 4 = 0$ is equivalent to $x^2 = -4$, which means that the square of a number is -4. There is no real number whose square is negative.

20. $x = $ a positive integer
$2x = $ an integer
$(2x)x + x + 2x = 119$
$2x^2 + 3x - 119 = 0$
$(2x + 17)(x - 7) = 0$
$x = -\frac{17}{2}$ $\quad$ $x = 7$

Since x is a positive integer, then $x = 7$ and $2x = 14$

21. $1728x^2 = (x + 2)^2 \cdot 432$
$1728x^2 = 432(x^2 + 4x + 4)$
$1728x^2 = 432x^2 + 1728x + 1728$
$1296x^2 - 1728x - 1728 = 0$
$432(3x^2 - 4x - 4) = 0$
$432(3x + 2)(x - 2) = 0$

$x = -\dfrac{2}{3}$ and 2

Since x represents the length, it must be positive. So
$x = 2$ in.

22.

$(2x - 4)$

Volume $= L \cdot H \cdot W$
Volume $= 896$ in.^{3}

$896 = 2 \cdot (x - 4) \cdot (2x - 4)$
$896 = (2x - 8)(2x - 4)$
$0 = 4x^2 - 24x + 32 - 896$
$0 = 4x^2 - 24x - 864$
$0 = 4(x^2 - 6x - 216)$
$0 = 4(x - 18)(x + 12)$
$x = -12, 18$

Since x represents the length of the original box, it cannot be negative. So 18 in. and 36 in. are the original dimensions.

23. $x^3 + x^2 - 36x - 36 = 0$
$x^2(x + 1) - 36(x + 1) = 0$
$(x^2 - 36)(x + 1) = 0$
$(x - 6)(x + 6)(x + 1) = 0$
$x = -6, -1, 6$
Check: $6^3 + 6^2 - 36 \cdot 6 - 36 = 0$
$216 + 36 - 216 - 36 = 0$
$0 = 0$ ✔
$(-6)^3 + (-6)^2 - 36 \cdot -6 - 36 = 0$
$-216 + 36 + 216 - 36 = 0$
$0 = 0$ ✔
$(-1)^3 + (-1)^2 - 36 \cdot -1 - 36 = 0$
$-1 + 1 + 36 - 36 = 0$
$0 = 0$ ✔

24. $d^3 - d^2 - 9d + 9 = 0$
$d^2(d - 1) - 9(d - 1) = 0$
$(d^2 - 9)(d - 1) = 0$
$(d - 3)(d + 3)(d - 1) = 0$
$d = -3, 1, 3$
Check: $3^3 - 3^2 - 9 \cdot 3 + 9 = 0$
$27 - 9 - 27 + 9 = 0$
$0 = 0$ ✔
$(-3)^3 - (-3)^2 - 9 \cdot -3 + 9 = 0$
$-27 - 9 + 27 + 9 = 0$
$0 = 0$ ✔
$(1)^3 - (1)^2 - 9(1) + 9 = 0$
$1 - 1 - 9 + 9 = 0$
$0 = 0$ ✔

25. $p^3 + 12p^2 + 36p = 0$
$p(p^2 + 12p + 36) = 0$
$p(p + 6)(p + 6) = 0$
$p = -6, -6, 0$
Check:
$0^3 + 12(0)^2 + 36(0) = 0$
$0 = 0$ ✔
$(-6)^3 + 12(-6)^2 + 36(-6) = 0$
$-216 + 432 - 216 = 0$
$0 = 0$ ✔

26. $x^3 - 14x^2 + 49x = 0$
$x(x^2 - 14x + 49) = 0$
$x(x - 7)(x - 7) = 0$
$x = 0, 7$
Check:
$0^3 - 14(0)^2 + 49(0) = 0$
$0 = 0$ ✔
$7^3 - 14(7)^2 + 49(7) = 0$
$343 - 686 + 343 = 0$
$0 = 0$ ✔

27. $-3x^3 + 3x^2 + 36x = 0$
$-3x(x^2 - x - 12) = 0$
$-3x(x - 4)(x + 3) = 0$
$x = -3, 0, 4$
Check:
$-3(0)^3 + 3(0)^2 + 36(0) = 0$
$0 = 0$ ✔
$-3(4)^3 + 3(4)^2 + 36(4) = 0$
$-192 + 48 + 144 = 0$
$0 = 0$ ✔
$-3(-3)^3 + 3(-3)^2 + 36(-3) = 0$
$81 + 27 - 108 = 0$
$0 = 0$ ✔

28. $5n^3 - 20n^2 - 25n = 0$

$5n(n^2 - 4n - 5) = 0$

$5n(n - 5)(n + 1) = 0$

$n = -1, 0, 5$

Check:

$$5(0)^3 - 20(0)^2 - 25(0) = 0$$
$$0 = 0\checkmark$$
$$5(5)^3 - 20(5)^2 - 25(5) = 0$$
$$625 - 500 - 125 = 0$$
$$0 = 0\checkmark$$
$$5(-1)^3 - 20(-1)^2 - 25(-1) = 0$$
$$-5 - 20 + 25 = 0$$
$$0 = 0\checkmark$$

29. Let $x = $ the length of one piece of the wire. Then $52 - x = $ the length of the other piece of wire.

$$\left(\frac{x}{4}\right)^2 + \left(\frac{52 - x}{4}\right)^2 = 97$$

$$\frac{x^2}{16} + \frac{(52 - x)^2}{16} = 97$$

$$x^2 + (52 - x)^2 = 1552$$

$$x^2 + 2704 - 104x + x^2 - 1552 = 0$$

$$2x^2 - 104x + 1152 = 0$$

$$(2x - 72)(x - 16) - 0$$

$$x = 16 \text{ in.}, 36 \text{ in.}$$

30a. $P(x) = x^2 + 60x - 945 - (2x^2 + 10x - 420)$

$\quad = x^2 + 60x - 945 - 2x^2 - 10x + 420$

$\quad = -x^2 + 50x - 525$

30b. $-x^2 + 50x - 525 = 0$

$x^2 - 50x + 525 = 0$

$(x - 15)(x - 35) = 0$

$x = 15 \quad x = 35$

15,000 or 35,000 figurines

31. $(x - 0)(x - 3) = 0$

$x(x - 3) = 0$

$x^2 - 3x = 0$

32. $(x - 2)(x - 5) - 0$

$x^2 - 5x - 2x + 10 - 0$

$x^2 - 7x + 10 = 0$

33. $(x - 7)(x + 2) = 0$

$x^2 - 5x - 14 = 0$

34. $(x + 3)(x + 5) = 0$

$x^2 + 8x + 15 = 0$

35. $(x - 4)(x - 4) = 0$

$x^2 - 8x + 16 = 0$

36. $(x + 5)(x + 5) = 0$

$x^2 + 10x + 25 = 0$

37. $(x - 0)(x - 6)(x - 6) = 0$

$x(x^2 - 12x + 36) = 0$

$x^3 - 12x^2 + 36x = 0$

38. $(x - 0)(x - 3)(x + 4) = 0$

$x(x^2 + x - 12) = 0$

$x^3 + x^2 - 12x = 0$

39. $(x - 2)(x - 5)(x - 5) = 0$

$(x - 2)(x^2 - 10x + 25) = 0$

$x^3 - 10x^2 + 25x - 2x^2 + 20x + 50 = 0$

$x^3 - 12x^2 + 45x - 50 = 0$

40. $(x - 3)(x + 2)(x + 2) = 0$

$(x - 3)(x^2 + 4x + 4) = 0$

$x^3 + 4x^2 + 4x - 3x^2 - 12x - 12 = 0$

$x^3 + x^2 - 8x - 12 = 0$

41. $(x - 2)(x - 5)(x - 8) = 0$

$(x^2 - 7x + 10)(x - 8) = 0$

$x^3 - 8x^2 - 7x^2 + 56x + 10x - 80 = 0$

$x^3 - 15x^2 + 66x - 80 = 0$

42. $(x + 3)(x - 4)(x + 6) = 0$

$(x^2 - x - 12)(x + 6) = 0$

$x^3 + 6x^2 - x^2 - 6x \quad 12x - 72 = 0$

$x^3 + 5x^2 - 18x - 72 = 0$

EXTEND

43. $(3x + 1)^2 + 5(3x + 1) = 0$

$(3x + 1 + 5)(3x + 1) = 0$

$(3x + 6)(3x + 1) = 0$

$x = -2, -\frac{1}{3}$

Check: $\quad (3 \cdot -2 + 1)^2 + 5(3 \cdot -2 + 1) \overset{?}{=} 0$

$\qquad\qquad\qquad (-5)^2 + 5(-5) \overset{?}{=} 0$

$\qquad\qquad\qquad\qquad\quad 25 \quad 25 \overset{?}{=} 0$

$\qquad\qquad\qquad\qquad\qquad\qquad 0 = 0\checkmark$

$\left(\frac{3}{1} \cdot -\frac{1}{3} + 1\right)^2 + 5\left(\frac{3}{1} \cdot -\frac{1}{3} + 1\right) \overset{?}{=} 0$

$\qquad\qquad\qquad 0^2 + 5(0) \overset{?}{=} 0$

$\qquad\qquad\qquad\qquad\quad 0 = 0\checkmark$

44. $(x - 2)^2 - 4(x - 2) - 12 = 0$

$x^2 - 4x + 4 - 4x + 8 - 12 = 0$

$x^2 - 8x = 0$

$x(x - 8) = 0$

$x = 0, 8$

Check: $\quad (0 - 2)^2 - 4(0 - 2) - 12 \overset{?}{=} 0$

$\qquad\qquad\qquad\qquad 4 + 8 - 12 \overset{?}{=} 0$

$\qquad\qquad\qquad\qquad\qquad\quad 0 = 0\checkmark$

$\qquad (8 - 2)^2 - 4(8 - 2) - 12 \overset{?}{=} 0$

$\qquad\qquad\qquad\quad 36 - 4(6) - 12 \overset{?}{=} 0$

$\qquad\qquad\qquad\quad 36 - 24 - 12 \overset{?}{=} 0$

$\qquad\qquad\qquad\qquad\qquad\quad 0 = 0\checkmark$

45. $x^4 - 13x^2 + 36 = 0$
$(x^2 - 4)(x^2 - 9) = 0$
$(x - 2)(x + 2)(x - 3)(x + 3) = 0$
$x = -3, -2, 2, 3$
Check: $(2)^4 - 13(2)^2 + 36 \stackrel{?}{=} 0$
$16 - 52 + 36 \stackrel{?}{=} 0$
$0 = 0$ ✓
$(-2)^4 - 13(-2)^2 + 36 \stackrel{?}{=} 0$
$16 - 52 + 36 \stackrel{?}{=} 0$
$0 = 0$ ✓
$(3)^4 - 13(3)^2 + 36 \stackrel{?}{=} 0$
$81 - 117 + 36 \stackrel{?}{=} 0$
$0 = 0$ ✓
$(-3)^4 - 13(-3)^2 + 36 \stackrel{?}{=} 0$
$81 - 117 + 36 \stackrel{?}{=} 0$
$0 = 0$ ✓

46. $x^4 - 29x^2 + 100 = 0$
$(x^2 - 25)(x^2 - 4) = 0$
$(x - 5)(x + 5)(x - 2)(x + 2) = 0$
$x = -5, -2, 2, 5$
Check: $(5)^4 - 29(5)^2 + 100 \stackrel{?}{=} 0$
$625 - 725 + 100 \stackrel{?}{=} 0$
$0 = 0$ ✓
$(-5)^4 - 29(-5)^2 + 100 \stackrel{?}{=} 0$
$625 - 725 + 100 \stackrel{?}{=} 0$
$0 = 0$ ✓
$(2)^4 - 29(2)^2 + 100 \stackrel{?}{=} 0$
$16 - 116 + 100 \stackrel{?}{=} 0$
$0 = 0$ ✓
$(-2)^4 - 29(-2)^2 + 100 \stackrel{?}{=} 0$
$16 - 116 + 100 \stackrel{?}{=} 0$
$0 = 0$ ✓

47.

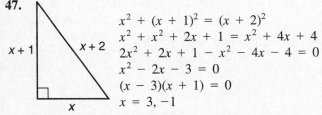

$x^2 + (x + 1)^2 = (x + 2)^2$
$x^2 + x^2 + 2x + 1 = x^2 + 4x + 4$
$2x^2 + 2x + 1 - x^2 - 4x - 4 = 0$
$x^2 - 2x - 3 = 0$
$(x - 3)(x + 1) = 0$
$x = 3, -1$

Since x represents the length of a leg of a right triangle it cannot be negative. So $x = 3$ in., 4 in., 5 in.

48. $x = $ tens' digit
$y = $ units digit
$10x + y = $ number
$y^2 = 10x + y + 2$
$y - x = 3$
$y = x + 3$
$x = y - 3$
$y^2 = 10(y - 3) + y + 2$
$y^2 = 10y - 30 + y + 2$
$y^2 = 11y - 28$
$y^2 - 11y + 28 = 0$

$(y - 4)(y - 7) = 0$
$y = 4, 7$
$x = 1, 4$
The two possible numbers are 14 and 47.

49. $x = $ one positive integer
$x + 2 = $ consecutive odd integer
$x^2 + (x + 2)^2 = 1570$
$x^2 + x^2 + 4x + 4 = 1570$
$2x^2 + 4x - 1566 = 0$
$2(x^2 + 2x - 783) = 0$
$2(x - 27)(x - 29) = 0$
$x = 27, 29$

THINK CRITICALLY

50. $\begin{cases} x^2 + y^2 = 18 \\ x = y \end{cases}$
$x^2 + x^2 = 18$
$2x^2 = 18$
$2x^2 - 18 = 0$
$2(x^2 - 9) = 0$
$2(x - 3)(x + 3) = 0$
$x = -3, 3$
$(3, 3); (-3, -3)$

51. $\begin{cases} x^2 + 3 = y \\ y = 5 - x \end{cases}$ $\qquad x = -2, 1$
$x^2 + 3 = 5 - x$ $\qquad y = 5 + 2 = 7$
$x^2 + x - 2 = 0$ $\qquad y = 5 - 1 = 4$
$(x + 2)(x - 1) = 0$ $\qquad (-2, 7); (1, 4)$

52. $\begin{cases} y + 3 = 2x \\ 3x^2 = 4 + xy \end{cases}$

from the first equation: $\dfrac{y + 3}{2} = x$

Substituting $\dfrac{y + 3}{2}$ into x in the second equation, we have

$3\left(\dfrac{y + 3}{2}\right)^2 = 4 + \left(\dfrac{y + 3}{2}\right)y$

$4\left[\dfrac{3}{4}(y + 3)^2 = 4 + \dfrac{y^2 + 3y}{2}\right]$

$3(y + 3)^2 = 16 + 2(y^2 + 3y)$
$3(y^2 + 6y + 9) = 16 + 2y^2 + 6y$
$3y^2 + 18y + 27 - 16 - 2y^2 - 6y = 0$
$y^2 + 12y + 11 = 0$
$(y + 11)(y + 1) = 0$
$y = -11, -1$

$-11 + 3 = 2x$ and $-1 + 3 = 2x$
$-8 = 2x$ $\qquad\qquad 2 = 2x$
$x = -4$ $\qquad\qquad x = 1$
$(-4, -11)$ $\qquad\qquad (1, -1)$

53. The graph touches the x-axis at 5.

54. $\begin{bmatrix} 4 & 3 \\ -2 & 0 \\ 1 & 5 \end{bmatrix} + \begin{bmatrix} 0 & -7 \\ 1 & 8 \\ 6 & -4 \end{bmatrix} = \begin{bmatrix} 4 & -4 \\ -1 & 8 \\ 7 & 1 \end{bmatrix}$

55. $\begin{bmatrix} 3 & -8 \\ 1 & -1 \end{bmatrix} \begin{bmatrix} 2 & 0 \\ 3 & -5 \end{bmatrix} = \begin{bmatrix} -18 & 40 \\ -1 & 5 \end{bmatrix}$

56. C; mean = 6; mode = 6; median = 6

57. no; $a \not> a$ **58.** yes   **59.** B

ALGEBRAWORKS

1.
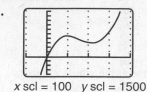
x scl = 100 y scl = 1500

2. The graph increases to a maximum value of 104 items where profit is $7307.22. Then the profit begins decreasing.

3. The graph decreased to a minimum profit of $4507.60 for 230 items. Then profits rise again.

4. maximum profit = $11,743.75. Must make 325 items to reach this profit.

5. The owner should make at least 295 items, and try to make 325 items.

Lesson 5.6, pages 262–265

EXPLORE THE PROBLEM

1. Each expression on the left is the sum of three consecutive whole numbers.

2. Each number on the right has a factor of 3.

3. Examples may vary. Some examples are conjecture: the sum of three consecutive whole numbers always has a factor of 3.
21 + 22 + 23 = 66 and
47 + 48 + 49 = 144.

4. $m + (m + 1) + (m + 2) = 3m + 3$

5. $3m + 3 = 3(m + 1)$; the sum is 3 times the middle number.

6. $1023 + 1024 + 1025 = 3(1024)$
$= 3072$

7. Answers will vary. Students may suggest the sum of 4 consecutive whole numbers has a factor of 4 (not true) or the sum of five consecutive whole numbers has a factor of 5 (true). In general, the sum of n consecutive **odd** integers has a factor of n.

INVESTIGATE FURTHER

8. 49, 79, 128, 207, 335

9. sum = 869

10. The seventh number is a factor of the sum.
869 = 11(79).

11. The sum is always 11 times the seventh number

12. $m, n, m + n, m + 2n, 2m + 3n, 3m + 5n,$
$5m + 8n, 8m + 13n, 13m + 21n, 21m + 34n$, the seventh number is $5m + 8n$

13. $55m + 88n = 11(5m + 8n)$; the sum of 11 times $5m + 8n$, which is the seventh number.

14. seventh number = 97
11(97) = 1067
5, 9, 14, 23, 37, 60, 97, 157, 254, 411
sum = 1067

15. Answers will vary.

APPLY THE STRATEGY

16. perimeter = 12 ft; area = 5 ft^2

17. perimeter = 16 ft; area = 10 ft^2

18. perimeter = 20 ft; area = 10 ft^2

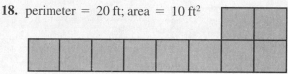

19. perimeter = 24 ft; area = 20 ft^2

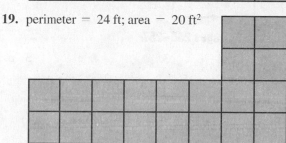

20. Doubling both dimensions doubles the perimeter; doubling either dimension doubles the area.

21. $p = z + r, z = 8, r = 4$

22. vertical doubling: $p = z + 2r, p = 8 + 2(4) = 16$ ft
horizontal doubling:
$p = 2z + r, p = 2(8) + 4 = 20$ ft
doubling both dimensions:
$p = 2z + 2r = 2(8) + 2(4) = 24$ ft

23. $z + 2r + 2z + r = 3z + 3r = 3(z + r)$;
$3(8 + 4) = 3(12) = 36$ ft

24. perimeter would triple; perimeter would be multiplied by n.

25. area was multiplied by 4 or 2^2; area will be multiplied by 9 or 3^2; area will be multiplied by n^2.

26. $abc = abc + 1000abc = abc(1 + 1000) =$
$1001(abc)$ and 11 is a factor of 1001. The other factors are 7 and 13.

27. $a^2 + (1 - a) = (1 - a)^2 + a$; if you multiply each side of the equation, you get $a^2 - a + 1$.

28. Result is a perfect square.
$(x)(x + 1)(x + 2)(x + 3) + 1 = (x^2 + 3x + 1)^2$

REVIEW PROBLEM SOLVING STRATEGIES

1a. equations; four variables.
  I. $c + s = p$
  II. $s = c + t$
  III. $2p = 3t$

1b. Add c to each side of II to get $s + c = 2c + t$.

1c. By I, $s + c = p$ so $p = 2c + t$. Multiply by 2 to get $2p = 4c + 2t$; substitute in III to get $4c + 2t = 3t$, so $4c = t$. Substitute in II to find $s = 5c$. So, 5 calculators balance a stapler.

2. 557 pages; pages 1–10 use 11 digits, pages 11–90 use 160 digits, pages 91–100 use 21 digits, pages 101–200 use 300 digits, 301–400 use 300 digits, 401–500 use 300 digits for a total of 1392 digits. The remaining 171 digits will number $171 \div 3 = 57$ pages

3. Yes;

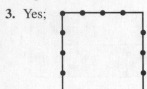

Chapter Review, pages 266–267

1. c   **2.** d   **3.** a   **4.** e   **5.** b

6. $x^3 - 2x^2 + 2x - 5$ **7.** $x^3 + 2x^2 - 4x + 4$

8. $2x^3 - 2x - 1$

9. $(x^3 - 2x^2 + 2x - 5) - (-x^3 - 2x^2 + 4x - 4)$

10. Subtraction is the addition of an opposite

11. $(3x - 2)(2x - 1)$ **12.** $6x^2 - 7x + 2$

13. $(6x^2 - 7x + 2) \div (3x - 2) = 2x - 1$
  $(6x^2 - 7x + 2) \div (2x - 1) = 3x - 2$

14. factoring $6x^2 - 7x + 2$ **15.** $-6x^7$ **16.** $16y^6$

17. $4m^3$ **18.** 3   **19.** $\frac{1}{4}$ **20.** $4x^2 - 6x - 9$

21. $-2x^3 - 3x^2 + x - 4$ **22.** $8x^2 + 10x - 3$

23. $6k^4 - 15k^3 + 6k^2$ **24.** $1 - 4x^2$

25. $8xy(3x^2 - y^3)$ **26.** $(2x + 5)(2x - 5)$

27. $(y - 5)^2$ **28.** $(x - 1)(2x + 3)$

29. $3a(x - 3)(2x + 1)$ **30.** $(z^2 + 4)(z + 2)(z - 2)$

31. $(a + 1)(b - 2)$ **32.** $(x - 4)(x^2 + 4x + 16)$

33. $x^2 - 5x = 0$ **34.** $4a^2 - 64 = 0$
  $x(x - 5) = 0$ $4(a^2 - 16) = 0$
  $x = 0, 5$ $4(a - 4)(a + 4) = 0$
                        $a = \pm 4$

35. $z^2 - 6z + 9 = 0$ **36.** $r^2 - 3r + 6 = 4$
  $(z - 3)(z - 3) = 0$ $r^2 - 3r + 2 = 0$
  $z = 3, 3$ $(r - 2)(r - 1) = 0$
                              $r = 1, 2$

37. $3x^2 - 10x + 8 = 0$ **38.** $2x^3 - 6x^2 - 20x = 0$
  $(3x - 4)(x - 2) = 0$ $x(2x^2 - 6x - 20) = 0$
  $x = \frac{4}{3}, 2$ $x(2x - 10)(x + 2) = 0$
                                 $x = -2, 0, 5$

39. $x^3 + x^2 - 4x - 4 = 0$
  $x^2(x + 1) - 4(x + 1) = 0$
  $(x^2 - 4)(x + 1) = 0$
  $(x - 2)(x + 2)(x + 1) = 0$
  $x = -2, -1, 2$

40. $y^5 - 4y^3 = 0$
  $y^3(y^2 - 4) = 0$
  $y^3(y - 2)(y + 2) = 0$
  $y = -2, 0, 2$

41. The sum of any seven consecutive whole numbers always has a factor of 7.
  $m + (m + 1) + (m + 2) + (m + 3) + (m + 4) + (m + 5) + (m + 6) = 7m + 21$
  $= 7(m + 3)$

Chapter Assessment, pages 268–269

CHAPTER TEST

1. After removing zero pairs of x-blocks, the product modeled is $2x^2 + x - 6$. This model represents
  $(x + 2)(2x - 3) = 2x^2 + x - 6$,
  $(2x^2 + x - 6) \div (x + 2) = 2x - 3$,
  $(2x^2 + x - 6) \div (2x - 3) = x + 2$, or
  $2x^2 + x - 6 = (x + 2)(2x - 3)$.

2. y^4 **3.** $-2m^7$ **4.** $27x^6$ **5.** 4   **6.** $10z^4$

7. $\frac{2}{16} = \frac{1}{8}$ **8.** C   **9.** $-x^3 - 3x^2 - x + 1$

10. $-5y^2 - 2y - 2$ **11.** $12w^4z^3 - 8w^5z^2 + 8w^6z$

12. $6x^2 + 8x - 8$ **13.** $6q^3 - 19q^2 + 17q - 3$

14. $\frac{(4x + 2)(x - 2)}{2} = \frac{4x^2 - 6x + 4}{2} = 2x^2 - 3x + 2$

15. $\frac{1}{2}(4x)(x + 3 + x + 7)$
  $= (2x)(2x + 10)$
  $= 4x^2 + 20x$

16. $7ab(3a - 1)$ **17.** $(5 + 3x)(5 - 3x)$

18. $(y + 2)(y - 6)$ **19.** $(2m + 5)^2$

20. $(3z - 2)(2z - 3)$ **21.** $2a(3x + 1)(2x - 5)$

22. $2(m - 5)(m + 5)$ **23.** $(x - 2)(x - 4y)$

24. $(a + 2)(a + 2)(a - 2)$ **25.** $(m - 5)(m^2 + 5m + 25)$

26. $(r + 4)(r^2 - 4r + 16)$

27. Subtract 9 from each side: $x^2 + x - 2 = 0$.
  Factor the trinomial: $(x + 2)(x - 1) = 0$.
  Apply the zero product property: $x + 2 = 0$ or $x - 1 = 0$.
  Then solve each equation: $x = -2$ and $x = 1$.
  Check by substituting each of the values -2 and 1 into the original equation $x^2 + x + 7 = 9$.

28. $x^2 + 7x = 0$ Check: $0^2 + 7(0) \overset{?}{=} 0 + 0 = 0$ ✓
     $x(x + 7) = 0$ $7^2 + 7(-7) \overset{?}{=} 49 - 49 = 0$ ✓
     $x = -7, 0$

29. $3y^2 - 27 = 0$ Check: $3(-3)^2 - 27 \overset{?}{=} 0$
     $3(y + 3)(y - 3) = 0$ $27 - 27 = 0$ ✓
     $y = -3, 3$

$$3(3)^2 - 27 \overset{?}{=} 0$$
$$27 - 27 = 0 ✓$$

30. $(3a - 2)(a - 1) = 0$

     $a = \dfrac{2}{3}, 1$ Check: $(3 \cdot 1 - 2)(1 - 1) = 0$

$$1 \cdot 0 = 0 ✓$$
$$\left(3 \cdot \frac{2}{3} - 2\right)\left(\frac{2}{3} - 1\right) = 0$$
$$0 \cdot -\frac{1}{3} = 0 ✓$$

31. $2y^2 + 5y - 3 = 0$ Check: $2(-3)^2 + 5(-3) + 5 \overset{?}{=} 8$
     $(2y - 1)(y + 3) = 0$ $18 - 15 + 5 \overset{?}{=} 8$

     $y = -3, \dfrac{1}{2}$ $3 + 5 = 8$ ✓

$$2\left(\frac{1}{2}\right)^2 + 5\left(\frac{1}{2}\right) + 5 \overset{?}{=} 8$$
$$\frac{1}{2} + \frac{5}{2} + 5 \overset{?}{=} 8$$
$$3 + 5 = 8 ✓$$

32. $(x - 7)(x - 7) = 0$ Check: $7^2 - 14(7) + 49 \overset{?}{=} 0$
     $x = 7, 7$ $49 - 98 + 49 \overset{?}{=} 0$
                         $0 = 0$ ✓

33. $x(x - 4)(x + 3) = 0$ Check: $0^3 - 0^2 - 12(0) \overset{?}{=} 0$
     $x = -3, 0, 4$ $0 = 0$ ✓

$$4^3 - 4^2 - 12(4) \overset{?}{=} 0$$
$$64 - 16 - 48 \overset{?}{=} 0$$
$$0 = 0 ✓$$
$$(-3)^3 - (-3)^2 - 12(-3) \overset{?}{=} 0$$
$$-27 - 9 + 36 \overset{?}{=} 0$$
$$0 = 0 ✓$$

34. $m^2(m + 1) - 9(m + 1) = 0$
     $(m - 3)(m + 3)(m + 1) = 0$
     $m = -3, -1, 3$

$$\text{Check: } 3^3 + 3^2 - 9(3) - 9 \overset{?}{=} 0$$
$$27 + 9 - 27 - 9 \overset{?}{=} 0$$
$$0 = 0 ✓$$
$$(-1)^3 + (-1)^2 - 9(-1) - 9 \overset{?}{=} 0$$
$$-1 + 1 + 9 - 9 \overset{?}{=} 0$$
$$0 = 0 ✓$$
$$(-3)^3 + (-3)^2 - 9(-3) - 9 \overset{?}{=} 0$$
$$-27 + 9 + 27 - 9 \overset{?}{=} 0$$
$$0 = 0 ✓$$

35. $z^2(z^2 - 5) = 0$ Check: $(\sqrt{5})^4 - 5(\sqrt{5})^2 \overset{?}{=} 0$
     $z = 0, 0, \pm\sqrt{5}$ $25 - 25 \overset{?}{=} 0$
                            $0 = 0$ ✓
$$(-\sqrt{5})^4 - 5(-\sqrt{5})^2 \overset{?}{=} 0$$
$$25 - 25 \overset{?}{=} 0$$
$$0 = 0 ✓$$
$$0^4 - 5(0)^2 \overset{?}{=} 0$$
$$0 = 0 ✓$$

36. $2x^2 = (x + 2)(x + 4) - 1$
            $2x^2 = x^2 + 6x + 7$
       $x^2 - 6x - 7 = 0$
   $(x - 7)(x + 1) = 0$
             $x = -1, 7$
        $7, 9, 11$ or $-1, 1, 3$

37.

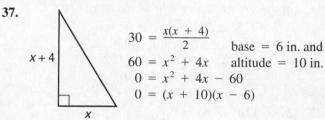

$30 = \dfrac{x(x + 4)}{2}$ base = 6 in. and
$60 = x^2 + 4x$ altitude = 10 in.
$0 = x^2 + 4x - 60$
$0 = (x + 10)(x - 6)$

Cumulative Review, page 270

1. D   **2.** $-2a^2 - 8a$

3. $8x^2 + x + 9$ **4.** $-8m^7 - 28m^5 + 4m^3$

5. $12z^2 + 7z - 12$

6. The greatest common factor of a polynomial is the greatest common factor of all the coefficients and the lowest power of the variables in every term.

7. $3m(m + 2)(m - 2)$ **8.** $(3z - 1)^2$

9. $(y + 3)(x - 2)$ **10.** $(y - 3)(y^2 + 3y + 9)$

11. $3(2a - 3)(a + 7)$ **12.** $6xy^2(3x - 5y)(2x + 7y)$

13. $3(q^2 - 81) = 0$ Check: $3(9)^2 - 243 \overset{?}{=} 0$
     $3(q - 9)(q + 9) = 0$ $243 - 243 = 0$ ✓
     $q = \pm 9$ $3(-9)^2 - 243 \overset{?}{=} 0$
                          $243 - 243 = 0$ ✓

14. $(3x - 2)(x + 4) = 0$ Check: $3\left(\frac{2}{3}\right)^2 + 10\left(\frac{2}{3}\right) - 8 = 0$

     $x = -4, \dfrac{2}{3}$ $\dfrac{4}{3} + \dfrac{20}{8} - 8 = 0$

$$\frac{24}{3} - 8 = 0$$
$$0 = 0 ✓$$
$$3(-4)^2 + 10(-4) - 8 = 0$$
$$48 - 40 - 8 = 0$$
$$0 = 0 ✓$$

15. $z^2(z + 2) - 25(z + 2) = 0$
$(z - 5)(z + 5)(z + 2) = 0$
$z = -5, -2, 5$
Check: $(5)^3 + 2(5)^2 - 25(5) - 50 = 0$
$125 + 50 - 125 - 50 = 0$
$0 = 0$ ✓
$(-5)^3 + 2(-5)^2 - 25(-5) - 50 = 0$
$-125 + 50 + 125 - 50 = 0$
$0 = 0$ ✓
$(-2)^3 + 2(-2)^2 - 25(-2) - 50 = 0$
$-8 + 8 + 50 - 50 = 0$
$0 = 0$ ✓

16. $(2x + 16)(x + 12) = 0$
$x = -12, -8$
Check: $2(-8)^2 + 40(-8) + 192 = 0$
$128 - 320 + 192 = 0$
$0 = 0$ ✓
$2(-12)^2 + 40(-12) + 192 = 0$
$288 - 480 + 192 = 0$
$0 = 0$ ✓

17. $(3y - 12)(y - 4) = 0$
$y = 4, 4$
Check: $3(4)^2 - 24(4) + 48 = 0$
$48 - 96 + 48 = 0$
$0 = 0$ ✓

18. Write *abba* in expanded form:
$a(10^3) + b(10^2) + b(10) + a$
Group terms and factors: $a(10^3 + 1) + b(10^2 + 10)$
$= a(1001) + b(110)$
$= 11a(91) + 11b(10)$
$= 11(91a + 10b)$

19. Set up a system of 3 equations and solve.
$10A + 12B + 15C = 327$
$24A + 18B + 20C = 576$
$16A \qquad + 14C = 318$

Multiply the first equation by 3 and the second by 2.
$30A + 36B + 45C = 981$
$\underline{-(48A + 36B + 40C = 1152)}$
$-18A \qquad + 5C = -171$

Multiply the above equation by 8 and the third equation at the top by 9.
$-144A + 40C = -1368$
$\underline{+(144A + 126C = 2862)}$
$\qquad\qquad 166C = 1494$
$\qquad\qquad\quad C = 9$

$16A + 14(9) = 318$
$\qquad 16A = 192$
$\qquad\quad A = 12$

$10(12) + 12B + 15(9) = 327$
$\qquad 12B + 255 = 327$
$\qquad\qquad 12B = 72$
$\qquad\qquad\quad B = 6$

A:12; B:6; C:9

20. In Cramer's Rule, the value of each variable is obtained from a fraction that uses a determinant as the numerator and as the denominator. If the value of the coefficient determinant is 0, it is not possible to apply Cramer's rule even though the system may have a solution.

21. Let H = horsepower of steam engine
R = number of revolutions per minute.
$H = KR$
$1280 = K160$
$K = 8$
So $\quad H = 8R$
$H = 8(220)$
$H = 1760$ horsepower.

22a. $A = p + prt$
$A - p = prt$
$\dfrac{A - p}{pr} = t$
$t = \dfrac{A - p}{pr}$

22b. $A = p(1 + rt)$
$\dfrac{A}{1 + rt} = p$
$p = \dfrac{A}{1 + rt}$

Standardized Test, page 271

1. D; $(2x^2)^3 = 8x^6$

2. C; $49 - 14x - x^2$
$= (7 - x)(7 - x)$
$= (7 - x)^2$

3. E; $3(x^2 - 16)$
$= 3(x - 4)(x + 4)$

4. D

5. E; $\frac{1}{2}x + 2 = y$
$2x + 2 = 4y - 6$

6. C

7. D; $\begin{vmatrix} 12 & -4 \\ -8 & -2 \end{vmatrix} = -24 - 32 = -56$

8. B

9. E; $f(n + 1) = \dfrac{n + 1 - 2}{n + 1 + 1} = \dfrac{n - 1}{n + 2}$

10. A; $2y = 6x + 1 \qquad\qquad -\dfrac{8}{3} = b$
$\quad y = 3x + 1 \qquad\qquad y = -\dfrac{1}{3}x - \dfrac{8}{3}$
$\text{slope} = 3 \qquad\qquad 3y = -x - 8$
$-3 = -\dfrac{1}{3}(1) + b \qquad 3y + x + 8 = 0$
$-3 = -\dfrac{1}{3} + b$
$-3 + \dfrac{1}{3} = b$

11. B

Chapter 6 Quadratic Functions and Equations

Data Activity, pages 272–273

1. Answers will vary; $300 + 400 + 400 + 200 +$
   $100 + 200 + 100 + 100 + 100 + 200 =$
   $2100; 2,100,000$

2. Let x represent the percent of departures from the ten
   busiest airports; $7,200,000x = 2,100,000x = 0.2916\overline{6}$,
   $x \approx 29\%$

3. Chicago, Dallas/Ft. Worth, Atlanta, Los Angeles,
   Denver, San Francisco, Phoenix, Detroit, Newark,
   Minneapolis/St. Paul

4. Answers will vary. Possible answer: A bar graph of
   pictograph would effectively display the data and allow
   for visual comparison.

5. Let x represent the number of passengers in 1988;
   $11,295 = x + 2.34x, x = 11,295 \div 3.34$,
   $x \approx 3381.736527; 3,382,000$ passengers

6. Answers will vary.

Lesson 6.1, pages 275–281

EXPLORE

1. $x^2 + 6x = -3$

2a.
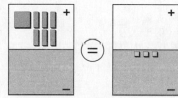

2b.
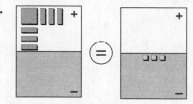

3. Since the model represents an equation, you must also
   add 9 unit blocks to the right side of the mat.

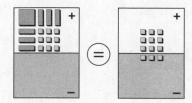

4.
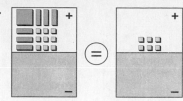

$x^2 + 6x + 9 = 6$

5. $(x + 3)^2 = 6, x + 3 = \pm\sqrt{6}, x = -3 \pm \sqrt{6}$

6. $x^2 + 4x = -1$

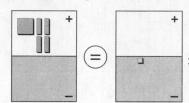

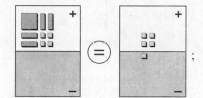

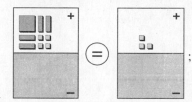

$(x + 2)^2 = 3, x + 2 = \pm\sqrt{3}, x = -2 \pm \sqrt{3}$

7. $x^2 + 2x - 2$

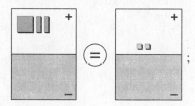

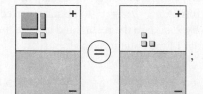

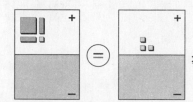

$(x + 1)^2 = 3, x + 1 = \pm\sqrt{3}, x = -1 \pm \sqrt{3}$

8. $x^2 + 8x = -4$

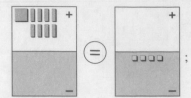

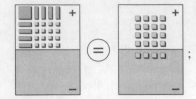

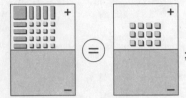

$(x + 4)^2 = 12, x + 4 = \pm\sqrt{12},$
$x + 4 = \pm 2\sqrt{3}, x = -4 \pm 2\sqrt{3}$

9. $\left(\dfrac{b}{2}\right)^2$

TRY THESE

1. $x^2 + 10x + 25$ **2.** $x^2 - 20x + 100$

3. $x^2 - 13x + \dfrac{169}{4}$

4. Let n represent the number. $n^2 = 2n + 120$,
$n^2 - 2n - 120, (n - 12)(n + 10); n = 12$ and
$n = -10$; The numbers are -10 and 12.

5. $\left(\dfrac{b}{2}\right)^2 = 36$ **6.** $\left(\dfrac{b}{2}\right)^2 = 16$ **7.** $\left(\dfrac{b}{2}\right)^2 = 49$

$\dfrac{b}{2} = \pm 6$ $\dfrac{b}{2} = \pm 4$ $\dfrac{b}{2} = \pm 7$

$b = \pm 12$ $b = \pm 8$ $b = \pm 14$

8. Answers will vary. Possible answer: In a quadratic
equation of the form $x^2 + bx = c$, to obtain a perfect
square divide b by 2, square the result, and add the final
result to both sides of the equation. Example:
$x^2 + 6x = -41$ $x^2 + 6x + 9 = -32$

9. $x^2 - 8x = -15$ $5^2 - 8(5) \stackrel{?}{=} -15$
$x^2 - 8x + 16 = -15 + 16$ $-15 = -15$
$(x - 4)^2 = 1$ $3^2 - 8(3) \stackrel{?}{=} -15$
$x - 4 = \pm 1$ $-15 = -15$
$x = 5$ or $x = 3$

10. $x^2 + 14x + 3 = 0$
$x^2 + 14x = -3$
$x^2 + 14x + 49 = -3 + 49$
$(x + 7)^2 = 46$
$x + 7 = \pm\sqrt{46}$
$x = -7 + \sqrt{46}$
$(-7 + \sqrt{46})^2 + 14(-7 - \sqrt{46}) + 33 \stackrel{?}{=} 0, 0 = 0$
$(-7 - \sqrt{46})^2 + 14(-7 - \sqrt{46}) + 33 \stackrel{?}{=} 0, 0 = 0$

11. $x^2 + 8x - 2 = 0$
$x^2 + 8x + 16 = 2 + 16$
$(x + 4)^2 = 18$
$x + 4 = \pm\sqrt{18}$
$x = -4 \pm 3\sqrt{2}$
$(-4 + 3\sqrt{2})^2 + 8(-4 + 3\sqrt{2}) -2 \stackrel{?}{=} 0, 0 = 0;$
$(-4 - 3\sqrt{2})^2 + 8(-4 - 3\sqrt{2}) -2 \stackrel{?}{=} 0, 0 = 0$

12. $x^2 - 12x - 4 = 5$
$x^2 - 12x + 36 = 9 + 36$
$(x - 6)^2 = 45$
$x - 6 = \pm\sqrt{45}$
$x = 6 \pm\sqrt{45}$
$x = 6 \pm 3\sqrt{5}$
$(6 + 3\sqrt{5})^2 - 12(6 + 3\sqrt{5}) - 4 \stackrel{?}{=} 5, 5 = 5$
$(6 - 3\sqrt{5})^2 - 12(6 - 3\sqrt{5}) - 4 \stackrel{?}{=} 5, 5 = 5$

13. $x^2 - 18x - 7 = 3$
$x^2 - 18x + 81 = 10 + 81$
$(x - 9)^2 = 91$
$x - 9 = \pm\sqrt{91}$
$x = 9 \pm\sqrt{91}$
$(9 + \sqrt{91})^2 - 18(9 + \sqrt{91}) - 7 = 3, 3 = 3$
$(9 - \sqrt{91})^2 - 18(9 - \sqrt{91}) - 7 = 3, 3 = 3$

14. $2x^2 + 5 + 6 = 8$
$x^2 + \dfrac{5}{2}x + 3 = 4$
$x + \dfrac{5}{2}x + \dfrac{25}{16} = 1 + \dfrac{25}{16}$
$\left(x + \dfrac{5}{4}\right)^2 = \dfrac{41}{16}$
$x + \dfrac{5}{4} = \dfrac{\pm\sqrt{41}}{4}$
$x = \dfrac{-5 \pm\sqrt{41}}{4}$

$2\left(\dfrac{-5 + \sqrt{41}}{4}\right)^2 + 5\left(\dfrac{-5 + \sqrt{41}}{4}\right) + 6 \stackrel{?}{=} 8. 8 = 8$

$2\left(\dfrac{-5 - \sqrt{41}}{4}\right)^2 + 5\left(\dfrac{-5 - \sqrt{41}}{4}\right) + 6 \stackrel{?}{=} 8, 8 = 8$

15. $\dfrac{x^2}{2} + 4x = 10$
$x^2 + 8x = 20$
$x^2 + 8x + 16 = 20 + 16$
$(x + 4)^2 = 36$
$x + 4 = \pm 6$
$x = -10$ or $x = 2$

$\frac{(-10)^2}{2} + 4(-10) \overset{?}{=} 10, 10 = 10$

$\frac{2^2}{2} + 4(2) \overset{?}{=} 10, 10 = 10$

16. $\frac{x^2}{2} + 5x = -1$

$x^2 + 10x = -2$

$x^2 + 10x + 25 = -2 + 25$

$(x + 5)^2 = 23$

$x + 5 = \pm\sqrt{23}$

$x = -5 \pm \sqrt{23}$

$\frac{(-5 + \sqrt{23})^2}{2} + 5(-5 + \sqrt{23}) \overset{?}{=} -1, -1 = -1$

$\frac{(-5 - \sqrt{23})^2}{2} + 5(-5 - \sqrt{23}) \overset{?}{=} -1, -1 = -1$

17. $\frac{x^2}{2} - 4x - 5 = 0$

$x^2 - 8x = 10$

$x^2 - 8x + 16 = 10 + 16$

$(x - 4)^2 = 26$

$x - 4 = \pm\sqrt{26}$

$x = 4 \pm \sqrt{26}$

$\frac{(4 + \sqrt{26})^2}{2} - 4(4 + \sqrt{26}) - 5 \overset{?}{=} 0, 0 = 0$

$\frac{(4 - \sqrt{26})^2}{2} - 4(4 - \sqrt{26}) - 5 \overset{?}{=} 0, 0 = 0$

18. Let w represent the width. Let $2w - 2$ represent the length.

$w(2w - 2) = 180$

$2w^2 - 2w = 180$

$2w - w = 90$

$w^2 - w + \frac{1}{4} = 90 + \frac{1}{4}$

$\left(w - \frac{1}{2}\right)^2 = \frac{361}{4}$

$w - \frac{1}{2} = \pm\sqrt{\frac{361}{4}}$

$w = \frac{1}{2} \pm \frac{19}{2}$

$w = -9 \text{ or } w = 10$

Disregard $w = -9$ since measurement must be positive.

Length $= 2w - 2 = 2(10) - 2 = 18$

The dimensions of the carpet are 18ft $\times$ 10ft

19. Let w represent the width. Let $w + 2$ represent the length.

$w(w + 2) = 15$

$w^2 + 2w = 15$

$w^2 + 2 + 1 = 15 + 1$

$(w + 1)^2 = 16$

$w + 1 = \pm 4$

$w = 3 \text{ or } w = -5$

Disregard $w = -5$ since measurement must be positive.

Length $= w + 2 = 3 + 2 = 5$

The dimensions of the rectangle are 3 yd by 5 yd.

PRACTICE

1. $x^2 + 40x + 400$ **2.** $x^2 - 2x + 1$

3. $x^2 - 11x + \frac{121}{4}$ **4.** $x^2 + 5x + \frac{25}{4}$

5. $x^2 - \frac{10}{3}x + \frac{25}{9}$ **6.** $x^2 + \frac{5}{2}x + \frac{25}{16}$

7. Let x represent the first positive integer, then $x + 2$ represents the next consecutive positive odd integer.

$x(x + 2) = 143$

$x^2 + 2x = 143$

$x^2 + 2x + 1 = 143 + 1$

$(x + 1)^2 = 144$

$x + 1 = \pm 12$

$x = -13 \text{ or } x = 11$

Disregard -13 since it is not positive.

$x + 2 = 11 + 2 = 13;$

$x + (x + 2) = 11 + 13 = 24$

8. $\left(\frac{b}{2}\right)^2 = 81$ **9.** $\left(\frac{b}{2}\right)^2 = 144$ **10.** $\left(\frac{b}{2}\right)^2 = 225$

$\frac{b}{2} = \pm 9$ $\frac{b}{2} = \pm 12$ $\frac{b}{2} = \pm 15$

$b = \pm 18$ $b = \pm 24$ $b = \pm 30$

11. Manuel's work is not correct. His first step should be to multiply both sides of the equation by -1 to get $x^2 - 4x = -5$.

12. $x^2 + 10x + 9 = 0$

$x^2 + 10x + 25 = -9 + 25$

$(x + 5)^2 = 16$

$x + 5 = \pm 4$

$x = -9 \text{ or } x = -1$

$(-9)^2 + 10(-9) + 9 \overset{?}{=} 0, 0 = 0$

$(-1)^2 + 10(-1) + 9 \overset{?}{=} 0, 0 = 0$

13. $x^2 + 4x - 8 = 0$

$x^2 + 4x + 4 = 8 + 4$

$(x + 2)^2 = 12$

$x + 2 = \pm 2\sqrt{3}$

$x = -2 \pm 2\sqrt{3}$

$(-2 + 2\sqrt{3})^2 + 4(-2 + 2\sqrt{3}) - 8 \overset{?}{=} 0, 0 = 0$

$(-2 - 2\sqrt{3})^2 + 4(-2 - 2\sqrt{3}) - 8 \overset{?}{=} 0, 0 = 0$

14. $-x^2 + 14x + 7 = -9$

$x^2 - 14x - 7 = 9$

$x^2 - 14x + 49 = 16 + 49$

$(x - 7)^2 = 65$

$x - 7 = \pm\sqrt{65}$

$x = 7 \pm \sqrt{65}$

$-(7 + \sqrt{65})^2 + 14(7 + \sqrt{65}) + 7 \overset{?}{=} -9, -9 = -9$

$-(7 - \sqrt{65})^2 + 14(7 - \sqrt{65}) + 7 \overset{?}{=} -9, -9 = -9$

15.
$$x^2 - 24x - 2 = 1$$
$$x^2 - 24x + 144 = 3 + 144$$
$$(x - 12)^2 = 147$$
$$x - 12 = \pm\sqrt{147}$$
$$x = 12 \pm \sqrt{147}$$
$$(12 + \sqrt{147})^2 - 24(12 + \sqrt{147}) - 2 \stackrel{?}{=} 1$$
$$1 = 1$$
$$(12 - \sqrt{147})^2 - 24(12 - \sqrt{147}) - 2 \stackrel{?}{=} 1$$
$$1 = 1$$

16.
$$4x^2 - 10x = 6$$
$$x^2 - \frac{10}{4}x = \frac{6}{4}$$
$$x^2 - \frac{10}{4}x + \frac{100}{64} = \frac{6}{4} + \frac{100}{64}$$
$$\left(x - \frac{10}{8}\right)^2 = \frac{196}{64}$$
$$x - \frac{10}{8} = \pm\frac{14}{8}$$
$$x = \frac{10}{8} \pm \frac{14}{8}$$
$$x = \frac{1}{2} \text{ or } x = 3$$
$$4\left(-\frac{1}{2}\right)^2 - 10\left(-\frac{1}{2}\right)^2 \stackrel{?}{=} 6, 6 = 6$$
$$4(3)^2 - 10(3) \stackrel{?}{=} 6, 6 = 6$$

17.
$$6x + 6 = -x^2$$
$$x^2 + 6x = -6$$
$$x^2 + 6x + 9 = -6 + 9$$
$$(x + 3)^2 = 3$$
$$x + 3 = \pm\sqrt{3}$$
$$x = -3 \pm \sqrt{3}$$
$$6(-3 + \sqrt{3}) + 6 \stackrel{?}{=} -(-3 + \sqrt{3})^2$$
$$-12 + 6\sqrt{3} = -12 + 6\sqrt{3}$$
$$6(-3 - \sqrt{3}) + 6 \stackrel{?}{=} -(-3 - \sqrt{3})^2$$
$$-12 - 6\sqrt{3} = -12 - 6\sqrt{3}$$

18.
$$\frac{x^2}{4} + 2x = 6$$
$$x^2 + 8x = 24$$
$$x^2 + 8x + 16 = 24 + 16$$
$$(x + 4)^2 = 40$$
$$x + 4 = \pm 2\sqrt{10}$$
$$x = -4 \pm 2\sqrt{10}$$
$$\frac{(-4 + 2\sqrt{10})^2}{4} + 2(-4 + 2\sqrt{10}) \stackrel{?}{=} 6, 6 = 6$$
$$\frac{(-4 - 2\sqrt{10})^2}{4} + 2(-4 - 2\sqrt{10}) \stackrel{?}{=} 6, 6 = 6$$

19.
$$\frac{x^2}{5} + 2x = -2$$
$$x^2 + 10x = -10$$
$$x^2 + 10x + 25 = -10 + 25$$
$$(x + 5)^2 = 15$$
$$x + 5 = \pm\sqrt{15}$$
$$x = -5 \pm \sqrt{15}$$
$$\frac{(-5 + \sqrt{15})^2}{5} + 2(-5 + \sqrt{15}) \stackrel{?}{=} -2, -2 = -2$$

$$\frac{(-5 - \sqrt{15})^2}{5} + 2(-5 - \sqrt{15}) \stackrel{?}{=} -2, -2 = -2$$

20.
$$\frac{x^2}{2} - 6x - 8 = 0$$
$$x^2 - 12x - 16 = 0$$
$$x^2 - 12x + 36 = 16 + 36$$
$$(x - 6)^2 = 52$$
$$x - 6 = \pm 2\sqrt{13}$$
$$x = 6 \pm 2\sqrt{13}$$
$$\frac{(6 + 2\sqrt{13})^2}{2} - 6(6 + 2\sqrt{13}) - 8 \stackrel{?}{=} 0, 0 = 0$$
$$\frac{(6 - 2\sqrt{13})^2}{2} - 6(6 - 2\sqrt{13}) - 8 \stackrel{?}{=} 0, 0 = 0$$

21. Let l represent the length, then $\frac{3}{4}l$ represents the width.
$$\left(\frac{3}{4}l - 1\right)(l - 2) = 80$$
$$\frac{3}{4}l^2 - 1l - \frac{6}{4}l + 2 = 80$$
$$\frac{3}{4}l^2 - \frac{10}{4}l + 2 = 80$$
$$\frac{3}{4}l^2 - \frac{10}{4}l = 78$$
$$l^2 - \frac{10}{3}l = 104$$
$$l^2 - \frac{10}{3}l + \frac{100}{36} = 104 + \frac{100}{36}$$
$$\left(l - \frac{10}{6}\right)^2 = \frac{3844}{36}$$
$$l - \frac{10}{6} = \pm\frac{62}{6}$$
$$l = \pm\frac{62}{6} + \frac{10}{6}$$
$$l = -\frac{52}{6} \text{ or } l = \frac{72}{6} = 12$$

Disregard $-\frac{52}{6}$ since measurement must be positive.

$l = 12; \frac{3}{4}l = \frac{3}{4}(12) = 9;$ The dimensions of the page are 12in. $\times$ 9 in.

22a.
$$-x^2 + 68x = 900$$
$$x^2 - 68x = -900$$
$$x^2 - 68x + 1156 = -900 + 1156$$
$$(x - 34)^2 = 256$$
$$x - 34 = \pm 16$$
$$x = 34 \pm 16$$
$$x = 18 \text{ or } x = 50$$

22b.
$$-x^2 + 68x = 1075$$
$$x^2 - 68x = -1075$$
$$x^2 - 68x + 1156 = -1075 + 1156$$
$$(x - 34)^2 = 81$$
$$x - 34 = \pm 9$$
$$x = 34 \pm 9$$
$$x = 25 \text{ or } x = 43$$

22c. In step 4, they did not add 1 to both sides of the equation.
$$3x^2 - 6x - 15 = 0$$
$$x - 2x - 5 = 0$$
$$x - 2x = 5$$
$$x - 2x + 1 = 5 + 1$$
$$(x - 1)^2 = 6$$
$$x - 1 = \pm\sqrt{6}$$
$$x = 1 \pm \sqrt{6}$$

EXTEND

23. $x^2 - 4ax - 21a^2 = 0$
$$x^2 - 4ax + 4a^2 = 21a^2 + 4a^2$$
$$(x - 2a)^2 = 25a^2$$
$$x - 2a = \pm 5a$$
$$x = 2a \pm 5a$$
$$x - -3a \text{ or } x = 7a$$
$$(-3a)^2 - 4a(-3a) - 21a^2 \stackrel{?}{=} 0, 0 = 0$$
$$(7a)^2 - 4a(7a) - 21a^2 \stackrel{?}{=} 0, 0 = 0$$

25. $x^2 + 7ax - 18a^2 = 0$
$$x^2 + 7ax + \frac{49}{4}a^2 = 18a^2 + \frac{49}{4}a^2$$
$$\left(x + \frac{7}{2}a\right)^2 = \frac{121}{4}a^2$$
$$x + \frac{7}{2}a = \pm\frac{11}{2}$$
$$x = -\frac{7}{2}a \pm \frac{11}{2}$$
$$x = -9a \text{ or } x = 2a$$
$$(-9a)^2 + 7a(-9a) - 18a^2 \stackrel{?}{=} 0, 0 = 0$$
$$(2a)^2 + 7a(2a) - 18a^2 \stackrel{?}{=} 0, 0 = 0$$

26.
$$x^2 - \frac{8a}{3}x = a^2$$
$$x^2 - \frac{8a}{3}x + \frac{64a^2}{36} = a^2 + \frac{64a^2}{36}$$
$$\left(x - \frac{8a}{6}\right)^2 = \frac{100a^2}{36}$$
$$x - \frac{8a}{6} = \pm\frac{10a}{6}$$
$$x = \frac{8a}{6} \pm \frac{10a}{6}$$
$$x = -\frac{2a}{6} = -\frac{1}{3}a \text{ or } x = \frac{18a}{6} = 3a$$

$$\left(-\frac{1}{3}a\right)^2 - \frac{8}{3}\left(-\frac{1}{3}a\right) \stackrel{?}{=} a^2, a^2 = a^2$$
$$(3a)^2 - \frac{8}{3}a(3a) \stackrel{?}{=} a^2, a^2 = a^2$$

27.
$$x^2 + \frac{15a}{4}x = a^2$$
$$x^2 + \frac{15a}{4}x + \frac{225a^2}{64} = a^2 + \frac{225a^2}{64}$$
$$\left(x + \frac{15a}{8}\right)^2 = \frac{289a^2}{64}$$
$$x + \frac{15a}{8} = \pm\frac{17a}{8}$$
$$x = -\frac{15a}{8} \pm \frac{17a}{8}$$
$$x = -\frac{32a}{8} = -4a \text{ or } x = \frac{2}{8}a = \frac{1}{4}a$$
$$(-4a)^2 + \frac{15a}{4}(-4a) \stackrel{?}{=} a^2, a^2 = a^2$$
$$\left(\frac{1}{4}a\right)^2 + \frac{15a}{4}\left(\frac{1}{4}a\right) \stackrel{?}{=} a^2, a^2 = a^2$$

28. Let x represent the length of a side of a square using the Pythagorean theorem:
$$x^2 + x^2 = (x + 12)^2$$
$$2x^2 = x^2 + 24x + 144$$
$$x^2 - 24x = 144$$
$$x^2 - 24x + 144 = 144 + 144$$
$$(x - 12)^2 = 288$$
$$x - 12 = \pm\sqrt{288}$$
$$x = 12 \pm \sqrt{288}$$
$$x \approx -4.97 \text{ or } x \approx 28.97$$
Disregard -4.97 since measurement is positive. The length of a side of a square is about 28.97 ft.

29.
$$(x^2 + 4x) + (y^2 + 6y) = 5$$
$$(x^2 + 4x + 4) + (y^2 + 6y + 9) = 5 + 4 + 9$$
$$(x + 2) + (y + 3)^2 = 18$$

30.
$$(x^2 + 6x) + (y^2 - 8y) + 10 = 0$$
$$(x^2 + 6x + 9) + (y^2 - 8y + 16) = -10 + 9 + 16$$
$$(x + 3)^2 + (y - 4)^2 = 15$$

THINK CRITICALLY

31. Answers will vary. Possible example:
$$x^2 + 2x - 35 = 0$$

32. Answers will vary. Possible example:
$$x^2 - 22x + 4 = 0.$$

33. Answers will vary.

34. No. By completing the square, you obtain the equation $(x - 1)^2 = \pm\sqrt{-5}$, which has no real solution.

35. $4x^2 + \frac{4a}{3}x - \frac{4a^2}{3} = 0$
$$x^2 + \frac{a}{3}x = \frac{a^2}{3}$$
$$x^2 + \frac{a}{3}x + \frac{a^2}{36} = \frac{a^2}{3} + \frac{a^2}{36}$$
$$\left(x + \frac{a}{6}\right)^2 = \frac{13a^2}{36}$$
$$x + \frac{a}{6} = \pm\frac{a\sqrt{13}}{6}$$
$$x = \frac{-a \pm a\sqrt{13}}{6}$$

MIXED REVIEW

36. C

37. $(6x + 10)(6x - 10)$
$4(3x + 5)(3x - 5)$

38. $2x(x - 44)$

39. $x = \frac{1}{2}y - 8$ **40.** $x = \frac{1}{4}y + 2$

$-\frac{1}{2}y = -x - 8$ $-\frac{1}{4}y = -x + 2$

$y = 2x + 16$ $y = 4x - 8$

$f^{-1}(x) = 2x + 16$ $f^{-1}(x) = 4x - 8$

PROJECT

1–5. Answers will vary.

ALGEBRAWORKS

1. radius $= 0.5 \times$ diameter; $r = 0.5b$

2. Volume of a cylinder $= \pi r^2 h$
$V = \pi(0.5b)^2 s$, $V = 0.25\pi b^2 s$

3. $V = 8(0.25)\pi b^2 s$, $V = 2\pi b^2 s$

4. $V = 6(0.25)(\pi)(3.026)^2(4.245) \approx 183$; 183 in.³

5a. $184 = 6\pi[(0.5)(3.026 + x)]^2(4.245)$

5b. ≈ 0.007 in.

5c. $3.026 + 0.007 = 3.033$; 3.033 in.

Lesson 6.2, pages 282–286

THINK BACK

1. reflection **2.** dilation **3.** translation

4. translate up 4 units **5.** translate left 2 units

6. translate right 5 units and down 6 units

EXPLORE

7. up

8. The x values are squared producing positive y-values. The y-values are positive only in Quadrants I and II.

9.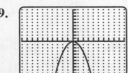

10. down **11.** quadrants III and IV

12. reflect it over the x-axis

13.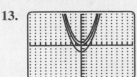

14. up 1 unit **15.** down 2 units

16. When K is positive, the graph of m is located K units up from the group of f. When K is negative, the graph of m is located K units down from the graph of f.

17.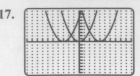

18. left 3 units **19.** right 4 units

20. When h is positive, the graph of n is located h units to the left of the graph of f. When h is negative, the graph of n is located h units to the right of the graph of f.

21. Answers will vary.

22.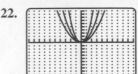

23. Similarities: Both graphs open up. Both graphs are located in quadrants I and II.

Difference: The graph of p is wider than f.

24. Similarities: Both graphs open up. Both graphs are located in quadrants I and II.

Difference: The graph of q is narrower than f.

25. They are reflections of each other over the x-axis.

26. $x^2 + x - 10 = 2$; $x^2 + x - 12 = 0$;
$(x + 4)(x - 3) = 0$; $x = -4$ or $x = 3$

27. $x^2 + x - 10 = 2$; $x^2 + x + \frac{1}{4} = 2 + 10 + \frac{1}{4}$;
$\left(x + \frac{1}{2}\right)^2 = 12\frac{1}{4}$; $x + \frac{1}{2} = \pm 3.5$; $x = -4$ or $x = 3$

28. $x = -4$ or $x = 3$; The results are the same.

MAKE CONNECTIONS

29a. $g(x) = -3x^2$

29b. $g(x) = 2x^2$

29c. $g(x) = \frac{1}{4}x^2$

29d. $g(x) = 2.4x^2$

30a. left 5 units and up $\frac{1}{2}$ unit

30b. down 8.8 units

30c. right 9 units and down 3.33 units

30d. right $\frac{1}{2}$ unit and up $\frac{1}{4}$ unit

31. a, d, b, c

32a. II **32b.** I **32c.** III

33a. II **33b.** III **33c.** I

34a. $g(x) = (x + 5)^2 - 1$ **34b.** $g(x) = (x - 6)^2 + 2$

34c. $g(x) = (x - 7)^2 - 3$ **34d.** $g(x) = (x + 8)^2 + 4$

35a. $x = -2$ or $x = 5$ **35b.** $x = -2$ or $x = 5$

36. Yes. Finding the intersection of the graphs of $y_1 = 2x^2 - 6x - 16$ and $y_2 = 4$ is equivalent to finding the intersection of the graphs of $y_1 = 2x^2 - 6x - 20$ and $y_2 = 0$.

37. The real solutions are located at the point or points at which the graph intersects the x-axis.

SUMMARIZE

38. The graph of g is a translation up or down by $|k|$ units and left or right by $|h|$ units from f.

39. The graph of g is a reflection of f over the x-axis.

40. Graph $y_1 = ax^2 + bx + c$ and $y_2 = d$ and then determine their points of intersection.

41. Stacey; Jacob will obtain the graph of $g(x) = -(x + 2)^2 - 1$.

42. Answers will vary. Graphs should not cross x-axis. Possible graph:

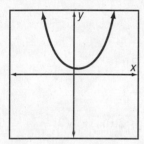

PROJECT CONNECTION

1–4. Answers will vary.

ALGEBRAWORKS

1. $7926 \div 2 = 3963$ mi

2. $S = 4\pi(3963)^2 - 197,359,488$; $197,359,488$ mi^2

3. Let x represent the percentage of the Earth's surface that is shaded. $360x = 75x \approx 0.2083$; 20.83%

4. $197,359,488 (0.2083) \approx 41,109,981$; $41,109,981$ mi^2 or, using unrounded values, $4\pi(3963)^2 \times \dfrac{75}{360} = 41,116,560$ mi^2

5. $C = 2\pi r = 2\pi(3963)24900$; $\dfrac{24,900}{360} = \dfrac{x}{75}$

$x = 5188$ mi

6. $24,900 \div 2 = 12,450$ mi

7. rectangle area: $5188 (12450) = 64,590,600$
$64,590,600 - 41,109,981$
$23,480,619$ mi^2 or using unrounded values,
$64,590,600 - 41,116,560 = 23,474,040$ mi^2

Lesson 6.3, pages 287–295

EXPLORE

1. $j(x) = 3(x - 1)^2 + 2 = 3(x^2 - 2x + 1) + 2 = 3x^2 - 6x + 5$

2. $(x^2 - 2x + 1) + 2 = 3x^2 - 6x + 5$; Yes. You are graphing the same function expressed in two different forms.

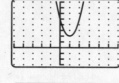

3.

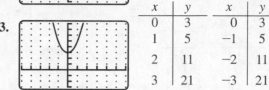

x	y	x	y
0	3	0	3
1	5	−1	5
2	11	−2	11
3	21	−3	21

4. You obtain the same y-value for x and $-x$.

5. b

6. Since $f(x) = f(-x)$ $f(x)$ is an even function.

TRY THESE

1. $-\dfrac{b}{2a} = -\dfrac{12}{2(-1)} = 6$;
$f(6) = -(6)^2 + 12(6) - 2 = 34$; $(6, 34)$; $x = 6$; downward

2. $-\dfrac{b}{2a} = -\dfrac{4}{2(-2)} = -1$; $f(-1) = -2(-1)^2 - 4(-1) + 5 = 7$; $(-1, 7)$; $x = -1$; downward

3. $-\dfrac{b}{2a} = -\dfrac{0}{2(4)} = 0$; $f(0) = 4(0)^2 + 10 = 10$; $(0, 10)$; $x = 0$; upward

4. $-\dfrac{b}{2a} = -\dfrac{0}{2(7)} = 0$; $f(0) = 7(0)^2 + 14 = 14$; $(0, 14)$; $x = 0$; upward

5. $-\dfrac{b}{2a} = -\dfrac{6}{2(-6)} = 0.5$; $f(0.5) = 6(0.5) - 1 - 6(0.5)^2 = 0.5$; $(0.5, -0.5)$; $x = 0.5$; downward

6. $-\dfrac{b}{2a} = -\dfrac{-10}{2(5)} = 1$; $f(1) = -10(1) + 9 + 5(1)^2 = 4$; $(1, 4)$; $x = 1$; upward

7. Answers will vary. Possible answer: First determine the axis of symmetry for $f(x)$. To find the point of reflection for (x, y), determine how many units x is from the line of symmetry. Let c represent the number of units, then (x, y)'s reflection point is $(x + 2c, y)$. For example see Example 2 in text.

8. $-\dfrac{b}{2a} = -\dfrac{2}{2(1)} = -1;$

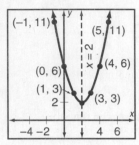

point	reflection
$(-5, 10)$	$(3, 10)$
$(0, -5)$	$(-2, -5)$
$(2, 3)$	$(-4, 3)$

9. $-\dfrac{b}{2a} = -\dfrac{-4}{2(1)} = 2; f(2) = 2^2 - 4(2) + 6 = 2$

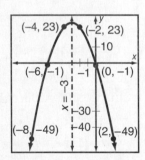

point	reflection
$(-1, 11)$	$(5, 11)$
$(0, 6)$	$(4, 6)$
$(3, 3)$	$(1, 3)$

10. $-\dfrac{b}{2a} = -\dfrac{-18}{2(-3)} = -3; f(-3) =$
$-3(-3)^2 - 18(-3) - 1 = 26$

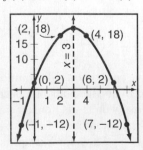

point	reflection
$(-4, 23)$	$(-2, 23)$
$(0, -1)$	$(-6, -1)$
$(2, -49)$	$(-8, -49)$

11. $-\dfrac{b}{2a} = -\dfrac{12}{2(-2)} = 3;$

$f(3) = -2(3)^2 + 12(3) + 2 = 20$

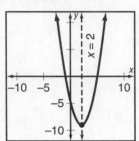

point	reflection
$(-1, -12)$	$(7, -12)$
$(0, 2)$	$(6, 2)$
$(4, 18)$	$(2, 18)$

12a. $h(t) = -16t^2 + 160t$

12b. 400 ft

12c. $-\dfrac{b}{2a} = -\dfrac{160}{(-16)} = 5; f(5) = -16(5)^2 + 160(5) =$

400; max. at (5, 400); The rocket reached maximum height at 5s.

12d. The rocket will hit the ground after 10s.

13. $-\dfrac{b}{2a} = -\dfrac{-4}{2(1)} = 2$

$f(2) = 2^2 - 4(2) - 5 = -9; x = 2; (2, -9)$

$f(0) = 0^2 - 4(0) - 5 = -5$

$0 = x^2 - 4x - 5$

$0 = (x - 5)(x + 1)$

$x = 5 \text{ or } x = -1$

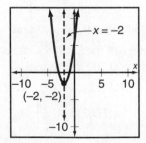

14. $-\dfrac{b}{2a} = -\dfrac{8}{2(2)} = -2; f(-2) =$

$2(-2)^2 + 8(-2) + 6 = -2$

$x = -2; (-2, -2); f(0) = 2(0)^2 + 8(0) + 6 = 6$

$0 = 2x^2 + 8x + 6$

$0 = x^2 + 4x + 3$

$0 = (x + 3)(x + 1)$

$x = -3 \text{ or } x = -1$

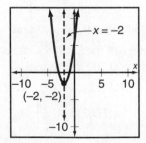

15. $-\dfrac{b}{2a} = -\dfrac{-4}{2(-1)} = -2$

$f(-2) = -(-2)^2 - 4(-2) + 12 = 16$
$x = -2; (-2, 16); f(0) = -0^2 - 4(0) + 12 = 12$
$0 = -x^2 - 4x + 12$
$0 = x^2 + 4x - 12$
$0 = (x + 6)(x - 2)$
$x = -6$ or $x = 2$

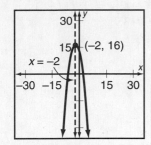

16. For $(0, -2)$;
For $(-2, -6)$;
For $(-6, 8)$:

$c = -2$
$4a - 2b + c = -6$
$36a - 6b + c = 10$

$4a - 2b - 2 = -6 \quad \rightarrow \quad -12a + 6b + 6 = 18$
$36a \quad 6b \quad 2 = 10 \quad \rightarrow \quad \dfrac{36a \quad 6b \quad 2 = 10}{24a \qquad\quad 4 = 28}$

$24a \qquad\qquad = 24$
$a \qquad\qquad = 1$

$4a - 2b - 2 = -6$
$4(1) - 2b - 2 = -6$
$-2b = -8$
$b = 4$
$y = x^2 + 4x - 2$

17. For $(0, 3)$:
For $(1, 4)$:
For $(3, -4)$

$c = 3$
$a + b + c = 4$
$9a + 3b + c = -4$

$a + b + 3 = 4 \quad \rightarrow \quad -3a - 3b - 9 = -12$
$9a + 3b + 3 = -4 \quad \rightarrow \quad \dfrac{9a + 3b + 3 = -4}{6a \qquad -6 = -16}$

$6a \qquad\qquad = -10$

$a = \dfrac{-5}{3}$

$a + b + 3 = 4 \qquad a + b + 3 = 4$
$-\dfrac{5}{3} + b + 3 = 4 \qquad -\dfrac{5}{3}$
$b + \dfrac{4}{3} = 4$
$b = \dfrac{8}{3}$
$y = -\dfrac{5}{3}x^2 + \dfrac{8}{3}x + 3$

18. For $(0, -16)$:
For $(-2, 0)$:
For $(2, 0)$:

$c = -16$
$4a - 2b + c = 0$
$4a + 2b + c = 0$

$4a - 2b - 16 = 0 \qquad 4a - 2b - 16 = 0$
$\dfrac{4a + 2b - 16 = 0}{8a \qquad - 32 = 0} \qquad 4(4) - 2b - 16 = 0$

$8a = 32 \qquad\qquad -2b = 0$
$a = 4 \qquad\qquad b = 0$
$y = 4x^2 - 16$

19. For $(0, -9)$:
For $(-3, 0)$:
For $(3, 0)$:

$c = -9$
$9a - 3b + c = 0$
$9a + 3b + c = 0$

$9a - 3b - 9 = 0 \qquad 9a - 3b - 9 = 0$
$\dfrac{9a + 3b - 9 = 0}{18a \qquad - 18 = 0} \qquad 9(1) - 3b - 9 = 0$

$a \qquad\qquad = 1 \qquad 3b = 0$
$y = x^2 - 9 \qquad\qquad b = 0$

20. $-5x^2 + 370x + 19{,}800; \; x^2 - 74x - 3960;$
$(x - 100)(x + 36) = 0; \; x = 100,$ or $x = -36$
Disregard $x = 100$ since the fair must decrease;
$1.10 - 0.36 = 0.73$

PRACTICE

1. $-\dfrac{b}{2a} = -\dfrac{4}{2(-1)} = 2; f(2) = -(2)^2 + 4(2) + 1 = 5;$
$(2, 5); x = 2;$ downward

2. $-\dfrac{b}{2a} = -\dfrac{-8}{2(-2)} = -2;$
$f(-2) = -2(-2)^2 - 8(-2) + 3 = 11;$
$(-2, 11); x = -2;$ downward

3. $-\dfrac{b}{2a} = -\dfrac{0}{2(6)} = 0; f(0) = 6(0)^2 + 18 = 18;$
$(0, 18); x = 0;$ upward

4. $-\dfrac{b}{2a} = \dfrac{0}{2(5)} = 0; f(0) = 5(0)^2 + 15 = 15;$
$(0, 15); x = 0;$ upward

5. $-\dfrac{b}{2a} = -\dfrac{2}{2(-3)} = \dfrac{2}{6} = \dfrac{1}{3};$
$f\left(\dfrac{1}{3}\right) = 2\left(\dfrac{1}{3}\right) - 1 - 3\left(\dfrac{1}{3}\right)^2 = -\dfrac{2}{3};$
$\left(\dfrac{1}{3}, -\dfrac{2}{3}\right); x = \dfrac{1}{3};$ downward

6. $-\dfrac{b}{2a} = -\dfrac{-5}{2(2)} = \dfrac{5}{4};$
$f\left(\dfrac{5}{4}\right) = -5\left(\dfrac{5}{4}\right) + 3 + 2\left(\dfrac{5}{4}\right)^2 = -\dfrac{1}{8};$
$\left(\dfrac{5}{4}, -\dfrac{1}{8}\right); x = \dfrac{5}{4};$ upward

7. If a is positive, the graph opens upward, and the vertex is a minimum. If a is negative, the graph opens downward, and the vertex is a maximum.

8. $-\dfrac{b}{2a} = -\dfrac{8}{2(1)} = -4;$

$f(-4) = -4^2 + 8(-4) - 2 = -18;$

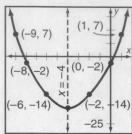

point	reflection
$(-6, -14)$	$(-2, -14)$
$(0, -2)$	$(-8, -2)$
$(1, 7)$	$(-9, 7)$

9. $-\dfrac{b}{2a} = -\dfrac{-12}{2(1)} = 6; f(6) = 6^2 - 12(6) + 1 = -35;$

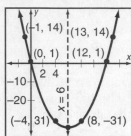

point	reflection
$(-1, 14)$	$(13, 14)$
$(0, 1)$	$(12, 1)$
$(4, -31)$	$(8, -31)$

10. $-\dfrac{b}{2a} = -\dfrac{-20}{2(-5)} = -2;$

$f(-2) = -5(-2)^2 - 20(-2) - 3 = 17;$

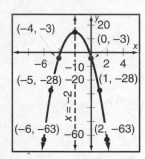

point	reflection
$(-5, -28)$	$(1, -28)$
$(0, -3)$	$(-4, -3)$
$(2, -63)$	$(-6, -63)$

11. $-\dfrac{b}{2a} = -\dfrac{24}{2(-6)} = 2;$

$f(2) = -6(2)^2 + 24(2) + 2 = 26;$

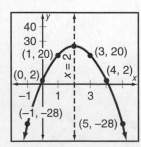

point	reflection
$(-1, -28)$	$(5, -28)$
$(0, 2)$	$(4, 2)$
$(3, 20)$	$(1, 20)$

12. Let x represent the length of one of the sides of the bottom. Since the perimeter of the bottom is 64, the length of the adjacent side must be $\dfrac{64 - 2x}{2} = 32 - x.$

Volume of the box: $12(x)(32 - x) =$
$12x(32 - x) = 384x - 12x^2 =$
$-12x^2 + 384 = -12x(x - 16); x = 0$ or $x = 16;$
Disregard 0; $32 - x = 32 - 16 = 16;$ The
dimensions of the bottom are 16cm $\times$ 16cm $\times$ 12cm.
Volume of the box: $16 \cdot 16 \cdot 12 = 3072;$ 3072 cm^3

13. $-\dfrac{b}{2a} = -\dfrac{-2}{2(1)} = 1;$

$f(1) = 1^2 - 2(1) - 3 = -4; x = 1; (1, -4);$
$f(0) = 0^2 - 2(0) - 3 = -3;$
$0 = x^2 - 2x - 3$
$0 = (x - 3)(x + 1)$
$x = 3$ or $x = -1$

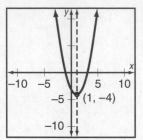

14. $-\dfrac{b}{2a} = -\dfrac{-6}{2(1)} = 3;$

$f(3) = 3^2 - 6(3) - 16 = -25; x = 3;$
$(3, -25); f(0) = 0^2 - 6(0) - 16 = -16;$
$0 = x^2 - 6x - 16$
$0 = (x - 8)(x + 2)$
$x = 8$ or $x = -2$

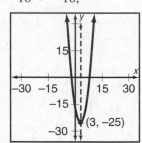

15. $-\dfrac{b}{2a} = -\dfrac{18}{2(4)} = -\dfrac{9}{4};$

$f\left(-\dfrac{9}{4}\right) = 4\left(-\dfrac{9}{4}\right)^2 + 18\left(-\dfrac{9}{4}\right) + 8 = -12;$

$x = -\dfrac{9}{4}; \left(-\dfrac{9}{4}, -12\right); f(0) = 4(0)^2 + 18(0) + 8 = 8;$

$0 = 4x^2 + 18x + 8$
$0 = (2x + 8)(2x + 1)$

$x = -4$ or $x = -\dfrac{1}{2}$

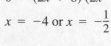

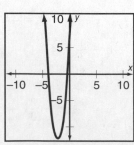

16. $-\dfrac{b}{2a} = -\dfrac{-8}{2(2)} = 2;$

$f(2) = 2(2)^2 - 8(2) + 12 = 4;\ x = 2;$

$(2, 4);\ f(0) = 2(0)^2 - 8(0) + 12 = 12;$

$0 = 2x^2 - 8x + 12$

$0 = x^2 - 4x + 6$

no real sol.

17. $-\dfrac{b}{2a} = -\dfrac{-4}{2(-1)} = -2;$

$f(-2) = -(-2)^2 - 4(-2) + 3 = 7;$

$x = -2;\ (-2, 7);\ f(0) = -(0)^2 - 4(0) + 3 = 3;$

$0 = -x^2 - 4x + 3$

$0 = x^2 + 4x - 3$

$3 = x^2 + 4x$

$7 = x^2 + 4x + 4$

$7 = (x + 2)^2$

$\pm\sqrt{7} = x + 2$

$x = -2 \pm \sqrt{7}$

$x \approx -4.6\text{ or } x \approx 0.6$

18. $-\dfrac{b}{2a} = -\dfrac{2}{2(-1)} = 1;$

$f(1) = -(1)^2 + 2(1) + 8 = 9;\ x = 1;$

$(1, 9);\ f(0) = -(0)^2 + 2(0) + 8 = 8;$

$0 = -x^2 + 2x + 8$

$0 = x^2 - 2x - 8$

$0 = (x - 4)(x + 2)$

$x = 4\text{ or } x = -2$

19. For $(0, -1)$: $\quad c = -1$

For $(-2, -5)$: $\quad 4a - 2b + c = -5$

For $(1, 4)$: $\quad a + b + c = 4$

$4a - 2b - 1 = -5 \rightarrow \quad 4a - 2b - 1 = -5$

$a + b - 1 = 4 \quad\rightarrow \quad \underline{2a + 2b - 2 = 8}$

$\qquad\qquad\qquad\qquad 6a \qquad\quad -3 = 3$

$\qquad\qquad\qquad\qquad\qquad 6a = 6$

$\qquad\qquad\qquad\qquad\qquad\ a = 1$

$a + b - 1 = 4$

$1 + b - 1 = 4$

$\qquad\quad b = 4$

$y = x^2 + 4x - 1$

20. For $(0, -3)$: $\qquad c = -3$

For $(2, 5)$: $\qquad 4a + 2b + c = 5$

For $(3, 3)$: $\qquad 9a + 3b + c = 3$

$4a + 2b - 3 = 5 \quad\rightarrow\quad -12a - 6b + 9 = -15$

$9a + 3b - 3 = 3 \quad\rightarrow\quad \underline{18a + 6b - 6 = 6}$

$\qquad\qquad\qquad\qquad\qquad 6a \qquad\ + 3 = -9$

$\qquad\qquad\qquad\qquad\qquad\qquad 6a = -12$

$\qquad\qquad\qquad\qquad\qquad\qquad\ \ a = -2$

$4a + 2b - 3 = 5$

$4(-2) + 2b - 3 = 5$

$\qquad\qquad 2b = 16$

$\qquad\qquad\ b = 8$

$y = -2x^2 + 8x - 3$

21. For $(0, -6)$ $\qquad c = -6$

For $(-2, 2)$: $\qquad 4a - 2b + c = 2$

For $(-4, -6)$: $\qquad 16a - 4b + c = -6$

$4a - 2b - 6 = 2 \quad\rightarrow\quad -8a + 4b + 12 = -4$

$16a - 4b - 6 = -6 \rightarrow \underline{16a - 4b - 6 = -6}$

$\qquad\qquad\qquad\qquad\quad 8a \qquad\quad + 6 = -10$

$\qquad\qquad\qquad\qquad\qquad\quad 6a = -12$

$\qquad\qquad\qquad\qquad\qquad\quad\ \ a = -2$

$4a - 2b - 6 = 2$

$4(-2) - 2b - 6 = 2$

$\qquad\qquad -2b = 16$

$\qquad\qquad\ \ b = -8$

$y = -2x^2 + 8x - 6$

22. For $(0, 0)$: $\qquad c = 0$

For $(-2, -4)$: $\qquad 4a - 2b + c = -4$

For $(4, 0)$: $\qquad 16a - 4b + c = 0$

$4a - 2b = -4 \quad\rightarrow\quad -8a + 4b = 8$

$16a - 4b = 0 \quad\rightarrow\quad \underline{16a - 4b - 0}$

$\qquad\qquad\qquad\qquad\quad 8a \qquad\quad = 8$

$\qquad\qquad\qquad\qquad\qquad\ a = 1$

$4a - 2b = -4$

$4(1) - 2b = -4$

$\qquad -2b = -8$

$\qquad\quad b = 4$

$y = x^2 + 4x$

23. $-\dfrac{b}{2a} = -\dfrac{8}{2(-1)} = 4;$

$f(4) = -(4)^2 + 8(4) + 78 = 94;$ max. at $(4, 94);$

Ross should study 4 h to achieve his maximum grade of 94.

EXTEND

24. $y = x^2 + 3x - 40$ **25.** $y = x^2 + 5x + 10$

26. $y = 2x^2 + 8x + 6$ **27.** $y = 3x^2 + 4x - 7$

28. $h(t) = -16t^2 + 40t + 8;$

$-\dfrac{b}{2a} = -\dfrac{40}{2(-16)} = \dfrac{40}{32} = \dfrac{5}{4};$

$f\!\left(\dfrac{5}{4}\right) = -16\!\left(\dfrac{5}{4}\right)^2 + 40\!\left(\dfrac{5}{4}\right) + 8 = 33;$ max. at $\left(\dfrac{5}{4}, 33\right);$

The maximum height attained by the ball is 33 ft.

29. Yes. The function for this situation is
$h(t) = -16t^2 + 40t + 7$, which is translated down by 1 from $h(t)$. The ball will reach a maximum height of 32 ft.

30. Revenue, $R(x)$, is the price per mug times the number of mugs sold. If x is the number of times that the price is decreased by $0.05 then
$R(x) = (1.5 - 0.05x)(200 + 10x)$
$R(x) = -0.5x^2 + 5x + 300$

The maximum is at the vertex, $-\dfrac{b}{2a}$.

$-\dfrac{b}{2a} = -\dfrac{5}{2(-0.5)} = 5$

$R(5) = (1.5 - 0.05(5))(200 + 10(5))$
$R(5) = 312.5$
$P(5) = (1.5 - 0.05(5))$
$P(5) = 1.25$

The price per mug should be $1.25 for a maximum revenue of $312.50.

THINK CRITICALLY

31. Since $f(-x) = f(x)$, the axis of symmetry must be $x = 0$.

32. For $f(-x) = f(x)$, $ax^2 + bx + c = ax^2 - bx + c$, thus, $b = 0$.

33. No. Two points do not determine a unique parabola. It requires at least three non-collinear points.

34. Yes. As long as there are at least three non-collinear points, it will provide a quadratic equation.

35. $\begin{cases} -39 = 16a - 4b - 7 \\ -4 = -\dfrac{b}{2a} \end{cases} \rightarrow \begin{cases} 16a - 4b = -32 \\ -8a + b = 0 \end{cases}$

$\begin{array}{l} 16a - 4b = -32 \\ \underline{-32a + 4b = 0} \\ -16a = -32 \\ a = 2 \end{array}$

$\begin{array}{l} -8a + b = 0 \\ \\ -8(2) + b = 0 \\ \\ b = 16 \end{array}$

36. $\begin{cases} 7 = a + b + 6 \\ 1 = -\dfrac{b}{2a} \end{cases} \rightarrow \begin{cases} a + b = 1 \\ 2a + b = 0 \end{cases}$

$\begin{array}{l} a + b = 1 \\ \underline{-2a - b = 1} \\ -a = 1 \\ a = -1 \end{array}$

$\begin{array}{l} a + b = 1 \\ \\ (-1) + b = 1 \\ \\ b = 2 \end{array}$

MIXED REVIEW

37. 4, frequency 3

38. 0, frequency 3

39. C; $y - 4 = -3(x - 2)$, $y - 4 = -3x + 6$,
$y = -3x + 10$

40. 11.7898

41. 23.5584

42. $-\dfrac{b}{2a} = -\dfrac{-24}{2(3)} = 4$;
$f(4) = 3(4)^2 - 24(4) + 2 = -46$; $(4, -46)$; $x = 4$

43. $-\dfrac{b}{2a} = -\dfrac{-16}{2(-2)} = -4$;
$f(-4) = -2(-4)^2 - 16(-4) - 10 = 22$;
$(-4, 22)$; $x = -4$

44. $x = -3$ or $x = 3$

45. $x - 6 = 5$ or $x - 6 = -5$
$ x = 11$ or $x = 1$

46. no solution

47. $x + 3 < -1$ or $x + 3 > 1$
$ x < -4$ or $ x > -2$

48. $x = 4y$, $y = \dfrac{1}{4}x$

49. $x = 3y + 6$, $3y = x - 6$, $y = \dfrac{1}{3}x - 2$

50. $x = y^2$, $y = \sqrt{x}$, if $x > 0$.

51. $x = \dfrac{1}{y + 1}$, $x(y + 1) = 1$, $xy + x = 1$, $xy = 1 - x$,

$y = \dfrac{1}{x} - 1$

ALGEBRAWORKS

1. $-\dfrac{b}{2a} = -\dfrac{0}{2\left(\dfrac{1}{9000}\right)} = 0$;

$f(0) = \dfrac{1}{9000}(0)^2 + 5 = 5$; $(0, 5)$

2. $\dfrac{1}{9000}(1600)^2 + 5 = 289.\overline{44}$; $289\,\text{ft}$

3. 4

4. $4200 \div 2 = 2100$; $\dfrac{1}{9000}(2100)^2 + 5 = 495$; 495 ft

5. $0 = -0.05x^2 + 5.25x - 47.5$
$0 = x^2 - 105x + 950$
$0 = (x - 10)(x - 95)$
$x = 10$ or $x = 95$
$95 - 10 = 85$; The width of the road is 85 ft.

6. $-\dfrac{b}{2a} = -\dfrac{5.25}{2(-0.05)} = 52.5$;

$f(52.5) = -0.05(52.5)^2 + 5.25(52.5) - 47.5 = 90.3125$; $90.3125 + 12 = 102.3125$; $y = 102.3125$

Lesson 6.4, pages 296–302

EXPLORE

1. $\dfrac{c}{a}$; Subtract $\dfrac{c}{a}$ from both sides.

2. $\left(\dfrac{b}{2a}\right)^2$; $\left(\dfrac{b}{2a}\right)^2$; Add $\left(\dfrac{b}{2a}\right)^2$ to both sides.

3. $\dfrac{b^2}{4a^2}$; Simplify.

4. $\dfrac{b}{2a}$; Factor the left side.

5. $\dfrac{b^2 - 4ac}{4a^2}$; Simplify.

6. Take the square root of both sides.

7. $-\dfrac{b}{2a}$; Subtract $\dfrac{b}{2a}$ from both sides.

8. 2a; Simplify.

9. 2a; Add fractions.

TRY THESE

1. $x = \dfrac{-b \pm \sqrt{b^2 - 4ac}}{2a} = \dfrac{-4 \pm \sqrt{4^2 - 4(2)(-1)}}{2(2)} =$

$\dfrac{-4 \pm \sqrt{24}}{4} =$

$\dfrac{-4 \pm 2\sqrt{6}}{4} = -1 \pm \dfrac{\sqrt{6}}{2}$;

$2x^2 + 4x - 1 = 0: 2\left(-1 + \dfrac{\sqrt{6}}{2}\right)^2 +$

$4\left(-4 + \dfrac{\sqrt{6}}{2}\right) - 1 \overset{?}{=} 0, 0 = 0; 2\left(-1 - \dfrac{\sqrt{6}}{2}\right)^2 +$

$4\left(-1 - \dfrac{\sqrt{6}}{2}\right) - 1 \overset{?}{=} 0, 0 = 0$

2. $y = \dfrac{-b \pm \sqrt{b^2 - 4ac}}{2a} = \dfrac{-2 \pm \sqrt{2^2 - 4(4)(-8)}}{2(4)} =$

$\dfrac{-2 \pm \sqrt{132}}{8} = \dfrac{-2 \pm 2\sqrt{33}}{8} = \dfrac{1 \pm \sqrt{33}}{4}$;

$4y^2 + 2y - 8 = 0: 4\left(\dfrac{-1 + \sqrt{33}}{4}\right)^2 +$

$2\left(\dfrac{-1 + \sqrt{33}}{4}\right) - 8 \overset{?}{=} 0, 0 = 0; 4\left(\dfrac{-1 - \sqrt{33}}{4}\right)^2 +$

$2\left(\dfrac{-1 + \sqrt{33}}{4}\right) - 8 \overset{?}{=} 0, 0 = 0$

3. $z = \dfrac{-b \pm \sqrt{b^2 - 4ac}}{2a} =$

$\dfrac{-(-30) \pm \sqrt{(-30)^2 - 4(9)(25)}}{2(9)} = \dfrac{30}{18} = \dfrac{5}{3}$;

$9z^2 - 30z + 25 = 0: 9\left(\dfrac{5}{3}\right)^2 - 30\left(\dfrac{5}{3}\right) + 25 \overset{?}{=} 0, 0 = 0$

4. Yes. If s_1 and s_2 are solutions to a quadratic equation, then $(x - s_1)(x - s_2) = x^2 - s_2x - s_1x + s_1s_2 = x^2 - (s_1 + s_2)x + s_1s_2$ is the quadratic equation.

5. $b^2 - 4ac = 6^2 - 4(2)(8) = -28$; no real solutions

6. $b^2 - 4ac = 12^2 - 4(4)(9) = 0$; one real, rational solution

7. $b^2 - 4ac = (-2)^2 - 4(3)(-10) = 124$; two real irrational solutions

8. 1   **9.** 0   **10.** 2

11. $20^2 - 4(k)(25) = 0, -100k = -400, k = 4$

12. $(-k)^2 - 4(36)(1) = 0, (-k)^2 = 144, k = \pm 12$

13. $24^2 - 4(9)(k) = 0, 576 = 36k, k = 16$

14. $P(x) = 8x^2 + 85x + 20{,}222 = 8x^2 + 85x + 20,$

$0 = 8x^2 + 85x - 202; x = \dfrac{-b \pm \sqrt{b^2 - 4ac}}{2a} =$

$\dfrac{-85 \pm \sqrt{85^2 - 4(8)(-202)}}{2(8)} = \dfrac{-85 \pm \sqrt{13{,}689}}{16} =$

$\dfrac{-85 \pm 117}{16} = -12.625$ or 2; Disregard -12.625 since number of items must be positive. $2 \cdot 100 = 200$; The company sold 200 items.

15. $-\dfrac{b}{a} = -\dfrac{-12}{1} = 12; \dfrac{c}{a} = \dfrac{4}{1} = 4$

16. $2g^2 - 8g - 4 = 0; -\dfrac{b}{a} = -\dfrac{-8}{2} = 4;$

$\dfrac{c}{a} = \dfrac{-4}{2} = -2$

17. $3h^2 - 4h - 7 = 0; -\dfrac{b}{a} = -\dfrac{-4}{3} = \dfrac{4}{3};$

$\dfrac{c}{a} = \dfrac{-7}{3} = -\dfrac{7}{3}$

18. $s_1 + s_2 = 4 + \left(-\dfrac{1}{2}\right) = \dfrac{7}{2};$

$s_1s_2 = 4\left(-\dfrac{1}{2}\right) = -2; x^2 - \dfrac{7}{2}x + (-2) = 0,$

$x^2 - \dfrac{7}{2}x - 2 = 0, 2x^2 - 7x - 4 = 0$

19. $s_1 + s_2 = -6 - 7 = -13; s_1s_2 = -6(-7) = 42$
$x^2 - (-13)x + 42 = 0, x^2 + 13x + 42 = 0$

20. $s_1 + s_2 = -5 + \left(-\dfrac{1}{4}\right) = -\dfrac{21}{4}; s_1s_2 = -5\left(-\dfrac{1}{4}\right) = \dfrac{5}{4};$

$x^2 - \left(\dfrac{21}{4}\right) + \dfrac{5}{4} = 0, x^2 + \dfrac{21}{4} + \dfrac{5}{4} = 0,$

$4x^2 + 21x + 5 = 0$

21. Let x represent the number. $x^2 = 6x - 135,$
$x^2 - 6x + 135 = 0, (x + 9)(x - 15) = 0, x = -9$
or $x = 15$; The number is either -9 or 15.

PRACTICE

1. $x = \dfrac{-b \pm \sqrt{b^2 - 4ac}}{2a} =$

$\dfrac{(9) \pm \sqrt{(9)^2 - 4(3)(1)}}{2(3)} = \dfrac{9 \pm \sqrt{69}}{6} = \dfrac{3}{2} \pm \dfrac{\sqrt{69}}{6};$

$3x^2 - 9x + 1 = 0: 3\left(\dfrac{3}{2} + \dfrac{\sqrt{69}}{6}\right)^2 -$

$9\left(\dfrac{3}{2} + \dfrac{\sqrt{69}}{6}\right) + 1 \overset{?}{=} 0, 0 = 0; 3\left(\dfrac{3}{2} - \dfrac{\sqrt{69}}{6}\right)^2 -$

$9\left(\dfrac{3}{2} - \dfrac{\sqrt{69}}{6}\right) + 1 \overset{?}{=} 0, 0 = 0$

2. $y = \dfrac{-b \pm \sqrt{b^2 - 4ac}}{2a} =$

$\dfrac{-(-7) \pm \sqrt{(-7)^2 - 4(4)(2)}}{2(4)} = \dfrac{7 \pm \sqrt{17}}{8};$

$4y^2 - 7y + 2 = 0: 4\left(\dfrac{7 + \sqrt{17}}{8}\right)^2 - 7\left(\dfrac{7 + \sqrt{17}}{8}\right) +$

$2 \overset{?}{=} 0, 0 = 0; 4\left(\dfrac{7 - \sqrt{17}}{8}\right)^2 - 7\left(\dfrac{7 - \sqrt{17}}{8}\right) + 2 \overset{?}{=} 0,$

$0 = 0$

3. $z = \dfrac{-b \pm \sqrt{b^2 - 4ac}}{2a} = \dfrac{-16 \pm \sqrt{16^2 - 4(3)(5)}}{2(3)} =$

$\dfrac{-16 \pm \sqrt{196}}{6} = \dfrac{-16 \pm 14}{6} = -5$ or $-\dfrac{2}{6} = -\dfrac{1}{3};$

$3z^2 + 16z + 5 = 0: 3(-5)^2 + 16(-5) + 5 \overset{?}{=} 0,$

$0 = 0; 3\left(-\dfrac{1}{3}\right)^2 + 16\left(-\dfrac{1}{3}\right) + 5 \overset{?}{=} 0, 0 = 0$

4. $p = \dfrac{-b \pm \sqrt{b^2 - 4ac}}{2a} = \dfrac{-(-2) \pm \sqrt{(-2)^2 - 4(5)(-6)}}{2(5)} =$

$\dfrac{2 \pm \sqrt{124}}{10} = \dfrac{2 \pm 2\sqrt{31}}{10} = \dfrac{1 \pm \sqrt{31}}{5};$

$5p^2 - 2p - 6 = 0: 5\left(\dfrac{1 \pm \sqrt{31}}{5}\right)^2 -$

$2\left(\dfrac{1 + \sqrt{31}}{5}\right) - 6 \stackrel{?}{=} 0, 0 = 0;$

$5\left(\dfrac{1 - \sqrt{31}}{5}\right)^2 - 2\left(\dfrac{1 - \sqrt{31}}{5}\right) - 6 \stackrel{?}{=} 0, 0 = 0$

5. $q = \dfrac{-b \pm \sqrt{b^2 - 4ac}}{2a} =$

$\dfrac{-(-5) \pm \sqrt{(-5)^2 - 4(-3)(8)}}{2(-3)} =$

$\dfrac{5 \pm \sqrt{121}}{-6} = \dfrac{5 \pm 11}{-6} = -\dfrac{16}{6} = -\dfrac{8}{3}$ or $\dfrac{-6}{-6} = 1;$

$-3q^2 - 5q + 8 = 0; -3\left(-\dfrac{8}{3}\right)^2 - 5\left(-\dfrac{8}{3}\right) + 8 \stackrel{?}{=} 0,$

$0 = 0; -3(1)^2 - 5(1) + 8 \stackrel{?}{=} 0, 0 = 0$

6. $r = \dfrac{-b \pm \sqrt{b^2 - 4ac}}{2a} =$

$\dfrac{-(-24) \pm \sqrt{(-24)^2 - 4(-16)(9)}}{2(-16)} = \dfrac{24 \pm \sqrt{1152}}{-32} =$

$\dfrac{24 \pm 6\sqrt{32}}{-32} = \dfrac{-3 \pm 3\sqrt{2}}{4}; -16r^2 - 24r + 9 = 0:$

$-16\left(\dfrac{-3 + 3\sqrt{2}}{4}\right)^2 - 24\left(\dfrac{-3 + 3\sqrt{2}}{4}\right) + 9 \stackrel{?}{=} 0, 0 = 0;$

$-16\left(\dfrac{-3 - 3\sqrt{2}}{4}\right)^2 - 24\left(\dfrac{-3 - 3\sqrt{2}}{4}\right) + 9 \stackrel{?}{=} 0, 0 = 0$

7. $-48 = -16t^2, 16t^2 - 48 = 0, 16(t^2 - 3) = 0,$
$t^2 - 3 = 0, t^2 = 3, t = \sqrt{3}$; The ride is in free fall for $\sqrt{3}$ s.

8. $b^2 - 4ac = 2^2 - 4(4)(7) = -108$; no real solutions

9. $b^2 - 4ac = (-5)^2 - 4(2)(-5) = 65$; two real, irrational solutions

10. $b^2 - 4ac = 12^2 - 4(-36)(-1) = 0$; one real rational solution

11. $b^2 - 4ac = 24^2 - 4(16)(9) = 0$; one real rational solution

12. $b^2 - 4ac = 7^2 - 4(2)(-30) = 289$; two real rational solutions

13. $b^2 - 4ac = 24^2 - 4(5)(-5) = 676$; two real rational solutions

14. 1 **15.** 2 **16.** 0

17. 0 **18.** 2 **19.** 2

20. $(-40)^2 - 4(k)(25) = 0, 1600 = 100k, k = 16$

21. $k^2 - 4(25)(9) = 0, k^2 = 900, k = \pm 30$

22. $(-42)^2 - 4(9)(k) = 0, 1764 = 36k, k = 49$

23. The graph is a parabola that does not intersect the x-axis.

24. $-\dfrac{b}{a} = -\dfrac{-15}{1} = 15; \dfrac{c}{a} = \dfrac{5}{1} = 5$

25. $-\dfrac{b}{a} = -\dfrac{8}{1} = -8; \dfrac{c}{a} = \dfrac{-12}{1} = -12$

26. $3h^2 - 7h + 7 = 0; -\dfrac{b}{a} = -\dfrac{-7}{3} = \dfrac{7}{3}; \dfrac{c}{a} = \dfrac{7}{3}$

27. $4m^2 - 5m - 9 = 0; -\dfrac{b}{a} = -\dfrac{-5}{4} = \dfrac{5}{4}; \dfrac{c}{a} = \dfrac{-9}{4} = -\dfrac{9}{4}$

28. $\dfrac{1}{2}n^2 - \dfrac{1}{4}n - 3 = 0; -\dfrac{b}{a} = \dfrac{-\frac{1}{4}}{\frac{1}{2}} = \dfrac{1}{4} \cdot \dfrac{2}{1} = \dfrac{2}{4} =$

$\dfrac{1}{2}; \dfrac{c}{a} = \dfrac{-3}{\frac{1}{2}} = -3 \cdot 2 = -6$

29. $\dfrac{1}{2}p^2 - \dfrac{1}{2}p + 3 = 0; -\dfrac{b}{a} = \dfrac{-\frac{1}{2}}{\frac{1}{2}} = 1;$

$\dfrac{c}{a} = \dfrac{3}{\frac{1}{2}} = 3 \cdot 2 = 6$

30. $s_1 + s_2 = 5 + (-12) = -7; s_1 s_2 5(-12) = -60;$
$x - (-7)x + (-60) = 0, x^2 + 7x - 60 = 0$

31. $s_1 + s_2 = -10 + \dfrac{9}{11} = -\dfrac{101}{11}; s_1 s_2 (-10)$

$\left(\dfrac{9}{11}\right) = -\dfrac{90}{11}; x^2 - \left(-\dfrac{101}{11}\right)x + \left(-\dfrac{90}{11}\right) = 0,$

$x^2 + \dfrac{101}{11}x - \dfrac{90}{11} = 0, 11x^2 + 101x - 90 = 0$

32. $s_1 + s_2 = -8 + \dfrac{11}{12} = -\dfrac{85}{12}; s_1 s_2 = (-8)\left(\dfrac{11}{12}\right) = -\dfrac{88}{12};$

$x^2 - \left(-\dfrac{85}{12}\right)x + \left(-\dfrac{88}{12}\right) = 0, x^2 + \dfrac{85}{12}x - \dfrac{88}{12} = 0,$

$12x^2 + 85x - 88 = 0$

33. $s_1 + s_2 = -\dfrac{2}{3} + \left(-\dfrac{4}{9}\right) = -\dfrac{30}{27};$

$s_1 s_2 = \left(-\dfrac{2}{3}\right)\left(-\dfrac{4}{9}\right) = \dfrac{8}{27};$

$x^2 - \left(-\dfrac{30}{27}\right)x + \dfrac{8}{27} = 0, x^2 + \dfrac{30}{27}x + \dfrac{8}{27} = 0,$

$27x^2 + 30x + 8 = 0$

34. Let x represent the equal amount on all sides that the wallhanging is to be increased. Then $2x + 24$ represents one side and $2x + 12$ represents the other side.

$(2x + 24)(2x + 12) = (24 \times 12) + 160,$
$4x^2 + 72x + 288 = 448, 4x^2 + 72x - 160 = 0,$
$x^2 + 18x - 40 = 0, (x + 20)(x - 2) = 0,$
$x = -20$ or $x = 2$; Disregard -20 since measurement must be positive. $2x + 24 = 2(2) + 24 = 28;$
$2x + 12 = 2(2) + 12 = 16;$ The dimensions of the wallhanging is 28 in. by 16 in.

35. $x^2 - (-3)x + (-40) = 0, x^2 + 3x - 40 = 0;$

$\dfrac{-b \pm \sqrt{b^2 - 4ac}}{2a} = \dfrac{-3 \pm \sqrt{3^2 - 4(1)(-40)}}{2(1)} =$

$\dfrac{-3 \pm \sqrt{169}}{2} = \dfrac{-3 \pm 13}{2}; x = \dfrac{-16}{2} =$

-8 or $x = \dfrac{10}{2} = 5$

36. $x^2 - 11x + (-152) = 0, x^2 - 11x - 152 = 0;$

$$\frac{-b \pm \sqrt{b^2 - 4ac}}{2a} = \frac{(-11) \pm \sqrt{(-11)^2 - 4(1)(-152)}}{2(1)}$$

$$\frac{11 \pm \sqrt{729}}{2} = \frac{11 \pm 27}{2};$$

$$x = \frac{-16}{2} = -8 \text{ or } x = \frac{38}{2} = 19$$

37. $x^2 - 35x + (-750) = 0, x^2 - 35x - 750 = 0;$

$$\frac{-b \pm \sqrt{b^2 - 4ac}}{2a} =$$

$$\frac{-(-35) \pm \sqrt{(-35)^2 - 4(1)(-750)}}{2(1)} =$$

$$\frac{35 \pm \sqrt{4225}}{2} = \frac{35 \pm 65}{2};$$

$$x = \frac{-30}{2} = -15 \text{ or } x = \frac{100}{2} = 50$$

EXTEND

38. $\dfrac{-b \pm \sqrt{b^2 - 4ac}}{2a} = \dfrac{-q \pm \sqrt{q^2 - 4(1)(-30q^2)}}{2(1)} =$

$$\frac{-q \pm \sqrt{121q^2}}{2} = \frac{-q \pm 11q}{2};$$

$$x = \frac{-12q}{2} = -6q \text{ or } x = \frac{10q}{2} = 5q$$

39. $\dfrac{-b \pm \sqrt{b^2 - 4ac}}{2a} = \dfrac{-q \pm \sqrt{q^2 - 4(6)(-2q^2)}}{2(6)} =$

$$\frac{-q \pm \sqrt{49q^2}}{12} = \frac{-q \pm 7q}{12}; x = \frac{-8q}{12} =$$

$$-\frac{2}{3}q \text{ or } x = \frac{6q}{12} = \frac{1}{2}q$$

40. $\dfrac{-b \pm \sqrt{b^2 - 4ac}}{2a} = \dfrac{-14q \pm \sqrt{14^2 - 4(5)(-3q^2)}}{2(5)} =$

$$\frac{-14q + \sqrt{256q^2}}{10} = \frac{-14q \pm 16q}{10};$$

$$x = \frac{-30q}{10} = -3q \text{ or } x = \frac{2q}{18} = \frac{1}{5}q$$

41. $s_1 + s_2 = 6f + 2f = 8f; s_1 s_2 = 6f(2f) = 12f^2;$
$x^2 - 8fx + 12f^2 = 0$

42. $s_1 + s_2 = 5g + (-8g) = -3g;$
$s_1 s_2 = 5g(-8g) = -40g^2;$
$x^2 - (-3g)x + (-40g^2) = 0, x^2 + 3gx - 40g^2 = 0$

43. $s_1 + s_2 = -7h + (-3h) = -10h;$
$s_1 s_2 = -7h(-3h) = 21h^2;$
$x^2 - (-10h)x + 21h^2 = 0, x^2 + 10hx + 21h^2 = 0$

THINK CRITICALLY

44. a,c,d **45.** $(-8)^2 - 4(-1)(k) < 0, 4k < -64, k < -16$

MIXED REVIEW

46.
$$\begin{array}{rr} -2a - b & = 4 \\ 2a + 2b + 2c & = 2 \\ \hline b + 2c & = 6 \end{array} \qquad \begin{array}{rr} 4b + 6c & = 16 \\ -4b - 8c & = -24 \\ \hline -2c & = -8 \\ c & = 4 \end{array}$$

$$\begin{array}{rl} b + 2c & = 6 \\ b + 2(4) & = 6 \\ b & = -2 \end{array} \qquad \begin{array}{rl} -2a - b & = 4 \\ -2a - (-2) & = 4 \\ -2a & = 2 \\ a & = -1 \end{array}$$

$$\begin{array}{rl} 4b + 6c & = 16 \\ 4(-2) + 6(4) & \overset{?}{=} 16 \\ 16 & = 16 \end{array} \qquad \begin{array}{rl} -2a - b & = 4 \\ -2(-1) - (-2) & \overset{?}{=} 4 \\ 4 & = 4 \end{array}$$

$$\begin{array}{rl} a + b + c & = 1 \\ -1 + (-2) + 4 & \overset{?}{=} 1 \\ 1 & = 1 \end{array}$$

The solution is $(-1, -2, 4)$.

47.
$$\begin{array}{rr} 2a + b - c & = -4 \\ a - 2b + c & = 13 \\ \hline 3a - b & = 9 \end{array} \qquad \begin{array}{rr} 2a + b - c & = -4 \\ a - b + c & = 4 \\ \hline 3a - 2b & = 0 \end{array}$$

$$\begin{array}{rll} 3a - b = 9 & \to & 6a - 2b = 18 \\ 3a - 2b = 0 & \to & 3a + 2b = 0 \\ & & \hline 9a = 18 \\ & & a = -2 \end{array}$$

$$\begin{array}{rl} 3a - b & = 9 \\ 3(2) - b & = 9 \\ -b & = 3 \\ b & = -3 \end{array} \qquad \begin{array}{rl} a + b + c & = 4 \\ 2 + (-3) + c & = 4 \\ c & = 5 \end{array}$$

$$\begin{array}{rl} 2a + b - c & = -4 \\ 2(2) + (-3) - 5 & \overset{?}{=} -4 \\ -4 & = -4 \end{array} \qquad \begin{array}{rl} a - 2b + c & = 13 \\ 2 - 2(-3) + 5 & \overset{?}{=} 13 \\ 13 & = 13 \end{array}$$

$$\begin{array}{rl} a + b + c & = 4 \\ 2 + (-3) + 5 & \overset{?}{=} 4 \\ 4 & = 4 \end{array}$$

The solution is $(2, -3, 5)$.

48. $x^2 + x - 20 = 0, (x + 5)(x - 4) = 0,$

$x = -5 \text{ or } x - 4,$
$x^2 + x - 20 = 0: (-5)^2 + (-5) - 20 \overset{?}{=} 0,$
$0 = 0; 4^2 + 4 - 20 \overset{?}{=} 0, 0 = 0$

49. $2x^2 - 7x - 30 = 0, (2x + 5)(x - 6) = 0,$
$x = -2.5 \text{ or } x = 6;$
$2x^2 - 7x - 30 = 0: 2(2.5)^2 - 7(2.5) - 30 \overset{?}{=} 0,$
$0 = 0; 2(6)^2 - 7(6) - 30 \overset{?}{=} 0, 0 = 0$

50. B; $1^2 - 4(5)(-9) = 181$

Lesson 6.5, pages 303–307

EXPLORE

1. $y^2 - 7y + 10 = 0$ **2.** quadratic

3. $z^2 - 9z - 18 = 0$ **4.** quadratic

5. $x^3 + 11x^2 - 6x = 0, x(x^2 + 11x - 6) = 0;$ yes

1. no   **2.** no   **3.** $(x^2)^2 + 4(x^2) - 5 = 0$

4. $(x^2)^2 - 8(x^2) + 12 = 0$

5. $(\sqrt{x})^2 - 7(\sqrt{x}) + 10 = 0$

6. $(\sqrt{x})^2 - 6(\sqrt{x}) - 72 = 0$

7. He correctly let $x = m + 3$ and then solved the equation $x^2 - 10x + 24 = 0$. He obtained the answers $x = 4$ and $x = 6$. However, he forgot to substitute back in the equation $x = m + 3$ to obtain $m = 1$ and $m = 3$.

8. $x^3 - 7x = 0, x(x^2 - 7) = 0; x^2 - 7$

9. $x^3 + 5x = 0, x(x^2 + 5) = 0; x^2 + 5$

10. $x^5 - 4x^4 + 9x^3 = 0, x^3(x^2 - 4x + 9) = 0;$
$x^2 - 4x + 9$

11. $x^5 + 8x^4 - 9x^3 = 0, x^3(x^2 + 8x - 9) = 0;$
$x^2 + 8x - 9$

12. $x^3 - 2x^2 - 5 = 0;$ no

13. $x^3 + 3x^2 - 10 = 0;$ no

14. Let x represent the shorter leg, then $3\sqrt{x}$ represents the longer leg. $x^2 + (3\sqrt{x})^2 = 6^2, x^2 + 9x - 36 = 0,$ $(x - 3)(x + 12) = 0, x = 3$ or $x = 12$; Disregard -12 since measurement must be positive. The shorter leg is 3 cm.

15.
$$x^4 - 14x^2 + 45 = 0$$
$$(x^2)^2 - 14(x^2) + 45 = 0$$
$$(x^2 - 9)(x^2 - 5) = 0$$
$$x^2 = 9 \text{ or } x^2 = 5$$
$$x = \pm 3 \text{ or } x = \pm\sqrt{5}$$
$$x^4 - 14x^2 + 45 = 0$$
$$(\pm 3)^4 - 14(\pm 3)^2 + 45 \stackrel{?}{=} 0$$
$$x^4 - 14x^2 + 45 = 0$$
$$(\pm\sqrt{5})^4 - 14(\pm\sqrt{5})^2 + 45 \stackrel{?}{=} 0$$
$$0 = 0$$

The solutions are 3, -3, $-\sqrt{5}$ and $\sqrt{5}$.

16.
$$x^4 - 15x^2 + 36 = 0$$
$$(x^2)^2 - 15(x^2) + 36 = 0$$
$$(x^2 - 3)(x^2 - 12) = 0$$
$$x^2 = 3 \text{ or } x^2 = 12$$
$$x = \pm\sqrt{3} \text{ or } x = \pm\sqrt{12} = \pm 2\sqrt{3}$$
$$x^4 - 15x^2 + 36 = 0$$
$$(\pm\sqrt{3})^4 - 15(\pm\sqrt{3})^2 + 36 \stackrel{?}{=} 0$$
$$0 = 0$$
$$x^4 - 15x^2 + 36 = 0$$
$$(\pm 2\sqrt{3})^4 - 15(\pm 2\sqrt{3})^2 + 36 \stackrel{?}{=} 0$$
$$0 = 0$$

The solutions are $\sqrt{3}, -\sqrt{3}, 2\sqrt{3},$ and $-2\sqrt{3}$.

17.
$$x - 4\sqrt{x} - 5 = 0$$
$$(\sqrt{x})^2 - 4(\sqrt{x}) - 5 = 0$$
$$(\sqrt{x} + 1)(\sqrt{x} - 5) = 0$$
$$\sqrt{x} = -1 \text{ or } \sqrt{x} = 5$$

$$x = 25$$
$$x - 4\sqrt{x} - 5 = 0$$
$$25 - 4\sqrt{25} - 5 \stackrel{?}{=} 0$$
$$0 = 0$$

The solution is 25.

18.
$$(x - 5)^2 - 5(x - 5) + 6 = 0$$
$$p^2 - 5p + 6 = 0$$
$$(p - 3)(p - 2) = 0$$
$$p = 3 \text{ or } p = 2$$
$$x - 5 = 3 \text{ or } x - 5 = 2$$
$$x = 8 \text{ or } x = 7$$
$$(x - 5)^2 - 5(x - 5) + 6 = 0$$
$$(8 - 5)^2 - 5(8 - 5) + 6 \stackrel{?}{=} 0$$
$$0 = 0$$
$$(x - 5)^2 - 5(x - 5) + 6 = 0$$
$$(7 - 5)^2 - 5(7 - 5) + 6 \stackrel{?}{=} 0$$
$$0 = 0$$

The solutions are 7 and 8.

19.
$$(x - 4)^2 - 10(x - 4) + 9 = 0$$
$$p^2 - 10p + 9 = 0$$
$$(p - 1)(p - 9) = 0$$
$$p = 1 \text{ or } p = 9$$
$$x - 4 = 1 \text{ or } x - 4 = 9$$
$$x = 5 \text{ or } x = 13$$
$$(x - 4)^2 - 10(x - 4) + 9 = 0$$
$$(5 - 4)^2 - 10(5 - 4) + 9 \stackrel{?}{=} 0$$
$$0 = 0$$
$$(x - 4)^2 - 10(x - 4) + 9 = 0$$
$$(13 - 4)^2 - 10(13 - 4) + 9 \stackrel{?}{=} 0$$
$$0 = 0$$

The solutions are 5 and 13.

20. Let x represent the number.
$$x^4 + 36 = 12x^2$$
$$x^4 - 12x^2 + 36 = 0$$
$$(x^2)^2 - 12(x^2) + 36 = 0$$
$$(x^2 - 6)(x^2 - 6) = 0$$
$$x^2 = 6$$
$$x = \pm\sqrt{6}$$

The number is $\pm\sqrt{6}$.

PRACTICE

1. $(x^2)^2 - 5(x^2) + 15 = 0$

2. $(x^2)^2 + 6(x^2) - 9 = 0$ **3.** no   **4.** no

5. $(\sqrt{x})^2 - 3(\sqrt{x}) + 1 = 0$

6. $(\sqrt{x})^2 - 9(\sqrt{x}) + 3 = 0$

7. Answers will vary.

8. $3x^3 + 6x = 0, 3x(x^2 + 2) = 0; x^2 + 2$

9. $4x^3 + 2x = 0, 2x(2x^2 + 1) = 0; 2x^2 + 1$

10. $x^5 - 12x^4 + 2x^3 = 0, x^3(x^2 - 12x + 2) = 0;$
$x^2 - 12x + 2$

11. $x^5 + 6x^4 - 3x^3 = 0$, $x^3(x^2 + 6x - 3) = 0$;
$x^2 + 6x - 3$

12. $2x^5 - 2x - 5 = 0$; no

13. $5x^3 + 5x^2 - 5 = 0$; no

14. Let x represent the width, then $2\sqrt{x}$ represents the length.
$x^2 + (2\sqrt{x})^2 = 10^2$, $x^2 + 4x - 100 = 0$;
$$\frac{-b \pm \sqrt{b^2 - 4ac}}{2a} = \frac{-4 \pm \sqrt{4^2 - 4(1)(-100)}}{2(1)} =$$
$$\frac{-4 \pm \sqrt{416}}{2} = \frac{-4 \pm 4\sqrt{26}}{2} = -2 \pm 2\sqrt{26};$$ The
width of the rectangle is $-2 + 2\sqrt{26}$

15.
$$x^4 - 3x^2 - 40 = 0$$
$$(x^2)^2 - 3(x^2) - 40 = 0$$
$$(x^2 + 5)(x^2 - 8) = 0$$
$$x^2 = -5 \text{ or } x^2 = 8$$
$$x = \pm\sqrt{8}$$
$$x = \pm 2\sqrt{2}$$
$$x^4 - 3x^2 - 40 = 0$$
$$(\pm 2\sqrt{2})^4 - 3(\pm\sqrt{2})^2 - 40 \overset{?}{=} 0$$
$$0 = 0$$

16.
$$x^4 - 14x^2 - 32 = 0$$
$$(x^2)^2 - 14(x^2) - 32 = 0$$
$$(x^2 + 2)(x^2 - 16) = 0$$
$$x^2 = -2 \text{ or } x^2 = 16$$
$$x = \pm 4$$
$$x^4 - 14x^2 - 32 = 0$$
$$(\pm 4)^4 - 14(\pm 4)^2 - 32 \overset{?}{=} 0$$
$$0 = 0$$

17.
$$9x^4 - 18x^2 + 8 = 0$$
$$9(x^2)^2 - 18(x^2) + 8 = 0$$
$$(3x^2 - 2)(3x^2 - 4) = 0$$
$$3x^2 = 2 \text{ or } 3x^2 = 4$$
$$x = \pm\sqrt{\frac{2}{3}} \text{ or } x \pm \sqrt{\frac{4}{3}}$$
$$9x^4 - 18x^2 + 8 = 0$$
$$9\left(\pm\sqrt{\frac{2}{3}}\right)^4 - 18\left(\pm\sqrt{\frac{2}{3}}\right)^2 + 8 \overset{?}{=} 0$$
$$0 = 0$$
$$9x^4 - 18x^2 + 8 = 0$$
$$9\left(\pm\sqrt{\frac{4}{3}}\right)^4 - 18\left(\pm\sqrt{\frac{4}{3}}\right)^2 + 8 \overset{?}{=} 0$$
$$0 = 0$$

18.
$$12x^4 + 5x^2 - 2 = 0$$
$$12(x^2)^2 + 5(x^2) - 2 = 0$$
$$(3x^2 + 2)(4x^2 - 1) = 0$$
$$3x^2 = -2 \text{ or } 4x^2 = 1$$
$$x = \pm\sqrt{\frac{1}{4}}$$
$$x = \pm\frac{1}{2}$$
$$12x^4 + 5x^2 - 2 = 0$$
$$12\left(\pm\frac{1}{2}\right)^4 + 5\left(\pm\frac{1}{2}\right)^2 - 2 \overset{?}{=} 0$$

$$0 = 0$$

19.
$$x - 3\sqrt{x} - 4 = 0$$
$$(\sqrt{x})^2 - 3(\sqrt{x}) - 4 = 0$$
$$(\sqrt{x} + 1)(\sqrt{x} - 4) = 0$$
$$\sqrt{x} = -1 \text{ or } \sqrt{x} = 4$$
$$x = 16$$
$$x - 3\sqrt{x} - 4 = 0$$
$$16 - 3\sqrt{16} - 4 \overset{?}{=} 0$$
$$0 = 0$$

20.
$$x - 8\sqrt{x} + 15 = 0$$
$$(\sqrt{x})^2 - 8(\sqrt{x}) + 15 = 0$$
$$(\sqrt{x} - 3)(\sqrt{x} - 5) = 0$$
$$\sqrt{x} = 3 \text{ or } \sqrt{x} = 5$$
$$x = 9 \text{ or } x = 25$$
$$x - 8\sqrt{x} + 15 = 0$$
$$9 - 8\sqrt{9} + 15 \overset{?}{=} 0$$
$$0 = 0$$
$$x - 8\sqrt{x} + 15 = 0$$
$$25 - 8\sqrt{25} + 15 \overset{?}{=} 0$$
$$0 = 0$$

21.
$$x - 12\sqrt{x} + 32 = 0$$
$$(\sqrt{x})^2 - 12(\sqrt{x}) + 32 = 0$$
$$(\sqrt{x} - 4)(\sqrt{x} - 8) = 0$$
$$\sqrt{x} = 4 \text{ or } \sqrt{x} = 8$$
$$x = 16 \text{ or } x = 64$$
$$x - 12\sqrt{x} + 32 = 0$$
$$16 - 12\sqrt{16} + 32 \overset{?}{=} 0$$
$$0 = 0$$
$$x - 12\sqrt{x} + 32 = 0$$
$$64 - 12\sqrt{64} + 32 \overset{?}{=} 0$$
$$0 = 0$$

22.
$$x - 14\sqrt{x} + 24 = 0$$
$$(\sqrt{x})^2 - 14(\sqrt{x}) + 24 = 0$$
$$(\sqrt{x} - 2)(\sqrt{x} - 12) = 0$$
$$\sqrt{x} = 2 \text{ or } \sqrt{x} = 12$$
$$x = 4 \quad x = 144$$
$$x - 14\sqrt{x} + 24 = 0$$
$$4 - 14\sqrt{4} + 24 \overset{?}{=} 0$$
$$0 = 0$$
$$x - 14\sqrt{x} + 24 = 0$$
$$144 - 14\sqrt{144} + 24 \overset{?}{=} 0$$
$$0 = 0$$

23.
$$(x - 6)^2 - 17(x - 6) + 70 = 0$$
$$p^2 - 17p + 70 = 0$$
$$(p - 10)(p - 7) = 0$$
$$p = 10 \text{ or } p = 7$$
$$x - 6 = 10 \text{ or } x - 6 = 7$$
$$x = 16 \text{ or } x = 13$$
$$(x - 6)^2 - 17(x - 6) + 70 = 0$$
$$(16 - 6)^2 - 17(16 - 6) + 70 \overset{?}{=} 0$$
$$0 = 0$$
$$(x - 6)^2 - 17(x - 6) + 70 = 0$$
$$(13 - 6)^2 - 17(13 - 6) + 70 \overset{?}{=} 0$$
$$0 = 0$$

24.
$$(x - 1)^2 - 16(x - 1) + 60 = 0$$
$$p^2 - 16p + 60 = 0$$
$$(p - 10)(p - 6) = 0$$
$$p = 10 \text{ or } p = 6$$
$$x - 1 = 10 \text{ or } x - 1 = 6$$
$$x = 11 \text{ or } x = 7$$
$$(x - 1)^2 - 16(x - 1) + 60 = 0$$
$$(11 - 1)^2 - 16(11 - 1) + 60 \overset{?}{=} 0$$
$$0 = 0$$
$$(x - 1)^2 - 16(x - 1) + 60 = 0$$
$$(7 - 1)^2 - 16(7 - 1) + 60 \overset{?}{=} 0$$
$$0 = 0$$

25.
$$(x + 2)^2 - 4(x + 2) - 32 = 0$$
$$p^2 - 4p - 32 = 0$$
$$(p + 4)(p - 8) = 0$$
$$p = -4 \text{ or } p = 8$$
$$x + 2 = -4 \text{ or } x + 2 = 8$$
$$x = -6 \text{ or } x = 6$$
$$(x + 2)^2 - 4(x + 2) - 32 = 0$$
$$(-6 + 2)^2 - 4(-6 + 2) - 32 \overset{?}{=} 0$$
$$0 = 0$$
$$(x + 2)^2 - 4(x + 2) - 32 = 0$$
$$(6 + 2)^2 - 4(6 + 2) - 32 \overset{?}{=} 0$$
$$0 = 0$$

26.
$$(x + 1)^2 - (x + 1) - 20 = 0$$
$$p^2 - p - 20 = 0$$
$$(p + 4)(p - 5) = 0$$
$$x = -4 \text{ or } p = 5$$
$$x + 1 = -4 \text{ or } x + 1 = 5$$
$$x = -5 \text{ or } x = 4$$
$$(x + 1)^2 - (x + 1) - 20 = 0$$
$$(-5 + 1)^2 - (-5 + 1) - 20 \overset{?}{=} 0$$
$$0 = 0$$
$$(x + 1)^2 - (x + 1) - 20 = 0$$
$$(4 + 1)^2 - (4 + 1) - 20 \overset{?}{=} 0$$
$$0 = 0$$

27. Let x represent the width, then $x + 12$ represents the length, $x + 6$ represents the height, and 2016 represents the volume.
$$x(x + 12)(x + 6) = 2016x$$
$$x^3 + 12x^2 + 6x^2 + 72x = 2016x$$
$$x^3 + 18x^2 - 1944x = 0$$
$$x(x^2 + 18x - 1944) = 0$$
$$x(x + 54)(x - 36) = 0$$
$$x = 0 \text{ or } x = -54 \text{ or } x = 36$$
Disregard 0 and -54 since measurement must be positive; $x + 12 = 36 + 12 = 48$; $x + 6 = 36 + 6 = 42$; The dimensions of the box are 48 in. $\times$ 36 in. $\times$ 42 in.

EXTEND

28.
$$\left(\frac{x + 1}{x - 1}\right)^2 + \frac{(x + 1)}{(x - 1)} - 6 = 0$$
$$p^2 + p - 6 = 0$$

$$(p - 2)(p + 3) = 0$$
$$p = 2 \text{ or } p = -3$$
$$\frac{x + 1}{x - 1} = 2 \text{ or } \frac{x + 1}{x - 1} = -3$$
$$x + 1 = 2(x - 1) \text{ or } x + 1 = -3(x - 1)$$
$$x + 1 = 2x - 2 \text{ or } x + 1 = -3x + 3$$
$$-x = -3 \text{ or } 4x = 2$$
$$x = 3 \text{ or } x = \frac{2}{4} = \frac{1}{2}$$
$$\left(\frac{x + 1}{x - 1}\right)^2 + \left(\frac{x + 1}{x - 1}\right) - 6 = 0$$
$$\left(\frac{3 + 1}{3 - 1}\right)^2 + \left(\frac{3 + 1}{3 - 1}\right) - 6 \overset{?}{=} 0$$
$$0 = 0$$
$$\left(\frac{x + 1}{x - 1}\right)^2 + \left(\frac{x + 1}{x - 1}\right) - 6 = 0$$
$$\left(\frac{\frac{1}{2} + 1}{\frac{1}{2} - 1}\right)^2 + \left(\frac{\frac{1}{2} + 1}{\frac{1}{2} - 1}\right) - 6 \overset{?}{=} 0$$
$$0 = 0$$

29.
$$2\left(\frac{x + 3}{x - 3}\right)^2 - 9\left(\frac{x + 3}{x - 3}\right) + 10 = 0$$
$$2p^2 - 9p + 10 = 0$$
$$(p - 2)(2p - 5) = 0$$
$$p = 2 \text{ or } 2p = 5$$
$$\frac{x + 3}{x - 3} = 2 \text{ or } p = \frac{5}{2}$$
$$x + 3 = 2(x - 3) \text{ or } \frac{x + 3}{x - 3} = \frac{5}{2}$$
$$x + 3 = 2x - 6 \text{ or } 2(x + 3) = 5(x - 3)$$
$$-x = -9 \text{ or } 2x + 6 = 5x - 15$$
$$x = 9 \text{ or } -3x = -21$$
$$x = 7$$
$$2\left(\frac{x + 3}{x - 3}\right)^2 - 9\left(\frac{x + 3}{x - 3}\right) + 10 = 0$$
$$2\left(\frac{9 + 3}{9 - 3}\right)^2 - 9\left(\frac{9 + 3}{9 - 3}\right) + 10 \overset{?}{=} 0$$
$$0 = 0$$
$$2\left(\frac{x + 3}{x - 3}\right)^2 - 9\left(\frac{x + 3}{x - 3}\right) + 10 = 0$$
$$2\left(\frac{7 + 3}{7 - 3}\right)^2 - 9\left(\frac{7 + 3}{7 - 3}\right) + 10 \overset{?}{=} 0$$
$$0 = 0$$

30a.
$$\frac{\sqrt{d}}{4} + \frac{d}{1100} = \frac{59}{11}$$
$$275\sqrt{d} + d = 5900$$
$$d + 275\sqrt{d} - 5900 = 0$$
$$(\sqrt{d})^2 + 275(\sqrt{d}) - 5900 = 0$$
$$(\sqrt{d} + 295)(\sqrt{d} - 20) = 0$$
$$\sqrt{d} = -295 \text{ or } \sqrt{d} = 20$$
$$d = 400$$
Disregard -295 since distance must be positive. The drop is 400 ft.

30b.
$$\frac{\sqrt{d}}{4} + \frac{d}{1100} = 10$$
$$275\sqrt{d} + d = 11{,}000$$
$$d + 275\sqrt{d} - 11{,}000 = 0$$
$$(\sqrt{d})^2 + 275(\sqrt{d}) - 11{,}000 = 0$$
$$\sqrt{d} = \frac{-b \pm \sqrt{b^2 - 4ac}}{2a} =$$
$$\frac{-275 \pm \sqrt{275^2 - 4(1)(-11000)}}{2(1)} \approx$$
-310 or 35.434; $\sqrt{d} = 35.434$; $d \approx 1256$
The drop is about 1256 ft.

31.
$$1000D = 108d^2 + 803d + 620$$
$$1000(115) = 108d^2 + 803d + 620$$
$$0 = 108d^2 + 803d - 114{,}380$$
Graph $y = 108d^2 + 803d - 114{,}380$. The positive
x-intercept is about 29.037, so the depth is about 29m.

THINK CRITICALLY

32.
$$|a - 2|^2 - 9|a - 2| = -18$$
$$|a - 2|^2 - 9|a - 2| + 18 = 0$$
$$p^2 - 9p + 18 = 0$$
$$(p - 3)(p - 6) = 0$$
$$p = 6$$
$|a - 2| = 3$ or $|a - 2| = 6$
$a - 2 = 3$ or $a - 2 = -3$ or $a - 2 = 6$ or $a - 2 = -6$
$a = 5$ or $a = -1$ or $a = 8$ or $a = -4$
$$|a - 2|^2 - 9|a - 2| = -18$$
$$|a - 2|^2 - 9|a - 2| = -18$$
$$|5 - 2|^2 - 9|5 - 2| \stackrel{?}{=} -18$$
$$|-1 - 2|^2 - 9|-1 - 2| \stackrel{?}{=} -18$$
$$-18 = -18 \qquad -18 = -18$$
$$|a - 2|^2 - 9|a - 32| = -18$$
$$|A - 2|^2 - 9|A - 2| = -18$$
$$|8 - 2|^2 - 9|8 - 2| \stackrel{?}{=} -18$$
$$|-4 - 2|^2 - 9|-4 - 2| \stackrel{?}{=} -18$$
$$-18 = -18 \qquad -18 = -18$$

33.
$$|b + 1|^2 - 10|b + 1| = -16$$
$$|b + 1|^2 - 10|b + 1| + 16 = 0$$
$$p^2 - 10p + 16 = 0$$
$$(p - 8)(p - 2) = 0$$
$$p = 2$$
or $|b + 1| = 2$
$b + 1 = -8$ or $b + 1 = 8$ or $b + 1 = 2$ or $b + 1 = -2$
$b = -9$ or $b = 7$ or $b = 1$ or $b = -3$
$$|b + 1|^2 - 10|b + 1| = -16$$
$$|b + 1|^2 - 10|b + 1| = -16$$
$$|-9 + 1|^2 - 10|-9 + 1| \stackrel{?}{=} -16$$
$$|1 + 1|^2 - 10|1 + 1| \stackrel{?}{=} -16$$
$$-16 = -16 \qquad -16 = -16$$
$$|b + 1|^2 - 10|b + 1| = -16$$
$$|b + 1|^2 - 10|b + 1| = -16$$
$$|7 + 1|^2 - 10|7 + 1| \stackrel{?}{=} -16$$
$$|-3 + 1|^2 - 10|-3 + 1| \stackrel{?}{=} -16$$
$$-16 = -16 \qquad -16 = -16$$

34.
$$2 - \frac{11}{x} - \frac{6}{x^2} = 0$$
$$-6p^2 - 11p + 2 = 0$$
$$(-6p + 1)(p + 2) = 0$$
$$-6p = -1 \text{ or } p = -2$$
$$p = \frac{1}{6}$$
$$x = \frac{1}{\frac{1}{6}} = 6 \text{ or } x = -\frac{1}{2}$$
$$2 - \frac{11}{x} - \frac{6}{x^2} = 0$$
$$2 - \frac{11}{6} - \frac{6}{6^2} \stackrel{?}{=} 0$$
$$0 = 0$$
$$2 - \frac{11}{x} - \frac{6}{x^2} = 0$$
$$2 - \frac{11}{-\frac{1}{2}} - \frac{6}{\left(-\frac{1}{2}\right)^2} \stackrel{?}{=} 0$$
$$0 = 0$$

35.
$$3 - \frac{7}{x} - \frac{20}{x^2} = 0$$
$$-20p^2 - 7p + 3 = 0$$
$$(-4p + 1)(5p + 3) = 0$$
$-4p = -1$ or $5p = -3$
$$p = \frac{1}{4} \qquad \text{or} \qquad p = -\frac{3}{5}$$
$$x = \frac{1}{\frac{1}{4}} = 4 \text{ or } x = \frac{1}{-\frac{3}{5}} = -\frac{5}{3}$$
$$3 - \frac{7}{x} - \frac{20}{x^2} = 0$$
$$3 - \frac{7}{4} - \frac{20}{4^2} \stackrel{?}{=} 0$$
$$0 = 0$$
$$3 - \frac{7}{x} - \frac{20}{x^2} = 0$$
$$3 - \frac{7}{-\frac{3}{5}} - \frac{20}{-\frac{3}{5}} \stackrel{?}{=} 0$$
$$0 = 0$$

36. $x^2 - 3x - 4 = 0$ can be factored to
$(x + 4)(x - 1) = 0$, so -4 and 1 are solutions.
$x + 3\sqrt{x} - 4 = 0$ can be factored to
$(\sqrt{x} + 4)(\sqrt{x} - 1) = 0$. However, $\sqrt{x} + 4 = 0$ if
$\sqrt{x} = -4$ is not a real solution. So, $\sqrt{x} - 1 = 0$ if
$x = 1$ is the only solution.

MIXED REVIEW

37. $|x - 5| = 10$
$x - 5 = 10$ or $x - 5 = -10$
$x = 15$ $\qquad$ $x = -5$
$|x - 5| = 10$ $\qquad$ $|x - 5| = 10$

$|15 - 5| \overset{?}{=} 10$ $|-5 - 5| \overset{?}{=} 10$
$\quad 10 = 10$ $\qquad 10 = 10$

38. $|x + 8| = 10$
$\quad x + 8 = 10$ or $\quad x + 8 = -10$
$\qquad x = 2$ or $\qquad x = -18$
$\quad |x + 8| = 10$ $\quad |x + 8| = 10$
$\quad |2 + 8| \overset{?}{=} 10$ $|-18 + 8| \overset{?}{=} 10$
$\qquad 10 = 10$ $\qquad 10 = 10$

39. $C; \frac{1}{2} \cdot \frac{1}{2} \cdot \frac{1}{2} = \frac{1}{8}$

40. $x^3 - 125, (x - 5)(x^2 + 5x + 25)$

41. $x^3 - 1000, (x - 10)(x^2 + 10x + 100)$

42.
$$x^4 - 15x^2 + 54 = 0$$
$$(x^2)^2 - 15(x^2) + 54 = 0$$
$$(x^2 - 6)(x^2 - 9) = 0$$
$$x^2 = 6 \text{ or } x^2 = 9$$
$$x = \pm\sqrt{6} \qquad x = \pm 3$$
$$x^4 - 15x^2 + 54 = 0$$
$$(\pm\sqrt{6})^4 - 15(\pm\sqrt{6})^2 + 54 \overset{?}{=} 0$$
$$0 = 0$$
$$x^4 - x^2 + 54 = 0$$
$$(\pm 3)^4 - 15(\pm 3)^2 + 54 \overset{?}{=} 0$$
$$0 = 0$$

43.
$$x - 3\sqrt{x} - 40 = 0$$
$$(\sqrt{x})^2 - 3(\sqrt{x}) - 40 = 0$$
$$(\sqrt{x} + 5)(\sqrt{x} - 8) = 0$$
$$\sqrt{x} = -5 \text{ or } \sqrt{x} = 8$$
$$x = 64$$
$$x - 3\sqrt{x} - 40 = 0$$
$$64 - 3\sqrt{64} - 40 \overset{?}{=} 0$$
$$0 = 0$$

Lesson 6.6, pages 308–311

EXPLORE THE PROBLEM

1. $y_0 = f(t_0); y = f(t)$

2. $\Delta y = f(t_0 + \Delta t) - f(t_0)$

3. $\frac{\Delta y}{\Delta x}$ is the ratio of distance traveled to elapsed time.

4. $\frac{\Delta y}{\Delta t} = v = \frac{y - y_0}{t - t_0}; v(t - t_0) = y - y_0,$
$y = y_0 + v(t - t_0)$

5. $y = y_0 + v(t - t_0), y = y_0 + vt - vt_0; y = mt + b;$
$y = mt + b$ is of the slope-intercept form.

6. $\frac{\Delta y}{\Delta t} = \frac{f(t_0 + \Delta t) - f(t_0)}{\Delta t}$
$= \frac{m(t_0 + \Delta t) + b - [m(t_0) + b]}{\Delta t}$
$= \frac{mt_0 + m\Delta t + b - mt_0 - b}{\Delta t}$
$= \frac{m\Delta t}{\Delta t}$
$= m$

7. $y = 4t - 3$ is a linearfunction with $m = 4$; so $v = 4$ at all times and in particular at $t = 5$.

INVESTIGATE FURTHER

8. If $\Delta t = 0$ then $\Delta y = 0$ and $\frac{\Delta y}{\Delta t} = \frac{0}{0}$ which is meaningless.

9. $(3^2 + 1)$
$9 + 6\Delta t + (\Delta t)^2 + 1 - 10$
$6\Delta t + (\Delta t)^2$
$6 + \Delta t$

10. Yes, as Δt gets close to 0, $6 + \Delta t$ gets close to 6.

11. 6ft/sec

APPLY THE STRATEGY

12. uniform; 4 ft/sec

13. uniform; 7 ft/sec

14. nonuniform; 8 ft/sec

15. nonuniform; 24 ft/sec

16. nonuniform; 4 ft/sec

17. This method becomes unwieldly if the function describing motion is complicated.

18. $v = 8t; v = 8 \cdot 11, v = 88$; 88 ft/sec

REVIEW PROBLEM SOLVING STRATEGIES

1a.

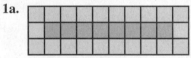

1b.

1c.

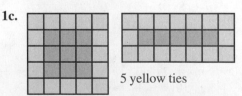

5 yellow ties

9 yellow ties

1d. Answers may vary. $G = 2(l + w) + 4; Y = lw$

1e. Many designs are possible. One way to determine the dimensions is to pick an even whole number for l, substitute in the equation $lw = 2(l + w) + 4$, and solve for w. One solution is $l = 6, w = 4, Y = G = 24$, and another is $l = 10, w = 3, Y = G = 30$.

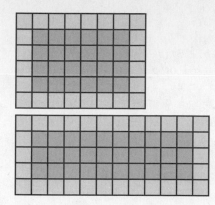

2a. If they arrived home 30 minutes earlier, Mr. Lee must have spent 30 minutes less driving. The trip must have taken 15 minutes less in each direction.

2b. If the trip one way took 15 minutes less, Mr. Lee must have met Mrs. Lee 15 minutes before the usual time, at 5:15 rather than 5:30. Mrs. Lee must have walked from 4:30 to 5:15, for 45 minutes.

2c. If Mrs. Lee had walked 5 minutes longer, for 50 minutes, she would have met her husband at 5:20. The trip one way would have taken 10 minutes less, so Mr. Lee would spend 20 minutes less driving the entire trip. Mrs. Lee would arrive home 20 minutes earlier than usual.

3. Let x represent her entire life in years.

$$\frac{1}{4}x + \frac{1}{6}x + \frac{1}{2}x + 7 - x$$

$$\frac{11}{12}x + 7 - x$$

$$7 = \frac{1}{12}x$$

$$84 = x$$

$$x = 84$$

Roy's aunt was 84 years old when she died.

Chapter Review, pages 312–313

1. e **2.** d **3.** a **4.** c **5.** b

6. $\left(\frac{b}{2}\right)^2 = 256$ **7.** $\left(\frac{b}{2}\right)^2 = 400$ **8.** $\left(\frac{b}{2}\right)^2 = \frac{25}{36}$

$\frac{b}{2} = \pm16$ $\quad\quad \frac{b}{2} = \pm20$ $\quad\quad \frac{b}{2} = \pm\frac{5}{6}$

$b = \pm32$ $\quad\quad b = \pm40$ $\quad\quad b = \pm\frac{10}{6} = \pm\frac{5}{3}$

9. $x^2 + 9x + 18 = 0$

$$x^2 + 9x + \frac{81}{4} = -18 + \frac{81}{4}$$

$$\left(x + \frac{9}{2}\right)^2 = \frac{9}{4}$$

$$x + \frac{9}{2} = \pm\frac{3}{2}$$

$$x = -\frac{9}{2} \pm \frac{3}{2}$$

$$x = -\frac{12}{2} = -6 \text{ or } x = -\frac{6}{2} = -3$$

10. $\quad x^2 - 10x = 2$

$$x^2 - 10x + 25 = 2 + 25$$

$$(x - 5)^2 = 27$$

$$x - 5 = \pm\sqrt{27}$$

$$x = 5 \pm \sqrt{27}$$

11. $\quad 4x^2 - 6x = 5$

$$x^2 - \frac{6}{4}x = \frac{5}{4}$$

$$x^2 - \frac{6}{4}x + \frac{36}{64} = \frac{5}{4} + \frac{36}{64}$$

$$\left(x - \frac{6}{8}\right)^2 = \frac{116}{64}$$

$$x - \frac{6}{8} = \pm\sqrt{\frac{116}{64}}$$

$$x = \frac{6}{8} \pm \sqrt{\frac{116}{64}}$$

$$x = \frac{3}{4} \pm \sqrt{\frac{29}{16}}$$

$$x = \frac{3 \pm \sqrt{29}}{4}$$

12. right 3 units and down 5 units

13. left 6 units

14. left 4 units and down 8 units

15. c, a, b, d

16. $-\dfrac{b}{2a} = -\dfrac{2}{2(1)} = -1;$

$f(-1) = (-1)^2 + 2(-1) - 3 = -4;$

$x = -1; (-1, -4);$
$f(0) = 0^2 + 2(0) - 3;$
$0 = x^2 + 2x - 3$
$0 = (x + 3)(x - 1)$
$x = -3 \quad\text{or}\quad x = 1$

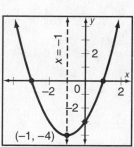

17. $-\dfrac{b}{2a} = -\dfrac{-4}{2(1)} = 2; g(2) = 2^2 - 4(2) + 4 = 0;$

$x = 2; (2, 0);$

$g(0) = 0^2 - 4(0) + 4 = 4;$
$0 = x^2 - 4x + 4$
$0 = (x - 2)^2$
$x = 2$

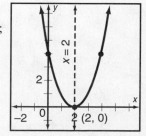

18. $-\dfrac{b}{2a} = -\dfrac{-4}{2(-1)} = -2;$

$h(-2) = -(-2)^2 - 4(-2) + 5 = 9;$

$x = -2; (-2, 9); h(0) = -0^2 - 4(0) + 5 = 5;$

$0 = -x^2 - 4x + 5$

$0 = (-x + 1)(x + 5)$

$x = 1 \quad \text{or} \quad x = -5$

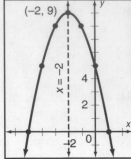

19. 324ft

20. $\dfrac{-b \pm \sqrt{b^2 - 4ac}}{2a} = \dfrac{-(-4) \pm \sqrt{(-4)^2 - 4(2)(-5)}}{2(2)} =$

$\dfrac{4 \pm \sqrt{56}}{4} = \dfrac{4 \pm 2\sqrt{14}}{4} = \dfrac{2 \pm \sqrt{14}}{2}$

21. $\dfrac{-b \pm \sqrt{b^2 - 4ac}}{2a} = \dfrac{-1 \pm \sqrt{1^2 - 4(5)(-7)}}{2(5)} =$

$\dfrac{-1 \pm \sqrt{141}}{10}$

22. $\dfrac{-b \pm \sqrt{b^2 - 4ac}}{2a} =$

$\dfrac{-(-24) \pm \sqrt{(-24)^2 - 4(2)(10)}}{2(2)} =$

$\dfrac{24 \pm \sqrt{496}}{4} = \dfrac{24 \pm 4\sqrt{31}}{4} = 6 \pm \sqrt{31}$

23. $s_1 + s_2 = -8 + 6 = -2; s_1 s_2 = -8(6) = -48;$

$x^2 - (-2x) + (-48) = 0, x^2 + 2x - 48 = 0$

24. $s_1 + s_2 = 3 + \dfrac{1}{4} = \dfrac{13}{4}; s_1 s_2 = 3\left(\dfrac{1}{4}\right) = \dfrac{3}{4};$

$x^2 - \dfrac{13}{4}x + \dfrac{3}{4} = 0, 4x^2 - 13x + 3 = 0$

25. $s_1 + s_2 = 4 + \left(-\dfrac{1}{2}\right) = \dfrac{7}{2};$

$s_1 s_2 = 4\left(-\dfrac{1}{2}\right) = -\dfrac{4}{2} = -2;$

$x^2 - \dfrac{7}{2}x + (-2) = 0, 2x^2 - 7x - 4 = 0$

26. $x^4 - 16x^2 + 48 = 0$

$(x^2)^2 - 16(x^2) + 48 = 0$

$(x^2 - 12)(x^2 - 4) = 0$

$x^2 = 12 \quad \text{or} \quad x^2 = 4$

$x = \pm 2\sqrt{3} \quad \text{or} \quad x = \pm 2$

27. $x + 11\sqrt{x} - 26 = 0$

$(\sqrt{x})^2 + 11(\sqrt{x}) - 26 = 0$

$(\sqrt{x} + 13)(\sqrt{x} - 2) = 0$

$\sqrt{x} = -13 \quad \text{or} \quad \sqrt{x} = 2$

$x = 4$

28. $(x - 2)^2 + 7(x - 2) + 12 = 0$

$p^2 + 7p + 12 = 0$

$(p + 3)(p + 4) = 0$

$p = -3 \quad \text{or} \quad p = -4$

$x - 2 = -3 \quad \text{or} \quad x - 2 = -4$

$x = -1 \quad \text{or} \quad x = -2$

29. uniform; 6 ft/sec

30. nonuniform; 20 ft/sec

Chapter Assessment, pages 314–315

CHAPTER TEST

1. $\left(\dfrac{b}{2}\right)^2 = 169$

$\dfrac{b}{2} = \pm 13$

$b = \pm 26$

2. $\left(\dfrac{b}{2}\right)^2 = \dfrac{121}{196}$

$\dfrac{b}{2} = \pm \dfrac{11}{14}$

$b = \pm \dfrac{22}{14} = \pm \dfrac{11}{7}$

3. Answers will vary. Possible example: To solve the quadratic equation, $x^2 - 8x - 50 = 0$, by completing the square, first add 50 to both sides of the equation. Then add $\left(\dfrac{8}{2}\right)^2$ to both sides of the equation to get $x^2 - 8x + 16 = 50 + 16$. Factor the left-hand side and solve for x.

To solve the quadratic equation, $x^2 - 8x - 50 = 0$, by using the quadratic formula, let $a = 1$ $b = -8$ and $c = -50$.

Then plug these values into $\dfrac{-b \pm \sqrt{b^2 - 4ac}}{2a}$ and solve.

4. $x^2 + 12x + 11 = 0$

$x^2 + 12x = -11$

$x^2 + 12x + 36 = -11 + 36$

$(x + 6)^2 = 25$

$x + 6 = \pm 5$

$x = -1 \quad \text{or} \quad x = -11$

5. $x^2 - 4x - 1 = 7$

$x^2 - 4x = 8$

$x^2 - 4x + 4 = 8 + 4$

$(x - 2)^2 = 12$

$x - 2 = \pm 2\sqrt{3}$

$x = 2 \pm 2\sqrt{3}$

6. $2x^2 + 8x - 16 = 0$

$x^2 + 4x - 8 = 0$

$x^2 + 4x = 8$

$x^2 + 4x + 4 = 8 + 4$

$(x + 2)^2 = 12$

$x + 2 = \pm 2\sqrt{3}$

$x = -2 \pm 2\sqrt{3}$

7. $\dfrac{x^2}{5} - 4x + 3 = 0$

$x^2 - 20x + 15 = 0$

$x^2 - 20x = -15$

$x^2 - 20x + 100 = -15 + 100$

$(x - 10)^2 = 85$

$x - 10 = \pm\sqrt{85}$

$x = 10 \pm \sqrt{85}$

8. Let n represent the number. $\dfrac{1}{n}$ is its reciprocal.

$$n + \dfrac{1}{n} = \dfrac{130}{63}$$

$$n^2 + 1 = \dfrac{130}{63}n$$

$$63n^2 + 63 = 130n$$

$$63n^2 - 130n + 63 = 0$$

$$(9n - 7)(7n - 9) = 0$$

$9n - 7 = 0 \quad$ or $\quad 7n - 9 = 0$

$9n = 7 \qquad\qquad 7n = 9$

$n = \dfrac{7}{9} \quad$ or $\qquad n = \dfrac{9}{7}$

9. $g(x) = (x + 9)^2 + 3$ **10.** $g(x) = (x + 1)^2 - 10$

11. a, b, c, d

12. $-\dfrac{b}{2a} = -\dfrac{-10}{2(1)} = 5; f(5) = 5^2 - 10(5) + 6 = -19;$

$(5, -19); x = 5;$ upward

13. $-\dfrac{b}{2a} = -\dfrac{5}{2\left(-\frac{1}{4}\right)} = 10; f(10) =$

$\dfrac{1}{4}(10) + 5(10) - 7 - 18; (10, 18); x = 10;$

downward

14. $\dfrac{-b \pm \sqrt{b^2 - 4ac}}{2a} = \dfrac{-(-6) \pm \sqrt{(-6)^2 - 4(3)(2)}}{2(3)} =$

$\dfrac{6 + \sqrt{12}}{6} = \dfrac{6 + 2\sqrt{3}}{6} = \dfrac{3 + \sqrt{3}}{3}$

15. $\dfrac{-b \pm \sqrt{b^2 - 4ac}}{2a} = \dfrac{-(-60) \pm \sqrt{(-60)^2 - 4(25)(36)}}{2(25)}$

$\dfrac{60 \pm \sqrt{0}}{50} = \dfrac{60}{50} = \dfrac{6}{5}$

16. $\dfrac{-b \pm \sqrt{b^2 - 4ac}}{2a} = \dfrac{-(-3) \pm \sqrt{(-3)^2 - 4(7)(-1)}}{2(7)} =$

$\dfrac{3 \pm \sqrt{37}}{14}$

17. $b^2 - 4ac = 5^2 - 4(1)(6) = 1;$ two real rational solutions

18. $b^2 - 4ac = (-12)^2 - 4(1)(36) = 0;$ one real rational solution

19. $b^2 - 4ac = 5^2 - 4(1)(7) = -3;$ no real solutions

20. $b^2 - 4ac = 4^2 - 4(1)(2) = 8;$ two real irrational solutions

21. $-\dfrac{b}{a} = -\dfrac{4}{3}; \dfrac{c}{a} = \dfrac{-9}{3} = -3$

22. $-\dfrac{b}{a} = -\dfrac{-\frac{1}{2}}{\frac{1}{4}} = 2; \dfrac{c}{a} = \dfrac{-8}{\frac{1}{4}} = -32$

23. $s_1 + s_2 = 6 + -3 = 3; s_1 s_2 = 6(-3) = -18;$

$x^2 - 3x + (-18) = 0, x^2 - 3x - 18 = 0$

24. $s_1 + s_2 = 11 + -12 = -1;$

$s_1 s_2 = 11(-12) = -132;$

$x^2 - (-x) + (-132) = 0, x^2 + x - 132 = 0$

25. D

26. $x^3 + 5x + 3;$ no

27. $x^3 - 6x^2 - 4x = 0, x(x^2 - 6x - 4) = 0;$

$x^2 - 6x - 4$

28. $\quad x^4 + x^2 - 30 = 0$

$(x^2)^2 + (x^2) - 30 = 0$

$(x^2 - 5)(x^2 + 6) = 0$

$x^2 = 5 \quad$ or $\quad x^2 = -6$

$x = \pm\sqrt{5}$

29. $\quad x + 3\sqrt{x} - 108 = 0$

$(\sqrt{x})^2 + 3(\sqrt{x}) - 108 = 0$

$(\sqrt{x} - 9)(\sqrt{x} + 12) = 0$

$\sqrt{x} = 9 \qquad$ or $\qquad \sqrt{x} = -12$

$x = 81$

Cumulative Review, page 316

1. Answers will vary. Possible example: Remove 2, the greatest common factor. Then add the square of one-half the coefficient of x. The perfect square trinomial is $2(x^2 - 30x + 225)$ or $2(x - 15)^2$.

2. 3 units left

3. 3 units up

4. $-\dfrac{b}{2a} = -\dfrac{4}{2(-1)} = 2; f(2) = -(2^2) + 4(2) - 6 = -2; (2, -2); x = -2;$ downward

5. $-\dfrac{b}{2a} = -\dfrac{-8}{2(2)} = 2; f(2) = 3 - 8(2) + 2(2^2) = -5; (2, -5); x = 2;$ upward

6. $\quad x^4 - 13x^2 + 36 = 0$

$(x^2)^2 - 13(x^2) + 36 = 0$

$(x^2 - 9)(x^2 - 4) = 0$

$x^2 = 9 \quad$ or $\quad x^2 = 4$

$x = \pm 3 \qquad\qquad x = \pm 2$

7. $(x - 3)^2 + 2(x - 3) - 8 = 0$

$p^2 + 2p - 8 = 0$

$(p - 2)(p + 4) = 0$

$p = 2 \quad$ or $\quad p = -4$

$x - 3 = 2 \quad$ or $\quad x - 3 = -4$

$x = 5 \quad$ or $\quad x = -1$

8. Let x represent the height, then $x + 6$ represents the base.

$$\frac{1}{2}bh = A$$

$$\frac{1}{2}(x + 6)(x) = 80$$
$$(x + 6)(x) = 160$$
$$x^2 + 6x = 160$$
$$x^2 + 6x - 160 = 0$$
$$(x - 10)(x + 16) = 0$$
$$x = 10 \text{ or } x = -16$$

Disregard -16 since measurement must be positive.
$x + 6 = 10 + 6 = 16$; The base of the triangle is 16 in. and the height of the triangle is 10 in.

9. $3x^4 - 48$, $3(x^4 - 16)$, $3(x^2 + 4)(x^2 - 4)$,

$3(x^2 + 4)(x + 2)(x - 2)$

10. $y^3 + 125$, $(y + 5)(y^2 - 5y + 25)$

11. $2x^2 + 3x - 2$

$$\begin{array}{r} 2x - 3 \\ \hline 4x^3 + 6x^2 - 4x \\ -6x^2 - 9x + 6 \\ \hline 4x^3 \qquad - 13x + 6 \end{array}$$

12. $m^2 - 3m + 5$

$$\begin{array}{r} 2m + 1 \\ \hline 2m^3 - 6m^2 + 10m \\ m^2 - 3m + 5 \\ \hline 2m^3 - 5m + 7m + 5 \end{array}$$

13. Mode would be the most useful in determing stock needed. Especially when considering most frequently sold sizes, styles, and colors.

14. C ; $(2 - x) + (3x^2 + 3x + 4) = 3x^2 + 2x + 6$; $(3x^2 + 2x + 6) - (6 + 2x) = 3x^2$

15. Let B = black, W = white, F = fabric, R = braiding, and D = beads.

$$\begin{array}{cc} F \ R \ D \\ \begin{array}{c} B \\ W \end{array}\begin{bmatrix} 6 & 5 & 2 \\ 5 & 4 & 3 \end{bmatrix} \cdot \begin{array}{c} F \\ R \\ D \end{array}\begin{bmatrix} 40 \\ 20 \\ 15 \end{bmatrix} = \begin{array}{c} B \\ W \end{array}\begin{bmatrix} 370 \\ 325 \end{bmatrix} \end{array}$$

The total cost for the black design is \$370 and for the white design is \$325.

16. Let c = the amount invested in CDs and $b = 5$ the amount invested in bonds.

$c + b = 25{,}000$
$0.045c + 0.06b = 1300$

$$\begin{bmatrix} 1 & 1 \\ 0.04 & 0.06 \end{bmatrix}\begin{bmatrix} c \\ b \end{bmatrix} = \begin{bmatrix} 25{,}000 \\ 1300 \end{bmatrix}$$

$$\begin{bmatrix} 3 & -50 \\ -2 & 50 \end{bmatrix}\begin{bmatrix} 1 & 1 \\ 0.04 & 0.06 \end{bmatrix}\begin{bmatrix} c \\ b \end{bmatrix} = \begin{bmatrix} 3 & -50 \\ -2 & 50 \end{bmatrix}\begin{bmatrix} 25{,}000 \\ 1300 \end{bmatrix}$$

$$\begin{bmatrix} 1 & 0 \\ 0 & 1 \end{bmatrix}\begin{bmatrix} c \\ b \end{bmatrix} = \begin{bmatrix} 10{,}000 \\ 15{,}000 \end{bmatrix}$$

Ms. Paul invested \$10,000 in CDs and \$15,000 in bonds.

17. D

18. $d(p) = -60p + 12{,}000$

19. $-3 \leq \frac{2x + 3}{4} < 7$
$-12 \leq 2x + 3 \leq 21$
$-15 \leq 2x \leq 18$
$-\frac{15}{2} < x < 9$

Standardize Test, page 317

1. $k = \left(\frac{9}{2}\right)^2 = \frac{81}{4}$

2. 3 units

3. $p(x) = -x^2 + 350x - 15{,}000$
Maximum at $\frac{-b}{2a} = 175$ units

4. The break even point is when $P(x) = 0$.
$0 = -x^2 + 350x - 15{,}000$
$0 = x^2 - 350x + 15{,}000$
Use quadratic formula.

$$x = \frac{-(-350) \pm \sqrt{350^2 - 4(15{,}000)}}{2} =$$

$$\frac{350 \pm 250}{2} = 300 \text{ or } 50$$

The greatest number of items the company can produce and break even is 300.

5. $s_1 + s_2 = \frac{1}{2} + \frac{1}{2} = 1$; $s_1 s_2 = \frac{1}{2}\left(\frac{1}{2}\right) = \frac{1}{4}$;
$x^2 - x + \left(\frac{1}{4}\right) = 0$, $4x^2 - 4x + 1 = 0$; $a = 4$

or a multiple of 4

6. $(x - 2)^2 - 4(x - 2) - 5 = 0$
$p^2 - 4p - 5 = 0$
$(p + 1)(p - 5) = 0$
$p = -1 \text{ or } p = 5$
$x - 2 = -1 \quad x - 2 = 5$
$x = 1 \text{ or } x = 7$
or $x^2 - 4x + 4 - 4x + 8 - 5 = 0$
$x^2 - 8x + 7 = 0$
$(x - 7)(x - 1) = 0$
$x = 7 \text{ or } x = 1$
The product of the solutions is $7(1) = 7$

7. $(4x^2)^3 = 64x^6$; $n = 6$

8. $4y^2 - 12y + 9$, $(2y - 3)^2$, $(2y + (-3))^2$; $b = -3$

9. Factors are $(x + 2)(x - 3)(x + 1)$
$(x^2 - 3x + 2x - 6)(x + 1)$
$(x^2 - x - 6)(x + 12)$
$x^3 + x^2 - x^2 - x - 6x - 6$
$x^3 - 7x - 6$ Constant term is -6.

10. In reflecting point $P(x, y)$ about the line $y = -x$ the new point is $P'(-y, -x)$. The point is $P'(-3, 1)$. The x-coordinate is -3.

11.
$$\begin{cases} 4B + 6C = 1 \\ 5A + 2B + 3C = 1 \rightarrow \\ 5A \quad\quad + 6C = 1 \rightarrow \end{cases}$$

$$\begin{array}{r} 5A + 2B + 3C = 1 \\ -5A \quad\quad\quad\; 6C = -1 \\ \hline 2B - 3C = 0 \end{array}$$

$$\begin{cases} 4B + 6C = 1 \rightarrow 4B + 6C = 1 \\ 2B - 3C = 0 \rightarrow \underline{4B - 6C = 0} \\ \quad\quad\quad\quad\quad\quad\quad 8B = 1 \end{cases}$$

$$B = \frac{1}{8}$$

$$4B + 6C = 1 \quad\quad\quad 5A + 6C = 1$$

$$4\left(\frac{1}{8}\right) + 6C = 1 \quad\quad 5A + 6\left(\frac{1}{12}\right) = 1$$

$$6C = \frac{1}{2} \quad\quad\quad 5A = \frac{1}{2}$$

$$C = \frac{1}{12} \quad\quad\quad A = \frac{1}{10}$$

In one hour, pipe A could fill $\frac{1}{10}$ of the pool, therefore in 10h, it could fill the entire pool.

12. 5

13. $\dfrac{3(x - 1)}{4} = -12x$

$$3x - 3 = -48x$$
$$51x = 3$$
$$x = \frac{3}{51} = \frac{1}{17}$$

14.
$$\begin{aligned}
\text{total monthly expenses} &= x \\
\text{rent} &= 0.4x \\
\text{cleaning and repairs} &= \tfrac{1}{3}(x - 0.4x) = \tfrac{1}{3}(0.6x) = 0.2x \\
\text{advertising} &= 0.25(0.6x - 0.2x) = 0.25(0.4x) = 0.1x \\
\text{utilities} &= 0.8(0.4x - 0.1x) = 0.8(0.3x) = 0.24x \\
\text{office supplies} &= 96 \\
0.4x + 0.2x + 0.1x + 0.24x + 96 &= x \\
0.94x + 96 &= x \\
96 &= 0.06x \\
1600 &= x \\
x &= 1600 \\
\text{total monthly expenses} &= \$1600
\end{aligned}$$

15. 4

16. $0.50(30) = 15$

Chapter 7 Inequalities and Linear Programming

Data Activity, pages 318–319

1. Individuals with a grade school education.

2. The number of female Republican women is 37. The total number of females is 100. The probability is $\frac{37}{100}$.

3. The number of black Democrats is 81. The number of non-Democratic black individuals is 19. The odds are 81 to 19.

4. 50% of the college graduates said they were Republican.

   $(0.5)(1500) = 750$ Republicans
   43% of the college graduates said they were Democrats.
   $(0.43)(1500) = 645$ Democrats
   $750 - 645 = 105$ more Republicans

5.

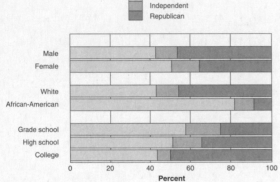

Lesson 7.1, pages 321–323

THINK BACK

1. $2x - 1 \leq -11$
   $2x \leq -1.0$
   $x \leq -5$

2. $-3x < 15$ Change direction of the inequality
   $x > -5$ when dividing by a negative

   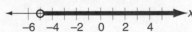

3. $3x + 2 \geq 8$
   $3x \geq 6$
   $x \geq 2$

4. Answers will vary. Students should choose points in the shaded area. They should substitute this value in the inequality to produce a true statement.

5. Answers will vary. Students should pick points in the non-shaded part of the graph. They should substitute this value in the inequality to produce a false statement.

9., 11., 13., 14., Answers will vary.

6. Answers will vary. Students should write and draw a graph of a linear equation.

7. Students should replace the equal symbol with an inequality symbol.

8. Answers will vary. Students must name four ordered pairs that are not on their line but ones which satisfy their inequality. They should justify their answers by substitution into the inequality.

10. Students will repeat their activity from exercise 6 finding ordered pairs that do not satisfy their inequality.

12. Students will choose two points on their line to substitute into their equation and inequality students must indicate if these ordered pairs are solutions to the inequality.

15. Students may note that all points that are solutions lie in the region on one side of the line and all points that are not solutions lie in the region on the other side of the line.

MAKE CONNECTIONS

16. The points students choose to test will vary.

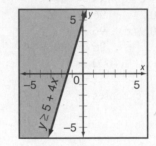

   $(0, 0); 0 \overset{?}{\geq} 5 + 4(0);$
   $0 \overset{?}{\geq} 5;$ no
   $(-2, -2);$
   $-2 \overset{?}{\geq} 5 + 4(-2);$
   $-2 \overset{?}{\geq} 5 - 8(-2);$
   $-2 \overset{?}{\geq} -3$ yes

17. The points students choose to test will vary.

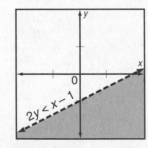

   $(0, 0); 2(0) \overset{?}{<} 0 - 1;$
   $0 \overset{?}{<} 1;$ no
   $(4, -8); 2(-8) \overset{?}{<} 4 - 1;$
   $-16 < 3;$ yes

18. The points students choose to test will vary.

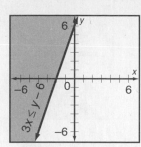

$(0, 0); 3(0) \overset{?}{\leq} 0 - 6;$

$0 \overset{?}{\leq} -6$ no

$(-3, 3); 3(-3) \overset{?}{\leq} 3 - 6;$

$-9 \overset{?}{\leq} 3$ yes

19. Shade above the boundary line for $y \geq 5 + 4x$ and $3x \leq y - 6$; shade below for $2y < x - 1$.

SUMMARIZE

20.

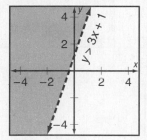

Test points on both sides of the line $y = 3x + 1$ to determine which half-plane to shade. Since the inequality symbol is $>$, the line is dashed.

Specific points given will vary.

21.

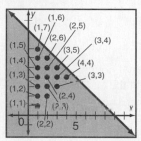

The linear inequality is $x + y < 9$. There are 16 different pairs.

22.

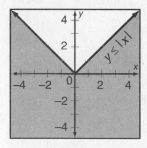

Students points will vary. Points Y_1 to Y_{10} should be placed in the shaded region of the graph and points N_1 to N_{10} should be in the non-shaded portion.

The solution of the inequality is one of the regions bounded by the graph of the equation. In this case, the solution is the closed region below the graph of $y = |x|$.

23. It has no solutions since the absolute value on the left side of the equation cannot be less than any negative number.

24.

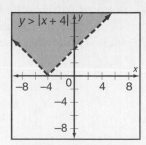

$y > |x + 4|$ implies $x + 4 < y$ or $x + 4 > -y$ Graph the equation $x + 4 = y$ and $x + 4 = -y$ only including the portions where $y > 0$.

25.

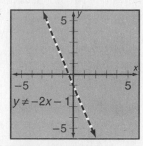

The $\neq$ symbol means "not equal to". The graph of $y \neq -2x - 1$ is the entire plane except for the dashed line itself.

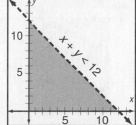

The $\not<$ symbol means "is not less than" which can be interpreted as "is greater than or equal to." The graph of $y \not< x + 2$ is the graph of $y \geq x + 2$.

Lesson 7.2, pages 324–331

EXPLORE

1. There are two acceptable forms for the first statement: $x + y < 12, x \geq 0, y \geq 0$ or $y + x < 12$, $x \geq 0, y \geq 0$.

For the second statement: $y - x > 3, x \geq 0, y \geq 0$.

2.

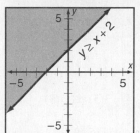

Convert each equation to $y = mx + b$ form. Then graph each line and shade according to the inequality. The graphs are limited to the first quadrant because of the restriction given that x and y are positive.

3.

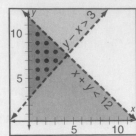

Answers will vary. Students should name points that are located in the shaded area.

4. Answers will vary but should show two arithmetic examples for each set of numbers from Question 3.

5. 12; They are (1, 5), (1, 6), (1, 7), (1, 8), (1, 9), (1, 10), (2, 6), (2, 7), (2, 8), (2, 9), (3, 7), (3, 8). These points are located in the double shaded region.

6.–8. Answers will vary.

TRY THESE

1a. yes; $2(1) + (10) \overset{?}{\geq} 10$ $3(1) \overset{?}{\leq} 5(10) - 1$
$2 + (10) \overset{?}{\geq} 10$ $3 \overset{?}{\leq} 50 - 1$
$12 > 10 \checkmark$ $3 \leq 49 \checkmark$

1b. no; $2(0) + 0 \overset{?}{\geq} 10$
$0 + 0 \overset{?}{\geq} 10$
$0 \overset{?}{\geq} 10$ no

1c. no; $2(5) + 3 \overset{?}{\geq} 10$ $3(5) \overset{?}{\leq} 5(3) - 1$
$10 + 3 \overset{?}{\geq} 10$ $15 \overset{?}{\leq} 15 - 1$
$13 > 10 \checkmark$ $15 \overset{?}{\leq} 14$ no

1d. no; $2(0) + 2 \overset{?}{\geq} 10$
$0 + 2 \overset{?}{\geq} 10$
$2 \overset{?}{\geq} 10$ no

1e. no; $2(-3) + 13 \overset{?}{\geq} 10$
$-6 + 13 \overset{?}{\geq} 10$
$7 \overset{?}{\geq} 10$ no

1f. no; $2(-3) + 14 \overset{?}{\geq} 10$
$-6 + 14 \overset{?}{\geq} 10$
$8 \overset{?}{\geq} 10$ no

2.

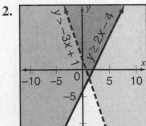

3.

4.

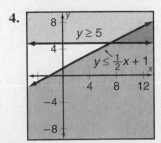

5.

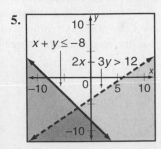

6.

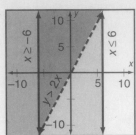

7.

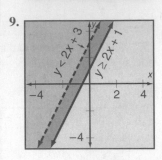

8.

9.

10.

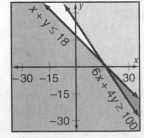

11. Let x represent the number of hours Joelle works at the golf course. Let y represent the number of hours Joelle works as a Mother's helper. The number of hours Joelle wants to work each week is modeled by the inequality $x + y \leq 18$. The amount of money Joelle needs to earn can be modeled by the inequality $6x + 4y \leq 100$. The system is:

$$\begin{cases} x + y \leq 18 \\ 6x + 4y \geq 100 \end{cases}$$

12.

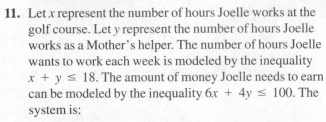

13. Answers will vary but should include ordered pairs in the double-shaded area of the first quadrant only.

14. $\begin{cases} y < 3x - 4 \\ 2y - x \geq 4 \end{cases}$ or equivalent inequalities

15.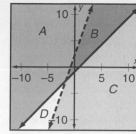

A. This region satisfies $y \geq x - 1$.

B. This region satisfies both $y \geq x - 1$ and $y < 3x + 2$. This is shown by the double-shading on the graph.

C. This region satisfies $y < 3x + 2$.

D. There is no shading in this region. It does not satisfy either inequality.

PRACTICE

1a. both; this is a double-shaded region

1b. one; $y < 3x - 5$ **1c.** one; $x + 2y \geq 7$

1d. neither; there is no shading in this region.

2. 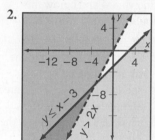   **3.**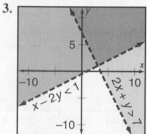

4. 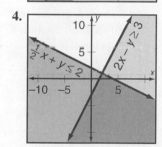   **5.**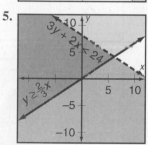

6. 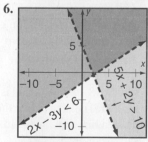   **7.**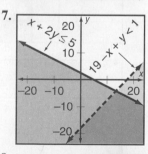

8. 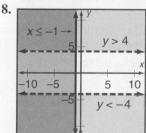   **9.**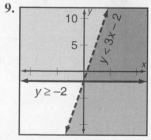

10a. Let x represent one number and y represent the other. The inequality $x + y < 8$ represents "the sum of two positive numbers less than 8". Since it is unknown if x or y is greater, the second inequality can either be $x - y > 6$ or $y - x > 6$. The two numbers are specified as positive numbers, therefore $x > 0$, and $y > 0$.

10b. Answers will vary. Students should list five pairs of numbers which satisfy both equations. The numbers must be non-integers.

11a. Let x represent age and y represent pulse rate. The equation for minimum desired pulse rate is $y = 0.72(220 - x)$. The equation for maximum desired pulse rate is $y = 0.87(220 - x)$. The system representing the range for recommended pulse rate is:
$$\begin{cases} y > 0.72(220 - x) \\ y < 0.87(220 - x) \end{cases}$$

11b.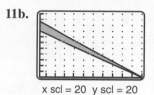

x scl = 20   y scl = 20

11c. Let $x = 40$ in each equation given in part a and solve for y.

$y\,(\text{minimum}) = 0.72(220 - x)$
$= 0.72(220 - 40) = 129.6 \approx 130$

$y\,(\text{maximum}) = 0.87(220 - x)$
$= 0.87(220 - 40)$
$= 156.6 \approx 157$
$130 < y < 157$

12. Answers will vary. Both systems of inequalities and systems of equations can be solved by graphing. The graph of a system of equations consists of lines that may or may not intersect. The graph of a system of inequalities consists of half-planes that may or may not intersect. A system of equations can have a single point as a solution and never has a whole region as a solution.

13a. Let x represent minutes of television and y represent minutes of radio. The candidate wants no more than thirty minutes of advertising. This is modeled by the inequality $x + y \leq 30$. The candidate wants to reach at least two million voters. This is modeled by the inequality $0.2x + 0.5y \geq 2$. The system that models the candidate's need is
$$\begin{cases} x + y \leq 30 & \quad x \geq 0 \\ 0.2x + 0.05y \geq 2 & \quad y \geq 2 \end{cases}$$

13b.

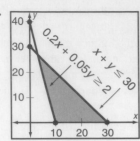

13c. Answers will vary. Students should give integer pairs that are located in the double shaded region in quadrant I.

13d. All radio is (30, 0) and is in the solution but all TV (0, 30) is not.

13e. Let N represent the number of voters reached in millions. Substitute $x = 20$ and $y = 10$ into the following equation.

$$N = 0.2x + 0.05y$$
$$= 0.2(20) + 0.05(10)$$
$$= 4.5 \text{ million voters}$$

Now, substitute $x = 30$ and $y = 0$ into the same equation.

$$N = 0.2x + 0.05y$$
$$= 0.2(30) + 0.05(0)$$
$$= 6 \text{ million voters}$$

EXTEND

14. First, determine the equations of the two lines. The y-intercept of the first line is $(0, 0)$ and its slope is 1. This yields $y = x$. The y-intercept of the second line is -6 and its slope is 5. The equation for this line is $y = 5x - 6$. The points to the right of the first line are shaded and the line is dashed indicating that this portion of the graph is represented by $x > y$. The points to the left of the second line are shaded and the line is dashed indicating that this portion of the graph is represented by $y > 5x - 6$.

The system is $\begin{cases} y > x \\ y > 5x - 6 \end{cases}$

or equivalent inequalities.

15. Determine the equations of the two lines. The first line has a y-intercept $(0, 2)$ and a slope equal to 3. This yields $y = 3x + 2$. The y-intercept of the second line is $(0, -4)$ and the slope is 1. The equation for this line is $y = x - 4$. The points to the right of the first line are shaded and the line is dashed indicating that this portion of the graph is represented by $y > x - 4$. The points to the left of the second line are shaded and the line is dashed indicating that this portion of the graph is represented by $y < 3x + 2$. The system is

$\begin{cases} y > x - 4 \\ y < 3x + 2 \end{cases}$ or equivalent inequalities.

16.

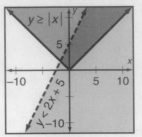

The area above the line $y = |x|$ satisfies $y \geq |x|$ and is shaded. The area to the right of the line $y = 2x + 5$ satisfies $y < 2x + 5$ and is shaded. The double-shaded area is the solution to the system.

17.

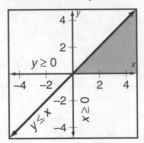

$|x + 2| < 3$ can be written as the inequality $-3 < x + 2 < 3$. Solve for x: $-5 < x < 1$. Shade the area between the lines $x = 1$ and $x = -5$. The equation $|y| > 4$ can be written $y > 4$ and $y < -4$. Shade the area above the dashed line $y = 4$ and below the dashed line $y = -4$. The double-shaded area is the solution to the system.

18.

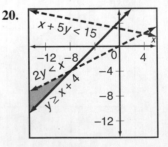

For $x \geq 0$, the area to the right of the line $x = 0$ is shaded. For $y \geq 0$, the area above the line $y = 0$ is shaded. The area below the line $y = x$ satisfies $y \leq x$ and is shaded. The triple-shaded area is the solution to the system.

19.

For $x > 5$ the area to the right of the line $x = 5$ is shaded. For $y < 3x + 2$ the area to the right of the line $y = 3x + 2$ is shaded. The area below the line $y = 10$ satisfies $y \leq 10$. The triple-shaded region is the solution to the system.

20.

The area to the left of the solid line $y = x + 4$ satisfies $y \geq x + 4$ and is shaded. The area below the dashed line $x + 5y = 15$ satisfies $x + 5y < 15$ and is shaded. For the inequality $2y < x$ the area below the dashed line $2y = x$ is shaded. The triple-shaded region is the solution to the system.

21.

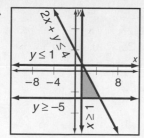

For the inequality $x \geq 1$ the area to the right of the solid line $x = 1$ is shaded. The area above the solid line $y = -5$ satisfies $y \geq -5$ and is shaded. For the inequality $y \leq 1$ the area below the solid line $y = 1$ is shaded. The area to the left of the solid line $2x + y = 4$ satisfies $2x + y \leq 4$ and is shaded. The overlap of all four of these regions is the solution to the system.

22.

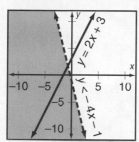

Use substitution to determine the intersection of the two lines.
$$y = -4x - 1$$
$$y = 2x + 3$$
$$-4x - 1 = 2x + 3$$
$$-6x = 4$$
$$x = -\frac{2}{3}$$

Any point on the line $y = 2x + 3$ where $x < -\frac{2}{3}$ is an acceptable answer. Answers will vary and students should list four points.

THINK CRITICALLY

23. Since absolute value is never negative $|3y - 4x| \neq -4$ for any x and y. The entire coordinate plane is the region of solutions.

24. The system has no solution. The expression cannot simultaneously be greater than -15 and less than -25.

25.

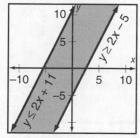

Marilyn is correct. As long as the parallel boundary lines do not have half-planes shaded in opposite directions there will be an overlap to the solution of the two linear inequalities.

26. False. For positive real numbers less than 1 this is not true. Let $x = \frac{1}{4}$, $\sqrt{x} = \frac{1}{2}$, $\frac{1}{2} > \frac{1}{4}$.

MIXED REVIEW

27. 17, $1.\overline{32}$, 1.3, $\sqrt{64}$, $\frac{3}{8}$, and 0 are rational numbers. When written as decimals, they either terminate or repeat. $-\sqrt{2}$, $\sqrt{17}$, and $\sqrt{\frac{5}{2}}$ do not repeat or terminate. When written as decimals, these are irrational numbers.

28. B; if $x > 1$ then $x^2 > 1$.
Incorrect answers:
A. $x > 2$ counter example: $x = 1.1$
C. $x^3 \geq 8$ counter example: $x = 1.5$
$$x^3 = 3.375$$
D. $-x > 0$ counter example: $x = 5$
$$-x = -5$$

29.

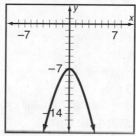

$$2x - 7 > 5$$
$$2x > 12$$
$$x > 6$$

Since 6 is not included in the solution, use an open circle. All real numbers to the right of six satisfy this inequality.

30. To determine the x-intercept let $y = 0$ and solve for x.
$$y = 4x - 9$$
$$0 = 4x - 9$$
$$9 = 4x$$
$$x = \frac{9}{4} \left(\frac{9}{4}, 0\right)$$

31. Factor the left side of the equation to determine roots.
$$x^2 - 8x + 15 = 0$$
$$(x - 5)(x - 3) = 0$$
$$x - 5 = 0 \quad \text{or} \quad x - 3 = 0$$
$$x = 5 \quad \text{or} \quad x = 3$$

32.

False; the graph lies in Quadrants III and IV

33. False; $(2x + 5)^2 = (2x + 5)(2x + 5)$
$$= 4x^2 + 10x + 10x + 25$$
$$= 4x^2 + 20x + 25$$

ALGEBRAWORKS

1. $Y_4 = 0.36(x - 140{,}000) + 35{,}704.50$
if $140{,}000 < x \leq 250{,}000$.
$Y_5 = 0.396(x - 250{,}000) + 75{,}304.50$
if $x > 250{,}000$.

2. $(0, 0)$; $(38000, 5700)$; $(91850, 20778)$; $(140000, 35704.50)$ $(250000, 75304.50)$

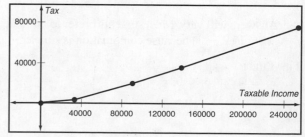

3. $f(x) = 0.24(x - 9000)$

4.

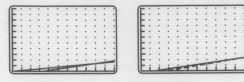

$f(x)$ intersects Y_1 at (240,000, 3600)

$f(x)$ intersects Y_2 at (69,500, 14,520)

5.

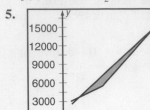

Since $f(x) > t(x)$, people in this income range pay more tax under the flat tax, and they should be against the above stated flat tax.

Lesson 7.3, pages 332–338

EXPLORE

1. Answers will vary; $x < -7$ or $x > 1$.

2. Answers will vary; $-7 < x < 1$

3. $x^2 + 6x - 7 = 0$
   $(x + 7)(x - 1) = 0$
   $x + 7 = 0$ or $x - 1 = 0$
   $x = -7$ or $x = 1$

4. Answers will vary; $x < -7$ or $x > 1$.
   Students may explain that they looked for numbers to make both factors positive or negative.

5. Answers will vary; $-7 < x < 1$.

6. $(x + 7)(x - 1) = 0$
   $x + 7 = 0$ or $x - 1 = 0$
   $x = -7$ or $x = 1$

7.

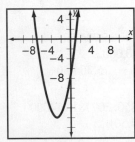

They have the same graph.
$y = x^2 + 6x - 7$

8. Answers will vary. Most students prefer to evaluate $(x + 7)(x - 1)$ because computation is simpler.

TRY THESE

1. $x^2 - 4x + 6 < 0$; $x = 3$
   $(3)^2 - 4(3) + 6 \overset{?}{<} 0$
   $9 - 12 + 6 \overset{?}{<} 0$
   $3 \overset{?}{<} 0$; no

2. $x^2 + 5x - 3 > 0$; $x = 6$
   $(6)^2 + 5(6) - 3 \overset{?}{>} 0$
   $36 + 30 - 3 \overset{?}{>} 0$
   $63 \overset{?}{>} 0$; yes

3. $x^2 + 1.5x - 6.25 < 0$
   $(-4.5)^2 + 1.5(-4.5) - 6.25 \overset{?}{<} 0$
   $20.25 - 6.75 - 6.25 \overset{?}{<} 0$
   $7.25 \overset{?}{<} 0$; no

4. $2x^2 - 8x \geq 0$; $x = 4$
   $2(4)^2 - 8(4) \overset{?}{\geq} 0$
   $32 - 32 \overset{?}{\geq} 0$
   $0 \overset{?}{\geq} 0$; yes

5. both ≥ 0 or both ≤ 0

6. one ≤ 0 and one ≥ 0

7. one < 0 and one > 0

8. both > 0 or both < 0

9. $x^2 - 2x - 24 < 0$
   $x^2 - 2x - 24 = 0$
   $(x - 6)(x + 4) = 0$
   $x = 6$ or $x = -4$

$x - 6$	$-$	$-$	$+$
$x + 4$	$-$	$+$	$+$
$(x-6)(x+4)$	$+$	$-$	$+$

$-4 < x < 6$

Check: $x = 0$; $(0)^2 - 2(0) - 24 \overset{?}{<} 0$; $-24 \overset{?}{<} 0$; yes

10. $x^2 + x - 12 \leq 0$
    $x^2 + x - 12 = 0$
    $(x + 4)(x - 3) = 0$
    $x + 4 = 0$ or $x - 3 = 0$
    $x = -4$ or $x = 3$

$x + 4$	$-$	$+$	$+$
$x - 3$	$-$	$-$	$+$
$(x+4)(x-3)$	$+$	$-$	$+$

$-4 \leq x \leq 3$

Check: $x = 0$; $(0)^2 + (0) - 12 \overset{?}{\leq} 0$
$-12 \overset{?}{\leq} 0$; yes

11. $x^2 - 4x - 5 > 0$
    $x^2 - 4x - 5 = 0$
    $(x - 5)(x + 1) = 0$
    $x - 5 = 0$ or $x + 1 = 0$
    $x = 5$ or $x = -1$

$x - 5$	$-$	$-$	$+$
$x + 1$	$-$	$+$	$+$
$(x-5)(x+1)$	$+$	$-$	$+$

$x < -1$ or $x > 5$

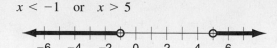

Check: $x = -3$; $(-3)^2 - 4(-3) - 5 \overset{?}{>} 0$
$$9 + 12 - 5 \overset{?}{>} 0$$
$$16 \overset{?}{>} 0; \text{ yes}$$
$x = 6$; $(6)^2 - 4(6) - 5 \overset{?}{>} 0$
$$36 - 24 - 5 \overset{?}{>} 0$$
$$7 \overset{?}{>} 0; \text{ yes}$$

12. $2x^2 - 5x - 3 \geq 0$
$2x^2 - 5x - 3 = 0$
$(2x + 1)(x - 3) = 0$
$2x + 1 = 0$ or $x - 3 = 0$
$x = -\dfrac{1}{2}$ or $x = 3$

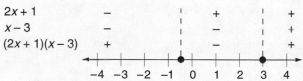

$x \leq -\dfrac{1}{2}$ or $x \geq 3$

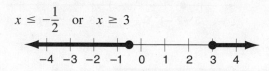

Check: $x = -2$; $2(-2)^2 - 5(-2) - 3 \overset{?}{\geq} 0$
$$8 + 10 - 3 \overset{?}{\geq} 0$$
$$15 \overset{?}{\geq} 0; \text{ yes}$$
$x = 4$; $2(4)^2 - 5(4) - 3 \overset{?}{\geq} 0$
$$32 - 20 - 3 \overset{?}{\geq} 0$$
$$9 \overset{?}{\geq} 0; \text{ yes}$$

13. $3x^2 + 2x - 5 < 0$
$3x^2 + 2x - 5 = 0$
$(3x + 5)(x - 1) = 0$
$3x + 5 = 0$ or $x - 1 = 0$
$x = -\dfrac{5}{3}$ or $x = 1$

$3x + 5$	$-$	$+$	$+$
$x - 1$	$-$	$-$	$+$
$(3x + 5)(x - 1)$	$+$	$-$	$+$

$-\dfrac{5}{3} < x < 1$

Check: $x = 0$; $3(0)^2 + 2(0) - 5 \overset{?}{<} 0$
$$0 + 0 - 5 \overset{?}{<} 0$$
$$-5 < 0; \text{ yes}$$

14. $x^2 - 6x + 9 > 0$
$x^2 - 6x + 9 = 0$
$(x - 3)(x - 3) = 0$
$x - 3 = 0$ $x = 3$

x is all reals except $x = -3$.

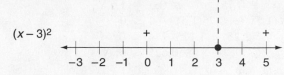

Check: $x = 0$; $(0)^2 - 6(0) + 9 > 0$
$$9 \overset{?}{>} 0; \text{ yes}$$
$x = 5$; $(5)^2 - 6(5) + 9 \overset{?}{>} 0$
$$25 - 30 + 9 \overset{?}{>} 0$$
$$4 \overset{?}{>} 0; \text{ yes}$$

15. $h = -16t^2 + vt + s$
$s = 80$ $v = 48$
$-16t^2 + 48t + 80 \geq 16$
$-16t^2 + 48t + 64 \geq 0$
$-16(t^2 - 3t - 4) \geq 0$
$-16(t - 4)(t + 1) \geq 0$
The three intervals are $t < -1$; $-1 < t < 4$ and $t > 4$.

$t - 4$	$-$	$-$	$+$
$t + 1$	$-$	$+$	$+$
$-16(t - 4)(t + 1)$	$-$	$+$	$-$

Since time must be positive, the solution is
$0 \leq t \leq 4 \text{ sec}$

16. $h = -16t^2 + vt + s$
$s = 80$ $v = 48$
$-16t^2 + 48t + 80 \leq 112$
$16t^2 + 48t - 32 \leq 0$
$-16(t^2 - 3t + 2) \leq 0$
$-16(t - 2)(t - 1) \leq 0$
The three intervals are $t < 1$; $1 < t < 2$; and $t > 2$.

$t - 2$	$-$	$-$	$+$
$t - 1$	$-$	$+$	$+$
$-16(t - 2)(t - 1)$	$-$	$+$	$-$

Since time must be positive, the solution is $0 \leq t \leq 1$ and $t \geq 2$.

17. Since x represents dollars spent, the interval $x < -3$ makes no sense in this context. The solution is $x > 1$.

PRACTICE

1. $x^2 + 8x - 1 < 0$; $x = 0$
$(0)^2 + 8(0) - 1 \overset{?}{<} 0$
$$-1 \overset{?}{<} 0; \text{ yes}$$

2. $3x^2 - 2x + 5 < 0; x \overset{?}{=} -1$

$3(-1)^2 - 2(-1) + 5 \overset{?}{<} 0$

$3 + 2 + 5 \overset{?}{<} 0$

$10 \overset{?}{<} 0;$ no

3. $x^2 + 0.75x - 4 \leq 0; x \overset{?}{=} 1.5$

$(1.5)^2 + 0.75(1.5) - 4 \overset{?}{\leq} 0$

$2.25 + 1.125 - 4 \overset{?}{\leq} 0$

$-0.625 \leq 0;$ yes

4. $x^2 + \dfrac{3}{2} < 0; x = -\dfrac{2}{3}$

$\left(-\dfrac{2}{3}\right)^2 + \dfrac{3}{2} \overset{?}{<} 0$

$\dfrac{4}{9} + \dfrac{3}{2} \overset{?}{<} 0$

$\dfrac{35}{18} \overset{?}{<} 0;$ no

5. $x^2 - 4x + 3 < 0$

$x^2 - 4x + 3 = 0$

$(x - 3)(x - 1) = 0$

$x = 3 \ \text{ or } \ x = 1$

$x - 3$	−	−	+
$x - 1$	−	+	+
$(x-3)(x-1)$	+	−	+

(number line: −1 0 1 2 3 4 5)

$1 < x < 3$

(number line: −1 0 1 2 3 4 5)

Check: $x = 2;$ $(2)^2 - 4(2) + 3 \overset{?}{\leq} 0$

$4 - 8 + 3 \overset{?}{\leq} 0$

$-1 \overset{?}{\leq} 0;$ yes

6. $x^2 - 6x < 0$

$x^2 - 6x = 0$

$x(x - 6) = 0$

$x = 0 \ \text{ or } \ x = 6$

x	−	+	+
$x - 6$	−	−	+
$x(x-6)$	+	−	+

(number line: −2 −1 0 1 2 3 4 5 6)

$0 < x < 6$

(number line: −2 −1 0 1 2 3 4 5 6)

Check: $x = 3;$ $(3)^2 - 6(3) \overset{?}{\leq} 0$

$-9 \overset{?}{\leq} 0;$ yes

7. $x^2 - 5x + 4 > 0$

$x^2 - 5x + 4 = 0$

$(x - 4)(x - 1) = 0$

$x = 4 \text{ or } x = 1$

$x - 4$	−	−	+
$x - 1$	−	+	+
$(x-4)(x-1)$	+	−	+

(number line: 0 1 2 3 4 5 6 7)

$x < 1 \text{ or } x > 4$

(number line: 0 1 2 3 4 5 6 7)

Check: $x = 0;$

$(0)^2 - 5(0) + 4 \overset{?}{>} 0$

$4 \overset{?}{>} 0;$ yes

$x = 5; (5)^2 - 5(5) + 4 \overset{?}{>} 0$

$4 \overset{?}{>} 0;$ yes

8. $x^2 - 3x - 10 < 0$

$x^2 - 3x - 10 = 0$

$(x - 5)(x + 2) = 0$

$x = 5 \ \text{ or } \ x = -2$

$x - 5$	−	−	+
$x + 2$	−	+	+
$(x-5)(x+2)$	+	−	+

(number line: −4 −3 −2 −1 0 1 2 3 4 5 6 7)

$-2 < x < 5$

(number line: −4 −3 −2 −1 0 1 2 3 4 5 6 7)

Check: $x = 0;$ $(0)^2 - 3(0) - 10 \overset{?}{<} 0$

$-1 \overset{?}{<} 0;$ yes

9. $2x^2 + 3x - 2 \geq 0$

$2x^2 + 3x - 2 = 0$

$(2x - 1)(x + 2) = 0$

$x = \dfrac{1}{2} \ \text{ or } \ x = -2$

$2x - 1$	−	−	+
$x + 2$	−	+	+
$(2x-1)(x+2)$	+	−	+

(number line: −4 −3 −2 −1 0 1 2 3)

$x \leq -2 \ \text{ or } \ x \geq \dfrac{1}{2}$

(number line: −4 −3 −2 −1 0 1 2 3)

Check: $x = -3;$ $2(-3)^2 + 3(-3) - 2 \overset{?}{\geq} 0$

$18 - 9 - 2 \overset{?}{\geq} 0$

$7 \overset{?}{\geq} 0;$ yes

$x = 2;$ $2(2)^2 + 3(2) - 2 \overset{?}{\geq} 0$

$8 + 6 - 2 \overset{?}{\geq} 0$

$12 \overset{?}{\geq} 0;$ yes

10. $3x^2 - 27 > 0$
$3x^2 - 27 = 0$
$3(x^2 - 9) = 0$
$3(x - 3)(x + 3) = 0$
$x = 3$ or $x = -3$

$x - 3$	$-$	$-$	$+$
$x + 3$	$-$	$+$	$+$
$(x-3)(x+3)$	$+$	$-$	$+$

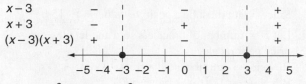

$x < -3$ or $x > 3$

Check: $x = -4;\ 3(-4)^2 - 27 \overset{?}{>} 0$
$21 \overset{?}{>} 0;$ yes
$x = 4;\quad 3(4)^2 - 27 \overset{?}{>} 0$
$21 \overset{?}{>} 0;$ yes

11. $x^2 + 8x + 16 \le 0$
$x^2 + 8x + 16 = 0$
$(x + 4)(x + 4) = 0$
$x = -4$

$(x + 4)^2$ $+$ | $+$

$x = -4$

Check: $x = -4;\ (-4)^2 + 8(-4) + 16 \overset{?}{\le} 0$
$0 \overset{?}{\le} 0;$ yes

12. $3x^2 - 7x + 2 \le 0$
$3x^2 - 7x + 2 = 0$
$(3x - 1)(x - 2) = 0$
$x = \dfrac{1}{3}$ or $x = 2$

$3x - 1$	$-$	$+$	$+$
$x - 2$	$-$	$-$	$+$
$(3x-1)(x-2)$	$+$	$-$	$+$

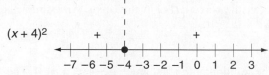

$\dfrac{1}{3} \le x \le 2$

Check: $x = 1;\ 3(1)^2 - 7(1) + 2 \overset{?}{\le} 0$
$-2 \overset{?}{\le} 0;$ yes

13. $-2x^2 - 5x + 3 > 0$
$-2x^2 - 5x + 3 = 0$
$(-2x + 1)(x + 3) = 0$

$x = \dfrac{1}{2}$ or $x = -3$

$-2x + 1$	$+$	$+$	$-$
$x + 3$	$-$	$+$	$+$
$(-2x+1)(x+3)$	$-$	$+$	$-$

$-3 < x < \dfrac{1}{2}$

Check: $x = 0;\qquad -2(0)^2 - 5(0) + 3 \overset{?}{>} 0$
$0 - 0 + 3 \overset{?}{>} 0$
$3 > 0$ ✓

14. $h = -16t^2 + 208t$
$-16t^2 + 208t \ge 480$
$-16t^2 + 208t - 480 \ge 0$
$-16(t^2 - 13t + 30) \ge 0$
$-16(t - 10)(t - 3) \ge 0$
The three intervals are $t < 3;\ 3 < t < 10;\ t > 10$.

$t - 10$	$-$	$-$	$+$
$t - 3$	$-$	$+$	$+$
$-16\,(t-10)(t-3)$	$-$	$+$	$-$

$3 \le t \le 10$.

15. Let x represent the width. The length is 30 in. longer than the width therefore the length is $x + 30$.
$A = l \times w$
$A = x(x + 30) \ge 1800$
$x^2 + 30x \ge 1800$
$x^2 + 30x - 1800 \ge 0$
$(x + 60)(x - 30) \ge 0$
The three intervals are $x < -60;\ -60 < x < 30;$ and $x > 30$.

$x + 60$	$-$	$+$	$+$
$x - 30$	$-$	$-$	$+$
$(x+60)(x-30)$	$+$	$-$	$+$

Since width must be a positive value, $x \ge 30$ in.

16. $x > -2.75$ **17.** $x < -2.75$

18. $x \ge -2.75$ **19.** $x \le -2.75$

20. Let $y_1 = x^2 - x - 30$ and $y_2 = x^2 + 7x - 8$
Solve $y_1 \le y_2$
$x^2 - x - 30 \le x^2 + 7x - 8$
$-8x \le 22$
$x \ge -2.75$
Solve $y_1 \ge y_2$
$x^2 - x - 30 \ge x^2 + 7x - 8$
$-8x \ge 22$
$x \le -2.75$

These results correspond exactly with the results determined by graphing.

21. $0.5 < x < 4$ **22.** $x < 0.5$ or $x > 4$

23. $0.5 \le x \le 4$ **24.** $x \le 0.5$ or $x \ge 4$

25. Let $y_1 = x^2 - 4x + 4$ and $y_2 = -x^2 + 5x$
Solve $y_1 \le y_2$
$$x^2 - 4x + 4 \le -x^2 + 5x$$
$$2x^2 - 9x + 4 \le 0$$
$$(2x - 1)(x - 4) \le 0$$
The intervals to consider are $x \le 0.5, 0.5 \le x \le 4$; and $x \ge 4$.

$y_1 \le y_2$ when $0.5 \le x \le 4$
Solve $y_1 \ge y_2$
$$x^2 - 4x + 4 \ge -x^2 + 5x$$
$$2x^2 - 9x + 4 \ge 0$$
$$(2x - 1)(x - 4) \ge 0$$
$y_1 \ge y_2$ when $x \le 0.5$ or $x \ge 4$
These results correspond exactly with the results determined by graphing.

26. Solve by graphing.
$x^2 - x - 12 \le x^2 + 6x$

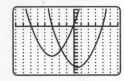

Solve algebraically.
$$x^2 - x - 12 \le x^2 + 6x$$
$$-7x \le 12$$
$$x \ge -\frac{12}{7}$$

27. Solve by graphing.
$x^2 + 4x - 5 > x^2 - x - 6$

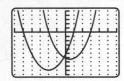

Solve algebraically.
$$x^2 + 4x - 5 > x^2 - x - 6$$
$$5x > -1$$
$$x > -0.2$$

28. Let width = x and length = $2x - 10$.
$$(2x - 10)(x) \ge 300$$
$$2x^2 - 10x - 300 \ge 0$$
$$2(x^2 - 5x - 150) \ge 0$$
$$2(x - 15)(x + 10) \ge 0$$
The three intervals are $x < -10; -10 < x < 15$; $x > 15$.

$x - 15$	$-$	$-$	$+$
$x + 10$	$-$	$+$	$+$
$2(x - 15)(x + 10)$	$+$	$-$	$+$

Since width must be a positive value, $x \ge 15$ ft.

29. Let x = number of bracelets sold and let $60 - x$ = price per bracelet.
Revenue = number sold $\cdot$ price
Revenue = $(x)(60 - x) \ge 500$
$$-x^2 + 60x - 500 \ge 0$$
$$(x - 10)(-x + 50) \ge 0$$
The three intervals are $x < 10; 10 < x < 50$; and $x > 50$.

$x - 10$	$-$	$+$	$+$
$-x + 50$	$+$	$+$	$-$
$(x - 10)(-x + 50)$	$-$	$+$	$-$

$10 \le x \le 50$ bracelets per week

30. If there is no solution, then the graph might fall entirely above the x-axis or below the x-axis. An example is $2x^2 + 4 \le 0$.

EXTEND

31. The solution of the first graph is $-4 \le x \le 4$. The solution of the second graph is $x \ge 1$. If these inequalities are graphed simultaneously the solution is $1 \le x \le 4$.

32. The solution of the first graph is $-3 \le x \le 2$. The solution of the second graph is $-2 < x < 4$. If these inequalities are graphed simultaneously the solution is $-2 < x \le 2$.

33. $x^2 - x \ge 0$
  $x^2 - x = 0$
 $x(x - 1) = 0$
  $x = 0$ or $x = 1$

x	$-$	$+$	$+$
$x - 1$	$-$	$-$	$+$
$x(x - 1)$	$+$	$-$	$+$

$x \le 0$ or $x \ge 1$

$x^2 - 1 \ge 0$
$x^2 - 1 = 0$
$(x - 1)(x + 1) = 0$
$x = 1$ or $x = -1$

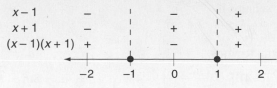

$x \le -1$ or $x \ge 1$.

If these 2 inequalities are graphed simultaneously the solution is $x \le -1$ and $x \ge 1$.

34. $x^2 + x - 6 \le 0$
$x^2 + x - 6 = 0$
$(x + 3)(x - 2) = 0$
$x = -3$ or $x = 2$

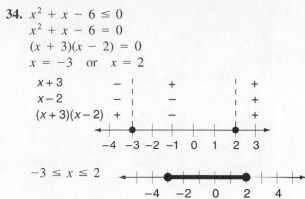

$-3 \le x \le 2$

$x^2 - 5x + 4 \ge 0$
$x^2 - 5x + 4 = 0$
$(x - 4)(x - 1) = 0$
$x = 4$ or $x = 1$

$x \le 1$ or $x \ge 4$

If these 2 inequalities are graphed simultaneously the solution is $-3 \le x \le 1$.

35. $-4 \le x^2 - 4x \le 12$
$x^2 - 4x \ge -4$ and $x^2 - 4x \le 12$
$x^2 - 4x + 4 = 0$ $x^2 - 4x - 12 = 0$
$(x - 2)(x - 2) = 0$ $(x - 6)(x + 2) = 0$
$x = 2$ $x = 6$ or $x = -2$

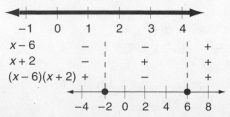

$x =$ all real numbers

36. $-7 < x^2 + 8x < 20$
$x^2 + 8x > -7$ and $x^2 + 8x < 20$
$x^2 + 8x + 7 = 0$ $x^2 + 8x - 20 = 0$
$(x + 7)(x + 1) = 0$ $(x + 10)(x - 2) = 0$
$x = -7$ or $x = -1$ $x = -10$ or $x = 2$

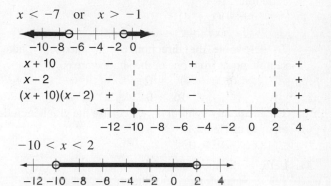

$x < -7$ or $x > -1$

$-10 < x < 2$

If the two inequalities are graphed simultaneously the solution is $-10 < x < -7$ or $-1 < x < 2$.

THINK CRITICALLY

37. $x = -2$ and $x = 2$ are separation points for the three intervals on the graph. These are solutions of the needed equation.
$(x - 2)(x + 2) = 0$
$x^2 - 4 = 0$
To determine the direction of the inequality take a sample point from one of the shaded areas.
$x = -3; (-3)^2 - 4 \stackrel{?}{\underline{\quad}} 0$
$5 > 0$
The inequality symbol is $\ge$ because the graph includes closed circles on -2 and 2.
$x^2 - 4 \ge 0$

38. $x = -2$ and $x = 2$ are separation points for the three intervals on the graph. These are the solutions of the needed equation.
$(x + 2)(x - 2) = 0$
$x^2 - 4 = 0$
To determine the direction of the inequality, take a sample point from the shaded area.
$x = 0; (0)^2 - 4 \stackrel{?}{\underline{\quad}} 0$
$-4 < 0$

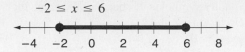

The inequality symbol is $\leq$ because the graph includes closed circles on -2 and 2.
$$x^2 - 4 \leq 0$$

39. $x = -3$ and $x = 0$ are separation points for the three intervals on the graph. These are the solutions of the needed equation.
$$(x + 3)(x - 0) = 0$$
$$x^2 + 3x = 0$$
To determine the direction of the inequality take a sample point from one of the shaded areas.
$$x = 2; (2)^2 + 3(2) \overset{?}{\_} 0$$
$$10 > 0$$
The inequality symbol is $>$ because the graph includes open circles on -3 and 0.
$$x^2 + 3x > 0$$

40. $x = 4$ is the only separation point for the two intervals on the graph. This is the solution of the needed equation.
$$(x - 4)(x - 4) = 0$$
$$(x - 4)^2 = 0$$
To determine the direction of the inequality take a sample point from one of the shaded areas.
$$x = 0; (0 - 4)^2 \overset{?}{\_} 0$$
$$16 > 0$$
The inequality symbol is $>$ because the graph includes an open circle on 4.
$$(x - 4)^2 > 0$$

41. Let $x =$ the short side of the chip
$x + 2 =$ the long side of the chip
$$(x + 2)(x) = x^2 + 2x$$
$$15 \leq x^2 + 2x \leq 24$$
$$x^2 + 2x - 15 \geq 0 \quad \text{and} \quad x^2 + 2x - 24 \leq 0$$
$$x^2 + 2x - 15 = 0 \qquad x^2 + 2x - 24 = 0$$
$$(x + 5)(x - 3) = 0 \qquad (x + 6)(x - 4) = 0$$
$$x = -5 \quad \text{or} \quad x = 3 \qquad x = -6 \quad \text{or} \quad x = 4$$

$$x \leq -5 \quad \text{or} \quad x \geq 3$$

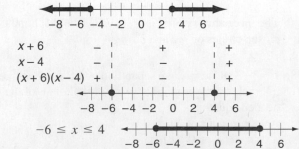

$$-6 \leq x \leq 4$$

If the two inequalities are graphed simultaneously, the solution is $-6 \leq x \leq -5$ or $3 \leq x \leq 4$.
Since length is a positive value the only sensible

solution is
$$3 \leq x \leq 4 \text{ cm}$$

MIXED REVIEW

42. $|(-4)^3| - \dfrac{(36 + 8)}{2}$
$$= |64| - \frac{44}{2}$$
$$= 64 - 22$$
$$= 42$$

43. $-5t^3 + 4t - 8t^2 + t - 8$
$$-5t^3 - 8t^2 + (4t + t) - 8$$
$$-5t^3 - 8t^2 + 5t - 8$$

44. $\dfrac{1.5 \text{ cm}}{30 \text{ km}} = \dfrac{9 \text{ cm}}{x \text{ km}}$
$$1.5x = 9(30)$$
$$1.5x = 270$$
$$x = 180 \text{ km}$$

45.
$$-2 < 2y + 8 \qquad < 18$$
$$-2 - 8 < 2y + 8 - 8 < 18 - 8$$
$$-10 < 2y \qquad < 10$$
$$-\frac{10}{2} < \frac{2y}{2} \qquad < \frac{10}{2}$$
$$-5 < y < 5$$

46.
$$|-4 + 3k| \leq 5$$
$$-4 + 3k \leq 5 \quad \text{and} \quad -4 + 3k \geq 5$$
$$3k \leq 9 \qquad\qquad 3k \geq -1$$
$$k \leq 3 \quad \text{and} \qquad k \geq -\frac{1}{3}$$
$$-\frac{1}{3} \leq k \leq 3$$

47. Let width $= w$
length $= 2w^2$
depth $= 2w^2 - 4$
$V = wld$
$$= w(2w^2)(2w^2 - 4)$$
$$= 2w^3(2w^2 - 4)$$
$$= 4w^5 - 8w^3$$

48. C; $2x^2 + kx - 12 \leq 0$ Determine k if $-1.5 \leq x \leq 4$.
$x = -1.5$ and $x = 4$ are solutions for the equation:
$$(x + 1.5)(x - 4) = 0$$
$$x^2 - 2.5x - 6 = 0$$
Multiplying the left side of this equation by 2 results in the left side of the given inequality.
$$2x^2 - 5x - 12 \leq 0; k = -5$$

Lesson 7.4, pages 339–345

EXPLORE

1a. $y < x^2 - 8x + 7; (0, 8)$
$$8 \overset{?}{<} (0)^2 - 8(0) + 7$$
$$8 \overset{?}{<} 7; \text{ no}$$

1b. $y < x^2 - 8x + 7; (-3, 1)$
$$1 \overset{?}{<} (-3)^2 - 8(-3) + 7$$
$$1 \overset{?}{<} 40; \text{ yes}$$

1c. $y < x^2 - 8x + 7; (4, -10)$
$$-10 < (4)^2 - 8(4) + 7$$
$$-10 < -9; \text{ yes}$$

1d. $y < x^2 - 8x + 7$; $(8, 9)$

$9 < (8)^2 - 8(8) + 7$

$9 < 7$; no

1e. $y < x^2 - 8x + 7$; $(10, 2)$

$2 < (10)^2 - 8(10) + 7$

$2 < 27$; yes

1f. $y < x^2 - 8x + 7$; $(5, 0)$

$0 < 25 - 40 + 7$

$0 < -8$; no

2. Answers will vary but should satisfy $y < x^2 - 8x + 7$.

3. Answers will vary but should not satisfy $y < x^2 - 8x + 7$.

4.

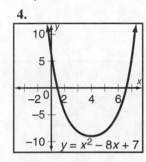

$y = x^2 - 8x + 7$

$y = (x^2 - 8x + 16) + 7 - 16$

$\quad = (x - 4)^2 - 9$

Vertex; $(4, -9)$

Find x-intercepts

Let $y = 0$.

$0 = x^2 - 8x + 7$

$0 = (x - 7)(x - 1)$

$x = 7 \quad$ or $\quad x = 1$

The axis of symmetry is $x = 4$. To find the y-intercept, let $x = 0$.

$y = (0)^2 - 8(0) + 7 = 7$; $y = 7$

5.

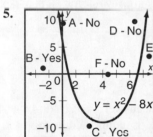

(points G–L will vary)

6. Answers will vary, but should include the fact that points that do satisfy and points that don't satisfy lie in non-intersecting regions of which the parabola is a boundary curve.

TRY THESE

1. $y \leq x^2 - 8x + 2$; $(0, 0)$

$0 \overset{?}{\leq} (0)^2 - 8(0) + 2$

$0 \overset{?}{\leq} 2$; yes

2. $y \leq x^2 - 8x + 2$; $(5, 3)$

$3 \overset{?}{\leq} (5)^2 - 8(5) + 2$

$3 \overset{?}{\leq} -13$; no

3. $y \leq x^2 - 8x + 2$; $(-3, 10)$

$10 \overset{?}{\leq} (-3)^2 - 8(-3) + 2$

$10 \overset{?}{\leq} 35$; yes

4. $y \leq x^2 - 8x + 2$; $(1, -5)$

$-5 \overset{?}{\leq} (1)^2 - 8(1) + 2$

$-5 \overset{?}{\leq} -5$; yes

5. $y \leq x^2 - 8x + 2$; $(-2, -2)$

$-2 \leq (-2)^2 - 8(-2) + 2$

$-2 \leq 22$; yes

6. $y \leq x^2 - 8x + 2$; $(-1, 0)$

$0 \overset{?}{\leq} (-1)^2 - 8(-1) + 2$

$0 \overset{?}{\leq} 11$; yes

7. $y \leq x^2 - 8x + 2$; $(-10, 192)$

$192 \overset{?}{\leq} (-10)^2 - 8(-10) + 2$

$192 \overset{?}{\leq} 182$; no

8. $y \leq x^2 - 8x + 2$; $(0, 2)$

$2 \overset{?}{\leq} (0)^2 - 8(0) + 2$

$2 \overset{?}{\leq} 2$; yes

9. $y < x^2 + 1$

$y = x^2 + 1$

$y = (x - 0)^2 + 1$

vertex $= (0, 1)$

Let $x = 0$

$y = (0)^2 + 1$

$y = 1$

y-intercept $= (0, 1)$

Let $y = 0$

$0 = x^2 + 1$

$-1 = x^2$

$x = \pm \sqrt{-1}$

no x-intercepts

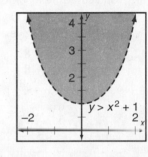

The boundary parabola is a dashed curve. Test points in both regions.

$(0, 4)$; $4 \overset{?}{>} (0)^2 + 1 \qquad (0, 0)$; $0 \overset{?}{>} (0)^2 + 1$

$\quad\quad 4 > 1$; yes $\qquad\qquad\quad 0 > 1$; no

10. $y \leq x^2 - 7x + 12$

$y = x^2 - 7x + 12$

$y = \left(x^2 - 7x + \dfrac{49}{4}\right) + 12 - \dfrac{49}{4}$

$y = \left(x - \dfrac{7}{2}\right)^2 - \dfrac{1}{4}$

Vertex $= (3.5, -0.25)$

Let $x = 0$

$y = (0)^2 - 7(0) + 12$

$y = 12$

y-intercept $= (0, 12)$

Let $y = 0$

$0 = x^2 - 7x + 12$

$0 = (x - 3)(x - 4)$

$x = 3 \quad$ or $\quad x = 4$

x-intercepts $= (3, 0)$

$\qquad\qquad\qquad (4, 0)$

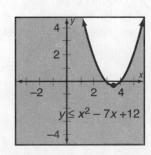

The boundary parabola is a solid curve. Test points in both regions.

$(0, 0); 0 \overset{?}{\leq} (0)^2 - 7(0) + 12$
$0 \overset{?}{\leq} 12;$ yes
$(3, 6); 6 \overset{?}{\leq} (3)^2 - 7(3) + 12$
$6 \overset{?}{\leq} 0;$ no

11. $y < -x^2 + x$

$y = -x^2 + x$

$y = -1\left(x^2 - x + \dfrac{1}{4}\right) + \dfrac{1}{4}$

$y = -1\left(x - \dfrac{1}{2}\right)^2 + \dfrac{1}{4}$

Vertex $= (0.5, 0.25)$

Let $x = 0$

$y = -(0)^2 + 0$

$y = 0; y\text{-intercept} = (0, 0)$

Let $y = 0$

$0 = x^2 + x$

$-x(x - 1) = 0$

$x = 0 \quad$ or $\quad x = 1;$

$x\text{-intercepts} = (0, 0); (1, 0)$

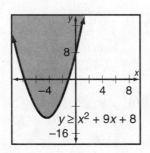

The boundary parabola is a
solid curve. Test points in both regions.
$(1, -1); -1 \overset{?}{<} -(1)^2 + 1 \qquad (0, 1); 1 \overset{?}{<} -(0)^2 + 0$
$\quad -1 \overset{?}{<} 0;$ yes $\qquad\qquad\qquad 1 \overset{?}{<} 0;$ no

12. $y \geq x^2 + 9x + 8$

$y = x^2 + 9x + 8$

$y = \left(x^2 + 9x + \dfrac{81}{4}\right) + 8 - \dfrac{81}{4}$

$y = \left(x + \dfrac{9}{2}\right)^2 - \dfrac{49}{4}$

Vertex $= (-4.5, -12.25)$

Let $x = 0$

$y = (0)^2 + 9(0) + 8$

$y = 8; y\text{-intercept} = (0, 8)$

Let $y = 0$

$0 = x^2 + 9x + 8$

$0 = (x + 8)(x + 1)$

$x = -8 \quad$ or $\quad x = -1$

$x\text{-intercepts} = (-8, 0);$
$(-1, 0)$

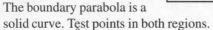

The boundary parabola is a solid curve. Test points in
both regions.
$(0, 0); 0 \overset{?}{\geq} (0)^2 + 9(0) + 8$
$0 \overset{?}{\geq} 8;$ no
$(-4, 0); 0 \overset{?}{\geq} (-4)^2 + 9(-4) + 8$
$0 \overset{?}{\geq} -12;$ yes

13. $y > (x - 7)^2$

$y = (x - 7)^2$

$y = (x - 7)^2 + 0$

Vertex $= (7, 0)$

Let $x = 0$

$y = (0 - 7)^2; y = 49$

$y\text{-intercept} = (0, 49)$

Let $y = 0$

$0 = (x - 7)^2$

$x = 7; x\text{-intercept} = (7, 0)$

The boundary parabola is a
dashed curve. Test points in
both regions.

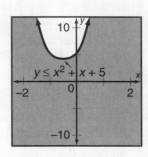

$(0, 0); 0 \overset{?}{>} (0 - 7)^2 \qquad (7, 10); 10 \overset{?}{>} (7 - 7)^2$
$\quad 0 \overset{?}{>} 49;$ no $\qquad\qquad\qquad 10 \overset{?}{>} 0;$ yes

14. $y \leq x^2 + x + 5$

$y = x^2 + x + 5$

$y = \left(x^2 + x + \dfrac{1}{4}\right) + 5 - \dfrac{1}{4}$

$y = \left(x + \dfrac{1}{2}\right)^2 + \dfrac{19}{4}$

Vertex $= (-0.5, 4.75)$

Let $x = 0$

$y = (0)^2 + (0) + 5$

$y = 5; y\text{-intercept} = (0, 5)$

Let $y = 0$

$0 = x^2 + x + 5$

$x = \dfrac{-1 \pm \sqrt{1 - 4(1)(5)}}{2(1)}$

$x = \dfrac{-1 \pm \sqrt{-19}}{2};$ no real

$x\text{-intercepts}$

The boundary parabola is a solid curve. Test points in
both regions.
$(0, 0); 0 \overset{?}{\leq} (0)^2 + (0) + 5$
$0 \overset{?}{\leq} 5;$ yes
$(-1, 10); 10 \overset{?}{\leq} (-1)^2 + (-1) + 5$
$10 \overset{?}{\leq} 5;$ no

15. Answers will vary. Possible answer: $y \geq x^2 - 4x + 4$.

16. Changes in driving conditions require additional
braking distance. The distance will be at least 0.05S, or
$b \geq 0.05S$.

PRACTICE

1. $y > 2x^2 - x - 3; (1, 5)$
$5 \overset{?}{>} 2(1)^2 - (1) - 3$
$5 \overset{?}{>} -4;$ yes

2. $y > 2x^2 - x - 3; (-1, 5)$
$5 \overset{?}{>} 2(-1)^2 - (-1) - 3$
$5 \overset{?}{>} 0;$ yes

3. $y > 2x^2 - x - 3; (12, 0)$
$0 \overset{?}{>} 2(12)^2 - 12 - 3$
$0 \overset{?}{>} 273;$ no

4. $y > 2x^2 - x - 3; (0, -4)$
$-4 \overset{?}{>} 2(0)^2 - 0 - 3$
$-4 \overset{?}{>} -3;$ no

5. $y > 2x^2 - x - 3; (2, 3)$

$3 \overset{?}{>} 2(2)^2 - 2 - 3$
$3 \overset{?}{>} 3$; no

6. $y \le 2x^2 + 7$
$y = 2x^2 + 7$
$y = 2(x + 0)^2 + 7$
Vertex $= (0, 7)$

Let $x = 0$; $y = 2(0)^2 + 7$
$y = 7$ y-intercept $= (0, 7)$

Let $y = 0$; $0 = 2x^2 + 7$
$x = \dfrac{0 \pm \sqrt{0^2 - 4(2)(7)}}{2(2)}$
$= \dfrac{0 \pm \sqrt{-56}}{4}$; no real

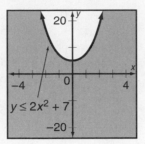

x-intercepts
The boundary parabola is a
solid curve. Test points in both regions.
$(0, 0)$; $0 \overset{?}{\le} 2(0)^2 + 7$
$0 \overset{?}{\le} 7$; yes
$(0, 10)$; $10 \overset{?}{\le} 2(0)^2 + 7$
$10 \overset{?}{\le} 7$; no

7. $y > -4x^2 + x$
$y = -4x^2 + x$
$y = -4\left(x^2 - \dfrac{1}{4}x + \dfrac{1}{64}\right) + \dfrac{1}{16}$
$y = -4\left(x - \dfrac{1}{8}\right)^2 + \dfrac{1}{16}$
Vertex $= \left(\dfrac{1}{8}, \dfrac{1}{16}\right)$

Let $x = 0$
$y = -4(0)^2 + 0$
$y = 0$; y-intercept $= (0, 0)$

Let $y = 0$
$0 = -4x^2 + x$
$0 = -x(4x - 1)$
$x = 0$ or $x = \dfrac{1}{4}$;
x-intercepts $= (0, 0)$; $\left(\dfrac{1}{4}, 0\right)$

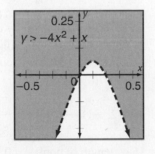

The boundary parabola is a dashed curve. Check points
in both regions.
$(0, -1)$; $-1 \overset{?}{>} -4(0)^2 + 0$
$-1 \overset{?}{>} 0$; no
$(0, 1)$; $1 \overset{?}{>} -4(0)^2 + 0$
$1 \overset{?}{>} 0$; yes

8. $y < x^2 + 9x + 10$
$y = x^2 + 9x + 10$
$y = \left(x^2 + 9x + \dfrac{81}{4}\right) + 10 - \dfrac{81}{4}$
$y = \left(x + \dfrac{9}{2}\right)^2 - \dfrac{41}{4}$
Vertex $= (-4.5, -10.25)$

Let $x = 0$
$y = (0)^2 + 9(0) + 10$
$y = 10$; y-intercept $= (0, 10)$

Let $y = 0$
$0 = x^2 + 9x + 10$
$x = \dfrac{-9 \pm \sqrt{81 - 4(1)(10)}}{2(1)}$
$x = \dfrac{-9 \pm \sqrt{41}}{2} = -7.702$;
-1.298

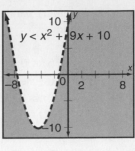

x intercepts $= (-7.7, 0)$; $(-1.3, 0)$
The boundary parabola is a dashed curve. Test points in
both regions.
$(0, 0)$; $0 \overset{?}{<} (0)^2 + 9(0) + 10$
$0 \overset{?}{<} 10$; yes
$(-3, 5)$; $5 \overset{?}{<} (-3)^2 + 9(-3) + 10$
$5 \overset{?}{<} -8$; no

9. $y \ge 2x^2 + 7x + 3$
$y = 2x^2 + 7x + 3$
$y = 2\left(x^2 + \dfrac{7}{2}x + \dfrac{49}{16}\right) + 3 - \dfrac{49}{8}$
$y = 2\left(x + \dfrac{7}{4}\right)^2 - \dfrac{25}{8}$
Vertex $= (-1.75, -3.125)$

Let $x = 0$
$y = 2(0)^2 + 7(0) + 3$
$y = 3$; y-intercept $= (0, 3)$

Let $y = 0$
$0 = 2x^2 + 7x + 3$
$0 = (2x + 1)(x + 3)$;
$x = -\dfrac{1}{2}, -3$
x-intercepts $= \left(-\dfrac{1}{2}, 0\right)$;
$(-3, 0)$

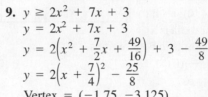

The boundary parabola is a solid curve. Test points in
both regions.
$(0, 0)$; $0 \overset{?}{\ge} 2(0)^2 + 7(0) + 3$
$0 \overset{?}{\ge} 3$; no
$(-2, 2)$; $0 \overset{?}{\ge} 2(-2)^2 + 7(-2) + 3$
$2 \overset{?}{\ge} -3$; yes

10. $y \le -0.5x^2$
$y = -0.5x^2$
$y = -0.5(x - 0)^2 + 0$
Vertex $= (0, 0)$

Let $x = 0$
$y = -0.5(0)^2$
$y = 0$
y-intercept $= (0, 0)$

Let $y = 0$
$0 = -0.5x^2$
$x = 0$; x-intercept $= (0, 0)$

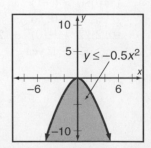

The boundary parabola is a solid curve. Test points in both regions.

$(0, -5); -5 \overset{?}{\le} -0.5(0)^2$

$\qquad -5 \le 0;$ yes

$(5, 0); 0 \overset{?}{\le} -0.5(5)^2$

$\qquad 0 \overset{?}{\le} 12.5;$ no

11. $y \le 16 - x^2$

$y = -x^2 + 16$

$y = -(x - 0)^2 + 16$

Vertex $= (0, 16)$

Let $x = 0$

$y = 16 - (0)^2$

$y = 16;$ y-intercept $= (0, 16)$

Let $y = 0$

$0 = -x^2 + 16$

$x^2 = 16$

$x = \pm 4$

x-intercepts $= (4, 0); (-4, 0)$

The boundary parabola is a solid curve. Test points in both regions.

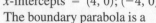

$(0, 0); 0 \overset{?}{\le} 16 - (0)^2$

$\qquad 0 \overset{?}{\le} 16;$ yes

$(0, 18); 18 \overset{?}{\le} 16 - (0)^2$

$\qquad 18 \overset{?}{\le} 16;$ no

12. $y > -x^2 + 2x + 1$

$y = -x^2 + 2x + 1$

$y = -(x^2 - 2x + 1) + 1 + 1$

$y = -(x - 1)^2 + 2$

Vertex $= (1, 2)$

Let $x = 0$

$y = -(0)^2 + 2(0) + 1$

$y = 1;$ y-intercept $= (0, 1)$

Let $y = 0$

$0 = -x^2 + 2x + 1$

$x = \dfrac{-2 \pm \sqrt{2^2 - 4(-1)(1)}}{2(-1)}$

$x = \dfrac{-2 \pm \sqrt{8}}{-2} = 1 - \sqrt{2}$

$1 + \sqrt{2}$

x-intercepts $= (-0.4, 0); (2.4, 0)$

The boundary parabola is a dashed curve. Test points in both regions.

$(1, 0); 0 \overset{?}{>} (-1)^2 + 2(-1) + 1$

$\qquad 0 \overset{?}{>} 0;$ no

$(0, 2); 2 \overset{?}{>} -(0)^2 + 2(0) + 1$

$\qquad 2 \overset{?}{>} 1;$ yes

13. $y > (x - 5)^2$

$y = (x - 5)^2 + 0$

Vertex $= (5, 0)$

Let $x = 0$

$y = (0 - 5)^2; y = 25$

y-intercept $= (0, 25)$

Let $y = 0$

$0 = (x - 5)^2; x = 5$

x-intercept $= (5, 0)$

The boundary parabola is a dashed curve. Test points in both regions.

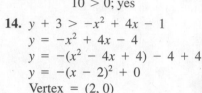

$(0, 0); 0 \overset{?}{>} (0 - 5)^2$

$\qquad 0 \overset{?}{>} 25;$ no

$(5, 10); 10 \overset{?}{>} (5 - 5)^2$

$\qquad 10 \overset{?}{>} 0;$ yes

14. $y + 3 > -x^2 + 4x - 1$

$y = -x^2 + 4x - 4$

$y = -(x^2 - 4x + 4) - 4 + 4$

$y = -(x - 2)^2 + 0$

Vertex $= (2, 0)$

Let $x = 0$

$y = -(0)^2 + 4(0) - 4$

$y = -4;$

y-intercept $= (0, -4)$

Let $y = 0$

$0 = -x^2 + 4x - 4$

$0 = -(x - 2)(x - 2)$

$x = 2;$ x-intercept $= (2, 0)$

The boundary parabola is a dashed curve. Test points in both regions.

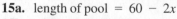

$(0, 0); 0 + 3 \overset{?}{>} -(0)^2 + 4(0) - 1$

$\qquad 3 \overset{?}{>} -1;$ yes

$(2, -1); -1 + 3 \overset{?}{>} -(2)^2 + 4(2) - 1$

$\qquad 2 \overset{?}{>} 3;$ no

15a. length of pool $= 60 - 2x$

15b. width of pool $= 50 - 2x$

15c. Area of pool and surrounding patio

$= l \times w$

$= 60 \cdot 50 = 3000$ ft^2

15d. Area of pool $= l \times w$

$= (60 - 2x)(50 - 2x)$

15e. amount of concrete $<$ area of pool and patio $-$ area of pool

$y < 3000 - (60 - 2x)(50 - 2x)$

15f. No. If $x = 6$, the pool's area is $48 \cdot 38 = 1824$. The patio including the stones, measures 1176 ft^2. Since 200 ft^2 are covered by stones, 976 ft^2 of concrete is needed; 900 ft^2 is not enough.

16a. Area covered ≤ number of heads · area covered per head

$$A \leq 10\pi r^2$$

16b.

r	2	4	6	8	10	12
A	≈ 126	≈ 503	≈ 1131	≈ 2011	≈ 3142	≈ 4524

Check: $(0, 1000)$

$1000 \overset{?}{\leq} 10\pi(0)^2$; no

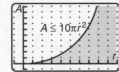

16c. $A = 10 \cdot \pi \cdot 14^2$

$A = 1960\pi = 6158 \text{ ft}^2$

17. $\dfrac{3000 \text{ mg}}{5 \text{ days}} = 600$ mg per day

Find w such that $600 \leq 0.05w^2 + 2w$

Graph the parabola $y = 0.05w^2 + 2w - 600$

$y = 0.05(w^2 + 40w + 400) - 600 - 20$

$y = 0.05(w + 20)^2 - 620$

Vertex = $(-20, -620)$

Let $w = 0$

$y = 0.05(0 + 20)^2 - 620$

$y = -600$;

y-intercept = $(0, -600)$

Let $y = 0$

$0 = 0.05w^2 + 2w - 600$

$w = \dfrac{-2 \pm \sqrt{2^2 - 4(0.05)(-600)}}{2(0.05)}$

$w = 91.4$ or -131.4

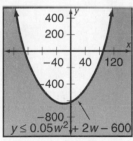

Test points in both regions.

$y \leq 0.05w^2 + 2w - 600$

$(0, 0)$; $600 \overset{?}{\leq} 0.05(0)^2 + 2(0)$

$600 \overset{?}{\leq} 0$; no

$(100, 0)$; $600 \overset{?}{\leq} 0.05(100)^2 + 2(100)$

$600 \overset{?}{\leq} 700$; yes

$w \geq 91.4$ lb

18. Answers will vary. Similarities may include determining if region separations are dashed or solid and testing points on either side of the boundary separation. Differences may include that in quadratic inequalities one must determine a vertex and the direction of opening, and in a linear inequality one determines slope and y-intercept.

19a.

$y \leq 6$ includes the area below the solid line $y = 6$.

$y > x^2 + 2$

Vertex = $(0, 2)$

There are no x-intercepts.

The curve is dashed.

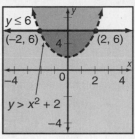

Check $(0, 6)$; $6 \overset{?}{>} (0)^2 + 2, 6 > 2$

Shade inside the parabola. The solution to the system is the double shaded area.

19b. Substitute $y = 6$ into the equation $y = x^2 + 2$.

$y = x^2 + 2$, $6 = x^2 + 2$, $4 = x^2$, $x = 2$ or $x = -2$

The boundary lines intersect at $(2, 6)$ and $(-2, 6)$

19c. It is not; the second inequality is not a true statement when $x = 1$ and $y = 2$.

19d. Answers will vary. Possible solutions are:

$(1, 5)$; check: $y \leq 6$, $5 \leq 6$; $y > x^2 + 2$;

$5 \overset{?}{>} (1)^2 + 2, 5 > 3$

$(-1, 5)$; check: $y \leq 6$, $5 \leq 6$; $y > x^2 + 2$;

$5 \overset{?}{>} (-1)^2 + 2, 5 > 3$

$(0, 3)$; check: $y \leq 6$, $3 \leq 6$; $y > x^2 + 2$;

$3 \overset{?}{>} (0)^2 + 2, 3 > 2$

20a. Check points from each region:

A; $(3, 20)$; check: $y < x^2 - 8x + 15$,

$20 \overset{?}{<} (3)^2 - 8(3) + 15, 20 \not< 0$

B; $(-6, 8)$; check: $y < x^2 - 8x + 15$,

$8 \overset{?}{<} (-6)^2 - 8(-6) + 15, 8 < 99$

C; $(0, 0)$; check: $y < x^2 - 8x + 15$,

$0 \overset{?}{<} (0)^2 - 8(0) + 15, 0 < 15$

D; $(4, 1)$; check: $y < x^2 - 8x + 15$,

$1 \overset{?}{<} (4)^2 - 8(4) + 15, 1 \not< -1$

E; $(10, 20)$; check: $y < x^2 - 8x + 15$,

$20 \overset{?}{<} (10)^2 - 8(10) + 15, 20 < 35$

B, C, and E have solutions to the first inequality.

20b. Check points from each region to see which have solutions to $y \geq x + 7$.

A; $(3, 20)$; check: $y \geq x + 7$, $20 \overset{?}{\geq} 3 + 7$,

$20 \geq 10$

B; $(-6, 8)$; check: $y \geq x + 7$, $8 \overset{?}{\geq} -6 + 7$,

$8 \geq 1$

C; $(0, 0)$; check: $y \geq x + 7$, $0 \overset{?}{\geq} 0 + 7$,

$0 \not\geq 7$

D; $(4, 1)$; check: $y \geq x + 7$, $1 \overset{?}{\geq} 4 + 7$,

$1 \not\geq 11$

E; $(10, 20)$; check: $y \geq x + 7$, $20 \overset{?}{\geq} 10 + 7$,

$20 \geq 17$

B and E have solutions to both inequalities.

20c. A **20d.** C

21. $y < -x^2 + 4$

$y = -(x + 0)^2 + 4$

Vertex $= (0, 4)$

Let $x = 0$

$y = -(0)^2 + 4 = 0 + 4 = 4$

x-intercept $= (0, 4)$

Let $y = 0$

$0 = -x^2 + 4 = (-x + 2)(x + 2)$

$x = -2$ or $x = 2$

y-intercepts $= (-2, 0)$ and $(2, 0)$

Check: $(0, 0)$: $0 < -x^2 + 4$, $0 \overset{?}{<} -(0)^2 + 4$, $0 < 4$

Shade inside the parabola.

$y > |x|$

$(0, 2)$: $y > |x|$, $2 \overset{?}{>} |0|$, $2 > 0$

Graph above the dashed lines

$y = x$ and $y = -x$. The solution to the system is the double shaded area.

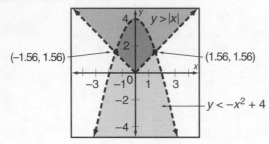

22a. $y = x^2 - 9x + 8$

$= \left(x^2 - 9x + \dfrac{81}{4}\right) + 8 - \dfrac{81}{4}$

$= \left(x - \dfrac{9}{2}\right)^2 - \dfrac{49}{4}$

Vertex $= (4.5, -12.25)$

Let $x = 0$

$y = x^2 - 9x + 8 = (0)^2 - 9(0) + 8 = 8$

y-intercept $= (0, 8)$

Let $y = 0$

$0 = x^2 - 9x + 8 = (x - 8)(x - 1); x = 8$ or

$x = 1$

x-intercepts $= (8, 0)$ or $(1, 0)$

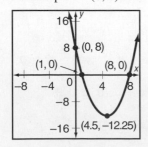

22b.

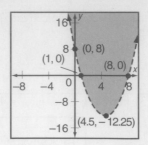

Check $(4, 0)$: $y > x^2 - 9x + 8$.

$0 \overset{?}{>} (4)^2 - 9(4) + 8, 0 > -12$

Check $(0, 0)$: $y > x^2 - 9x + 8$,

$0 \overset{?}{>} (0)^2 - 9(0) + 8, 0 \not> 8$

22c. $y = |x^2 - 9x + 8|$

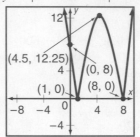

The portion of the graph $y = x^2 - 9x + 8$ that was below the x-axis is now reflected in the x-axis.

22d. $y > |x^2 - 9x + 8|$

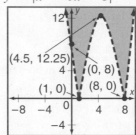

Test points in all regions:

$(0, 0); 0 \overset{?}{>} |0^2 - 9(0) + 8|, 0 \not> 8$

$(1, 8); 8 \overset{?}{>} |(1)^2 - 9(1) + 8|, 8 > 0$

$(4, 4); 4 \overset{?}{>} |(4)^2 - 9(4) + 8|, 4 \not> 12$

$(8, 4); 4 \overset{?}{>} |(8)^2 - 9(8) + 8|, 4 > 0$

$(9, 1); 1 \overset{?}{>} |(9)^2 - 9(9) + 8|, 1 \not> 8$

THINK CRITICALLY

23. b; the triangle inequality is violated:

$x + 1 + x + 5 < 3x + 6$.

24. The solution is all points except the origin. Answers will vary; the graph should be displayed by shading the entire plane except for an open circle at the origin.

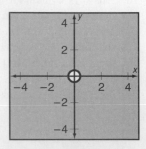

25a. Any of the following are correct:

$y < x^2$ and $y > x^2 + 10$
$y \le x^2$ and $y \ge x^2 + 10$
$y < x^2$ and $y \ge x^2 + 10$
$y \le x^2$ and $y > x^2 + 10$

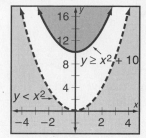

25b. Any of the following are correct:

$y > x^2$ and $y < x^2 + 10$
$y \ge x^2$ and $y \le x^2 + 10$
$y \ge x^2$ and $y < x^2 + 10$
$y > x^2$ and $y \le x^2 + 10$

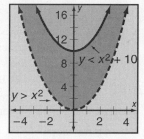

26.

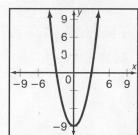

Graph $y = x^2 - 9$
Vertex $= (0, -9)$
x-intercepts $=$
  $(3, 0); (-3, 0)$
y-intecept $= (0, -9)$

To graph $y = |x^2 - 9|$, take the portion of the graph $y = x^2 - 9$ that is below the x-axis and reflect it in the x-axis. Test points in all regions and shade accordingly.

$(-6, 0); 0 \overset{?}{\le} |(-6)^2 - 9|, \ 0 \le 27$
$(-3, 9); 9 \overset{?}{\le} |(-3)^2 - 9|, \ 9 \ne 0$
$(0, 3); 3 \overset{?}{\le} |(0)^2 - 9|, \ 3 \le 9$
$(3, 9); 9 \overset{?}{\le} |(3)^2 - 9|, \ 9 \ne 0$
$(6, 0); 0 \overset{?}{\le} |(6)^2 - 9|, \ 0 \le 27$

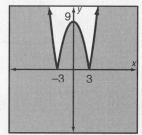

27. Graph $y \ne x^2 - 4$

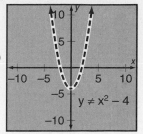

$y = x^2 - 4$
Vertex $= (0, -4)$
x-intercepts $= (2, 0); (-2, 0)$
y-intercept $= (0, -4)$
Shade entire plane except for the dashed parabola
$y = x^2 - 4$

28. Answers will vary, but should include the fact that a negative discriminant implies no real roots, no x-intercepts. The x^2 coefficient implies the parabola opens up.

MIXED REVIEW

29. B; 1.7329 is the largest number.

30. The fraction $3x$ is undefined
$x^2 - 5x + 4$
when the denominator is zero.
$(x^2 - 5x + 4) = 0, (x - 4)(x - 1) = 0; x = 4$ or $x - 1$.

31. The multiplicative inverse of $\dfrac{1}{x - 7}$ is $\dfrac{x - 7}{1} = x - 7$.

32. $\dfrac{\sqrt{x - 4}}{3}$ is defined when $x - 4 \ge 0; x \ge 4$
$\{x \,|\, x \ge 4\}$

33.

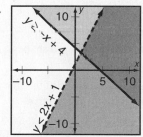

$y < 2x + 1$; Test $(0, 0)$;
$0 < 2(0) + 1; \ 0 < 1$
Test $(0, 2); 2 < 2(0) + 1$;
$2 \not< 1$
Shade below the dashed line $y = 2x + 1$

$y \ge -x + 4$; Test $(0, 0)$;
$0 \overset{?}{\ge} -(0) + 4; 0 \not\ge 4$
Test $(6, 0); 0 \overset{?}{\ge} -(6) + 4$;
$0 \ge -2$

Shade above the solid line $y = -x + 4$
The double shaded region is the solution to the inequality.

34.

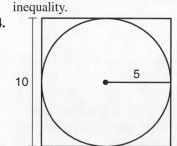

$A = \pi r^2 = 3.14(5)^2$
$\approx 78.5 \ \text{ft}^2$

35. Substitute $x = 5$ in the equation $y = 2x + 3$.
$y = 2(5) + 3 = 13$; The graphs intersect at $(5, 13)$

36. $0.71 = \dfrac{71}{100}$

37. Let n represent $0.71717\ldots$, then $100n = 71.71717\ldots$

$$100n = 71.71717\ldots$$
$$\underline{-\quad n = 0.71717}$$
$$99n = 71$$
$$n = \frac{71}{99}$$

38.

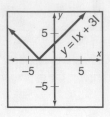

$y = |x + 3|$

x	y
-10	7
-5	2
-3	0
0	3
5	8

39.

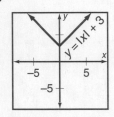

$y = |x| + 3$

x	y
-10	13
-5	8
0	3
5	8
10	13

40. $0 < x < 1$; An example is $\frac{1}{2}$; $\frac{1}{2} \stackrel{?}{>} \left(\frac{1}{2}\right)^2; \frac{1}{2} > \frac{1}{4}$

Lesson 7.5 pages 346–349

THINK BACK/WORKING TOGETHER

1. A and B are polygons since they are closed shapes whose sides are line segments.

2. The polygon is a right triangle. The slope of the line that forms the hypotenuse of the triangle is negative.

3. The polygon is a quadrilateral.

EXPLORE

4. The graph on the left is the solution to a system of four inequalities. The graph on the right is the solution to a system of three inequalities.

5-8. Answers will vary but should match the steps outlined in the activity.

MAKE CONNECTIONS

9. 1, 3, and 4 since points in A are positive x and y values and in the region below 13.

10. All inequalities since points in B are in the shaded region formed by the intersection of the graphs.

11. 1, 2, and 4 since points in C are positive x and y values and in the region above 12.

12. Answers will vary. One solution is $x \geq 0, y \geq 0,$ $x + y \leq 5, x + y \leq 10$. The inequalities $x \geq 0$ and $y \geq 0$ restrict the solution to the first quadrant since all coordinates are positive in that quadrant.

SUMMARIZE

13.

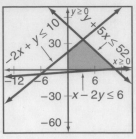

Yes, all of the inequalities include all points on the boundaries.

14. The shading includes only positive x and y values, therefore two of the inequalities are $x \geq 0$ and $y \geq 0$.

The line with a y-intercept of 6 has slope $\frac{2}{6} = \frac{1}{3}$; $y = \frac{1}{3}x + 6$. The points that are shaded below this line satisfy the inequality $y \leq \frac{1}{3}x + 6$.

The line with an x-intercept of 8 has slope $\frac{8}{-2} = -4$; $y = -4x + 32$. The points that are shaded to the left of this line satisfy the inequality $y \leq -4x + 32$.

15.

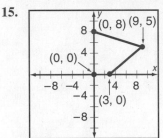

The polygon includes only positive x and y values, therefore two of the inequalities are $x \geq 0$ and $y \geq 0$.

The line with a y-intercept of 8 has slope $\frac{3}{-9} = -\frac{1}{3}$; $y = -\frac{1}{3}x + 8$. The points within the polygon satisfy the inequality $y \leq -\frac{1}{3}x + 8$.

The line with an x-intercept of 3 has slope $\frac{5}{6}$; $y = \frac{5}{6}x - \frac{5}{2}$. The points within the polygon satisfy the inequality $y \geq \frac{5}{6}x - \frac{5}{2}$.

16. Answers will vary. One possible solution is $x \geq 0$, $y \geq 0, y \leq x + 5, y \geq x - 5, y \leq -x + 8$.

ALGEBRAWORKS

1. The year 1912 is represented by 12.

2. $y = 0.9963x - 69.1818$. It is a good predictor based on the strong positive correlation 0.99. $y = 0.9963x - 69.1818$

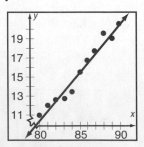

3. $y = 0.9963x - 69.1818$
$y = 0.9963(103) - 69.1818 = 33.4371$
The percent generated by nuclear energy will be 33.4% in 2003.

4. $y = 0.9963x - 69.1818$

$50 = 0.9963x - 69.1818$

$119.1818 = 0.9963x; x = 119.62$

50% of our power will come from nuclear energy in the year 2020.

5. $y < 0.9963x - 69.1818$ **6.** $y \le 20.5$

7. The coordinate $(120, 39)$ indicates that in the year 2020, 39% of power comes from nuclear energy.

8. At the current rate, the year 2020 would have 50% power from nuclear sources so congresswoman A views 39% as a cutback. Congressman B sees it as an increase because it is above the 1990 level he wants to maintain.

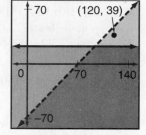

Lesson 7.6, pages 350–357

EXPLORE

1.

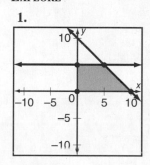

2.

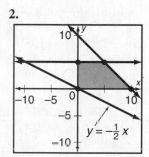

3. $(5, 5)$

4. Answers will vary, but will be one of the four vertices; $(5, 5), (0, 5), (9, 0), (0, 0)$

5. It is one of the vertices of the polygonal region.

TRY THESE

1. no   **2.** yes   **3.** yes   **4.** no   **5.** yes

6. $P = 15x + 12y$

$(0, 8); 15(0) + 12(8) = 96$

$(0, 0); 15(0) + 12(0) = 0$

$(15, 0); 15(15) + 12(0) = 225$

maximum 225 at $(15, 0)$

7. $C = 15x + 12y$

$(3, 3); 15(3) + 12(3) = 81$

$(3, 10); 15(3) + 12(10) = 165$

$(8, 10); 15(8) + 12(10) = 240$

$(12, 4); 15(12) + 12(4) = 228$

maximum 240 at $(8, 10)$

8. $C = 3x + 2y$

$(0, 6); 3(0) + 2(6) = 12$

$(11, 0); 3(11) + 2(0) = 33$

$(8, 0); 3(8) + 2(0) = 24$

minimum 12 at $(0, 6)$

9. $C = 3x + 2y$

$(2, 8); 3(2) + 2(8) = 22$

$(12, 10); 3(12) + 2(10) = 56$

$(8, 5); 3(8) + 2(5) = 34$

$(16, 5); 3(16) + 2(5) = 58$

minimum 22 at $(2, 8)$

10. $P = 10x + 6y$

$(0, 10); 10(0) + 6(10) = 60$

$(5, 15); 10(5) + 6(15) = 140$

$(8, 8); 10(8) + 6(8) = 128$

$(12, 0); 10(12) + 6(0) = 120$

minimum 60 at $(0, 10)$

maximum 140 at $(5, 15)$

11. $P = 4x + 5y$

$(2, 5); 4(2) + 5(5) = 33$

$(2, 9); 4(2) + 5(9) = 53$

$(6, 11); 4(6) + 5(11) = 79$

$(8, 5); 4(8) + 5(5) = 57$

minimum 33 at $(2, 5)$

maximum 79 at $(6, 11)$

12. $C = 1.25x + 0.75y$

$(0, 4); 1.25(0) + 0.75(4) = 3$

$(9, 15); 1.25(9) + 0.75(15) = 22.5$

$(20, 2); 1.25(20) + 0.75(2) = 26.5$

minimum 3 at $(0, 4)$

maximum 26.5 at $(20, 2)$

13. $C = 120x + 180y$

$(6, 6); 120(6) + 180(6) = 1800$

$(6, 10); 120(6) + 180(10) = 2520$

$(10, 12); 120(10) + 180(12) = 3360$

$(13, 11); 120(13) + 180(11) = 3540$

$(13, 6); 120(13) + 180(6) = 2640$

minimum 1800 at $(6, 6)$

maximum 3540 at $(13, 11)$

14. Let x represent the number of short-sleeved shirts and y represent the number of long-sleeved shirts.

Cost $= C = 3x + 5y$

15. $x \ge 15$ (at least 15 short-sleeved shirts)

$y \ge 10$ (at least 10 long-sleeved shirts)

$x + y \le 50$ (will take up to a total of 50 shirts)

16.

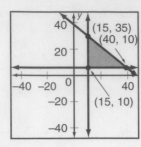

17. $C = 3x + 5y$
$(15, 10); 3(15) + 5(10) = 95$ min
$(15, 35); 3(15) + 5(35) = 220$ max
$(40, 10); 3(40) + 5(10) = 170$
They should sell 15 short- and 10 long-sleeved shirts.

18. Answers will vary. One possible solution is
$y \leq -\frac{2}{3}x + 10$ and $y \leq -\frac{3}{2}x + 15$.

19. When x and y represent numbers of real items, values of x and y usually must be positive.

PRACTICE

1. 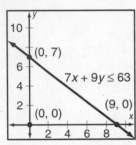 $(0, 0), (0, 7), (9, 0)$

2. 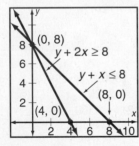 $(0, 8), (4, 0), (8, 0)$

3. 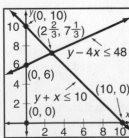 $(0, 6), \left(2\frac{2}{3}, 7\frac{1}{3}\right),$
$(10, 0), (0, 0)$

4. $A = 2x + 3y$
$(0, 0); 2(0) + 3(0) = 0$
$(12, 0); 2(12) + 3(0) = 24$
$(0, 16); 2(0) + 3(16) = 48$
maximum 48 at $(0, 16)$;
minimum 0 at $(0, 0)$

5. $A = 3x - 2y$
$(0, 0); 3(0) - 2(0) = 0$
$(12, 0); 3(12) - 2(0) = 36$
$(0, 16); 3(0) - 2(16) = -32$
maximum 36 at $(12, 0)$;
minimum -32 at $(0, 16)$

6. $A = 0.5x + 0.75y$
$(6, 6); 0.5(6) + 0.75(6) = 7.5$
$(6, 12); 0.5(6) + 0.75(12) = 12$
$(10, 12); 0.5(10) + 0.75(12) = 14$
$(15, 6); 0.5(15) + 0.75(6) = 12$
maximum 14 at $(10, 12)$;
minimum 7.5 at $(6, 6)$

7. $A = 10x + 12y$
$(4, 9); 10(4) + 12(9) = 148$
$(6, 14); 10(6) + 12(14) = 228$
$(12, 12); 10(12) + 12(12) = 264$
$(14, 2); 10(14) + 12(2) = 164$
maximum 264 at $(12, 12)$;
minimum 148 at $(4, 9)$

8. $P = -4x + 5y$
$(0, 5); -4(0) + 5(5) = 25$
$(0, 14); -4(0) + 5(14) = 70$
$\left(\frac{55}{19}, \frac{40}{19}\right); -4\left(\frac{55}{19}\right) + 5\left(\frac{40}{19}\right) = \frac{-20}{19}$
$\left(\frac{380}{23}, \frac{132}{23}\right); -4\left(\frac{380}{23}\right) + 5\left(\frac{132}{23}\right) = -\frac{860}{23} = -37.39$
maximum 70 at $(0, 14)$

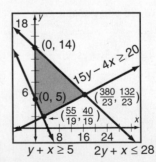

9. $C = 5x + 2y$
$(2, 4); 5(2) + 2(4) = 18$
$(2, 8); 5(2) + 2(8) = 26$
$(6, 8); 5(6) + 2(8) = 46$
$(18, 4); 5(18) + 2(4) = 98$
minimum 18 at $(2, 4)$

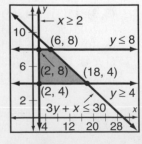

10. Answers will vary. One possible solution is
$P = 12x + 10y$ and $R = 8x + 2y$.

11. $P = 5x + 6y$
$(0, 5); 5(0) + 6(5) = 30$
$(1, 3); 5(1) + 6(3) = 23$
$(2, 0); 5(2) + 6(0) = 10$
maximum 30 at $(0, 5)$

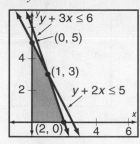

12. $P = 2x + y$
$(0, 1); 2(0) + 1 = 1$
$(3, 1); 2(3) + 1 = 7$
$(12, 4); 2(12) + 4 = 28$
$(0, 16); 2(0) + 16 = 16$
maximum 28 at $(12, 4)$

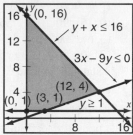

13. $P = -2x + 3y$
$(1, 0); -2(1) + 3(0) = -2$
$(4, 0); -2(4) + 3(0) = -8$
$(4, 4); -2(4) + 3(4) = 4$
$(1, 1); -2(1) + 3(1) = 1$
maximum 4 at $(4, 4)$

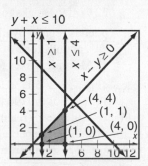

14. $P = -2x + 4y$
$(0, 0); -2(0) + 4(0) = 0$
$(2, 2); -2(2) + 4(2) = 4$
$(0, 4); -2(0) + 4(4) = 16$
maximum 16 at $(0, 4)$

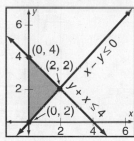

15. $C = x + 7y$
$\left(0, \dfrac{9}{4}\right); 0 + 7\left(\dfrac{9}{4}\right) = 15.75$
$(0, 0); 0 + 7(0) = 0$
$(3, 3); 3 + 7(3) = 24$
minimum 0 at $(0, 0)$

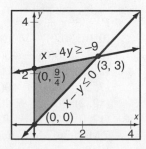

16.

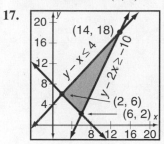

$C = 1.5x + 2.5y$
$(9, 5); 1.5(9) + 2.5(5) = 26$
$(9, 15); 1.5(9) + 2.5(15) = 51$
$(15, 5); 1.5(15) + 2.5(5) = 35$
$(15, 9); 1.5(15) + 2.5(9) = 45$
minimum 26 at $(9, 5)$

17.

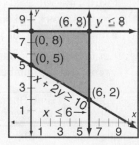

$C = 3x + 4y$
$(2, 6); 3(2) + 4(6) = 30$
$(6, 2); 3(6) + 4(2) = 26$
$(14, 18); 3(14) + 4(18) = 114$
minimum 26 at $(6, 2)$

18. $C = -x + 3y$
$(0, 5); -0 + 3(5) = 15$
$(0, 8); -0 + 3(8) = 24$
$(6, 2); -6 + 3(2) = 0$
$(6, 8); -6 + 3(8) = 18$
minimum 0 at $(6, 2)$

19. Let x represent iceberg lettuce and y represent romaine.
Profit $= P = 200x + 250y$
$x \leq 3000$ (no more than 3000)
$y \leq 2500$ (no more than 2500)
$x + y \leq 3600$ (3600-acre field available)

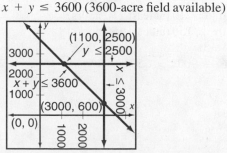

$P = 200x + 250y$
$(0, 0); 200(0) + 250(0) = 0;$ minimum
$(0, 3600); 200(0) + 250(3600) = 900,000$
$(1100, 2500); 200(1100) + 250(2500) = 845,000;$
maximum
$(3000, 600); 200(3000) + 250(600) = 750,000$
Plant 1100 acres of iceberg and 2500 acres of romaine
to yield a maximum profit of \$845,000.

20. Let x represent the number of pairs of adult skates and
    let y represent the number of pairs of child skates.
    Profit = $P = 40x + 20y$
    $x + y \leq 80$ (can stock at most 80)
    $x \leq 50$ (can order up to 50
    $y \leq 50$ pairs of each type)

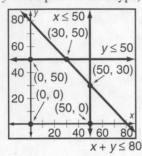

$P = 40x + 20y$
$(0, 0); 40(0) + 20(0) = 0$
$(0, 50); 40(0) + 20(50) = 1000$
$(30, 50); 40(30) + 20(50) = 2200$
$(50, 30); 40(50) + 20(30) = 2600$
$(50, 0); 40(50) + 20(0) = 2000$
She must stock and sell 50 adult and 30 child skates for
a maximum profit of \$2600.

21. Let x represent the number of statisticians and let
    y represent the number of interviewers.

Type of Worker	Time Spent Collecting	Time Spent Analyzing	Cost
Statistician	$10x$	$30x$	$900x$
Interviewer	$30y$	$10y$	$550y$
Totals	$10x + 30y$	$30x + 10y$	$900x +$ $550y$

$10x + 30y \geq 210$ (at least 210 hours)
$30x + 10y \geq 150$ (at least 150 hours)
$x \geq 0; y \geq 0$
Cost = $C = 900x + 550y$

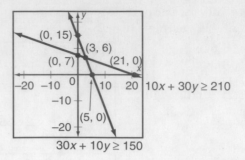

$C = 900x + 550y$
$(0, 15); 900(0) + 550(15) = 8250$
$(3, 6); 900(3) + 550(6) = 6000$
$(21, 0); 900(21) + 550(0) = 18,900$
3 statisticians and 6 interviewers for a cost of \$6000.

22. Let x represent number of Morning Mix packages and
    y represent number of Coffee Break Mix packages.

22a. $\dfrac{\text{pounds decaffeinated available}}{\text{pounds for mix}} = \dfrac{90}{4} = 22.5$
    22 Coffee Breaks for a profit of $22(40) = \$880$

22b. $\dfrac{\text{pounds regular available}}{\text{pounds for mix}} = \dfrac{50}{4} = 12.5$
    12 Morning Mixes for a profit of $12(30) = \$360$

22c. $4x + 1y \leq 50$ (50 lbs. regular total)
    $1x + 4y \leq 90$ (90 lbs. decaf. total)
    $x \geq 0, y \geq 0$ (reality constraints)
    $y \leq 22$ (from part a)
    $x \leq 12$ (from part b)

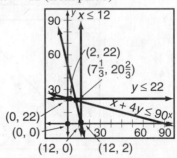

$P = 30x + 40y$
$(0, 0); 30(0) + 40(0) = 0$
$(0, 22); 30(0) + 40(22) = 880$
$(2, 22); 30(2) + 40(22) = 940$
$(7, 20); 30(7) + 40(20) = 1010$
$(12, 2); 30(12) + 40(2) = 440$
$(12, 0); 30(12) + 40(0) = 360$
7 Morning Mixes and 20 Coffee Breaks for a profit of
\$1010.

22d. All the regular coffee will be used but 10 lbs of
    decaffeinated will be left.

23. The region does not have an upper boundary.

24–30. Let x represent the number of tons of ore A and y represent the number of tons of ore B.

Ore	Tons of Iron	Tons of Copper	Tons of Zinc
Grade A	$0.15x$	$0.05x$	$0.02x$
Grade B	$0.10y$	$0.10y$	$0.08y$
TOTAL	$0.15x + 0.10y$	$0.05x + 0.10y$	$0.02x + 0.08y$

24. $0.15x + 0.10y \geq 20$

25. $0.05x + 0.10y \geq 15$

26. $0.02x + 0.08y \geq 8$

27.

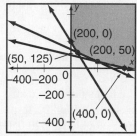

28. $(0, 200)$; $(50, 125)$; $(200, 50)$; $(400, 0)$

29. $C(\text{cost}) = 30x + 40y$

30. $(0, 200)$; $30(0) + 40(200) = 8000$
$(50, 125)$; $30(50) + 40(125) = 6500$
$(200, 50)$; $30(200 + 40(50) = 8000$
$(400, 0)$; $30(400) + 40(0) - 12000$.
$(50, 125)$; 50 tons of ore A and 125 tons of ore B for a cost of $6500

MIXED REVIEW

31. B; $\dfrac{98 + 89 + 88 + 80 + x}{5} = 90$; $\dfrac{355 + x}{5} = 90$;
$355 + x = 450$; $x = 95$

32. $2x + 8 = -x - 1$; $2x + x + 8 = -x + x - 1$;
$3x + 8 = -1$; $3x + 8 - 8 = -1 - 8$; $3x = -9$;
$x = -3$

33. $-3y + 4 \leq 25$; $-3y + 4 - 4 \leq 25 - 4$; $-3y \leq 21$;
$y \geq -7$

34. $|2x + 3| = 9$
$2x + 3 = 9$ or $2x + 3 = -9$
$2x + 3 = 9$; $2x = 6$; $x = 3$ or
$2x + 3 = -9$; $2x = -12$; $x = -6$
$x = 3$ or $x = -6$

35. $3x + 2y = 16$
$-4x + y = -3$
Multiply the second equation by -2 and add the two equations.
$\quad 3x + 2y = 16$
$\quad 8x - 2y = 6$
$\quad 11x = 22$
$\quad\quad x = 2$
$3(2) + 2y = 16$; $6 + 2y = 16$; $2y = 10$; $y = 5$
$(2, 5)$

36. $-2x + 3y = 11$
$\quad x - y = -4$
Multiply equation two by 2 and add the two equations.
$\quad -2x + 3y = 11$
$\quad\ \ 2x - 2y = -8$
$\quad\quad\quad y = 3$
$x - (3) = -4$; $x = -1$; $(-1, 3)$

37. $y = 5 - 2x$
$x = 5 - 2y$; $x - 5 = -2y$; $\dfrac{x - 5}{-2} = y$;
$y = -\dfrac{1}{2}x + \dfrac{5}{2}$

ALGEBRAWORKS

1. $I = C + 3D$; Total income equals one dollar times the number of pounds of Product C plus three dollars times the number of pounds of product D.

2. This inequality expresses the amount of product A that would be needed to make both C and D. No more than one pound of product A is to be used per batch.

3. This inequality expresses the amount of product B that would be needed to make both C and D. No more than one pound of product B is to be used per batch.

4. $C \geq 0$, $D \geq 0$; these are the reality constraints

5.

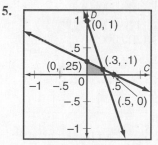

$I = C + 3D$
$(0, 0)$; $0 + 3(0) = 0$
$(0, 0.25)$; $0 + 3(0.25) = 0.75$
$(0.\overline{33}, 0)$; $0.\overline{33} + 3(0) = 0.\overline{33}$
$(0.3, 0.1)$; $0.3 + 3(0.1) = 0.6$

6. Produce 0 pounds of Product C and 0.25 pounds of Product D to maximize the income.

EXPLORE THE PROBLEM

Let x represent the amount invested in Bond A, and let y represent the amount invested in Bond B.

1.

Investment	Amount Invested	Interest Rate
Bond A	x	9.5%
Bond B	y	11%

2. $x + y \leq 200{,}000$ (investment of up to \$200,000)
   $x + y \geq 100{,}000$ (total investment no less than \$100,000)
   $x \geq 0; y \geq 0$ (reality constraints)

3. $y \leq 0.35(x + y)$ (at most 35% of the total)

4. $y \leq \dfrac{35}{100}x + \dfrac{35}{100}y; y - \dfrac{35}{100}y \leq \dfrac{35}{100}x;$
   $\dfrac{65}{100}y \leq \dfrac{35}{100}x; y \leq \dfrac{35}{65}x; y \leq \dfrac{7}{13}x$

5. $9.5\% = 0.095$
   $0.095x$

6. $11\% = 0.11$
   $0.11y$

7. $R = 0.095x + 0.11y$

8. The return is combined interest earned by each investment.

9. $X_{min} = 0; X_{max} = 200{,}000; X_{scl} = 20{,}000;$
   $Y_{min} = 0; Y_{max} = 200{,}000; Y_{scl} = 20{,}000$
   This window will include all of the vertices of the feasible region.

10.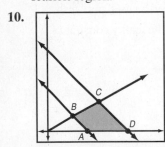
   Vertices
   (65,000, 35,000)
   (100,000, 0)
   (130,000, 70,000)
   (200,000, 0)

11. $R = 0.095x + 0.11y$
   (100,000, 0); $0.095(100{,}000) + 0.11(0) = 9500$
   (200,000, 0); $0.095(200{,}000) + (0.11)(0) = 19{,}000$
   (65,000, 35,000);
   $0.095(65{,}000) + 0.11(35{,}000) = 10{,}025$
   (130,000, 70,000);
   $0.095(130{,}000) + 0.11(70{,}000) = 20{,}050$
   maximum 20,050 at (130,000, 70,000)

12. \$130,000 of Bond A and \$70,000 of Bond B

13. The city will earn \$20,050.

INVESTIGATE FURTHER

14. $\dfrac{3}{4}\% = 0.0075$
   $0.0075x$

15. $\dfrac{1}{2}\% = 0.005$
   $0.005y$

16. $F = 0.0075x + 0.005y$

17. (100,000, 0); $0.0075(100{,}000) + (0.005)(0) = 750$
   (200,000, 0); $0.0075(200{,}000) + (0.005)(0) = 1500$
   (65,000, 35,000)
   $= (0.0075)(65{,}000) + (0.005)(35{,}000) = 662.50$
   (130,000, 70,000)
   $= (0.0075)(130{,}000) + (0.005)(70{,}000) = 1325$
   minimum \$662.50 at (65,000, 35,000)

18. There is a fee of \$662.50 if \$65,000 of Bond A and \$35,000 of Bond B are purchased.

19. No. The maximum value on the return of the investment and the minimum value on the fees occur at two different points in the feasible region.

20. Answers will vary. While the fee is minimized at (65,000, 35,000), the return is only \$10,025. At (130,000, 70,000), the return is doubled at \$20,050 and the fee is also doubled to \$1325. Students may conclude the extra income justifies the higher fee expense.

APPLY THE STRATEGY

21a. Let m represent money invested in money market funds and b represent money invested in bonds.
   $m + b \leq 31{,}500$ (has \$31,500 to invest)

21b. $10{,}000 \leq m \leq 20{,}000$ (no less than \$10,000 and no more than \$20,000)
   $b \leq 0.75m$ (will invest up to 75% of the money market amount in bonds)
   $b \geq 0, m \geq 0$ (reality constraints)

21c. Return on money market $= 0.075m$
   Return on bonds $= 0.105b$
   Total return $= R = 0.075m + 0.105b$

21d.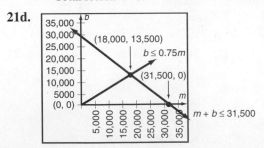

   $R = 0.075m + 0.105b$
   (0, 0); $0.075(0) + 0.105(0) = 0$
   (31,500, 0); $0.075(31{,}500 + 0.105(0) = 2362.5$
   (18,000, 13,500);
   $0.075(18{,}000) + 0.105(13{,}500) = 2767.5$
   Rhea should invest \$18,000 in money markets and \$13,500 in bonds for a maximum return of \$2767.50.

22. Let x represent the amount of money invested at the bank and let y represent the amount of money invested in a bond fund.

$x + y \le 100,000$ (Their savings total \$100,000)
$y \le 0.4(100,000)$ (no more than 40% in bonds)
$x \ge 0, y \ge 0$ (reality constraints)

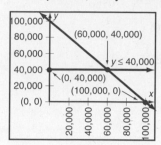

$R = 0.04x + 0.1y$
$(0, 0); 0.04(0) + 0.1(0) = 0$
$(0, 40,000); 0.04(0) + 0.1(40,000) = 4000$
$(100,000, 0); 0.04(100,000) + (0.1)(0) = 4000$
$(60,000, 40,000); 0.04(60,000) + (0.1)(40,000) = 6400$
They should invest \$60,000 in the bank account and \$40,000 in the bond fund for a return of \$6400.

23. Let a represent the amount invested in Bond A and let b represent the amount invested in Bond B.

$a + b \le 60,000$ (has \$60,000 to invest)
$a + b \ge 30,000$ (total investment at least \$30,000)
$b \ge 15,000$ (at least $\frac{1}{4}$ of her money in Bond B)
$a > 30,000$ (no less than $\frac{1}{2}$ in Bond A)
$a \ge 0; b \ge 0$ (reality constraints)

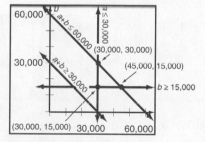

23a. $R = 0.05a + 0.1b$
$(30,000, 15,000); 0.05(30,000) + 0.1(15,000) = 3000$
$(45,000, 15,000); 0.05(45,000) + 0.1(15,000) = 3750$
$(30,000, 30,000); 0.05(30,000) + 0.1(30,000) = 4500$
She should invest \$30,000 in Bond A and \$30,000 in Bond B to maximize her return of \$4500.

23b. $F = 0.01a + 0.008b$
$(30,000, 15,000); 0.01(30,000) + 0.008(15,000) = 420$
$(45,000, 15,000); 0.01(45,000) + 0.008(15,000) = 570$
$(30,000, 30,000); 0.01(30,000) + 0.008(30,000) = 540$
She should invest \$30,000 in Bond A and \$15,000 in Bond B to minimize her fees of \$420.

24. Activity; answers will vary.

REVIEW PROBLEM SOLVING STRATEGIES

1a. The new number has six digits; the new number is $10N + 1$

1b. The new number is $100,000 + N$.

1c. $10N + 1 = 3(100,000 + N)$

1d. $10N + 1 = 3(100,000 + N)$
$10N + 1 = 300,000 + 3N$
$7N + 1 = 300,000; 7N = 299999; N = 42,857$

2a. Let w represent digit one, x represent digit two, y represent digit three and z represent digit four. From the given information we can derive four equations:

1) $w + x + y + z = 14$ 2) $\dfrac{\begin{array}{c}zxyw\\-wxyz\end{array}}{2997}$

3) $\dfrac{\begin{array}{c}wyxz\\-wxyz\end{array}}{90}$ 4) $\dfrac{\begin{array}{c}wyxz\\+wxyz\end{array}}{4780}$

From **4)** $z + z = 0$ *or* $z + z = 10$

If $z + z = 0$;

From **2)** $\dfrac{\begin{array}{c}0xyw\\-wxy0\end{array}}{2997}$ $0 - w \ne 2$, so $z \ne 0$

So, $z + z = 10; z = 5$

From **2)** $\dfrac{\begin{array}{c}5xyw\\-wxy5\end{array}}{2997}$ is equivalent to $\dfrac{\begin{array}{c}2997\\+wxy5\end{array}}{5xyw}$

Since $7 + 5 = 12, w = 2$.

From **3)** $\dfrac{\begin{array}{c}2yx5\\-2xy5\end{array}}{90}$ due to borrowing
$\downarrow$
$(10 + x) - y = 9$
$x - y = -1$
$x = y - 1$

From **1)** $w + x + y + z - 2 + x + y + 5 = 14$;
$x + y = 7$

Solve the system: $\begin{cases} x = y - 1 \\ x + y = 7 \end{cases}$

$(y - 1) + y = 7; 2y = 8; y = 4; x = 3$
The number is 2345.

2b. Answers will vary.

3. Kim; the order is Hayley, Kim, Terry, Pat, Erin

Chapter Review, pages 352–353

1. e **2.** f **3.** b **4.** c **5.** a **6.** d

7. $y \le 2x + 6; (1, 10)$
$10 \le 2(1) + 6; 10 \not\le 8$; no

8. $4y \ge 3x; (0, 0)$
$4(0) \ge 3(0); 0 \ge 0$; yes

9. $5x - y > 3; (1, 3)$
$5(1) - 3 \overset{?}{<} 3; 2 \not> 3;$ no

10.
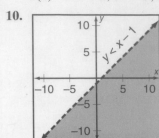

$y < x - 1$
Check: $(0, 0); 0 \overset{?}{<} 0 - 1;$
$0 \leq -1$
$(2, -2); -2 \overset{?}{<} 2 - 1;$
$-2 < 1;$ yes
The solution to $y < x - 1$
is the shaded area below the
dashed line $y = x - 1$.

11.
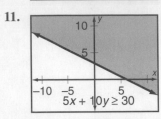

$5x + 10y \geq 30$
Check: $(0, 0);$
$5(0) + 10(0) \overset{?}{\geq} 30;$
$30 \not> 30$
$(5, 5); 5(5) + 10(5) \overset{?}{\geq} 30;$
$75 \geq 30;$ yes

12.
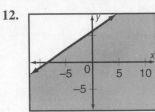

$4y - 3x \leq 24$
Check: $(0, 0);$
$4(0) - 3(0) \overset{?}{\leq} 24;$
$0 \leq 24;$ yes
$(0, 10); 4(10) - 3(0) \overset{?}{\leq} 24;$
$40 \not\leq 24$

13.
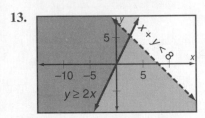

14.
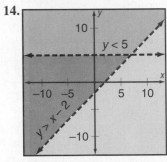

15.
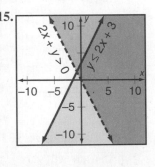

16. Let x represent the larger integer and let y represent the smaller integer.

$\begin{cases} x + y < 10 \\ x - y > 4 \end{cases}$

x and y must be
positive integers
so we only consider
points from the
double shaded area
in the first quadrant.
$(6, 1), (7, 1), (8, 1), (7, 2)$

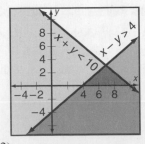

17. $x^2 + 2x - 80 \geq 0; x = 9$
$(x + 10)(x - 8) \geq 0$
$(19)(1) \overset{?}{\geq} 0$
$19 \geq 0;$ yes

18. $x^2 + 6x - 2 > 0; x = 0$
$(0)^2 + 6(0) - 2 \overset{?}{>} 0; -2 \not> 0;$ no

19. $4x^2 - 8x - 2.25 < 0; x = -\frac{1}{4}$
$4\left(-\frac{1}{4}\right)^2 - 8\left(-\frac{1}{4}\right) - 2.25 \overset{?}{<} 0$
$0 \not< 0;$ no

20. $3x^2 + 2x - 5 \leq 0; x = \frac{1}{2}$
$(3x + 5)(x - 1) \leq 0$
$\left[3\left(\frac{1}{2}\right) + 5\right]\left[\left(\frac{1}{2}\right) - 1\right] \overset{?}{\leq} 0$
$-\frac{13}{4} \leq 0;$ yes

21. $y > 2x^2 - 7x - 3; (3, 1)$
$1 \overset{?}{>} 2(3)^2 - 7(3) - 3$
$1 > -6;$ yes

22. $y > 2x^2 - 7x - 3; (0, -3)$
$-3 \overset{?}{>} 2(0)^2 - 7(0) - 3$
$-3 \not> -3;$ no

23. $y > 2x^2 - 7x - 3; (5, 15)$
$15 \overset{?}{>} 2(5)^2 - 7(5) - 3$
$15 > 12;$ yes

24. $y > 2x^2 - 7x - 3; (8, -1)$
$-1 \overset{?}{>} 2(8)^2 - 7(8) - 3$
$-1 \not> 69;$ no

25. $y \leq x^2 - 4x + 1$
$y = x^2 - 4x + 1$
$y = (x^2 - 4x + 4) + 1 - 4$
$y = (x - 2)^2 - 3$
Vertex $= (2, -3)$
Let $x = 0$
$y = (0)^2 - 4(0) + 1$
$y = 1; (0, 1)$
Let $y = 0$
$0 = x^2 - 4x + 1$
$x = \dfrac{-(-4) \pm \sqrt{(-4)^2 - 4(1)(1)}}{2(1)}$
$= \dfrac{4 \pm \sqrt{12}}{2} = 2 \pm \sqrt{3}$
x-intercepts $= (2 + \sqrt{3}, 0); (2 - \sqrt{3}, 0)$
Test $(0, 0): y \leq x^2 - 4x + 1$
$0 \overset{?}{\leq} (0)^2 - 4(0) + 1$
$0 \leq 1;$ yes

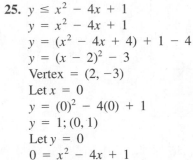

26. $y > -0.25x^2$
$y = -0.25x^2$
$y = -0.25(x - 0)^2 + 0$
Vertex $= (0, 0)$

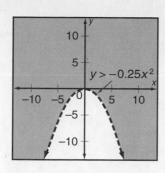

Let $x = 0$
$y = -0.25(0^2)$
$y = 0$
x- and y-intercept $= (0, 0)$
Check $(0, -5)$:
$-5 \overset{?}{>} -0.25(0)^2; -5 \not> 0$
$(0, 5): 5 \overset{?}{>} -0.25(0)^2;$
$5 > 0$; yes

27. $y < 9 - 16x^2$
$y = -16(x - 0)^2 + 9$
Vertex $= (0, 9)$
Let $x = 0; y = 9 - 16(0)^2; y = 9$
y-intercept: $(0, 9)$
Let $y = 0; 0 = 9 - 16x^2;$
$0 = (3 - 4x)(3 + 4x)$
$3 - 4x = 0$ or $3 + 4x = 0$
$x = 0.75$ or $x = -0.75$
x-intercepts: $(0.75, 0); (-0.75, 0)$

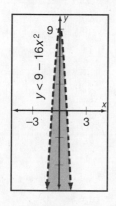

Check $(0, 0): 0 \overset{?}{<} 9 - 16(0)^2$
$0 < 9$; yes
$(3, 0): 0 \overset{?}{<} 9 - 16(3)^2$
$0 \not< -135$

28. $x \geq 0, y \geq 0, y \geq x - 4, 7y \leq -2x + 35$
$(0, 0), (0, 5), (7, 3), (4, 0)$

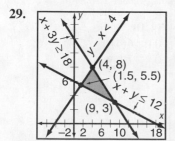

29.

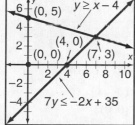

$P = -3x + 4y$
$(4, 8); -3(4) + 4(8) = 20$

$(9, 3); -3(9) + 4(3) = -15$
$(1.5, 5.5); -3(1.5) + 4(5.5) = 17.5$
maximum 20 at $(4, 8)$

30. Let a represent the amount of money invested in Bond A and let b represent the amount invested in Bond B.
$a + b \leq 90,000$ (has \$90,000 to invest)

$b \geq 30,000$ (at least $\frac{1}{3}$ of his money in Bond B)

$a \leq 45,000$ (no more than $\frac{1}{2}$ of his money in Bond A)

$a + b \geq 50,000$ (total investment at least \$50,000)
$a \geq 0; b \geq 0$ (reality constraints)

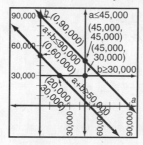

$R = 0.06a + 0.12b$
$(0, 90,000) = 0.06(0) + 0.12(90,000) = 10,800$
$(0, 60,000) = 0.06(0) + 0.12(60,000) = 7200$
$(20,000, 30,000 = 0.06(20,000) +$
$(0.12(30,000) = 4800$
$(45,000, 30,000) = 0.06(45,000) +$
$(0.12(30,000) = 6300$
$(45,000, 45,000) = 0.06(45,000) +$
$(0.12(45,000) = 8100$
To maximize his return he should invest \$0 in Bond A and \$90,000 in Bond B.

Chapter Assessment, pages 364–365

CHAPTER TEST

1. $y > 3x + 1; (3, 5)$
$5 \overset{?}{>} 3(3) + 1; 5 \not> 10$; no

2. $y < 2x - 5; (2, -4)$
$-4 \overset{?}{<} 2(2) - 5; -4 < -1$; yes

3. 0; the absolute value of an expression is never negative

4.

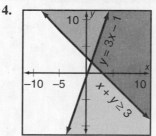

5.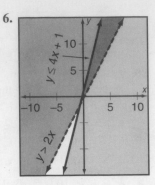

6.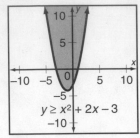

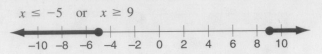
$$x \leq -5 \quad \text{or} \quad x \geq 9$$

7. C; $x^2 + 2x - 1 < 0; x = 0$
$(0) + 2(0) - 1 \overset{?}{<} 0; -1 < 0;$ yes
$x^2 + 2x - 1 < 0; x = -1$
$(-1)^2 + 2(-1) - 1 \overset{?}{<} 0; -2 < 0;$ yes
$x^2 + 2x - 1 < 0; x = -3$
$(-3)^2 + 2(-3) - 1 \overset{?}{<} 0; 2 \not< 0;$ no

8. $(x + 4)(x - 3) \leq (x + 2)(x - 5)$
$x^2 + x - 12 \leq 2 - 3x - 10$
$x - 12 \leq -3x - 10$
$4x - 12 \leq -10$
$4x \leq 2; x \leq \dfrac{1}{2}$

9. $(x + 6)(x - 2) \geq (x + 3)(x - 1)$
$x^2 + 4x - 12 \geq x^2 + 2x - 3$
$4x - 12 \geq 2x - 3$
$2x - 12 \geq -3$
$2x \geq 9; x \geq \dfrac{9}{2}$

10. $x \geq 0, y \geq 0, y \geq 2x - 6, y \leq -0.5x - 7$

11. $x^2 + 3x - 40 \leq 0$
$(x + 8)(x - 5) = 0$
$x + 8 = 0 \quad \text{or} \quad x - 5 = 0$
$x = -8 \quad \text{or} \quad x = 5$

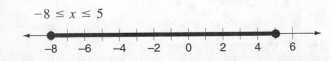

$-8 \leq x \leq 5$

12. $x^2 - 4x - 45 \geq 0$
$(x - 9)(x + 5) = 0$
$x - 9 = 0 \quad \text{or} \quad x + 5 = 0$
$x = 9 \quad \text{or} \quad x = -5$

13. $y \geq x^2 + 2x - 3$
$y = x^2 + 2x - 3$
$y = (x^2 + 2x + 1) - 3 - 1$
$y = (x + 1)^2 - 4$
Vertex $= (-1, -4)$

Let $x = 0; y = (0)^2 + 2(0) - 3; y = -3$
y-intercept $= (0, -3)$
Let $y = 0; 0 = x^2 + 2x - 3; 0 = (x + 3)(x - 1);$
$x = -3 \quad \text{or} \quad 1$
x-intercepts $= (-3, 0) \quad \text{and} \quad (1, 0)$
Check $(0, 0); 0 \overset{?}{\geq} (0)^2 + 2(0) + 3; 0 \not\geq 3;$ no
$(2, 0); 0 \overset{?}{\geq} (2)^2 + 2(2) + 3; 11 \geq 3;$ yes

14. $y \leq 6x^2 - 5x - 6$
$y = 6x^2 - 5x - 6$
$y = 6\left(x^2 - \dfrac{5}{6}x + \dfrac{25}{144}\right) - 6 - \dfrac{25}{24}$
$y = 6\left(x - \dfrac{5}{12}\right)^2 - \dfrac{169}{24}$

Vertex $= \left(\dfrac{5}{12}, -\dfrac{169}{24}\right)$
Let $x = 0; y = 6(0)^2 - 5(0) - 6; y = -6$
y-intercept $= (0, -6)$

Let $y = 0; 0 = 6x^2 - 5x - 6 = (3x + 2)(2x - 3);$

$3x + 2 = 0 \quad \text{or} \quad 2x - 3 = 0; x = -\dfrac{2}{3} \quad \text{or} \quad x = \dfrac{3}{2}$

x-intercepts $= \left(-\dfrac{2}{3}, 0\right) \quad \text{and} \quad \left(\dfrac{3}{2}, 0\right)$

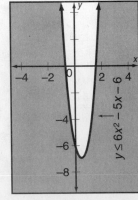

Check:
$(0, 0);$
$0 \overset{?}{\leq} 6(0)^2 - 5(0) - 6;$
$0 \not\leq -6;$ no
$(4, 0);$
$0 \overset{?}{\leq} 6(4)^2 - 5(4) - 6;$
$0 \leq 70;$ yes

15.

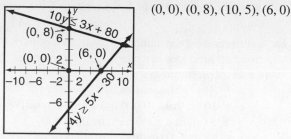

$(0, 0), (0, 8), (10, 5), (6, 0)$

16. Answers will vary. Students should mention assigning variables, identifying constraints, writing the objective function and graphing the system of inequalities resulting from the constraints. The student should discuss identifying the vertices of the feasible region, evaluating the function at each vertex, and determining the maximum and minimum values.

17.

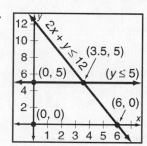

$P = 3x + 2y$
$(0, 0); 3(0) + 2(0) = 0$
$(0, 5); 3(0) + 2(5) = 10$
$(6, 0); 3(6) + 2(0) = 18$
$(3.5, 5); 3(3.5) + 2(5) = 20.5$
maximum 20.5 at $(3.5, 5)$

18.

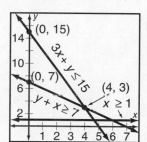

$C = 20x + 30y$
$(0, 15);$
$20(0) + 30(15) = 450$
$(0, 7); 20(0) + 30(7) = 210$
$(4, 3); 20(4) + 30(3) = 170$
minimum 170 at $(4, 3)$

19. Since the region is not bounded in the first quadrant, there will be no maximum.

20. Let a represent amount invested in AAA bonds and let b represent amount invested in B bonds.
$a + b \leq 12{,}000$ (has $12,000 to invest)
$a \geq 2b$ (at least twice as much AAA bonds as B bonds)
$a \geq 0; b \geq 0$ (reality constraints)

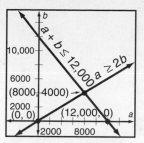

$R = 0.06a + 0.09b$
$(0, 0); 0.06(0) + 0.09(0) = 0$
$(12{,}000, 0); 0.06(12{,}000) + 0.09(0) = 720$
$(8000, 4000); 0.06(8000) + 0.09(4000) = 840$
Stacey should invest $8000 in AAA bonds and $4000 in B bonds for a return of $840.

21. Let b represent bumper stickers and let h represent hankies.
$b \leq 300$ (no more than 300)
$h < 150$ (no more than 150)
$b + h \geq 200$ (must sell at least 200 items)
$b \geq 0, h \geq 0$ (reality constraints)

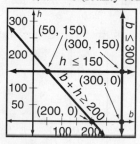

$C(\text{cost}) = b + 1.25h$
$(50, 150) = 50 + 1.25(150) = 237.5$
$(200, 0) = 200 + 1.25(0) = 200$
$(300, 0) = 300 + 1.25(0) = 300$
$(300, 150) = 300 + 1.25(150) = 487.5$
To minimize costs the students should sell 200 bumper stickers and 0 hankies.
$P(\text{profit}) = (2 - 1)b + (2.5 - 1.25)h$
$\qquad P = b - 1.25h$
This is the same as the cost function which had a maximum value of 487.5 at $(300, 150)$.
They should sell 300 bumper stickers and 150 hankies to maximize profit.

PERFORMANCE ASSESSMENT

3. Answers will vary. An example is
$$x \geq 1, x \leq 7, y \leq 4, y \leq -\frac{2}{3}x + \frac{14}{3}.$$

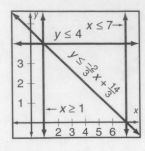

1.

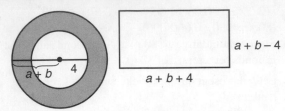

$$\frac{A_{\text{shaded region of circle}}}{A_{\text{rectangle}}} = \frac{\pi(a + b)^2 - 16\pi}{(a + b + 4)(a + b - 4)} =$$

$$\frac{\pi[(a + b)^2 - 16]}{a^2 + 2ab + b^2 - 16} = \frac{\pi(a^2 + 2ab + b^2 - 16)}{a^2 + 2ab + b^2 - 16} = \pi$$

2.

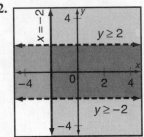

3.

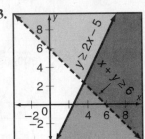

4. To find the difference in cost we only need to consider row 3 in the Miles Driven matrix and the Cost Per Mile matrix. The only change will come from Jan's driving habits (row 3 of the Miles Driven matrix)
Find the cost if Jan drove both cars:
Multiply Row 3 by Cost Matrix:

$$[1790 \quad 1928] \times \begin{bmatrix} 0.23 \\ 0.31 \end{bmatrix} = (3718)(0.23) \times$$

$(1790)(0.23) + (1928)(0.31) = \$1009.38$

Now determine the cost if Jan always drove Car 1.

$$[3718 \quad 0] \times \begin{bmatrix} 0.23 \\ 0.31 \end{bmatrix} = (3718)(0.23) \times$$

$(0)(0.31) = \$855.14$
The difference between these two costs is the savings:
$1009.38 - 855.14 = \$154.24$

5. Graph the line $y = 3x + 7$ to divide the plane into two half-planes. Since the given inequality does not contain the equality relation, draw the line as dotted to indicate that it is not to be included in the solution.
Test a point from one of the half-planes in the inequality. If the point satisfies the inequality, shade that half-plane for the solution set. If the test point does not

satisfy the inequality, shade the other half-plane for the solution.

6a. Let c represent the number of thousands of barrels of crude oil produced in a week and let d represent the diesel oil produced in a week.

$c + d \leq 30$ (at most 30,000)
$c \geq 10$ (at least 10,000 must be crude oil)
$d \geq 5$ (at least 5000 must be diesel oil)
$c \geq 0, \ d \geq 0$ (reality constraints)

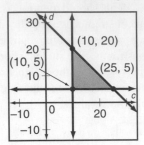

6b. $P = 18.50c + 20.75d$
$(10, 20); \ 18.50(10) + 20.75(20) = 600$
$(10, 5); \ 18.50(10) + 20.75(5) = 288.75$
$(25, 5); \ 18.50(25) + 20.75(5) = 566.25$
maximum 600 at (10, 20)
The production output that will yield a maximum revenue is 10,000 barrels of crude oil an 20,000 barrels of diesel oil.

7. Let g represent gallons of gas and let m represent miles driven.
$g = km; \ 11.5 = k(150); \ k = 0.0\overline{77}$
$g = km; \ g = (0.0\overline{77})(375) = 28.75$

8. $x^2 - 5x \geq 6$
$x^2 - 5x - 6 \geq 0$
$x^2 - 5x - 6 = 0$
$(x - 6)(x + 1) = 0$
$x = 6 \quad \text{or} \quad x = -1$

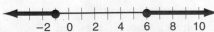

$x \leq -1 \quad \text{or} \quad x \geq 6$

9. $y \geq 0$ (the triangle is bounded by the solid line $y = 0$)
$4y \leq -5x + 20$ (the triangle is located to the left of the solid line $y = \frac{-5}{4}x + 5$)
$2y \leq 5x + 10$ (the triangle is located to the right of the solid line $y = \frac{5}{2}x + 5$)

10. D; $4x^2 - 12x + 9 = 0$
Factor: $(2x - 3)(2x - 3) = 0$
$$2x - 3 = 0 \qquad 2x - 3 = 0$$
$$x = \frac{3}{2}$$
The sum of its solutions is not 12.
Find the discriminant
$$d = \sqrt{(-12)^2 - 4(4)(9)}$$
$$d = \sqrt{144 - 144} = 0$$
The x-intercept is $\left(\frac{3}{2}, 0\right)$

11. Jon has a total score of 1876 in his first 8 games.
Let x represent the total of his last 4 games.
$$\frac{x + 1876}{12} = 240; x + 1876 = 2880; x = 1004$$
$$\text{average} = \frac{x}{4} = \frac{1004}{4} = 251$$

12. $y \le -x^2 + 5x - 4$
$$y = -\left(x^2 - 5x + \frac{25}{4}\right) - 4 + \frac{25}{4}$$
$$y = -\left(x - \frac{5}{2}\right)^2 + \frac{9}{4}$$

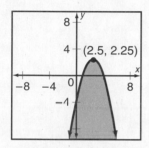

Vertex = (2.5, 2.25)
Let $x = 0; y = -(0)^2 + 5(0) - 4 = -4$
y-intercept = (0, -4)
Let $y = 0; 0 = -x^2 + 5x - 4 = (-x + 1)(x - 4)$;
$x = 1$ or $x = 4$; x-intercepts = (1, 0), (4, 0)
Test: $(0, 0); 0 \overset{?}{\le} -(0)^2 + 5(0) - 4; 0 \not\le -4$
$(2, 0); 0 \overset{?}{\le} -(2)^2 + 5(2) - 4; 0 \le 2;$ yes

Standardized Test, page 367

1. B; Let x represent the number of blue marbles and the number of yellow marbles. Then:

red	white	blue	yellow
$\overline{2x + 8}$	$\overline{2x}$	$\overline{x}$	$\overline{x}$

$(2x + 8) + (2x) + x + x = 56 = $ total marbles
$6x + 8 = 56; 6x = 48; x = 8$
8 yellow, 8 blue, 16 white, 24 red
Probability $= \frac{8}{56} = \frac{1}{7};$ 1:7
marble is blue

2. E; $-21 \cdot 4 - (-7 \cdot 12) = -84 + 84 = 0$

3. C; The region between the solid lines $y = 1$ and $y = -1$ is shaded.

This area satisfies the inequality $|y| \le 1$.
The region between the solid lines $x = 1$ and $x = -1$ is shaded.
This area satisfies the inequality $|x| \le 1$.

4. C; $\dfrac{\text{area } ABEF}{\text{area } FECD} = \dfrac{(x + 3)(x + 1)}{x(x + 3)} = \dfrac{x + 1}{x} = \dfrac{4}{1};$
$4(x) = 1(x + 1); 4x = x + 1; 3x = 1; x = \dfrac{1}{3}$
$DC = x + 3 = 3\dfrac{1}{3};$
$BC = x + 1 + x = 2x + 1 = 1\dfrac{2}{3}$
The dimensions are $3\dfrac{1}{3} \times 1\dfrac{2}{3}$

5. E; Consider the graph $f(x) = x^3$

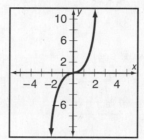

I. The graph is one-to-one. No horizontal line hits the graph more than once.
II. $F^{-1}(x)$ exists and is a function: $y = x^3$
$$x = y^3$$
$$y = \sqrt[3]{x} = x^{\frac{1}{3}};$$
$$f^{-1}(x) = x^{\frac{1}{3}}$$
III. $f(x)$ has point symmetry. $f(x)$ is symmetric about the point $(0, 0)$

6. C; $2.1 \ne [2]; [2] = 2$

7. A;
$$2x^2 + 5x - 3 \le 0$$
$$2x^2 + 5x - 3 = 0$$
$$(2x - 1)(x + 3) = 0$$
$$2x - 1 = 0 \quad \text{or} \quad x + 3 = 0$$
$$x = \frac{1}{2} \quad \text{or} \quad x = -3$$

$2x - 1$	$-$	$-$	$+$
$x + 3$	$-$	$+$	$+$
$(2x - 1)(x + 3)$	$+$	$-$	$+$

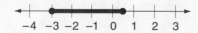

8. B; $h = -16t^2 + v_0 t + h_0; h_0 = 0; v_0 = 48$
$h = -16t^2 + 48t + 0$
$$h = -16\left(t^2 - 3t + \frac{9}{4}\right) + 36$$
$$h = -16\left(t - \frac{3}{2}\right)^2 + 36$$

Vertex $= \left(\frac{3}{2}, 36\right)$

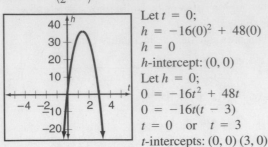

Let $t = 0$;
$h = -16(0)^2 + 48(0)$
$h = 0$
h-intercept: $(0, 0)$
Let $h = 0$;
$0 = -16t^2 + 48t$
$0 = -16t(t - 3)$
$t = 0$ or $t = 3$
t-intercepts: $(0, 0)$ $(3, 0)$

The ball reaches its maximum height of 36 feet at $t = 1\frac{1}{2}$s.

9. D; $(x - 1)8x^2 + (x - 1)(-2x) + (x - 1)(-3) =$
$8x^3 - 8x^2 - 2x^2 + 2x - 3x + 3 =$
$8x^3 - 10x^2 - x + 3$

List the possible solutions of

$f(x) = 8x^3 - 10x^2 - x + 3, \frac{p}{q}$, where p is an integral

factor of 3 and q is an integral factor of 8.

$q \in (\pm 1, \ \pm 2, \ \pm 4, \ \pm 8)$
$p \in (\pm 1, \ \pm 3)$
possible solutions:

$\pm 1, \ \pm\frac{1}{2}, \ \pm\frac{1}{4}, \ \pm\frac{1}{8}, \ \pm 3, \ \pm\frac{3}{2}, \ \pm\frac{3}{4}, \ \pm\frac{3}{8}$

$f(x) = 8x^3 - 10x^2 - x + 3$

$f(1) = 0; f(-1) = -14; f\left(\frac{1}{2}\right) = 1; f\left(-\frac{1}{2}\right) = 0;$

$f\left(\frac{1}{4}\right) = 2.25; f\left(-\frac{1}{4}\right) = 2.5; f\left(\frac{1}{8}\right) = 2.73475;$

$f\left(-\frac{1}{8}\right) = 2.9531; f(3) = 123; f(-3) = -300;$

$f\left(\frac{3}{2}\right) = 6; f\left(-\frac{3}{2}\right) = -45; f\left(\frac{3}{4}\right) = 0; f\left(-\frac{3}{4}\right) = -5.25;$

$f\left(\frac{3}{8}\right) = 1.640625; f\left(-\frac{3}{8}\right) = 1.54685$

The solution is $\left\{1, -\frac{1}{2}, \frac{3}{4}\right\}$

Data Activity, page 369

1. $1993: 66.5 + 8.1 + 28.9 + 20.7 = 124.2$;
$\$1.242 \times 10^{11}$
$1994: 58.2 + 7.3 + 28.5 + 18.8 = 112.8$;
$\$1.128 \times 10^{11}$
$1995: 56.7 + 6.6 + 27.8 + 18.2 = 109.3$;
$\$109.3 \times 10^{11}$

2. $58.2x = 32.0, x \approx 0.5498; 55.0\%$

3. $7.3 - 8.1 = -0.8; 6.6 - 7.3 = -0.7$;
$\dfrac{-0.8 + -0.7}{2} = -0.75; \$-0.75$ billion

4. $\dfrac{18.2}{14.5} = \dfrac{27.8}{x}, x = \dfrac{(27.8)(14.5)}{18.2} \approx 22.14; \$22.1$ billion

5. Answers will vary.

Lesson 8.1, pages 371–373

THINK BACK

1. $y^4 \cdot y^5 = y^{4+5} = y^9$ 2. $\dfrac{n^6}{n} = n^{6-1} - n^5$

3. $\dfrac{k^7 \cdot k^9}{k^2} = k^{(7+9-2)} - k^{14}$

4. $x^2(x + 3)^6(x + 3)^4 = x^2(x + 3)^{6+4} = x^2(x + 3)^{10}$

5. $(5m - 7)^0 = 1$ 6. $\dfrac{14p^3q^5}{2n^3pq^2} = \dfrac{7p^{3-1}q^{5-2}}{n^3} = \dfrac{7p^2q^3}{n^3}$

7. $b^{-7} = \dfrac{1}{b^7}$ 8. $\dfrac{4x^{-3}y^4}{6x^{-5}y^{10}} = \dfrac{2x^{-3+5}}{3y^{10-4}} = \dfrac{2x^2}{3y^6}$

9. $\dfrac{a^5(a + 1)^{-9}}{(a + 1)^{-3}} = \dfrac{a^5}{(a + 1)^{-3+9}} - \dfrac{a^5}{(a + 1)^6}$

10. $(2h)^4 = 2^4h^4 = 16h^4$ 11. $5(l^3)^4 = 5l^{3\cdot 4} = 5e^{12}$

12. $(a^5b^3)^4 = a^{5\cdot 4}b^{3\cdot 4} = a^{20}b^{12}$

13. $\left(\dfrac{d^3z^0}{w^4}\right)^5 \dfrac{d^{3\cdot 5}z^{0\cdot 5}z^{0\cdot 5}}{w^{4\cdot 5}} = \dfrac{d^{15}z^0}{w^{20}} = \dfrac{d^{15}}{w^{20}}$

14. $\left(\dfrac{x^{-3}}{x^2}\right)^4\left(\dfrac{x^6}{x^{-5}}\right)^3 = \left(\dfrac{x^{-3\cdot 4}}{x^{2\cdot 4}}\right)\left(\dfrac{x^{6\cdot 3}}{x^{-5\cdot 3}}\right) = \left(\dfrac{x^{-12}}{x^8}\right)\left(\dfrac{x^{18}}{x^{-15}}\right) =$
$(x^{(-12-8)})(x^{[18-(-15)]}) = (x^{-20})(x^{33}) = x^{-20+33} = x^{13}$

15. $\left(\dfrac{a^{-2}}{b^{-3}}\right)^{-4} = \dfrac{a^{-2(-4)}}{b^{-3(-4)}} = \dfrac{a^8}{b^{12}}$

EXPLORE

16. $(\overline{AE})^2 + (\overline{DE})^2 = (\overline{AD})^2, 2^2 + 2^2 = (\overline{AD})^2$,
$8 = (\overline{AD})^2, \sqrt{8} = \overline{AD}$

17. $(\overline{DF})^2 + (\overline{CF})^2 = (\overline{CD})^2, 1^2 + 1^2 = (\overline{CD})^2$,
$\sqrt{2} = \overline{CD}$

18. 4 square units

19. area $ABCD$ = length $\times$ width, $\overline{AD} \cdot \overline{CD} = 4$,
$\sqrt{8} \cdot \sqrt{2} = \sqrt{16}$

20. Students may conjecture that
$\sqrt{8} \cdot \sqrt{18} = \sqrt{144}$.
They can confirm this with a sketch like the figure shown. By the Pythagorean theorem, $AD = \sqrt{8}$ and $DC = \sqrt{18}$. By counting squares, area $ABCD$ is 12 square units. Area $ABCD$ = length $\times$ width, $AD \cdot DC = 12$, $\sqrt{8} \cdot \sqrt{18} = \sqrt{144}$

21. $\sqrt{5} \cdot \sqrt{5} = \sqrt{25}$; Conjecture results from the idea $\sqrt{a} \cdot \sqrt{b} = \sqrt{ab}$. To check by calculator, solve left-hand side by $\boxed{\sqrt{}}$ 5 $\boxed{\times}$ $\boxed{\sqrt{}}$ 5 $\boxed{=}$ 5, then solve right-hand side by $\boxed{\sqrt{}}$ 25 $\boxed{=}$ 5.

On some calculators, students will obtain a value very close to 5, like 4.9999996, which they should recognize as an approximation resulting from rounding.

22. $(\overline{AG})^2 + (\overline{EG})^2 = (\overline{AE})^2, 1^2 + 2^2 = (\overline{AE})^2$,
$5 = (\overline{AE})^2, \sqrt{5} = (\overline{AE}); (\overline{EF})^2 + (\overline{CF})^2 = (\overline{EC})^2$,
$1^2 + 2^2 = (\overline{EC})^2, \sqrt{5} = \overline{EC}; \overline{AE} + \overline{EC} =$
$\sqrt{5} + \sqrt{5} = \overline{AC}; 1$
Explanation: Using the Pythagorean theorem in $\triangle AEG$ and $\triangle ECF$, determine that $\overline{AE} = \overline{EC} = \sqrt{5}$. Furthermore $\overline{AE} + \overline{EC} = \sqrt{5} + \sqrt{5} = \overline{AC}$.

23. $(\overline{AB})^2 + (\overline{BC})^2 = (\overline{AC})^2, 2^2 + 4^2 = (\overline{AC})^2$,
$20 = (\overline{AC})^2, \sqrt{20} = \overline{AC}$; No, $\sqrt{5} + \sqrt{5} \neq \sqrt{10}$ since from Question 22 $\sqrt{5} + \sqrt{5} = \overline{AC}$ and by substitution $\sqrt{5} + \sqrt{5} = \sqrt{20}$.

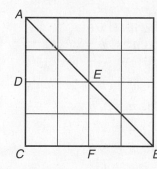

24. Students may conjecture that $\sqrt{8} + \sqrt{8} = \sqrt{32}$. They can confirm this with a sketch like the figure shown. By the Pythagorean theorem in $\triangle AED$ and $\triangle EBF$, $\overline{AE} = \overline{EB} = \sqrt{8}$. By the Pythagorean theorem in $\triangle ABC, \overline{AB} = \sqrt{32}$. Since $\overline{AE} + \overline{EB} = \overline{AB}$, $\sqrt{8} + \sqrt{8} = \sqrt{32}$.

25. true 26. false 27. true

28. true 29. false 30. true

31. 1, 2, 3, 4, 5, 6 32. 1, 2, 3, 4, 5, 6

33. Students should conjecture that raising a number to the $\frac{1}{2}$ power is the same as taking the positive square root of the number.

34. Test numbers will vary. Possible test numbers: $\sqrt{64} = (64)^{\frac{1}{2}} = 8$; $\sqrt{100} = (100)^{\frac{1}{2}} = 10$; yes

35. 1, 2, 3, 4, 5   **36.** 1, 2, 3, 4, 5

37. Students should conjecture that raising a number to the $\frac{1}{3}$ power is the same as taking the cube root of the number.

38. Test numbers will vary. Possible test numbers: $\sqrt[3]{512} = (512)^{\frac{1}{3}} = 8$; $\sqrt[3]{1000} = (10)^{\frac{1}{3}} = 10$; yes.

39. $(81)^{\frac{1}{4}} = 3$; $(32)^{\frac{1}{5}} = 2$

40. To raise a number to the $\frac{1}{4}$ power is the same as taking the fourth root of the number. To raise a number to the $\frac{1}{5}$ power is the same as taking the fifth root of the number.

41a. Each expression equals 27.

41b. Each expression equals 32.

41c. Each expression equals 64.

42. $(128^5)^{\frac{1}{7}}$; $(128^{\frac{1}{7}})^5$; Each equals 32.

MAKE CONNECTIONS

43. false; $\sqrt{9} + \sqrt{9} = 3 + 3 = 6$; $\sqrt{9 + 9} = \sqrt{18} = 3\sqrt{2}$

44. false; $\sqrt{25} - \sqrt{16} = 5 - 4 = 1$; $\sqrt{25 - 16} = \sqrt{9} = 3$

45. true; $\sqrt{4}\sqrt{9} = 2 \cdot 3 = 6$; $\sqrt{9 \cdot 4} = \sqrt{36} = 6$

46. true; $\dfrac{\sqrt{36}}{\sqrt{9}} = \dfrac{6}{3} = 2$; $\sqrt{\dfrac{36}{9}} = \sqrt{4} = 2$

47a. $a = \sqrt{b}$ $a^2 = b$ **47b.** $a = \sqrt[3]{b}$ $a^3 = b$

47c. $a = \sqrt[4]{b}$ $a^4 = b$ **47d.** $a = \sqrt[n]{b}$ $a^n = b$

48a. $a = \sqrt{b}$ $a = b^{\frac{1}{2}}$ **48b.** $a = \sqrt[3]{b}$ $a = b^{\frac{1}{3}}$

48c. $a = \sqrt[4]{b}$ $a = b^{\frac{1}{4}}$ **48d.** $a = \sqrt[n]{b}$ $a = b^{\frac{1}{n}}$

49. $k^{\frac{1}{2}}$ **50.** $7^{\frac{1}{3}}$ **51.** $5^{\frac{1}{6}}$

52. $11.2^{\frac{1}{9}}$ **53.** $(x^2)^{\frac{1}{3}} = x^{\frac{2}{3}}$ **54.** $(15^5)^{\frac{1}{11}} = 15^{\frac{5}{11}}$

SUMMARIZE

55. volume $= lwh$; $l^2 = 1^2 + \sqrt{7}^2$, $l^2 = 8$, $l = \sqrt{8}$; $w^2 = (\sqrt{2})^2 + 1^2$, $w^2 = 3$, $w = \sqrt{3}$; $h^2 = 1^2 + \sqrt{5}^2$, $h^2 = 6$, $h = \sqrt{6}$; $lwh = (\sqrt{8})(\sqrt{3})(\sqrt{6}) = \sqrt{144} = 12$

56. Yes. They are equal if $a = 0$ and $b \geq 0$, or $b = 0$ and $a \geq 0$. Examples will vary. Possible example: let $a = 0$ and $b = 4$, then $\sqrt{0 + 4} = 2$ and $\sqrt{0} + \sqrt{4} = 2$.

57. $(x^{\frac{1}{2}})^{\frac{1}{3}} = 2$, $x^{\frac{1}{2}} = 2^3$, $x = (2^3)^2$, $x = 2^6 = 64$

58. $(n^{\frac{1}{4}})^{\frac{1}{5}} = n^{\frac{1}{20}} = \sqrt[20]{n}$

59. Students should graph $y = x^{\frac{1}{5}} = x^{0.2}$ and trace to find the value of y when $x = 2$. To the nearest hundred-thousandth, the value is 1.14870.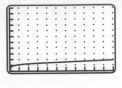

60. Explanations will vary. Assume that $n = x^{\frac{1}{2}}$. Then $n^2 = x$. For any real number n, positive or negative, n^2 is a nonnegative number. Therefore, x must be nonnegative.

61. Assume that $n = x^{\frac{1}{3}}$. Then $n^3 = x$. If n is positive, then n^3 is positive. If n is negative, then n^3 is negative. Therefore, x may be positive or negative, and no values must be excluded from the domain of the function.

Lesson 8.2, pages 374–380

EXPLORE

1. 1, 2, 4, 8, 16. Answers will vary. Students should see that $2^{2.5}$ would be midway between $2^{2.0}$ and $2^{3.0}$ only if the function $y = 2^x$ were linear, that is, if equal increases in x resulted in equal increases in y. However, for integer values of x from 0 to 4, the corresponding values of y are 1, 2, 4, 8, 16. Equal increases of 1 in x result in increases of 1, 2, 4, and 8 in y. Therefore, $y = 2^x$ is not linear, and students should not expect $2^{2.5}$ to be midway between $2^{2.0}$ and $2^{3.0}$.

2.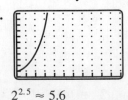

$2^{2.5} \approx 5.6$

3. The results confirm that $2^{2.5}$ is not equal to 6, the value halfway between $2^2(4)$ and $2^3(8)$, but is less than 6.

4. $\sqrt{2^5} = (\sqrt{2})^5 \approx 5.66$, approximately the value of $2^{2.5}$ obtained graphically.

5. The patterns suggest that
$3^{1.5} = 3^{\frac{3}{2}} = \sqrt{3^3} = \left(\sqrt{3}\right)^3 \approx 5.2$
The graph of $y = 3^x$ is shown below. Tracing to $x = 1.5$ confirms that $3^{1.5} \approx 5.2$.

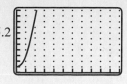

TRY THESE

1. -9 **2.** -2 **3.** no real sol. **4.** -1

5. $\sqrt{45} = \sqrt{3 \cdot 3 \cdot 5} = 3\sqrt{5}$

6. $-2\sqrt{200} = -2\sqrt{10 \cdot 10 \cdot 2} = -20\sqrt{2}$

7. $-\sqrt{24x^2} = -\sqrt{2 \cdot 2 \cdot 6 \cdot x \cdot x} = -2|x|\sqrt{6}$

8. $\sqrt[3]{-108x^3} = \sqrt[3]{(-3)^3 \cdot 4 \cdot x^3} = -3x\sqrt[3]{4}$

9. $16^{\frac{1}{2}} = \sqrt{16} = 4$ **10.** $125^{-\frac{1}{3}} = \frac{1}{125^{\frac{1}{3}}} = \frac{1}{\sqrt[3]{125}} = \frac{1}{5}$

11. $7.233^{\frac{5}{4}} \approx 11.85558$ **12.** $-(64)^{\frac{5}{6}} = -32$

13. $\sqrt[4]{4} = \sqrt[4]{2^2} = 2^{\frac{2}{4}} = 2^{\frac{1}{2}} = \sqrt{2}$

14. $\sqrt[9]{(-20)^3} = (-20)^{\frac{3}{9}} = (-20)^{\frac{1}{3}} = \sqrt[3]{(-20)}$

15. $\sqrt{5}\sqrt[3]{5} - 5^{\frac{1}{2}} \cdot 5^{\frac{1}{3}} = 5^{\frac{1}{2}+\frac{1}{3}} = 5^{\frac{3}{6}+\frac{2}{6}} = 5^{\frac{5}{6}} = \sqrt[6]{5^5}$

16. $a^{\frac{2}{5}}b^{\frac{1}{2}} = a^{\frac{4}{10}}b^{\frac{5}{10}} = (a^4b^5)^{\frac{1}{10}} = \sqrt[10]{a^4b^5}$

17. $\sqrt{gr} = \sqrt{(32)(192)} = \sqrt{6144} = \sqrt{1024 \cdot 6} = 32\sqrt{6}$; $32\sqrt{6}$ ft/s

18. Answers will vary. If $x^2 = y^2$, then, taking the square root of both sides of the equation yields $|x| = |y|$. Therefore, x may or may not equal y. For example, $5^2 = 5^2$, and $5 = 5$. But $(-3)^2 = (3)^2$, and -3 does not equal 3.

PRACTICE

1. $\sqrt{3}$

2. $\sqrt[3]{8^5}$

3. $\sqrt[10]{(x + y)^3}$

4. $p^{-\frac{2}{5}} = \frac{1}{p^{\frac{2}{5}}} = \frac{1}{\sqrt[5]{p^2}}$

5. $15^{\frac{1}{4}}$

6. $e^{\frac{9}{8}}$

7. $\frac{1}{\sqrt[5]{n}} = \frac{1}{n^{\frac{1}{5}}} = n^{-\frac{1}{5}}$

8. $\sqrt[9]{(3x^5 - y)^2} = (3x^5 - y)^{\frac{2}{9}}$

9. $-\sqrt{121} = -\sqrt{11 \cdot 11} = -11$

10. $\sqrt[4]{1} = \sqrt[4]{1^4} = 1^{\frac{4}{4}} = 1$

11. $\sqrt{0.09} = \sqrt{0.3 \cdot 0.3} = 0.3$

12. $\sqrt[5]{-243} = \sqrt[5]{(-3)^5} = (-3)^{\frac{5}{5}} = -3$

13. $\sqrt[7]{0} = \sqrt[7]{0^7} = 0^{\frac{7}{7}} = 0$

14. $\sqrt{-0.25}$; no real solution

15. $\sqrt[5]{6^{10}} = 6^{\frac{10}{5}} = 6^2 = 36$

16. $\sqrt[8]{(-10)^{24}}$; no real solution

17. $\sqrt{98} = \sqrt{7 \cdot 7 \cdot 2} = 7\sqrt{2}$

18. $-3\sqrt{375} = -3\sqrt{5 \cdot 5 \cdot 15} = -15\sqrt{15}$

19. $\sqrt{2h^2} = \sqrt{2 \cdot h \cdot h} = |h|\sqrt{2}$

20. $-\sqrt[4]{32} = -\sqrt[4]{2^4 \cdot 2} = -2\sqrt[4]{2}$

21. $\sqrt{0.64x^6} = \sqrt{0.8 \cdot 0.8 \cdot x^3 \cdot x^3} = 0.8|x^3|$

22. $\sqrt[4]{10,000x^8} = \sqrt[4]{(10)^4 \cdot (x^2)^4} = 10x^2$

23. $\sqrt{16a^2b^4} = \sqrt{4 \cdot 4 \cdot a \cdot a \cdot b^2 \cdot b^2} = 4|a|b^2$

24. $-\sqrt[3]{-64k^{12}} = -\sqrt[3]{(-4) \cdot (-4) \cdot (-4) \cdot k^4 \cdot k^4 \cdot k^4} = 4k^4$

25. $\sqrt[5]{32x^{25}} = \sqrt[5]{2^5(x^5)^5} = 2x^5$

26. $\sqrt{(c - 2)^2} = |c - 2|$

27. $\sqrt[7]{(x + 4)^7} = x + 4$

28. $\sqrt{y^2 - 4y + 4} = \sqrt{(y - 2)^2} = |y - 2|$

29. $1024^{0.1} = 1024^{\frac{1}{10}} = \sqrt[10]{1024} = \sqrt[10]{2^{10}} = 2$

30. $(2^{k-2})(2^{2-k}) = 2^{k-2+2-k} = 2^0 = 1$

31. $\sqrt[3]{b^{12n-15}} = \sqrt[3]{(b^{4n-5})^3} = b^{4n-5}$

32. No; Students may give an example such as $\sqrt{-s}$, which is not a real number if $s > 0$.

33. $4^{\frac{1}{2}} = \sqrt{4} = \sqrt{2^2} = 2$

34. $27^{-\frac{1}{3}} = \frac{1}{27^{\frac{1}{3}}} = \frac{1}{\sqrt[3]{27}} = \frac{1}{\sqrt[3]{3^3}} = \frac{1}{3}$

35. $16^{\frac{3}{4}} = (\sqrt[4]{16})^3 = 2^3 = 8$

36. $64^{-\frac{2}{3}} = \frac{1}{64^{\frac{2}{3}}} = \frac{1}{(\sqrt[3]{64})^2} = \frac{1}{4^2} = \frac{1}{16}$

37. $49^{\frac{3}{2}} = (\sqrt{49})^3 = 7^3 = 343$

38. $9^{\frac{5}{2}} = (\sqrt{9})^5 = 3^5 = 243$

39. $-125^{\frac{4}{3}} = -(\sqrt[3]{125})^4 = -(5)^4 = -625$

40. $32^{\frac{2}{5}} = (\sqrt[5]{32})^2 = 2^2 = 4$

41. $(9^{\frac{4}{3}})^{\frac{9}{8}} = 9^{\frac{36}{24}} = 9^{\frac{3}{2}} = (\sqrt{9})^3 = 3^3 = 27$

42. $81^{\frac{3}{8}} 81^{\frac{7}{8}} = 81^{\frac{3}{8}+\frac{7}{8}} = 81^{\frac{10}{8}} = 81^{\frac{5}{4}} = (\sqrt[4]{81})^5 = 3^5 = 243$

43. $12.08^{\frac{7}{2}} \approx 6126.808$

44. $\left(\frac{2}{3}\right)^{\frac{2}{3}} \approx 0.763143$

45. $\sqrt[6]{7^3} = 7^{\frac{3}{6}} = 7^{\frac{1}{2}} = \sqrt{7}$

46. $36^{\frac{1}{4}} = \sqrt[4]{36} = \sqrt[4]{6^2} = 6^{\frac{2}{4}} = 6^{\frac{1}{2}} = \sqrt{6}$

47. $\sqrt[16]{3^4} = 3^{\frac{4}{16}} = 3^{\frac{1}{4}} = \sqrt[4]{3}$

48. $\sqrt[6]{8} = \sqrt[6]{2^3} = 2^{\frac{3}{6}} = 2^{\frac{1}{2}} = \sqrt{2}$

49. $\sqrt[3]{2} \sqrt[4]{2^3} = 2^{\frac{1}{3}} 2^{\frac{3}{4}} = 2^{\frac{1}{3}+\frac{3}{4}} = 2^{\frac{13}{12}} = \sqrt[12]{2^{13}}$

50. $\sqrt[20]{x^8} \sqrt[5]{x^2} = x^{\frac{8}{20}} x^{\frac{2}{5}} = x^{\frac{8}{20}+\frac{2}{5}} = x^{\frac{16}{20}} = x^{\frac{4}{5}} = \sqrt[5]{x^4}$

51. $m^{\frac{3}{5}} n^{\frac{2}{3}} = m^{\frac{9}{15}} n^{\frac{10}{15}} = \sqrt[15]{m^9 n^{10}}$

52. $\sqrt{\sqrt[3]{5}} = (\sqrt[3]{5})^{\frac{1}{2}} = 5^{\frac{1}{2}\cdot\frac{1}{3}} = 5^{\frac{1}{6}} = \sqrt[6]{5}$

53a. $d = 16t^2, t^2 = \frac{d}{16}, t = \left(\frac{d}{16}\right)^{\frac{1}{2}}$

53b. $t = \sqrt{\frac{d}{16}}; t = \sqrt{\frac{864}{16}}, t = \sqrt{54}, t = 3\sqrt{6}; 3\sqrt{6}\text{ s}$

54. $\frac{(\sqrt{24})^3 \sqrt{2}}{3} = \frac{(\sqrt{24})^2(\sqrt{24})\sqrt{2}}{3} = \frac{(24)(2\sqrt{6})\sqrt{2}}{3} =$
$8(2\sqrt{6})\sqrt{2} = 16\sqrt{12} = 32\sqrt{3}\text{ cm}^3$

55a. $P = d(\sqrt[16]{2.7})^t, P = d(2.7)^{\frac{t}{16}}$

55b. $P = 2400(2.7)^{\frac{28}{16}} \approx 13,649; \$13,649.00$

56. Explanations will vary. Students should point out that when a real number is squared, the result is always positive. Therefore, only positive numbers have real number square roots. On the other hand, when a number is cubed, the result is positive if the number is positive and negative if the number is negative. Therefore, negative numbers have cube roots, each of them the result of cubing a negative number.

EXTEND

57. $x \geq 0$ **58.** all x

59. $2x + 6 \geq 0, 2x \geq -6, x \geq -3$ **60.** no solutions

61. $x \geq -5$; Students should explain that the equation is true if $x + 5 \geq 0$.

62. $2x^5 - 10 = 0, 2x^5 = 10, x^5 = 5, x = 5^{\frac{1}{5}}, x = \sqrt[5]{5}$

63. $6x^{\frac{2}{3}} = 150, x^{\frac{2}{3}} = 25, x = 25^{\frac{3}{2}} = (\sqrt{25})^3 = 5^3$
$= 125$

64. $\sqrt[4]{x^5} = 32, x^{\frac{5}{4}} = 32, x = 32^{\frac{4}{5}} = (\sqrt[5]{32})^4, x = 2^4,$
$x = 16$

65. $x^9 + x^2 = 0, x^9 = -x^2; x = 0 \text{ or } x = -1$

THINK CRITICALLY

66. $x + 5 \geq 0, x \geq -5; x^2 - 5x + 6 \neq 0,$
$(x - 2)(x - 3) \neq 0, x \neq 2 \text{ or } x \neq 3$

67. $\frac{1}{a^{-\frac{1}{n}}} = a^{\frac{1}{n}} = \sqrt[n]{a}$

68. $\sqrt[n]{\sqrt[m]{a}} = (\sqrt[m]{a})^{\frac{1}{n}} = a^{\frac{1}{n}\cdot\frac{1}{m}} = a^{\frac{1}{mn}} = \sqrt[mn]{a}$

69. $a^{\frac{x}{z}} \cdot a^{\frac{y}{z}} = a^{\frac{x+y}{z}} = \sqrt[z]{a^{x+y}}$ **70.** $\left(a^{\frac{w}{x}}\right)^{\frac{y}{z}} = a^{\frac{wy}{xz}} = \sqrt[xz]{a^{wy}}$

71. 0 and 1

72. $(y + 9)(y - 9)$ **73.** $(3a + 2)(3a - 2)$

74. $2d(c^4 - 16), 2d(c^2 + 4)(c^2 - 4),$
$2d(c^2 + 4)(c + 2)(c - 2)$

75. $(x - 1)(x^2 + x + 1)$

76. $B; 16^{-\frac{3}{4}} = \frac{1}{16^{\frac{3}{4}}} = \frac{1}{(\sqrt[4]{16})^3} = \frac{1}{2^3} = \frac{1}{8}$

PROJECT CONNECTION

1. $V = \pi r^2 h + \frac{4}{3}\pi r^2 = \pi r^2(h + \frac{4}{3}r)$

2–4. Answers will vary.

ALGEBRAWORKS

1. 3.67 billion $= 3.67 \times 10^9$ mi $\times \frac{5280\text{ ft}}{1\text{ mi}} \approx$
1.94×10^{13} ft

$184 = 1.84 \times 10^2$ ft $\frac{1.94 \times 10^{13}\text{ ft}}{1.84 \times 10^2\text{ ft}} \approx 1.1 \times 10^{11}$
times as far

2a. $t = \left(\frac{d}{6}\right)^{\frac{3}{2}}, t = \sqrt{\left(\frac{d}{6}\right)^3}, t = \sqrt{\frac{d^3}{6^3}}$

2b. $t = \sqrt{\frac{24^3}{6^3}}, t = \sqrt{64}, t = 8$ h

2c. about 4.2 mi

3a. $t = 49.6\sqrt[25]{D^6}, t = 49.6(D)^{\frac{6}{25}}, t = 49.6D^{0.24}$

3b. $t = 49.6(6.0)^{0.24} \approx 76.24; 76.3°$

3c. $t = 49.6(3.29)^{0.24} \approx 66.00; 66.0°$

4. $T_{wc} = 0.045(5.27\sqrt{36} + 10.45 - 0.28(36))$
$(0 - 33) + 33 \approx -14.50; -14.5°C$

Lesson 8.3, pages 381–384

THINK BACK

1.

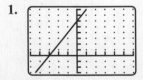

2. No vertical line intersects the graph in more than one point.

3. Exactly one value of y corresponds to each value of x.

4. $x = 2y + 6$, $2y = x - 6$, $y = \frac{1}{2}x - 3$;

$f^{-1}(x) = \frac{1}{2}x - 3$

5.

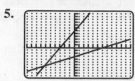

6. $f^{-1}(x)$ consists of ordered pairs obtained by interchanging the domain and range of ordered pairs in $f(x)$.

7. The graph of $f^{-1}(x)$ is a reflection of the graph of $f(x) = 2x + 6$ over the line $y = x$.

8. The graph of $y = x^2$ is a parabola with vertex $(0, 0)$, opening upward and symmetric to the y-axis.

9a. translated up 5 units

9b. translated down 7 units

9c. translated left 3 units

9d. translated right 4 units

9e. closed tighter around the y-axis

9f. opened wider around the y-axis and reflected over the x-axis so that it opens downward

EXPLORE

10. 16; 9; 4; 1; 0; 1; 4; 9; 16

11. Yes. For each value of x there is exactly one value of y.

12. 0; ±1; ±2; ±3; ±4

13. No. For each value of x other than 0, there are two values for y.

Graph for Questions 14 and 15:

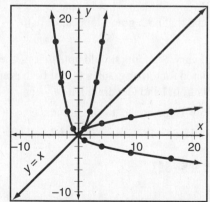

14. The graphs confirm that the relation $y = x^2$ is a function because no vertical line intersects the graph in more than one point. The graphs confirm that the inverse of the relation $y = x^2$ is not a function because any vertical line that intersects the graph at one point (except $(0, 0)$) intersects it at two points.

15. The graph of an inverse relation is a reflection of the original relation across the line $y = x$, so each of the curves graphed here is the reflection of the other across

the line. Students may confirm this by finding the midpoints (see Lesson 12.1 for formula) of line segments connecting corresponding points on the two graphs. Each midpoint lies on the line $y = x$. For example, the midpoint of the segment joining $(4, 16)$ and $(16, 4)$ is $(10, 10)$.

16. Answers will vary. Students may show that $y = \pm \sqrt{x}$ generates the values in the table in Question 12. They may also apply the method for finding the inverse of a relation, solving the equation $y = x^2$ for x and then interchanging x and y.

17.

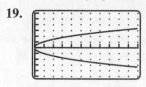 To obtain the graph, students may enter $y = x^{0.5}$ or $x^{(1 \div 2)}$ or $\sqrt{x}$

18. For each value of x in the graph in Question 14, the two points $(x, \sqrt{x})$ and $(x, -\sqrt{x})$ are plotted. For each value of x in the graph here, only the point $(x, \sqrt{x})$ is plotted.

19.

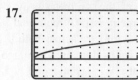 To obtain the graph, students should enter $y_2 = -(x^{0.5})$ or $-(x^{(1 \div 2)})$, along with the equation entered earlier for y_1.

20. 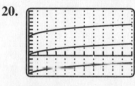 **21.**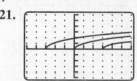

22. 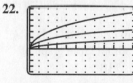 **23.**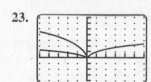

24.

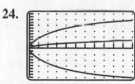

MAKE CONNECTIONS

25. The graph is the graph of $y = \sqrt{x}$ translated up c units (for positve c) or translated down $|c|$ units (for negative c).

26. The graph is the graph of $y = \sqrt{x}$ translated to the left c units (for positive c) or to the right $|c|$ units (for negative c).

27. The smaller the value of c, the closer to the positive x-axis the curve will appear. For $c > 1$, the graph of $y = \sqrt{x}$ is between the curve and the positive x-axis. For $0 < c < 1$, the curve is between the graph of $y = \sqrt{x}$ and the positive x-axis.

28. The graph is the graph of $y = c\sqrt{x}$ reflected over the y-axis.

29. The graph is the graph of $y = c\sqrt{x}$ reflected over the x-axis.

30. b **31.** f **32.** c **33.** e **34.** a **35.** d

36. Answers will vary. In the first quadrant, both graphs increase at a continually greater rate as x increases, with $y = x^3$ increasing at a greater rate than $y = x^2$. Both graphs contain (0, 0) and (1, 1). The section of each graph that is not in Quadrant I is symmetric to the section in Quadrant I. For $y = x^2$, the symmetry is over the y-axis; for $y = x^3$, the symmetry is 180° rotational symmetry about the orrigin (or symmetry across both axes). Each graph has all real x for its domain. The range of $y = x^2$ is $y \geq 0$; the range of $y = x^3$ is all real y.

37. Answers will vary. Students may show that $y = x^3$ and $y = \sqrt[3]{x}$ generate ordered pairs with their domains and ranges interchanged. They may also apply the method for finding the inverse of a relation, solving the equation $y = x^3$ for x and then interchanging x and y.

38. Answers will vary. Students should mention that the graph will be a reflection of the graph of $y = x^3$ over the line $y = x$.

39.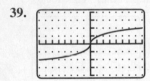

40. Answers will vary. Both graphs increase for increasing x in the first quadrant, with $y = \sqrt{x}$ increasing at a greater rate. Both graphs contain (0, 0) and (1, 1). The graph of $y = \sqrt{x}$ is only in Quadrant I; the graph of $y = \sqrt[3]{x}$ exists in Quadrants I and III.

41. Compared to the graph of $y = \sqrt[3]{x}$, the graph increases at a greater rate in quadrant I and decreases at a greater rate in Quadrant III.

42. The graph is the graph of $y = \sqrt[3]{x}$ translated up 9 units.

43. The graph is the graph of $y = \sqrt[3]{x}$ translated to the right 4 units.

44. The graph is the graph of $y = \sqrt[3]{x}$ reflected over the x-axis.

45. Compared to the graph of $y = \sqrt[3]{x}$, the graph increases at a lesser rate in Quadrant I and decreases at a lesser rate in Quadrant III.

46. The graph is the graph of $y = \sqrt[3]{x}$ reflected over the y-axis.

47. Answers will vary. $y = \pm\sqrt{x}$ is not a function because for each value of x there are two corresponding values of y. Graphically this can be seen because any vertical line that intersects the graph of the relation in one point (except (0, 0)) intersects it in two points. The other two relations are functions because for each value of x there is a unique corresponding value of y. Graphically this can be seen because no vertical line intersects either graph in more than one point.

48. Increase the steepness of the graph of $y = \sqrt{x}$ by a factor of 3. Reflect the new graph over the x-axis. Translate the resulting graph 9 units down and translate it to the left 6 units.

49a.
$$\sqrt{x} = \pm\sqrt[3]{x}$$
$$x^3 = x^2$$
$$x^3 - x^2 = 0$$
$$x^2(x - 1) = 0$$
$$x = 0 \text{ or } x = 1$$
Or graph
$y_1 = \sqrt{x}, y_2 = \sqrt[3]{x}, y_3 = -\sqrt[3]{x}$ and find points of intersection $x = 0$ and $x = 1$.

49b.
$$\sqrt[3]{x} = \sqrt[5]{x}$$
$$x^5 = x^3$$
$$x^5 - x^3 = 0$$
$$x^3(x^2 - 1) = 0$$
$$x^3 = 0 \text{ or } x^2 = 1$$
$$x = 0, \pm 1$$
Or graph $y_1 = \sqrt[3]{x}$ and $y_2 = \sqrt[5]{x}$ and find points of intersection $x = -1$, $x = 0$, and $x = 1$.

50a. Answers will vary. Students should note that $x \geq 0$ for both graphs and that both graphs appear only in Quadrant I.

50b. Answers will vary. Students should note that neither x nor y is restricted for either graph and that both graphs appear in Quadrant I and Quadrant III.

51. The graph of $\sqrt[4]{x}$ is the one that is higher up.

52a. 7.6 **52b.** -48.8 **52c.** 62.4, 24.2

Lesson 8.4, pages 385–391

EXPLORE

1–4. In each group, a and b are false, and c and d are true.

TRY THESE

1. $2 - \sqrt{6}$ **2.** $3\sqrt{11} + 8\sqrt{16}$ **3.** $-\sqrt{3.5} - 5$

4. $7 + \sqrt{21}$ **5.** $\dfrac{\sqrt{6}}{\sqrt{6}}$ **6.** $\dfrac{\sqrt{7} + 3\sqrt{3}}{\sqrt{7} + 3\sqrt{3}}$ **7.** $\dfrac{\sqrt[3]{100n}}{\sqrt[3]{100n}}$

8. $\dfrac{\sqrt[5]{108x^4}}{\sqrt[5]{108x^4}}$ **9.** $\sqrt{3} \cdot \sqrt{5} = \sqrt{3 \cdot 5} = \sqrt{15}$

10. $2\sqrt[3]{9} \cdot \sqrt[3]{3} = 2\sqrt[3]{9 \cdot 3} = 2\sqrt[3]{27} = 2 \cdot 3 = 6$

11. $8^{-\frac{2}{3}} = \frac{1}{8^{\frac{2}{3}}} = \frac{1}{(\sqrt[3]{8})^2} = \frac{1}{2^2} = \frac{1}{4}$

12. $\frac{\sqrt{5}}{\sqrt{7}} = \frac{\sqrt{5}}{\sqrt{7}} \cdot \frac{\sqrt{7}}{\sqrt{7}} = \frac{\sqrt{35}}{7}$

13. $\sqrt[3]{\frac{8}{9n}} = \frac{\sqrt[3]{8}}{\sqrt[3]{9n}} = \frac{\sqrt[3]{8}}{\sqrt[3]{9n}} \cdot \frac{\sqrt[3]{3n^2}}{\sqrt[3]{3n^2}} = \frac{\sqrt[3]{24n^2}}{\sqrt[3]{27n^3}} = \frac{2\sqrt[3]{3n^2}}{3n}$

14. $\frac{3x}{\sqrt{20x}} = \frac{3x}{\sqrt{20x}} \cdot \frac{\sqrt{20x}}{\sqrt{20x}} = \frac{3x\sqrt{20x}}{20x} = \frac{6x\sqrt{5x}}{20x} = \frac{3\sqrt{5x}}{10}$

15. $\sqrt{10} + \sqrt{40} - 3\sqrt{250} = \sqrt{10} + 2\sqrt{10} - 15\sqrt{10} = -12\sqrt{10}$

16. $\sqrt[4]{2k} - \sqrt[4]{162k} = \sqrt[4]{2k} - 3\sqrt[4]{2k} = -2\sqrt[4]{2k}$

17. $\frac{1 + \sqrt{2}}{1 - \sqrt{2}} = \frac{1 + \sqrt{2}}{1 - \sqrt{2}} \cdot \frac{1 + \sqrt{2}}{1 + \sqrt{2}} =$
    $\frac{1 + \sqrt{2} + \sqrt{2} + 2}{1 - \sqrt{2} + \sqrt{2} - 2} =$
    $\frac{3 + 2\sqrt{2}}{-1} = -3 - 2\sqrt{2}$

18. perimeter $= 2(\sqrt{600} + \sqrt{384}) = 2(10\sqrt{6} + 8\sqrt{6}) = 2(18\sqrt{6}) = 36\sqrt{6}$; area $= \sqrt{600} \cdot \sqrt{384} = (10\sqrt{6} \cdot 8\sqrt{6}) = 80 \cdot 6 = 480$; The frame's perimeter is $36\sqrt{6}$ in. and the frame's area is 480 in.²

19. The radical must be expressed using the least possible index. The radical should not contain a fraction or a perfect nth power of any factor. No denominator in the expression should contain a radical.

PRACTICE

1. $\sqrt{11} \cdot \sqrt{7} = \sqrt{77}$

2. $\sqrt[3]{4} \cdot 5\sqrt[3]{6} = 5\sqrt[3]{24} = 10\sqrt[3]{3}$

3. $\sqrt[6]{3 \cdot 2} \cdot \sqrt[6]{20} = \sqrt[6]{64} = 2$

4. $\sqrt[3]{9} \cdot \sqrt[3]{-9} = \sqrt[3]{-81} = -3\sqrt[3]{3}$

5. $\sqrt{18x^2y} \cdot \sqrt{2yz} = \sqrt{36x^2y^2z} = 6|x|y\sqrt{z}$

6. $\sqrt{5ab} \cdot \sqrt{10ab} = \sqrt{50a^2b^2} = 5ab\sqrt{2}$

7. $\sqrt{147} \div \sqrt{3} = \sqrt{49} = 7$

8. $8\sqrt[3]{56} \div 4\sqrt[3]{7} = 2\sqrt[3]{8} = 4$

9. $64^{-\frac{1}{2}} = \frac{1}{64^{\frac{1}{2}}} = \frac{1}{\sqrt{64}} = \frac{1}{8}$

10. $243^{-\frac{4}{5}} = \frac{1}{243^{\frac{4}{5}}} = \frac{1}{(\sqrt[5]{243})^4} = \frac{1}{3^4} = \frac{1}{81}$

11. $\left(\frac{1}{8}\right)^{-\frac{3}{4}} = \frac{1}{\left(\frac{1}{8}\right)^{\frac{3}{4}}} = 8^{\frac{3}{4}} = (2^3)^{\frac{3}{4}} = 2^{\frac{9}{4}}$

12. $\frac{1}{x^{-\frac{8}{10}}} = x^{\frac{8}{10}} = x^{\frac{4}{5}}$ 13. $\frac{\sqrt{3}}{\sqrt{5}} = \frac{\sqrt{3}}{\sqrt{5}} \cdot \frac{\sqrt{5}}{\sqrt{5}} = \frac{\sqrt{15}}{5}$

14. $\frac{6\sqrt[3]{32}}{\sqrt[3]{2}} = \frac{6\sqrt[3]{32}}{\sqrt[3]{2}} \cdot \frac{\sqrt[3]{4}}{\sqrt[3]{4}} = \frac{6\sqrt[3]{128}}{2} = \frac{24\sqrt[3]{2}}{2} = 12\sqrt[3]{2}$

15. $\frac{\sqrt[5]{-24x^{20}}}{\sqrt[5]{12x^5}} = \frac{\sqrt[5]{-24x^{20}}}{\sqrt[5]{12x^5}} \cdot \frac{\sqrt[5]{648}}{\sqrt[5]{648}} = \frac{\sqrt[5]{-15,552x^{20}}}{\sqrt[5]{7776x^5}} =$
    $\frac{6x^4\sqrt[5]{-2}}{6x} = x^3\sqrt[5]{-2}$

16. $\sqrt{\frac{7}{12b}} = \frac{\sqrt{7}}{\sqrt{12b}} \cdot \frac{\sqrt{3b}}{\sqrt{3b}} = \frac{\sqrt{21b}}{\sqrt{36b^2}} = \frac{\sqrt{21b}}{6b}$

17. $\sqrt{\frac{25x^3}{y^2}} = \frac{\sqrt{25x^3}}{\sqrt{y^2}} = \frac{5x\sqrt{x}}{|y|}$

18. $\sqrt[3]{\frac{162x^5y^7}{6x^2y^2}} = \frac{\sqrt[3]{162x^5y^7}}{\sqrt[3]{6x^2y^2}} \cdot \frac{\sqrt[3]{36xy}}{\sqrt[3]{36xy}} = \frac{\sqrt[3]{5832x^6y^8}}{\sqrt[3]{216x^3y^3}} =$
    $\frac{18x^2y^2\sqrt[3]{y^2}}{6xy} = 3xy\sqrt[3]{y^2}$

19. $\frac{\sqrt{10a^2b}}{\sqrt{18a^5b^3}} = \frac{\sqrt{10a^2b}}{\sqrt{18a^5b^3}} \cdot \frac{\sqrt{18a^5b^3}}{\sqrt{18a^5b^3}} = \frac{\sqrt{180a^7b^4}}{\sqrt{324a^{10}b^6}} =$
    $\frac{6a^3b^2\sqrt{5a}}{18a^5b^3} = \frac{\sqrt{5a}}{3a^2b}$

20. $\frac{12h}{\sqrt[4]{18h}} = \frac{12h}{\sqrt[4]{18h}} \cdot \frac{\sqrt[4]{72h^3}}{\sqrt[4]{72h^3}} = \frac{12h\sqrt[4]{72h^3}}{\sqrt[4]{1296h^4}} = \frac{12h\sqrt[4]{72h^3}}{6h} =$
    $2\sqrt[4]{72h^3}$

21. $v = \sqrt{\frac{2k}{m}}, v = \frac{\sqrt{2k}}{\sqrt{m}}, v = \frac{\sqrt{2k}}{\sqrt{m}} \cdot \frac{\sqrt{m}}{\sqrt{m}}, v = \frac{\sqrt{2km}}{m}$

22. $6\sqrt{3} + 8\sqrt{3} = 14\sqrt{3}$

23. $8\sqrt{45} + 7\sqrt{20} = 8\sqrt{9 \cdot 5} + 7\sqrt{4 \cdot 5} = 24\sqrt{5} + 14\sqrt{5} = 38\sqrt{5}$

24. $\sqrt[3]{54} - \sqrt[3]{128} = \sqrt[3]{27 \cdot 2} - \sqrt[3]{64 \cdot 2} = 3\sqrt[3]{2} - 4\sqrt[3]{2} = -\sqrt[3]{2}$

25. $\sqrt[4]{48m^5} - \sqrt[4]{768m} =$
    $\sqrt[4]{16 \cdot 3 \cdot m^4 \cdot m} - \sqrt[4]{256 \cdot 3 \cdot m} =$
    $2m\sqrt[4]{3m} - 4\sqrt[4]{3m} =$
    $(2m - 4)\sqrt[4]{3m} = 2(m - 2)\sqrt[4]{3m}$

26. $(\sqrt{2} + \sqrt{3})(\sqrt{2} + \sqrt{3}) =$
    $\sqrt{4} + \sqrt{6} + \sqrt{6} + \sqrt{9} =$
    $2 + 2\sqrt{6} + 3 = 5 + 2\sqrt{6}$

27. $(\sqrt{10} - \sqrt{6})(\sqrt{10} - \sqrt{6}) =$
    $\sqrt{100} - \sqrt{60} - \sqrt{60} + \sqrt{36} =$
    $10 - 2\sqrt{60} + 6 = 16 - 2\sqrt{4 \cdot 15} = 16 - 4\sqrt{15}$

28. $(\sqrt{7} - \sqrt{5})(\sqrt{7} - \sqrt{5}) =$
    $\sqrt{49} - \sqrt{35} + \sqrt{35} - \sqrt{25} =$
    $7 - 5 = 2$

29. $(2\sqrt{x} - 6\sqrt{y})(2\sqrt{x} - 6\sqrt{y}) =$
    $4\sqrt{x^2} - 12\sqrt{xy} - 12\sqrt{xy} +$
    $36\sqrt{y^2} = 4x + 36y - 24\sqrt{xy}$

30. $\dfrac{3}{3+\sqrt{3}} = \dfrac{3}{3+\sqrt{3}} \cdot \dfrac{3-\sqrt{3}}{3-\sqrt{3}} =$

$\dfrac{9-3\sqrt{3}}{9+3\sqrt{3}-3\sqrt{3}-\sqrt{9}} =$

$\dfrac{9-3\sqrt{3}}{6} =$

$\dfrac{3-\sqrt{3}}{2}$

31. $\dfrac{5-\sqrt{8}}{\sqrt{2}} = \dfrac{5-\sqrt{8}}{\sqrt{2}} \cdot \dfrac{\sqrt{2}}{\sqrt{2}} = \dfrac{5\sqrt{2}-\sqrt{16}}{\sqrt{4}} =$

$\dfrac{5\sqrt{2}-4}{2}$

32. $\dfrac{\sqrt{k+1}}{\sqrt{k-1}} = \dfrac{\sqrt{k+1}}{\sqrt{k-1}} \cdot \dfrac{\sqrt{k-1}}{\sqrt{k-1}} =$

$\dfrac{\sqrt{k^2-k+k-1}}{\sqrt{k^2-2k+1}} = \dfrac{\sqrt{k^2-1}}{\sqrt{(k-1)^2}} =$

$\dfrac{\sqrt{k^2-1}}{|k-1|}$

33. $\dfrac{2\sqrt{5}-3\sqrt{2}}{3\sqrt{6}-4\sqrt{3}} = \dfrac{(2\sqrt{5}-3\sqrt{2})}{(3\sqrt{6}-4\sqrt{3})} \cdot \dfrac{(3\sqrt{6}+4\sqrt{3})}{(3\sqrt{6}+4\sqrt{3})} =$

$\dfrac{6\sqrt{30}+8\sqrt{15}-9\sqrt{12}-12\sqrt{6}}{9\sqrt{36}-12\sqrt{18}+12\sqrt{18}-16\sqrt{9}} =$

$\dfrac{6\sqrt{30}+8\sqrt{15}-18\sqrt{3}-12\sqrt{6}}{54-48} =$

$\dfrac{3\sqrt{30}+4\sqrt{15}-9\sqrt{3}-6\sqrt{6}}{3}$

34. $\sqrt{1440} - \sqrt{90} = \sqrt{144 \cdot 10} - \sqrt{9 \cdot 10} =$

$12\sqrt{10} - 3\sqrt{10} = 9\sqrt{10}$ mi

35a. $m_v = 100\left(1 - \dfrac{299{,}000^2}{299{,}793^2}\right)^{-\frac{1}{2}} \approx 1375.77;\ 1376$ kg

35b. $m_v = m\left(1 - \dfrac{v^2}{c^2}\right)^{-\frac{1}{2}} = m\left(\dfrac{c^2-v^2}{c^2}\right)^{-\frac{1}{2}} = m\left(\dfrac{c^2}{c^2-v^2}\right)^{\frac{1}{2}} =$

$\dfrac{mc}{(c^2-v^2)^{\frac{1}{2}}} = \dfrac{mc(c^2-v^2)^{\frac{1}{2}}}{c^2-v^2}$

36. Answers will vary. Students might prove the result in general by showing that the product of $a\sqrt{b} + c\sqrt{d}$ and $a\sqrt{b} - c\sqrt{d}$ is $a^2b - c^2d$, which does not contain a radical. Or they might state that when two conjugates are multiplied, the result is a difference of two squares, and that the two squarings together remove all the radicals appearing in the conjugates.

EXTEND

37. $\sqrt{\dfrac{2}{3}} + \sqrt{\dfrac{3}{2}} = \dfrac{\sqrt{2}}{\sqrt{3}} + \dfrac{\sqrt{3}}{\sqrt{2}} = \dfrac{\sqrt{2}}{\sqrt{3}} \cdot \dfrac{\sqrt{2}}{\sqrt{2}} + \dfrac{\sqrt{3}}{\sqrt{2}} \cdot \dfrac{\sqrt{3}}{\sqrt{3}} =$

$\dfrac{\sqrt{4}}{\sqrt{6}} + \dfrac{\sqrt{9}}{\sqrt{6}} = \dfrac{2+3}{\sqrt{6}} = \dfrac{5}{\sqrt{6}} = \dfrac{5\sqrt{6}}{6}$

38. $\sqrt[4]{9} - \sqrt{\dfrac{1}{3}} = \sqrt[4]{3^2} - \dfrac{\sqrt{1}}{\sqrt{3}} = (3^2)^{\frac{1}{4}} -$

$\dfrac{1}{\sqrt{3}} = 3^{\frac{1}{2}} - \dfrac{1}{\sqrt{3}} = \sqrt{3} - \dfrac{1}{\sqrt{3}} =$

$\sqrt{3} \cdot \dfrac{\sqrt{3}}{\sqrt{3}} - \dfrac{1}{\sqrt{3}} = \dfrac{3-1}{\sqrt{3}} = \dfrac{2}{\sqrt{3}} = \dfrac{2\sqrt{3}}{3}$

39. $\sqrt[4]{144} + 3\sqrt[4]{9} - 5\sqrt{48} =$

$\sqrt[4]{12^2} + 3\sqrt[4]{3^2} - 5\sqrt{16 \cdot 3} =$

$(12^2)^{\frac{1}{4}} + 3(3^2)^{\frac{1}{4}} - 20\sqrt{3} = 12^{\frac{1}{2}} + 3(3)^{\frac{1}{2}} - 20\sqrt{3} =$

$\sqrt{12} + 3\sqrt{3} - 20\sqrt{3} = 2\sqrt{3} + 3\sqrt{3} - 20\sqrt{3} =$

$-15\sqrt{3}$

40. $\left(\dfrac{x^{-2}y^6}{9}\right)^{-\frac{1}{2}} = \left(\dfrac{y^6}{9x^2}\right)^{-\frac{1}{2}} = \left(\dfrac{9x^2}{y^6}\right)^{\frac{1}{2}} = \sqrt{\dfrac{9x^2}{y^6}} = \dfrac{\sqrt{9x^2}}{\sqrt{y^6}} = \dfrac{3|x|}{y^3}$

41. $\dfrac{x^{\frac{1}{3}}}{x^{-\frac{1}{2}}+x^{-\frac{3}{2}}} = \dfrac{x^{\frac{1}{3}}}{x^{-\frac{3}{2}}(x+1)} = \dfrac{x^{\frac{3}{2}} \cdot x^{\frac{1}{2}}}{x+1} = \dfrac{x^{\frac{4}{2}}}{x+1} = \dfrac{x^2}{x+1}$

42. $\dfrac{m^{-\frac{2}{3}}n^{\frac{1}{2}}}{n^{-\frac{3}{2}}\sqrt[3]{m}} = \dfrac{m^{-\frac{2}{3}}n^{\frac{1}{2}}}{n^{-\frac{3}{2}}m^{\frac{1}{3}}} = \dfrac{n^{\frac{3}{2}} \cdot n^{\frac{1}{2}}}{m^{\frac{2}{3}} \cdot m^{\frac{1}{3}}} = \dfrac{n^{\frac{4}{2}}}{m^{\frac{3}{3}}} = \dfrac{n^2}{m}$

THINK CRITICALLY

43. Reason that: $\dfrac{3}{\sqrt{3}} = \sqrt{3};\ \dfrac{5}{\sqrt{5}} = \sqrt{5};$ and $\sqrt{5} > \sqrt{3}$,

thus $\dfrac{5}{\sqrt{5}} > \dfrac{3}{\sqrt{3}}$.

44. $(\sqrt{7}-\sqrt{2})^2 = (\sqrt{7}-\sqrt{2})(\sqrt{7}-\sqrt{2}) = 7 - \sqrt{14} - \sqrt{14} + 2 = 9 - 2\sqrt{14}$

45. $(\sqrt{5}+\sqrt{3})^2 = (\sqrt{5}+\sqrt{3})(\sqrt{5}+\sqrt{3}) = 5 + 2\sqrt{15} + 3 = 8 + 2\sqrt{15}$, so $\sqrt{5} + \sqrt{3} = \sqrt{8+2\sqrt{15}}$

46. $(\sqrt{11}-\sqrt{3})^2 = (\sqrt{11}-\sqrt{3})(\sqrt{11}-\sqrt{3}) = 11 - 2\sqrt{33} + 3 = 14 - 2\sqrt{33}$, so $\sqrt{11} - \sqrt{3} = \sqrt{14-2\sqrt{33}}$

47. $\sqrt{6} + \sqrt{5}$

48. $\sqrt{10} - \sqrt{7}$

49. $\sqrt{7} - \sqrt{6}$

50. $n - 1 = n^{\frac{3}{3}} - 1 = \left(n^{\frac{1}{3}} - 1\right)\left(n^{\frac{2}{3}} + n^{\frac{1}{3}} + 1\right)$. The first

term of the factorization is the denominator of $\dfrac{1}{\sqrt[3]{n}-1} \cdot$ in exponential form. This suggests that by multiplying the denominator by the second term of the factorization, a term without radicals $(n-1)$ will be obtained.

$\dfrac{1}{\sqrt[3]{n}-1} \cdot \dfrac{\sqrt[3]{n^2}+\sqrt[3]{n}+1}{\sqrt[3]{n^2}+\sqrt[3]{n}+1} = \dfrac{\sqrt[3]{n^2}+\sqrt[3]{n}+1}{n-1}$

51. $\begin{cases} x + y + z = 5 \\ 2x + 3y - z = 9 \\ -4x - y + 6z = -4 \end{cases}$ $\rightarrow$ $\begin{array}{l} x + y + z = 5 \\ \underline{2x + 3y - z = 9} \\ 3x + 4y = 14 \end{array}$

$\begin{cases} 2x + 3y - z = 9 \\ -4x - y + 6z = -4 \end{cases}$ $\rightarrow$ $\begin{array}{l} 12x + 18y - 6z = 54 \\ \underline{-4x - y + 6z = -4} \\ 8x + 17y = 50 \end{array}$

$\begin{cases} 3x + 4y = 14 \\ 8x + 17y = 50 \end{cases}$ $\rightarrow$ $\begin{array}{l} 24x + 32y = 112 \\ \underline{-24x - 51y = -150} \\ -19y = -38 \\ y = 2 \end{array}$

$\begin{array}{ll} 3x + 4y = 14 & x + y + z = 5 \\ 3x + 4(2) = 14 & 2 + 2 + z = 5 \\ 3x = 6 & z = 1 \\ x = 2 & \end{array}$

$(2, 2, 1)$

52. $\begin{cases} 2x + y + 2z = 0 \\ 3x - 2y - 3z = 1 \\ -x + 3y + z = -5 \end{cases}$ $\rightarrow$ $\begin{array}{l} 6x + 3y + 6z = 0 \\ \underline{6x - 4y - 6z = 2} \\ 12x - y = 2 \end{array}$

$\begin{cases} 3x - 2y - 3z = 1 \\ -x + 3y + z = -5 \end{cases}$ $\rightarrow$ $\begin{array}{l} 3x - 2y - 3z = 1 \\ \underline{-3x + 9y + 3z = -15} \\ 7y = -14 \\ y = -2 \end{array}$

$\begin{array}{ll} 12x - y = 2 & 2x + y + 2z = 0 \\ 12x - (-2) = 2 & 2(0) + (-2) + 2z = 0 \\ 12x = 0 & 2z = 2 \\ x = 0 & z = 1 \end{array}$

$(0, -2, 1)$

53. $\begin{array}{l} x^2 - 2x - 15 = 0 \\ (x + 3)(x - 5) = 0 \\ x = -3 \text{ or } x = 5 \end{array}$

Check: $\begin{array}{l} (-3)^2 - 2(-3) - 15 \stackrel{?}{=} 0 \\ 0 = 0 \\ 5^2 - 2(5) - 15 \stackrel{?}{=} 0 \\ 0 = 0 \end{array}$

54. $\dfrac{-b \pm \sqrt{b^2 - 4ac}}{2a} = \dfrac{-5 \pm \sqrt{5^2 - 4(2)(1)}}{2(2)} = \dfrac{-5 \pm \sqrt{17}}{4}$;

Check: $2\left(\dfrac{-5 \pm \sqrt{17}}{4}\right)^2 + 5\left(\dfrac{-5 \pm \sqrt{17}}{4}\right) + 1 \stackrel{?}{=} 0$,

$0 = 0$

55. $2x^2 = 5, x^2 = \dfrac{5}{2}, x = \pm\sqrt{\dfrac{5}{2}}, x = \dfrac{\sqrt{5}}{\sqrt{2}} \cdot \dfrac{\sqrt{2}}{\sqrt{2}}$,

$x = \pm\dfrac{\sqrt{10}}{2}; 2x^2 = 5, 2\left(\dfrac{\pm\sqrt{10}}{2}\right)^2 \stackrel{?}{=} 5, 5 = 5$

56. D; $\sqrt{\dfrac{3}{8}} = \dfrac{\sqrt{3}}{\sqrt{8}} = \dfrac{\sqrt{3}}{\sqrt{8}} \cdot \dfrac{\sqrt{8}}{\sqrt{8}} = \dfrac{\sqrt{24}}{8} = \dfrac{2\sqrt{6}}{8} - \dfrac{\sqrt{6}}{4}$

PROJECT CONNECTION

1–4. Answers will vary.

ALGEBRA WORKS, PAGE 391

1. $v = (3.7 \times 10^{-10})\sqrt{\dfrac{(5.98 \times 10^{24})}{6376}} \approx 11.3312$;

11.3 km/s

2a. $\dfrac{1}{2}(24 + 24 + 24) = 36$; 36 cm

2b. $A = \sqrt{36(36 - 24)(36 - 24)(36 - 24)} =$
$\sqrt{(3 \cdot 12)(12)(12)(12)} = \sqrt{3(144)^2} = 144\sqrt{3}$ cm^2

2c. $144\sqrt{3} \cdot \sqrt{0.12} = 144 \times 0.6 = 86.4$; 86.4 watts

3. Use the Pythagorean theorem.

$\begin{array}{l} \text{Legs} = r, d \\ \text{Hypotenuse} = r + h \\ r^2 + d^2 = (r + h)^2 \\ r^2 + d^2 = r^2 + 2rh + h^2 \\ d^2 = 2rh + h^2 \\ d = \sqrt{2rh + h^2} \end{array}$

Lesson 8.5, pages 392–398

EXPLORE

1. $x + 2 = -2x - 1, 3x = -3, x = -1$

2. Set $y_1 = x + 2$ and $y_2 = -2x - 1$. Then graph the two equations on the same set of axes.
The x-coordinate of the point of intersection $(-1, 1)$ is the solution of the equation, $x = -1$.

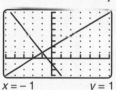

$x = -1 \qquad y = 1$

3. $x + 2 = -2x - 1, (x + 2)^2 = (-2x - 1)^2, x^2 + 2x + 2x + 4 = 4x^2 + 2x + 2x + 1, x^2 + 4x + 4 = 4x^2 + 4x + 1$

4. $x^2 + 4x + 4 = 4x^2 + 4x + 1, 3x^2 - 3 = 0, 3(x^2 - 1) = 0, (x + 1)(x - 1) = 0, x = -1$ or $x = 1$

5. The graphs of trhe two equations show points of intersection at $x = -1$ and $x = 1$.

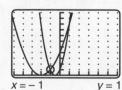 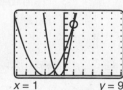

$x = -1 \qquad y = 1 \qquad\quad x = 1 \qquad y = 9$

6. Answers will vary. Students should note that the solution of the linear equation, -1, was also a solution of the quadratic. However, the quadratic had an additional solution. Squaring appears to preserve the solutions of an equation but also to produce new solutions.

TRY THESE

1. $\sqrt{x-4} = 7, (\sqrt{x-4})^2 = 7^2, x - 4 = 49, x = 53$;
Check: $\sqrt{53-4} \stackrel{?}{=} 7, 7 = 7$; The solution is 53.

2. $3 = \sqrt{2x-7} - 2, 3 + 2 = \sqrt{2x-7}, (3+2)^2 = (\sqrt{2x-7})^2, 25 = 2x - 7, 32 = 2x, x = 16$;
Check: $3 \stackrel{?}{=} \sqrt{2(16)-7} - 2, 3 = 3$; The solution is 16.

3. $\sqrt[3]{x} = -4, (\sqrt[3]{x})^3 = -4^3, x = -64$;
Check: $\sqrt[3]{-64} \stackrel{?}{=} -4, -4 = -4$; The solution is -64.

4. $\sqrt[4]{3x-5} + 11 = 18, \sqrt[4]{3x-5} = 7,$
$(\sqrt[4]{3x-5})^4 = 7^4, 3x - 5 = 2401, 3x = 2406,$
$x = 802$; Check: $\sqrt[4]{3(802)-5} + 11 \stackrel{?}{=} 18, 18 = 18$;
The solution is 802.

5. $x^{\frac{1}{7}} = -3, x = -3^7, x = -2187$; Check: $-2187^{\frac{1}{7}} \stackrel{?}{=} -3,$
$-3 = -3$; The solution is -2187.

6. $x^{\frac{3}{5}} = 1.728, x = 1.728^{\frac{5}{3}}, x = 2.48832$;
Check: $2.48832^{\frac{3}{5}} \stackrel{?}{=} 1.728, 1.728 = 1.728$;
The solution is 2.48832.

7. $\sqrt{x-5} = \sqrt{4x-29}, (\sqrt{x-5})^2 = (\sqrt{4x-29})^2,$
$x - 5 = 4x - 29, 3x = 24, x = 8$;
Check: $\sqrt{8-5} \stackrel{?}{=} \sqrt{4(8)-29}, \sqrt{3} = \sqrt{3}$;
The solution is 8.

8. $\sqrt{x+7} - x = 1, \sqrt{x+7} = x + 1,$
$(\sqrt{x+7})^2 = (x+1)^2, x + 7 = x^2 + 2x + 1,$
$x^2 + x - 6 = 0, (x+3)(x-2) = 0, x = -3$ or
$x = 2$;
Check: $\sqrt{-3+7} - (-3) \stackrel{?}{=} 1; 5 \neq 1$;
$\sqrt{2+7} - 2 \stackrel{?}{=} 1, 1 = 1$; The solution is 2.

9. $\sqrt{x+12} - \sqrt{x} = 2, \sqrt{x+12} = \sqrt{x} + 2,$
$(\sqrt{x+12})^2 = (\sqrt{x}+2)^2, x + 12 = x + 4\sqrt{x} + 4,$
$8 = 4\sqrt{x}, 8^2 = (4\sqrt{x})^2, 64 = 16x, x = 4$;
Check: $\sqrt{4+12} - \sqrt{4} \stackrel{?}{=} 2, 2 = 2$;
The solution is 4.

10. $r = \sqrt[3]{\frac{GMt^2}{4\pi^2}}, r^3 = \left(\sqrt[3]{\frac{GMt^2}{4\pi^2}}\right)^3, r^3 = \frac{GMt^2}{4\pi^2}, \frac{4\pi^2 r^3}{Gt^2} = M$

11. $r = \left(\frac{3V}{4\pi}\right)^{\frac{1}{3}}, r^3 = \left[\left(\frac{3V}{4\pi}\right)^{\frac{1}{3}}\right]^3, r^3 = \frac{3V}{4\pi}, 4\pi r^3 = 3V,$
$V = \frac{4\pi r^3}{3}; V = \frac{4\pi r^3}{3} = \frac{4\pi(0.145)^3}{3} \approx 0.0128;$
0.0128 cm^3

12. The solution to the equation $x = 3$ is 3. To obtain the second equation, both sides of the first equation are squared. The new equation $x^2 = 9$ has 3 and -3 as its solutions. Therefore, an additional solution $x = -3$ has been introduced by squaring the first equation.

PRACTICE

1. $\sqrt{x} = 11, (\sqrt{x})^2 = 11^2, x = 121$;
$\sqrt{121} = 11, 11 = 11$; The solution is 121.

2. $-6 = -\sqrt{x-9}, (-6)^2 = (-\sqrt{x-9})^2, 36 = x - 9, x = 45; -6 = -\sqrt{45-9}, -6 = -6$;
The solution is 45.

3. $\sqrt{x+11} - 7 = 3, \sqrt{x+11} = 10,$
$(\sqrt{x+11})^2 = 10^2, x + 11 = 100, x = 89;$
$\sqrt{89+11} - 7 = 3, 3 = 3$; The solution is 89.

4. $-\sqrt{3x-2} + 6 = 1, -\sqrt{3x-2} = -5,$
$-(\sqrt{3x-2})^2 = (-5)^2,$
$3x - 2 = 25, 3x = 27, x = 9;$
$-\sqrt{3(9)-2} + 6 = 1, 1 = 1$; The solution is 9.

5. $8 - \sqrt{\frac{x}{4}} = -1, -\sqrt{\frac{x}{4}} = -9, \left(-\sqrt{\frac{x}{4}}\right)^2 = (-9)^2,$
$\frac{x}{4} = 81, x = 324; 8 - \sqrt{\frac{324}{4}} = -1, -1 = -1;$
The solution is 324.

6. $16 = 9 + \sqrt{\frac{9x-1}{2}}, 7 = \sqrt{\frac{9x-1}{2}}, 7^2 = \left(\sqrt{\frac{9x-1}{2}}\right)^2, 49 = \frac{9x-1}{2}, 98 = 9x - 1, 9x = 99,$
$x = 11; 16 = 9 + \sqrt{\frac{9(11)-1}{2}}, 16 = 16;$
The solution is 11.

7. $-3 = \sqrt[3]{x}, -3^3 = (\sqrt[3]{x})^3, -27 = x;$
$-3 \stackrel{?}{=} \sqrt[3]{-27}, -3 = -3$; The solution is -27.

8. $\sqrt[3]{-x-6} - 3 = -9, \sqrt[3]{-x-6} = -6,$
$(\sqrt[3]{-x-6})^3 = -6^3, -x - 6 = -216, -x = -210,$
$x = 210; \sqrt[3]{-210-6} - 3 \stackrel{?}{=} -9,$
$-9 = -9$; The solution is 210.

9. $\sqrt[4]{5x+16} + 11 = 14, \sqrt[4]{5x+16} = 3,$
$(\sqrt[4]{5x+16})^4 = 3^4, 5x + 16 = 81, 5x = 65, x = 13;$
$\sqrt[4]{5(13)+16} + 11 \stackrel{?}{=} 14, 14 = 14$; The solution is 13.

10. $2 - \sqrt[4]{\frac{5x+8}{3}} = 0, 2 = \sqrt[4]{\frac{5x+8}{3}}, 2^4 = \left(\sqrt[4]{\frac{5x+8}{3}}\right)^4,$
$16 = \frac{5x+8}{3}, 48 = 5x + 8, 40 = 5x, x = 8;$
$2 - \sqrt[4]{\frac{5(8)+8}{3}} \stackrel{?}{=} 0, 0 = 0$; The solution is 8.

11. $\sqrt[5]{\frac{1}{x}} - 1 = -0.8, \sqrt[5]{\frac{1}{x}} = 0.2, \left(\sqrt[5]{\frac{1}{x}}\right)^5 = 0.2^5,$
$\frac{1}{x} = 0.00032, x = 3125; \sqrt[5]{\frac{1}{3125}} - 1 \stackrel{?}{=} -0.8;$
$-0.8 = -0.8$. The solution is 3125.

12. $\sqrt[7]{x^7} = 3.62, x = 3.62; \sqrt[7]{3.62^7} \stackrel{?}{=} 3.62,$
$3.62 = 3.62$; The answer is 3.62.

13. $x^{\frac{1}{2}} - 8 = 0$, $\sqrt{x} = 8$, $(\sqrt{x})^2 = 8^2$, $x = 64$;

$64^{\frac{1}{2}} - 8 \stackrel{?}{=} 0$, $0 = 0$; The solution is 64.

14. $x^{\frac{1}{5}} + 3 = 1$, $\sqrt[5]{x} = -2$, $(\sqrt[5]{x})^5 = -2^5$, $x = -32$;

$(-32)^{\frac{1}{5}} + 3 \stackrel{?}{=} 1$, $1 = 1$; The solution is -32.

15. $1000 = x^{\frac{3}{2}}$, $1000^{\frac{2}{3}} = \left(x^{\frac{3}{2}}\right)^{\frac{2}{3}}$, $100 = x$;

$100^{\frac{3}{2}} \stackrel{?}{=} 1000$, $1000 = 1000$;
The solution is 100.

16. $(3x + 5)^{\frac{2}{3}} = 25$, $\left[(3x + 5)^{\frac{2}{3}}\right]^{\frac{3}{2}} = 25^{\frac{3}{2}}$,

$3x + 5 = 125$, $3x = 120$, $x = 40$;

$[3(40) + 5]^{\frac{2}{3}} \stackrel{?}{=} 25$, $25 = 25$; The solution is 40.

17. $(4x + 3.713)^{\frac{3}{4}} = 6.859$, $\left[(4x + 3.713)^{\frac{3}{4}}\right]^{\frac{4}{3}} = 6.859^{\frac{4}{3}}$,

$4x + 3.713 = 13.0321$, $4x = 9.3191$, $x = 2.329775$;

$(4(2.329775) + 3.713)^{\frac{3}{4}} \stackrel{?}{=} 6.859$; $6.895 = 6.859$; The
solution is 2.329775.

18. $\left(\frac{1}{x} - 0.00007\right)^{0.8} = 0.0081$, $\left[\left(\frac{1}{x} - 0.00007\right)^{0.8}\right]^{1.25} =$

$0.0081^{1.25}$, $\frac{1}{x} - 0.00007 = 0.00243$, $\frac{1}{x} = 0.0025$,

$0.0025x = 1$, $x = 400$; $\left(\frac{1}{400} - 0.00007\right)^{0.8} \stackrel{?}{=} 0.0081$,

$0.0081 = 0.0081$; The solution is 400.

19a. $R = \sqrt{F_1^2 + F_2^2}$, $R^2 = F_1^2 + F_2^2$, $F_1^2 = R^2 - F_2^2$,

$F_1 = \sqrt{R^2 - F_2^2}$

19b. $F_1 = \sqrt{1000^2 - 800^2}$, $F_1 = 600$ lb

20. There is no value of x that makes both sides of the
equation 0, and $\sqrt{x - 4}$ cannot equal a negative
number.

21. $\sqrt{4x + 4} = \sqrt{5x - 4}$, $(\sqrt{4x + 4})^2 = (\sqrt{5x - 4})^2$,

$4x + 4 = 5x - 4$, $x = 8$; $\sqrt{4(8) + 4} \stackrel{?}{=}$
$\sqrt{5(8) - 4}$, $6 = 6$; The solution is 8.

22. $\sqrt{\frac{1}{x + 1}} = \sqrt{\frac{x - 1}{8}}$, $\left(\sqrt{\frac{1}{x + 1}}\right)^2 = \left(\sqrt{\frac{x - 1}{8}}\right)^2$,

$\frac{1}{x + 1} = \frac{x - 1}{8}$, $8 = (x + 1)(x - 1)$,

$8 = x^2 + x - x - 1$, $8 = x^2 - 1$, $9 = x^2$,

$x = \pm 3$; $\sqrt{\frac{1}{-3 + 1}} \stackrel{?}{=} \sqrt{\frac{-3 - 1}{8}}$, $\sqrt{-\frac{1}{2}} = \sqrt{-\frac{4}{8}}$,

not real; $\sqrt{\frac{1}{-3 + 1}} \stackrel{?}{=} \sqrt{\frac{3 - 1}{8}}$, $\frac{1}{2} = \frac{1}{2}$;

The solution is 3.

23. $3\sqrt{2x + 3} = \sqrt{x + 10}$, $(3\sqrt{2x + 3})^2 = (\sqrt{x + 10})^2$,

$9(2x + 3) = x + 10$, $18x + 27 = x + 10$,

$17x = -17$, $x = -1$; $3\sqrt{2(-1) + 3} \stackrel{?}{=}$
$\sqrt{-1 + 10}$, $3 = 3$; The solution is -1.

24. $\sqrt{x + 4} = x - 2$, $(\sqrt{x + 4})^2 = (x - 2)^2$,

$x + 4 = x^2 - 4x + 4$, $x^2 - 5x = 0$, $x(x - 5) = 0$,

$x = 0$ or $x = 5$; $\sqrt{0 + 4} \stackrel{?}{=} 0 - 2$, $2 \neq -2$;

$\sqrt{5 + 4} \stackrel{?}{=} 5 - 2$, $3 = 3$; The solution is 5.

25. $\sqrt{x + 16} - 4 = x + 6$, $\sqrt{x + 16} = x + 10$,

$(\sqrt{x + 16})^2 = (x + 10)^2$, $x + 16 =$

$x^2 + 20x + 100$, $x^2 + 19x + 84 = 0$,

$(x + 12)(x + 7) = 0$, $x = -12$ or $x = -7$;

$\sqrt{-12 + 16} - 4 \stackrel{?}{=} -12 + 6$, $-2 \neq -6$;

$\sqrt{-7 + 16} - 4 \stackrel{?}{=} -7 + 6$, $-1 = -1$; The solution
is -7.

26. $x - \sqrt{3x + 1} = 3$, $x - 3 = \sqrt{3x + 1}$, $(x - 3)^2 =$

$(\sqrt{3x + 1})^2$, $x^2 - 6x + 9 = 3x + 1$,

$x^2 - 9x + 8 = 0$, $(x - 1)(x - 8) = 0$, $x = 1$ or
$x = 8$; $1 - \sqrt{3(1) + 1} \stackrel{?}{=} 3$, $-1 \neq 3$;

$8 - \sqrt{3(8) + 1} = 3$, $3 = 3$; The solution is 8.

27. $\sqrt{x + 10} = 8 - \sqrt{x - 6}$, $(\sqrt{x + 10})^2 =$

$(8 - \sqrt{x - 6})^2$, $x + 10 =$

$64 - 16\sqrt{x - 6} + x - 6$, $-48 = -16\sqrt{x - 6}$,

$48 = 16\sqrt{x - 6}$, $48^2 = (16\sqrt{x - 6})^2$, $2304 =$

$256(x - 6)$, $2304 = 256x - 1536$, $3840 = 256x$,

$x = 15$; $\sqrt{15 + 10} \stackrel{?}{=} 8 - \sqrt{15 - 6}$, $5 = 5$;
The solution is 15.

28. $\sqrt{3x + 3} - \sqrt{6x + 7} = -1$, $\sqrt{3x + 3} =$

$\sqrt{6x + 7} - 1$, $(\sqrt{3x + 3})^2 = (\sqrt{6x + 7} - 1)^2$,

$3x + 3 = 6x + 7 - 2\sqrt{6x + 7} + 1$,

$-3x - 5 = 2\sqrt{6x + 7}$, $(-3x - 5)^2 =$

$(2\sqrt{6x + 7})^2$, $9x^2 + 30x + 25 = 4(6x + 7)$,

$9x^2 + 30x + 25 = 24x + 28$, $9x^2 + 6x - 3 = 0$,

$3x^2 + 2x - 1 = 0$, $(3x - 1)(x + 1) = 0$, $x = \frac{1}{3}$

or $x = -1$;

$\sqrt{3\left(\frac{1}{3}\right) + 3} - \sqrt{6\left(\frac{1}{3}\right) + 7} \stackrel{?}{=} -1$, $-1 = -1$;

$\sqrt{3(-1) + 3} - \sqrt{6(-1) + 7} \stackrel{?}{=} -1$, $-1 = -1$;
The solutions are $\frac{1}{3}$ and -1.

29. $\sqrt{x + 5} + \sqrt{1 - 2x} = 4$,
$\sqrt{x + 5} = 4 - \sqrt{1 - 2x}$,
$(\sqrt{x + 5})^2 = (4 - \sqrt{1 - 2x})^2$,
$x + 5 = 16 - 8\sqrt{1 - 2x} + 1 - 2x$,
$3x - 12 = -8\sqrt{1 - 2x}$, $(3x - 12)^2 =$
$(-8\sqrt{1 - 2x})^2$, $9x^2 - 72x + 144 = 64(1 - 2x)$,
$9x^2 - 72x + 144 =$
$64 - 128x$, $9x^2 + 56x + 80 = 0$,

$(x + 4)(9x + 20) = 0, x = -4 \text{ or } x = -\dfrac{20}{9};$

$\sqrt{-4 + 5} + \sqrt{1 - 2(-4)} \overset{?}{=} 4, 4 = 4;$

$\sqrt{-\dfrac{20}{9} + 5} + \sqrt{1 - 2\left(-\dfrac{20}{9}\right)} \overset{?}{=} 4, 4 = 4;$

The solutions are -4 and $-\dfrac{20}{9}$.

30. $1 - \sqrt{3x + 1} + \sqrt{x + 4} = 0,$
$1 - \sqrt{3x + 1} = -\sqrt{x + 4};$
$(1 - \sqrt{3x + 1})^2 = (-\sqrt{x + 4})^2, 1 - 2\sqrt{3x + 1} +$
$3x + 1 = x + 4, 2x - 2 = 2\sqrt{3x + 1},$
$(2x - 2)^2 = (2\sqrt{3x + 1})^2, 4x^2 - 8x + 4 =$
$4(3x + 1), 4x^2 - 8x + 4 = 12x + 4,$
$4x^2 - 20x = 0, 4x(x - 5) = 0, x = 0 \text{ or } x = 5;$
$1 - \sqrt{3(0) + 1} + \sqrt{0 + 4} \overset{?}{=} 0, 2 \neq 0;$
$1 - \sqrt{3(5) + 1} + \sqrt{5 + 4} \overset{?}{=} 0, 0 = 0;$
The solution is 5.

31. $f = \sqrt{\dfrac{\pi F^2}{4A}}, f^2 = \dfrac{\pi F^2}{4A}, F^2 = \dfrac{f^2 4A}{\pi}, F = \sqrt{\dfrac{4AF^2}{\pi}}$

32. $r = \sqrt[3]{\dfrac{A}{P}} - 1, r + 1 = \sqrt[3]{\dfrac{A}{P}}, (r + 1)^3 = \dfrac{A}{P},$
$A = P(r + 1)^3$

33. $a = \sqrt[3]{p^2}, a^3 = p^2, p = \sqrt{a^3}$

34. $E = Z\dfrac{\sigma}{\sqrt{n}}, E\sqrt{n} = Z\sigma, \sqrt{n} = \dfrac{Z\sigma}{E}, n = \left(\dfrac{Z\sigma}{E}\right)^2,$
$n = \dfrac{Z^2\sigma^2}{E^2}$

35. $1260 = \left(\dfrac{3h}{2}\right)^{\frac{1}{2}}, 1260^2 = \dfrac{3h}{2}, \dfrac{2 \cdot 1260^2}{3} = h,$
$h = 1{,}058{,}400; 1{,}058{,}400 \div 5280 \approx 200;$
200 mi

36a. $13 = \sqrt{(-1 - 4)^2 + (y - (-2)^2},$
$13 = \sqrt{(-5)^2 + (y + 2)^2},$
$13 = \sqrt{25 + y^2 + 4y + 4}, 13 = \sqrt{y^2 + 4y + 29},$
$169 = y^2 + 4y + 29, y^2 + 4y - 140 = 0,$
$(y + 14)(y - 10) = 0, y = -14 \text{ or } y = 10$

36b. $37 = \sqrt{(x - (-15)^2 + (-11 - 24)^2},$
$37 = \sqrt{(x + 15)^2 + (-35)^2},$
$37 = \sqrt{x^2 + 30x + 225 + 1225},$
$37^2 = (\sqrt{x^2 + 30x + 1450})^2,$
$1369 = x^2 + 30x + 1450,$
$x^2 + 30x + 81 = 0, (x + 27)(x + 3) = 0, x = -27$
or $x = -3$.

36c. $d = \sqrt{(x_2 - x_1)^2 + (y_2 - y_1)^2}$
$d^2 = (\sqrt{(x_2 - x_1)^2 + (y_2 - y_1)^2})^2,$
$d^2 = (x_2 - x_1)^2 + (y_2 - y_1)^2,$
$(x_2 - x_1)^2 = d^2 - (y_2 - y_1)^2,$
$\sqrt{(x_2 - x_1)^2} = \sqrt{d^2 - (y_2 - y_1)^2},$
$x_2 - x_1 = \sqrt{d^2 - (y_2 - y_1)^2},$
$x_2 = x_1 + \sqrt{d^2 - (y_2 - y_1)^2}$

37. $2.8 = \dfrac{d^{\frac{9}{4}}}{400(1.2)}, (2.8 \cdot 400 \cdot 1.2)^{\frac{4}{9}} = d, d \approx 24.6;$
24.6 m

38a. $T = 2\pi\sqrt{\dfrac{L}{32}}, T^2 = \left(2\pi\sqrt{\dfrac{L}{32}}\right)^2, T^2 = 4\pi^2\left(\dfrac{L}{32}\right),$
$L = \dfrac{32T^2}{4\pi^2}, L = \dfrac{8T^2}{\pi^2}$

38b. $L = \dfrac{8(3^2)}{\pi^2} \approx 7.30; 7.30$ ft

39. $x - 7\sqrt{x} + 12 = 0, (\sqrt{x} - 3)(\sqrt{x} - 4) = 0,$
$\sqrt{x} = 3 \text{ or } \sqrt{x} = 4, x = 9 \text{ or } x = 16;$
$9 - 7\sqrt{9} + 12 \overset{?}{=} 0, 0 = 0; 16 - 7\sqrt{16} + 12 \overset{?}{=} 0,$
$0 = 0;$ The solutions are 9 and 16.

40. $x - 11\sqrt{x} + 30 = 0, (\sqrt{x} - 6)(\sqrt{x} - 5) = 0,$
$\sqrt{x} = 6 \text{ or } \sqrt{x} = 5, x = 36 \text{ or } x = 25;$
$36 - 11\sqrt{36} + 30 \overset{?}{=} 0, 0 = 0;$
$25 - 11\sqrt{25} + 30 \overset{?}{=} 0, 0 = 0;$
The solutions are 25 and 36.

41. $x - 18 = -7\sqrt{x}, x + 7\sqrt{x} - 18 = 0,$
$(\sqrt{x} + 9)(\sqrt{x} - 2) = 0, \sqrt{x} = -9 \text{ or }$
$\sqrt{x} = 2, x = 81 \text{ or } x = 4; 81 - 18 = -7\sqrt{81},$
$63 \neq -63; 4 - 18 \overset{?}{=} -7\sqrt{4}, -14 = -14;$
The solution is 4.

42. $2x - 5\sqrt{x} - 12 = 0, (2\sqrt{x} + 3)(\sqrt{x} - 4) = 0;$
$2\sqrt{x} = -3 \text{ or } \sqrt{x} = 4, x = \left(-\dfrac{3}{2}\right)^2 = \dfrac{9}{4} \text{ or } x = 16;$
$2\left(\dfrac{9}{4}\right) - 5\sqrt{\dfrac{9}{4}} - 12 \overset{?}{=} 0, -15 \neq 0;$
$2(16) - 5\sqrt{16} - 12 \overset{?}{=} 0, 0 = 0;$ The solution is 16.

43. $\sqrt{x + 1} - \sqrt{x - 2} - \sqrt{4x - 11} = 0,$
$\sqrt{x + 1} - \sqrt{x - 2} = \sqrt{4x - 11},$
$(\sqrt{x + 1} - \sqrt{x - 2})^2 = (\sqrt{4x - 11})^2,$
$x + 1 - 2\sqrt{(x + 1)(x - 2)} + x - 2 = 4x - 11,$
$2x - 10 = 2\sqrt{(x + 1)(x - 2)},$
$(2x - 10)^2 = (2\sqrt{(x + 1)(x - 2)})^2,$
$4x^2 - 40x + 100 = 4(x + 1)(x - 2),$
$4x^2 - 40x + 100 = 4x^2 - 4x - 8,$
$-36x = -108, x = 3;$
$\sqrt{3 + 1} - \sqrt{3 - 2} - \sqrt{4(3) - 11} \overset{?}{=} 0, 0 = 0;$
The solution is 3.

44. $\sqrt{\sqrt{x - 4} + x} = 2, \left(\sqrt{\sqrt{x - 4} + x}\right)^2 = 2^2,$
$\sqrt{x - 4} + x = 4, \sqrt{x - 4} = 4 - x, (\sqrt{x - 4})^2 =$
$(4 - x)^2, x - 4 = 16 - 8x + x^2,$
$x^2 - 9x + 20 = 0, (x - 5)(x - 4) = 0, x = 5 \text{ or }$
$x = 4; \sqrt{\sqrt{5 - 4} + 5} \overset{?}{=} 2, \sqrt{6} \neq 2;$
$\sqrt{\sqrt{4 - 4} + 4} \overset{?}{=} 2; 2 = 2;$ The solution is 4.

45. 0 is the only solution. For any other real number the left side is positive and the right side is negative.

46. orbital period in seconds: $365 \times 24 \times 60 \times 60 = 31{,}536{,}000$; orbital radius in meters:
$(9.3 \times 10^7) \times 1609 \approx 1.5 \times 10^{11}$;
$M = \dfrac{4\pi^2 r^3}{Gt^2} = \dfrac{4\pi^2(1.5 \times 10^{11})^3}{(6.67 \times 10^{-11})(31{,}536{,}000^2)} \approx$
2×10^{30} kg; $(5.98 \times 10^{24})x = 2 \times 10^{30}$;
$x \approx 3.3 \times 10^5$; The sun is about 330,000 times as massive as the Earth.

MIXED REVIEW

47. $D; \dfrac{5}{6}$

48. $P = 2l + 2w, 2w = P - 2l, w = \dfrac{P - 2l}{2}$

49. $C = \dfrac{5}{9}(F - 32), \dfrac{9}{5}C = F - 32, F = \dfrac{9}{5}C + 32$

50. $A = \dfrac{1}{2}h(a + b), 2A = h(a + b),$
$\dfrac{2A}{h} = a + b, a = \dfrac{2A}{h} - b$

51. $g(x) = -3x + 2; f(g(x)) = 2(-3x + 2) = -6x - 4$

52. $g(x) = x + 1; f(g(x)) = (x + 1)^2 = x^2 + 2x + 1$

53. $g(x) = \sqrt{x}; f(g(x)) = \dfrac{1}{\sqrt{x}} = \dfrac{1}{\sqrt{x}} \cdot \dfrac{\sqrt{x}}{\sqrt{x}} = \dfrac{\sqrt{x}}{x}$

54. $g(x) = -\dfrac{3}{4}; f(g(x)) = 16^{-\frac{3}{4}} = \dfrac{1}{16^{\frac{3}{4}}} = \dfrac{1}{(\sqrt[4]{16})^3} =$
$\dfrac{1}{2^3} = \dfrac{1}{8}$

55. $\sqrt{x + 3} - \sqrt{x - 2} = 1, \sqrt{x + 3} = \sqrt{x - 2} + 1,$
$(\sqrt{x + 3})^2 = (\sqrt{x - 2} + 1)^2,$
$x + 3 = x - 2 + 2\sqrt{x - 2} + 1,$
$4 = 2\sqrt{x - 2}, 2 = \sqrt{x - 2}, 2^2 = (\sqrt{x - 2})^2,$
$4 = x - 2, x = 6$

PROJECT CONNECTION

1–2. Answers will vary.

Lesson 8.6, pages 399–404

EXPLORE

1. $x^2 - x - 6 = 0, (x - 3)(x + 2) = 0, x = 3$
or $x = -2$

2.

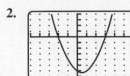

The solutions of the equation are the x-intercepts 3 and -2, the x-coordinates of the points where the graph crosses the x-axis.

3.

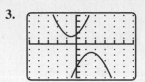

4. The method doesn't work because the graphs do not intersect the x-axis.

5. $25; -3; -7$; Students should note that the discriminant of the equation whose graph intersects the x-axis and which has real number solutions is positive, while the discriminant of the equations whose graphs do not intersect the x-axis and which have no real number solutions is negative.

TRY THESE

1. $\sqrt{-25} = \sqrt{25}\sqrt{-1} = i\sqrt{5}$

2. $3\sqrt{-200} = 3\sqrt{100}\sqrt{2}\sqrt{-1} = 30i\sqrt{2}$

3. $\sqrt{-\dfrac{1}{2}} = \dfrac{\sqrt{-1}}{\sqrt{2}} = \dfrac{i}{\sqrt{2}} = \dfrac{i}{\sqrt{2}} \cdot \dfrac{\sqrt{2}}{\sqrt{2}} = \dfrac{i\sqrt{2}}{2}$

4. $(\sqrt{-15})^2 = -15$

5. $i^{17} = (i^8)^2 \cdot i = 1 \cdot i = i$

6. $-5i^{46} = -5 \cdot (i^6)^7 \cdot i^4 = -5 \cdot (-1) \cdot 1 = 5$

7. $14i - 9i + 3i = 8i$

8. $10i \cdot (-3i) = -30i^2 = -30 \cdot (-1) = 30$

9. $20 \div 5i = \dfrac{20}{5i} \cdot \dfrac{i}{i} = \dfrac{20i}{5i^2} = \dfrac{20i}{-5} = -4i$

10. $\sqrt{-2} + \sqrt{-18} = i\sqrt{2} + 3i\sqrt{2} = 4i\sqrt{2}$

11. $\sqrt{-3}\sqrt{-5} = i\sqrt{3} \cdot i\sqrt{5} = i^2\sqrt{15} =$
$-1\sqrt{15} = -\sqrt{15}$

12. $\dfrac{10i}{\sqrt{-5}} = \dfrac{10i}{i\sqrt{5}} = \dfrac{10}{\sqrt{5}} = \dfrac{10}{\sqrt{5}} \cdot \dfrac{\sqrt{5}}{\sqrt{5}} = \dfrac{10\sqrt{5}}{5} = 2\sqrt{5}$

13. $\dfrac{-b \pm \sqrt{b^2 - 4ac}}{2a} = \dfrac{-0 \pm \sqrt{0^2 - 4(1)(81)}}{2(1)} =$
$\dfrac{\pm\sqrt{-324}}{2} = \dfrac{\pm 18i}{2} = \pm 9i$

14. $\dfrac{-b \pm \sqrt{b^2 - 4ac}}{2a} = \dfrac{-0 \pm \sqrt{0^2 - 4(6)(4)}}{2(6)} =$
$\dfrac{\pm\sqrt{-96}}{12} = \dfrac{\pm 4i\sqrt{6}}{12} = \dfrac{\pm i\sqrt{6}}{3}$

15. $\dfrac{-b \pm \sqrt{b^2 - 4ac}}{2a} = \dfrac{-(-6) \pm \sqrt{(-6)^2 - 4(3)(4)}}{2(3)} =$
$\dfrac{6 \pm \sqrt{-12}}{6} = \dfrac{6 \pm 2i\sqrt{3}}{6} = \dfrac{3 \pm i\sqrt{3}}{3}$

16. $4x^2 - 12x + 9 = 0;$
$b^2 - 4ac = (-12)^2 - 4(4)(9) = 0$, so the solutions are real and equal.

17. $3x^2 - 2x + 1 = 0;$
$b^2 - 4ac = (-2)^2 - 4(3)(1) = -8$, so the solutions are imaginary.

18. $5x^2 + 2x - 3 = 0; b^2 - 4ac =$
$2^2 - 4(5)(-3) = 64$, so the solutions are real
and unequal.

19. $X_T = X_L - X_C, X_T = 11.61i - 9.45i = 2.16i$;
$2.16i$ ohms

20. Since $\sqrt[3]{64} = 4 = 4 + 0i$, the number belongs to the
complex numbers, real numbers, rational numbers.
Since $\sqrt{-64} = i\sqrt{64} = 8i = 0 + 8i$, the number
belongs to the complex numbers, the imaginary
numbers, and the pure imaginary numbers. The number
$2 + i\sqrt{2}$ belongs to the complex numbers and
imaginary numbers.

Practice

1. $\sqrt{-98} = \sqrt{49}\sqrt{2}\sqrt{-1} = 7i\sqrt{2}$

2. $\sqrt{-800} = \sqrt{400}\sqrt{2}\sqrt{-1} = 20i\sqrt{2}$

3. $-n\sqrt{-36} = -n\sqrt{36}\sqrt{-1} = -6ni$

4. $i\sqrt{-45} = i\sqrt{9}\sqrt{5}\sqrt{-1} = 3i^2\sqrt{5} = -3\sqrt{5}$

5. $3i(2i\sqrt{-16}) = 3i(2i\sqrt{16}\sqrt{-1}) = 3i(8i^2) = $
$24i^3 = -24i$

6. $-i(8i)\sqrt{-243} = -i(8i)\sqrt{81}\sqrt{3}\sqrt{-1} = $
$-i(8i)9i\sqrt{3} = -72i^3\sqrt{3} = 72i\sqrt{3}$

7. $(\sqrt{-11})^2 = -11$

8. $i(\sqrt{-18})^2 = -18i$

9. $\sqrt{-\dfrac{4}{9}} = \dfrac{\sqrt{-4}}{\sqrt{9}} = \dfrac{2}{3}i$

10. $\sqrt{-\dfrac{15}{16}} = \dfrac{\sqrt{-15}}{\sqrt{16}} = \dfrac{i\sqrt{15}}{4}$

11. $\dfrac{2\sqrt{5}}{\sqrt{-3}} = \dfrac{2\sqrt{5}}{\sqrt{-3}} \cdot \dfrac{\sqrt{-3}}{\sqrt{-3}} = \dfrac{2\sqrt{5}i\sqrt{3}}{-3} = -\dfrac{2i\sqrt{15}}{3}$

12. $\dfrac{3\sqrt{-6}}{4\sqrt{-12}} = \dfrac{3\sqrt{-6}}{4\sqrt{-12}} \cdot \dfrac{\sqrt{-12}}{\sqrt{-12}} = \dfrac{3\sqrt{-6}\sqrt{-12}}{4(-12)} = $
$\dfrac{3i\sqrt{6}\,2i\sqrt{3}}{4(-12)} = \dfrac{6i^2\sqrt{18}}{-48} = \dfrac{18i^2\sqrt{2}}{-48} = \dfrac{18\sqrt{2}}{48} = \dfrac{3\sqrt{2}}{8}$

13. $i^{13} = (i^6)^2 \cdot i = (-1)^2 \cdot i = i$

14. $i^{104} = (i^8)^8 \cdot (i^8)^5 = 1^8 \cdot 1^5 = 1$

15. $i^{67} = (i^8)^8 \cdot i^3 = 1^8 \cdot (-i) = -i$

16. $i^{50} = (i^8)^5 \cdot (i^2)^5 = 1^8 \cdot (-1)^5 = -1$

17. $3i(i^9) = 3i \cdot i^8 \cdot i = 3i \cdot 1 \cdot i = 3i^2 = -3$

18. $-9i(i^{19}) = -9i \cdot (i^8)^2 \cdot i^3 = -9i \cdot 1^2 \cdot i^3 = $
$-9i^4 = -9$

19. $-5i + 14i = 9i$

20. $6.9i + 8.3i + 0.47i = 15.67i$

21. $778i + 697i = 1475i$

22. $0.091i - 0.77i = -0.679i$

23. $23i^{87} + 16i^{33} = 23 \cdot (i^8)^8 \cdot (i^7)^3 \cdot i^2 + $
$16 \cdot (i^8)^4 \cdot i = 23 \cdot 1^8 \cdot (-i)^3 \cdot i^2 + 16 \cdot 1^4 \cdot i = $
$23 \cdot [-(-i)] \cdot (-1) + 16 \cdot i = -23i + 16i = -7i$

24. $-8.1i^{31} + 11.5i^{-55} = -8.1 \cdot (i^6)^5 \cdot i + $
$11.5 \cdot (i^7)^7 \cdot i^6 = -8.1 \cdot (-1)^5 \cdot i + $
$11.5 \cdot (-i)^7 \cdot i^6 = 8.1i + 11.5 \cdot [-(-i)] \cdot -1 = $
$8.1i - 11.5i = -3.4i$

25. $(-8i)(9i) = -72i^2 = -72(-1) = 72$

26. $2.7i(-4.1i) = -11.07i^2 = -11.07(-1) = 11.07$

27. $3i^{13} \cdot 4i^{23} = 12i^{36} = 12(i^6)^6 = 12(-1)^6 = 12$

28. $6 \cdot 4i^{34} = 24i^{34} = 24 \cdot (i^6)^5 \cdot i^4 = $
$= 24 \cdot (-1)^5 \cdot 1 = -24$

29. No. Every imaginary number bi is a multiple of i, the
square root of -1. Since the square root of -1 is not a
real number, an imaginary number cannot be real.
Similarly, since no real number a contains an imaginary
part i, a real number cannot be imaginary. Therefore,
the real and imaginary numbers are mutually exclusive
sets. A correct statement would be that a real number is
also a complex number $(a + 0i)$ and a complex
number can be a real number.

30. $2\sqrt{-3} + 5\sqrt{-3} = 2i\sqrt{3} + 5i\sqrt{3} = 7i\sqrt{3}$

31. $\sqrt{-18} - \sqrt{-32} = 3i\sqrt{2} - 4i\sqrt{2} = -i\sqrt{2}$

32. $5i - \sqrt{-5} = 5i - i\sqrt{5}$

33. $i\sqrt{2} + 3\sqrt{-200} = i\sqrt{2} + 30i\sqrt{2} = 31i\sqrt{2}$

34. $\sqrt{-6}\sqrt{-6} = i\sqrt{6} \cdot i\sqrt{6} = 6i^2 = 6 \cdot (-1) = -6$

35. $-4\sqrt{-8}\sqrt{-4.5} = -4 \cdot i\sqrt{8} \cdot i\sqrt{4.5} = $
$-4i^2\sqrt{36} = -24i^2 = -24 \cdot (-1) = 24$

36. $\sqrt{-5}(2\sqrt{-9}) = i\sqrt{5}(6i) = 6i^2\sqrt{5} = 6(-1)\sqrt{5} = $
$-6\sqrt{5}$

37. $\sqrt{-10}\sqrt{-10}\sqrt{-10} = i\sqrt{10} \cdot i\sqrt{10} \cdot i\sqrt{10} = $
$i^3\sqrt{1000} = 10i^3\sqrt{10} = 10 \cdot (-i)\sqrt{10} = -10i\sqrt{10}$

38. $\dfrac{8}{i} = \dfrac{8}{i} \cdot \dfrac{i}{i} = \dfrac{8i}{i^2} = \dfrac{8i}{-1} = -8i$

39. $\dfrac{15}{9i} = \dfrac{15}{9i} \cdot \dfrac{i}{i} = \dfrac{15i}{9i^2} = \dfrac{15i}{-9} = -\dfrac{5i}{3}$

40. $\dfrac{3}{\sqrt{-3}} = \dfrac{3}{i\sqrt{3}} = \dfrac{3}{i\sqrt{3}} \cdot \dfrac{i\sqrt{3}}{i\sqrt{3}} = \dfrac{3i\sqrt{3}}{3i^2} = \dfrac{3i\sqrt{3}}{3(-1)} = $
$-i\sqrt{3}$

41. $\dfrac{12i}{3\sqrt{-32}} = \dfrac{12i}{12i\sqrt{2}} = \dfrac{12i}{12i\sqrt{2}} \cdot \dfrac{i\sqrt{2}}{i\sqrt{2}} = \dfrac{12i^2\sqrt{2}}{12i^2 \cdot 2} = $
$\dfrac{12(-1)\sqrt{2}}{12(-1) \cdot 2} = \dfrac{\sqrt{2}}{2}$

42. $\dfrac{-b \pm \sqrt{b^2 - 4ac}}{2a} = \dfrac{-0 \pm \sqrt{0^2 - 4(1)(16)}}{2(1)} = $
$\dfrac{\pm\sqrt{-64}}{2} = \dfrac{\pm 8i}{2} = \pm 4i$

43. $\dfrac{-b \pm \sqrt{b^2 - 4ac}}{2a} = \dfrac{-0 \pm \sqrt{0^2 - 4(1)(11)}}{2(1)} =$

$\dfrac{\pm\sqrt{-44}}{2} = \dfrac{\pm 2i\sqrt{11}}{2} = \pm i\sqrt{11}$

44. $4x^2 + 800 = 0; \dfrac{-b \pm \sqrt{b^2 - 4ac}}{2a} =$

$\dfrac{-0 \pm \sqrt{0^2 - 4(4)(800)}}{2(4)} = \dfrac{\pm\sqrt{-12800}}{8} =$

$\dfrac{\pm 80i\sqrt{2}}{8} = \pm 10i\sqrt{2}$

45. $\dfrac{-b \pm \sqrt{b^2 - 4ac}}{2a} = \dfrac{-1 \pm \sqrt{1^2 - 4(1)(1)}}{2(1)} =$

$\dfrac{-1 \pm \sqrt{-3}}{2} = \dfrac{-1 \pm i\sqrt{3}}{2}$

46. $\dfrac{-b \pm \sqrt{b^2 - 4ac}}{2a} = \dfrac{-(-6) \pm \sqrt{(-6)^2 - 4(5)(5)}}{2(5)} =$

$\dfrac{6 \pm \sqrt{-64}}{10} = \dfrac{6 \pm 8i}{10} = \dfrac{3 \pm 4i}{5}$

47. $3(x - 1) = 2x^2 + 1, 3x - 3 - 2x^2 + 1,$

$2x^2 - 3x + 4 = 0; \dfrac{-b \pm \sqrt{b^2 - 4ac}}{2a} =$

$\dfrac{-(-3) \pm \sqrt{(-3)^2 - 4(2)(4)}}{2(2)} =$

$\dfrac{3 \pm \sqrt{-23}}{4} = \dfrac{3 \pm i\sqrt{23}}{4}$

48. $b^2 - 4ac = 5^2 - 4(1)(7) = -3$, so the solutions are imaginary.

49. $b^2 - 4ac = (-3)^2 - 4(2)(-4) = 41$, so the solutions are real and unequal.

50. $b^2 - 4ac = 5^2 - 4(6)(3) = -47$, so the solutions are imaginary.

51. $b^2 - 4ac = (-6)^2 - 4(9)(1) = 0$, so the solutions are real and equal.

52. $b^2 - 4ac = 20^2 - 4(4)(25) = 0$, so the solutions are real and equal.

53. $-3(x^2 - 2) = 4, -3x^2 + 6 = 4, -3x^2 + 2 = 0;$
$b^2 - 4ac = 0^2 - 4(-3)(2) = 24$, so the solutions are real and unequal.

54. $X_T = X_L - X_C, 8.08i = X_L - 17.85i, X_L = 25.93i;$
25.93i ohms

EXTEND

55. Examples will vary for the first two questions. π or $2 + 3i$ are not rational numbers; 7 or $\sqrt{10}$ are not imaginary numbers. There is no number that is not a complex number since real numbers can be written $a + 0 \cdot i$ and pure imaginaries can be written $0 + bi$.

56. $i^{-1} = \dfrac{1}{i} = \dfrac{1}{i} \cdot \dfrac{i}{i} = \dfrac{i}{i^2} = \dfrac{i}{-1} = -i$

57. $i^{-2} = \dfrac{1}{i^2} = \dfrac{1}{-1} = -1$

58. $i^{-3} = \dfrac{1}{i^3} = \dfrac{1}{-i} = \dfrac{1}{-i} \cdot \dfrac{-i}{-i} = \dfrac{-i}{i^2} = \dfrac{-i}{-1} = i$

59. $i^{-4} = \dfrac{1}{i^4} = \dfrac{1}{1} = 1$　　**60.** $i^{-2} = i^2; i^{-4} = i^4$

61. The pattern consists of the continuously repeating sequence $-i, -1, i, 1$.

62. $4x = 20, x = 5; 2yi = 6i, y = 3$

63. $5x = 35, x = 7; 3yi = -18i, y = -6$

64. $ai + bi = (a + b)i$. Since the real numbers are closed under addition, $a + b$ is a real number. Therefore, $(a + b)i$ is a pure imaginary number.

65. $ai - bi = (a - b)i$. Since the real numbers are closed under subtraction, $a - b$ is a real number. Therefore, $(a - b)i$ is a pure imaginary number.

66. $ai \cdot bi = ab(i^2) = -ab$, a real number. Therefore, the set of pure imaginary numbers is not closed under multiplication.

67. $\dfrac{ai}{bi} = \dfrac{a}{b}$, a real number. Therefore, the set of pure imaginary numbers is not closed under division.

68. As the Problem Solving Tip for Example 3b points out, neither the product nor the quotient property holds for imaginary numbers. Step c follows from the quotient property and leads to the fallacy $1 = -1$.

MIXED REVIEW

69. $5A = 5\begin{bmatrix} 1 & 2 \\ 4 & 3 \end{bmatrix} = \begin{bmatrix} 5 & 10 \\ 20 & 15 \end{bmatrix}$

70. $-3C = -3\begin{bmatrix} 1 & -1 \\ -1 & 1 \end{bmatrix} = \begin{bmatrix} -3 & 3 \\ 3 & -3 \end{bmatrix}$

71. $AB = \begin{bmatrix} 1 & 2 \\ 4 & 3 \end{bmatrix}\begin{bmatrix} -3 & 5 \\ 2 & -1 \end{bmatrix} =$

$\begin{bmatrix} 1(-3) + 2(2) & 1(5) + 2(-1) \\ 4(-3) + 3(2) & 4(5) + 3(-1) \end{bmatrix} = \begin{bmatrix} 1 & 3 \\ -6 & 17 \end{bmatrix}$

72. $BC = \begin{bmatrix} -3 & 5 \\ 2 & -1 \end{bmatrix}\begin{bmatrix} 1 & -1 \\ -1 & 1 \end{bmatrix} =$

$\begin{bmatrix} -3(1) + 5(-1) & -3(-1) + 5(1) \\ 2(1) + (-1)(-1) & 2(-1) + (-1)(1) \end{bmatrix} = \begin{bmatrix} -8 & 8 \\ 3 & -3 \end{bmatrix}$

73. $MN = (2 + \sqrt{2})(-3 - 2\sqrt{2}) = -6 - 7\sqrt{2} - 4 = -10 - 7\sqrt{2}$

74. $p^2 = (2 - \sqrt{2})^2 = 4 - 4\sqrt{2} + 2 = 6 - 4\sqrt{2}$

75. $MP = (2 + \sqrt{2})(2 - \sqrt{2}) = 4 - 2 = 2$

76. $M + N + P = 2 + \sqrt{2} + (-3 - 2\sqrt{2}) + 2 - \sqrt{2} = 2 + \sqrt{2} - 3 - 2\sqrt{2} + 2 - \sqrt{2} = 1 - 2\sqrt{2}$

77. B; $6 \div \sqrt{-2} = \dfrac{6}{\sqrt{-2}} = \dfrac{6}{i\sqrt{2}} = \dfrac{6}{i\sqrt{2}} \cdot \dfrac{i\sqrt{2}}{i\sqrt{2}} = \dfrac{6i\sqrt{2}}{2i^2} = -3i\sqrt{2}$

EXPLORE

1a. commutative **1b.** associative **1c.** commutative

1d. associative **1e.** distributive

2. $(4 + 2i) + (3 - 5i) = (2i + 4) + (3 - 5i)$
$= 2i + (4 + 3) - 5i$
$= (4 + 3) + 2i - 5i$
$= (4 + 3) + (2i - 5i)$
$= (4 + 3) + (2 - 5)i$
$= 7 - 3i$

3. Answers will vary. Students may conjecture
$(a + bi) + (c + di) = (a + c) + (b + d)i$ and
$(a + bi) - (c + di) = (a - c) + (b - d)i$, which,
as they will learn in this lesson, are the correct methods
to use to add and subtract complex numbers. However,
some students may question, correctly, whether the
commutative, associative, and distributive properties
can validly be applied to complex numbers, and answer
that they do not have sufficient information to add and
subtract the given complex numbers.

TRY THESE

1. $(3 - 2i) + (-4 - i) = (3 - 4) + (-2i - i) =$
$-1 - 3i$

2. $(9 - 9i) - (12 - 3i) = (9 - 12) +$
$(-9i - (-3i)) = -3 - 6i$

3. $(5 + i)(2 - 2i) = 10 - 10i + 2i - 2i^2 =$
$10 - 10i + 2i - 2(-1) = (10 + 2) +$
$(-10 + 2)i = 12 - 8i$

4. $\dfrac{-1 + 4i}{2 + 6i} = \dfrac{-1 + 4i}{2 + 6i} \cdot \dfrac{2 - 6i}{2 - 6i} =$

$\dfrac{-2 + 6i + 8i - 24i^2}{4 - 12i + 12i - 36i^2} = \dfrac{22 + 14i}{40} = \dfrac{11}{20} + \dfrac{7}{20}i$

5. $(-2 + i)(-2 - i) = 4 + 2i - 2i - i^2 = 5$

6.–9.

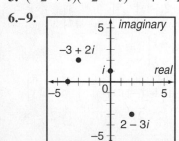

10. $|6 - 4i|; \sqrt{6^2 + 4^2} = \sqrt{52} = 2\sqrt{13}$

11.

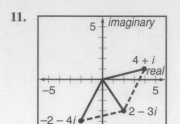

The sum is $2 - 3i$.

12. $f(-3 + i) = -3 + i + 5 + 2i = 2 + 3i$;
$f(2 + 3i) = 2 + 3i + 5 + 2i = 7 + 5i$;
$f(7 + 5i) = 7 + 5i + 5 + 2i = 12 + 7i$;
$f(12 + 7i) = 12 + 7i + 5 + 2i = 17 + 9i$; The
first four iterates of $f(z)$ are $2 + 3i$, $7 + 5i$, $12 + 7i$,
and $17 + 9i$.

13. $E = I \cdot Z$; $E = (4 + 2i)(-3 + 2i) =$
$-12 + 8i - 6i + 4i^2 = -16 + 2i$; $|3 + 2i| =$
$\sqrt{3^2 + 2^2} = \sqrt{13}$; The voltage is $-16 + 2i$ volts and
$|z|$ is $\sqrt{13}$ ohms.

14. Answers will vary. Since $z_1 - z_2 = z_1 + (-z_2)$,
determine $-z_2$ and add using the parallelogram method.
Another method is to reverse the process of adding. To
find the difference of $z_1 - z_2$, plot z_1 and z_2. Draw a
segment connecting z_2 to the origin. Draw a segment
connecting z_1 to z_2. Complete the parallelogram. The
fourth vertex represents the difference.

PRACTICE

1. $(2 + 7i) + (6 + 4i) = (2 + 6) + (7i + 4i) =$
$8 + 11i$

2. $(-5 + 5i) + (5 - 2i) = (-5 + 5) + (5i - 2i) =$
$0 + 3i = 3i$

3. $(8i) + (-1) = -1 + 8i$

4. $(14 - 7i) - (9 + 6i) = (14 - 9) + (-7i - 6i) =$
$5 - 13i$

5. $(-6 + i) - (13 - 2i) = (-6 - 13) + (i + 2i) =$
$-19 + 3i$

6. $(21.4 - 16.8i) - (32.9 + 11.4i) =$
$(21.4 - 32.9) + (-16.8i - 11.4i) = -11.5 - 28.2i$

7. $(3 + 5i)(2 + 6i) = 6 + 18i + 10i + 30i^2 =$
$-24 + 28i$

8. $(10 + i)(4 - 3i) = 40 - 30i + 4i - 3i^2 =$
$43 - 26i$

9. $(-2.5 + 12i)(6 - 14i) = -15 + 35i +$
$72i - 168i^2 = 153 + 107i$

10. $\dfrac{2 + 2i}{4 + 3i} = \dfrac{2 + 2i}{4 + 3i} \cdot \dfrac{4 - 3i}{4 - 3i} =$

$\dfrac{8 - 6i + 8i - 6i^2}{16 - 12i + 12i - 9i^2} = \dfrac{14 + 2i}{25} = \dfrac{14}{25} + \dfrac{2}{25}i$

11. $\dfrac{7-i}{7+i} = \dfrac{7-i}{7+i} \cdot \dfrac{7-i}{7-i} = \dfrac{49 - 7i - 7i + i^2}{49 - 7i + 7i - i^2} =$

$\dfrac{48 - 14i}{50} = \dfrac{24}{25} - \dfrac{7}{25}i$

12. $\dfrac{6}{5 - 8i} = \dfrac{6}{5 - 8i} \cdot \dfrac{5 + 8i}{5 + 8i} =$

$\dfrac{30 + 48i}{25 + 40i - 40i - 64i^2} = \dfrac{30 + 48i}{89} = \dfrac{30}{89} + \dfrac{48}{89}i$

13. $\dfrac{1}{5 - 2i} = \dfrac{1}{5 - 2i} \cdot \dfrac{5 + 2i}{5 + 2i} = \dfrac{5}{29} + \dfrac{2}{29}i;$

$(5 - 2i)\left(\dfrac{5}{29} + \dfrac{2}{29}i\right) = (5 - 2i)\left(\dfrac{5 + 2i}{29}\right) =$

$\dfrac{(5 - 2i)(5 + 2i)}{29} = \dfrac{29}{29} = 1$

14–17.

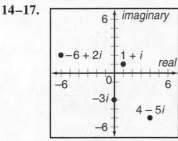

18. $|8 + 6i| = \sqrt{8^2 + 6^2} = \sqrt{100} = 10$

19. $|3 - 7i| = \sqrt{3^2 + 7^2} = \sqrt{58}$

20. $|-6 - 3i| = \sqrt{6^2 + 3^2} = \sqrt{45} = 3\sqrt{5}$

21. $|20 + 21i| = \sqrt{20^2 + 21^2} = \sqrt{841} = 29$

22.

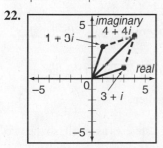

The sum is $4 + 4i$.

23.

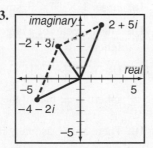

The sum is $-2 + 3i$.

24.

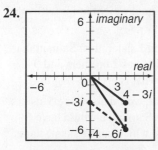

The sum is $4 - 6i$.

25.

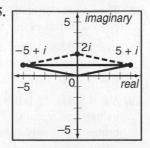

The sum is $2i$.

26. $i - 1; -1 - i; -i + 1; 1 + i;$ The next four iterates will be the same as the first four. That is because the fourth iterate equals the initial value. Therefore, the iterates will repeat the initial sequence $i - 1, -1 - i,$ $-i + 1, 1 + i,$ and will continue to do so every cycle of four iterates.

27. $f(1 - i) = (1 - i)^2 = 1 - 2i + i^2 = -2i;$
$f(-2i) = (-2i)^2 = 4i^2 = -4; f(4) = 4^2 = 16;$
$f(16) = 16^2 = 256;$ The first four iterates of $f(z)$ are $-2i, -4, 16,$ and $256.$

28. $E = I \cdot Z, 3 + 2i = 1.6 + 0.2i \cdot Z, Z = \dfrac{3 + 2i}{1.6 + 0.2i},$

$Z = \dfrac{3 + 2i}{1.6 + 0.2i} \cdot \dfrac{1.6 - 0.2i}{1.6 - 0.2i} =$

$\dfrac{4.8 - 0.6i + 3.2i - 0.4i^2}{2.56 - 0.32i + 0.32i - 0.04i^2} = \dfrac{5.2 + 2.6i}{2.6} = 2 + i;$

$2 + i$ ohms

29. Answers will vary. When graphed, the two numbers are collinear with the origin. Therefore, when the points are connected to the origin, the resulting line segments are part of a line instead of adjacent sides of a parallelogram. Since the parallelogram cannot be completed, the numbers cannot be added graphically. This will be true for any two numbers that, when graphed, are collinear with the origin. An example is $5 + 7i$ and $15 + 21i$. Students may note that one number is a constant multiple of the other.

EXTEND

30. $s_1 + s_2 = 4 + 3i + 4 - 3i = 8;$
$s_1 s_2 = (4 + 3i)(4 - 3i) =$
$16 - 12i + 12i - 9i^2 = 25; x^2 - 8x + 25 = 0$

31. $s_1 + s_2 = 1 + 2i + 1 - 2i = 2;$
$s_1 s_2 = (1 + 2i)(1 - 2i) = 1 - 2i + 2i - 4i^2 = 5;$
$x^2 - 2x + 5 = 0$

32. $s_1 + s_2 = \dfrac{2 + i\sqrt{2}}{2} + \dfrac{2 - i\sqrt{2}}{2} = \dfrac{4}{2} = 2; s_1 s_2 =$
$\left(\dfrac{2 + i\sqrt{2}}{2}\right)\left(\dfrac{2 - i\sqrt{2}}{2}\right) = \dfrac{4 - 2i\sqrt{2} + 2i\sqrt{2} - 2i^2}{4} =$
$\dfrac{6}{4} = \dfrac{3}{2}; x^2 - 2x + \dfrac{3}{2} = 0, 2x^2 - 4x + 3 = 0$

33. $s_1 + s_2 = \dfrac{-5 + i\sqrt{7}}{4} + \dfrac{-5 - i\sqrt{7}}{4} = \dfrac{-10}{4} = -\dfrac{5}{2};$
$s_1 s_2 = \left(\dfrac{-5 + i\sqrt{7}}{4}\right)\left(\dfrac{-5 - i\sqrt{7}}{4}\right) =$
$\dfrac{25 + 5i\sqrt{7} - 5i\sqrt{7} - 7i^2}{16} = \dfrac{32}{16} = 2;$
$x^2 - \left(-\dfrac{5}{2}\right)x + 2 = 0, x^2 + \dfrac{5}{2}x + 2 = 0,$
$2x^2 + 5x + 4 = 0$

34. $|a + 16i| = 34, \sqrt{a^2 + 16^2} = 34,$
$a^2 + 16^2 = 1156, a^2 = 900, a = \pm 30$

35. $\sqrt{-5 - 12i} = 2 - 3i$
$(\sqrt{-5 - 12i})^2 = (2 - 3i)^2$
$-5 - 12i = 4 - 6i - 6i + 9i^2$
$-5 - 12i = -5 - 12i$

36. $\left(\dfrac{-1 + i\sqrt{3}}{2}\right)^3 = \dfrac{(-1 + i\sqrt{3})^3}{2^3} = \dfrac{8}{8} = 1$

37. $|a + bi| = \sqrt{a^2 + b^2}$

38. $\dfrac{1}{a + bi} \cdot \dfrac{a - bi}{a - bi} = \dfrac{a - bi}{a^2 - (bi)^2} = \dfrac{a - bi}{a^2 + b^2} = \dfrac{a}{a^2 + b^2} - \dfrac{b}{a^2 + b^2}i$

39. $a^2 + b^2 = a^2 - b^2 i^2 = (a + bi)(a - bi)$

40.

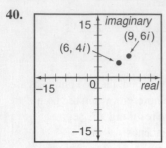

The sum is $6 + 4i$.
The answers are the same.

MIXED REVIEW

41. no   **42.** yes   **43.** no

44. $6x + 3y = 12, 3y = -6x + 12, y = -2x + 6$; $m = -2$

45. $x + y = 9, y = -x + 9$; $m = -1$

46. $\dfrac{y_1 - y}{x_1 - x} = \dfrac{8 - 4}{0 - 1} = \dfrac{4}{-1} = -4$; $m = -4$

47. $\dfrac{3 \cdot 5}{8.5 \cdot 11} = \dfrac{15}{93.5} \approx 0.1604$; 16.0%

48. A; $(3 - 4i)(3 + 4i) = 9 + 12i - 12i - 16i^2 = 25$

Lesson 8.8, pages 410–413

EXPLORE THE PROBLEM

1. $D = 1000; i - 0.06; n = 10$

2. $F = 1000 \cdot \dfrac{(1 + 0.06)^{10} - 1}{0.06} \approx \$13,180.79$

3. $D = 500; i = 0.06 \div 2 = 0.03; n = 10 \cdot 2 = 20$

4. $F = 500 \cdot \dfrac{(1 + 0.03)^{20} - 1}{0.03} \approx 13,435.19$;
$13,435.19 - 13,180.79 = 254.40$; Roberto would have \$254.40 more.

5. You can multiply \$14.49 by \$750 to find the future value.

INVESTIGATE FURTHER

6a. $D = 1000; n = 10; i = 0.06$

6b. $p = 1000 \cdot \dfrac{1 - (1.06)^{-10}}{0.06} \approx 7360.09$; \$7360.90

7a. $D = 750; n = 3 \cdot 2 = 6; i = 0.08 \div 2 = 0.04$

7b. $p = 750 \cdot \dfrac{1 - (1.04)^{-6}}{0.04} \approx 3931.60$; \$3931.60

8. $p = 1000 \cdot \dfrac{(1.04)^{20} - 1}{0.04} \approx 29,778.08$; \$29,778.08

9a. $30(1500) = 45,000$; \$45,000

9b. $p = 1500 \cdot \dfrac{(1.08)^{30} - 1}{0.08} \approx 169,924.81$; \$169,924.81

10. $p = 200 \cdot \left(\dfrac{1 - (1.005)^{-60}}{0.005} \right) \approx 10,345.11$; \$10,345.11

11. $p = 2000 \cdot \left(\dfrac{1 - (1.02)^{-16}}{0.02} \right) \approx 27,155.42$; \$27,155.42

12a. In other problems, the amount of each payment was known and the account total was to be determined. In this problem, the final amount is known and the payment must be determined.

12b. Substitute the known values for F, n, and i. Solve for D.

12c. $100,000 = D\left(\dfrac{(1.03)^{40} - 1}{0.03}\right), D = \dfrac{100,000(0.03)}{(1.03)^{40} - 1}$,
$D \approx 1326.24$; \$1326.24

13. $20,000 = D\left(\dfrac{(1.0125)^{20} - 1}{0.0125}\right), D = \dfrac{0.0125(20,000)}{(1.0125)^{20} - 1} \approx$ 886.41; \$886.41

14. For a present value problem, you generally determine a lump sum payment. For a sinking fund problem, you determine the amount of a payment to be made periodically for the term of the annuity.

REVIEW PROBLEM SOLVING STRATEGIES

1a. large: $3 \times 3 \times 3$, small: $4 \times 4 \times 4$

1b. The diameter of a large sphere is $\dfrac{4}{3}$ that of the small sphere.

1c. Use $V = \dfrac{4}{3}\pi r^3$. The volume (weight) of the larger sphere is $\dfrac{64}{27}$ that of the smaller.

1d. There are $\dfrac{27}{64}$ as many large spheres as small, so the two cartons weigh the same.

1e. The same relationship of equal weight holds with other pairs of numbers that are cubes.

2a. No; if Sid had 1 brother and 1 sister, then Samantha is the only sister. So, Samantha has 2 brothers and 0 sisters, which contradicts what is given.

2b. No; if Sid had 5 brothers, he would also have 5 sisters, including Samantha. Then Samantha would have 6 brothers, but only 4 sisters, which contradicts what is given.

2c. There are 7 children, 4 boys and 3 girls. Explanations may vary, but if you make a table and look for a pattern, the difference between the number of brothers and the number of sisters Samantha has is always 2; the only numbers b and s such that $b = 2s$ and $b - s = 2$ are $b = 4, s = 2$.

3a. Since $(ab)^2 = a^2b^2$ and $3^2 = 9$, the expression must be divisible by 9.

3b. The number on the left must also be divisible by 9, so the sum of the digits must be a multiple of 9. The sum of the digits of the number on the right is 23, so $y = 4$ (to give 27). Then solve $(241 + x)^2 = 66,049$ to get $241 + x = 257$ and $x = 16$.

Chapter Review, pages 414–415

1. c   **2.** d   **3.** a   **4.** b

5. $\sqrt[3]{k} = k^{\frac{1}{3}}$

6. $(\sqrt[6]{9})^5 = (9^{\frac{1}{6}})^5 = 9^{\frac{5}{6}}$

7. $b^{\frac{1}{2}} = \sqrt{b}$

8. $(m + 4)^{\frac{2}{9}} = \sqrt[9]{(m + 4)^2}$

9. $\sqrt[3]{x^3} = (x^3)^{\frac{1}{3}} = x$

10. $\sqrt[5]{64x^{10}} = \sqrt[5]{2^5 \cdot 2 \cdot x^{10}} = 2x^2\sqrt[5]{2}$

11. $125^{-\frac{1}{3}} = \dfrac{1}{125^{\frac{1}{3}}} = \dfrac{1}{(5^3)^{\frac{1}{3}}} = \dfrac{1}{5}$

12. $\sqrt[6]{11^3} = 11^{\frac{3}{6}} = 11^{\frac{1}{2}}$

13.
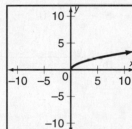

14.
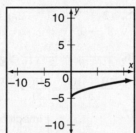

15.

16. $\sqrt{6} + \sqrt{24} = \sqrt{6} + 2\sqrt{6} = 3\sqrt{6}$

17. $2\sqrt[3]{5} \cdot \sqrt[3]{25} = 2\sqrt[3]{125} = 2 \cdot 5 = 10$

18. $\sqrt{\dfrac{1}{3}} = \dfrac{\sqrt{1}}{\sqrt{3}} = \dfrac{\sqrt{1}}{\sqrt{3}} \cdot \dfrac{\sqrt{3}}{\sqrt{3}} = \dfrac{\sqrt{3}}{3}$

19. $\dfrac{\sqrt[3]{3}}{\sqrt[3]{24}} = \dfrac{\sqrt[3]{3}}{\sqrt[3]{8} \cdot \sqrt[3]{3}} = \dfrac{1}{\sqrt[3]{2^3}} = \dfrac{1}{2}$

20. $\dfrac{2}{2 + \sqrt{2}} = \dfrac{2}{2 + \sqrt{2}} \cdot \dfrac{2 - \sqrt{2}}{2 - \sqrt{2}} =$
$\dfrac{4 - 2\sqrt{2}}{4 - 2\sqrt{2} + 2\sqrt{2} - 2} = \dfrac{4 - 2\sqrt{2}}{2} = 2 - \sqrt{2}$

21. $\sqrt{n + 6} = 8, (\sqrt{n + 6})^2 = 8^2, n + 6 = 64, n = 58;$
$\sqrt{58 + 6} = 8, 8 = 8;$ The solution is 58.

22. $\sqrt[3]{x} - 5 = -8, \sqrt[3]{x} = -3, (\sqrt[3]{x})^3 = -3^3, x = -27;$

$\sqrt[3]{-27} - 5 \overset{?}{=} -8, -8 = -8;$ The solution is -27.

23. $y^{\frac{3}{4}} = 8, y = 8^{\frac{4}{3}}, y = (\sqrt[3]{8})^4, y = 2^4, y = 16;$
$16^{\frac{3}{4}} \overset{?}{=} 8, 8 = 8;$ The solution is 16.

24. $\sqrt{5x - 1} + 3 = x, \sqrt{5x - 1} = x - 3,$
$(\sqrt{5x - 1})^2 = (x - 3)^2, 5x - 1 = x^2 - 6x + 9,$
$x^2 - 11x + 10 = 0, (x - 10)(x - 1) = 0, x = 10$
or $x = 1; \sqrt{5(10) - 1} + 3 \overset{?}{=} 10, 10 = 10;$
$\sqrt{5(1) - 1} + 3 \overset{?}{=} 1, 5 \neq 1;$ The solution is 10.

25. $\sqrt{-49} = \sqrt{49}\sqrt{-1} = 7i$

26. $\sqrt{-20} = \sqrt{4}\sqrt{5}\sqrt{-1} = 2i\sqrt{5}$

27. $i^{21} = (i^4)^5 \cdot i = 1^5 \cdot i = i$

28. $\sqrt{-3} + \sqrt{-75} = i\sqrt{3} + 5i\sqrt{3} = 6i\sqrt{3}$

29. $5i(2i) = 10i^2 = -10$

30. $12i - 10i = 2i$

31. $\sqrt{-2}\sqrt{-3} = i\sqrt{2} \cdot i\sqrt{3} = i^2\sqrt{6} = -\sqrt{6}$

32. $\dfrac{15}{5i} = \dfrac{15}{5i} \cdot \dfrac{i}{i} = \dfrac{15i}{i^2} = \dfrac{15i}{-5} = -3i$

33. $(4 + i) + (11 - 3i) = (4 + 11) + (i - 3i) =$
$15 - 2i$

34. $(2 - 5i) - (1 - 8i) = (2 - 1) + (-5i + 8i) =$
$1 + 3i$

35. $(-3 + i)(2 - 3i) = -6 + 9i + 2i - 3i^2 =$
$-6 + 11i + 3 = -3 + 11i$

36. $\dfrac{4 - 3i}{4 + 3i} = \dfrac{4 - 3i}{4 + 3i} \cdot \dfrac{4 - 3i}{4 - 3i} =$

$\dfrac{16 - 12i - 12i + 9i^2}{16 - 12i + 12i - 9i^2} = \dfrac{16 - 24i - 9}{16 + 9} =$

$\dfrac{7 - 24i}{25} = \dfrac{7}{25} - \dfrac{24}{25}i$

37.

imaginary

$-2 + 4i$ $1 + 6i$ $3 + 2i$ real

The sum is $1 + 6i$.

38. $F = 3000\left(\dfrac{(1.05)^{45} - 1}{0.05}\right) \approx 479,100;$ \$479,100

39. $10,000 = D\left(\dfrac{1 - (1.06)^{-4}}{0.06}\right), D = \dfrac{(0.06)(10,000)}{(1 - (1.06)^{-4})} \approx$
2886; \$2886

Chapter Assessment, pages 416–417

CHAPTER TEST

1. $\sqrt[4]{n^3} = n^{\frac{3}{4}}$

2. $(\sqrt{2p})^5 = \left[(12p)^{\frac{1}{2}} = (2p)^{\frac{5}{2}}\right]^5$

3. $k^{\frac{4}{7}} = \sqrt[7]{k^4}$

4. $(m + 5)^{\frac{1}{2}} = \sqrt{m + 5}$

5. $\sqrt{a^2} = |a|$

6. $32^{\frac{4}{5}} = (\sqrt[5]{32})^4 = (\sqrt[5]{2^5})^4 = 2^4 = 16$

7. $-\sqrt[4]{81} = -\sqrt[4]{3^4} = -3$

8. $\sqrt[4]{32x^4} = \sqrt[4]{2^4 \cdot 2 \cdot x^4} = 2|x|\sqrt[4]{2}$

9. $\sqrt[8]{100^4} = 100^{\frac{4}{8}} = 100^{\frac{1}{2}} = \sqrt{100} = 10$

10. $\sqrt{5} + \sqrt{45} = \sqrt{5} + 3\sqrt{5} = 4\sqrt{5}$

11. $\sqrt[3]{16y^6} - y^2\sqrt[3]{2} = \sqrt[3]{2 \cdot 2^3 \cdot y^6} - y^2\sqrt[3]{2} = 2y^2\sqrt[3]{2} - y^2\sqrt[3]{2} = y^2\sqrt[3]{2}$

12. $\sqrt{6}(3\sqrt{6} + 2\sqrt{7}) = 3 \cdot 6 + 2\sqrt{42} = 18 + 2\sqrt{42}$

13. $\sqrt{18}\sqrt{8} = \sqrt{144} = 12$

14. $\frac{\sqrt{35}}{\sqrt{7}} = \sqrt{\frac{35}{7}} = \sqrt{5}$

15. $\sqrt{\frac{9}{16}} = \frac{\sqrt{9}}{\sqrt{16}} = \frac{3}{4}$

16. $\frac{3}{\sqrt[3]{y}} = \frac{3}{\sqrt[3]{y}} \cdot \frac{\sqrt[3]{y^2}}{\sqrt[3]{y^2}} = \frac{3\sqrt[3]{y^2}}{y}$

17. $\frac{11}{\sqrt{11}} = \frac{11}{\sqrt{11}} \cdot \frac{\sqrt{11}}{\sqrt{11}} = \frac{11\sqrt{11}}{11} = \sqrt{11}$

18. $\frac{5x}{2 - \sqrt{5}} = \frac{5x}{2 - \sqrt{5}} \cdot \frac{2 + \sqrt{5}}{2 + \sqrt{5}} = \frac{10x + 5x\sqrt{5}}{4 + 2\sqrt{5} - 2\sqrt{5} - 5} = \frac{10x + 5x\sqrt{5}}{-1} = -10x - 5x\sqrt{5}$

19.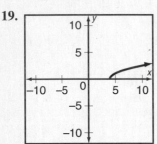

20.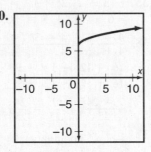

21. Answers will vary. Students should point out tht the square root of any quantity is always positive. By writing $\sqrt{x^2} = |x|$, we guarantee a positive solution, since $|x|$ is always positive.

22. $2\sqrt{p} = 18, \sqrt{p} = 9, (\sqrt{p})^2 = 9^2, p = 81$; $2\sqrt{81} \stackrel{?}{=} 18, 18 = 18$; The solution is 81.

23. $\sqrt{x - 3} = 5, (\sqrt{x - 3})^2 = 5^2, x - 3 = 25, x = 28$; $\sqrt{28 - 3} \stackrel{?}{=} 5, 5 = 5$; The solution is 28.

24. $\sqrt[3]{n} + 5 = 9, \sqrt[3]{n} = 4, (\sqrt[3]{n})^3 = 4^3, n = 64$; $\sqrt[3]{64} + 5 \stackrel{?}{=} 9, 9 = 9$; The solution is 64.

25. $k^{\frac{2}{3}} - 1 = 3, k^{\frac{2}{3}} = 4, (k^{\frac{2}{3}})^{\frac{3}{2}} = 4^{\frac{3}{2}}, k = \pm(\sqrt{4})^3$, $k = \pm(\sqrt{4})^3, k = \pm2^3, k = \pm8$; $8^{\frac{2}{3}} - 1 \stackrel{?}{=} 3$, $3 = 3; (-8)^{\frac{2}{3}} - 1 \stackrel{?}{=} 3, 3 = 3$; The solution is 8, -8.

26. $\sqrt{x - 4} = \sqrt{x + 5} - 3$, $(\sqrt{x - 4})^2 = (\sqrt{x + 5} - 3)^2$, $x - 4 = x + 5 - 6\sqrt{x + 5} + 9$,

$-18 = -6\sqrt{x + 5}, 18 = 6\sqrt{x + 5}$, $18^2 = (6\sqrt{x + 5})^2, 324 = 36(x + 5)$, $324 = 36x + 180, 144 = 36x, x = 4$; $\sqrt{4 - 4} \stackrel{?}{=} \sqrt{4 + 5} - 3, 0 = 0$; The solution is 4.

27. $\sqrt{3x - 3} = x - 7, (\sqrt{3x - 3})^2 = (x - 7)^2$, $3x - 3 = x^2 - 14x + 49, x^2 - 17x + 52 = 0$, $(x - 13)(x - 4) = 0, x = 13, x = 13$ or $x = 4$; $\sqrt{3(13) - 3} \stackrel{?}{=} 13 - 7, 6 = 6; \sqrt{3(4) - 3} \stackrel{?}{=} 4 - 7$, $3 \neq -3$; The solution is 13.

28. $\sqrt{-25} = \sqrt{25}\sqrt{-1} = 5i$

29. $i^{35} = (i^6)^5 \cdot i^5 = (-1)^5 \cdot i = -i$

30. $-6i(3i) = -18i^2 = 18$

31. $\sqrt{-5}\sqrt{-6} = i\sqrt{5} \cdot i\sqrt{6} = i^2\sqrt{30} = -\sqrt{30}$

32. $\sqrt{-8n} + \sqrt{-18n} - i\sqrt{2n}$ $= 2i\sqrt{2n} + 3i\sqrt{2n} - i\sqrt{2n} = 4i\sqrt{2n}$

33. $\frac{6}{i} = \frac{6}{i} \cdot \frac{i}{i} = \frac{6i}{i^2} = -6i$

34. $(8 + 11i) - (9 - 4i) = (8 - 9) + (11i + 4i) = -1 + 15i$

35. $(2 + 5i)(-1 - 2i) = -2 - 4i - 5i - 10i^2 = 8 - 9i$

36. $\frac{3 + i}{4 - 2i} = \frac{3 + i}{4 - 2i} \cdot \frac{4 + 2i}{4 + 2i} = \frac{12 + 6i + 4i + 2i^2}{16 + 8i - 8i - 4i^2} = \frac{10 + 10i}{20} = \frac{10}{20} + \frac{10}{20}i = \frac{1}{2} + \frac{1}{2}i$

37.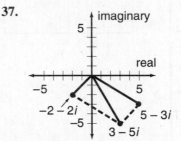

The sum is $3 - 5i$.

38. B; $b^2 - 4ac = 12^2 - 4(-1)(-36) = 0$, so the solutions are real and equal.

39. $X_T = X_L - X_C, 9.7i = 15.5i - X_C, X_C = 5.8i$; $5.8i$ ohms

40. perimeter $= 2(1 + w) = 2(\sqrt{12} + \sqrt{3}) = 2(2\sqrt{3} + \sqrt{3}) = 2(3\sqrt{3}) = 6\sqrt{3}$; area $= l \cdot w = \sqrt{12}\sqrt{3} = \sqrt{36} = 6$; The perimeter is $6\sqrt{3}$ cm and the area is 6 cm^2.

41. $V = \frac{4}{3}\pi r^3, \frac{9\pi}{2} = \frac{4}{3}\pi r^3, r^3 = \frac{3}{4\pi} \cdot \frac{9\pi}{2}, r^3 = \frac{27\pi}{8\pi}$, $r^3 = \frac{27}{8}, r = \sqrt[3]{\frac{27}{8}}, r = \frac{\sqrt[3]{27}}{\sqrt[3]{8}}, r = \frac{3}{2}$, the radius is $\frac{3}{2}$ in.

42. $F = 1000\left(\frac{(1.05)^{18} - 1}{0.05}\right) \approx 28{,}132.38$; \$28,132.38

ROOT MODEL

Possible models: (a) the side of a square with an area of 5; (b) the side of a cube with a volume of 5; (c) the area of a face of a cube with a volume of 5.

COMPLEX CIRCLE

Students should find that the two iterates lie on the circle. Using initial value $0.6 + 0.8i$, for example, the first two iterates are $-0.28 + 0.96i$ and $-0.8432 - 0.5376i$.

Cumulative Review, page 418

1. c

2. $x^2 - 30x + 225$

3. $21x^2 - 29x - 10, (7x + 2)(3x - 5)$

4. $x^3 + 5x^2 - 16x - 80, (x + 5)(x^2 - 16),$
$(x + 5)(x + 4)(x - 4)$

5. $\sqrt{y + 1} = 5, (\sqrt{y + 1})^2 = 5^2, y + 1 = 25, y = 24$
$\sqrt{24 + 1} = 5, 5 = 5;$ The solution is 24.

6. $\sqrt{2x + 1} - 1 = -4, \sqrt{2x + 1} = 5,$
$(\sqrt{2x + 1})^2 = 5^2, 2x + 1 = 25, 2x = 24, x = 12;$
$\sqrt{2(12) + 1} - 1 = -4; 4 \neq -4;$ There is no solution.

7. $x^2 + 625 = 0, x^2 = -625, x = \sqrt{-625}, x = \pm 25i;$
$(\pm 25i)^2 + 625 = 0, 0 = 0$

8. $y^2 - 4y + 13 = 0; \dfrac{-b \pm \sqrt{b^2 - 4ac}}{2a} =$
$\dfrac{-(-4) \pm \sqrt{(-4)^2 - 4(1)(13)}}{2(1)} = \dfrac{4 \pm \sqrt{-36}}{2} =$
$\dfrac{4 \pm 6i}{2} = 2 \pm 3i$

9. A; $x^2 + 4x + 3 = 0, (x + 3)(x + 1) = 0, x = -3$
or $x = -1; -3 < x < -1$

10. For $(0, 0)$: $c = 0$
For $(-2, -6)$: $4a - 2b + c = -6$
For $(4, 0)$: $16a + 4b + c = 0$
$\begin{cases} 4a - 2b = -6 & \to & -16a + 8b = 24 \\ 16a + 4b = 0 & \to & 16a + 4b = 0 \end{cases}$
$ \quad 12b = 24$
$ \quad b = 2$
$4a - 2b = -6$
$4a - 2(2) = -6$
$4a = -2$
$a = -\dfrac{1}{2}$
$y = -\dfrac{1}{2}x^2 + 2x$

11. $64^{-\frac{1}{2}} = \dfrac{1}{64^{\frac{1}{2}}} = \dfrac{1}{\sqrt{64}} = \dfrac{1}{8}$

12. $\sqrt[4]{-16};$ no real solution

13. $p = 1500 - \left(\dfrac{1 - (1.07)^{-15}}{0.07}\right) \approx 13,661.87; \$13,661.87$

14.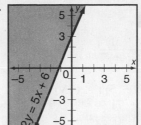

15.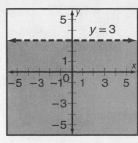

16. The graph of $y = 3\sqrt{x}$ shaes the same vertex, $(0, 0)$ as the original curve $y = \sqrt{x}$ and then follows the direction of the original but each ordinate is 3 times that of the original, thus placing th enew curve above the original for $x > 0$.
The graph of $y = \sqrt{x} + 3$ is the graph of $y = \sqrt{x}$ translated up 3 units.
The graph of $y = \sqrt{x + 3}$ is the graph of $y = \sqrt{x}$ translated 3 units to the left.

17. Let x represent the uniform width of the mat.
$2(8 + 2x) + 2(10 + 2x) \leq 48,$
$16 + 4x + 20 + 4x \leq 48, 8x \leq 12, x \leq 1.5;$ The width of the mat must be 1.5 in or less.

18. Let d represent the number of doorways widened and let r represent the number of ramps installed.
$\begin{cases} d + r = 34 \\ 350d + 540r = 12,850 \end{cases}$
$d = 34 - r; 350(34 - r) + 540r = 12,850,$
$11,900 - 350r + 540r = 12,850,$
$190r = 950, r = 5;$ There were 5 ramps installed.

19. $(30, 500), (90, 600)$
$m = \dfrac{600 - 500}{90 - 30} = \dfrac{100}{60} = \dfrac{5}{3}$
$y - 500 = \dfrac{5}{3}(x - 30)$
$y - 500 = \dfrac{5}{3}x - 50$
$y = \dfrac{5}{3}x + 450$
$V = \dfrac{5}{3}t + 450;$ when the temperature is above $-270°$ C, there is some quantity of this gas.

Standardized Test, page 419

1. C; $10x + 10y = 60, x + y = 6;$ mean $= 6 \div 2 = 3$

2. D

3. B; $x + y = z, x + y = 2y - 1, y = x + 1; x < y$

4. A; $x = -4$ and $y = -6;$
$\dfrac{x + y}{x \cdot y} = \dfrac{-4 + (-6)}{-4(-6)} = \dfrac{-10}{24} = -\dfrac{5}{12};$
$\dfrac{x \cdot y}{x + y} = \dfrac{-4(-6)}{-4 + (-6)} = \dfrac{24}{-10} = -\dfrac{12}{5}; -\dfrac{5}{12} > -\dfrac{12}{5}$

5. A; $\sqrt{3x + 3} - \sqrt{6x + 7} + 1 = 0$,
$\sqrt{3x + 3} = \sqrt{6x + 7} - 1$,
$(\sqrt{3x + 3})^2 = (\sqrt{6x + 7} - 1)^2$,
$3x + 3 = 6x + 6 - 2\sqrt{6x + 7} + 1$,
$-3x - 5 = -2\sqrt{6x + 7}$,
$3x + 5 = 2\sqrt{6x + 7}$, $(3x + 5)^2 = (2\sqrt{6x + 7})^2$,
$9x^2 + 30x + 25 = 4(6x + 7)$, $9x^2 + 30x + 25 = 24x + 28$, $9x^2 + 6x - 3 = 0$, $3x^2 + 2x - 1 = 0$,

$(3x - 1)(x + 1) = 0$, $x = \frac{1}{3}$ or $x = -1$;

$\sqrt{3\left(\frac{1}{3}\right) + 3} - \sqrt{6\left(\frac{1}{3}\right) + 7} + 1 = 0$, $0 = 0$;

$\sqrt{3(-1) + 3} - \sqrt{6(-1) + 7} + 1 = 0$, $0 = 0$;

The solutions are $\frac{1}{3}$ and -1. $s_1 s_2 = \frac{1}{3} \cdot -1 = -\frac{1}{3}$;

$s_1 + s_2 = \frac{1}{3} + (-1) = -\frac{2}{3}$; $-\frac{1}{3} > -\frac{2}{3}$

6. C; $x^{\frac{3}{2}} y^{\frac{1}{2}} = \sqrt{x^3}\sqrt{y} = \sqrt{x^3 y}$; $\sqrt{x^3 y} = \sqrt{x^3 y}$

7. A; $x^2 + 6x + 8 = 0$, $(x + 4)(x + 2) = 0$, $x = -4$ or $x = -2$; $-4 < x < -2$; $-6 < -4$

8. B; $[5.2] + [-5.2] = 5 + (-6) = -1$; $-1 < 0$

9. B; $1 - \sqrt{3} < -1 + \sqrt{3}$

10. C; $3x^2 - 2x + 2 = 0$; $\dfrac{-b \pm \sqrt{b^2 - 4ac}}{2a} =$

$\dfrac{-(-2) \pm \sqrt{(-2)^2 - 4(3)(2)}}{2(3)} = \dfrac{2 \pm \sqrt{-20}}{6} =$

$\dfrac{2 + 2i\sqrt{5}}{6} = \dfrac{1 \pm i\sqrt{5}}{3}$;

$s_1 + s_2 = \dfrac{1 + i\sqrt{5}}{3} + \dfrac{1 - i\sqrt{5}}{3} = \dfrac{2}{3}$;

$s_1 + s_2 = \left(\dfrac{1 + i\sqrt{5}}{3}\right)\left(\dfrac{1 - i\sqrt{5}}{3}\right) =$

$\dfrac{1 - i\sqrt{5} + \sqrt{5} - 5i^2}{9} = \dfrac{6}{9} = \dfrac{2}{3}$; $\dfrac{2}{3} = \dfrac{2}{3}$

11. B; $x = 0.10(81) \approx 8$; $y = 0.02(441) \approx 9$

12. D

13. C; $\dfrac{1}{1 - i} = \dfrac{1}{1 - i} \cdot \dfrac{1 + i}{1 + i} = \ = \dfrac{1 + i}{1 + i - i - i^2} =$

$\dfrac{1 + i}{2} = \dfrac{1}{2} + \dfrac{1}{2}i$; $\dfrac{1}{2} = \dfrac{1}{2}$

14. A; $10x^2 + 11x - 6 = (2x + 3)(5x - 2)$; $a = 3$; $b = -2$; $3 > -2$

15. B; Vertex of $y = 3|x - 3| - 3$ is $(3, -3)$, so product is -9; Vertex of $y = 3|x + 3| - 3$ is $(-3, -3)$, so product is 9.

Chapter 9 Exponential and Logarithmic Functions

Data Activity, pages 420–421

1.
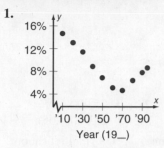

2. Answers will vary. A possible quadratic function is $y = 0.003x^2 - 0.407x + 19.668$ where x is the last two digits of the year.

3. Answers will vary. For 2000, the predicted value is about 9.0%.

4. Mexico, 39.8%; China, 3.6%.

5. Answers will vary. Students must decide on scales to use, how to round numbers, how many to represent by a pictograph symbol, and whether to group countries for the circle graph.

Lesson 9.1, pages 423–427

THINK BACK

1. $g(x) = -f(x)$ **2.** $g(x) = -3x$ **3.** $g(x) = -(x^2)$

4. $g(x) = |x|$ **5.** $3^{-4} = \dfrac{1}{3^4}$ **6.** $\dfrac{m^{-2}h^3}{p^{-5}} = \dfrac{p^5 n^3}{m^2}$

7. $4x^{-2}y = \dfrac{4y}{x^2}$

EXPLORE

8. $s(x) = 3^x$ and $h(x) = \left(\dfrac{1}{2}\right)^x$ are exponential functions; $t(x) = x^4$ is not because x is the base instead of the exponent.

9.
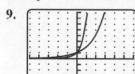
Df: {all real numbers},
Rf: {all positive real numbers}
Dg: {all real numbers},
Rg: {all positive real numbers}

10. As the value of x increases, the values of $f(x)$ and $g(x)$ increase exponentially.

11. As the value of x decreases, the values of $f(x)$ and $g(x)$ approach, yet never reach zero.

12. Both graphs approach zero for small values of x, and approach infinity for large values of x. However $g(x)$ approaches zero and infinity much faster than $f(x)$.

13. The point $(0, f(x))$ is common to both graphs; this point is present on all graphs of the form $y = b^x$ since b^0 is always equal to one.

14. f and g pass the horizontal line test and therefore are one-to-one functions.

15. The graph of $h(x) = 15^x$ would approach zero for decreasing x and infinity for increasing x even faster than $g(x) = 6^x$.

16.
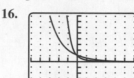
$D_j = D_k = $ {all real numbers}
$R_j = R_k = $
{all positive real numbers}

17. The horizontal line test determines that j and k are one to one functions.

18. As the value of x increases, j and k approach zero.

19. As the value of x decreases, j and k approach infinity.

20. the y axis; the y axis.   **21.** $j(x) = \dfrac{1}{2^x}$; $k(x) = \dfrac{1}{6^x}$

22. The graphs of q and r are reflections of each other over the y-axis.

23. Suppose $b = 0.5$, then $f(x) = 0.5^x$

x	$f(x)$
-2	4
-1	2
0	1
1	0.5
2	0.25

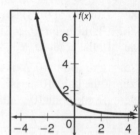

therefore, the x-axis is a horizontal asymptote of $f(x)$.

24. $f(x) = -2^x$; $k(x) = -6^{-x}$; $p(x) = -\left(\dfrac{1}{4}\right)^{-x}$

25.
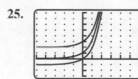
Df: {all real numbers},
Rf: {all positive real numbers},
Dg: {all real numbers},
Rg: {$y: y > 2$},
Dh: {all real numbers},
Rh: {$y: y > -1$}

26. $f(x); f(x) = 0$ $g(x); g(x) = 2$ $h(x); h(x) = -1$

27. The graph of g is above the graph of $f(x)$. The graph of h is below the graph of $f(x)$.

28. If c is a positive number, the graph of n is translated upward from the graph of m. If c is a negative number, the graph of n is translated downward from the graph of m.

29.
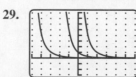
$D_f = D_g = D_h = $ {all real numbers},

$R_f = R_g = R_h = $ {all positive real numbers}

30. The horizontal asymptote for all three equations is the same; $y = 0$; $y = 0$; $y = 0$.

31. The graph of g is to the left of f. The graph of h is to the right of f.

32. If d is a positive number, the graph of n is translated right from the graph of m. If d is a negative number, the graph of n is translated left from the graph of m.

33. Answers will vary.

MAKE CONNECTIONS
34a. $g(x) = -f(x) = -5^x$; $h(x) = f(-x) = 5^{(-x)}$
34b. $g(x) = -f(x) = -7^{-x}$; $h(x) = f(-x) = 7^{(x)}$
34c. $g(x) = -f(x) = -\left(\frac{1}{11}\right)^x$; $h(x) = f(-x) = \left(\frac{1}{11}\right)^{-x}$
34d. $g(x) = -f(x) = -\left(\frac{1}{3}\right)^{-x}$; $h(x) = f(-x) = \left(\frac{1}{3}\right)^x$

35a. To the right and down   **36a.** steeper
35b. To the left and up   **36b.** less steep
35c. To the right and up   **36c.** less steep

37a. II, the point $(0, 0.25)$ is a solution.
37b. III, the point $(0, 3)$ is a solution.
37c. I, the point $(0, 4)$ is a solution.

38a. I, the point $(0, 8)$ is a solution.
38b. III, the point $(0, 242)$ is a solution.
38c. II, the point $(0, 1.004)$ is a solution.

39a. II, the point $(0, 4)$ is a solution.
39b. I, the point $(0, 3)$ is a solution.
39c. III, the point $(0, -2)$ is a solution.

40a. $g(x) = b^{x+8} - 1$
40b. $g(x) = b^{x-7} + 2$
40c. $g(x) = b^{x-6} - 3$
40d. $g(x) = b^{x+5} + 4$

41a. $y = 1$ is a horizontal asymptote
41b. $y = -1$ is a horizontal asymptote
41c. $y = \frac{1}{3}$ is a horizontal asymptote

SUMMARIZE
42. $y = 2^x$ and $y = \left(\frac{1}{2}\right)^x$ are mirrored around the y-axis;
$y = \left(\frac{1}{2}\right)^x$ and $y = 2^{-x}$ are identical

43. $g(x)$ is 2 units up and 5 units right of $f(x)$.

44. $h(x)$ and $f(x)$ are mirrored about y-axis

45. Square roots of negative numbers are not real.

46. If $b < 0$ and $x = \frac{1}{n}$ where n is an even integer, then b^x is not a real number.

47. The graphs of $f(x)$ and $g(x)$ are the same since:
$3^2 = 9 \therefore 3^{2x} = 9^x = f(x) = g(x)$

48. Answers will vary.

49.  $f(x) = e^x$, when
$x = 1$, $e^x = 2.71828$
$y = 2^x$ and $y = 3^x$

Lesson 9.2, pages 428–435

EXPLORE
1. $y = 25,000(1.05)$ **2.** $y = 25,000(1.05)^3$
$y = 25,000(1.05)^5$

3. \$31,907.04   **4.** addition   **5.** multiplication

TRY THESE
1. 278.378   **2.** 36.555   **3.** 0.012   **4.** 0.064

5. $y = a3^x10p$ **6.** $y = a3^x$ **7.** $y = a3^x$
$3 = a3^1$ $6 = a3^0$ $3 = a3^2$
$a = 1$ $6 = a^1$ $3 = a9$
　　　　$a = 6$ $a = \frac{1}{3}$

8. $y = a3^x$
$9 = a3^4$
$9 = a81$
$a = \frac{1}{9}$

9. 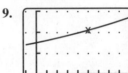 $A = P(1 + r)^n$
$A = 1500(1 + 0.07)^n$
$A = 1500(1.07)^n$
After 5 years, $A = \$2104$.

10. $7 = 49^x$ **11.** $2 = 16^x$ **12.** $4^x = \frac{1}{64}$
$7 = 7^{2x}$ $2 = 2^{4x}$ $4^x = \frac{1}{4^3}$
$1 = 2x$ $1 = 4x$ $4^x = 4^{-3}$
$x = \frac{1}{2}$ $x = \frac{1}{4}$ $x = -3$

13. $3^x = \frac{1}{81}$ **14.** $5^{-x+2} = 625$ **15.** $2^{-2x} = 64$
$3^x = \frac{1}{3^4}$ $5^{-x+2} = 5^4$ $2^{-2x} = 2^6$
$3^x = 3^{-4}$ $-x + 2 = 4$ $-2x = 6$
$x = -4$ $-x = 2$ $x = -3$
　　　　$x = -2$

16. $4^4 = 2^{x^2}$ **17.** $8^6 = 4^{x^2}$
$2^8 = 2^{x^2}$ $2^{18} = 2^{2x^2}$
$8 = x^2$ $18 = 2x^2$
$x = \pm 2\sqrt{2}$ $9 = x^2$
　　　　$x = \pm 3$

18. $D = 1300(0.94)^P$

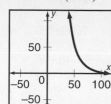

$D = 1300(0.94)^{25}$
$D = 276.783$ thousand bars.

19. $x = 2.096$ **20.** $x = 1.302$ **21.** $x = 0.774$ **22.** $x = 0.747$

23. The method assumes an average rate. If an early estimate is wrong, all future estimates are exponentially wrong.

PRACTICE

1. 39.874 **2.** 500.041 **3.** 0.001 **4.** 0.004

5. $y = a\left(\frac{1}{2}\right)^x$
$4 = a\left(\frac{1}{2}\right)^2$
$4 = 0.25a$
$a = 16$

6. $y = a\left(\frac{1}{2}\right)^x$
$9 = a\left(\frac{1}{2}\right)^0$
$a = 9$

7. $y = a\left(\frac{1}{2}\right)^x$
$4 = a\left(\frac{1}{2}\right)^{-3}$
$4 = 8a$
$a = \frac{1}{2}$

8. $y = a\left(\frac{1}{2}\right)^x$
$32 = a\left(\frac{1}{2}\right)^{-2}$
$32 = 4a$
$a = 8$

9. y gets exponentially larger as x increases.

10. $8 = 4^x$
$2^3 = 2^{2x}$
$3 = 2x$
$x = \frac{3}{2}$

11. $128 = 4^x$
$2^7 = 2^{2x}$
$7 = 2x$
$x = \frac{7}{2}$

12. $\frac{1}{2} = 4^x$
$\frac{1}{2^1} = \frac{1}{2^{-2x}}$
$1 = -2x$
$x = -\frac{1}{2}$

13. $\frac{1}{4} = 16^x$
$\frac{1}{4} = \frac{1}{4^{-2x}}$
$1 = -2x$
$x = -\frac{1}{2}$

14. $6^{3x-5} = 36^{4x+10}$
$6^{3x-5} = 6^{8x+20}$
$3x - 5 = 8x + 20$
$-5x = 25$
$x = -5$

15. $4^{3x-2} = 64^{3x+2}$
$4^{3x-2} = 4^{9x+6}$
$3x - 2 = 9x + 6$
$-6x = 8$
$x = -\frac{4}{3}$

16. $9^8 = 3^{x^2}$
$3^{16} = 3^{x^2}$
$x^2 = 16$
$x = \pm 4$

17. $\left(\frac{1}{2}\right)^6 = \left(\frac{1}{4}\right)^{x^2}$
$\left(\frac{1}{4}\right)^3 = \left(\frac{1}{4}\right)^{x^2}$
$x^2 = 3$
$x = \pm\sqrt{3}$

18. $100^x = 0.01$
$0.01^{-x} = 0.01^1$
$-x = 1$
$x = -1$

19. $1000^x = 0.1$
$0.1^{-3x} = 0.1^1$
$-3x = 1$
$x = -\frac{1}{3}$

20. $25^{2x+1} = 125^{x+2}$
$5^{4x+2} = 5^{3x+6}$
$4x + 2 = 3x + 6$
$x = 4$

21. $9x = 81^{2x+1}$
$3^{2x} = 3^{8x+4}$
$2x = 8x + 4$
$-6x = 4$
$x = -\frac{2}{3}$

22a. $y = 8(2)^t$ **22b.** $y = 8(2)^t$

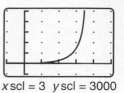

$x\,\text{scl} = 3 \quad y\,\text{scl} = 3000$

22c. $5(2) = 10; y = 8(2)^{10}; y = 8{,}192$

23. $x = 1.292$ **24.** $x = 2.631$

25. $x = 0.356$ **26.** $x = 0.540$

27a. $A = A_i(1 + r)^t$ **27b.**
$A = 18{,}000(1 - 0.15)^t$
$A = 18{,}000(0.85)^t$

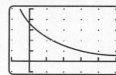

27c. \$9,396.11

28. For exponential growth, you *multiply* the y-value by the same number for a given change in x. For linear growth, you *add* the same number to the y-value for a given change in x.

EXTEND

29a. $y = (33.39)(1.076)^x$

29b. The curve fits the "trend", but will most likely not be exact. $r = 0.997$

29c. \$300.59 billion dollars **30a.** $y = 144.522(1.013)^x$

30b. Very accurate; $r = 0.999$

30c. 212.9 per 100,000 females; $y = 144.522(1.013)^{30}$

30d. No because of future developments in research.

31. An exponential curve is not appropriate because the population increased for a while and then decreased.

32a. $h = 1.7320$ **32b.** $a^h < a^x < a^k$
$k = 1.7321$
$5^{1.732} < a^x < 5^{1.7321}$
$16.24112 < a^x < 16.24374$
$a^x = 5^{\sqrt{3}} = 16.24245$ ✓

32c. The approximation of $5^{\sqrt{3}}$ can be improved by expanding on the decimal points used for h and k.

THINK CRITICALLY

33. Answers will vary.

34. $y = ab^x$ becomes steeper.

35. $y = ab^x$ decays toward the x-axis steeper.

36. The result would be a horizontal line at $y = a$.

37. $(5c)^x = (25c^2)^{x+4}$
$(5c)^x = (5c)^{2x+8}$
$x = 2x + 8$
$6 = 2x$

38. $(2c)^{x+3} = (8c^3)^{x-1}$
$(2c)^{x+3} = (2c)^{3x-3}$
$x + 3 = 3x - 3$
$x = -8$
$x = 3$

MIXED REVIEW

39. $x = 2$ **40.** $x = -8$

41. C; $(x - 5)(x + 2) = x^2 - 5x + 2x - 10 = x^2 - 3x - 10$

42. A., $(x - 5)(x^2 + 5x + 25) = x^3 + 5x^2 + 25x - 5x^2 - 25x - 125 = x^3 = 125$.

43. $x = 4y - 6$
$y = \frac{1}{4}x + \frac{3}{2}$

44. $x = -y - 3$
$y = -x - 3$

45. $x = 1.338$ **46.** $x = 1.456$

47. $\begin{bmatrix} 5+0 & 7+6 & 1+-2 & 2+1 \\ 3+3 & 6+7 & 4+4 & 8+9 \end{bmatrix} = \begin{bmatrix} 5 & 13 & -1 & 3 \\ 6 & 13 & 8 & 17 \end{bmatrix}$

48. $\begin{bmatrix} 23 & 39 \\ 35 & 119 \end{bmatrix}$

ALGEBRAWORKS

1. $y = 2^x$

2.
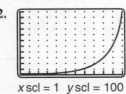
x scl = 1 y scl = 100

3. $y = 2^{30} = 1,073,741,824$ cells

4. $(1.1 \times 10^{12})(0.7) = 7.7 \times 10^{11}$

5. $(4.3 \times 10^{13})(0.3) = 1.29 \times 10^{13}$

Lesson 9.3, pages 436–442

EXPLORE

1. $y = 10^x$ is a function. $b > 1$ and is constant; $y = 10^x$ is an exponential function.

2. The inverse is a function. Use the horizontal line test.

3. The inverse of $y = 10^x$ is $x = 10^y$.

4.
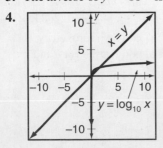

1. $7^2 = 49$ **2.** $10^{-1} = 0.1$ **3.** $3^{-3} = \frac{1}{27}$ **4.** $2^{-t} = i$

5. $\log_3 243 = 5$ **6.** $\log_{25} 5 = \frac{1}{2}$ **7.** $\log 0.001 = -3$

8. $\log_8 \frac{1}{2} = -\frac{1}{3}$ **9.** Answers will vary.

10. Let $x = \log_6 216$
$6^x = 216$
$6^x = 6^3$
$x = 3$
$\log_6 216 = 3$

11. Let $x = \log_3 \frac{1}{81}$
$3^x = \frac{1}{81}$
$3^x = 3^{-4}$
$x = -4$
$\log_3 \frac{1}{81} = -4$

12. Let $x = \log_{12} 12$
$12^x = 12$
$x = 1;$
$\log_{12} 12 = 1$

13. Let $x = \log_8 1$
$8^x = 1$
$x = 0;$
$\log_8 1 = 0$

14. Let $x = \log 62$
$10^x = 62$
$x = 1.7924;$
$\log 62 = 1.7924$

15. Let $x = \log 39$
$10^x = 39$
$x = 1.5911;$
$\log 39 = 1.5911$

16. Let $x = \log_5 10$
$5^x = 10$
$x = 1.4307$
$\log_5 10 = 1.4307$

17. Let $x = \log_7 2$
$7^x = 2$
$x = 0.3562$
$\log_7 2 = 0.3562$

18. $\log_{10} x = 3.2121$
$10^{3.2121} = 1629.6712$

19. $\log_{10} x = -1.4242$
$10^{-1.4242} = 0.0377$

20. $\log_{10} x = 15.5555$
$10^{15.5555} = 3.5934 \times 10^{15}$

21. $d = 2.5 \log r$
$d = 2.5 \log 1000$
$d = 7.5$

22. $\log_4 x = 4$
$4^4 = x = 256$

23. $\log_2 x = 8$
$2^8 = x = 256$

24. $\log_x 81 = 4$
$x^4 = 81$
$x = 3$

25. $\log_x 512 = 9$
$x^9 = 512$
$x = 2$

26. $\log_5 25 = 2x - 2$
$5^{2x-2} = 25$
$5^{2x-2} = 5^2$
$2x - 2 = 2$
$2x = 4$
$x = 2$

27. $\log_6 \frac{1}{6} = \frac{x}{9}$
$6^{\frac{x}{9}} = \frac{1}{6}$
$6^{\frac{x}{9}} = 6^{-1}$
$\frac{x}{9} = -1$
$x = -9$

PRACTICE

1. $\log_9 729 = 3$
$9^3 = 729$

2. $\log_{10} 0.0001 = -4$
$10^{-4} = 0.0001$

3. $\log_4 \frac{1}{256} = -4$
$4^{-4} = \frac{1}{256}$

4. $\log_8 1 = 0$
$8^0 = 1$

5. $\log_\pi \pi = 1$
$\pi^1 = \pi$

6. $\log_5 5^x = x$
$5^x = 5^x$

7. $t = \log_2 N$
 $t = \log_2 2048$
 $2^t = 2048$
 $t = 11$ hours

8. $\log_7 343 = 3$

9. $\log_{625} 5 = \frac{1}{4}$

10. $\log_2 \frac{1}{8} = -3$

11. $\log_{32} -\frac{1}{2} = -\frac{1}{5}$

12. $y = 5^x$ and $x = 5^y$

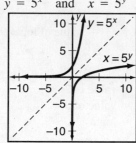

Answers will vary but can include: x and y axes are asymptotes; exponential curves; mirrored about $y = x$ line.

13. $\log_4 1024 = x$
 $4^x = 1024$
 $4^x = 4^5$
 $x = 5$
 $\log_4 1024 = 5$

14. $\log_6 \frac{1}{216} = x$
 $6^x = \frac{1}{216}$
 $6^x = 6^{-3}$
 $x = -3$
 $\log_6 \frac{1}{216} = -3$

15. $\log_9 1 = x$
 $9^x = 1$
 $x = 0$
 $\log_9 1 = 0$

16. $\log_8 8 = x$
 $8^x = 8$
 $x = 1$

17. $\log_{10} 119 = x$
 $10^x = 119$
 $x = 2.0755$

18. $\log_{10} 24 = x$
 $10^x = 24$
 $x = 1.3802$

19. $\log_6 6\sqrt{6} = x$ or $\log_6 14.69 = x$
 $6^x = 6\sqrt{6}$ $\frac{\log 14.69}{\log 6} = \frac{1.167}{0.778} = \frac{3}{2}$
 $6^x = 6(6)^{\frac{1}{2}}$
 $x = \frac{3}{2}$ $x = \frac{3}{2}$

20. $\log_{\frac{1}{2}} 32 = x$
 $\left(\frac{1}{2}\right)^x = 32$
 $2^{-x} = 2^5$
 $x = -5 = \log_{\frac{1}{2}} 32$

21. $\log_8 10 = x$
 $8^x = 10$
 $= \frac{\log 10}{\log 8} = \frac{1}{0.9031} = 1.1073$

22. $\log_9 13 = x$
 $\frac{\log 13}{\log 9} = \frac{1.11394}{0.95424} = 1.1674$

23. $\log_7 5 = x$
 $\frac{\log 5}{\log 7} = \frac{0.69897}{0.84510} = 0.8271$

24. $\log_{12} 8 = x$
 $\frac{\log 8}{\log 12} = \frac{0.90309}{1.07918} = 0.8368$

25. $x = 215.4269$

26. $x = 0.0016$

27. $x = 7.6243 \times 10^{11}$

28a. $D(I) = 10 \log \frac{I}{I_0}$
 $D(I) = 10 \log \frac{10^{-6}\,\text{w/cm}^2}{10^{-16}\,\text{w/cm}^2} = 100\,\text{dB}$

28b. $D(I) = 10 \log \frac{10^{-12}\,\text{w/cm}^2}{10^{-16}\,\text{w/cm}^2} = 40\,\text{dB}$

28c. $D(I) = 10 \log \frac{10^{-9}\,\text{w/cm}^2}{10^{-16}\,\text{w/cm}^2} = 70\,\text{dB}$

28d. $D(I) = 10 \log \frac{10^{-14}\,\text{w/cm}^2}{10^{-16}\,\text{w/cm}^2} = 20\,\text{dB}$

29. $\log_5 x = 5$
 $x = 3125$

30. $x = 729$

31. $x = 4$

32. $\log_x 729 = 6$
 $x^6 = 729$
 $x = 3$

33. $\log_x 6 = \frac{1}{2}$
 $x^{\frac{1}{2}} = 6$
 $x = 36$

34. $\log_7 49 = 3x - 3$
 $7^{3x-3} = 7^2$
 $3x - 3 = 2$
 $3x = 5$
 $x = \frac{5}{3}$

35. $\log_5 \frac{1}{5} = \frac{x}{8}$
 $5^{\frac{x}{8}} = \frac{1}{5}$
 $5^{\frac{x}{8}} = 5^{-1}$
 $-1 = \frac{x}{8}$
 $x = -8$

36. $\log_8 x = \frac{4}{3}$
 $8^{\frac{4}{3}} = x$
 $x = 16$

37. $\log_{16} x = \frac{3}{2}$
 $16^{\frac{3}{2}} = x$
 $x = 64$

EXTEND

38a. the graph of g is 2 units up from the graph of f

38b. the graph of h is 1 unit down from the graph of f

39a. the graph of j is 2 units left of the graph of f

39b. the graph of k is 3 units right of the graph of f

40. $pH = -\log [H+]$
 $pH = -\log [10^{-3}] = 3.0$

41. $pH = -\log [1.585 \times 10^{-8}] = 7.8$

42. $pH = -\log [H+]$
 $7 = -\log [H+]$
 $\log [H+] = -7$
 $10^{-7} = [H+]$
 $[H+] = 10^{-7}$ moles per liter

THINK CRITICALLY

43. -0.9420

44. $\log 0$ is undefined

45. Negative numbers are excluded from the range of the exponential function because $b^x > 0$ for $b > 0$. Since the logarithm function is the inverse of the exponential function, negative numbers must be excluded from its domain.

46. $x = \log_b a$

$b^x = a$, suppose $b = 4$ and $a = 2$

$\quad b^x = a$

$\quad 4^x = 2$

$\quad\; x = 0.5$

x must lie between zero and one

$0 < x < 1$

47. $f(g(x)) = x$

$\quad f(x) = \log_b x$

$\quad g(x) = b^x$

$\quad f(b^x) = \log_b(b^x) = x$

$\quad\; b^x = b^x$

48. $y = \dfrac{\log x}{\log 4}$

MIXED REVIEW

49. $(f + g)(x) = (3x + x - 4) = 4x - 4$

50. $(f \cdot g)(x) = (3x)(x - 4) = 3x^2 - 12x$

51. $x^2 - 8x - 9 = 0$

$(x - 9)(x + 1) = 0$

$\qquad\qquad x = -1, 9$

52. $x^2 - 2x - 15 = 0$

$(x + 3)(x - 5) = 0$

$\qquad\qquad x = -3, 5$

53. C; $\log_4 x = 3$, $4^3 = x$, $x = 64$

54. A; $\log_x 64 = 6$, $x^6 = 64$, $x = 2$

ALGEBRAWORKS

1. $M(x) = \log \dfrac{x}{x_0} = \log \dfrac{1 \text{ mm}}{0.001 \text{ mm}} = 3$

2. No, an earthquake of 8 is 10,000 times more powerful than an earthquake of magnitude 4.

Intensity level of $4 \to 10^4$

Intensity level of $8 \to \dfrac{10^8}{10^4}$

3. $M = \log_{10} \dfrac{x}{x_0}$

$10^M = \dfrac{x}{x_0}$

4. $\qquad\qquad 10^{6.1} = \dfrac{x}{0.001}$

$(1.25 \times 10^6)(0.001) = x$

$\qquad\qquad\quad x = 1{,}258.9254$

Lesson 9.4, pages 443–448

EXPLORE

1.

x	2	3	4	5	8	10	15
$\log x$	0.30103	0.47712	0.60206	0.69897	0.90309	1	1.17609

2a. 1.0   **2b.** 0.90309   **2c.** 1.7609

3. the value of log 2 + log 5 is the value of log 10

$\qquad\quad 2 \times 5 = 10$

$\qquad\quad 2 \times 4 = 8 \quad \leftarrow$ value of log 8

$\qquad\quad 3 \times 5 = 15 \leftarrow$ value of log 15

$\quad \log a + \log b = \log ab$

4a. 0.30103   **4b.** 0.69897   **4c.** 0.47712

5. the value of log 8 − log 4 = value of log 2

$\qquad\qquad\qquad 8 \div 4 = 2$

$\qquad\qquad\qquad 10 \div 2 = 5 \leftarrow$ value of log 5

$\qquad\qquad\qquad 15 \div 5 = 3 \leftarrow$ value of log 3

$\qquad \log a - \log b = \log \dfrac{a}{b}$

6a. 0.60206   **6b.** 0.90309

7. $2^2 = 4 \leftarrow$ value of log 4 = value of $2 \times \log 2$

$\quad 2^3 = 8 \leftarrow$ value of log 8 = value of $3 \times \log 2$

$\qquad\qquad c \log a = \log a^c$

TRY THESE

1. $\log_7 3x = \log_7 3 + \log_7 x$

2. $\log_2 \dfrac{y}{5} = \log_2 y - \log_2 5$

3. $\log_8 x^7 y^3 = 7 \log_8 x + 3 \log_8 y$

4. $\log_9 (3x)^3 = \log_9 27 + 3 \log_9 x$

5. Answers will vary.   **6.** $\log_3 \dfrac{y^3}{z^6}$

7. $\log_8 3^4 x^5 = \log_8 81 x^5$ **8.** $\log_4 \dfrac{p^5 q^2}{r^3}$

9. $\log_b \dfrac{7^2 d^6}{e^4} = \log_b \dfrac{49 d^2}{e^4}$

10. $\log F = \log k + \log m_1 + \log m_2 - 2 \log d$

11. $\log_6 t = 5 \log_6 2 - \log_6 8$

$\qquad\quad = \log_6 2^5 - \log_6 8$

$\qquad\quad = \log_6 32 - \log_6 8$

$\log_6 t = \log_6 \dfrac{32}{8}$

$\log_6 t = \log_6 4$

$\qquad t = 4$

12. $\log_7 V = 4 \log_7 3 - \log_7 9$

$\qquad\quad = \log_7 3^4 - \log_7 9$

$\qquad\quad = \log_7 81 - \log_7 9$

$\log_7 V = \log_7 \dfrac{81}{9} = \log_7 9$

$\qquad V = 9$

13. $\log_3 (w^2 + 11) = 3$

$\qquad\qquad 3^3 = w^2 + 11$

$\qquad\qquad 0 = w^2 - 16$

$(w + 4)(w - 4) = 0$

$\qquad\qquad w = \pm 4$

14. $\log_2 (z^2 - 8) = 3$
$$2^3 = z^2 - 8$$
$$8 = z^2 - 8$$
$$0 = z^2 - 16$$
$$(z + 4)(z - 4) = 0$$
$$z = \pm 4$$

15. $\log_4 (x - 12) = 3 - \log_4 x$
$$3 = \log_4 (x - 12) + \log_4 x$$
$$3 = \log_4 (x^2 - 12x)$$
$$4^3 = x^2 - 12x$$
$$x^2 - 12x - 64 = 0$$
$$x + 4 = 0 \quad \text{or} \quad x - 16 = 0$$
$$x = 4 \qquad\qquad x = 16$$
Since $x = 4$ is an extraneous solution, the solution is $x = 16$.

16.
$$\log_5 x = 1 - \log_5 (6x - 7)$$
$$\log_5 x + \log_5 (6x - 7) = 1$$
$$\log_5 (6x^2 - 7x) = 1$$
$$5^1 = 6x^2 - 7x$$
$$6x^2 - 7x - 5 = 0$$
$$(3x - 5)(2x + 1) = 0$$
$$x = \frac{5}{3}$$

17. $\log_b \dfrac{m}{n} = \log_b m - \log_b n$
Let $\log_b m = u$ and $\log_b n = v$
then $b^u = m$ and $b^v = n$
$$\frac{m}{n} = \frac{b^u}{b^v} \qquad \text{divide, quotient prop of exp.}$$
$$\frac{m}{n} = b^{u-v}$$
$$\log_b \frac{m}{n} = u - v$$
$$\log_b \frac{m}{n} = \log_b m - \log_b n \quad \text{substitute}$$

18. $d = 2.5 \log r$
$d = \log r^{2.5}$

19. $d = \log (10{,}000)^{2.5} = 10$

Practice

1. $\log_4 5y = \log_4 5 + \log_4 y$

2. $\log_9 7z = \log_9 7 + \log_9 z$

3. $\log_8 \dfrac{x}{6} = \log_8 x - \log_8 6$

4. $\log_6 \dfrac{t}{5} = \log_6 t - \log_6 5$

5. $\log_2 x^6 y^7 = 6 \log_2 x + 7 \log_2 y$

6. $\log_5 p^2 q^3 = 2 \log_5 p + 3 \log_5 q$

7. $\log_7 \sqrt{x^4 y^2} = \log_7 (x^4 y^2)^{\frac{1}{2}} = \log_7 (x^2 y) = 2 \log_7 x + \log_7 y$

8. $\log_3 (x^6 y^9)^{\frac{1}{3}} = \log_3 (x^2 y^3) = 2 \log_3 x + 3 \log_3 y$

9a. $D(I) = 10 \log \dfrac{I}{I_0} = 10 \log I - 10 \log I_0$

9b. $D(I) = 10 \log \left(10^{-9} \dfrac{\text{w}}{\text{cm}^2}\right) - 10 \log \left(10^{-16} \dfrac{\text{w}}{\text{cm}^2}\right)$
$= 70\text{dB}$

10. $\log_5 \dfrac{y^2}{z^7}$ **11.** $\log_3 u^3 v^8$ **12.** $\log_7 \dfrac{8^3 e^6}{f^4} = \log_7 \dfrac{512 e^6}{f^4}$

13. $\log_9 \dfrac{q^2 r^6}{t^4}$ **14.** $\log_8 y^{\frac{4}{3}}$ **15.** $\log_5 w^{\frac{7}{2}}$

16a. $\text{pH} = \log [\text{H+}]^{-1}$

16b. $\text{pH} = \log [10^{-4}]^{-1} = \log 10^4 = 4$

17. $\log_9 t = 4 \log_9 5 - \log_9 5$
$$= \log_9 5^4 - \log_9 5$$
$$\log_9 t = \log_9 \frac{625}{5} = \log_9 125$$
$$t = 125$$

18. $\log_8 v = 3 \log_8 4 - \log_8 32$
$$= \log_8 4^3 - \log_8 32$$
$$\log_8 v = \log_8 \frac{64}{32} = \log_8 2$$
$$v = 2$$

19. $\log_4 (w^2 + 15) = 3$
$$4^3 = w^2 + 15$$
$$64 - 15 = w^2$$
$$w^2 = 49, \, w = \pm 7$$

20. $\log_5 (z^2 + 225) = 4$
$$5^4 = z^2 + 225$$
$$625 - 225 = z^2 = 400, \quad z = \pm 20$$

21. $\log_7 2x + \log_7 3x = \log_7 384$
$$\log_7 2x \cdot 3x = \log_7 384$$
$$6x^2 = 384$$
$$x^2 = 64, \, x = 8$$

22. $\log_3 7x + \log_3 8x = \log_3 224$
$$56x^2 = 224$$
$$x^2 - 4, \, x - 2$$

23. $\log \dfrac{x}{5} = \log x - \log 5$

$\dfrac{\log x}{\log 5}$

Suppose $x = 3$;
then $\log 3 - \log 5 = 0.477 - 0.6989 = -0.22185$
and $\dfrac{\log 3}{\log 5} = 0.68261$. These expressions are not equal.

24. $\log_b m^p = p \log_b m$
Let $\log_b m = u$
then $b^u = m$
$$u^p = p(u)$$
$$= p \log_b m$$
$$\log_b m^p = p \log_b m$$

Extend

25. $\log_4 (x + 8) - \log_4 (x - 1) = 2$
$$\log_4 \frac{(x + 8)}{x - 1} = 2$$
$$4^2 - \frac{x + 8}{x - 1} = 16$$
$$x + 8 = 16(x - 1) = 16x - 16$$
$$24 = 15x$$
$$x = 1.6$$

26. $\log_3 (2x + 5) - \log_3 (x - 3) = 1$

$\log_3 \dfrac{2x + 5}{x - 3} = 1$

$3^1 = \dfrac{2x + 5}{x - 3}$

$3(x - 3) = 2x + 5 = 3x - 9$

$x = 14$

27. $\log_5 (x + 5) - \log_5 (x - 5) = 1$

$\log_5 \dfrac{x + 5}{x - 5} = 1$

$5^1 = \dfrac{x + 5}{x - 5}$

$5x - 25 = x + 5$

$4x = 30$

$x = 7.5$

28. $\log_2 (x + 4) = \log_2 (x - 3) = 3$

$\log_2 \dfrac{x + 4}{x - 3} = 3$

$2^3 = \dfrac{x + 4}{x - 3}$

$8 = \dfrac{x + 4}{x - 3}, 8x - 24 = x + 4,$

$7x = 28, x = 4$

29. $x = \sqrt{\dfrac{2cd}{e}}$

$\log x = \log \left(\dfrac{2cd}{e}\right)^{\frac{1}{2}} = \frac{1}{2} \log 2c + \frac{1}{2} \log d - \frac{1}{2} \log e$

$= \frac{1}{2} [\log 2 + \log c + \log d - \log e]$

30. $v = K\sqrt{PV}$

$\log v = \log K(PV)^{\frac{1}{2}} = \log K + \frac{1}{2} \log P + \frac{1}{2} \log V$

31. $p = 2\pi \sqrt{\dfrac{m}{k}}$

$\log p = \log \left(2\pi \sqrt{\dfrac{m}{k}}\right)$

$\log p = \log 2 + \frac{1}{2} \log m - \frac{1}{2} \log k + \log \pi$

THINK CRITICALLY

32. $\log_b (b^x) = x; \log_b (t) = x; b^x = (t)$

$b^x = (b^x)$

33. $b^{\log_b x} = x; t = \log_b x; b^t = x \therefore b^{\log_b x} = x$

34. $-\log_b x$

$= \log_b x^{-1} = \log_b \dfrac{1}{x}$

35. No. The domain of y_1 is all real numbers except $x = 3$. The domain of y_2 is $x : x > 3$.

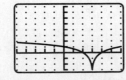

36. $\log x \div \log \dfrac{1}{x}$

$\dfrac{\log x}{\log \frac{1}{x}} = \dfrac{\log x}{\log x^{-1}} = \dfrac{\log x}{-\log x} = -1$

37. C

38. $[3 \quad 0 \ {-2}] \begin{bmatrix} 2 & 9 \\ 3 & -6 \\ 1 & 0 \end{bmatrix} = [4 \quad 27]$

39. $\begin{bmatrix} 3 & 4 & 5 \\ 1 & -1 & -6 \\ 0 & 4 & 2 \end{bmatrix} \begin{bmatrix} 1 & 1 & -2 \\ 4 & -2 & 0 \\ 5 & 1 & 6 \end{bmatrix} = \begin{bmatrix} 44 & -10 & 24 \\ -33 & 9 & -38 \\ 26 & -10 & 12 \end{bmatrix}$

40. $(5 + 3i)(4 - 2i) = 20 + 12i - 10i - 6i^2$

$= 20 + 2i - 6i^2 = 20 + 2i + 6 = 26 + 2i$

41. $(7 - 5i)(3 - 8i) = 21 - 15i - 56i + 40i^2$

$= 21 - 40 - 71i$

$= -(19 + 71i)$

42. $\log_8 4t = \log_8 4 + \log_8 t$

43. $\log_9 \dfrac{q}{2} = \log_9 q - \log_9 2$

44. $\log_3 4x^2 y^3 = \log_3 4 + 2 \log_3 x + 3 \log_3 y$

Lesson 9.5, pages 449–453

EXPLORE

1. $n \left(1 + \dfrac{1}{n}\right)^n$

1	2
10	2.59374
10^2	2.70481
10^3	2.71692
10^4	2.71815
10^6	2.71828
10^8	2.71828
10^{10}	2.71828

2. 2.71828

3. $1 + 1 + 0.5 + 0.16667 + 0.04167 + 0.00833 + 0.00139 + 1.98412 \times 10^{-4} + 2.48 \times 10^{-5}$
$= 2.71828$

4. They are the same.

TRY THESE

1. $e^5 = 148.4132$ **2.** $e^{-2.5} = 0.082$

3. $\ln 12.8 = 2.5494$ **4.** $4.2 \ln 8 + 6 = 14.7337$

5. e^3 **6.** e^3 **7.** $0.2e^{-4}$ **8.** $27e^{12}$

9. $\ln 650 = 6.47697$ **10.** $9 \ln a - 3 \ln b$
Thus $\ln 650 \neq 3.4770$

11. $\ln \sqrt{9u^8 v^{14}} = \ln (9u^8 v^{14})^{\frac{1}{2}} = \ln 3u^4 v^7$
$= \ln 3 + 4 \ln u + 7 \ln v$

12. $C = 5e^{-0.4t}$, if $t = 2.5h$, then
$C = 5e^{-0.4(2.5)} = 1.84$ mg

13. $\ln 64 \left(\dfrac{1}{2}\right)^4 x = \ln 4x$

14. $\ln \dfrac{80\left(\frac{1}{4}\right)^2}{y} = \ln \dfrac{5}{y}$

15. $N = 120 - 120e^{-0.09t}$, if $t = 3$ then
$N = 120 - 120e^{-0.09(3)} = 28.39$
$N = 28$ whole clocks

16. $\log_9 14 = \dfrac{\ln 14}{\ln 9} = 1.2011$

17. $\log_4 8 = \dfrac{\ln 8}{\ln 4} = 1.5000$

18. $\log_7 4 = \dfrac{\log 4}{\log 7} = 0.7124$

19. $\log_{10} 6 = \dfrac{\log 6}{\log 10} = 0.7782$

20. An approximation of the graph of e^x is the graph of $h(x) = b^x$ where $b = 2.7183$ or graph $y = 2^x$ and $y = 3^x$ and reason $y = e^x$ is between these two graphs.

PRACTICE

1. 1.2840   **2.** 60.3403   **3.** 0.6065   **4.** 1.6864

5. $6.227 + 2 = 8.2269$ **6.** 3.2958

7.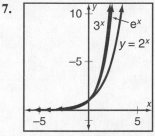

The graph of e^x lies between the graphs of 2^x and 3^x.
The graph of e^x is actually the graph of 2.7183^x.

8. $y = (T_1 - T_0)e^{-0.03t} + T_0$
$= (98.6 - 39)e^{-0.031(90)} + 39$
$y = 42.7°F$

9. e^9 **10.** e^4 **11.** $2e^5$ **12.** $256e^{20}$

13. $\ln 8 + \ln t$ **14.** $4 \ln u$

15. $\ln m^6 - \ln n^5 = 6 \ln m - 5 \ln n$

16. $\ln (4x^4y^{10})^{\frac{1}{2}} = \ln 2 + \ln x^2 + \ln y^5$
$= \ln 2 + 2 \ln x + 5 \ln y$

17. $\ln 10jk$ **18.** $\ln 10w$ **19.** $\ln \dfrac{7}{z}$

20. $p = 101352e^{-0.000122h}$; if $h = 2500$ m then
$p = 101352e^{-0.000122(2500)}$
$p = 74,000$ Pascals

21. $\log_8 10 = \dfrac{\ln 10}{\ln 8} = 1.1073$

22. $\log_3 11 = \dfrac{\ln 11}{\ln 3} = 2.1827$

23. $\log_9 6 = \dfrac{\ln 6}{\ln 9} = 0.8155$

24. $\log_{12} 2 = \dfrac{\ln 2}{\ln 12} = 0.2789$

EXTEND

25a.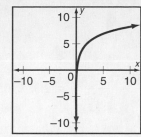

$y = -5.395 + 19.127 \ln x$

25b. $r = 0.998$, the curve fits the data very closely

25c. $y = -5.4 + 19.1 \ln (120) = 86\%$

26a. Logarithmic regression gives $r = 0.985$.
Linear regression gives $r = 0.979$.
Thus logarithmic regression is the best.

26b. $y = -15.306 + 2.634 \ln x$

26c. Very close; $r = 0.985$

27a. exponential regression

27b. $y - 3,732,113.235 \, (0.9931)^x$

27c. $r = -0.9707$; very close

28a. Quadratic   **28b.** $h = -16x^2 + 184x$

THINK CRITICALLY

29. $\ln e = 1 \quad \log_e e = 1; \therefore \ln e = 1$

30. Since: $\log_e x = \ln x$
$e^{\ln x} = x$

31. Prove $\ln e^x = x = x \ln e = x; x(1) = x; x = x$

32. $e^{0.25} = 1.284$
from calculator
$= 1 + 0.25 + 0.031 + 0.0026 + 0.0001$
$= 1.28376$

33. $\log 25$ is $\log_{10} 25$
$\ln 25$ is $\log_e 25$ or $\log_{2.7} 25$
therefore, $\ln 25$ is larger

MIXED REVIEW

34. B; $y = 5x + 35; y = mx + b$; where $m = $ slope
$10 = 5(-5) + 35 = -25 + 35$
$10 = 10$

35. $2x^2 + 6x - 5 = 0 \quad a = 2$
$b = 6$
$c = -5$

$x_1, x_2 = \dfrac{-b \pm \sqrt{b^2 - 4ac}}{2a}$

$= \dfrac{-6 \pm \sqrt{36 + 40}}{4} = \dfrac{-6 \pm 8.717}{4} = -3.679, 0.679$

or $-\dfrac{3}{2} \pm \dfrac{\sqrt{19}}{2}$

36. $3x^2 - 8x + 2 = 0$ $a = 3$
  $b = -8$
  $c = 2$

$$x = \frac{-b \pm \sqrt{b^2 - 4ac}}{2a}$$

$$x_1, x_2 = \frac{8 \pm \sqrt{64 - 24}}{6} = \frac{8 \pm 6.3}{6} = 2.387, \ 0.28$$

or $\frac{4}{3} \pm \frac{\sqrt{10}}{3}$

37. $\log_5 26 = \frac{\ln 26}{\ln 5} = 2.0244$

38. $\log_{12} 3 = \frac{\ln 3}{\ln 12} = 0.4421$

PROJECT CONNECTION

1. $l(t) = \dfrac{A}{1 + Be^0} = \dfrac{A}{1 + B}$

2. Be^{-kt} goes to 0 when t is very large.

3. $l(t) = \dfrac{A}{1 + 0} = A$

4. Answers will vary.

Lesson 9.6, pages 454–459

EXPLORE

1. From the graph of the two functions, I found
$x = 1.7712$

2. $\log_b b^x = x$, for this problem, the base 3 would be used
$3^x = 7$, $\log_3 3^x = \log_3 7$

3. You must take the logarithm of both sides

4. $\log_3 3^x = \log_3 7; x = \log_3 7$

5. $\log_3 7 = \dfrac{\ln 7}{\ln 3} = 1.7712$

6. Answers to 1 and 5 are the same.

7. Answers will vary.

TRY THESE

1. $5^x = 28; \log 5^x = \log 28, \ x \log 5 = \log 28$
$x = \dfrac{\log 28}{\log 5} = 2.0704$

2. $6^x = 3; x \log 6 = \log 3$
$x = \dfrac{\log 3}{\log 6} = 0.6131$

3. $e^x = 12; \ln e^x = \ln 12$
  $x = \ln 12 = 7.4849$

4. $e^x = 2; \ln e^x = \ln 2$
  $x = \ln 2 = 0.6931$

5. $b^x = y$ 1. Take log of both sides
      $\log b^x = \log y$
      $x \log b = \log y$
      $x = \dfrac{\log y}{\log b}$
    2. Graph both functions to find x intercept
    3. Take common base of both sides
      $\log_b b^x = y$
      $b^y = b^x$

6. $2e^x + 8 = 40$ Check:
   $2e^x = 32$ $2e^{(2.77)} + 8 \overset{?}{=} 40$
   $e^x = 16$ $40 = 40$ ✓
  $\ln e^x = \ln 16$
   $x = 2.7726$

7. $5e^x - 10 = 45$ Check:
   $5e^x = 55$ $5e^{(2.39789)} - 10 \overset{?}{=} 45$
   $e^x = 11$ $44.9 = 45$ ✓
  $\ln e^x = \ln 11$
   $x = 2.3979$

8. $7e^{2x} - 5 = 42$ Check:
   $7e^{2x} = 47$ $7e^{2(0.95212)} - 5 \overset{?}{=} 42$
   $e^{2x} = 6.71428$ $7(6.71478) - 5 \overset{?}{=} 42$
  $\ln e^{2x} = \ln 6.71428$ $47 - 5 \overset{?}{=} 42$
   $2x = 1.904$ $42 = 42$ ✓
   $x = 0.9521$

9. $10^x - 17 = 50$ Check:
   $10^x = 67$ $10^{1.82607} - 17 \overset{?}{=} 50$
  $\log 10^x = \log 67$ $49.999 = 50$ ✓
  $x \log 10 = \log 67$
   $x = \dfrac{\log 67}{\log 10} = 1.8261$

10. $10^{5x} + 12 = 60$ Check:
    $10^{5x} = 48$ $10^{5(0.33625)} + 12 \overset{?}{=} 60$
  $\log 10^{5x} = \log 48$ $60 = 60$ ✓
    $5x = 1.68$
    $x = 0.3362$

11. $10^{-3x} - 120 = 18$ Check:
    $10^{-3x} = 138$ $10^{-3(-0.71329)} - 120 \overset{?}{=} 18$
  $\log 10^{-3x} = \log 138$ $17.99 = 18$ ✓
    $-3x = 2.13987$
    $x = -0.7133$

12. $a = 300 - 300e^{-0.3t}$
   $175 - 300 = -300e^{-0.3t}$
   $\dfrac{-125}{-300} = e^{-0.3t}$
  $\ln 0.4167 = \ln e^{-0.3t}$
   $-0.8754 = -0.3t$
      $t = 2.91823$
      $t = 3$ days
Check:
   $175 \overset{?}{=} 300 - 300e^{-0.3(2.91)}$
   $175 \overset{?}{=} 300 - 124.99$
   $175 = 175$ ✓

13. $2 + 6 \log x = 10$ Check:
   $6 \log x = 8$ $2 + 6 \log 21.5 \overset{?}{=} 10$
   $\log x = 1.33$ $2 + 7.999 \overset{?}{=} 10$
  $10^{\log x} = 10^{1.33}$ $10 = 10$ ✓
   $x = 21.5443$

14.

$5 \log x - 8 = 4$

$5 \log x = 12$

$\log x = \dfrac{12}{5} = 2.4$

$10^{\log x} = 10^{2.4}$

$x = 251.1886$

Check:

$5 \log 251 - 8 \overset{?}{=} 4$

$11.99 - 8 \overset{?}{=} 4$

$3.9999 = 4$ ✓

15.

$3 \log 3x = 21$

$\log 3x = 7$

$\log 3 + \log x = 7$

$0.47712 + \log x = 7$

$\log x = 6.52287$

$10^{\log x} = 10^{6.52287}$

$x = 3{,}333{,}333.3333$

Check:

$3 \log 3(3 \times 10^6) \overset{?}{=} 21$

$3(6.99) \overset{?}{=} 21$

$20.999 = 21$ ✓

16.

$7 + 8 \ln x = 2$

$8 \ln x = -5$

$\ln x = -\dfrac{5}{8} = -0.625$

$e^{\ln x} = 3^{-0.625}$

$x = 0.53536$

Check:

$7 + 8 \ln (0.53526) \overset{?}{=} 2$

$7 + (-5) \overset{?}{=} 2$

$2 = 2$ ✓

17.

$9 - 3 \ln x = 6$

$-3 \ln x = -3$

$\ln x = +1$

$e^{\ln x} - e^1$

$x = 2.7183$

Check:

$9 - 3 \ln (2.71828) \overset{?}{=} 6$

$9 - 2.99999 \overset{?}{=} 6$

$6 = 6$ ✓

18.

$4 \ln 4x = 14$

$\ln 4x = 3.5$

$e^{\ln 4x} = e^{3.5}$

$4x = e^{3.5}$

$x = \dfrac{1}{4} e^{3.5}$

$x = 8.2789$

Check:

$4 \ln 4(8.27886) \overset{?}{=} 14$

$4(3.5) \overset{?}{=} 14$

$14 = 14$ ✓

PRACTICE

1.

$7^x = 35$

$\log 7^x = \log 35$

$x \log 7 = \log 35$

$x = \dfrac{\log 35}{\log 7}$

$x = 1.8271$

Check:

$7^{1.82708} \overset{?}{=} 35$

$34.9995 = 35$ ✓

2.

$8^{-x} = 3$

$\log 8^{-x} = \log 3$

$-x \log 8 = \log 3$

$-x = \dfrac{\log 3}{\log 8}$

$x = -0.5283$

Check:

$8^{-(-0.52832)} \overset{?}{=} 3$

$2.9999 = 3$ ✓

3.

$4^{-3x} = 13$

$\log 4^{-3x} = \log 13$

$-3x \log 4 = \log 13$

$-3x = \dfrac{\log 13}{\log 4}$

$-3x = 1.85022$

$x = -0.6167$

Check:

$4^{-3(-0.61674)} \overset{?}{=} 13$

$4^{1.85022} \overset{?}{=} 13$

$13 = 13$ ✓

4.

$5^{\frac{1}{2}x} = 12$

$\log 5^{\frac{1}{2}x} = \log 12$

$\dfrac{1}{2} x \log 5 = \log 12$

$\dfrac{1}{2} x = \dfrac{\log 12}{\log 5}$

$x = 3.0879$

Check:

$5^{\frac{1}{2}(3.08792)} \overset{?}{=} 12$

$12 = 12$ ✓

5.

$e^{-x} = 31$

$\ln e^{-x} = \ln 31$

$-x = 3.43399$

$x = -3.4340$

Check:

$e^{-(-3.43399)} \overset{?}{=} 31$

$30.999 = 31$ ✓

6.

$e^x = 18$

$\ln e^x = \ln 18$

$x = 2.8903$

Check:

$e^{2.89037} \overset{?}{=} 18$

$18 = 18$ ✓

7.

$e^{\frac{1}{4}x} = 55$

$\ln e^{\frac{1}{4}x} = \ln 55$

$\dfrac{1}{4} x = 4.00733$

$x = 16.0293$

Check:

$e^{\frac{1}{4}(16.02933)} \overset{?}{=} 55$

$54.999 = 55$ ✓

8.

$e^{-4x} - 5$

$\ln e^{-4x} = \ln 5$

$-4x = 1.60944$

$x = -0.4024$

Check:

$e^{-4(-0.40236)} \overset{?}{=} 5$

$4.999 = 5$ ✓

9. b is the correct set of steps.

Check: $3 \log (215.4435) \overset{?}{=} 7$

$7 = 7$ ✓

10.

$4e^x + 10 = 50$

$4e^x - 40$

$e^x = 10$

$\ln e^x = \ln 10$

$x = 2.3026$

Check:

$4e^{2.3025} + 10 \overset{?}{=} 50$

$40 + 10 \overset{?}{=} 50$

$50 = 50$ ✓

11.

$7e^x - 32 = 45$

$7e^x = 77$

$e^x = 11$

$\ln e^x = \ln 11$

$x = 2.3979$

Check:

$7e^{2.3979} - 32 \overset{?}{=} 45$

$77 - 32 \overset{?}{=} 45$

$45 = 45$ ✓

12.

$6e^{2x} - 6 = 66$

$6e^{2x} = 72$

$e^{2x} = 12$

$2x = \ln 12$

$2x = 2.4849$

$x = 1.2425$

Check:

$6e^{2(1.2425)} - 6 \overset{?}{=} 66$

$6(12) = 66 + 6 \overset{?}{=} 72$

$72 = 72$ ✓

13.

$10^x - 2 = 74$

$10^x = 76$

$\log 10^x = \log 76$

$x = 1.8808$

Check:

$10^{1.8808} - 2 \overset{?}{=} 74$

$75.99 - 2 \overset{?}{=} 74$

$74 = 74$ ✓

14. $10^{\frac{1}{2}x} + 32 = 111$

$10^{\frac{1}{2}x} = 79$

$\log 10^{\frac{1}{2}x} = \log 79$

$\frac{1}{2}x = 1.8976$

$x = 3.7953$

Check:

$10^{\frac{1}{2}(3.7953)} + 32 \stackrel{?}{=} 111$

$78.99 + 32 \stackrel{?}{=} 111$

$111 = 111$ ✓

15. $10^{-4x} - 88 = 4$

$10^{-4x} = 92$

$\log 10^{-4x} = \log 92$

$-4x = 1.9638$

$x = -0.4909$

Check:

$10^{-4(-0.4909)} - 88 \stackrel{?}{=} 4$

$91.99 - 88 \stackrel{?}{=} 4$

$4 = 4$ ✓

16. $p = 101352e$

$6000 = 101352e^{-0.000122h}$

$0.05919 = e^{-0.000122h}$

$\ln 0.05919 = \ln e^{-0.000122h}$

$-2.8268 = -0.000122h$

$h = 23{,}171 \text{ meters}$

Check:

$6000 \stackrel{?}{=} 101352e^{-(0.000122)(23.171)}$

$6000 \stackrel{?}{=} 101352(0.05919)$

$6000 = 5999.99$ ✓

17. $3 + 5 \log x = 18$

$5 \log x = 15$

$\log x = 3$

$10^{\log x} = 10^3$

$x = 1000$

Check:

$3 + 5 \log 1000 \stackrel{?}{=} 18$

$3 + 15 \stackrel{?}{=} 18$

$18 = 18$ ✓

18. $6 \log x - 1 = 11$

$6 \log x = 12$

$\log x = 2$

$10^{\log x} = 10^2$

$x = 100$

Check:

$6 \log 100 - 1 \stackrel{?}{=} 11$

$12 - 1 \stackrel{?}{=} 11$

$11 = 11$ ✓

19. $4 \log \frac{1}{2}x = 7$

$\log \frac{1}{2}x = 1.75$

$10^{\log \frac{1}{2}x} = 10^{1.75}$

$\frac{1}{2}x = 56.2341$

$x = 112.4683$

Check:

$4 \log \frac{1}{2}(112.4683) \stackrel{?}{=} 7$

$4(1.75) \stackrel{?}{=} 7$

$7 = 7$ ✓

20. $-9 + 2 \ln x = -5$

$2 \ln x = 4$

$\ln x = 2$

$e^{\ln x} = e^2$

$x = 7.3891$

Check:

$-9 + 2 \ln (7.3891) \stackrel{?}{=} -5$

$-9 + 4 \stackrel{?}{=} -5$

$-5 = -5$ ✓

21. $10 - 7 \ln x = 5$

$-7 \ln x = -5$

$\ln x = 0.71428$

$e^{\ln x} = e^{0.71428}$

$x = 2.0427$

Check:

$10 - 7 \ln (2.0427) \stackrel{?}{=} 5$

$10 - 4.99 \stackrel{?}{=} 5$

$5 = 5$ ✓

22. $5 \ln 5x = 23$

$\ln 5x = 4.6$

$e^{\ln 5x} = e^{4.6}$

$5x = \frac{e^{4.6}}{5}$

$x = 19.8969$

Check:

$5 \ln 5(19.8969) \stackrel{?}{=} 23$

$5(4.6) \stackrel{?}{=} 23$

$23 = 23$ ✓

23. $y = 600{,}744 \, (0.9847)^x$

$28{,}5000 = 600{,}744 \, (0.9847)^x$

$0.4744 = 0.9847^x$

$\log 0.4744 = \log 0.9847^x = x \log 0.9847$

$x = \dfrac{\log 0.4744}{\log 0.9847}$

$x = 48.365$

1998

Check:

$285{,}000 \stackrel{?}{=} 600{,}744 \, (0.9847)^{48.365}$

$285{,}000 \stackrel{?}{=} 600{,}744 \, (0.4744)$

$285{,}000 = 284{,}993$ ✓

24. $B = 800(2)^{\frac{t}{40}}$

$5000 = 800\left(2^{\frac{t}{40}}\right)$

$6.25 = 2^{\frac{t}{40}}$

$\log 6.25 = \log 2^{\frac{t}{40}}$

$\log 6.25 = \frac{t}{40} \log 2$

$\dfrac{\log 6.25}{\log 2} = \dfrac{t}{40}$

$t = 106 \text{ hours}$

Check:

$5000 \stackrel{?}{=} 800\left(2^{\frac{105.75}{40}}\right)$

$5000 \stackrel{?}{=} 800(6.233)$

$5000 = 4986$ ✓

EXTEND

25. $6^{x-2} = 3^{x+1}$

$\log 6^{x-2} = \log 3^{x+1}$

$(x - 2) \log 6 = (x + 1) \log 3$

$x \log 6 - 2 \log 6 = x \log 3 + \log 3$

$x \log 6 - x \log 3 = \log 3 + 2 \log 6$

$x (\log 6 - \log 3) = \log 3 + 2 \log 6$

$x = \dfrac{\log 3 + 2 \log 6}{\log 6 - \log 3}$

$x = 6.7549$

Check:

$6^{(6.7549 - 2)} \stackrel{?}{=} 3^{(6.7549 + 1)}$

$6^{4.7549} \stackrel{?}{=} 3^{7.7549}$

$5012.23 = 5012.19$ ✓

26.
$$12^x = 5^{x+5}$$
$$\log 12^x = \log 5^{x+5}$$
$$x \log 12 = (x + 5) \log 5$$
$$x \log 12 = x \log 5 + 5 \log 5$$
$$x \log 12 - x \log 5 = 5 \log 5$$
$$x (\log 12 - \log 5) = 5 \log 5$$
$$x = \frac{5 \log 5}{\log 12 - \log 5}$$
$$x = 9.1919$$

Check:
$$12^{9.1919} \stackrel{?}{=} 5^{9.1919+5}$$
$$8{,}312{,}393{,}528 = 8{,}312{,}131{,}486 \checkmark$$

27.
$$9^{x-4} = 4^{4-x}$$
$$\log 9^{x-4} = \log 4^{4-x}$$
$$(x - 4) \log 9 = (4 - x) \log 4$$
$$x \log 9 - 4 \log 9 = 4 \log 4 - x \log 4$$
$$x \log 9 + x \log 4 = 4 \log 4 + 4 \log 9$$
$$x (\log 9 + \log 4) = 4 (\log 4 + \log 9)$$
$$x = 4 \left(\frac{\log 4 + \log 9}{\log 4 + \log 9} \right)$$
$$x = 4$$

Check:
$$9^{x-4} \stackrel{?}{=} 4^{4-x}$$
$$9^{4-4} \stackrel{?}{=} 4^{4-4}$$
$$9^0 \stackrel{?}{=} 4^0$$
$$1 = 1 \checkmark$$

28.
$$7^{2+x} = 5^{2-x}$$
$$\log 7^{2+x} = \log 5^{2-x}$$
$$(2 + x) \log 7 = (2 - x) \log 5$$
$$2 \log 7 + x \log 7 = 2 \log 5 - x \log 5$$
$$x \log 7 + x \log 5 = 2 \log 5 - 2 \log 7$$
$$x (\log 7 + \log 5) = 2 (\log 5 - \log 7)$$
$$x = \frac{2(\log 5 - \log 7)}{\log 7 + \log 5} = \frac{-0.29226}{1.544} = -0.1893$$

Check:
$$7^{2-0.1893} \stackrel{?}{=} 5^{2+0.1893}$$
$$7^{1.8107} \stackrel{?}{=} 5^{2.1893}$$
$$33.9 - 33.9 \checkmark$$

29.
$$A = A_0 2^{\frac{-t}{5760}}$$
$$100 = 120 \left(2^{\frac{-t}{5760}} \right)$$
$$\frac{100}{120} = 2^{\frac{-t}{5760}}$$
$$0.833 = 2^{\frac{-t}{5760}}$$
$$\log 0.833 = \log 2^{\frac{-t}{5760}} = -\frac{t}{5760} \log 2$$
$$-\frac{t}{5760} = \frac{\log 0.833}{\log 2} = \frac{-0.07918}{0.301} = -0.2630$$
$$t = (0.2630)(5760) = 1515$$
$$t = 1500 \text{ years}$$

Check:
$$100 \stackrel{?}{=} 120 \left(2e^{-\frac{1515}{5760}} \right)$$
$$100 \stackrel{?}{=} 120(0.833)$$
$$100 \approx 99.84 \checkmark$$

30.
$$A_0 2^{-\frac{t}{5760}} = 0.75 A_0$$
$$\log 0.75 = -\frac{t}{5760} \log 2$$
$$-\frac{t}{5760} = \frac{\log 0.75}{\log 2}$$
$$t = 2390.6159$$
$$t \approx 2400 \text{ years}$$

Check:
$$0.75 \stackrel{?}{=} 2^{-\frac{2400}{5760}}$$
$$0.75 = 0.75 \checkmark$$

31.
$$A = A_0 \left(2e^{-\frac{t}{5760}} \right)$$
$$20 = 180 \left(2e^{-\frac{t}{5760}} \right)$$
$$\frac{20}{180} = 2e^{-\frac{t}{5760}}$$
$$\log 0.1111 = -\frac{t}{5760} \log 2$$
$$-\frac{t}{5760} = \frac{\log 0.111}{\log 2}$$
$$t = 18{,}259$$
$$t \approx 18{,}300 \text{ years}$$

Check:
$$20 \stackrel{?}{=} 180 \, 2e^{-\frac{18,300}{5760}}$$
$$20 \stackrel{?}{=} 180 \cdot 2(0.0417) = 15$$
$$20 \approx 15 \checkmark \text{ close due to approx}$$

THINK CRITICALLY

32.
$$A = A_0 2^{\frac{-t}{5760}}$$
$$60 = 80(2^{\frac{-t}{5760}})$$
$$0.75 = 2^{\frac{-t}{5760}}$$
$$\log 0.75 = \frac{-t}{5760} \log 2$$
$$\frac{-t}{5760} = \frac{\log 0.75}{\log 2}$$
$$t \sim 2391$$
$$30 = 40 \left(2^{\frac{-t}{5760}} \right)$$
$$0.75 = 2^{\frac{-t}{5760}}$$

they will take the same amount of time because
$$\frac{60}{80} = \frac{30}{40} = 0.75$$

33. Answers will vary.

34. Answers will vary.

35. $6^{3x-a} = 5^{a-2x}$
$$\log 6^{3x-a} = \log 5^{a-2x}$$
$$(3x - a)\log 6 = (a - 2x)\log 5$$
$$3x \log 6 - a \log 6 = a \log 5 - 2x \log 5$$
$$3x \log 6 + 2x \log 5 = a \log 5 + a \log 6$$
$$x(3 \log 6 + 2 \log 5) = a(\log 5 + \log 6)$$
$$x = \frac{a(\log 5 + \log 6)}{3 \log 6 + 2 \log 5}$$
$$x = 0.3958a$$

36.
$$4^{6x-a} = 3^{a-5x}$$
$$6x \log 4 - a \log 4 = a \log 3 - 5x \log 3$$
$$6x \log 4 + 5x \log 3 = a \log 3 + a \log 4$$
$$x(6 \log 4 + 5 \log 3) = a(\log 3 + \log 4)$$
$$x = a\left(\frac{\log 3 + \log 4}{6 \log 4 + 5 \log 3}\right)$$
$$x = 0.1799a$$

MIXED REVIEW

37. $x^2 + 6x + 25 = 0$
$$x_1, x_2 = \frac{-b \pm \sqrt{b^2 - 4ac}}{2a}$$
$$= \frac{-6 \pm \sqrt{36 - 100}}{2}$$
$$= \frac{-6 \pm 8i}{2}$$
$$= -3 \pm 4i$$

38. $x^2 + 2x + 5 = 0$
$$x_1, x_2 = 3 \pm 4i$$
$$= \frac{-2 \pm \sqrt{4 - 20}}{2}$$
$$= \frac{-2 \pm 4i}{2}$$
$$= -1 \pm 2i$$

39. $A - K = 13$ cards of those 3 are face cards
$$2 \times 3 = 6 \text{ red face cards}$$
$$\frac{6}{52} = 0.115$$

40. $F = \{(x, y): y = x + 3, x = 2, -2, 5\}$;
Domain: $\{2, -2, 5\}$; Range: $\{5, 1, 8\}$

41. $E = \{(x, y): y = 6x, x = 0, 1, -3\}$;
Domain: $\{0, 1, -3\}$; Range: $\{0, 6, -18\}$

42. D; $5^x = 44$
$$\log 5^x = \log 44$$
$$x \log 5 = \log 44$$
$$x = \frac{\log 44}{\log 5}$$
$$x = 2.3512$$

43. B; $3^{-5x} = 20$
$$\log 3^{-5x} = \log 20$$
$$-5x \log 3 = \log 20$$
$$-5x = \frac{\log 20}{\log 3}$$
$$x = -\frac{1}{5}(2.727)$$
$$x = -0.5454$$

ALGEBRAWORKS

1.
$$S = CA^z$$
$$9 = 3.09(35)^z$$
$$\frac{9}{3.09} = 35^z$$
$$2.9126 = 35^z$$
$$\log 2.9126 = z \log 35$$
$$z = \frac{\log 2.9126}{\log 35}$$
$$z = 0.301$$

2. $S = (3.09)(3980)^{0.301} = 3.09(12.1217) = 37$ species

3. $F = 100(1 - e^{-D})$
$F = 100(1 - e^{-4.8})$
$F = 99.177\%$

4. $(10,000) \dfrac{D^n}{n!} e^{-D}$

$$10,000\left(\frac{13^3}{3!} e^{-13}\right) = 10,000\left(\frac{2197}{6} e^{-13}\right) = 366 e^{-13}(10,000)$$
$$= (10,000)366(2.26 \times 10^{-6}) = 8.273$$

Lesson 9.7, pages 460–463

EXPLORE THE PROBLEM

1.

	Year 1	Year 2	Year 3
principal	10,000	11,000	12,100
interest	1,000	1,100	1,210
Amount	11,000	12,100	13,310

$A = P + I$

2. 1.1; 1.33   **3.** $A = P(1 + r)^n$

4. $A = 8000(1 + 0.06)^5 = \$10,705.80$;
My money earned $(10,705.80 - 8,000) = \$2,705.80$

INVESTIGATE FURTHER

5. $m = 4; \dfrac{r}{m} = \dfrac{0.06}{4} = 0.015$;
$A = 8000(1 + 0.015)^{20} = \$10,774.84$
More money is earned when compounded quarterly.

6.

	$\dfrac{\text{pds}}{\text{year}}$	$P\left(1 + \dfrac{r}{m}\right)^{mt}$	Amount
Annually	1	$A = 1\left(1 + \dfrac{1}{1}\right)^1 = 2$	2
Monthly	12	$A = 1\left(1 + \dfrac{1}{12}\right)^{12} = 2.61$	2.6130
Daily	365	$A = 1\left(1 + \dfrac{1}{365}\right)^{365} = 2.71$	2.7146
Hourly	8,760	$A = 1\left(1 + \dfrac{1}{8760}\right)^{8760} = 2.72$	2.7181
Each Minute	525,600	$A = 1\left(1 + \dfrac{1}{525,600}\right)^{525,600} = 2.72$	2.7183

7. $\$2.72 = e$ **8.** $A = 10,000e^{7(0.08)} = \$17,506.73$

9. How often your money is compounded will determine which deal is better.

10. $A = P\left(1 + \dfrac{r}{m}\right)^{mt}$ $i = \dfrac{0.066}{2} = 0.033$

    $= 9300\left(1 + \dfrac{0.066}{2}\right)^{2(4)} = 9300(1.296) = \$12,058.28$

    $12,058.28 - 9,300 = \$2,758$

11. $A = Pe^{rt}$

    $= 5800e^{5(0.09)} = \$9,096.21$

12. $A_1 = Pe^{rt}$

    $= 20,000e^{10(0.07)}$

    $= \$40,275.05$

    $A_2 = P\left(1 + \dfrac{r}{m}\right)^{mt} =$

    $20,000\left(1 + \dfrac{0.07}{4}\right)^{4(10)} = \$40,031.95$

    $A_d = A_1 - A_2 = \$243.10$

13. $A = Pe^{rt}$

    $19,257.39 = 11,000e^{0.08t}$

      $1.7507 = e^{0.08t}$

    $\ln 1.7507 = \ln e^{0.08t}$

      $0.5599 = 0.08t$

          $t = 7$ years

14. $A + P\left(1 + \dfrac{r}{m}\right)^{mt}$

    $15,687.50 = 8300\left(1 + \dfrac{0.072}{2}\right)^{2(t)}$

      $1.89 = 1.036^{2t}$

    $\dfrac{\log 1.89}{\log 1.036} = 2t = 18$

    $t = 9$ years

15. $E = \left(1 + \dfrac{0.054}{12}\right)^{12} - 1 = 5.54\%$

16. $0.0698 = \left(1 + \dfrac{r}{4}\right)^4 - 1$

    $r = 0.068 \rightarrow 6.8\%$

17. Use the formula $A = Pe^{rt}$ to determine the amount earned on $100 for 1 year. The difference between P and A is the effective annual rate. This number can then be used in the formula

$$EAR = \left(1 + \dfrac{r}{m}\right)^m - 1$$

there are many possible examples

18. $A = Pe^{rt}$

    $A = 1000e^{0.16t}$

19.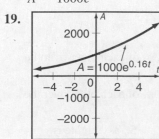

Use the formula when $A = 2P$

    $3200 = 1600e^{0.16t}$

      $2 = e^{0.16t}$

    $\ln 2 = \ln e^{0.16t}$

    $0.693 = 0.16t$

      $t = \dfrac{0.693}{0.16}$

      $t = 4.33$ years

REVIEW PROBLEM SOLVING STRATEGIES

1a. 8   **1b.** 9

1c. At each landing pad, for a total of 19 meetings.

1d. Daily, at noon and at midnight.

1e. Answers will vary. Possible diagram:

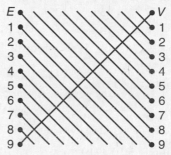

2. 3 candleholders; label the identical candleholders A, B, C; then AB, AC, and BC are three pairs.

3. Sandra

Chapter Review, pages 464–465

1. e   **2.** c   **3.** d   **4.** a   **5.** b

6. move f left 1 unit   **7.** move f down 2 units

8. reflect f over y-axis   **9.** reflect f over x-axis

10. $11 = 121^x$ Check:

    $11 = (11)^{2x}$ $11 \overset{?}{=} 121^{\frac{1}{2}}$

     $1 = 2x$ $11 = 11$ ✓

     $x = \dfrac{1}{2}$

11. $4^x = \dfrac{1}{64}$ Check:

    $4^x = 4^{-3}$ $4^{-3} \overset{?}{=} \dfrac{1}{64}$

     $x = -3$ $0.01562 = \dfrac{1}{64}$ ✓

12. $32 = 4^x$ Check:

    $2^5 = 2^{2x}$ $32 \overset{?}{=} 4^{\frac{5}{2}}$

     $5 = 2x$ $32 = 32$ ✓

     $x = \dfrac{5}{2}$

13. $16^8 = 4^x$ Check:

    $4^{16} = 4^x$ $16^8 \overset{?}{=} 4^{16}$

     $x = 16$ $4.29 \times 10^9 = 4.29 \times 10^9$ ✓

14. $3^{8x+2} = 9^{2x-3}$

$$\log 3^{8x+2} = \log 9^{2x-3}$$
$$(8x + 2)\log 3 = (2x - 3)\log 9$$
$$8x \log 3 + 2 \log 3 = 2x \log 9 - 3 \log 9$$
$$8x \log 3 - 2x \log 9 = -2 \log 3 - 3 \log 9$$
$$x (8 \log 3 - 2 \log 9) = -2 \log 3 - 3 \log 9$$
$$x = \frac{-2 \log 3 - 3 \log 9}{8 \log 3 - 2 \log 9}$$
$$x = -1.99 = -2$$

Check:

$$3^{8(-2)+2} \stackrel{?}{=} 9^{2(-2)-3}$$
$$3^{-14} \stackrel{?}{=} 9^{-7}$$
$$2.09 \times 10^{-7} = 2.09 \times 10^{-7} \checkmark$$

15. $\left(\frac{1}{5}\right)^6 = \left(\frac{1}{25}\right)^x$

Check

$$(0.2)^6 = (0.004)^x \qquad \left(\frac{1}{5}\right)^6 \stackrel{?}{=} \left(\frac{1}{25}\right)^3$$
$$0.2^6 = 0.2^{2x} \qquad 6.4 \times 10^{-5} = 6.4 \times 10^{-5} \checkmark$$
$$6 = 2x$$
$$x = 3$$

16. using a graphing utility $x = 1.3181$

Check:

$$13 = 7^x \stackrel{?}{=} 7^{1.3181}$$
$$13 = 12.999 \checkmark$$

17. using a graphing utility $x = 0.4719$

Check:

$$8 \stackrel{?}{=} 82^{0.4719}$$
$$8 = 8 \checkmark$$

18. $\log_9 31 = \dfrac{\log 31}{\log 9} = 1.5629$

19. $\log_8 15 = \dfrac{\log 15}{\log 8} = 1.3023$

20. $\log_{14} 7 = \dfrac{\log 7}{\log 14} = 0.7374$

21. $\log_6 5 = \dfrac{\log 5}{\log 6} = 0.8982$

22. $\log_x 729 = 3$
$$x^3 = 729$$
$$x = 9$$

23. $\log_x 5 = \dfrac{1}{4}$
$$x^{\frac{1}{4}} = 5$$
$$x = 5^4$$
$$x = 625$$

24. $\log_{16} x = \dfrac{5}{2}$
$$16^{\frac{5}{2}} = x$$
$$x = 1024$$

25. $\log_8 64 = 5x - 5$
$$8^{5x-5} = 64$$
$$8^{5x-5} = 8^2$$
$$5x - 5 = 2$$
$$5x = 7$$
$$x = \frac{7}{5}$$

26. $\log_6 \dfrac{j}{3} = \log_6 j - \log_6 3$

27. $\log_9 \left(\dfrac{x^4}{y^8}\right)^{\frac{1}{2}} = \log_9 \left(\dfrac{x^2}{y^4}\right) = 2 \log_9 x - 4 \log_9 y$

28. $3 \log_4 x - \log_4 5 = \log_4 \dfrac{x^3}{5}$

29. $\dfrac{1}{2} \log_a + \log_b = \log a^{\frac{1}{2}} b$

30. $\log_6 (x^2 - 9) = 3$ Check:
$$6^3 = x^2 - 9 \qquad \log_6 (x^2 - 9) = 3$$
$$216 + 9 = x^2 \qquad \log_6 (15^2 - 9) = 3$$
$$x = \pm 15 \qquad \log_6 216 = 3$$
$$6^3 = 216$$
$$216 = 216 \checkmark$$

31. $\log_5 2x + \log_5 3x = \log_5 600$
$$\log_5 6x^2 = \log_5 600$$
$$6x^2 = 600$$
$$x^2 = 100$$
$$x = 10$$

Check:

$$\log_5 2(10) + \log_5 3(10) \stackrel{?}{=} \log_5 600$$
$$\log_5 20(30) \stackrel{?}{=} \log_5 600$$
$$\log_5 600 = \log_5 600 \checkmark$$

32. $\ln 3hk = \ln 3 + \ln h + \ln k$

33. $\ln x^2 y^3 = 2 \ln x + 3 \ln y$

34. $4 \ln x - 2 \ln y + \ln 9 = \ln \dfrac{9x^4}{y^2}$

35. $3e^x + 12 = 56$ **36.** $8e^{2x} + 10 = 72$
$$3e^x = 44 \qquad\qquad 8e^{2x} = 62$$
$$e^x = 14.67 \qquad\qquad e^{2x} = 7.75$$
$$\ln e^x = \ln 14.67 \qquad \ln e^{2x} = \ln 7.75$$
$$x = 2.6856 \qquad\qquad 2x = 2.04769$$
$$x = 1.0238$$

37. $10^x - 7 = 77$ **38.** $8 + 5 \log x = 18$
$$10^x = 84 \qquad\qquad 5 \log x = 10$$
$$\log 10^x = \log 84 \qquad \log x = 2$$
$$x = 1.9243 \qquad\qquad 10^2 = x$$
$$x = 100$$

39. $4 \ln x - 1 = 11$ **40.** $3 \ln 9x = 11$
$$4 \ln x = 12 \qquad\qquad \ln 9x = 3.667$$
$$\ln x = 3 \qquad\qquad e^{\ln 9x} = e^{3.667}$$
$$e^{\ln x} = e^3 \qquad\qquad 9x = 39.1213$$
$$x = 20.0855 \qquad\qquad x = 4.3468$$

41. $A = P\left(1 + \dfrac{r}{m}\right)^{mt} = 4500\left(1 + \dfrac{0.08}{4}\right)^{5(4)} = \$6686.76$

42. $A = Pe^{rt} = 5200(e^{0.07(4)}) = \$6880.28$

Chapter Assessment, pages 466–467

CHAPTER TEST

1. D; $k(x) = 6^{x-4} + 3$

2. the two graphs are reflections of each other over the x-axis

3. the two graphs are reflections of each other over the y-axis

4. asymptote is $y = 2$ **5.** asymptote is $y = -3$

6. $3 = 27^x$
$3^1 = 3^{3x}$
$1 = 3x$
$x = \dfrac{1}{3}$

7. $2^{-2x} = 256$
$2^{-2x} = 2^8$
$-2x = 8$
$x = -4$

8. $9^6 = 3^{x^2}$
$3^{2 \cdot 6} = 3^{x^2}$
$\log 3^{12} = \log_3 3^{x^2}$
$12 = x^2$
$x = \pm 2\sqrt{3}$

9. $6^{2x} = 36^{3x-1}$
$6^{2x} = 6^{2(3x-1)}$
$\log_6 6^{2x} = \log_6 6^{2(3x-1)}$
$2x = 6x - 2$
$4x = 2$
$x = \dfrac{1}{2}$

10. Graph each side of the equation and find their intersection. Examples will vary.

11. $9 = 7^x$
$x = 1.129$

12. $19 = 5^x$
$x = 1.829$

13. $3 = 12^x$
$x = 0.442$

14. $7 = 13^x$
$x = 0.759$

15. $P = P_0 2^t$
$P = 12(2^8) = 3072$

16. $\log_4 6 = \dfrac{\log_6}{\log_4} = \dfrac{\log_{10} 6}{\log_{10} 4} = \dfrac{\ln 6}{\ln 4}$
II, I, and III

17. $\log_5 \sqrt{5} = x$
$5^x = \sqrt{5}$
$5^x = 5^{\frac{1}{2}}$
$x = \dfrac{1}{2}$

18. $\log_x 343 = 3$
$x^3 = 343$
$x^3 = 7^3$
$x = 7$

19. $\log_6 x = -3$
$6^{-3} = x$
$x = \dfrac{1}{216}$

20. $\log_5 625 = 2x - 1$
$5^{2x-1} = 625$
$5^{2x-1} = 5^4$
$2x - 1 = 4$
$2x = 5$
$x = \dfrac{5}{2}$

21. $\log_7 16 = \dfrac{\log 16}{\log 7} = 1.425$

22. $\log_9 2 = \dfrac{\log 2}{\log 9} = 0.315$

23. $\log ab^2 c^3 = \log a + \log b^2 + \log c^3 =$
$\log a + 2 \log b + 3 \log c$

24. $\log \sqrt{\dfrac{a^{10}}{b^{12}}} = \log \left(\dfrac{a^{10}}{b^{12}}\right)^{\frac{1}{2}} = \log \dfrac{a^5}{b^6} = \log a^5 - \log b^6 =$
$5 \log a - 6 \log b$

25. $3 \log_8 x + 2 \log_8 y = \log_8 x^3 y^2$

26. $4 \ln q - \ln 3 = \ln \dfrac{q^4}{3}$

27. $5e^x - 11 = 77$, $5e^x = 88$, $e^x = 17.6$
$\ln e^x = \ln 17.6$
$x = 2.8679$

28. $10^{2x} - 9 = 55$, $10^{2x} = 64$
$\log 10^{2x} = \log 64$
$2x = \log 64$
$x = \dfrac{1}{2} \log 64 = 0.9031$

29. $9 + 6 \log x = 25$
$\log x = \dfrac{25 - 9}{6} = \dfrac{8}{3}$
$x = 464.1589$

30. $8 \ln x - 4 = 19$
$\ln x = \dfrac{19 + 4}{8} = \dfrac{23}{8}$
$x = 17.7254$

31. $A = P\left(1 + \dfrac{r}{m}\right)^{mt}$
$A = 5000\left(1 + \dfrac{0.06}{12}\right)^{4(12)} = 5000(1.27) = \$6,352.45$

32. $A = Pe^{rt}$
$A = 3800e^{0.1(3)} = \$5,129.46$

33. $13,728.05 = 8000e^{0.09t}$
$1.716 = e^{0.09t}$
$\ln 1.716 = \ln e^{0.09t} = 0.09t$
$\dfrac{0.5399}{0.09} = t = 5.999 = 6$ years

Cumulative Review, page 468

1. $B \cdot A = \begin{bmatrix} 45 & 29 \\ -16 & -36 \end{bmatrix}$

2. $x = y + 1$
$2x + 3y - 12 = 0$
$2y + 2 + 3y - 12 = 0$
$5y = 10$
$y = 2$
$x = 3$

3. The curve has a horizontal asymptote at the x-axis and rises exponentially. Crosses the y-axis at $(0, 1)$. Domain is all real number $b > 0$; Range is all numbers greater than 0.

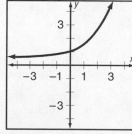

4. $x^2 + 4 = 6x$
$x^2 - 6x + 9 = 5$
$(x - 3)^2 = 5$
$x - 3 = \pm\sqrt{5}$
$x = 3 \pm \sqrt{5}$

5.
$$A_t = A_0 e^{rt}$$
$$2 = e^{0.08t}$$
$$\ln 2 = \ln e^{0.08t}$$
$$0.693 = 0.08t$$
$$t = 8.6643 \text{ years}$$

6. $f(x) = \dfrac{x-1}{2} = \dfrac{3-1}{2} = 1$

$g(x) = 2x + 1 = 2(3) + 1 = 7$

$\dfrac{7-1}{2} - (2+1) = 3 - 3 = 0, \text{A}$

7. $8^{\frac{2}{3}} + 8^0 + 8^{-1} = 4 + 1 + 0.125 = 5.125$

8. $125^x = 25$ Check:
$5^{3x} = 5^2$ $125^{\frac{2}{3}} = 25$
$3x = 2$ $25 = 25$ ✓
$x = \dfrac{2}{3}$

9. $\log_x 1331 = 3$ Check:
$x^3 = 1331$ $\log_{11} 1331 = 3$
$x^3 = 11^3$ $11^3 = 1331$ ✓
$x = 11$

10. $x = e^{2\ln 5}$ Check:
$\ln x = \ln e^{2\ln 5}$ $25 = e^{2\ln 5} = e^{3.219}$
$\ln x = 2\ln 5 = 3.219$ $25 = 25$ ✓
$\ln x = 3.219$
$e^{\ln x} = e^{3.219}$
$x = 25$

11. $x = 10^{\frac{1}{3}\log 8}$ Check
$\log x = \log 10^{\frac{1}{3}\log 8}$ $2 = 10^{\frac{1}{3}\log 8}$
$\log x = \dfrac{1}{3}\log 8$ $2 = 10^{0.30103}$
$\log x = 0.301$ $2 = 2$ ✓
$x = 10^{0.301} = 2$

12. $6 \cdot (4 + w) \cdot w \Rightarrow d \times l \times w = \text{volume}$
9 tons $\times$ 40 cubic ft/ton $= 360$ cu. ft.
$6w(4 + w) = 360 = 24w + 6w^2$
$6w^2 + 24w - 360 = 0, w^2 + 4w - 60 = 0,$
$(w + 10)(w - 6) = 0,$
$w = 6,$
$l = 4 + w = 6 \text{ ft}$
$w = 10 \text{ ft}$
$d = 6 \text{ ft}$
$6' \times 10' \times 6'$

13. $\log \dfrac{x^2}{y} = 2\log x - \log y$

14. Daily production should be
$60(28a) + 100(33b) + 40(33b) = 200$ units
for maximum profit.
$60a + 140b$
60 two-headed   140 four-headed

15. D; $2x^{-\frac{4}{3}} = 162$, x must be $\dfrac{1}{27}$

16.
$$x^2 - 3x > 0$$
$$x(x - 3) > 0$$

x	$-$	$+$	$+$
$x - 3$	$-$	$-$	$+$
$x(x-3)$	$+$	$-$	$+$

$-1 \quad 0 \quad 1 \quad 2 \quad 3 \quad 4$

$x < 0 \text{ or } x > 3$

17. $y = mx + b$ $(-2, 0) \ (0, -2)$
$m = \dfrac{-2 - 0}{0 - -2} = \dfrac{-2}{2} = -1$
$y = -x - 2$

18. E, $3x - 18 \geq -6 \text{ or } -4x \geq 16$
$2x \geq 8 \text{ or } 5x + 2 \leq -18$

Standardized Test, page 469

1. A; $3 + i$
$2 - 2i$
$\dfrac{a + bi}{1 + i}$ then:
$\dfrac{3 + 2 + a}{3} = 1$
$5 + a = 3$
$a = -2$
and $(1 - 2 + b) \div 3 = 1$
$(-1 + b) \div 3 = 1$
$b = 3 + 1 = 4$
therefore $x = -2 + 4i$

2. B; $x^2 + 2x \leq 3$
$x^2 + 2x - 3 \leq 0$
$(x + 3)(x - 1) \leq 0$

$x + 3$	$-$	$+$	$+$
$x - 1$	$-$	$-$	$+$
$(x+3)(x-1)$	$+$	$-$	$+$

$-4 \ -3 \ -2 \ -1 \ 0 \ 1 \ 2$

The solution of $x^2 + 2x \leq 3$ is $-3 \leq x \leq 1$

3. E; $\log_x \dfrac{2}{z} = -1, x^{-1} = \dfrac{2}{z} = \dfrac{1}{x}$
then, $x = \dfrac{z}{2}$

4. B; $(x^{\frac{1}{2}})^{\frac{1}{3}} = 3$
$\log x^{\frac{1}{10}} = \log 3$
$\dfrac{1}{10}\log x = \log 3$
$\log x = 10\log 3 = 4.771$
$x = 59.049 = 3^{10}$

5. C; $2x + 3y = 3$ $2x + 3y = 3$
$\dfrac{x + 2y = 1}{3x + 5y = 4}$ $\dfrac{-x - 2y = -1}{x + y = 2}$ 2 is 50% of 4.

6. C; $y = x^2 + bx + c$

$c = 3$

$y = x^2 + bx + c - 1$

$c = 2$

7. E; I is a true statement

The y-axis is an asymptote and the x-intercept is 1.

8. C; $\sqrt{\sqrt{y} + \sqrt{y} + \sqrt{y} + \sqrt{y}} = \sqrt{2}$

$\sqrt{\sqrt{\frac{1}{4}} + \sqrt{\frac{1}{4}} + \sqrt{\frac{1}{4}} + \sqrt{\frac{1}{4}}} = \sqrt{4\sqrt{\frac{1}{4}}} = \sqrt{2}$

9. D; If $2 \log x = \log 2x$ then x must be two;

$\log x^2 = \log 2x$; $x^2 = 2x$; $x = 2$

10. D; $\overline{PR} = 4x + 5$, $\overline{RS} = 2x + 3$ then $\overline{PS} = \overline{PR} + \overline{RS}$

$= 6x + 8$

$\overline{QS} = \dfrac{\overline{PS}}{2} = 3x + 4$, then $\overline{QR} = \overline{QS} - \overline{RS}$

$\overline{QR} = x + 1$

Data Activity, pages 470–471

1. From the table, the percent that visited is 44%, so the probability is 0.44

2. 51% of 2000 = (0.51)(2000) = 1020

3. Answers will vary. Attendance declines steeply as group age increases while attendance among those with at least a high school education is fairly consistent.

4.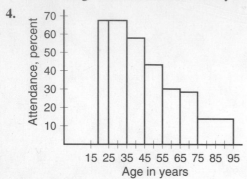

5. Answers will vary.

Lesson 10.1, pages 473–478

EXPLORE

1.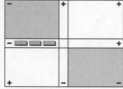

2.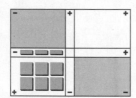

3.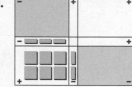

4. $(2x^2 - 6x) \div 2x = x - 3$

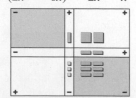

5. $(-3x^2 - 6x) \div (x + 2) = -3x$

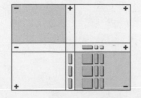

6. Multiply the quotient 4 by the divisor 2 and add the remainder 1.

7a.
$$\begin{array}{r} 698 \\ 38\overline{)26{,}537} \\ 228 \\ \hline 373 \\ 342 \\ \hline 317 \\ 304 \\ \hline 13 \end{array}$$

Divide 38 into 265 to get the first digit 6 of the quotient.
Multiply 6 by the divisor, 38.
Subtract 228 from 265. Bring down the next digit.
Divide 38 into 373 to get the next digit 9 of the quotient.
Multiply 9 by the divisor, 38.
Subtract 342 from 373. Bring down the next digit.
Divide 38 into 317 to get the next digit 8 of the quotient.
Multiply 8 by the divisor, 38.

$\dfrac{26{,}537}{38} = 698\dfrac{13}{38}$ Subtract 304 from 317. The remainder is 13.

8. The steps that are repeated in division are divide, multiply, subtract, and bring down the next digit.

TRY THESE

1. quotient = $2x^2 + x + 1$; remainder = 9;
dividend = $10x^3 + x^2 + 3x + 7$; divisor = $5x - 2$

2. Jon; the terms of the dividend and the divisor must be in the same order.

3.
$$\begin{array}{r} x - 7 \\ x - 2\overline{)x^2 - 9x + 7} \\ x^2 - 2x \\ \hline -7x + 7 \\ -7x + 14 \\ \hline -7 \end{array}$$

So, $(x^2 - 9x + 7) \div (x - 2) = x - 7 + \dfrac{-7}{x - 2}$

Check: dividend = quotient × divisor + remainder;
$(x - 7)(x - 2) + (-7) =$
$x^2 - 9x + 14 + (-7) = x^2 - 9x + 7$

4.
$$\begin{array}{r} z^2 - 6z - 18 \\ z - 2\overline{)z^3 - 8z^2 - 6z + 9} \\ z^3 - 2z^2 \\ \hline -6z^2 - 6z \\ -6z^2 + 12z \\ \hline -18z + 9 \\ -18z + 36 \\ \hline -27 \end{array}$$

So, $(z^3 - 8z^2 - 6z + 9) \div (z - 2) =$

$z^2 - 6z - 18 + \dfrac{-27}{z - 2}$

Check: $(z^2 - 6z - 18)(z - 2) + (-27) =$
$(z^2 - 6z - 18)z - (z^2 - 6z - 18)(2) + (-27) =$
$z^3 - 6z^2 - 18z - 2z^2 + 12z + 36 + (-27) =$
$z^3 - 8z^2 - 6z + 9$

5.

$$
\begin{array}{r}
a^2 - 4a + 2 \\
2a + 3 \overline{\smash{\big)}\ 2a^3 - 5a^2 - 8a + 2} \\
\underline{2a^3 + 3a^2} \\
-8a^2 - 8a \\
\underline{-8a^2 - 12a} \\
4a + 2 \\
\underline{4a + 6} \\
-4
\end{array}
$$

So, $\dfrac{-5a^2 - 8a + 2a^3 + 2}{2a + 3} = a^2 - 4a + 2 + \dfrac{-4}{2a + 3}$

Check: $(a^2 - 4a + 2)(2a + 3) + (-4) =$

$(a^2 - 4a + 2)2a + (a^2 - 4a + 2)3 + (-4)$

$= 2a^3 - 8a^2 + 4a + 3a^2 - 12a + 6 + (-4)$

$= 2a^3 - 5a^2 - 8a + 2$

6.

$$
\begin{array}{r}
x - 5 \\
x + 5 \overline{\smash{\big)}\ x^2 + 0x + 25} \\
\underline{x^2 + 5x} \\
-5x + 25 \\
\underline{-5x - 25} \\
50
\end{array}
$$

So, $(x^2 + 25) \div (x + 5) = x - 5 + \dfrac{50}{x + 5}$

Check: $(x - 5)(x + 5) + 50 =$

$x^2 - 25 + 50 = x^2 + 25$

7.

$$
\begin{array}{r}
x - 2 \\
x^2 + 4x + 6 \overline{\smash{\big)}\ x^3 + 2x^2 - 2x - 12} \\
\underline{x^3 + 4x^2 + 6x} \\
-2x^2 - 8x - 12 \\
\underline{-2x^2 - 8x - 12} \\
0
\end{array}
$$

So, $\dfrac{x^3 + 2x^2 - 2x - 12}{x^2 + 4x + 6} = x - 2$

Check: $(x - 2)(x^2 + 4x + 6) =$

$x(x^2 + 4x + 6) - 2(x^2 + 4x + 6) =$

$x^3 + 4x^2 + 6x - 2x^2 - 8x - 12$

$= x^3 + 2x^2 - 2x - 12$

8.

$$
\begin{array}{r}
x^2 - 8x + 2 \\
3x^2 + x - 5 \overline{\smash{\big)}\ 3x^4 - 23x^3 - 7x^2 + 42x - 10} \\
\underline{3x^4 + x^3 - 5x^2} \\
-24x^3 - 2x^2 + 42x \\
\underline{-24x^3 - 8x^2 + 40x} \\
6x^2 + 2x - 10 \\
\underline{6x^2 + 2x - 10} \\
0
\end{array}
$$

So, $\dfrac{3x^4 - 23x^3 - 7x^2 + 42x - 10}{3x^2 + x - 5} = x^2 - 8x + 2$

Check: $(x^2 - 8x + 2)(3x^2 + x - 5) =$

$(x^2 - 8x + 2)3x^2 + (x^2 - 8x + 2)x -$

$(x^2 - 8x + 2)5 = 3x^4 - 24x^3 + 6x^2 + x^3 -$

$8x^2 + 2x - 5x^2 + 40x - 10$

$= 3x^4 - 23x^3 - 7x^2 + 42x - 10$

9.

$$
\begin{array}{r}
x^2 \qquad + 4 \\
x - 2 \overline{\smash{\big)}\ x^3 - 2x^2 + 4x - 6} \\
\underline{x^3 - 2x^2} \\
0 + 4x - 6 \\
\underline{4x - 8} \\
2
\end{array}
$$

$x - 2$ is not a factor of $x^3 - 2x^2 + 4x - 6$ because the remainder is not zero.

10.

$$
\begin{array}{r}
2y^2 + 3y - 1 \\
3y + 1 \overline{\smash{\big)}\ 6y^3 + 11y^2 + 0y - 1} \\
\underline{6y^3 + 2y^2} \\
9y^2 + 0y \\
\underline{9y^2 + 3y} \\
-3y - 1 \\
\underline{-3y - 1} \\
0
\end{array}
$$

Vic's average speed is $(2y^2 + 3y - 1)$ mi/h.

11.

$V = lwh$

$x^3 + x^2 - 10x + 8 = (x - 2)(x - 1)h$

$x^3 + x^2 - 10x + 8 = (x^2 - 3x + 2)h$

$\dfrac{x^3 + x^2 - 10x + 8}{x^2 - 3x + 2} = h$

$$
\begin{array}{r}
x + 4 \\
x^2 - 3x + 2 \overline{\smash{\big)}\ x^3 + x^2 - 10x + 8} \\
\underline{x^3 - 3x^2 + 2x} \\
4x^2 - 12x + 8 \\
\underline{4x^2 - 12x + 8} \\
0
\end{array}
$$

So, $h = (x + 4)$ cm

PRACTICE

1.

$$
\begin{array}{r}
x + 2 \\
x - 1 \overline{\smash{\big)}\ x^2 + x - 1} \\
\underline{x^2 - x} \\
2x - 1 \\
\underline{2x - 2} \\
1
\end{array}
$$

Tanisha. In her answer, if you multiply the quotient by the divisor and add on the remainder, the result is the original division problem.

$x + 2 + \dfrac{1}{x - 1} = \dfrac{(x + 2)(x - 1) + 1}{x - 1}$

$= \dfrac{x^2 + x - 2 + 1}{x - 1} = \dfrac{x^2 + x - 1}{x - 1}$

2.

$$
\begin{array}{r}
2y + 8 \\
2y - 5 \overline{\smash{\big)}\ 4y^2 + 6y + 9} \\
\underline{4y^2 - 10y} \\
16y + 9 \\
\underline{16y - 40} \\
49
\end{array}
$$

So, $(4y^2 + 6y + 9) \div (2y - 5) = 2y + 8 + \dfrac{49}{2y - 5}$

Check: $(2y + 8)(2y - 5) + 49 =$

$4y^2 + 6y - 40 + 49 = 4y^2 + 6y + 9$

3.

$$
\begin{array}{r}
x + 2 \\
3x + 3 \overline{\smash{\big)}\ 3x^2 + 9x - 4} \\
\underline{3x^2 + 3x} \\
6x - 4 \\
\underline{6x + 6} \\
-10
\end{array}
$$

So, $(3x^2 + 9x - 4) \div (3x + 3) = x + 2 + \dfrac{-10}{3x + 3}$

Check: $(x + 2)(3x + 3) - 10 =$
$3x^2 + 9x + 6 - 10 = 3x^2 + 9x - 4$

4.

$$
\begin{array}{r}
3z^2 + 2z - 4 \\
z + 4 \overline{\smash{\big)}\ 3z^3 + 14z^2 + 4z - 4} \\
\underline{3z^3 + 12z^2} \\
2z^2 + 4z \\
\underline{2z^2 + 8z} \\
-4z - 4 \\
\underline{-4z - 16} \\
12
\end{array}
$$

So, $(3z^3 + 14z^2 + 4z - 4) \div (z + 4) =$
$3z^2 + 2z - 4 + \dfrac{12}{z + 4}$

Check: $(3z^2 + 2z - 4)(z + 4) + 12 =$
$3z^3 + 2z^2 - 4z + 12z^2 + 8z - 16 + 12 =$
$3z^3 + 14z^2 + 4z - 4$

5.

$$
\begin{array}{r}
2m^2 - 2m + 4 \\
2m + 1 \overline{\smash{\big)}\ 4m^3 - 2m^2 + 6m - 3} \\
\underline{4m^3 + 2m^2} \\
-4m^2 + 6m \\
\underline{-4m^2 - 2m} \\
8m - 3 \\
\underline{8m + 4} \\
-7
\end{array}
$$

So, $\dfrac{4m^3 - 2m^2 + 6m - 3}{2m + 1} =$

$2m^2 - 2m + 4 + \dfrac{-7}{2m + 1}$

Check: $(2m^2 - 2m + 4)(2m + 1) - 7 =$
$(2m^2 - 2m + 4)2m + (2m^2 - 2m + 4)1 - 7 =$
$4m^3 - 4m^2 + 8m + 2m^2 - 2m + 4 - 7 =$
$4m^3 - 2m^2 + 6m - 3$

6.

$$
\begin{array}{r}
3r^2 + 2r + 1 \\
5r + 1 \overline{\smash{\big)}\ 15r^3 + 13r^2 + 7r + 3} \\
\underline{15r^3 + 3r^2} \\
10r^2 + 7r \\
\underline{10r^2 + 2r} \\
5r + 3 \\
\underline{5r + 1} \\
2
\end{array}
$$

So, $\dfrac{7r + 3 + 13r^2 + 15r^3}{5r + 1} = 3r^2 + 2r + 1 + \dfrac{2}{5r + 1}$

Check: $(3r^2 + 2r + 1)(5r + 1) + 2 =$
$(3r^2 + 2r + 1)5r + (3r^2 + 2r + 1)1 + 2 =$
$15r^3 + 10r^2 + 5r + 3r^2 + 2r + 1 + 2 =$
$15r^3 + 13r^2 + 7r + 3$

7.

$$
\begin{array}{r}
x^2 - 2x + 4 \\
x + 2 \overline{\smash{\big)}\ x^3 + 0x^2 + 0x + 16} \\
\underline{x^3 + 2x^2} \\
-2x^2 + 0x \\
\underline{-2x^2 - 4x} \\
4x + 16 \\
\underline{4x + 8} \\
8
\end{array}
$$

So, $\dfrac{x^3 + 16}{x + 2} = x^2 - 2x + 4 + \dfrac{8}{x + 2}$

Check: $(x^2 - 2x + 4)(x + 2) + 8 =$
$(x^2 - 2x + 4)x + (x^2 - 2x + 4)2 + 8 =$
$x^3 - 2x^2 + 4x + 2x^2 - 4x + 8 + 8 = x^3 + 16$

8.

$$
\begin{array}{r}
16p^2 - 12p + 9 \\
4p + 3 \overline{\smash{\big)}\ 64p^3 + 0p^2 + 0p + 27} \\
\underline{64p^3 + 48p^2} \\
-48p^2 + 0p \\
\underline{-48p^2 - 36p} \\
36p + 27 \\
\underline{36p + 27} \\
0
\end{array}
$$

So, $(64p^3 + 27) \div (4p + 3) = 16p^2 - 12p + 9$
Check: $(16p^2 - 12p + 9)(4p + 3) =$
$(16p^2 - 12p + 9)(4p) + (16p^2 - 12p + 9)(3) =$
$64p^3 - 48p^2 + 36p + 48p^2 - 36p + 27 =$
$64p^3 + 27$

9.

$$
\begin{array}{r}
y^3 + y^2 + y + 1 \\
y - 1 \overline{\smash{\big)}\ y^4 + 0y^3 + 0y^2 + 0y - 1} \\
\underline{y^4 - y^3} \\
y^3 + 0y^2 \\
\underline{y^3 - y^2} \\
y^2 + 0y \\
\underline{y^2 - y} \\
y - 1 \\
\underline{y - 1} \\
0
\end{array}
$$

So, $(y^4 - 1) \div (y - 1) = y^3 + y^2 + y + 1$
Check: $(y^3 + y^2 + y + 1)(y - 1) =$
$(y^3 + y^2 + y + 1)y - (y^3 + y^2 + y + 1)1 =$
$y^4 + y^3 + y^2 + y - y^3 - y^2 - y - 1 = y^4 - 1$

10.

$$
\begin{array}{r}
y^2 \qquad - 2 \\
y^2 - 4 \overline{\smash{\big)}\ y^4 + 0y^3 - 6y^2 + 0y + 8} \\
\underline{y^4 \qquad - 4y^2} \\
-2y^2 + 0y + 8 \\
\underline{-2y^2 \qquad + 8} \\
0
\end{array}
$$

So, $(y^4 - 6y^2 + 8) \div (y^2 - 4) = y^2 - 2$
Check: $(y^2 - 2)(y^2 - 4) = y^4 - 6y^2 + 8$

11.

$$
\begin{array}{r}
2x^2 - 2x + 1 \\
2x^2 + 2x + 1\overline{)4x^4 + 0x^3 + 0x^2 + 0x + 1} \\
\underline{4x^4 + 4x^3 + 2x^2} \\
-4x^3 - 2x^2 + 0x \\
\underline{-4x^3 - 4x^2 - 2x} \\
2x^2 + 2x + 1 \\
\underline{2x^2 + 2x + 1} \\
0
\end{array}
$$

So, $(4x^4 + 1) \div (2x^2 + 2x + 1) = 2x^2 - 2x + 1$

Check: $(2x^2 - 2x + 1)(2x^2 + 2x + 1) = (2x^2 - 2x + 1)2x^2 + (2x^2 - 2x + 1)2x + (2x^2 - 2x + 1)1 = 4x^4 - 4x^3 + 2x^2 + 4x^3 - 4x^2 + 2x + 2x^2 - 2x + 1 = 4x^4 + 1$

12. Let p represent the amount Ms. Chen had invested.

First change $100(x + 2)\%$ to the fraction $\dfrac{100(x + 2)}{100}$.

Then
$$
\begin{aligned}
\frac{100(x + 2)}{100}(p) &= x^3 - 2x^2 - 5x + 6 \\
(x + 2)p &= x^3 - 2x^2 - 5x + 6 \\
p &= \frac{x^3 - 2x^2 - 5x + 6}{x + 2}
\end{aligned}
$$

$$
\begin{array}{r}
x^2 - 4x + 3 \\
x + 2\overline{)x^3 - 2x^2 - 5x + 6} \\
\underline{x^3 + 2x^2} \\
-4x^2 - 5x \\
\underline{-4x^2 - 8x} \\
3x + 6 \\
\underline{3x + 6} \\
0
\end{array}
$$

$$p = x^2 - 4x + 3$$

Ms. Chen had originally invested $(x^2 - 4x + 3)$ dollars.

13.

$$
\begin{array}{r}
y^2 + y - 2 \\
y + 3\overline{)y^3 + 4y^2 + y - 6} \\
\underline{y^3 + 3y^2} \\
y^2 + y \\
\underline{y^2 + 3y} \\
-2y - 6 \\
\underline{-2y - 6} \\
0
\end{array}
$$

$y^2 + y - 2 = (y + 2)(y - 1)$
$y^3 + 4y^2 + y - 6 = (y + 3)(y + 2)(y - 1)$

Check: $(y + 3)(y + 2)(y - 1) = (y + 3)(y^2 + y - 2) = y^3 + y^2 - 2y + 3y^2 + 3y - 6 = y^3 + 4y^2 + y - 6$

14.

$$
\begin{array}{r}
t^2 + 5t + 4 \\
t - 2\overline{)t^3 + 3t^2 - 6t - 8} \\
\underline{t^3 - 2t^2} \\
5t^2 - 6t \\
\underline{5t^2 - 10t} \\
4t - 8 \\
\underline{4t - 8} \\
0
\end{array}
$$

$t^2 + 5t + 4 = (t + 4)(t + 1)$
$t^3 + 3t^2 - 6t - 8 = (t - 2)(t + 4)(t + 1)$

Check: $(t - 2)(t + 4)(t + 1) = (t - 2)(t^2 + 5t + 4) = t^3 + 5t^2 + 4t - 2t^2 - 10t - 8 = t^3 + 3t^2 - 6t - 8$

15.

$$
\begin{array}{r}
x^2 - x - 20 \\
x + 1\overline{)x^3 + 0x^2 - 21x - 20} \\
\underline{x^3 + x^2} \\
-x^2 - 21x \\
\underline{-x^2 - x} \\
-20x - 20 \\
\underline{-20x - 20} \\
0
\end{array}
$$

$x^2 - x - 20 = (x - 5)(x + 4)$
$x^3 - 21x - 20 = (x + 1)(x - 5)(x + 4)$

Check: $(x + 1)(x - 5)(x + 4) = (x + 1)(x^2 - x - 20) = x^3 - x^2 - 20x + x^2 - x - 20 = x^3 - 21x - 20$

16.

$$
\begin{array}{r}
z^2 + z - 6 \\
z + 6\overline{)z^3 + 7z^2 + 0z - 36} \\
\underline{z^3 + 6z^2} \\
z^2 + 0z \\
\underline{z^2 + 6z} \\
-6z + 36 \\
\underline{-6z - 36} \\
0
\end{array}
$$

$z^2 + z - 6 = (z + 3)(z - 2)$
$z^3 + 7z^2 - 36 = (z + 6)(z + 3)(z - 2)$

Check: $(z + 6)(z + 3)(z - 2) = (z + 6)(z^2 + z - 6) = z^3 + z^2 - 6z + 6z^2 + 6z - 36 = z^3 + 7z^2 - 36$

17.

$$
\begin{array}{r}
-a^2 + 4a - 3 \\
a - 2\overline{)-a^3 + 6a^2 - 11a + 6} \\
\underline{-a^3 + 2a^2} \\
4a^2 - 11a \\
\underline{4a^2 - 8a} \\
-3a + 6 \\
\underline{-3a + 6} \\
0
\end{array}
$$

$(-a^2 + 4a - 3) = (-a + 1)(a - 3)$
$-a^3 + 6a^2 - 11a + 6 = (a - 2)(-a + 1)(a - 3)$

Check: $(a - 2)(-a + 1)(a - 3) = (a - 2)(-a^2 + 4a - 3) = -a^3 + 4a^2 - 3a + 2a^2 - 8a + 6 = -a^3 + 6a^2 - 11a + 6$

18.

$$
\begin{array}{r}
2x^2 + x - 3 \\
x + 2{\overline{\smash{\big)}\,2x^3 + 5x^2 - x - 6}} \\
\underline{2x^3 + 4x^2} \\
x^2 - x \\
\underline{x^2 + 2x} \\
-3x - 6 \\
\underline{-3x - 6} \\
0
\end{array}
$$

$2x^2 + x - 3 = (2x + 3)(x - 1)$
$2x^3 + 5x^2 - x - 6 = (x + 2)(2x + 3)(x - 1)$

Check: $(x + 2)(2x + 3)(x - 1) =$
$(x + 2)(2x^2 + x - 3) = 2x^3 + x^2 - 3x +$
$4x^2 + 2x - 6 = 2x^3 + 5x^2 - x - 6$

19.

$$V = \pi r^2 h$$
$$\pi(x^3 - 3x^2 + 4) = \pi r^2(x + 1)$$
$$\frac{\pi(x^3 - 3x^2 + 4)}{\pi(x + 1)} = r^2$$
$$r^2 = \frac{x^3 - 3x^2 + 4}{x + 1}$$

$$
\begin{array}{r}
x^2 - 4x + 4 \\
x + 1{\overline{\smash{\big)}\,x^3 - 3x^2 + 0x + 4}} \\
\underline{x^3 + x^2} \\
-4x^2 + 0x \\
\underline{-4x^2 - 4x} \\
4x + 4 \\
\underline{4x + 4} \\
0
\end{array}
$$

$$r^2 = x^2 - 4x + 4$$
$$r^2 = (x - 2)^2$$
$$r = x - 2$$

The radius of the cylinder is $(x - 2)$ cm.

20. $(x - 1)(x + 2) = x^2 + x - 2$

$$
\begin{array}{r}
x^2 - 7x + 12 \\
x^2 + x - 2{\overline{\smash{\big)}\,x^4 - 6x^3 + 3x^2 + 26x - 24}} \\
\underline{x^4 + x^3 - 2x^2} \\
-7x^3 + 5x^2 + 26x \\
\underline{-7x^3 - 7x^2 + 14x} \\
12x^2 + 12x - 24 \\
\underline{12x^2 + 12x - 24} \\
0
\end{array}
$$

$x^2 - 7x + 12 = (x - 4)(x - 3)$
$x^4 - 6x^3 + 3x^2 + 26x - 24 =$
$(x - 1)(x + 2)(x - 4)(x - 3)$

Check:
$(x - 1)(x + 2)(x - 4)(x - 3) = (x - 1)(x + 2)$
$(x^2 - 7x + 12) = (x - 1)(x^3 - 7x^2 + 12x +$
$2x^2 - 14x + 24) = (x - 1)$
$x^3 - 5x^2 - 2x + 24) =$
$x^4 - 5x^3 - 2x^2 + 24x - x^3 + 5x^2 + 2x - 24 =$
$x^4 - 6x^3 + 3x^2 + 26x - 24$

21. $(x + 4)(x - 1) = x^2 + 3x - 4$

$$
\begin{array}{r}
2x^2 - 3x - 2 \\
x^2 + 3x - 4{\overline{\smash{\big)}\,2x^4 + 3x^3 - 19x^2 + 6x + 8}} \\
\underline{2x^4 + 6x^3 - 8x^2} \\
-3x^3 - 11x^2 + 6x \\
\underline{-3x^3 - 9x^2 + 12x} \\
2x^2 - 6x + 8 \\
\underline{-2x^2 - 6x + 8} \\
0
\end{array}
$$

$2x^2 - 3x - 2 = (2x + 1)(x - 2)$
$2x^4 + 3x^3 - 19x^2 + 6x + 8 =$
$(x + 4)(x - 1)(2x + 1)(x - 2)$

Check: $(x + 4)(x - 1)(2x + 1)(x - 2) =$
$(x + 4)(x - 1)(2x^2 - 3x - 2) =$
$(x + 4)(2x^3 - 3x^2 - 2x - 2x^2 + 3x + 2) =$
$(x + 4)(2x^3 - 5x^2 + x + 2) =$
$2x^4 - 5x^3 + x^2 + 2x + 8x^3 - 20x^2 + 4x + 8 =$
$2x^4 + 3x^3 - 19x^2 + 6x + 8$

22. $(x + 1)^2 = x^2 + 2x + 1$

$$
\begin{array}{r}
-x^2 - x + 6 \\
x^2 + 2x + 1{\overline{\smash{\big)}\,-x^4 - 3x^3 + 3x^2 + 11x + 6}} \\
\underline{-x^4 - 2x^3 - x^2} \\
-x^3 + 4x^2 + 11x \\
\underline{-x^3 - 2x^2 - x} \\
6x^2 + 12x + 6 \\
\underline{6x^2 + 12x + 6} \\
0
\end{array}
$$

$-x^2 - x + 6 = (-x + 2)(x + 3)$
$-x^4 - 3x^3 + 3x^2 + 11x + 6 =$
$(x + 1)^2(-x + 2)(x + 3)$

Check: $(x + 1)^2(-x + 2)(x + 3) =$
$(x + 1)(x + 1)(-x^2 - x + 6) =$
$(x + 1)(-x^3 - x^2 + 6x - x^2 - x + 6) =$
$(x + 1)(-x^3 - 2x^2 + 5x + 6) =$
$-x^4 - 2x^3 + 5x^2 + 6x - x^3 - 2x^2 + 5x + 6 =$
$-x^4 - 3x^3 + 3x^2 + 11x + 6$

EXTEND

23.

$$
\begin{array}{r}
\frac{1}{2}x - 1 \\
3x + 1{\overline{\smash{\big)}\,\frac{3}{2}x^2 - \frac{5}{2}x - 1}} \\
\underline{\frac{3}{2}x^2 + \frac{1}{2}x} \\
-3x - 1 \\
\underline{-3x - 1} \\
0
\end{array}
$$

So, $\left(\frac{3}{2}x^2 - \frac{5}{2}x - 1\right) \div (3x + 1) = \frac{1}{2}x - 1$

Check: $\left(\frac{1}{2}x - 1\right)(3x + 1) = \frac{3}{2}x^2 - \frac{5}{2}x - 1$

24.

$$\begin{array}{r} \frac{3}{2}x - \frac{5}{4} \\ 2x + 1\overline{)3x^2 - x + \frac{3}{4}} \\ \underline{3x^2 + \frac{3}{2}x} \\ -\frac{5}{2}x + \frac{3}{4} \\ \underline{-\frac{5}{2}x - \frac{5}{4}} \\ 2 \end{array}$$

So, $\left(3x^2 - x + \frac{3}{4}\right) \div (2x + 1) = \frac{3}{2}x - \frac{5}{4} + \frac{2}{2x + 1}$

Check: $\left(\frac{3}{2}x - \frac{5}{4}\right)(2x + 1) + 2 =$

$3x^2 + \frac{3}{2}x - \frac{5}{2}x - \frac{5}{4} + 2 = 3x^2 - x + \frac{3}{4}$

25.

$$\begin{array}{r} 5x - 4y \\ 2x + y\overline{)10x^2 - 3xy + 9y^2} \\ \underline{10x^2 + 5xy} \\ -8xy + 9y^2 \\ \underline{-8xy - 4y^2} \\ 13y^2 \end{array}$$

So, $(10x^2 - 3xy + 9y^2) \div (2x + y) =$

$5x - 4y + \frac{13y^2}{2x + y}$

Check: $(5x - 4y)(2x + y) + 13y^2 =$
$10x^2 + 5xy - 8xy - 4y^2 + 13y^2 =$
$10x^2 - 3xy + 9y^2$

26.

$$\begin{array}{r} 2r^2 - 5rt - 3t^2 \\ r + t\overline{)2r^3 - 3r^2t - 8rt^2 - 3t^3} \\ \underline{2r^3 + 2r^2t} \\ -5r^2t - 8rt^2 \\ \underline{-5r^2t - 5rt^2} \\ -3rt^2 - 3t^3 \\ \underline{-3rt^2 - 3t^3} \\ 0 \end{array}$$

So, $(2r^3 - 3r^2t - 8rt^2 - 3t^2) \div (r + t) =$
$2r^2 - 5rt - 3t^2$

Check: $(2r^2 - 5rt - 3t^2)(r + t) =$
$(2r^2 - 5rt - 3t^2)r + (2r^2 - 5rt - 3t^2)t =$
$2r^3 - 5r^2t - 3rt^2 + 2r^2t - 5rt^2 - 3t^3 =$
$2r^3 - 3r^2t - 8rt^2 - 3t^3$

27.

$$\begin{array}{r} 2a^2 - 5b^2 \\ 2a^2 + 3b^2\overline{)4a^4 - 4a^2b^2 - 13b^4} \\ \underline{4a^4 + 6a^2b^2} \\ -10a^2b^2 - 13b^4 \\ \underline{-10a^2b^2 - 15b^4} \\ 2b^4 \end{array}$$

So, $(4a^4 - 4a^2b^2 - 13b^4) \div (2a^2 + 3b^2) =$

$2a^2 - 5b^2 + \frac{2b^4}{2a^2 + 3b^2}$

Check: $(2a^2 - 5b^2)(2a^2 + 3b^2) + 2b^4 =$
$4a^4 + 6a^2b^2 - 10a^2b^2 - 15b^4 + 2b^4 =$
$4a^4 - 4a^2b^2 - 13b^4$

28.

$$\begin{array}{r} 3x^2 - 2xy - y^2 \\ 2x - 3y\overline{)6x^3 - 13x^2y + 4xy^2 + 2y^3} \\ \underline{6x^3 - 9x^2y} \\ -4x^2y + 4xy^2 \\ \underline{-4x^2y + 6xy^2} \\ -2xy^2 + 2y^3 \\ \underline{-2xy^2 + 3y^3} \\ -y^3 \end{array}$$

So, $(6x^3 - 13x^2y + 4xy^2 + 2y^3) \div (2x - 3y) =$

$3x^2 - 2xy - y^2 + \frac{-y^3}{2x - 3y}$

Check: $(3x^2 - 2xy - y^2)(2x - 3y) - y^3 =$
$(3x^2 - 2xy - y^2)2x - (3x^2 - 2xy - y^2)$
$3y - y^3 = 6x^3 - 4x^2y - 2xy^2 - 9x^2y +$
$6xy^2 + 3y^3 - y^3 = 6x^3 - 13x^2y + 4xy^2 + 2y^3$

THINK CRITICALLY

29.

$$\begin{array}{r} x^2 - 2x + 4 \\ x + 2\overline{)x^3 + 0x^2 + 0x + 8} \\ \underline{x^3 + 2x^2} \\ -2x^2 + 0x \\ \underline{-2x^2 - 4x} \\ 4x + 8 \\ \underline{4x + 8} \\ 0 \quad \text{ycs; remainder} = 0 \end{array}$$

30.

$$\begin{array}{r} x^3 - 2x^2 + 4x - 8 \\ x + 2\overline{)x^4 + 0x^3 + 0x^2 + 0x + 16} \\ \underline{x^4 + 2x^3} \\ -2x^3 + 0x^2 \\ \underline{-2x^3 - 4x^2} \\ 4x^2 + 0x \\ \underline{4x^2 + 8x} \\ -8x + 16 \\ \underline{-8x - 16} \\ 32 \quad \text{no;} \\ \text{remainder} \ne 0 \end{array}$$

31.

$$\begin{array}{r} x^2 + 2x + 4 \\ x - 2\overline{)x^3 + 0x^2 + 0x + 8} \\ \underline{x^3 - 2x^2} \\ 2x^2 + 0x \\ \underline{2x^2 - 4x} \\ 4x + 8 \\ \underline{4x - 8} \\ 16 \quad \text{no; remainder} \ne 0 \end{array}$$

32.

$$\begin{array}{r} x^3 + 2x^2 + 4x + 8 \\ x - 2\overline{)x^4 + 0x^3 + 0x^2 + 0x + 16} \\ \underline{x^4 - 2x^3} \\ 2x^3 + 0x^2 \\ \underline{2x^3 - 4x^2} \\ 4x^2 + 0x \\ \underline{4x^2 - 8x} \\ 8x + 16 \\ \underline{8x - 16} \\ 32 \quad \text{no;} \\ \text{remainder} \ne 0 \end{array}$$

33.

$$
\begin{array}{r}
x^2 - 2x + 4 \\
x + 2 \overline{\smash{)}\ x^3 + 0x^2 + 0x - 8} \\
\underline{x^3 + 2x^2} \\
-2x^2 + 0x \\
\underline{-2x^2 - 4x} \\
4x - 8 \\
\underline{4x + 8} \\
-16 \quad \text{no; remainder} \neq 0
\end{array}
$$

34.

$$
\begin{array}{r}
x^3 - 2x^2 + 4x - 8 \\
x + 2 \overline{\smash{)}\ x^4 + 0x^3 + 0x^2 + 0x - 16} \\
\underline{x^4 + 2x^3} \\
-2x^3 + 0x^2 \\
\underline{-2x^3 - 4x^2} \\
4x^2 + 0x \\
\underline{4x^2 + 8x} \\
-8x - 16 \\
\underline{-8x - 16} \\
0 \quad \text{yes;} \\
\text{remainder} = 0
\end{array}
$$

35.

$$
\begin{array}{r}
x^2 + 2x + 4 \\
x - 2 \overline{\smash{)}\ x^3 + 0x^2 + 0x - 8} \\
\underline{x^3 - 2x^2} \\
2x^2 + 0x \\
\underline{2x^2 - 4x} \\
4x - 8 \\
\underline{4x - 8} \\
0 \quad \text{yes; remainder} = 0
\end{array}
$$

36.

$$
\begin{array}{r}
x^3 + 2x^2 + 4x + 8 \\
x - 2 \overline{\smash{)}\ x^4 + 0x^3 + 0x^2 + 0x - 16} \\
\underline{x^4 - 2x^3} \\
2x^3 + 0x^2 \\
\underline{2x^3 - 4x^2} \\
4x^2 + 0x \\
\underline{4x^2 - 8x} \\
8x - 16 \\
\underline{8x - 16} \\
0 \quad \text{yes;} \\
\text{remainder} = 0
\end{array}
$$

37. true, see Exercises 35 and 36

38. false, see Exercise 33

39. true, see Exercise 34

40. false, see Exercise 33

41. false, see Exercise 30

42. true, see Exercise 29

MIXED REVIEW

43.
$$
A = \frac{1}{2}h(a + b)
$$
$$
\frac{1}{2}h(a + b) = A
$$
$$
h(a + b) = 2A
$$
$$
a + b = \frac{2A}{h}
$$
$$
a = \frac{2A}{h} - b
$$
$$
a = \frac{2A - hb}{h}
$$

44.
$$
mv = Ft + mv_0
$$
$$
mv - mv_0 = Ft
$$
$$
m(v - v_0) = Ft
$$
$$
m = \frac{Ft}{v - v_0}
$$

45.
$$
\begin{cases} 4x + 3y = 27 \\ 2x - y = 1 \end{cases}
\begin{array}{l}
\to \ 4x + 3y = 27 \\
\to \ 6x - 3y = \ 3 \\
\hline
10x \qquad\ = 30 \\
x = 3 \\
2(3) - y = 1 \\
6 - y = 1 \\
-y = -5 \\
y = 5 \quad x = 3, y = 5
\end{array}
$$

Check: $4(3) + 3(5) \overset{?}{=} 27$ $\qquad$ $2(3) - 5 \overset{?}{=} 1$

$\qquad\quad 12 + 15 \overset{?}{=} 27$ $\qquad\qquad 6 - 5 \overset{?}{=} 1$

$\qquad\qquad\ \ 27 = 27 ✓$ $\qquad\qquad\qquad 1 = 1 ✓$

46.
$$
\begin{cases} 7x = 5 - 2y \\ 3y + 2x = 16 \end{cases}
$$

$$
\begin{cases} 7x + 2y = 5 \\ 2x + 3y = 16 \end{cases}
\begin{array}{l}
\to \ 21x + 6y = 15 \\
\to \ -4x - 6y = \ -32 \\
\hline
17x \qquad\ = -17 \\
x = -1 \\
2(-1) + 3y = 16 \\
-2 + 3y = 16 \\
3y = 18 \\
y = 6 \quad x = -1, y = 6
\end{array}
$$

Check: $7(-1) \overset{?}{=} 5 - 2(6)$ $\qquad$ $3(6) + 2(-1) \overset{?}{=} 16$

$\qquad\quad -7 \overset{?}{=} 5 - 12$ $\qquad\qquad\quad 18 - 2 \overset{?}{=} 16$

$\qquad\qquad -7 = -7 ✓$ $\qquad\qquad\qquad 16 = 16 ✓$

EXPLORE

1. $5x^3 + 2x^2 - 3x + 6 = (5x^2 + 2x - 3)x + 6 = ((5x + 2)x - 3)x + 6 = (((5)x + 2)x - 3)x + 6$

2. $3x^4 - 7x^3 + 2x^2 + 5x - 9 =$
$(3x^3 - 7x^2 + 2x + 5)x - 9 =$
$((3x^2 - 7x + 2)x + 5)x - 9 =$
$(((3x - 7)x + 2)x + 5)x - 9 =$
$((((3)x - 7)x + 2)x + 5)x - 9$

3. $F(3) = 82$

4. $F(3) = 82$

5. $F(x) = (((4)x - 3)x + 1)x - 2$
$= ((4x - 3)x + 1)x - 2$
$= 4x^3 - 3x^2 + x - 2$
$F(3) = 4(3)^3 - 3(3)^2 + 3 - 2$
$= 4(27) - 3(9) + 3 - 2$
$= 108 - 27 + 3 - 2$
$= 82$

6.
$$\begin{array}{r} 3x^2 + 5x + 8 + \dfrac{21}{x-2} \\ x - 2\overline{)3x^3 - x^2 - 2x + 5} \\ \underline{3x^3 - 6x^2} \\ 5x^2 \quad 2x \\ \underline{5x^2 - 10x} \\ 8x + 5 \\ \underline{8x - 16} \\ 21 \end{array}$$

7. $F(x) = 3x^3 + 2x^2 - 4x - 8$
$= (3x^2 + 2x - 4)x - 8$
$= ((3x + 2)x - 4)x - 8$
$= (((3)x + 2)x - 4)x - 8$
$F(3) = (((3)3 + 2)3 - 4)3 - 8$
$= ((11)3 - 4)3 - 8$
$= (29)3 - 8$
$= 79$
So, $(3x^3 + 2x^2 - 4x - 8) \div (x - 3) =$
$3x^2 + 11x + 29 + \dfrac{79}{x - 3}$

8. $F(x) = 5x^4 - 7x^3 - 2x^2 + 1$
$= (5x^3 - 7x^2 - 2x)x + 1$
$= ((5x^2 - 7x - 2)x)x + 1$
$= (((5x - 7)x - 2)x)x + 1$
$= ((((5)x - 7)x - 2)x)x + 1$
$F(1) = ((((5)1 - 7)1 - 2)1)1 + 1$
$= (((-2)1 - 2)1)1 + 1$
$= ((-4)1)1 + 1$
$= (-4)1 + 1$
$= -3$

So, $5x^4 - 7x^3 - 2x^2 + 1 \div (x - 1) =$
$5x^3 - 2x^2 - 4x - 4 + \dfrac{-3}{x - 1}$

TRY THESE

1. Answers will vary. Advantages include that you only have to work with coefficients and that you can do all the work on one line; a disadvantage is that the process is limited to divisors of the form $x - r$.

2.
$$\begin{array}{r|rrrr} 2 & 3 & 2 & -6 & -26 \\ & & 6 & 16 & 20 \\ \hline & 3 & 8 & 10 & -6 \end{array}$$
So, $(3x^3 + 2x^2 - 6x - 26) \div (x - 2) =$
$3x^2 + 8x + 10 + \dfrac{-6}{x - 2}$
Check: $(3x^2 + 8x + 10)(x - 2) + (-6) =$
$(3x^2 + 8x + 10)x - (3x^2 + 8x + 10)2 + (-6) =$
$3x^3 + 8x^2 + 10x - 6x^2 - 16x - 20 + (-6) =$
$3x^3 + 2x^2 - 6x - 26$

3.
$$\begin{array}{r|rrrr} -1 & 2 & -1 & 1 & -3 \\ & & -2 & 3 & -4 \\ \hline & 2 & -3 & 4 & -7 \end{array}$$
So, $(2y^3 - y^2 + y - 3) \div (y + 1) =$
$2y^2 - 3y + 4 + \dfrac{-7}{y + 1}$
Check: $(2y^2 - 3y + 4)(y + 1) + (-7) =$
$(2y^2 - 3y + 4)y + (2y^2 - 3y + 4)1 + (-7) =$
$2y^3 - 3y^2 + 4y + 2y^2 - 3y + 4 + (-7) =$
$2y^3 - y^2 + y - 3$

4.
$$\begin{array}{r|rrrrr} -2 & 1 & -2 & 0 & 5 & 13 \\ & & -2 & 8 & -16 & 22 \\ \hline & 1 & -4 & 8 & -11 & 35 \end{array}$$
So, $(z^4 - 2z^3 + 5z + 13) \div (z + 2) =$
$z^3 - 4z^2 + 8z - 11 + \dfrac{35}{z + 2}$
Check: $(z^3 - 4z^2 + 8z - 11)(z + 2) + 35 =$
$(z^3 - 4z^2 + 8z - 11)z +$
$(z^3 - 4z^2 + 8z - 11)2 + 35 = z^4 - 4z^3 + 8z^2 -$
$11z + 2z^3 - 8z^2 + 16z - 22 + 35 =$
$z^4 - 2z^3 + 5z + 13$

5.
$$\begin{array}{r|rrrrr} 1 & -2 & 1 & 0 & 2 & 1 \\ & & -2 & -1 & -1 & 1 \\ \hline & -2 & -1 & -1 & 1 & 2 \end{array}$$
So, $(2a + a^3 - 2a^4 + 1) \div (a - 1) =$
$-2a^3 - a^2 - a + 1 + \dfrac{2}{a - 1}$
Check: $(-2a^3 - a^2 - a + 1)(a - 1) + 2 =$
$(-2a^3 - a^2 - a + 1)a - (-2a^3 - a^2 - a + 1)1$
$+ 2 = -2a^4 - a^3 - a^2 + a + 2a^3 + a^2 + a - 1$
$+ 2 = 2a + a^3 - 2a^4 + 1$

6. $\frac{1}{2}\rfloor$

$$
\begin{array}{rrrr}
2 & 3 & 4 & 5 \\
 & 1 & 2 & 3 \\
\hline
2 & 4 & 6 & 8
\end{array}
$$

So, $(2m^3 + 3m^2 + 4m + 5) \div (m - \frac{1}{2}) =$

$2m^2 + 4m + 6 + \dfrac{8}{m - \frac{1}{2}}$

Check: $(2m^2 + 4m + 6)(m - \frac{1}{2}) + 8$

$(2m^2 + 4m + 6)m - (2m^2 + 4m + 6)\frac{1}{2} + 8 =$

$2m^3 + 4m^2 + 6m - m^2 - 2m - 3 + 8 =$

$2m^3 + 3m^2 + 4m + 5$

7. $\frac{-1}{3}\rfloor$

$$
\begin{array}{rrrrr}
6 & -10 & -4 & 3 & 3 \\
 & -2 & 4 & 0 & -1 \\
\hline
6 & -12 & 0 & 3 & 2
\end{array}
$$

So, $(6x^4 - 10x^3 - 4x^2 + 3x + 3) \div (x + \frac{1}{3}) =$

$6x^3 - 12x^2 + 3 + \dfrac{2}{x + \frac{1}{3}}$

Check: $(6x^3 - 12x^2 + 3)\left(x + \frac{1}{3}\right) + 2 =$

$(6x^3 - 12x^2 + 3)x + (6x^3 - 12x^2 + 3)\frac{1}{3} + 2 =$

$6x^4 - 12x^3 + 3x + 2x^3 - 4x^2 + 1 + 2 =$

$6x^4 - 10x^3 - 4x^2 + 3x + 3$

8. $\dfrac{6t^3 - 13t^2 - 12t + 4}{2t + 1} = \dfrac{1}{2}\left(\dfrac{6t^3 - 13t^2 - 12t + 4}{t + \frac{1}{2}}\right)$

$\frac{-1}{3}\rfloor$

$$
\begin{array}{rrrr}
6 & -13 & -12 & 4 \\
 & -3 & 8 & 2 \\
\hline
6 & -16 & -4 & 6
\end{array}
$$

$\dfrac{6t^3 - 13t^2 - 12t + 4}{2t + 1} = \dfrac{1}{2}\left(6t^2 - 16t - 4 + \dfrac{6}{t + \frac{1}{2}}\right)$

$= 3t^2 - 8t - 2 + \dfrac{6}{2t + 1}$

Check: $(3t^2 - 8t - 2)(2t + 1) + 6 =$

$(3t^2 - 8t - 2)2t + (3t^2 - 8t - 2)1 + 6 =$

$6t^3 - 16t^2 - 4t + 3t^2 - 8t - 2 + 6 =$

$6t^3 - 13t^2 - 12t + 4$

9. $\dfrac{3c^4 + 7c^3 - 10c + 5}{3c - 2} = \dfrac{1}{3}\left(\dfrac{3c^4 + 7c^3 - 10c + 5}{c - \frac{2}{3}}\right)$

$\frac{2}{3}\rfloor$

$$
\begin{array}{rrrrr}
3 & 7 & 0 & -10 & 5 \\
 & 2 & 6 & 4 & -4 \\
\hline
3 & 9 & 6 & -6 & 1
\end{array}
$$

$\dfrac{3c^4 + 7c^3 - 10c + 5}{3c - 2} =$

$\frac{1}{3}\left(3c^3 + 9c^2 + 6c - 6 + \dfrac{1}{c - \frac{2}{3}}\right) =$

$c^3 + 3c^2 + 2c - 2 + \dfrac{1}{3c - 2}$

Check: $(c^3 + 3c^2 + 2c - 2)(3c - 2) + 1 =$
$(c^3 + 3c^2 + 2c - 2)3c - (c^3 + 3c^2 + 2c - 2)2$
$+ 1 = 3c^4 + 9c^3 + 6c^2 - 6c - 2c^3 - 6c^2 - 4c$
$+ 4 + 1 = 3c^4 + 7c^3 - 10c + 5$

10. Let w represent width and h represent height
$$V = lwh$$
$2x^3 - 3x^2 - 11x + 6 = (x - 3)wh$
$$wh = \dfrac{2x^3 - 3x^2 - 11x + 6}{x - 3}$$

$3\rfloor$

$$
\begin{array}{rrrr}
2 & -3 & -11 & 6 \\
 & 6 & 9 & -6 \\
\hline
2 & 3 & -2 & 0
\end{array}
$$

$wh = 2x^2 + 3x - 2$
$wh = (2x - 1)(x + 2)$
The other dimensions are represented by $(2x - 1)$ cm and $(x + 2)$ cm.

11. $$d = rt$$
$6x^3 - 11x^2 - 46x - 24 = (2x + 3)t$
$$\dfrac{6x^3 - 11x^2 - 46x - 24}{2x + 3} = t$$
$$t = \frac{1}{2}\left(\dfrac{6x^3 - 11x^2 - 46x - 24}{x + \frac{3}{2}}\right)$$

$\frac{-3}{2}\rfloor$

$$
\begin{array}{rrrr}
6 & -11 & -46 & -24 \\
 & -9 & 30 & 24 \\
\hline
6 & -20 & -16 & 0
\end{array}
$$

$$t = \frac{1}{2}(6x^2 - 20x - 16)$$
$$t = (3x^2 - 10x - 8)\text{h}$$

12. $2\rfloor$

$$
\begin{array}{rrrr}
6 & 19 & 8 & -5 \\
 & 12 & 62 & 140 \\
\hline
6 & 31 & 70 & 135
\end{array}
$$

So, $F(2) = 135$

13. $-3\rfloor$

$$
\begin{array}{rrrr}
6 & 19 & 8 & -5 \\
 & -18 & -3 & -15 \\
\hline
6 & 1 & 5 & -20
\end{array}
$$

So, $F(-3) = -20$

14. $\frac{1}{3}\rfloor$

$$
\begin{array}{rrrr}
6 & 19 & 8 & -5 \\
 & 2 & 7 & 5 \\
\hline
6 & 21 & 15 & 0
\end{array}
$$

So, $F\left(\frac{1}{3}\right) = 0$

15. $\frac{-5}{2}\rfloor$

$$
\begin{array}{rrrr}
6 & 19 & 8 & -5 \\
 & -15 & -10 & 5 \\
\hline
6 & 4 & -2 & 0
\end{array}
$$

So, $F\left(-\frac{5}{2}\right) = 0$

16. $4\rfloor$

$$
\begin{array}{rrrr}
1 & 3 & -18 & 38 \\
 & 4 & 28 & 40 \\
\hline
1 & 7 & 10 & 78
\end{array}
$$
  no, is not a factor

17. $3\rfloor$

$$
\begin{array}{rrrr}
1 & -2 & 2 & -15 \\
 & 3 & 3 & 15 \\
\hline
1 & 1 & 5 & 0
\end{array}
$$
  yes, is a factor

18. $-2 \lfloor$ 1   0   -3   0   -4

       -2   4   -2   4
  ---
  1 -2   1   -2   0   yes, is a factor

19. $\dfrac{3x^3 - 2x^2 - 18x + 12}{3x - 2} = \dfrac{1}{3}\left(\dfrac{3x^3 - 2x^2 - 18x + 12}{x - \frac{2}{3}}\right)$

$\frac{2}{3}\lfloor$ 3   -2   -18   12

        2   0   -12
  ---
   3   0   -18   0   yes, is a factor

20. $-i\lfloor$ 2   3   2   3

        -2i -2-3i -3
  ---
   2 3-2i   -3i   0   yes, is a factor

21. $\sqrt{3}\lfloor$ 1 $-\sqrt{3}$ -2 $2\sqrt{3}$

          $\sqrt{3}$ 0 $-2\sqrt{3}$
  ---
   1   0   -2   0   yes, is a factor

PRACTICE

1. $3\lfloor$ 1   3   -8   -12

        3   18   30
  ---
   1   6   10   18

$\dfrac{x^3 + 3x^2 - 8x - 12}{x - 3} = x^2 + 6x + 10 + \dfrac{18}{x - 3}$

Check: $(x^2 + 6x + 10)(x - 3) + 18 =$
$(x^2 + 6x + 10)x - (x^2 + 6x + 10)3 + 18 =$
$x^3 + 6x^2 + 10x - 3x^2 - 18x - 30 + 18$
$= x^3 + 3x^2 - 8x - 12$

2. $-1\lfloor$ 1   6   2   3

        -1   -5   3
  ---
   1   5   -3   6

$\dfrac{a^3 + 6a^2 + 2a + 3}{a + 1} = a^2 + 5a - 3 + \dfrac{6}{a + 1}$

Check: $(a^2 + 5a - 3)(a + 1) + 6 =$
$(a^2 + 5a - 3)a + (a^2 + 5a - 3)1 + 6 =$
$a^3 + 5a^2 - 3a + a^2 + 5a - 3 + 6 =$
$a^3 + 6a^2 + 2a + 3$

3. $1\lfloor$ 1   2   3   1

        1   3   6
  ---
   1   3   6   5

$\dfrac{2y^2 + 3y + y^3 - 1}{y - 1} = y^2 + 3y + 6 + \dfrac{5}{y - 1}$

Check: $(y^2 + 3y + 6)(y - 1) + 5 =$
$(y^2 + 3y + 6)y - (y^2 + 3y + 6)1 + 5 =$
$y^3 + 3y^2 + 6y - y^2 - 3y - 6 + 5 =$
$2y^2 + 3y + y^3 - 1$

4. $2\lfloor$ 1   2   -7   -8   13

        2   8   2   -12
  ---
   1   4   1   -6   1

$\dfrac{b^4 - 8b - 7b^2 + 2b^3 + 13}{b - 2} =$
$b^3 + 4b^2 + b - 6 + \dfrac{1}{b - 2}$

Check: $(b^3 + 4b^2 + b - 6)(b - 2) + 1 =$
$(b^3 + 4b^2 + b - 6)b - (b^3 + 4b^2 + b - 6)2 + 1 =$
$b^4 + 4b^3 + b^2 - 6b - 2b^3 - 8b^2 - 2b + 12 +$
$1 = b^4 - 8b - 7b^2 + 2b^3 + 13$

5. $-2\lfloor$ 3   0   -2   2

        -6   12   -20
  ---
   3   -6   10   -18

$\dfrac{3z^3 - 2z + 2}{z + 2} = 3z^2 - 6z + 10 + \dfrac{-18}{z + 2}$

Check: $(3z^2 - 6z + 10)(z + 2) + (-18) =$
$(3z^2 - 6z + 10)z + (3z^2 - 6z + 10)2 + (-18) =$
$3z^3 - 6z^2 + 10z + 6z^2 - 12z + 20 + (-18) =$
$3z^3 - 2z + 2$

6. $2\lfloor$ 2   0   -7   0   -10

        4   8   2   4
  ---
   2   4   1   2   -6

$\dfrac{2c^4 - 7c^2 - 10}{c - 2} = 2c^3 + 4c^2 + c + 2 + \dfrac{-6}{c - 2}$

Check: $(2c^3 + 4c^2 + c + 2)(c - 2) + (-6) =$
$(2c^3 + 4c^2 + c + 2)c - (2c^3 + 4c^2 + c + 2)2 +$
$(-6) = 2c^4 + 4c^3 + c^2 + 2c - 4c^3 - 8c^2 - 2c$
$-4 + (-6) = 2c^4 - 7c^2 - 10$

7. $\dfrac{3w^3 + 5w^2 + 4w + 1}{3w - 1} = \dfrac{1}{3}\left(\dfrac{3w^3 + 5w^2 + 4w + 1}{w - \frac{1}{3}}\right)$

$\frac{1}{3}\lfloor$ 3   5   4   1

        1   2   2
  ---
   3   6   6   3

$\dfrac{3w^3 + 5w^2 + 4w + 1}{3w - 1} =$
$\dfrac{1}{3}\left(3w^2 + 6w + 6 + \dfrac{3}{w - \frac{1}{3}}\right) =$
$w^2 + 2w + 2 + \dfrac{3}{3w - 1}$

Check: $(w^2 + 2w + 2)(3w - 1) + 3 =$
$(w^2 + 2w + 2)3w - (w^2 + 2w + 2)1 + 3 =$
$3w^3 + 6w^2 + 6w - w^2 - 2w - 2 + 3 =$
$3w^3 + 5w^2 + 4w + 1$

8. $\dfrac{6v^4 - 3v^3 + 2v^2 + 3v - 3}{2v - 1} =$

$\dfrac{1}{2}\left(\dfrac{6v^4 - 3v^3 + 2v^2 + 3v - 3}{v - \frac{1}{2}}\right)$

$\frac{1}{2}\lfloor$ 6   -3   2   3   -3

        3   0   1   2
  ---
   6   0   2   4   -1

$\dfrac{6v^4 - 3v^3 + 2v^2 + 3v - 3}{2v - 1} =$
$\dfrac{1}{2}\left(6v^3 + 2v + 4 + \dfrac{1}{v - \frac{1}{2}}\right) =$
$3v^3 + v + 2 + \dfrac{-1}{2v - 1}$

Check: $(3v^3 + v + 2)(2v - 1) + (-1) =$
$(3v^3 + v + 2)2v - (3v^3 + v + 2)1 + (-1) =$
$6v^4 + 2v^2 + 4v - 3v^3 - v - 2 + (-1) =$
$6v^4 - 3v^3 + 2v^2 + 3v - 3$

9. $\dfrac{4t^3 + 4t^2 + 5t - 3}{2t + 1} = \dfrac{1}{2}\left(\dfrac{4t^3 + 4t^2 + 5t - 3}{t + \frac{1}{2}}\right)$

$$-\tfrac{1}{2}\,\big|\quad 4 \quad\ 4 \quad\ \ 5 \quad -3$$
$$\qquad\qquad\quad -2 \quad -1 \quad -2$$
$$\overline{\qquad\quad 4 \quad\ 2 \quad\ \ 4 \quad -5}$$

$\dfrac{4t^3 + 4t^2 + 5t - 3}{2t + 1} = \dfrac{1}{2}\left(4t^2 + 2t + 4 + \dfrac{-5}{t + \frac{1}{2}}\right)$

$= 2t^2 + t + 2 + \dfrac{-5}{2t + 1}$

Check: $(2t^2 + t + 2)(2t + 1) + (-5) =$
$(2t^2 + t + 2)2t + (2t^2 + t + 2)1 + (-5)$
$= 4t^3 + 2t^2 + 4t + 2t^2 + t + 2 + (-5)$
$= 4t^3 + 4t^2 + 5t - 3$

10. $\dfrac{4a^4 + 9a^2 + 21a + 9}{2a + 3} = \dfrac{1}{2}\left(\dfrac{4a^4 + 9a^2 + 21a + 9}{a + \frac{3}{2}}\right)$

$$-\tfrac{3}{2}\,\big|\quad 4 \quad\ \ 0 \quad\ \ 9 \quad\ \ 21 \quad\ 9$$
$$\qquad\qquad\quad -6 \quad\ \ 9 \quad -27 \quad\ 9$$
$$\overline{\qquad\quad 4 \quad -6 \quad\ 18 \quad -6 \quad 18}$$

$\dfrac{4a^4 + 9a^4 + 21a + 9}{2a + 3} =$

$\dfrac{1}{2}\left(4a^3 - 6a^2 + 18a - 6 + \dfrac{18}{a + \frac{3}{2}}\right)$

$= 2a^3 - 3a^2 + 9a - 3 + \dfrac{18}{2a + 1}$

Check: $(2a^3 + 3a^2 + 9a - 3)(2a + 3) + 18 =$
$(2a^3 - 3a^2 + 9a - 3)2a + (2a^3 - 3a^2 + 9a - 3)(3) +$
$18 = 4a^4 - 6a^3 + 18a^2 - 6a + 6a^3 - 9a^2 + 27a -$
$9 + 18 = 4a^4 + 9a^2 + 21a + 9$

11. Let n and m represent the other two numbers.
$nm(x)(2x - 1) = 6x^4 + 11x^3 - 3x^2 - 2x$
$nm(x)(2x - 1) = x(6x^3 + 11x^2 - 3x - 2)$
$nm(2x - 1) = 6x^3 + 11x^2 - 3x - 2$

$nm = \dfrac{6x^3 + 11x^2 - 3x - 2}{2x - 1}$

$nm = \dfrac{1}{2}\left(\dfrac{6x^3 + 11x^2 - 3x - 2}{x - \frac{1}{2}}\right)$

$$\tfrac{1}{2}\,\big|\quad 6 \quad\ 11 \quad -3 \quad -2$$
$$\qquad\qquad\quad\ \ 3 \quad\ \ 7 \quad\ \ 2$$
$$\overline{\qquad\quad 6 \quad 14 \quad\ \ 4 \quad\ \ 0}$$

$nm = \dfrac{1}{2}(6x^2 + 14x + 4)$

$nm = 3x^2 + 7x + 2$

$nm = (3x + 1)(x + 2)$

12. $V = \pi r^2 h$
$\pi(x - 2)^2 h = \pi(3x^3 - 8x^2 - 4x + 16)$
$(x - 2)^2 h = 3x^3 - 8x^2 - 4x + 16$
Divide by $(x - 2)$ twice using synthetic division

$$2\,\big|\quad 3 \quad -8 \quad -4 \quad\ \ 16$$
$$\qquad\qquad\quad\ \ 6 \quad -4 \quad -16$$
$$\overline{\qquad\quad 3 \quad -2 \quad -8 \quad\ \ \ 0}$$

$$2\,\big|\quad 3 \quad -2 \quad -8$$
$$\qquad\qquad\quad\ \ 6 \quad\ \ 8$$
$$\overline{\qquad\quad 3 \quad\ \ 4 \quad\ \ 0}$$

So, $h = 3x + 4$ or the representation of the height is $(3x + 4)$ cm.

13.
$$-3\,\big|\quad 1 \quad\ \ 4 \quad -3 \quad\ \ 7$$
$$\qquad\qquad\quad -3 \quad -3 \quad 18$$
$$\overline{\qquad\quad 1 \quad\ \ 1 \quad -6 \quad 25}$$

By the remainder theorem $F(-3) = 25$.

14.
$$-4\,\big|\quad \tfrac{1}{2} \quad -3 \quad\ \ 0 \quad\ \ \ 6 \quad -10$$
$$\qquad\qquad\qquad -2 \quad 20 \quad -80 \quad 296$$
$$\overline{\qquad\ \ \tfrac{1}{2} \quad -5 \quad 20 \quad -74 \quad 286}$$

By the remainder theorem $F(-4) = 286$.

15.
$$2\,\big|\quad 6 \quad\ \ 35 \quad\ \ 13 \quad -110 \quad\ 56$$
$$\qquad\qquad\quad\ 12 \quad\ \ 94 \quad\ \ 214 \quad 208$$
$$\overline{\qquad\quad 6 \quad\ \ 47 \quad 107 \quad\ \ 104 \quad 264}$$

By the remainder theorem $F(2) = 264$.

16.
$$-2\,\big|\quad 6 \quad\ \ 35 \quad\ \ 13 \quad -110 \quad\ 56$$
$$\qquad\qquad\quad -12 \quad -46 \quad\ \ 66 \quad\ \ 88$$
$$\overline{\qquad\quad 6 \quad\ \ 23 \quad -33 \quad -44 \quad 144}$$

By the remainder theorem $F(-2) = 144$.

17.
$$\tfrac{2}{3}\,\big|\quad 6 \quad\ \ 35 \quad 13 \quad -110 \quad\ 56$$
$$\qquad\qquad\quad\ \ 4 \quad\ \ 26 \quad\ \ 26 \quad -56$$
$$\overline{\qquad\quad 6 \quad\ \ 39 \quad 39 \quad -84 \quad\ \ 0}$$

By the remainder theorem $F\left(\dfrac{2}{3}\right) = 0$.

18.
$$-\tfrac{7}{2}\,\big|\quad 6 \quad\ \ 35 \quad\ \ 13 \quad -110 \quad\ 56$$
$$\qquad\qquad\quad -21 \quad -49 \quad\ \ 126 \quad -56$$
$$\overline{\qquad\quad 6 \quad\ \ 14 \quad -36 \quad\ \ 16 \quad\ \ 0}$$

By the remainder theorem $F\left(-\dfrac{7}{2}\right) = 0$

19. Synthetic division deals only with multiplication and addition, eliminating work with powers. If, however, $x^{10} + 3$ is to be divided by $x + 1$, substituting $x = -1$ into $F(x)$ is preferable to using synthetic division.

20. To find the total cost of producing 300 sailboats evaluate $C(300)$. Use the remainder theorem and synthetic division.

$C(x) = x^3 + 5x + 400$

$$
\begin{array}{r|rrrr}
-300 & 1 & 0 & 5 & 400 \\
 & & -300 & 90{,}000 & 27{,}001{,}500 \\
\hline
 & 1 & -300 & 90{,}005 & 27{,}001{,}900
\end{array}
$$

The cost is \$27,001,900.

21.
$$
\begin{array}{r|rrrr}
-1 & 1 & 3 & 3 & 1 \\
 & & -1 & -2 & -1 \\
\hline
 & 1 & 2 & 1 & 0
\end{array}
$$

yes, by the factor theorem, $x + 1$ is a factor.

22.
$$
\begin{array}{r|rrrr}
2 & 2 & -1 & -4 & 5 \\
 & & 4 & 6 & 4 \\
\hline
 & 2 & 3 & 2 & 9
\end{array}
$$

no, $x - 2$ is not a factor

23.
$$
\begin{array}{r|rrrrr}
-2 & 1 & 0 & 0 & 0 & 16 \\
 & & -2 & 4 & -8 & 16 \\
\hline
 & 1 & -2 & 4 & -8 & 32
\end{array}
$$

no, $x + 2$ is not a factor

24.
$$
\begin{array}{r|rrrrrr}
-3 & 1 & 0 & 0 & 0 & 0 & 243 \\
 & & -3 & 9 & -27 & 81 & -243 \\
\hline
 & 1 & -3 & 9 & -27 & 81 & 0
\end{array}
$$

yes, $x + 3$ is a factor

25. $\dfrac{3x^3 - 2x^2 - 19x - 6}{3x + 1} = \dfrac{1}{3}\left(\dfrac{3x^3 - 2x^2 - 19x - 6}{x + \frac{1}{3}}\right)$

$$
\begin{array}{r|rrrr}
-\frac{1}{3} & 3 & -2 & -19 & -6 \\
 & & -1 & 1 & 6 \\
\hline
 & 3 & -3 & -18 & 0
\end{array}
$$

yes, $x + \frac{1}{3}$ is a factor

26. $\dfrac{4x^3 - x^2 - 4x - 1}{4x - 1} = \dfrac{1}{4}\left(\dfrac{4x^3 - x^2 - 4x - 1}{x - \frac{1}{4}}\right)$

$$
\begin{array}{r|rrrr}
\frac{1}{4} & 4 & -1 & -4 & -1 \\
 & & 1 & 0 & -1 \\
\hline
 & 4 & 0 & -4 & -2
\end{array}
$$

no, $x - \frac{1}{4}$ is not a factor

27.
$$
\begin{array}{r|rrrr}
-\sqrt{5} & 1 & 1 & -5 & -5 \\
 & & -\sqrt{5} & -\sqrt{5} + 5 & 5 \\
\hline
 & 1 & 1 - \sqrt{5} & -\sqrt{5} & 0
\end{array}
$$

yes, $x + \sqrt{5}$ is a factor

28.
$$
\begin{array}{r|rrrr}
i & 1 & -3 & 1 & -3 \\
 & & i & -3i - 1 & 3 \\
\hline
 & 1 & -3 + i & -3i & 0
\end{array}
$$

yes, $x - i$ is a factor

29. Let p represent the product of the other two numbers.

$p(6x^2 + x - 2) = 6x^4 + 19x^3 - 23x^2 - 10x + 8$

$p = \dfrac{6x^4 + 19x^3 - 23x^2 - 10x + 8}{6x^2 + x - 2}$

$$
\require{enclose}
\begin{array}{r}
x^2 + 3x - 4 \\
6x^2 + x - 2 \enclose{longdiv}{6x^4 + 19x^3 - 23x^2 - 10x + 8} \\
\underline{6x^4 + x^3 - 2x^2} \\
18x^3 - 21x^2 - 10x \\
\underline{18x^3 + 3x^2 - 6x} \\
-24x^2 - 4x + 8 \\
\underline{-24x^2 - 4x + 8} \\
0
\end{array}
$$

a. The representation of the product of the other two numbers is $x^2 + 3x - 4$.

$x^2 + 3x - 4 = (x + 4)(x - 1)$

b. The representations of each of the four numbers are $(2x - 1)$, $(3x + 2)$, $(x + 4)$, and $(x - 1)$.

EXTEND

30. $\dfrac{x^4 - x^3 - 6x^2 + 1}{x + 2} = x^3 - 3x^2 + \dfrac{1}{x + 2}$

31. $\dfrac{4x^3 + 2x^2 - 5}{x - \frac{1}{2}} = 4x^2 + 4x + 2 + \dfrac{-4}{x - \frac{1}{2}}$

32.
$$
\begin{array}{r|rrrrr}
-1 & 2 & 0 & -5 & 2 & k \\
 & & -2 & 2 & 3 & -5 \\
\hline
 & 2 & -2 & -3 & 5 & k - 5
\end{array}
$$

$k - 5 = 0$

$k = 5$

33.
$$
\begin{array}{r|rrrr}
2 & 2 & 3 & -k & 6 \\
 & & 4 & 14 & -2k + 28 \\
\hline
 & 2 & 7 & -k + 14 & -2k + 34
\end{array}
$$

$-2k + 34 = 0$

$-2k = -34$

$k = 17$

34.
$$
\begin{array}{r|rrrr}
3 & k & -6 & 24 & -36 \\
 & & 3k & 9k - 18 & 27k + 18 \\
\hline
 & k & 3k - 6 & 9k + 6 & 27k - 18
\end{array}
$$

$27k - 18 = 0$

$27k = 18$

$k = \dfrac{18}{27}$

$k = \dfrac{2}{3}$

THINK CRITICALLY

35.
$$
\begin{array}{r|rrrr}
2 & 2 & -3 & 3 & -t \\
 & & 4 & 2 & 10 \\
\hline
 & 2 & 1 & 5 & -t + 10
\end{array}
$$

$-t + 10 = 6$

$-t = 6 - 10$

$-t = -4$

$t = 4$

36.

$$\begin{array}{r|rrrr} 1| & 3 & -2 & m & n \\ & & 3 & 1 & m+1 \\ \hline & 3 & 1 & m+1 & m+n+1 \end{array}$$

$m + n + 1 = 2$
$ m + n = 1$

$$\begin{array}{r|rrrr} -2| & 3 & -2 & m & n \\ & & -6 & 16 & -2m-32 \\ \hline & 3 & -8 & m+16 & -2m+n-32 \end{array}$$

$-2m + n - 32 = -37$
$ -2m + n = -5$

Solve the system of equations.

$$\begin{cases} -2m + n = -5 \\ m + n = 1 \end{cases} \rightarrow \begin{array}{r} -2m + n = -5 \\ \underline{2m + 2n = 2} \\ 3n = -3 \\ n = -1 \end{array}$$

$m + (-1) = 1$
$ m = 2$

So, $m = 2$ and $n = -1$

37.

$$\begin{array}{r|rrrrr} 3| & 2 & -7 & 0 & a & b \\ & & 6 & -3 & -9 & 3a-27 \\ \hline & 2 & -1 & -3 & a-9 & b+3a-27 \end{array}$$

$b + 3a - 27 = 0$
$ 3a + b = 27$

38. Yes. Find $F(1)$ by substituting $x = 1$ into $F(x)$. 1 to any power is 1. Synthetic division is inappropriate here because of the high powers.

39. Again synthetic division is inappropriate. Use the remainder theorem.
$F(-1) = (-1 + 4)^2 + (-1 + 3)^3 + (-1 + 2)^4 = (3)^2 + (2)^3 + (1)^4 = 9 + 8 + 1 = 18$

MIXED REVIEW

40. $3(2)^2 \cdot 4(2)^3 = 3(4) \cdot 4(8) = 12 \cdot 32 = 384$

41. $(4(2)^2)^3 = (4(4))^3 = (16)^3 = 4096$

42. $3(2)^0 = 3(1) = 3$ **43.** $8(2)^{-1} = 8\left(\frac{1}{2}\right) = 4$

44. $x^2 - 2x + 3 = 0$

$x = \dfrac{-b \pm \sqrt{b^2 - 4ac}}{2a}$

$x = \dfrac{2 \pm \sqrt{(-2)^2 - 4(1)(3)}}{2(1)} =$

$\dfrac{2 \pm \sqrt{4 - 12}}{2} = \dfrac{2 \pm \sqrt{-8}}{2} = \dfrac{2 \pm 2i\sqrt{2}}{2} =$

$\dfrac{2(1 \pm i\sqrt{2})}{2} = 1 \pm i\sqrt{2}$; B

45. $2^{x+1} = 16$
$2^{x+1} = 2^4$
$x + 1 = 4$
$ x = 3$

46. $4^{3x} = 8^{x+1}$
$(2^2)^{3x} = (2^3)^{x+1}$
$2^{6x} = 2^{3x+3}$
$6x = 3x + 3$
$3x = 3$
$ x = 1$

47. $7^{x^2+4x} = 7^{-3}$
$x^2 + 4x = -3$
$x^2 + 4x + 3 = 0$
$(x + 3)(x + 1) = 0$
$x + 3 = 0$ or $x + 1 = 0$
$ x = -3$ or $x = -1$

48. $\log x^2 - \log y = 2\log x - \log y$

49. $\log \sqrt{mn} = \frac{1}{2}\log mn = \frac{1}{2}(\log m + \log n) = \frac{1}{2}(x + y)$

Lesson 10.3, pages 486–490

THINK BACK

1a. $y = x^3 - 4x^2 + 8x - 8$ or
$f(x) = x^3 - 4x^2 + 8x - 8$

1b. $x^3 - 4x^2 + 8x - 8 = 0$ or
$x^3 - 4x^2 + 8x - 8 = a$, where a is a real number

2a. $x^2 + x - 6 = 0$
$(x + 3)(x - 2) = 0$
$x + 3 = 0$ or $x - 2 = 0$
$ x = -3$ or $x = 2$
So, the solutions are -3 and 2.

2b. Since $F(-3) = 0$ and $F(2) = 0$ the zeros of F are -3 and 2.

EXPLORE/WORKING TOGETHER

3.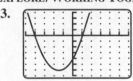
Xscl = 1   Yscl = 1
The zeros 1 and -4 are real and unequal.

4.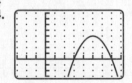
Xscl = 1   Yscl = 1
The zeros 3 and 7 are real and unequal.

5.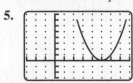
Xscl = 1   Yscl = 1
The zeros 5 and 5 are real and equal.

6.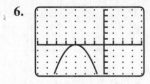
Xscl = 1   Yscl = 1
The zeros -3 and -3 are real and equal.

7.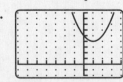
Xscl = 1   Yscl = 1
The zeros $1 \pm 2i$ are imaginary and unequal.

8.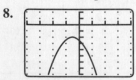
Xscl = 2   Yscl = 2
The zeros $-1 \pm 2i$ are imaginary and unequal.

9a. $a < 0$ **9b.** $a > 0$

10a. The y-value of the graph is a maximum.

10b. The y-value of the graph is a minimum.

11. two

12. The graph is tangent to the x-axis at the vertex or turning point of the hill or valley.

13. The graph never touches the x-axis.

14.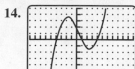

Xscl = 1 Yscl = 10
The zeros are -4, 1, and 4.

15.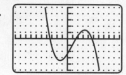

Xscl = 1 Yscl = 10
To the nearest tenth, the zeros are -3.3, 1.6, and 4.6.

16. The graph has both a hill and a valley.

17. When $a > 0$, the graph has a hill and then a valley. When $a < 0$, the graph has a valley and then a hill.

18.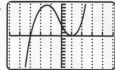

Xscl = 1 Yscl = 1
The point of tangency shows two equal zeros; a zero of multiplicity 2. The three real zeros are -3, 1, and 1.

19.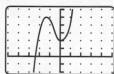

Xscl = 2 Yscl = 5
One; there is only one x-intercept.

20. All three zeros are equal to 5; 5 is a zero of multiplicity 3.

21.

Xscl = 1 Yscl = 1
The real zero of multiplicity 3 is not shown by a point of tangency. Rather, it is shown at the point where an "incomplete hill" becomes an "incomplete valley." That is, the curve rises to form the first portion of a hill and instead of falling to complete the hill, it continues to rise to form the second part of a valley. Although rising throughout, the nature of the rise changes at the zero of multiplicity 3.

22. 2 hills and a valley or 2 valleys and a hill

23a.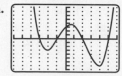

23b.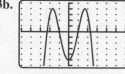

Xscl = 1 Yscl = 10   Xscl = 1   Yscl = 1

24a.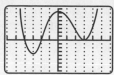

Xscl = 1   Yscl = 10

The zeros are 3, 3, -2, -4. The zero of multiplicity 2 will show as a point of tangency.

The graph has the characteristic quartic shape, 2 valleys and a hill. The minimum point of one valley, 3, is where the zero of multiplicity 2 occurs.

24b.

Xsc = 1   Yscl = 10

The zeros are 3, 3, 3, -2. The zero of multiplicity 3 will show as the point where an incomplete hill turns into an incomplete valley.

There is one distinct valley. The zero of multiplicity 3 is at the point where an incomplete hill turns into an incomplete valley.

24c.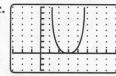

Xscl = 1   Yscl = 1

The zeros are 3, 3, 3, 3. The zero of multiplicity 4 will show as the point where two incomplete hills are trying to become two incomplete valleys.

Students will discover that the graph looks like the parabola $y = (x - 3)^2$. The difference between the two graphs is in the y-values of the points. For example, when $x = 5$, the point $(5, 2^2)$ is on the parabola while the point $(5, 2^4)$ is on the quartic function.

25. If the degree is n, the number of turns is at most $(n - 1)$.

26. 0 is the only zero of these functions, equal in multiplicity to the degree of the function.

27.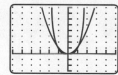

Xscl = 1   Yscl = 1
In both graphs, there's a single valley. The valley of the quartic is steeper than the valley of the quadratic.

$(0, 0)$, $(1, 1)$, $(1, 1)$. Both graphs are tangent to the x-axis at 0.

28.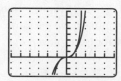

Xscl = 1   Yscl = 1
Both graphs have the same shape, continuing to rise from an incomplete hill to an incomplete valley. The graph of the fifth-degree curve is steeper than the graph of the cubic. $(0, 0)$, $(1, 1)$, $(-1, -1)$. Both graphs intersect the x-axis at the origin.

29. The graph of $y = x^6$ will have the same slope as the graph of $y = x^2$. The graph of $y = x^7$ will have the same shape as $y = x^3$.

Xscl = 1   Yscl = 1   Xscl = 1   Yscl = 1

30. If n is even, the graph is a single valley that is tangent to the x-axis at the origin.

If n is odd, the graph rises throughout, from an incomplete hill to an incomplete valley. The graph intersects the x-axis at the origin.

Both kinds of graphs increase in steepness as n increases.

31.    Xscl = 1   Yscl = 1
y_2 is y_1 translated 5 units to the right. y_3 is y_1 translated 5 units to the left.

32. $y = (x - a)^2$ is the result of translating $y = x^2$ in the horizontal direction by a distance of a units, to the right when $a > 0$ and to the left when $a < 0$.

33. 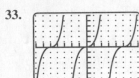 $y = (x - a)^3$ is the result of translating $y_1 = x^3$ in the horizontal direction by a distance of a units, to the right when $a > 0$ and to the left when $a < 0$.

34. 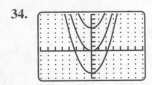   To translate $y = x^2$ up 5 units change it to $y = x^2 + 5$ and to translate it down 5 units change it to $y = x^2 - 5$.

MAKE CONNECTIONS

35. $x(x - 4)(x + 2)$

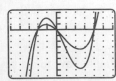

Xscl = 1   Yscl = 2
Both functions have the same zeros. Yes, an infinite number of functions have the same zeros; for example $a \cdot F(x)$ where a is any nonzero real number.

36. 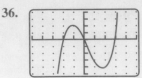 $f(x) = (x + 2)(x)(x - 3)$
$f(x) = x(x^2 - x - 6)$
$f(x) = x^3 - x^2 - 6x$
Any multiple of this polynomial would also be correct.

Xscl = 1   Yscl = 2

37. Values for x will vary but x will be consistently positive or negative in each interval.

$x < -2$

$$
\begin{array}{r|rrrr}
-3 & 1 & -1 & -6 & 0 \\
   &   & -3 & 12 & -18 \\
\hline
   & 1 & -4 & 6 & -18 \quad f(-3) = -18
\end{array}
$$

$-2 < x < 0$

$$
\begin{array}{r|rrrr}
-1 & 1 & -1 & -6 & 0 \\
   &   & -1 & 2 & 4 \\
\hline
   & 1 & -2 & -4 & 4 \quad f(-1) = 4
\end{array}
$$

$0 < x < 3$

$$
\begin{array}{r|rrrr}
1 & 1 & -1 & -6 & 0 \\
  &   & 1 & 0 & -6 \\
\hline
  & 1 & 0 & -6 & -6 \quad f(1) = -6
\end{array}
$$

$x > 3$

$$
\begin{array}{r|rrrr}
4 & 1 & -1 & -6 & 0 \\
  &   & 4 & 12 & 24 \\
\hline
  & 1 & 3 & 6 & 24 \quad f(4) = 24
\end{array}
$$

The first is negative; the second is positive; the third is negative; and the fourth is positive. The remainders change sign.

38. The graph changes position with respect to the x-axis before and after a zero, just as the remainders in synthetic division change sign.

39. Answers and descriptions for the functions will vary. Here are some possibilities.

39a. $y = -(x + 3)(x - 2)$; hill, points down at left, down right

39b. $y = -x(x - 4)(x + 1)$; valley, hill, points up left, down right

39c. $y = x(x + 1)(x - 1)(x + 2)$; valley, hill, valley, points up left and right

39d. $y = -x(x + 1)(x - 1)(x + 2)$; hill, valley, hill, points down left and right

39e. $y = x(x + 1)(x - 1)(x + 2)(x - 2)$; hill, valley, hill, valley, points down left, up right

39f. $y = -x(x + 1)(x - 1)(x + 2)(x - 2)$; valley, hill, valley, hill, points up left, down right

39g. $y = x(x + 1)(x - 1)(x + 2)(x - 2)(x + 3)$; valley, hill, valley, hill, valley, points up left and right

39h. $y = -x(x + 1)(x - 1)(x + 2)(x - 2)(x + 3)$; hill, valley, hill, valley, hill, points down left and right

SUMMARIZE

40. A polynomial is an expression in one variable (with exponents that are nonnegative integers). The related polynomial function has two variables. A related polynomial equation chooses a specific value for the second variable, such as 0. The zeros of the polynomial function $y = F(x)$ are the solutions of the polynomial equation $F(x) = 0$.

41. Changing the sign of the leading coefficient of a monomial function reflects the graph over the x-axis.

42. Answers will vary. Possible choices can differ, for example, in the multiplicity of zeros.
$y = (x + 2)(x - 3)(x - 5)$
$y = (x + 2)^2(x - 3)(x - 5)$
$y = (x + 2)(x - 3)^2(x - 5)$
Another possibility for difference can be in a constant multiplier.
$y = (x + 2)(x - 3)(x - 5)$
$y = 2(x + 2)(x - 3)(x - 5)$
$y = \frac{1}{2}(x + 2)(x - 3)(x - 5)$

43a.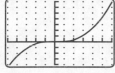
Xscl = 1   Yscl = 1

43b. The weight of a cubic foot of ice is approximately 57 lbs.

44a. $x \geq 0$

44b.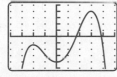
The company's profit is $0 at approximately 1.7 tons and 3.9 tons.
Xscl = 1   Yscl = 2

44c. Exporting 3 tons yields the company a profit of $7738.

45.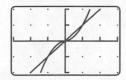
By subtracting y-distances of the two graphs, the graph of $y = x - x^3$ may be determined.

ALGEBRAWORKS

1. (0, 8.9), (10, 12.1), (20, 13.3), (30, 10.4), (40, 9.4), (50, 9.6), (60, 8.5), (70, 8.4)

2a. $y = -0.0015x^2 + 0.068x + 10.39$
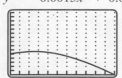
Find the y-value when $x = 39$.
Xscl = 10   Yscl = 2

2b. 10.76%

2c. No; in the table, the y-values are decreasing. Yet this y-value of 10.76 for the year 1999 is greater than the y-value in 1990.

3.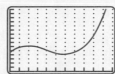
Xscl = 10   Yscl = 2

3a. when $x = 52$, $y = 8.79$; Yes; the value is consistent with the values in the table.

3b. When $x = 70$, $y = 9.17$; No, too high.

4. d; In general, the approximation improves as the degree of the polynomial increases.

Lesson 10.4, pages 491–499

EXPLORE/WORKING TOGETHER

1.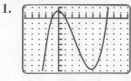
$-1, 1, 7$; The solution of the equation $x^3 - 7x^2 - x + 7 = 0$ are the same as the zeros of the function $y = x^3 - 7x^2 - x + 7$.
Xscl = 1   Yscl = 5

2a.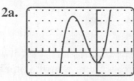
Zeros are $-5, -1, 1$; The solution of the equation $x^3 + 5x^2 - x - 5 = 0$ are the zeros of the function $y = x^3 + 5x^2 - x - 5$.
Xscl = 1   Yscl = 5

2b.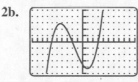
Zeros are $-6, -2, -3$; The solution of the equation $x^3 + 5x^2 - 12x - 36 = 0$ are the zeros of the function $y = x^3 + 5x^2 - 12x - 36$.
Xscl = 1   Yscl = 10

3.
7,	$\pm 7, \pm 1$	$-1, 1, 7$
-5,	$\pm 1, \pm 5$	$-1, 1, -5$
-36,	$\pm 1, \pm 2, \pm 3, \pm 4, \pm 6, \pm 9, \pm 12,$	$-2, 3, -6$
	$\pm 18, \pm 36$	

4. The solutions of a polynomial equation are factors of the constant term.

TRY THESE

1. Factors of 9 are $\pm 1, \pm 3, \pm 9$
Test factors using synthetic division.

```
1|     1   -1   -9    9
             1    0   -9
       _____
       1    0   -9    0
```

$x - 1$ is a factor

From the quotient determine the other factor is $x^2 - 9$.
Factor: $x^2 - 9 = (x - 3)(x + 3)$. Therefore, the
factors of $x^3 - x^2 - 9x + 9$ are $(x - 1)$, $(x - 3)$, and
$(x + 3)$. So $(x - 1)(x - 3)(x + 3) = 0$ and the
solutions are $x = 1$, $x = -3$, and $x = 3$.

2. $\frac{3}{5}$ cannot be zero of $3x^3 + 17x^2 - 5x + 5 = 0$; by the
rational zero theorem, the numerator must be a factor of
5 and the denominator a factor of 3.

3. Factors of 2 are ± 1, ± 2.
Test factors.

$$1\underline{|} \quad \begin{array}{rrrr} 3 & -2 & -7 & -2 \\ & 3 & 1 & -6 \\ \hline 3 & 1 & -6 & -8 \end{array} \quad 1 \text{ is not a zero}$$

$$-1\underline{|} \quad \begin{array}{rrrr} 3 & -2 & -7 & -7 \\ & -3 & 5 & 2 \\ \hline 3 & -5 & -2 & 0 \end{array}$$

-1 is a zero, and $x + 1$ is a factor. The other factor is
$3x^2 - 5x - 2$.

Factor: $3x^2 - 5x - 2 = (3x + 1)(x - 2)$.
$3x + 1 = 0$ yields the solution $x = -\frac{1}{3}$ and
$x - 2 = 0$ yields the solution $x = 2$ So, the rational
zeros of $3x^3 - 2x^2 - 7x - 2 = 0$ are $-\frac{1}{3}$, -1, 2.

4. Test factors of 4

$$1\underline{|} \quad \begin{array}{rrrr} 1 & -3 & -2 & 4 \\ & 1 & -2 & -4 \\ \hline 1 & -2 & -4 & 0 \end{array}$$

1 is one solution and $x - 1$ is one factor with the other
factor $x^2 - 2x - 4$.

Solve $x^2 - 2x - 4 = 0$ using quadratic formula
$$x = \frac{-b \pm \sqrt{b^2 - 4ac}}{2a}; x = \frac{2 \pm \sqrt{4 - 4(1)(-4)}}{2(1)}$$
$$= \frac{2 \pm \sqrt{4 + 16}}{2} = \frac{2 \pm \sqrt{20}}{2} =$$
$$= \frac{2 \pm 2\sqrt{5}}{2} = \frac{2(1 \pm \sqrt{5})}{2} = 1 \pm \sqrt{5}$$
Solutions are $x = 1, 1 \pm \sqrt{5}$
Check: $x = 1$
$(1)^3 - 3(1)^2 - 2(1) + 4 = 1 - 3 - 2 + 4 = 0$
Check the other two solutions by substituting and
evaluating with a calculator.

5. Test factors of 2.

$$1\underline{|} \quad \begin{array}{rrrr} 1 & -2 & 3 & -2 \\ & 1 & -1 & 2 \\ \hline 1 & -1 & 2 & 0 \end{array}$$

1 is one solution and $x - 1$ is a factor with
$(x^2 - x + 2)$ the other factor.

Solve $x^2 - x + 2 = 0$ using quadratic formula
$$x = \frac{-b \pm \sqrt{b^2 - 4ac}}{2a};$$
$$x = \frac{1 \pm \sqrt{1 - (4)(1)(2)}}{2(1)} = \frac{1 \pm \sqrt{1 - 8}}{2} = \frac{1 \pm i\sqrt{7}}{2}$$

Solutions are $x = 1, \dfrac{1 \pm i\sqrt{7}}{2}$
Check: $x = 1$
$(1)^3 - 2(1)^2 + 3(1) - 2 = 1 - 2 + 3 - 2 = 0$
Check: $x = \dfrac{1 + i\sqrt{7}}{2}$

$$\left(\frac{1 + i\sqrt{7}}{2}\right)^3 - 2\left(\frac{1 + i\sqrt{7}}{2}\right)^2 + 3\left(\frac{1 + i\sqrt{7}}{2}\right) - 2 =$$
$$\left(\frac{1 + i\sqrt{7}}{2}\right)\left(\frac{1 + i\sqrt{7}}{2}\right)\left(\frac{1 + i\sqrt{7}}{2}\right) -$$
$$2\left(\frac{1 + i\sqrt{7}}{2}\right)\left(\frac{1 + i\sqrt{7}}{2}\right) + 3\left(\frac{1 + i\sqrt{7}}{2}\right) - 2 =$$
$$\left(\frac{-6 + 2i\sqrt{7}}{4}\right)\left(\frac{1 + i\sqrt{7}}{2}\right) -$$
$$2\left(\frac{-6 + 2i\sqrt{7}}{4}\right) + 3\left(\frac{1 + i\sqrt{7}}{2}\right) - 2 =$$
$$\frac{-20 - 4i\sqrt{7}}{8} - \left(\frac{-6 + 2i\sqrt{7}}{2}\right) + \frac{3 + 3i\sqrt{7}}{2} - 2 =$$
$$-\frac{5}{2} - \frac{i\sqrt{7}}{2} + 3 - i\sqrt{7} + \frac{3}{2} + \frac{3i\sqrt{7}}{2} - 2 =$$
$$\left(-\frac{5}{2} + 3 + \frac{3}{2} - 2\right) + \left(\frac{-\sqrt{7}}{2} - \sqrt{7} + \frac{3\sqrt{7}}{2}\right)i =$$
$$0 + 0i = 0$$

Check: $x = \dfrac{1 - i\sqrt{7}}{2}$

$$\left(\frac{1 - i\sqrt{7}}{2}\right)^3 - 2\left(\frac{1 - i\sqrt{7}}{2}\right)^2 + 3\left(\frac{1 - i\sqrt{7}}{2}\right) - 2 =$$
$$\left(\frac{1 - i\sqrt{7}}{2}\right)\left(\frac{1 - i\sqrt{7}}{2}\right)\left(\frac{1 - i\sqrt{7}}{2}\right) -$$
$$2\left(\frac{1 - i\sqrt{7}}{2}\right)^2 + 3\left(\frac{1 - i\sqrt{7}}{2}\right) - 2 =$$
$$\left(\frac{-6 - 2i\sqrt{7}}{4}\right)\left(\frac{1 - i\sqrt{7}}{2}\right) -$$
$$2\left(\frac{-6 - 2i\sqrt{7}}{4}\right) + 3\left(\frac{1 - i\sqrt{7}}{2}\right) - 2 =$$
$$\left(\frac{-20 + 4i\sqrt{7}}{8}\right) + 6 + i\sqrt{7} + \frac{3}{2} - \frac{3i\sqrt{7}}{2} - 2 =$$
$$\frac{-5}{2} + \frac{i\sqrt{7}}{2} + 3 + i\sqrt{7} + \frac{3}{2} - \frac{3\sqrt{7}}{2}i - 2 =$$
$$\left(-\frac{5}{2} + 3 + \frac{3}{2} - 2\right) + \left(\frac{\sqrt{7}}{2} + \sqrt{7} - \frac{3\sqrt{7}}{2}\right)i =$$
$$0 + 0i = 0$$

6. Test factors of 8

x	1	2	-2	0	8
1	1	3	1	1	9
-1	1	1	-3	3	5
2	1	4	6	12	32
-2	1	0	-2	4	0

-2 is a solution. Test it again.

$$\begin{array}{r|rrrr}
-2] & 1 & 0 & -2 & 4 \\
& & -2 & 4 & -4 \\
\hline
& 1 & -2 & 2 & 0
\end{array}$$

-2 is a double solution. So, two of the factors are $(x + 2)$, $(x + 2)$. The other factor is $(x^2 - 2x + 2)$. Solve for the other solutions.

$x^2 - 2x + 2 = 0$

$x = \dfrac{-b \pm \sqrt{b^2 - 4ac}}{2a} = \dfrac{2 \pm \sqrt{4 - 4(1)(2)}}{2(1)} =$

$\dfrac{2 \pm \sqrt{4 - 8}}{2} = \dfrac{2 \pm \sqrt{-4}}{2} = \dfrac{2 \pm 2i}{2} =$

$\dfrac{2(1 \pm i)}{2} = 1 \pm i$

Therefore, the solutions are $x = -2, -2, 1 \pm i$

Check: $x = -2$

$(-2)^4 + 2(-2)^3 - 2(-2)^2 + 8 =$

$16 - 16 - 8 + 8 = 0$

Check: $1 + i$

$(1 + i)^2 = (1 + i)(1 + i) = 1 - 1 + 2i = 2i$

$(1 + i)^3 = (1 + i)2i = -2 + 2i$

$(1 + i)^4 = (1 + i)(-2 + 2i) =$

$(-2 - 2 + 2i - 2i) = -4$

$(1 + i)^4 + 2(1 + i)^3 - 2(1 + i)^2 + 8 =$

$-4 + 2(-2 + 2i) - 2(2i) + 8 =$

$-4 - 4 + 4i - 4i + 8 = 0$

Check: $1 - i$

$(1 - i)^2 = (1 - i)(1 - i) = 1 - 1 - 2i = -2i$

$(1 - i)^3 = (1 - i)(-2i) = -2 - 2i$

$(1 - i)^4 = (1 - i)(-2 - 2i) =$

$(-2 - 2 + 2i - 2i) = -4$

$(1 - i)^4 + 2(1 - i)^3 - 2(1 - i)^2 + 8 =$

$-4 + 2(-2 - 2i) - 2(-2i) + 8 =$

$-4 - 4 - 4i + 4i + 8 = 0$

7. Use synthetic division to evaluate function at various integers.

x	3	1	2	2	-6
2	3	7	16	34	62
1	3	4	6	8	2
0	3	1	2	2	-6
-1	3	-2	4	-2	-4
-2	3	-5	12	-22	38

By the location theorem, since $f(-1)$ and $f(-2)$ have opposite signs there must be at least one zero between -2 and -1 or between 0 and 1.

8. Apply Descartes' rule of signs.

$f(x) = x^3 - x^2 + 2x - 2$

Examine the signs of the coefficients of $f(x)$.

$\underset{1}{+} \underset{}{-} \underset{2}{+} \underset{}{-}$ 3 sign changes.

So, $f(x)$ can have 3 or 1 positive zeros.

$f(-x) = -x^3 - x^2 - 2x - 2$

Examine the signs of the coefficients of $f(-x)$.

$- \quad - \quad - \quad -$ No sign changes.

So, $f(x)$ has 0 negative zeros.

9. Apply Descartes' rule of signs.

$f(x) = x^5 - x^3 + 1$

Examine the signs of the coefficients of $f(x)$.

$\underset{1}{+} \underset{}{-} \underset{2}{-} +$ 2 sign changes.

So, $f(x)$ can have 2 or 0 positive coefficients.

$f(-x) = -x^5 + x^3 + 1$

Examine the signs of the coefficients of $f(-x)$.

$\underset{1}{-} \underset{}{+} \quad +$ 1 sign change.

So, $f(x)$ can have 1 negative zero.

10. Apply the upper and lower bound theorem. Use synthetic division to test positive integers.

x	1	0	0	-6	1
1	1	1	1	-5	-4
2	1	2	4	2	5	all positive

upper bound $- 2$

$f(-x) = x^4 + 6x + 1$ all positive

lower bound $= 0$

11. Let x represent the length of the original side of the cube. Therefore, $x - 1$ represents the length of one side of the cube after a slice 1cm thick is cut off. The formula for the volume of a cube is $V = s^3$ where s is the length of one side. Substitute and solve.

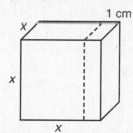
1 cm

$s^3 = V$

$x(x)(x - 1) = 48$

$x^3 - x^2 = 48$

$x^3 - x^2 - 48 = 0$

Test factors of 48.

x	1	-1	0	-48
1	1	0	0	-48
2	1	1	2	-44
3	1	2	6	-30
4	1	3	12	0

The length of the original side of the cube is 4 cm.

12.

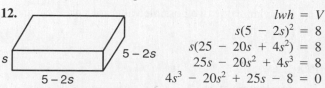

$lwh = V$

$s(5 - 2s)^2 = 8$

$s(25 - 20s + 4s^2) = 8$

$25s - 20s^2 + 4s^3 = 8$

$4s^3 - 20s^2 + 25s - 8 = 0$

Apply rational zero theorem to obtain possible zeros.

factors of 4: $\pm 1, \pm 2, \pm 4$

factors of 8: $\pm 1, \pm 2, \pm 4, \pm 8$

possible rational zeros:

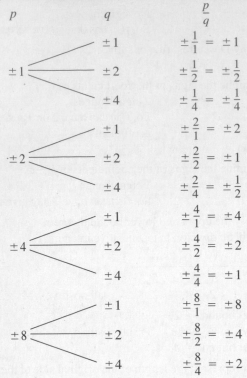

| p | | q | $\dfrac{p}{q}$ |

zeros to test: $\pm\dfrac{1}{4},\ \pm\dfrac{1}{2},\ \pm 1,\ \pm 2,\ \pm 4,\ \pm 8$

Since s is a distance the negative integers may be ruled out.

Also rule out 4 and 8 because 4 cm and 8 cm squares could not be cut from each corner. This leaves $\dfrac{1}{4}, \dfrac{1}{2}, 1,$ and 2 as possible solutions.

x	4	-20	25	-8
$\dfrac{1}{4}$	4	-19	$-\dfrac{81}{4}$	$-\dfrac{209}{16}$
$\dfrac{1}{2}$	4	-18	16	0

Solve: $4x^2 - 18x + 16 = 0$
$\qquad\ \ 2x^2 - 9x + 8 = 0$

$x = \dfrac{-b \pm \sqrt{b^2 - 4ac}}{2a} = \dfrac{9 \pm \sqrt{81 - 4(2)(8)}}{2(2)}$

$= \dfrac{9 \pm \sqrt{81 - 64}}{2} = \dfrac{9 \pm \sqrt{17}}{2}$ since $\dfrac{9 + \sqrt{17}}{2} \approx 6.6$

is too large, the two possible values of s are $\dfrac{1}{2}$ cm and

$\dfrac{9 - \sqrt{17}}{2} \approx 2.4$ cm

PRACTICE

1. Test factors of 5.

1⌋	1	-5	-1	5
		1	-4	-5
	1	-4	-5	0

$x - 1$ is a factor and so is $x^2 - 4x - 5$.

Solve $x^2 - 4x - 5 = 0$
$\quad (x + 1)(x - 5) = 0$
$\quad x + 1 = 0 \quad$ or $\quad x - 5 = 0$
$\qquad\quad x = -1 \quad$ or $\qquad x = 5$

The integer solutions are -1, 1, and 5.

2. Test factors of 16.

x	1	-3	-8	12	16
1	1	-2	-10	2	18
-1	1	-4	-4	16	0

$x + 1$ and $x^3 - 4x^2 - 4x + 16$ are factors.
Continue to test.

x	1	-4	-4	16
-1	1	-5	1	15
2	1	-2	-8	0

$x - 2$ and $x^2 - 2x - 8$ are factors.
Solve $x^2 - 2x - 8 = 0$
$\quad (x - 4)(x + 2) = 0$
$\quad\ x - 4 = 0 \quad$ or $\quad x + 2 = 0$
$\qquad\quad x = 4 \quad$ or $\qquad x = -2$

The integer solutions are -2, -1, 2, and 4.

3. Test factors of 8.

-1⌋	1	0	-11	-18	-8
		-1	1	10	8
	1	-1	-10	-8	0
-1⌋		-1	2	8	
	1	-2	-8	0	

Factors of $x^4 - 11x^2 - 18x$ are $(x + 1)$, $(x + 1)$, and $(x^2 - 2x - 8)$.
Solve $x^2 - 2x - 8 = 0$
$\quad (x - 4)(x + 2) = 0$
$\quad\ x - 4 = 0 \quad$ or $\quad x + 2 = 0$
$\qquad\quad x = 4 \quad$ or $\qquad x = -2$

The rational solutions are -2, -1, -1, and 4.

4. Factors of 5 are $\pm 1, \pm 5$
Factors of 3 are $\pm 1, \pm 3$
Possible rational zeros are

p		q	$\dfrac{p}{q}$
± 1		± 1	$\pm\dfrac{1}{1} = \pm 1$
		± 5	$\pm\dfrac{1}{5} = \pm\dfrac{1}{5}$
± 3		± 1	$\pm\dfrac{3}{1} = \pm 3$
		± 5	$\pm\dfrac{3}{5} = \pm\dfrac{3}{5}$

zeros to test: $\pm 1,\ \pm\dfrac{1}{5},\ \pm 3,\ \pm\dfrac{3}{5}$

Use synthetic division to test these.

-1⌋	5	-11	-13	3
		-5	16	-3
	5	-16	3	0

One factor is $x + 1$ and the other is $5x^2 - 16x + 3$.
Solve $5x^2 - 16x + 3 = 0$

$$(5x - 1)(x - 3) = 0$$
$$(5x - 1) = 0 \quad \text{or} \quad x - 3 = 0$$
$$5x = 1$$
$$x = \frac{1}{5} \quad \text{or} \quad x = 3$$

The rational solutions are $-1, \frac{1}{5}$, and 3.

5. The lowest possible degree is 5 because the conjugate zero theorem states that complex zeros and irrational zeros occur in pairs. So, $1 - \sqrt{2}$ and $2 + 3i$ are also zeros.

6. Test factors of 15.

$$
\begin{array}{r|rrrr}
-3 & 1 & 1 & -11 & -15 \\
   &   & -3 & 6 & 15 \\
\hline
   & 1 & -2 & -5 & 0
\end{array}
$$
-3 is a solution.

$x^2 - 2x - 5 = 0$

$x = \dfrac{-b \pm \sqrt{b^2 - 4ac}}{2a} = \dfrac{2 \pm \sqrt{4 - 4(1)(-5)}}{2(1)} =$

$\dfrac{2 \pm \sqrt{4 + 20}}{2} = \dfrac{2 \pm \sqrt{24}}{2} =$

$\dfrac{2 \pm 2\sqrt{6}}{2} = \dfrac{2(1 \pm \sqrt{6})}{2} = 1 \pm \sqrt{6}$

The solutions are $x = -3, 1 \pm \sqrt{6}$

Check: $x = -3$

$(-3)^3 + (-3)^2 - 11(-3) - 15 =$
$-27 + 9 + 33 - 15 = 42 - 42 = 0$

$x = 1 + \sqrt{6}$

$(1 + \sqrt{6})^3 + (1 + \sqrt{6})^2 - 11(1 + \sqrt{6}) - 15 =$
$19 + 9\sqrt{6} + 7 + 2\sqrt{6} - 11 - 11\sqrt{6} - 15 =$
$19 + 7 - 11 - 15 + 9\sqrt{6} + 2\sqrt{6} - 11\sqrt{6} = 0$

$x = 1 - \sqrt{6}$

$(1 - \sqrt{6})^3 + (1 - \sqrt{6})^2 - 11(1 - \sqrt{6}) - 15 =$
$19 - 9\sqrt{6} + 7 - 2\sqrt{6} - 11 + 11\sqrt{6} - 15 =$
$19 + 7 - 11 - 15 - 9\sqrt{6} - 2\sqrt{6} + 11\sqrt{6} = 0$

7. Test factors of 8.

$$
\begin{array}{r|rrrr}
-2 & 1 & 4 & 8 & 8 \\
   &   & -2 & -4 & -8 \\
\hline
   & 1 & 2 & 4 & 0
\end{array}
$$
-2 is a solution

$x^2 + 2x + 4 = 0$

$x = \dfrac{-b \pm \sqrt{b^2 - 4ac}}{2a} = \dfrac{-2 \pm \sqrt{4 - 4(1)(4)}}{2(1)} =$

$\dfrac{-2 \pm \sqrt{4 - 16}}{2} = \dfrac{-2 \pm \sqrt{-12}}{2} = \dfrac{-2 \pm 2i\sqrt{3}}{2} =$

$\dfrac{2(-1 \pm i\sqrt{3})}{2} = -1 \pm i\sqrt{3}$

The solutions are $x = -2, -1 \pm i\sqrt{3}$

Check: $x = -2$

$(-2)^3 + 4(-2)^2 + 8(-2) + 8 =$
$-8 + 4(4) - 16 + 8 = -8 + 16 - 16 + 8 = 0$

$x = -1 + i\sqrt{3}$

$(-1 + i\sqrt{3})^3 + 4(-1 + i\sqrt{3})^2 + 8(-1 + i\sqrt{3}) +$
$8 = 8 + 4(-2 - 2i\sqrt{3}) - 8 + 8i\sqrt{3} + 8 =$
$8 - 8 - 8i\sqrt{3} - 8 + 8i\sqrt{3} + 8 = 0$

$x = -1 - i\sqrt{3}$

$(-1 - i\sqrt{3})^3 + 4(-1 - i\sqrt{3})^2 + 8(-1 - i\sqrt{3}) +$
$8 = 8 + 4(-2 + 2i\sqrt{3}) - 8 - 8i\sqrt{3} + 8 =$
$8 - 8 + 8i\sqrt{3} - 8 - 8i\sqrt{3} + 8 = 0$

8. Test factors of 8.

$$
\begin{array}{r|rrrr}
-1 & 1 & -3 & 4 & 8 \\
   &   & -1 & 4 & -8 \\
\hline
   & 1 & -4 & 8 & 0
\end{array}
$$
-1 is a solution

$x^2 - 4x + 8 = 0$

$x = \dfrac{-b \pm \sqrt{b^2 - 4ac}}{2a} = \dfrac{4 \pm \sqrt{16 - 4(1)(8)}}{2(1)} =$

$\dfrac{4 \pm \sqrt{16 - 32}}{2} = \dfrac{4 \pm \sqrt{-16}}{2} = \dfrac{4 \pm 4i}{2} =$

$\dfrac{2(2 \pm 2i)}{2} = 2 \pm 2i$

The solutions are $x = -1, 2 \pm 2i$

Check: $x = -1$

$(-1)^3 - 3(-1)^2 + 4(-1) + 8 =$
$-1 - 3 - 4 + 8 = 0$

$x = 2 + 2i$

$(2 + 2i)^3 - 3(2 + 2i)^2 + 4(2 + 2i) + 8 =$
$-16 + 16i - 3(8i) + 4(2 + 2i) + 8 =$
$-16 + 16i - 24i + 8 + 8i + 8 = 0$

$x = 2 - 2i$

$(2 - 2i)^3 - 3(2 - 2i)^2 + 4(2 - 2i) + 8 =$
$-16 - 16i - 3(-8i) + 4(2 - 2i) + 8 =$
$-16 - 16i + 24i + 8 - 8i + 8 = 0$

9. Apply the location theorem.

x	4	4	-13	5	
-3	4	-8	11	-28	
-2	4	-4	-5	15	$\leftarrow$ sign change
-1	4	0	-13	18	
0	4	4	-13	5	
1	4	8	-5	0	
2	4	12	11	27	

A real zero exists between -3 and -2. Another zero exists at 1.

10. Apply the location theorem.

x	4	6	-23	-15	
-1	4	2	-25	10	
0	4	6	-23	-15	$\leftarrow$ sign change
1	4	10	-13	-28	
2	4	14	5	-5	
3	4	18	31	77	$\leftarrow$ sign change

A real zero exists between -1 and 0 and between 2 and 3

11. Apply Descartes' rule of signs.

$f(x) = x^3 - 4x^2 + x + 6$

$$+\underbrace{\quad}-\underbrace{\quad}+\quad+$$
$$\quad 1 \qquad 2$$

2 or 0 positive zeros

$f(-x) = -x^3 - 4x^2 - x + 6$

$$-\quad-\quad-\underbrace{\quad}+$$
$$\qquad\qquad 1$$

1 negative zero

12. Apply Descartes' rule of signs.

$f(x) = 2x^4 - 2x^3 + x^2 - 3$

$$+\underbrace{\quad}-\underbrace{\quad}+\underbrace{\quad}-$$
$$\quad 1 \qquad 2 \qquad 3$$

3 or 1 positive zeros

$f(-x) = 2x^4 + 2x^3 + x^2 - 3$

$$+\quad+\quad+\underbrace{\quad}-$$
$$\qquad\qquad 1$$

1 negative zero

13. $f(x) = x^3 - 2x^2 + 3x + 5$

x	1	-2	3	5
1	1	-1	2	7
2	1	0	3	11

upper bound is 2

$f(-x) = -x^3 - 2x^2 - 3x + 5$

x	-1	-2	-3	5
1	-1	-3	-6	-1

lower bound is -1

14. $f(x) = 2x^3 - 3x^2 - 3x + 3$

x	2	-3	-3	3
1	2	-1	-4	-1
2	2	1	-1	1
3	2	3	6	21

upper bound is 3

$f(-x) = -2x^3 - 3x^2 + 3x + 3$

x	-2	-3	3	3
1	-2	-5	-2	1
2	-2	-7	-11	-19

lower bound is -2

15. Use the volume formula, $V = lwh$ to find the volume of the box.

$V = (3)(4)(5) = 50$ in.^{3}

So, $2V = 120$ in.^{3}

If the same amount x is added to each dimension of the box, the new dimensions are $x + 3$, $x + 4$, and $x + 5$. Substitute into the volume formula.

$$(x + 3)(x + 4)(x + 5) = 120$$
$$(x^2 + 7x + 12)(x + 5) = 120$$
$$x^3 + 7x^2 + 12x + 5x^2 + 35x + 60 = 120$$
$$x^3 + 12x^2 + 47x - 60 = 0$$

Use synthetic division to show that $x = 1$ is a solution of this equation.

1⌋	1	12	47	-60
		1	13	60
	1	13	60	0

So, if $x = 1$ in. is added to each dimension the new dimensions are 4 in. $\times$ 5 in. $\times$ 6 in. and the volume is doubled. $V = (4)(5)(6) = 120$ in.^{3}

16. domain for x is $0 \le x \le 10$ because the beam is 10 m long

Find the zeros.

$$-x^4 + 26x^3 - 160x^2 = 0$$
$$-x^2(x^2 - 26x + 160) = 0$$
$$-x^2(x - 16)(x - 10) = 0$$
$$-x^2 = 0 \quad \text{or} \quad x - 16 = 0 \quad \text{or} \quad x - 10 = 0$$
$$\underbrace{x = 0}_{\substack{\text{double}\\\text{zero}}} \qquad\qquad x = 16 \qquad\qquad x = 10$$

The zeros are 0, 0, 10, and 16. There is no sag at 0 and 10, where the beam is supported. The zero at 16 is meaningless since it is out of the domain.

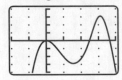

Window will vary. The graph is tangent to the x-axis at 0, indicating the double zero at 0. The graph intersects the x-axis at the other two real zeros, 10 and 16.

Xscl = 10   Yscl = 500

The maximum deflection of the beam is 14.42 mm and it occurs at $x = 5.9$ m from the wall.

EXTEND

17.
$$lwh = V$$
$$(x - 2)(x + 3)(x - 4) = 50$$
$$(x^2 + x - 6)(x - 4) = 50$$
$$x^3 + x^2 - 6x - 4x^2 - 4x + 24 = 50$$
$$x^3 - 3x^2 - 10x + 24 = 50$$
$$x^3 - 3x^2 - 10x - 26 = 0$$

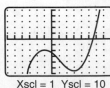

Xscl = 1   Yscl = 10

Graph the function $y = x^3 - 3x^2 - 10x - 26$ on a graphing utility and approximate the value of x where the graph crosses the x-axis. x is approximately 5.61 cm.

18. The volume of a cylinder is $V = \pi r^2 h$. Substitute into this formula.

$$\pi x^2(x + 9) = 100$$

$$x^2(x + 9) = \frac{100}{\pi}$$

$$x^3 + 9x^2 = \frac{100}{\pi}$$

$$x^3 + 9x^2 - \frac{100}{\pi} = 0$$

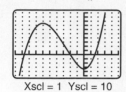

Xscl = 1 Yscl = 10

Graph the function
$$y = x^3 - 9x^2 - \frac{100}{\pi}$$

on a graphing utility and approximate the value of x where the graph crosses the x-axis. x is approximately 1.72 cm. The negative zeros are outside the domain.

THINK CRITICALLY

19. Since r_1 and r_2 are solutions of a quadratic equation then by the factor theorem $x - r_1$ and $x - r_2$ are factors. Therefore, the equation is $(x - r_1)(x - r_2) = 0$.

20.
$$(x - r_1)(x - r_2) = 0$$
$$x^2 - r_1x - r_2x + r_1r_2 = 0$$
$$x^2 - (r_1 + r_2)x + r_1r_2 = 0$$
When the leading coefficient is 1:
The coefficient of the x-term is the negative of the sum of the solutions.
The constant term is the product of the solutions.

21a. $\quad x^2 - (r_1 + r_2)x + r_1r_2 = 0 \quad r_1 = -2, r_2 = 3$
$$x^2 - (-2 + 3)x + (-2)(3) = 0$$
$$x^2 - x - 6 = 0$$

21b.
$$x^2 - (r_1 + r_2)x + r_1r_2 = 0$$
$$r_1 = 1 + \sqrt{5}, r_2 = 1 - \sqrt{5}$$
$$x^2 - (1 + \sqrt{5} + 1 - \sqrt{5})x + (1 + \sqrt{5})(1 - \sqrt{5}) = 0$$
$$x^2 - 2x + 1 - 5 = 0$$
$$x^2 - 2x - 4 = 0$$

21c.
$$x^2 - (r_1 + r_2)x + r_1r_2 = 0$$
$$r_1 = 1 + 2i, r_2 = 1 - 2i$$
$$x^2 - (1 + 2i + 1 - 2i)x + (1 + 2i)(1 - 2i) = 0$$
$$x^2 - 2x + 1 + 4 = 0$$
$$x^2 - 2x + 5 = 0$$

22. Since r_1, r_2, and r_3 are solutions of a cubic equation then by the factor theorem $x - r_1$, $x - r_2$, and $x - r_3$ are factors. Therefore the equation is $(x - r_1)(x - r_2)(x - r_3) = 0$.

23.
$$(x - r_1)(x - r_2)(x - r_3) = 0$$
$$(x^2 - r_1x - r_2x + r_1r_2)(x - r_3) = 0$$
$$(x^2 - (r_1 + r_2)x + r_1r_2)(x - r_3) = 0$$
$$(x^2 - (r_1 + r_2)x + r_1r_2)x -$$
$$(x^2 - (r_1 + r_2)x + r_1r_2)r_3 = 0$$
$$x^3 - (r_1 + r_2)x^2 + r_1r_2x - r_3x^2 +$$
$$(r_1 + r_2)r_3x - r_1r_2r_3 = 0$$
$$x^3 - (r_1 + r_2)x^2 - r_3x^2 + r_1r_2x +$$
$$r_1r_3x + r_2r_3x - r_1r_2r_3 = 0$$
$$x^3 - (r_1 + r_2 + r_3)x^2 + (r_1r_2 + r_1r_3 + r_2r_3)x -$$
$$r_1r_2r_3 = 0$$

When the leading coefficient is 1:
The coefficient of the x^2-term is the negative of the sum of the solutions.
The coefficient of the x-term is the sum of the products of the solutions taken two at a time.
The constant term is the negative of the product of the solutions.

24. $x^3 - (r_1 + r_2 + r_3)x^2 + (r_1r_2 + r_1r_3 + r_2r_3)x - r_1r_2r_3 = 0$

24a. $r_1 = -2, r_2 = 4, r_3 = 7$

$$x^3 - (-2 + 4 + 7)x^2 + ((-2)(4) + (-2)(7) + (4)(7))x - (-2)(4)(7) = 0$$
$$x^3 - 9x^2 + (-8 + (-14) + 28)x + 56 = 0$$
$$x^3 - 9x^2 + 6x + 56 = 0$$

24b. $r_1 = 3, r_2 = 1 + \sqrt{3}, r_3 = 1 - \sqrt{3}$
$$x^3 - (3 + 1 + \sqrt{3} + 1 - \sqrt{3})x^2 +$$
$$(3(1 + \sqrt{3}) + 3(1 - \sqrt{3}) + (1 + \sqrt{3})(1 - \sqrt{3}))x -$$
$$3(1 + \sqrt{3})(1 - \sqrt{3}) = 0$$
$$x^3 - 5x^2 + (3 + 3\sqrt{3} + 3 - 3\sqrt{3} + 1 - 3)x -$$
$$x - 3(1 - 3) = 0$$
$$x^3 - 5x^2 + 4x - 3(-2) = 0$$
$$x^3 - 5x^2 + 4x + 6 = 0$$

24c. $r_1 = 2, r_2 = i, r_3 = -i$
$$x^3 - (2 + i + (-i))x^2 + (2(i) + 2(-i) + i(-i))x - 2(i)(-i) = 0$$
$$x^3 - 2x^2 + x - 2 = 0$$

MIXED REVIEW

25. $(f \circ g)(x) = f(x + 1) = (x + 1)^2$

26. $(f \circ g)(3) = f(3 + 1) = f(4) = 4^2 = 16$

27. $(g \circ f)(x) = g(x^2) = x^2 + 1$

28. $(g \circ f)(3) = g(9) = 9 + 1 = 10$

29. c; because the discriminant, $b^2 - 4ac = 9 - 4(1)(1) = 5$, is positive and not a square number

1a. $d = 0.5at^2$

$0.5(5)t^2 = 40$

$2.5t^2 = 40$

$t^2 = 16$

$t = \pm 4$

$t = 4$ s disregard the negative value of time

1b. $0.5(5)t^2 = 45$

$2.5t^2 = 45$

$t^2 = 18$

$t = \pm \sqrt{18}$

$t = 4.24$ s again disregard negative time

1c. $0.5(5)t^2 = 50$

$2.5t^2 = 50$

$t^2 = 20$

$t = \pm\sqrt{20}$

$t = 4.47$ s

1d. $d = 0.5at^2 = 0.5(5)(6)^2 = 90$ m

1e. $d = 0.5at^2 = 0.5(5)(7)^2 = 122.5\ m$

2. $V_i = 0$ m/s

3. $V_f = V_i + at$ where $V_i = 0$

3a. $V_f = 5(4) = 20$ m/s

3b. $V_f = 5(4.24) = 21.2$ m/s

3c. $V_f = 5(4.47) = 22.35$ m/s

3d. $V_f = 5(6) = 30$ m/s

3e. $V_f = 5(7) = 35$ m/s

4. Answers will vary; to judge the impact on the rider when the car stops; to determine how the next segment of the ride should be designed

Lesson 10.5, pages 500–503

EXPLORE THE PROBLEM

1. Yes, they differ by 1.

2.
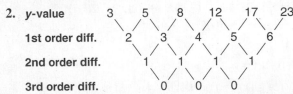

3. yes; since the third differences are zero, a polynomial function of degree 2 may be used. $y = ax^2 + bx + c$

4.
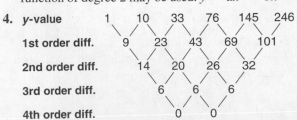

The 4th order differences are all 0, so the descriptive polynomial is a third-degree (cubic) function.

INVESTIGATE FURTHER

5. $2a = 1$, so $a = \frac{1}{2}$

6. $3a + b = 2$

$3\left(\frac{1}{2}\right) + b = 2$

$b = \frac{1}{2}$

7. $a + b + c = 3$

$\frac{1}{2} + \frac{1}{2} + c = 3$

$c = 2$

8. $y = \frac{1}{2}x^2 - \frac{1}{2}x + 2$

Check: $x = 3$

$y = \frac{1}{2}(3)^2 - \frac{1}{2}(3) + 2 = \frac{1}{2}(9) + \frac{3}{2} + 2 =$

$\frac{12}{2} + 2 = 6 + 2 = 8$

APPLY THE STRATEGY

9. Yes; the x-values increase in constant increments of 3.

10.

y-value	41	74	107	140
1st order diff.		33	33	33
2nd order diff.			0	0

11. By the given rule, since the 2nd order differences are 0, a polynomial of degree 1 can be used as a model. The general form is $y = ax + b$.

12. This is a linear function. Using two pairs of data points for (x, y), you can determine a and b by solving the system.

13. $x = 6$; $y = 74$

$y = ax + b$

$74 = a(6) + b \longrightarrow 6a + b = 74$

$x = 3, y = 41$

$y = ax + b$

$41 = a(3) + b \longrightarrow 3a + b = 41 \longrightarrow -3a - b = -41$

$\begin{cases} 6a + b = 74 \\ -3a - b = -41 \end{cases}$

$\overline{3a \qquad = 33}$

$a = 11$

$3(11) + b = 41$

$33 + b = 41$

$b = 8$

So, $y = 11x + 8$

14. $y = 11x + 8 = 200$

$11x = 192$

$x = \frac{192}{11}$

$x = 17\frac{5}{11}$

So, it will occur during week 17.

15. Students' answers should include

   a. y is a constant function (degree 0) of the form $y = c$.

   b. y is a linear function (degree 1) of the form
$y = ax + b$.

   c. y is a quadratic function (degree 2) of the form
$y = ax^2 + bx + c$.

   d. y is a cubic function (degree 3) of the form
$y = ax^3 + bx^2 + cx + d$.

16.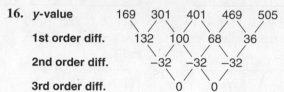

y-value	169		301		401		469		505
1st order diff.		132		100		68		36	
2nd order diff.			−32		−32		−32		
3rd order diff.				0		0			

17. $2a = -32$

   $a = -16$

      $3a + b = 132$

  $3(-16) + b = 132$

    $-48 + b = 132$

         $b = 180$

     $a + b + c = 169$

 $-16 + 180 + c = 169$

      $164 + c = 169$

          $c = 5$

 $y = -16x^2 + 180x + 5$

18a. $x = 7$

  $y = -16(7)^2 + 180(7) + 5 = 481$ ft

18b. $x = 10$

  $y = -16(10)^2 + 180(10) + 5 = 205$ ft

19. Set $y = 0$ and solve for x.

  $-16x^2 + 180x + 5 = 0$

  $x = \dfrac{-b \pm \sqrt{b^2 - 4ac}}{2a} =$

  $\dfrac{-180 \pm \sqrt{(180)^2 - 4(-16)(5)}}{2(-16)}$

  $\dfrac{-180 \pm \sqrt{32,720}}{-32} = 11.278$

It hits the ground between 11 and 12 seconds. (11.28 s)

20.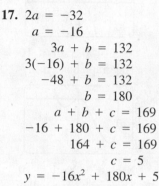

y-value	8		57		100		137		168
1st order diff.		49		43		37		31	
2nd order diff.			−6		−6		−6		
3rd order diff.				0		0			

  $2a = -6$

   $a = -3$

     $3a + b = 49$

  $3(-3) + b = 49$

    $-9 + b = 49$

        $b = 58$

      $a + b + c = 8$

    $-3 + 58 + c = 8$

       $55 + c = 8$

           $c = -47$

 $y = -3x^2 + 58x - 47$

21.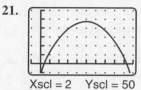

Xscl = 2   Yscl = 50

22. Plant should manufacture $9{,}667 \approx 10{,}000$ radios to maximize profit.

23. Maximum profit is $23,333.

REVIEW PROBLEM SOLVING STRATEGIES

2a. Let $x =$ number of years corresponding to 1 hour. Use a proportion.

  $\dfrac{x}{1} = \dfrac{2 \times 10^{10}}{24}$

  $x \approx 8.3 \times 10^8$ years

2b. One minute is $\dfrac{1}{60}$ of an hour. So, the number of years corresponding to one minute is approximately

  $\dfrac{1}{60}(8.3 \times 10^8) \approx 1.4 \times 10^7$ years.

2c. Let $y =$ number of hours that 4.6 billion years corresponds to

  $\dfrac{y}{4.6 \times 10^9} = \dfrac{24}{2.0 \times 10^{10}}$

  $y = \dfrac{24(4.6 \times 10^9)}{2.0 \times 10^{10}}$

  $y = 5.52$

Since $0.52h \times 60 = 31$ min, 5.52 h ≈ 5 h 31 min
So, 4.6 billion years ago corresponds to 18:29 o'clock.

2d. Let $w =$ number of hours that 65 million years correspond to.

  $\dfrac{w}{6.5 \times 10^7} = \dfrac{24}{2.0 \times 10^{10}}$

  $w = \dfrac{24(6.5 \times 10^7)}{2.0 \times 10^{10}}$

  $w = 0.078$

Since 0.078 h $\times 60 = 4.68$ min and 0.68
min $\times 60 = 41$ s, 0.078 h ≈ 4.68 min ≈ 5 min
So, 65 million years ago corresponds to approximately
24h − (5 min) or 23:55 o'clock.

2e. Let $z =$ number of hours that 4.4 million years correspond to.

  $\dfrac{z}{4.4 \times 10^6} = \dfrac{24}{2.0 \times 10^{10}}$

  $z = \dfrac{24(4.4 \times 10^6)}{2.0 \times 10^{10}}$

  $z = 0.00528$

Since $0.00528 \times 60 = 0.3168$ h and
0.3168 h $\times 60 = 195$, 0.00528 hours ≈ 0.3168
minutes ≈ 19 seconds
So, 4.4 million years ago corresponds to
approximately 24h $-$ 19s or 23:59:41 o'clock.

3. Sons A and B paddle across; son A stays on the other
side and son B comes back. Son B gets out and the
father paddles to the other side where he stays. Son A
comes back, picks up son B, and they both paddle to the
other side.

Chapter Review, pages 504–505

1. d　　2. a　　3. e　　4. c　　5. b

6.
$$3x + 2 \overline{)6x^3 + x^2 + 7x + 10}$$
quotient: $2x^2 - x + 3$

$$
\begin{array}{r}
6x^3 + 4x^2 \\ \hline
-3x^2 + 7x \\
-3x^2 - 2x \\ \hline
9x + 10 \\
9x + 6 \\ \hline
4
\end{array}
$$

So, $(6x^3 + x^2 + 7x + 10) \div (3x + 2) =$
$2x^2 - x + 3 + \dfrac{4}{3x + 2}$

Check: $(2x^2 - x + 3)(3x + 2) + 4 =$
$6x^3 + 4x^2 - 3x^2 - 2x + 9x + 6 + 4 =$
$6x^3 + x^2 + 7x + 10$

7.
$$x + 6 \overline{)x^2 + 0x + 36}$$
quotient: $x - 6$

$$
\begin{array}{r}
x^2 + 6x \\ \hline
-6x + 36 \\
-6x - 36 \\ \hline
72
\end{array}
$$

So, $(x^2 + 36) \div (x + 6) = x - 6 + \dfrac{72}{x + 6}$
Check: $(x - 6)(x + 6) + 72 =$
$x^2 - 6x + 6x - 36 + 72 = x^2 + 36$

8.
$$
\begin{array}{r|rrrr}
4 & 3 & -5 & -7 & -20 \\
& & 12 & 28 & 84 \\ \hline
& 3 & 7 & 21 & 64
\end{array}
$$

so, $(3x^3 - 5x^2 - 7x - 20) \div (x - 4) =$
$3x^2 + 7x + 21 + \dfrac{64}{x - 4}$

9.
$$
\begin{array}{r|rrrr}
-2 & 1 & 0 & 0 & -8 \\
& & -2 & 4 & -8 \\ \hline
& 1 & -2 & 4 & -16
\end{array}
$$

so, $(x^3 - 8) \div (x + 2) = x^2 - 2x + 4 - \dfrac{16}{x + 2}$

10.
$$
\begin{array}{r|rrrr}
-2 & 4 & 12 & 10 & -15 \\
& & -8 & -8 & -4 \\ \hline
& 4 & 4 & 2 & -19
\end{array}
$$

no, $(x + 2)$ is not a factor

11.
$$
\begin{array}{r|rrrr}
3 & 6 & -17 & -2 & -3 \\
& & 18 & 3 & 3 \\ \hline
& 6 & 1 & 1 & 0
\end{array}
$$
yes, $(x - 3)$ is a factor

12.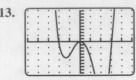
Xscl = 1 Yscl = 10
The zeros are $-1, -1,$
and 2.

13.
Xscl = 1 Yscl = 10
The zeros are $-2, 0, 0,$
and 3.

14.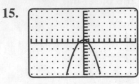
Xscl = 1 Yscl = 10
The graph has a valley.

15.
Xscl = 1 Yscl = 10
The graph has a hill.

16.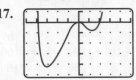
Xscl = 1 Yscl = 10
The graph has a hill,
then a valley.

17.
Xscl = 1 Yscl = 10
The graph has a valley,
then a hill, then a valley.

18. Test factors of 15.
$$
\begin{array}{r|rrrr}
-1 & 1 & -1 & -17 & -15 \\
& & -1 & 2 & 15 \\ \hline
& 1 & -2 & -15 & 0
\end{array}
$$

One factor is $x + 1$ and the other is $x^2 - 2x - 15$.
Solve $x^2 - 2x - 15 = 0$
$(x - 5)(x + 3) = 0$
$x - 5 = 0$　or　$x + 3 = 0$
$x = 5$　　　　$x = -3$
The rational zeros are $-1, -3,$ and 5.

19. Factors of 2 are $\pm 1, \pm 2$.
Possible rational zeros are:

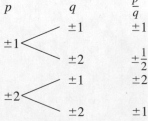

$$
\begin{array}{ccc}
p & q & \dfrac{p}{q} \\
& \pm 1 & \pm 1 \\
\pm 1 & & \\
& \pm 2 & \pm \dfrac{1}{2} \\
& \pm 1 & \pm 2 \\
\pm 2 & & \\
& \pm 2 & \pm 1
\end{array}
$$

zeros to test: $\pm 1, \pm \dfrac{1}{2}, \pm 2$

Use synthetic division to test these.

$$\begin{array}{r|rrrr} 1] & 2 & 1 & -5 & 2 \\ & & 2 & 3 & -2 \\ \hline & 2 & 3 & -2 & 0 \end{array}$$

One factor is $x - 1$ and the other is $2x^2 + 3x - 2$.

Solve $2x^2 + 3x - 2 = 0$

$(2x - 1)(x + 2) = 0$

$2x - 1 = 0$ or $x + 2 = 0$

$2x = 1$ $\qquad\qquad x = -2$

$x = \dfrac{1}{2}$

The rational zeros are $1, \dfrac{1}{2}, -2$.

20. Test factors of 2.

$$\begin{array}{r|rrrr} -1] & 1 & -1 & 0 & 2 \\ & & -1 & 2 & -2 \\ \hline & 1 & -2 & 2 & 0 \end{array}$$

One factor is $x + 1$ and the other is $x^2 - 2x + 2$.

Solve $x^2 - 2x + 2 = 0$

$x = \dfrac{-b \pm \sqrt{b^2 - 4ac}}{2a} =$

$\dfrac{-(-2) \pm \sqrt{(-2)^2 - 4(1)(2)}}{2(1)} =$

$\dfrac{2 \pm \sqrt{-4}}{2} = \dfrac{2 \pm 2i}{2} = 1 \pm i$

The solutions are $-1, 1 \pm i$.

21. Test factors of 8.

$$\begin{array}{r|rrrr} -2] & 1 & 0 & -8 & -8 \\ & & -2 & 4 & 8 \\ \hline & 1 & -2 & -4 & 0 \end{array}$$

One factor is $x + 2$ and the other is $x^2 - 2x - 4$.

Solve $x^2 - 2x - 4 = 0$

$x = \dfrac{-b \pm \sqrt{b^2 - 4ac}}{2a} = \dfrac{-(-2) \pm \sqrt{(-2)^2 - 4(1)(-4)}}{2(1)} =$

$\dfrac{2 \pm \sqrt{20}}{2} = \dfrac{2 \pm 2\sqrt{5}}{2} = 1 \pm \sqrt{5}$

The solutions are $-2, 1 \pm \sqrt{5}$.

22. Apply Descartes' rule of signs.

$F(x) = -3x^5 + 2x^4 - 3x^2 - 3x + 1$

2 or 0 positive zeros

$F(-x) = -3x^5 + 2x^4 - 3x^2 + 3x + 1$

3 or 1 negative zeros.

23. Apply Descartes' rule of signs.

$F(x) = 2x^3 - 4x^2 + x - 7$

3 or 1 positive zeros.

$F(-x) = -2x^3 - 4x^2 - x - 7$

0 negative zeros.

24. $F(x) = 3x^3 - 5x^2 + 7x + 4$

$$\begin{array}{r|rrrr} x & 3 & -5 & 7 & 4 \\ \hline 1 & 3 & -2 & 5 & 9 \\ 2 & 3 & 1 & 9 & 22 \end{array}$$

Upper Bound is 2.

$F(-x) = -3x^3 - 5x^2 - 7x + 4$

$$\begin{array}{r|rrrr} x & -3 & -5 & -7 & 4 \\ \hline 1 & -3 & -8 & -15 & -11 \end{array}$$

Lower Bound is -1.

25. $F(x) = x^3 - 4x^2 - 5x + 14$

$$\begin{array}{r|rrrr} x & 1 & -4 & -5 & 14 \\ \hline 1 & 1 & -3 & -8 & 6 \\ 2 & 1 & -2 & -9 & -4 \\ 3 & 1 & -1 & -8 & -10 \\ 4 & 1 & 0 & -5 & -6 \\ 5 & 1 & 1 & 0 & 14 \end{array}$$

Upper Bound is 5.

$F(-x) = -x^3 - 4x^2 + 5x + 14$

$$\begin{array}{r|rrrr} x & -1 & -4 & 5 & 14 \\ \hline 1 & -1 & -5 & 0 & 14 \\ 2 & -1 & -6 & -7 & 0 \\ 3 & -1 & -7 & -16 & -34 \end{array}$$

Lower Bound is -3.

26. Yes, the x-values increase in constant increments of 1.

y-value	2	9	22	41	66
1st order diff.		7	13	19	25
2nd order diff.			6	6	6
3rd order diff.				0	0

Since the third differences are zero, a polynomial function of degree 2 may be used.

27. No, the x-values do not increase in constant increments.

28.

y-value	2	5	10	17	26
1st order diff.		3	5	7	9
2nd order diff.			2	2	2
3rd order diff.				0	0

$2a = 2$

$a = 1$

$3a + b = 3$

$3(1) + b = 3$

$3 + b = 3$

$b = 0$

$a + b + c = 2$

$1 + 0 + c = 2$

$c = 1$

$y = x^2 + 1$

29.

| y-value | 0 | −1 | −4 | −9 | −16 |

1st order diff. −1 −3 −5 −7

2nd order diff. −2 −2 −2

3rd order diff. 0 0

$$2a = -2$$
$$a = -1$$
$$3a + b = -1$$
$$3(-1) + b = -1$$
$$-3 + b = -1$$
$$b = 2$$
$$a + b + c = 0$$
$$-1 + 2 + c = 0$$
$$1 + c = 0$$
$$c = -1$$
$$y = -x^2 + 2x - 1$$

Chapter Assessment, pages 506–507

CHAPTER TEST

1.

$$\require{enclose}\begin{array}{r} 2x^2 - 5x + 2 \\ 5x + 1 \enclose{longdiv}{10x^3 - 23x^2 + 5x + 5} \\ \underline{10x^3 + 2x^2} \\ -25x^2 + 5x \\ \underline{-25x^2 - 5x} \\ 10x + 5 \\ \underline{10x + 2} \\ 3 \end{array}$$

so, $(10x^3 - 23x^2 + 5x + 5) \div (5x + 1) =$
$$2x^2 - 5x + 2 + \frac{3}{5x + 1}$$

2.

$$\begin{array}{r} x^2 - xy + y^2 \\ x + y \enclose{longdiv}{x^3 + 0x^2y + 0xy^2 + y^3} \\ \underline{x^3 + x^2y} \\ -x^2y + 0xy^2 \\ \underline{-x^2y - xy^2} \\ xy^2 + y^3 \\ \underline{xy^2 + y^3} \\ 0 \end{array}$$

so, $(x^3 + y^3) \div (x + y) = x^2 - xy + y^2$

3.

$$\begin{array}{r|rrrr} 2 & 1 & -4 & 9 & -12 \\ & & 2 & -4 & 10 \\ \hline & 1 & -2 & 5 & -2 \end{array}$$

so, $(x^3 - 4x^2 + 9x - 12) \div (x - 2) =$
$$x^2 - 2x + 5 - \frac{2}{x - 2}$$

4. $\dfrac{4x^3 + 4x^2 - x - 7}{2x + 1} = \dfrac{1}{2}\left(\dfrac{4x^3 + 4x^2 - x - 7}{x + \frac{1}{2}}\right)$

$$\begin{array}{r|rrrr} -\frac{1}{2} & 4 & 4 & -1 & -7 \\ & & -2 & -1 & 1 \\ \hline & 4 & 2 & -2 & -6 \end{array}$$

so, $(4x^3 + 4x^2 - x - 7) \div (2x + 1) =$

$\dfrac{1}{2}\left(4x^2 + 2x - 2 - \dfrac{6}{x + \frac{1}{2}}\right) = 2x^2 + x - 1 - \dfrac{6}{2x + 1}$

5.

$$\begin{array}{r|rrrrr} -3 & 1 & 0 & 0 & 0 & 81 \\ & & -3 & 9 & -27 & 81 \\ \hline & 1 & -3 & 9 & -27 & 162 \end{array}$$

no, $(x + 3)$ is not a factor

6.

$$\begin{array}{r|rrrr} \frac{2}{3} & 9 & -27 & -4 & 12 \\ & & 6 & -14 & -12 \\ \hline & 9 & -21 & -18 & 0 \end{array}$$

yes, $(3m - 2)$ is a factor

7.

$$\begin{array}{r|rrrrr} -2 & -1 & 2 & 0 & -7 & 4 \\ & & 2 & -8 & 16 & -18 \\ \hline & -1 & 4 & -8 & 9 & -14 \end{array}$$

so, $F(-2) = -14$

8.

$$\begin{array}{r|rrrr} 3 & -\frac{1}{3} & -3 & -6 & 18 \\ & & -1 & -12 & -54 \\ \hline & -\frac{1}{3} & -4 & -18 & -36 \end{array}$$

so, $F(3) = -36$

9. Solve $x^2 - 4x - 5 = 0$
$$(x + 1)(x - 5) = 0$$
$$x + 1 = 0 \quad \text{or} \quad x - 5 = 0$$
$$x = -1 \qquad\qquad x = 5$$

The zeros are -1 and 5.

10. Solve $x^2 + 2x + 1 = 0$
$$(x + 1)(x + 1) = 0$$
$$x + 1 = 0 \quad \text{or} \quad x + 1 = 0$$
$$x = -1 \qquad\qquad x = -1$$
The zeros are -1 and -1.

11. C; Subtracting 3 from x moves the graph 3 units right.

12.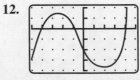

Xscl = 1 Yscl = 10
The real zeros are
$-4, -1,$ and 4.

13.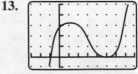

Xscl = 1 Yscl = 10
The real zeros are
$-1, 5,$ and 5.

14.

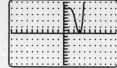

The real zeros are 3 and 3.

15. Test factors of 10.

$$\underline{1|}\ \ \begin{array}{rrrr} 1 & -4 & -7 & 10 \\ & 1 & -3 & -10 \\ \hline 1 & -3 & -10 & 0 \end{array}$$

One factor is $x - 1$ and the other is $x^2 - 3x - 10$.

Solve $x^2 - 3x - 10 = 0$

$(x - 5)(x + 2) = 0$

$x - 5 = 0\ \ $ or $\ \ x + 2 = 0$

$\ \ \ \ x = 5\ \ \ \ \ \ \ \ \ \ \ \ \ \ x = -2$

The rational zeros are -2, 1, and 5.

16. $F(x) = x(2x^3 - x^2 - 18x + 9)$

One factor is x and the other is $2x^3 - x^2 - 18x + 9$

Factors of 2 are $\pm1,\ \pm2$

Factors of 9 are $\pm1,\ \pm3,\ \pm9$

$$\begin{array}{ccc} p & q & \dfrac{p}{q} \\ & \pm1 & \pm1 \\ \pm1 & & \\ & \pm2 & \pm\dfrac{1}{2} \\ & \pm1 & \pm3 \\ \pm3 & & \\ & \pm2 & \pm\dfrac{3}{2} \\ & \pm1 & \pm9 \\ \pm9 & & \\ & \pm2 & \pm\dfrac{9}{2} \end{array}$$

zeros to test: $\pm1,\ \pm\dfrac{1}{2},\ \pm3,\ \pm\dfrac{3}{2},\ \pm9,\ \pm\dfrac{9}{2}$

Use synthetic division to test these.

$$\underline{3|}\ \ \begin{array}{rrrr} 2 & -1 & -18 & 9 \\ & 6 & 15 & -9 \\ \hline 2 & 5 & -3 & 0 \end{array}$$

Two of the factors are x and $x - 3$, and the other is $2x^2 + 5x - 3$.

Solve $2x^2 + 5x - 3 = 0$

$(2x - 1)(x + 3) = 0$

$2x - 1 = 0\ \ $ or $\ \ x + 3 = 0$

$\ \ \ \ 2x = 1\ \ \ \ \ \ \ \ \ \ \ \ \ x = -3$

$\ \ \ \ \ \ x = \dfrac{1}{2}$

The rational zeros are -3, 0, $\dfrac{1}{2}$, and 3.

17. Test factors of 16.

$$\underline{-1|}\ \ \begin{array}{rrrr} 1 & -7 & 8 & 16 \\ & -1 & 8 & -16 \\ \hline 1 & -8 & 16 & 0 \end{array}$$

One factor is $x + 1$ and the other is $x^2 - 8x + 16$.

Solve $x^2 - 8x + 16 = 0$

$(x - 4)(x - 4) = 0$

$x - 4 = 0\ \ $ or $\ \ x - 4 = 0$

$\ \ \ \ x = 4\ \ \ \ \ \ \ \ \ \ \ \ \ \ x = 4$

The solutions are -1, 4, and 4.

Check: $(-1)^3 - 7(-1)^2 + 8(-1) + 16 \stackrel{?}{=} 0$

$\ \ \ \ \ \ \ -1\ -\ 7\ \ \ \ \ \ -\ 8\ \ \ \ \ +\ 16 \stackrel{?}{=} 0$

$\ 0 = 0\ \checkmark$

$\ \ \ \ \ \ \ \ \ \ (4)^3 - 7(4)^2 + 8(4) + 16 \stackrel{?}{=} 0$

$\ \ \ \ \ \ \ \ \ \ 64\ -\ 112\ \ +\ 32\ \ +\ 16 \stackrel{?}{=} 0$

$\ 0 = 0\ \checkmark$

18. Test factors of 15.

$$\underline{-3|}\ \ \begin{array}{rrrr} 1 & -1 & -7 & 15 \\ & -3 & 12 & -15 \\ \hline 1 & -4 & 5 & 0 \end{array}$$

One factor is $x + 3$ and the other is $x^2 - 4x + 5$.

Solve $x^2 - 4x + 5 = 0$

$x = \dfrac{-b \pm \sqrt{b^2 - 4ac}}{2a} = \dfrac{-(-4) \pm \sqrt{(-4)^2 - 4(1)(5)}}{2(1)} =$

$\dfrac{4 \pm \sqrt{-4}}{2} = \dfrac{4 \pm 2i}{2} = 2 \pm i$

The solutions are -3, $2 + i$, and $2 - i$.

Check: $(-3)^3 - (-3)^2 - 7(-3) + 15 \stackrel{?}{=} 0$

$\ \ \ \ \ \ \ \ -27\ -\ 9\ \ \ \ \ +\ 21\ \ \ \ \ +\ 15 \stackrel{?}{=} 0$

$\ 0 \stackrel{?}{=} 0\ \checkmark$

$(2 + i)^2 = (2 + i)(2 + i) = 4 + 2i + 2i + i^2 =$

$\ \ \ \ \ 4 + 4i + -1 = 3 + 4i$

$(2 + i)^3 = (2 + i)^2(2 + i) = (3 + 4i)(2 + i) =$

$\ \ \ \ \ 6 + 3i + 8i + 4i^2 = 6 + 11i - 4 = 2 + 11i$

$(2 + i)^3 - (2 + i)^2 - 7(2 + i) + 15 \stackrel{?}{=} 0$

$2 + 11i - (3 + 4i) - 14 - 7i + 15 \stackrel{?}{=} 0$

$\ \ 2 + 11i - 3 - 4i - 14 - 7i + 15 \stackrel{?}{=} 0$

$\ 0 = 0\ \checkmark$

$(2 - i)^2 = (2 - i)(2 - i) = 4 - 2i - 2i + i^2 =$

$\ \ \ \ \ 4 - 4i - 1 = 3 - 4i$

$(2 - i)^3 = (2 - i)^2(2 - i) = (3 - 4i)(2 - i) =$

$\ \ \ \ \ 6 - 3i - 8i + 4i^2 = 6 - 11i - 4 = 2 - 11i$

$(2 - i)^3 - (2 - i)^2 - 7(2 - i) + 15 \stackrel{?}{=} 0$

$2 - 11i - (3 - 4i) - 14 + 7i + 15 \stackrel{?}{=} 0$

$\ \ 2 - 11i - 3 + 4i - 14 + 7i + 15 \stackrel{?}{=} 0$

$\ 0 = 0\ \checkmark$

19. Test factors of 10.

$$\underline{-5|}\quad \begin{array}{rrrr} 1 & 3 & -12 & -10 \\ & -5 & 10 & 10 \\ \hline 1 & -2 & -2 & 0 \end{array}$$

One factor is $x + 5$ and the other is $x^2 - 2x - 2$.

Solve $x^2 - 2x - 2 = 0$

$$x = \frac{-b \pm \sqrt{b^2 - 4ac}}{2a} = \frac{-(-2) \pm \sqrt{(-2)^2 - 4(1)(-2)}}{2(1)} =$$

$$\frac{2 \pm \sqrt{12}}{2} = \frac{2 \pm 2\sqrt{3}}{2} = 1 \pm \sqrt{3}$$

The solutions are -5, $1 + \sqrt{3}$, and $1 - \sqrt{3}$.

Check: $(-5)^3 + 3(-5)^2 - 12(-5) - 10 \overset{?}{=} 0$
$$-125 + 75 + 60 - 10 \overset{?}{=} 0$$
$$0 = 0 \checkmark$$

$(1 + \sqrt{3})^2 = (1 + \sqrt{3})(1 + \sqrt{3}) =$
$\quad 1 + \sqrt{3} + \sqrt{3} + 3 = 4 + 2\sqrt{3}$
$(1 + \sqrt{3})^3 = (1 + \sqrt{3})^2(1 + \sqrt{3}) =$
$\quad (4 + 2\sqrt{3})(1 + \sqrt{3}) =$
$\quad 4 + 4\sqrt{3} + 2\sqrt{3} + 6 = 10 + 6\sqrt{3}$

$(1 + \sqrt{3})^3 + 3(1 + \sqrt{3})^2 - 12(1 + \sqrt{3}) - 10 \overset{?}{=} 0$
$10 + 6\sqrt{3} + 3(4 + 2\sqrt{3}) - 12 - 12\sqrt{3} - 10 \overset{?}{=} 0$
$\quad 10 + 6\sqrt{3} + 12 + 6\sqrt{3} - 12 - 12\sqrt{3} - 10 \overset{?}{=} 0$
$$0 = 0 \checkmark$$

$(1 - \sqrt{3})^2 = (1 - \sqrt{3})(1 - \sqrt{3}) =$
$\quad 1 - \sqrt{3} - \sqrt{3} + 3 = 4 - 2\sqrt{3}$
$(1 - \sqrt{3})^3 = (1 - \sqrt{3})^2(1 - \sqrt{3}) =$
$\quad (4 - 2\sqrt{3})(1 - \sqrt{3}) = 4 - 4\sqrt{3} - 2\sqrt{3} + 6 =$
$\quad 10 - 6\sqrt{3}$
$(1 - \sqrt{3})^3 + 3(1 - \sqrt{3})^2 - 12(1 - \sqrt{3}) - 10 \overset{?}{=} 0$
$10 - 6\sqrt{3} + 3(4 - 2\sqrt{3}) - 12 + 12\sqrt{3} - 10 \overset{?}{=} 0$
$\quad 10 - 6\sqrt{3} + 12 - 6\sqrt{3} - 12 + 12\sqrt{3} - 10 \overset{?}{=} 0$
$$0 = 0 \checkmark$$

20. No. If one or more of the zeros has a multiplicity of 2 or greater, or if some of the zeros are imaginary, the graph will intersect the x-axis in fewer than n different points.

21. Apply the location theorem.

x	5	-13	31	-15	
-2	5	-23	77	-169	
-1	5	-18	49	-64	
0	5	-13	31	-15	
1	5	-8	23	8	← sign change

A real zero exists between 0 and 1.

22. Apply Descartes' rule of signs.
$$F(x) = 2x^4 + x^3 - x^2 + x - 7$$
$$\underset{1}{\underbrace{+\quad +}}\ \underset{2}{\underbrace{-}}\ \underset{3}{\underbrace{+\quad -}}$$
3 or 1 positive zeros.
$$F(-x) = 2x^4 - x^3 - x^2 - x - 7$$
$$\underset{1}{\underbrace{+\quad -}}\ \ -\ \ -\ \ -$$
1 negative zero.
4 real (3 pos. and 1 neg.) or 2 real (1 pos. and 1 neg.) and 2 imaginary (1 pair).

23. Apply Descartes' rule of signs.
$$F(x) = x^5 - 3x^4 + 5x^3 + x^2 - 4$$
$$\underset{1}{\underbrace{+\ -}}\ \underset{2}{\underbrace{-\ +}}\ \underset{3}{\underbrace{+\ -}}$$
3 or 1 positive zeros.
$$F(-x) = -x^5 - 3x^4 - 5x^3 + x^2 - 4$$
$$-\ \ -\ \ \underset{1}{\underbrace{-\ +}}\ \underset{2}{\underbrace{+\ -}}$$
2 or 0 negative zeros.
5 real (3 pos. and 2 neg.) or 3 real (3 pos. and 0 neg. or 1 pos. and 2 neg.) and 2 imaginary (1 pair), or 1 real (pos.) and 4 imaginary (2 pairs).

24. $F(x) = -x^4 + 6x - 1$

x	-1	0	0	6	-1
1	-1	-1	-1	5	4
2	-1	-2	-4	-2	-5

Upper Bound is 2.
$F(-x) = -x^4 - 6x - 1$

x	-1	0	0	-6	-1
1	-1	-1	-1	-7	-8
0	-1	0	0	-6	-1

Lower Bound is 0; A

25. $V = l \times w \times h = (10)(8)(6) = 480$
length of increase $= x$
new volume $= (10 + x)(8 + x)(6 + x) = 2(480)$
Solve $\quad (10 + x)(8 + x)(6 + x) = 960$
$\quad (80 + 10x + 8x + x^2)(6 + x) = 960$
$\quad (80 + 18x + x^2)(6 + x) = 960$
$480 + 80x + 108x + 18x^2 + 6x^2 + x^3 = 960$
$\quad 480 + 188x + 24x^2 + x^3 = 960$
$\quad x^3 + 24x^2 + 188x - 480 = 0$

Test factors of 480.

$$\underline{2|}\quad \begin{array}{rrrr} 1 & 24 & 188 & -480 \\ & 2 & 52 & 480 \\ \hline 1 & 26 & 240 & 0 \end{array}$$

One factor is $x - 2$ and the other is $x^2 + 26x + 240$.
The only real solution is 2 because solving
$x^2 + 26x + 240 = 0$ produces imaginary solutions.
Thus each dimension of the box should be increased 2"
to double the volume.

26.

y-value	-1		2		9		20		35		54
1st order diff.		3		7		11		15		19	
2nd order diff.			4		4		4		4		
3rd order diff.				0		0		0			

$2a = 4$
$a = 2$
$3a + b = 3$
$3(2) + b = 3$
$6 + b = 3$
$b = -3$

$$a + b + c = -1$$
$$2 + (-3) + c = -1$$
$$-1 + c = -1$$
$$c = 0$$
$$y = 2x^2 - 3x$$

27. Next 3 x values:

7    $y = 2(7)^2 - 3(7) = 98 - 21 = 77$

8    $y = 2(8)^2 - 3(8) = 128 - 24 = 104$

9    $y = 2(9)^2 - 3(9) = 162 - 27 = 135$

The next 3 pairs of data are $(7, 77)$, $(8, 104)$, and $(9, 135)$.

Cumulative Review, page 508

1. $\dfrac{(4 - 6i)(1 + i)}{(1 - i)(1 + i)} = \dfrac{4 - 6i + 4i - 6i^2}{1 - i^2} =$

$\dfrac{4 - 2i + 6}{1 + 1} = \dfrac{10 - 2i}{2} = 5 - i$

2. Apply Descartes' rule of signs

$$f(x) = x^5 - x^2 + 10$$

2 or 0 positive solutions

$$f(-x) = -x^5 - x^2 + 10$$

1 negative solution

Since the polynomial is of degree 5 there are 5 solutions, so at least 2 are imaginary. Therefore, the possibilities are 3 real solutions (2 positive and 1 negative) and 2 imaginary solutions (a conjugate pair) or 1 negative, 0 positive and 4 imaginary (2 conjugate pairs).

3. Let $x =$ the amount each dimension is increased.

$$A = lw$$
$$(x + 20)(x + 30) \leq 3000$$
$$x^2 + 50x + 600 \leq 3000$$
$$x^2 + 50x - 2400 \leq 0$$
$$(x - 30)(x + 80) \leq 0$$

$x - 30$	$-$		$-$		$+$
$x + 80$	$-$		$+$		$+$
$(x - 30)(x + 80)$	$+$		$-$		$+$

$$-80 \leq x \leq 30$$

However x must be positive, so $0 < x \leq 30$.
Therefore, the amount of increase could be any amount up to and including 30 m.

4.

	A	B	C	D
A	0	0	1	1
B	1	0	1	1
C	1	1	0	0
D	2	1	3	0

5. B; y-value

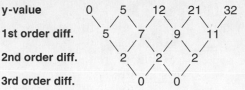

0	5	12	21	32
1st order diff.	5	7	9	11
2nd order diff.		2	2	2
3rd order diff.			0	0

$2a = 2$	$3a + b = 5$	$a + b + c = 0$
$a = 1$	$3(1) + b = 5$	$1 + 2 + c = 0$
	$b = 2$	$c = -3$

The descriptive polynomial is $y = x^2 + 2x - 3$.

6. Let $t = 0$ in 1970 when there were 12 prairie dogs.

So, $P_0 e^{k(0)} = 12$

$$P_0(1) = 12$$
$$P_0 = 12$$

For 1980 $t = 10$ and there were 90 prairie dogs.

So, $12e^{k(10)} = 90$

$$e^{10k} = \frac{90}{12} = \frac{15}{2}$$

$$\ln e^{10k} = \ln \frac{15}{2}$$

$$10k = \ln \frac{15}{2}$$

$$k = \frac{1}{10} \ln \frac{15}{2}$$

Therefore, in 2010 when $t = 40$ the expected prairie dog population is

$$P = 12e^{(\frac{1}{10} \ln \frac{15}{2})40} \approx 37{,}968$$

So, the expected population is approximately 37,970 prairie dogs.

7. For line $\overline{AB}$; $m = \dfrac{2 - (-2)}{3 - 0} = \dfrac{4}{3}, b = -2$

So, the equation is $y = \dfrac{4}{3}x - 2$.

For line $\overline{BC}$; $m = \dfrac{2 - 4}{3 - 0} = -\dfrac{2}{3}, b = 4$

So, the equation is $y = -\dfrac{2}{3}x + 4$.

For line $\overline{CD}$; $m = \dfrac{4 - 0}{0 - (-1)} = \dfrac{4}{1} = 4, b = 4$

So, the equation is $y = 4x + 4$.

For line $\overline{AD}$; $m = \dfrac{-2 - 0}{0 - (-1)} = \dfrac{-2}{1} = -2, b = -2$

So, the equation is $y = -2x - 2$.

So, the required system of inequalities is

$$y \geq \frac{4}{3}x - 2 \longrightarrow 3y \geq 4x - 6 \longrightarrow 3y - 4x \geq -6$$

$$y \leq -\frac{2}{3}x + 4 \longrightarrow 3y \leq -2x + 12 \longrightarrow 3y + 2x \leq 12$$

$$y \leq 4x + 4 \longrightarrow y \leq 4x + 4 \longrightarrow y - 4x \leq 4$$

$$y \geq -2x - 2 \longrightarrow y \geq -2x - 2 \longrightarrow y + 2x \geq -2$$

8.

$$
\begin{array}{r}
2x^2 + 5x - 10 \\
x + 2\overline{)2x^3 + 9x^2 + 0x - 10} \\
\underline{2x^3 + 4x^2} \\
5x^2 + 0x \\
\underline{5x^2 + 10x} \\
-10x - 10 \\
\underline{-10x - 20} \\
10
\end{array}
$$

$(9x^2 + 2x^3 - 10) \div (x + 2) =$

$2x^2 + 5x - 10 + \dfrac{10}{x + 2}$

9. $\dfrac{\log 2 + \log 18}{2} = \dfrac{\log (2)(18)}{2} = \dfrac{\log 36}{2} = \dfrac{1}{2}\log 36 =$

$\log 36^{\frac{1}{2}} = \log 6;$ D

10. $-\dfrac{3}{2}\bigg|$

$$
\begin{array}{r|rrrrr}
-\frac{3}{2} & 1 & 0 & -3 & 0 & -10 \\
& & -\frac{3}{2} & \frac{9}{4} & \frac{9}{8} & -\frac{27}{16} \\
\hline
& 1 & -\frac{3}{2} & -\frac{3}{4} & \frac{9}{8} & -\frac{187}{16}
\end{array}
$$

$F\!\left(-\dfrac{3}{2}\right) = -\dfrac{187}{16}$

11. $x^2 - 2x + 5 = 0$

$x = \dfrac{-b \pm \sqrt{b^2 - 4ac}}{2a} = \dfrac{2 \pm \sqrt{4 - 4(1)(5)}}{2(1)} =$

$\dfrac{2 \pm \sqrt{4 - 20}}{2} = \dfrac{2 \pm \sqrt{-16}}{2} = \dfrac{2 \pm 4i}{2} =$

$\dfrac{2(1 \pm 2i)}{2} = 1 \pm 2i$

Check:
$(1 + 2i)^2 - 2(1 + 2i) + 5 =$
$\quad 1 - 4 + 4i - 2 - 4i + 5 = 0$
$(1 - 2i)^2 - 2(1 - 2i) + 5 =$
$\quad 1 - 4 - 4i - 2 + 4i + 5 = 0$

12. If the zeros are real and unequal, the graph crosses the x-axis in two distinct places. If the zeros are real and equal, the graph is tangent to the x-axis. If the zeros are imaginary, the graph does not intersect the x-axis.

Standardized Test, page 509

1. -3

2. $3 + (x - 2)i = 3 + 7i$
$x - 2 = 7$
$\quad\quad x = 9$

3.

	y-value	1		12		27		46		69
	1st order diff.		11		15		19		23	
	2nd order diff.			4		4		4		
	3rd order diff.				0		0			

$2a = 4 \quad\quad 3a + b = 11 \quad\quad a + b + c = 1$
$\ a = 2 \quad\quad 3(2) + b = 11 \quad\quad 2 + 5 + c = 1$
$\quad\quad\quad\quad\quad 6 + b = 11 \quad\quad\quad 7 + c = 1$
$\quad\quad\quad\quad\quad\quad\quad b = 5 \quad\quad\quad\quad\quad c = -6$

$f(x) = 2x^2 + 5x - 6$
$f(12) = 2(12)^2 + 5(12) - 6 =$
$\quad\quad 2(144) + 60 - 6 = 342$

4. $2y = 3x - 5$
$\quad y = \dfrac{3}{2}x - \dfrac{5}{2}$

$\quad m = \dfrac{3}{2}$

Slopes of perpendicular lines are negative reciprocals. So, the slope of the other line is $-\dfrac{2}{3}$.

5. After 3000 years $n = \dfrac{3000}{5700} = \dfrac{10}{19}$

$A(n) = A_0(0.5)^n$

$A\!\left(\dfrac{10}{19}\right) = A_0(0.5)^{\frac{10}{19}} \approx 0.694A_0$

This means that after 3000 years the percent of carbon-14 that would be in the substance is 69.4% of the original amount.

6.

$$
\begin{array}{r}
x - 5 \\
x + 7\overline{)x^2 + 2x - 30} \\
\underline{x^2 + 7x} \\
-5x - 30 \\
\underline{-5x - 35} \\
5 \longleftarrow \text{Remainder}
\end{array}
$$

7. $3\text{ cm} = 30\text{ mm}$
$\quad 2^x > 30$
$\quad \ln 2^x > \ln 30$
$\quad x \ln 2 > \ln 30$
$\quad\quad x > \dfrac{\ln 30}{\ln 2}$
$\quad\quad x > 4.9$
After 5 cuts.

8. $\sqrt{2x + 3} - 5 = 1$
$\quad \sqrt{2x + 3} = 6$
$\quad\quad 2x + 3 = 36$
$\quad\quad\quad\quad 2x = 33$
$\quad\quad\quad\quad x = \dfrac{33}{2}$

9.

$$
\begin{array}{r|rrrr}
-3 & 1 & 4 & -3 & -12 \\
& & -3 & -3 & 18 \\
\hline
& 1 & 1 & -6 & 6
\end{array}
$$

$(x^3 + 4x^2 - 3x - 12) \div (x + 3) =$

$\quad x^2 + x - 6 + \dfrac{6}{x + 3};$ Remainder is 6

10. $\quad x^2 - 4x \le 21$
$\quad x^2 - 4x - 21 \le 0$
$\quad (x - 7)(x + 3) \le 0$

$x - 7$	$-$	$-$	$+$
$x + 3$	$-$	$+$	$+$
$(x - 7)(x + 3)$	$+$	$-$	$+$

So, $-3 \le x \le 7$.
$(-3)(7) = -21$

11. Let y represent the number of Model H engines and x represent the number of Model C engines and P the total profit.

$0 < y \leq 7$
$0 < x \leq 11$
$y + x \leq 12$
$y \leq 2x$
$100x + 200y > 1000$
  or $x + 2y > 10$

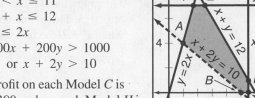

Profit on each Model C is $300 and on each Model H is $200.

Vertex	$P = 300x + 200y$
$A(2, 4)$	$P = 300(2) + 200(4) = 1400$
$B(10, 0)$	$P = 300(10) + 200(0) = 3000$
$C(11, 0)$	$P = 300(11) + 200(0) = 3300$
$D(11, 1)$	$P = 300(11) + 200(1) = 3500$
$E(5, 7)$	$P = 300(5) + 200(7) = 2900$
$F(3, 7)$	$P = 300(3) + 200(7) = 2300$

The greatest feasible daily profit is $3500.

12. $F(x) = x^4 + 12x^2 - 10x - 60$

x	1	0	12	-10	-60
1	1	1	13	3	-57
2	1	2	16	22	-16
3	1	3	21	53	99

3 is the upper bound.

13.
$$\frac{x + y + z}{3} = 15$$
$$x + y + z = 45$$
$$x + 10 + z = 45 \qquad \text{substitute 10 for } y.$$
$$x + z = 35$$

14. Let $x =$ the number of couples
$\dfrac{500}{x} =$ cost per couple for rental fee
$50 =$ total additional costs per couple (dinner, souvenirs, miscellaneous)

$$\frac{500}{x} + 50 \leq 70$$
$$\frac{500}{x} \leq 20$$
$$500 \leq 20x$$
$$25 \leq x$$

The minimum number of couples is 25.

15. D; Area of pool: 25π
Area of garden: 400π

Probability leaf lands in pool $= \dfrac{25\pi}{400\pi} = \dfrac{1}{16}$

Chapter 11 Rational Expressions and Equations

Data Activity, pages 510–511

1. $868 + 340 + 401 + 278 + 194 + 73 + 620 + 76 + 1831 = \$4681$ million

2. $346 + 172 + 338 + 384 + 206 + 324 + 3655 + 1170 + 2618 = \$9213$ million; no

3. $\dfrac{868}{4681} \cdot 100 = 18.54\%$

4. $\dfrac{3655}{9223} \cdot 100 = 39.63\%$

5. Exports − Imports = Balance of Payments to U.S.

Canada	Mex.	UK	Ger.	France	Italy	Japan	SK	OT/C
522	168	63	−106	−12	−251	−3035	−1094	−787

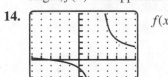

This graph has positive values when exports are greater than imports and negative for imports greater than exports; the balance is closest to zero for FRANCE

6. Answers will vary.

Lesson 11.1, pages 513–516

THINK BACK

1. A number that can be written as the ratio of two integers a and b, where $b \neq 0$.

2a. rational   2b. rational   2c. rational

2d. no; $\sqrt{3}$ cannot be written as the ratio of two integers

2e. rational   2f. rational

3. $n \neq 0$. The denominator of a fraction cannot equal zero because division by zero is undefined.

4a. $x \neq -2$ 4b. $x \neq 4$ 4c. $x \neq 3$ or $x \neq -3$

EXPLORE

5. Answers will vary. Both are ratios or quotients. Neither can have a denominator equal to zero.

6a. rational   6b. rational

6c. irrational. The fraction cannot be written as $\dfrac{p(x)}{q(x)}$ where $p(x)$ and $q(x)$ are polynomials in x.

6d. rational   6e. rational

6f. irrational. The function cannot be written as $\dfrac{p(x)}{q(x)}$ where $p(x)$ and $q(x)$ are polynomials in x.

7. $x \neq 5$ because the denominator will equal zero.

8. $\dfrac{x^2 - 25}{x - 5} = \dfrac{(x - 5)(x + 5)}{x - 5} = x + 5$

The graph will be a straight line with slope of 1 and y-intercept of 5.

9.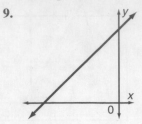

The graph has a discontinuity at $x = 5$. This is because when $x = 0$ the function equals $\dfrac{0}{0}$ which is undefined.

10.

x	0	1	1.5	1.8	1.9	1.99	1.999
$f(x)$	−1.5	−4	−9	−24	−49	−499	−4999

11.

x	4	3	2.5	2.2	2.1	2.01	2.001
$f(x)$	3.5	6	11	26	51	501	5001

12. As x approaches 2 from the left $f(x)$ decreases more and more rapidly. As x approaches 2 from the right $f(x)$ increases more and more rapidly. There will be a discontinuity at $x = 2$.

13. when $x = 1000, f(x) = 1.005$
when $x = -1000, f(x) = 0.995$
In regions to the far right of the origin, $f(x)$ will approach 1 from above. In regions to the far left of the origin, $f(x)$ will approach 1 from below.

14. $f(x) = \dfrac{x + 3}{x - 2}$

15. the horizontal asymptote of $f(x) = \dfrac{x + 3}{x - 2}$ is $y = 1$ and the vertical asymptote of $f(x)$ is $x = 2$

MAKE CONNECTIONS

16. $x = 3$ 17. none   18. $x = -2$

19. $x^2 + 3x - 4 = (x - 4)(x + 1); x = -4; x = 1$

20. $y = \dfrac{x^2 - 1}{x + 1} = \dfrac{(x - 1)(x + 1)}{x + 1}$; c

21. $y = x^3 - 2x^2 - x + 2$; a

22. $y = \dfrac{x - 1}{x - 2}$; d   23. $y = x - 1$; b   24. $x = 3$

25. $x = -5$

26. $x^2 - 2x - 3 = (x + 3)(x - 1); x = 1$ and $x = -3$

SUMMARIZE

27. Answers will vary; possible example
$$f(x) = \dfrac{1}{(x - 3)(x - 4)}$$

The domain will be all values of x except $x = 3$ and $x = 4$. Vertical asymptotes will exist at discontinuities.

28. $c = \dfrac{3.45 + 0.33p}{p}$

29. Yes, because p is percent and must be between 0 and 100.

30a. Answers will vary. The function will have $(x - 6)$ in the denominator

30b. Answers will vary. The function will have $(x - 8)$ in the denominator

30c. Answers will vary. The function will have $\left(x - \dfrac{2}{3}\right)$ in the denominator.

31. $f(x) = \dfrac{2x + 1}{x - 1}$

31a. When x is very large, the $+1$, and -1 have very little influence since they will be insignificant in relation to $2x$ and x.

31b. $f(x) = \dfrac{2x}{x}$

31c. horizontal asymptote is $f(x) = \dfrac{2}{1} = 2$. Since the exponent of the numerator and denominator is the same, $y = \dfrac{bx^n}{cx^m}$, $n = m$, then the horizontal asymptote is $y = \dfrac{b}{c}$

31d. from above because for large positive $|x|$ the function is slightly greater than 2.

31e. from below because for large negative $|x|$ the function is slightly less than 2.

31f. $x = 1$, vertical asymptote

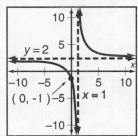

31g. $f(x) = \dfrac{5x - 1}{2x + 1}$, horizontal asymptote is $y = \dfrac{5}{2}$

Lesson 11.2, pages 517–523

EXPLORE

1. $M = \dfrac{I}{S}$

Tables will vary.

S	M
2	0.5
3	0.33
4	0.25
5	0.2

2.

3. As S increases, M decreases. The graph gets closer and closer to the s-axis

4. As S decreases, M increases. The graph gets closer and closer to the M-axis.

5. $M = \dfrac{I}{S}$ is not defined for $S = 0$.

6. $S = 0$ and $M = 0$; the s-axis and the m-axis

TRY THESE

1. The domain is restricted to values of x not equal to -1.

2. The domain is all values of x.

3. $x^2 - 16 = 0$
   $\quad x^2 = 16$
   $\quad x = \pm 4$; therefore the domain is restricted to values of x not equal to ± 4.

4. $\quad x^2 + x - 6 = 0$
   $(x + 3)(x - 2) = 0$
   $\quad\quad\quad\quad x = -3, 2$
   The domain is restricted to values of x not equal to -3 and 2.

5. The domain is restricted to values of x not equal to 0.

6. $3x - 5 = 0$
   $\quad\quad 3x = 5$
   $\quad\quad\; x = \dfrac{5}{3}$

   The domain is restricted to values of x not equal to $\dfrac{5}{3}$.

7. $x = 4$ 8. $x = -4$

9. $f(x) = \dfrac{(x^2 + 4x - 5)(x - 3)}{x^2 + 2x - 15} =$

   $\dfrac{(x + 5)(x - 1)(x - 3)}{(x + 5)(x - 3)}$

   The function will have discontinuities at $x = -5$ and $x = 3$. The graph will look like the graph of $y = x - 1$ with discontinuities at $(-5, -6)$ and $(3, 2)$.

10.

x	1	1.5	1.9	1.99	1.999	2.001	2.01	2.1	2.5	3
f(x)	0	−1	−9	−99	−999	1001	101	11	3	2

11. horizontal asymptote at $f(x) = 0$
    vertical asymptote at $x = 3$

12. horizontal asymptote at $f(x) = 0$
    vertical asymptotes at $x = 2$ and $x = 3$

13. horizontal asymptote at $f(x) = \dfrac{3}{1} = 3$
    vertical asymptote at $x = -2$

14. $f(x) = \dfrac{1}{x - 1}$

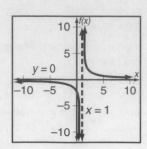

15. $f(x) = \dfrac{(x + 3)(x - 1)}{(x + 3)} = (x - 1)$

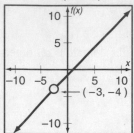

16. $f(x) = \dfrac{6x}{2x - 5}$

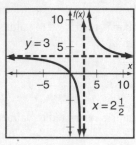

17a. $C_T = 10,000 + 20,000 + 2x$
$x = 5000$
$C_T = \$40,000$

17b. Cost/book $= \dfrac{40,000}{5000} = \$8$

17c. $C_T = 30,000 + 2n$
$C_B = \dfrac{30,000 + 2n}{n}$

17d. As n increases, the cost per book decreases. As n approaches zero, the cost per book increases.

18. Discontinuous means the graph has breaks or undefined range values. Rational functions are often discontinuous because the denominator cannot be equal to zero, therefore the value of x that makes the denominator equal to zero is usually a point of discontinuity.

PRACTICE

1. $f(x) = \dfrac{4}{x}$
$x = 0$
The domain is restricted to values of x not equal to 0.

2. $f(x) = \dfrac{x^2 - 16}{8}$
$8 \neq 0$
The domain includes all values of x.

3. $f(x) = \dfrac{x + 1}{x - 5}$
$x - 5 = 0$
$x = 5$
The domain is restricted to values of x not equal to 5.

4. $f(x) = \dfrac{2x + 3}{2x - 3}$
$2x - 3 = 0$
$2x = 3$
$x = \dfrac{3}{2}$
The domain is restricted to values of x not equal to $\dfrac{3}{2}$.

5. $f(x) = \dfrac{5x - 4}{x^2 - 100}$
$x^2 - 100 = 0$
$x^2 = 100$
$x = \pm 10$
The domain is restricted to values of x not equal to 10 and -10.

6. $f(x) = \dfrac{12x^2 - 5x}{x^2 - 6x + 8}$
$x^2 - 6x + 8 = 0$
$(x - 4)(x - 2) = 0$
$x = 4, 2$
The domain is restricted to values of x not equal to 2 and 4.

7. $f(x) = \dfrac{x^2 - 5x - 14}{x - 7}$
$x - 7 = 0$
$x = 7$
The domain is restricted to values of x not equal to 7.

8. $f(x) = \dfrac{x^3 - 5x + 2}{6x^2 + x - 2}$
$6x^2 + x - 2 = 0$
$(3x + 2)(2x - 1) = 0$
$x = -\dfrac{2}{3}, \dfrac{1}{2}$
The domain is restricted to values of x not equal to $-\dfrac{2}{3}$ and $\dfrac{1}{2}$.

9. $f(x) = \dfrac{1}{x^4 - 13x^2 + 36}$
$x^4 - 13x^2 + 36 = 0$
$(x^2 - 9)(x^2 - 4) = 0$
$(x - 3)(x + 3)(x - 2)(x + 2) = 0$
$x = -3, -2, 2, 3$
The domain is restricted to values of x not equal to $-3, -2, 2$, and 3.

10. $f(x) = \dfrac{x^2 - 1}{x + 1} = \dfrac{(x - 1)(x + 1)}{x + 1}$
A discontinuity will occur at $(-1, -2)$

11. $f(x) = \dfrac{x^2 - x - 12}{x - 2} = \dfrac{(x + 3)(x - 4)}{x - 2}$
There are no points of discontinuity.

12. $f(x) = \dfrac{(x^2 - 3x - 18)}{(x + 3)} = \dfrac{(x - 6)(x + 3)}{(x + 3)}$

A discontinuity will occur at $(-3, -9)$.

13.

x	2	2.5	2.9	2.99	2.999	3.001	3.01	3.1	3.5	4
$f(x)$	−1	−3	−19	−199	−1999	2001	201	21	5	3

14. horizontal asymptote at $f(x) = 0$
vertical asymptote at $x = 5$

15. horizontal asymptote at $f(x) = 0$
vertical asymptotes at $x = -5$ and $x = 5$

16. horizontal asymptote at $f(x) = 0$
vertical asymptotes at $x = 2$ and $x = 5$

17. horizontal asymptote at $f(x) = 6$
vertical asymptote at $x = 3$

18. horizontal asymptotes, NONE

vertical asymptote at $x = \dfrac{3}{2}$

19. horizontal asymptote at $f(x) = 0$
vertical asymptotes at $x = 2$, $x = -2$, and $x = 4$

20. For very large $|x|$; $\dfrac{2x + 3}{x - 1}$ approaches 2.

If $|x|$ is very large, the 3 and 1 are insignificant, leaving

$\dfrac{2x}{x} = 2$

21. $f(x) = \dfrac{1}{x + 1}$ **22.** $f(x) = \dfrac{4}{x^2 - 9}$

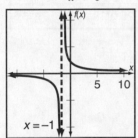   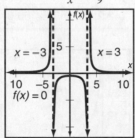

23. $f(x) = \dfrac{(x + 4)(x - 1)}{x - 1}$ **24.** $f(x) = \dfrac{6x}{3x + 5}$

$= x + 4$

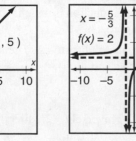

25. $f(x) = \dfrac{x + 3}{x^2 - 6x + 5}$ **26.** $f(x) = \dfrac{x - 3}{x - 4}$

$= \dfrac{x + 3}{(x - 5)(x - 1)}$

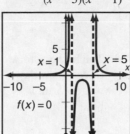   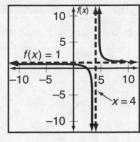

27. $f(x) = \dfrac{1}{x - 2}$ **28.** $f(x) = \dfrac{4x - 2}{2x + 1}$

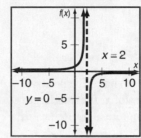   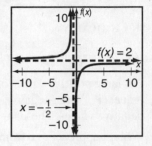

29. $f(x) = \dfrac{1}{(x - 2)^2}$ **30a.** $M = \dfrac{I}{S}$ $M = \dfrac{4}{S}$

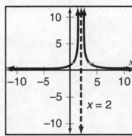   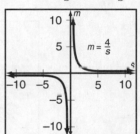

30b. $0.25 = \dfrac{4}{S}$

$S = \dfrac{4}{0.25} = 16\ \text{cm}$

EXTEND

31. $f(x) = \dfrac{2x^2}{(x - 5)(x + 2)}$ Answers will vary.

32a. $D_T = v_1 t_1 + v_2 t_2$ **32b.** $V_A = \dfrac{v_1 t_1 + v_2 t_2}{t_1 + t_2}$

$T_T = t_1 + t_2$

32c. $v = \dfrac{50t_1 + 60(4)}{t_1 + 4} = \dfrac{50t_1 + 240}{t_1 + 4}$

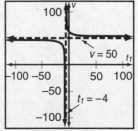

33a. $p(h) = \dfrac{9h^2 + h + 9}{h^2 + 1}$

$h = 0$

$p(0) = \dfrac{9}{1} = \$9.00;\ p(8) = \$9\dfrac{1}{8}$

33b. horizontal asymptote at $p(h) = 9$

33c.

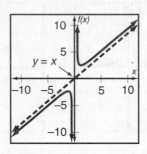

33d. The highest stock price for the day was \$9.50 at $h = 1$ hour

THINK CRITICALLY

34. $f(x) = \dfrac{x^2 + 1}{x} = \dfrac{x^2}{x} + \dfrac{1}{x} = x + \dfrac{1}{x}$

35. For large positive values of x, $f(x)$ has large positive values. For large negative values of x, $f(x)$ has large negative values. The slant asymptote is $f(x) = x$.

36. When x approaches 0 from the right $f(x)$ has very large positive values. As x approaches 0 from the left $f(x)$ has very large negative values. The vertical asymptote is $x = 0$.

37. $f(x) = \dfrac{x^2 + 1}{x}$

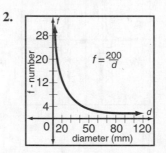

38. $W(x) = \begin{cases} 25 & 0 \le x \le 30 \\ 25 + 0.6(x - 30) & x > 30 \end{cases}$

38a. $W(28) = \$25$

$W(32) = 25 + 0.6(32 - 30) = \$26.20$

38b. As x approaches 30 from the left $W(x)$ approaches 25.

38c. As x approaches 30 from the right, $W(x)$ approaches 25.

38d. $W(x)$ is continuous

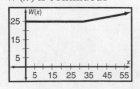

MIXED REVIEW

39. $\begin{bmatrix} 4 + -5 & -7 + 13 & 12 + 11 \\ 8 + -8 & 3 + -9 & -11 + 11 \end{bmatrix} = \begin{bmatrix} -1 & 6 & 23 \\ 0 & -6 & 0 \end{bmatrix}$

40. $\begin{bmatrix} -5 - 4 & 13 - (-7) & 11 - 12 \\ -8 - 8 & -9 - 3 & 11 - (-11) \end{bmatrix} =$

$\begin{bmatrix} -9 & 20 & -1 \\ -16 & -12 & 22 \end{bmatrix}$

41. $\begin{bmatrix} 3 \cdot (-5) & 3 \cdot 13 & 3 \cdot 11 \\ 3 \cdot (-8) & 3 \cdot (-9) & 3 \cdot 11 \end{bmatrix} = \begin{bmatrix} -15 & 39 & 33 \\ -24 & -27 & 33 \end{bmatrix}$

42. $\begin{bmatrix} 2(4) - 2(-5) & 2(-7) - 2(13) & 2(12) - 2(11) \\ 2(8) - 2(-8) & 2(3) - 2(-9) & 2(-11) - 2(11) \end{bmatrix} =$

$\begin{bmatrix} 18 & -40 & 2 \\ 32 & 24 & -44 \end{bmatrix}$

43. no, because for $x = 2$ there are two values in the range: 3 and 4.

44. yes, because each value in the domain has one value in the range.

45. yes.   **46.** c

Lesson 11.3, pages 524–529

EXPLORE

1. $f = \dfrac{F}{d} = \dfrac{200}{d}$ $d = \dfrac{200}{f}$

f	1.4	2	2.8	4	5.6	8	11	16	22	32
d	142.9	100.0	71.4	50.0	35.7	25.0	18.2	12.5	9.1	6.3

2.

f-number vs diameter (mm) graph of $f = \dfrac{200}{d}$

3. The graph has asymptotes of $f = 0$ and $d = 0$ because when $d = 0$ the function is undefined. When d gets very large f approaches 0

4. f and d are related indirectly. As f gets larger, d gets smaller (inverse variation).

TRY THESE

1. $d = vt$, $v = \dfrac{d}{t}$; inversely   **2.** directly   **3.** inversely

4. inversely   **5.** directly   **6.** directly; $k = 6$

7. jointly; $k = 0.5$ **8.** inversely; $k = 1$

9. directly; $k = 12$

10. $x = \frac{1}{y}(k)$

$16 = \frac{1}{9}(k), k = 144$

then $x = \frac{144}{y} = \frac{144}{24} = 6$

11. $x = kyz$
$56.7 = k(1.5)(9)$

$k = \frac{56.7}{(1.5)(9)} = 4.2$

then $x = 4.2yz = 4.2(3)(4.4) = 55.44$

12. $y = \frac{k}{x^2}$

$4 = \frac{k}{(3.2)^2}, k = 40.96$

then $y = \frac{40.96}{(4)^2} = 2.56$

13. $f = \frac{k}{d}$

$150 = \frac{k}{6}$

$k = 900$

then $120 = \frac{900}{d}, d = 7.5$ ft

14. $v = klw$
$16.2 = k(3.6)(2.5)$
$k = 1.8$
$v = (1.8)(4)(1.5) = 10.8$ cm³

15. $S_I = \frac{k}{d^2}$

$9\ wm^2 = \frac{k}{(6km)^2} = \frac{k}{(6 \times 10^3)^2 m}$

$k = 3.24 \times 10^8$

$S_I = \frac{3.24 \times 10^8}{(2\ km)^2} = 81\ wm^2$

16. Answers will vary.

PRACTICE

1. directly　　**2.** inversely　　**3.** inversely

4. inversely　　**5.** directly　　**6.** jointly; $k = 1.72$

7. directly; $k = 7.99$　　**8.** inversely; $k = 3.25$

9. inversely; $k = 20$

10. $x = \frac{k}{y}; 0.4 = \frac{k}{12}, k = 4.8$

$y = \frac{4.8}{5} = 0.96$

11. $a = \frac{k}{b^2}$

$100 = \frac{k}{36}, k = 3600$

$a = \frac{3600}{(15)^2} = 16$

12. $w = ksz$

$75 = k(10)(5)$

$k = 1.5$

$z = \frac{w}{ks} = \frac{115.5}{7(1.5)} = 11$

13. $t = \frac{k}{a}, 3 = \frac{k}{8}, k = 24$

$t = \frac{24}{10} = 2.4$ hours

14. $L_1 = kdwr$
$9072 = k(30)(36)(8)$
$k = 1.05$
$L_2 = 1.05(24)(20)(5) = 2520$ in.

15.

m	5	10	15	20
n	144	36	16	9

As m increases n decreases so the relation is inverse. Since the product of m and n is not constant it is not a simple inverse relation.

$5 = \frac{k}{\sqrt{144}}; \ 10 = \frac{k}{\sqrt{36}}; \ 15 = \frac{k}{\sqrt{16}}; \ 20 = \frac{k}{\sqrt{9}}$

$k = 60$

$m = \frac{60}{\sqrt{n}}$

EXTEND

16. $g = \frac{khl}{m}$

$40 = \frac{k(10)(20)}{0.4}$

$k = 0.08$

$g = \frac{(0.08)(5)(10)}{0.2} = 20$

17. $P = KAV^2$
$1 = KA(15)^2$
$KA = 0.004$
$5 = (0.00444)V^2$
$1125 = V^2$
$V = 33.5$ mi/h
The equation is solved for KA since the area of the building is unknown. This value of KA is used to find the wind velocity on the same building.

18. $T = K\sqrt{L}$
$1 = K\sqrt{39}$
$K = 0.1601$
$T = 0.1601\sqrt{39} + 3 = 1.04$ seconds

19. $L = \dfrac{KW}{l}$

$5 = \dfrac{K(1.5)}{6}$

$K = 20$

$L = \dfrac{20(2)}{4} = 10$ tons

20. $y = -x$

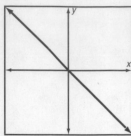

21. $y = \dfrac{k}{x}$

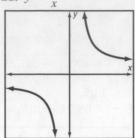

22. $y = kx^2$

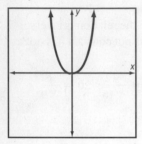

23. $n = \dfrac{k}{s^2 + t^2}$

$3 = \dfrac{k}{10^2 + 10^2}$; $k = 600$

$h = \dfrac{600}{20^2 + 20^2} = 0.75$

24. $F_1 = \dfrac{km_1 m_2}{d^2}$

$F_2 = \dfrac{k(2m_1)(3m_2)}{\left(\frac{1}{2}d\right)^2} = \dfrac{6km_1 m_2}{\frac{1}{4}d^2} = \dfrac{24km_1 m_2}{d^2}$

$F_2 = 24F_1$

ALGEBRAWORKS

1. f/5.6 lets in less light

2. 45; The sequence is $\sqrt{2}, \sqrt{4}, \sqrt{8}, \sqrt{16} \ldots$

3. $A = \pi r^2 \rightarrow A = \pi\left(\dfrac{d}{2}\right)^2 = \dfrac{\pi d^2}{4}$

4. $d = \dfrac{F}{f}$

$A = \pi\dfrac{\left(\frac{F}{f}\right)^2}{4} = \dfrac{\pi F^2}{4f^2}$

5. Approximate areas: 1826 mm², 895 mm², 456 mm², 224 mm². Each increase of one f-number halves the area. When standard f-numbers are used in the formula, the aperture areas are only approximate and the "area-halving" relationship may not be immediately obvious. When the exact terms are used ($\sqrt{2}, \sqrt{4}, \sqrt{8}, \sqrt{16}, \sqrt{32}, \ldots$), the areas are exact and the relationship clear.

Lesson 11.4, pages 531–535

EXPLORE

1. b; the answer is in simplest form.

2. a; all terms must be divided by the same factor

3. a; for the same reason as **2.**

4. b; when multiplying, multiply the numerators and multiply the denominators.

TRY THESE

1. $\dfrac{6x^5}{4x^2} = \dfrac{3}{2}x, \ x \neq 0$

2. $\dfrac{15(x+3)}{5x^2(x+3)} = \dfrac{3}{x^2}, \ x \neq 0, \ x \neq -3$

3. $\dfrac{x^2 - 3x - 4}{4x - 16} = \dfrac{(x-4)(x+1)}{4(x-4)} = \dfrac{x+1}{4}, \ x \neq 4$

4. $\dfrac{5x^2}{3y^3} \cdot \dfrac{2xy^2}{x^3} = \dfrac{10x^3 y^2}{3x^3 y^3} = \dfrac{10}{3y}$

5. $\dfrac{2x+2}{3} \cdot \dfrac{9x}{4x+4} = \dfrac{2(x+1) \cdot 9x}{3 \cdot 4(x+1)} = \dfrac{3x}{2}$

6. $\dfrac{2x^5}{5y^2} \div \dfrac{4x^2}{15y^3} = \dfrac{2x^5}{5y^2} \cdot \dfrac{15y^3}{4x^2} = \dfrac{3x^3 y}{2}$

7. $\dfrac{2x-10}{x+10} \cdot \dfrac{3x+3}{x-5}$

$\dfrac{3 \cdot 2(x-5)(x+1)}{(x+10)(x-5)} = \dfrac{6(x+1)}{(x+10)}$

8. $\dfrac{x^2 - 64}{x^2 - 9} \cdot \dfrac{x+3}{x+8}$

$\dfrac{(x-8)(x+8)(x+3)}{(x-3)(x+3)(x+8)} = \dfrac{x-8}{x-3}$

9. $\dfrac{2x^2 + x - 1}{2 - x} \cdot \dfrac{x^2 - 5x + 6}{2x^2 - x}$

$\dfrac{(2x-1)(x+1)(x-2)(x-3)}{(2-x)(2x-1)(x)}$

$= \dfrac{(x+1)(x-2)(x-3)}{-(x-2)(x)} = \dfrac{-(x+1)(x-3)}{x}$

10. $A = lw$

$l = \dfrac{A}{w} = \dfrac{2x^2 + 2xy}{4xy + 4y^2} = \dfrac{2x(x+y)}{4y(x+y)} = \dfrac{x}{2y}$

11. $\dfrac{\text{hinges}}{\text{hour}} = \dfrac{b}{a^2 - 1}$

$\text{profit} = \dfrac{a^2 - a}{b} \dfrac{\text{cents}}{\text{hinge}}$

$\text{total profit} = \text{hours worked} \cdot \dfrac{\text{hinges}}{\text{hour}} \cdot \dfrac{\text{cents}}{\text{hinge}}$

$= a \cdot \dfrac{b}{a^2 - 1} \cdot \dfrac{a^2 - a}{b}$

$= a \cdot \dfrac{b}{(a + 1)(a - 1)} \cdot \dfrac{a(a - 1)}{b} = \dfrac{a^2}{a + 1}$

12. $\dfrac{x^2 - 2x - 8}{x + 2} = \dfrac{(x - 4)(x + 2)}{(x + 2)} = x - 4$

Answers will vary but should all simplify to the same expression.

PRACTICE

1. $\dfrac{2abc}{6a^2c^2} = \dfrac{b}{3ac}, a \neq 0, c \neq 0$

2. $\dfrac{20xyz^3}{25x^3yz} = \dfrac{z^2}{5x^2}, x \neq 0, y \neq 0, z \neq 0$

3. $\dfrac{x^2 + 4x}{x + 4} = \dfrac{x(x + 4)}{x + 4} = x, x \neq -4$

4. $\dfrac{x^2 - 25}{x - 5} = \dfrac{(x - 5)(x + 5)}{(x - 5)} = x + 5, x \neq 5$

5. $\dfrac{20x^2 - 32x}{4x} = 5x - 8, x \neq 0$

6. $\dfrac{x^2 - 7x + 12}{x^2 - 5x + 6} = \dfrac{(x - 4)(x - 3)}{(x - 3)(x - 2)} = \dfrac{x - 4}{x - 2}$,
$x \neq 3, x \neq 2$

7. $\dfrac{a^2 - b^2}{a^2 - ab} = \dfrac{(a - b)(a + b)}{a(a - b)} = \dfrac{a + b}{a}, a \neq 0, a \neq b$

8. $\dfrac{x^2 - 9x + 20}{x^2 + x - 20} = \dfrac{(x - 4)(x - 5)}{(x + 5)(x - 4)} = \dfrac{x - 5}{x + 5}$,
$x \neq -5, x \neq 4$

9. $\dfrac{2xy^2}{3z^2} \cdot \dfrac{6x^2z^2}{5y^3} = \dfrac{4x^3}{5y}$

10. $\dfrac{40mn}{18p^2} \cdot \dfrac{27p^3q}{25n} = \dfrac{12mpq}{5}$

11. $\dfrac{x^2 + 5x - 6}{2x^2} \cdot \dfrac{6x}{x - 1}$

$\dfrac{(x + 6)(x - 1)}{2x^2} \cdot \dfrac{6x}{(x - 1)} = \dfrac{3(x + 6)}{x}$

12. $\dfrac{7x + 7y}{y - x} \cdot \dfrac{4x - 4y}{21} = \dfrac{-7(x + y)}{x - y} \cdot \dfrac{4(x - y)}{21}$

$= \dfrac{-4(x + y)}{3}$

13. $\dfrac{x^2 - 16}{2x + 6} \cdot \dfrac{x + 3}{x - 4} = \dfrac{(x + 4)(x - 4)(x + 3)}{2(x + 3)(x - 4)} = \dfrac{x + 4}{2}$

14. $\dfrac{x + 2}{x^3} \cdot \dfrac{x}{6x + 12} = \dfrac{(x + 2)x}{x^3 \cdot 6(x + 2)} = \dfrac{1}{6x^2}$

15. $xy \div \dfrac{x}{y + 1} = xy \cdot \dfrac{y + 1}{x} = y(y + 1)$

16. $\dfrac{2x + 4y}{9} \cdot \dfrac{27}{6x + 8y} = \dfrac{2(x + 2y) \cdot 27}{9 \cdot 2(3x + 4y)} = \dfrac{3(x + 2y)}{3x + 4y}$

17. $\dfrac{2a^2 - 8}{a + 2} \div (2a - 4) = \dfrac{2a^2 - 8}{a + 2} \cdot \dfrac{1}{2a - 4} =$

$\dfrac{2(a^2 - 4)}{(a + 2)2(a - 2)} = \dfrac{2(a - 2)(a + 2)}{(a + 2)2(a - 2)} = 1$

18. $\dfrac{a^2 - b^2}{a^2 + 2ab + b^2} \cdot \dfrac{ab + b^2}{a^2 - ab} =$

$\dfrac{(a - b)(a + b)b(a + b)}{(a + b)(a + b)a(a - b)} = \dfrac{b}{a}$

19. $\dfrac{x^2 + 8x + 16}{y^2 - 6y + 9} \div \dfrac{2x + 8}{3y - 9} = \dfrac{x^2 + 8x + 16}{y^2 - 6y + 9} \cdot \dfrac{3y - 9}{2x + 8}$

$= \dfrac{(x + 4)(x + 4)3(y - 3)}{(y - 3)(y - 3)2(x + 4)} = \dfrac{3(x + 4)}{2(y - 3)}$

20. $\dfrac{9x + 81}{x^2 - 9} \cdot \dfrac{2x^2 - 6x}{x^2 + 18x + 81} =$

$\dfrac{9(x + 9)2x(x - 3)}{(x + 3)(x - 3)(x + 9)(x + 9)} = \dfrac{18x}{(x + 3)(x + 9)}$

21. $\dfrac{x^3 - 27}{x^2 - 9} \cdot \dfrac{x^2 - 6x + 9}{x^2 + 3x + 9} =$

$\dfrac{(x - 3)(x^2 + 3x + 9)(x - 3)(x - 3)}{(x - 3)(x + 3)(x^2 + 3x + 9)} = \dfrac{(x - 3)^2}{x + 3}$

22. $\dfrac{3x + 12}{20 - x - x^2} \cdot \dfrac{x^2 + 7x + 10}{4x + 12} =$

$\dfrac{3(x - 4)(x + 5)(x + 2)}{(x + 5)(x - 4)4(x + 3)} = \dfrac{-3(x + 2)}{4(x + 3)}$

23. $\dfrac{x^2 - 10x + 21}{x^2 - 12x + 27} \cdot \dfrac{x^2 - 14x + 45}{x^2 - 11x + 30} =$

$\dfrac{(x - 7)(x - 3)(x - 9)(x - 5)}{(x - 3)(x - 9)(x - 5)(x - 6)} = \dfrac{x - 7}{x - 6}$

24. $\dfrac{n^2 - n - 6}{(n^2 - 3n - 10)} \cdot \dfrac{n - 5}{1} = \dfrac{(n - 3)(n + 2)(n - 5)}{(n - 5)(n + 2)}$

$= n - 3 \text{ g/cm}^3$

25. Answers will vary but should mention that the rules for operating on both are the same.

26. $\dfrac{3x^2 + 3x - 36}{x} \cdot \dfrac{x^2 + x - 6}{x + 4} \cdot \dfrac{2x}{x + 3} =$

$\dfrac{2 \cdot 3(x + 4)(x - 3)}{x} \cdot \dfrac{(x + 3)(x - 2)}{x + 4} \cdot \dfrac{x}{x + 3} =$

$6(x - 3)(x - 2)$

27a. $F_t = \dfrac{m}{300}, \quad F_t = \dfrac{m}{400}$

27b. $F_t = \dfrac{m}{300} + \dfrac{m}{400} = \dfrac{4m}{1200} + \dfrac{3m}{1200} = \dfrac{7m}{1200}$

27c. average rate $= \dfrac{\text{distance}}{\text{time}} = \dfrac{m + m}{\dfrac{7m}{1200}} = 2m \cdot \dfrac{1200}{7m}$

$$= \dfrac{2400}{7}$$

$$\approx 342.9 \text{ mi/h}$$

27d. The average rate does not depend on the distance m between the two cities. It is also not equal to what might be guessed as the average rate when a plane flies at 300 mi/h in one direction and 400 mi/h in the other direction.

28. window: $h = \dfrac{w}{2}$

door: $H = 4w$

$$W = \dfrac{2}{3}H$$

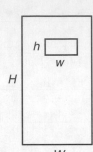

door area $= H \cdot W$ window area $= h \cdot w$

$\dfrac{\text{window area}}{\text{door area}} = \dfrac{h \cdot w}{H \cdot W} = \dfrac{\dfrac{w}{2} \cdot w}{4w \cdot \dfrac{2}{3}4w} = \dfrac{\dfrac{w^2}{2}}{\dfrac{32w^2}{3}} =$

$\dfrac{w^2}{2} \cdot \dfrac{3}{32w^2} = \dfrac{3}{64}$

EXTEND

29. $\dfrac{x^{10m}y^{8n}}{x^{5m}y^{4n}} = x^{5m}y^{4n}$

30. $\dfrac{x^{2a} + 10x^a + 24}{x^{2a} - x^a - 20} = \dfrac{(x^a + 6)(x^a + 4)}{(x^a - 5)(x^a + 4)} = \dfrac{x^a + 6}{x^a - 5}$

31. $\dfrac{x^{4m} - 9x^{2m} + 14}{x^{4m} - 8x^{2m} + 12} = \dfrac{(x^{2m} - 7)(x^{2m} - 2)}{(x^{2m} - 6)(x^{2m} - 2)} = \dfrac{x^{2m} - 7}{x^{2m} - 6}$

32. $f(x) = 5x \quad f(x + h) = 5(x + h)$

$\dfrac{f(x + h) - f(x)}{h} = \dfrac{5(x + h) - 5x}{h} =$

$\dfrac{5x + 5h - 5x}{h} = \dfrac{5h}{h} = 5$

33. $f(x) = x^2 \quad f(x + h) = (x + h)^2$

$\dfrac{f(x + h) - f(x)}{h} = \dfrac{(x + h)^2 - x^2}{h} =$

$\dfrac{x^2 + 2xh + h^2 - x^2}{h} = \dfrac{2xh + h^2}{h} = \dfrac{h(2x + h)}{h} =$

$2x + h$

34. $f(x) = x^2 - 4x \quad f(x + h) = (x + h)^2 - 4(x + h)$

$\dfrac{f(x + h) - f(x)}{h} = \dfrac{(x + h)^2 - 4(x + h) - (x^2 - 4x)}{h}$

$= \dfrac{x^2 + 2xh + h^2 - 4x - 4h - x^2 + 4x}{h}$

$= \dfrac{2xh + h^2 - 4h}{h} = \dfrac{h(2x + h - 4)}{h}$

$= 2x + h - 4$

THINK CRITICALLY

35. $\dfrac{x^2 - 2x - 15}{x^2 - 3x - 18} \cdot \dfrac{P}{x^2 - 3x - 10} = 1$

$P = \dfrac{(x^2 - 3x - 10)(x^2 - 3x - 18)}{x^2 - 2x - 15} =$

$\dfrac{(x - 5)(x + 2)(x - 6)(x + 3)}{(x - 5)(x + 3)}$

$= (x + 2)(x - 6) \quad \text{or} \quad x^2 - 4x - 12$

36. $\dfrac{x^2 - 16}{2x^2} \div \dfrac{2x^2 - 11x + 12}{P} = \dfrac{x + 4}{2x - 1}$

$\dfrac{x^2 - 16}{2x^2} \cdot \dfrac{P}{2x^2 - 11x + 12} = \dfrac{x + 4}{2x - 1}$

$P = \dfrac{x + 4}{2x - 1} \cdot \dfrac{(2x^2 - 11x + 12)2x^2}{x^2 - 16}$

$= \dfrac{x + 4}{2x - 1} \cdot \dfrac{(2x - 3)(x - 4)2x^2}{(x + 4)(x - 4)} = \dfrac{2x^2(2x - 3)}{2x - 1}$

MIXED REVIEW

37. $\begin{cases} y = 2x \\ 3x + 2y = 21 \end{cases}$

use substitution: $3x + 2(2x) = 21$

$3x + 4x = 21$

$7x = 21$

$x = 3$

solve for y: $\quad y = 2x \qquad$ Check: $y = 2x$

$\qquad\qquad y = 2(3) \qquad\qquad\quad 6 = 2(3)$

$\qquad\qquad y = 6 \qquad\qquad\qquad\quad 6 = 6 ✓$

$\qquad\qquad\qquad\qquad\qquad\quad 3x + 2y = 21$

$\qquad\qquad\qquad\qquad\quad 3(3) + 2(6) = 21$

$\qquad\qquad\qquad\qquad\qquad 9 + 12 = 21$

$\qquad\qquad\qquad\qquad\qquad\qquad 21 = 21 ✓$

38. $\begin{cases} x + 2y = 7 \\ 3x - 2y = -11 \end{cases}$

solve for x: $x = 7 - 2y$

substitute: $3(7 - 2y) - 2y = -11$

$21 - 6y - 2y = -11$

$21 - 8y = -11$

$-8y = -32$

$y = 4$

solve for x: $x = 7 - 2y \quad$ Check: $x + 2y = 7$

$\qquad\qquad x = 7 - 2(4) \qquad -1 + 2(4) = 7$

$\qquad\qquad x = 7 - 8 \qquad\qquad -1 + 8 = 7$

$\qquad\qquad x = -1 \qquad\qquad\qquad\quad 7 = 7 ✓$

$\qquad\qquad\qquad\qquad\qquad\quad 3x - 2y = -11$

$\qquad\qquad\qquad\qquad\quad 3(-1) - 2(4) = -11$

$\qquad\qquad\qquad\qquad\qquad -3 - 8 = -11$

$\qquad\qquad\qquad\qquad\qquad\quad -11 = -11 ✓$

39. $\begin{cases} 4x - 3y = -7 \\ 3x + 2y = 16 \end{cases}$

use elimination:

$4x - 3y = -7 \rightarrow 8x - 6y = -14$ multiply by 2
$3x + 2y = 16 \rightarrow \underline{9x + 6y = 48}$ multiply by 3
$\qquad\qquad\qquad\qquad 17x \quad\;\;\; = 34$ add
$\qquad\qquad\qquad\qquad\qquad x = 2$ solve

solve for y: $3x + 2y = 16$
$\qquad\qquad\quad 3(2) + 2y = 16$
$\qquad\qquad\qquad 6 + 2y = 16$
$\qquad\qquad\qquad\qquad 2y = 10$
$\qquad\qquad\qquad\qquad\; y = 5$

Check: $\;4x - 3y = -7 \qquad\quad 3x + 2y = 16$
$\qquad\;\; 4(2) - 3(5) = -7 \quad\; 3(2) + 2(5) = 16$
$\qquad\qquad 8 - 15 = -7 \qquad\quad 6 + 10 = 16$
$\qquad\qquad\qquad -7 = -7 \;\checkmark \qquad\quad 16 = 16 \;\checkmark$

40. $\begin{cases} 4(1 - x) = 8x + 7y \\ 6x + y + 8 = 0 \end{cases}$

solve for y: $6x + y + 8 = 0$
$\qquad\qquad\qquad\quad y = -8 - 6x$

substitute: $4(1 - x) = 8x + 7y$
$\qquad\qquad 4(1 - x) = 8x + 7(-8 - 6x)$
$\qquad\qquad\; 4 - 4x = 8x - 56 - 42x$
$\qquad\qquad\; 4 - 4x = -34x - 56$
$\qquad\qquad 4 + 30x = -56$
$\qquad\qquad\quad\;\; 30x = -60$
$\qquad\qquad\qquad\;\; x = -2$
solve for y: $y = -8 - 6x$
$\qquad\qquad\qquad\; = -8 - 6(-2)$
$\qquad\qquad\qquad\; = -8 + 12$
$\qquad\qquad\qquad\; = 4$

Check: $4(1 - x) = 8x + 7y \qquad\qquad 6x + y + 8 = 0$
$\quad 4(1 - (-2)) = 8(-2) + 7(4) \;\; 6(-2) + 4 + 8 = 0$
$\qquad\quad 4(3) = -16 + 28 \qquad\; -12 + 4 + 8 = 0$
$\qquad\quad\; 12 = 12 \;\checkmark \qquad\qquad\qquad\quad 0 = 0 \checkmark$

41. $x^4 - 13x^2 + 36 = 0$
let $m = x^2$, then $m^2 = x^4$
substitute: $m^2 - 13m + 36 = 0$
factor: $(m - 4)(m - 9) = 0$
solve: $m - 4 = 0 \qquad m - 9 = 0$
$\qquad\qquad m = 4 \qquad\qquad m = 9$
replace m with x^2: $x^2 = 4 \qquad x^2 = 9$
solve for x: $x = 2 \quad x = -2 \qquad\quad x = 3 \quad x = -3$

42. $x = 7\sqrt{x} - 8 = 0$
let $n = \sqrt{x}$ then $n^2 = x$
substitute: $n^2 - 7n - 8 = 0$
factor: $(n - 8)(n + 1) = 0$
solve: $n - 8 = 0 \qquad\qquad n + 1 = 0$
$\qquad\qquad n = 8 \qquad\qquad\qquad n = -1$

replace z with x: $\sqrt{x} = 8 \qquad \sqrt{x} = -1$
solve for x: $\qquad\quad x = 64$
since there is no real value of x, with $\sqrt{x} = -1$ reject
that solution.

43. $x^{\frac{4}{3}} - 29x^{\frac{2}{3}} + 100 = 0$
let $p = x^{\frac{2}{3}}$, then $p^2 = x^{\frac{4}{3}}$
substitute: $\;\; p^2 - 29p + 100 = 0$
factor: $\qquad (p - 4)(p - 25) = 0$
solve: $\qquad\;\; p - 4 = 0 \qquad p - 25 = 0$
$\qquad\qquad\qquad\qquad p = 4 \qquad\qquad p = 25$

replace with x: $x^{\frac{2}{3}} = 4 \qquad x^{\frac{2}{3}} = 25$
solve: $\qquad\qquad\;\; x = 8 \qquad\;\; x = 125$

44. B; $\dfrac{3x - 3}{3} \cdot \dfrac{x + 1}{x^2 - 1} = \dfrac{3(x - 1)(x + 1)}{3(x - 1)(x + 1)} = 1$

Lesson 11.5, page 536–541

EXPLORE

1. The image distance will decrease $\dfrac{1}{5} + \dfrac{1}{I} = \dfrac{1}{F}$,
If F is constant and S increases, then I must decrease.

2. C; $\dfrac{1}{S} + \dfrac{1}{I} = \dfrac{I}{SI} + \dfrac{S}{SI} = \dfrac{S + I}{SI}$

3. $\dfrac{1}{3} + \dfrac{1}{2} = \dfrac{2}{2}\left(\dfrac{1}{3}\right) + \dfrac{3}{3}\left(\dfrac{1}{2}\right) = \dfrac{5}{6}$
To add fractions, they must have a common
denominator.

TRY THESE

1. LCD $= x^2 y^3$

2. $36abc^3 d$

3. $4(x - y)(x + y)$

4. $\dfrac{7x}{y}$

5. $\dfrac{-3 + 2 - k}{m} = \dfrac{-1 - k}{m}$

6. $\dfrac{x - y}{x - y} = 1$

7. $\dfrac{3y - 1}{y - 2} + \dfrac{y + 3}{2 - y} = \dfrac{3y - 1}{y - 2} + \dfrac{y + 3}{-(y - 2)}$

$= \dfrac{3y - 1}{y - 2} - \dfrac{y + 3}{y - 2}$

$= \dfrac{3y - 1 - (y + 3)}{y - 2}$

$= \dfrac{3y - 1 - y - 3}{y - 2}$

$= \dfrac{2y - 4}{y - 2} = \dfrac{2(y - 2)}{(y - 2)} = 2$

8. $\dfrac{3x+2}{3x-6} - \dfrac{x+2}{x^2-4} = \dfrac{3x+2}{3(x-2)} - \dfrac{x+2}{(x+2)(x-2)}$

$= \dfrac{(3x+2)(x+2)}{3(x-2)(x+2)} - \dfrac{(x+2)3}{(x+2)(x-2)3}$

$= \dfrac{(3x+2)(x+2) - (x+2)3}{3(x+2)(x-2)}$

$= \dfrac{3x^2+8x+4-3x-6}{3(x+2)(x-2)} = \dfrac{3x^2+5x-2}{3(x+2)(x-2)}$

$= \dfrac{(3x-1)(x+2)}{3(x+2)(x-2)}$

$= \dfrac{3x-1}{3x-6}$

9. $\dfrac{n}{n^2-n-6} - \dfrac{n+2}{n^2+5n+6} =$

$\dfrac{n}{(n-3)(n+2)} - \dfrac{(n+2)}{(n+3)(n+2)}$

$= \dfrac{n(n+3)}{(n+3)(n-3)(n+2)} - \dfrac{(n+2)(n-3)}{(n+3)(n+2)(n-3)}$

$= \dfrac{n^2+3n-n^2+n+6}{(n+3)(n+2)(n-3)} = \dfrac{4n+6}{(n+3)(n+2)(n-3)}$

10a. $\dfrac{1}{S} + \dfrac{1}{I} = \dfrac{I}{I}\left(\dfrac{1}{S}\right) + \dfrac{S}{S}\left(\dfrac{1}{I}\right) = \dfrac{I}{IS} + \dfrac{S}{IS} = \dfrac{I+S}{IS}$

10b. $F = \dfrac{I+S}{IS}$

11. You multiply the numerator & denominator by the same quantity which does not change the value of the fraction

12a. $d = vt$

$t_1 = \dfrac{x}{v_1} = \dfrac{x}{15}$

$t_2 = \dfrac{x}{v_2} = \dfrac{x}{10}$

12b. $t_1 = \dfrac{12}{15} = 0.8$ hour

$t_2 = \dfrac{12}{10} = 1.2$ hours

$t_2 - t_1 = 1.2 - 0.8 = 0.4\text{h} = 24$ minutes

PRACTICE

1. $24x^3y$

2. $108a^4bc^2$

3. $m(m+n)$

4. $10(x-2y)$

5. $3(x+4)(x+5)$

6. $x(x+1)(x-2)^4$

7. $\dfrac{5x}{x-3} - \dfrac{2x+1}{x-3} = \dfrac{5x-2x-1}{x+3} = \dfrac{3x-1}{x+3}$

8. $\dfrac{x}{4} + \dfrac{2x}{5} = \dfrac{5x}{20} + \dfrac{8x}{20} = \dfrac{5x+8x}{20} = \dfrac{13x}{20}$

9. $\dfrac{3}{p+1} - \dfrac{5}{p+1} = \dfrac{3-5}{p+1} = \dfrac{-2}{p+1}$

10. $\dfrac{4}{x^2} + \dfrac{3}{xy} = \dfrac{4y}{x^2y} + \dfrac{3x}{x^2y} = \dfrac{4y+3x}{x^2y}$

11. $\dfrac{2x}{2-x} + \dfrac{x}{2} = \dfrac{2(2x)}{2(2-x)} + \dfrac{(2-x)x}{(2-x)2} =$

$\dfrac{4x+2x-x^2}{2(2-x)} = \dfrac{6x-x^2}{2(2-x)}$

12. $\dfrac{2}{3x} + \dfrac{3}{4x} - \dfrac{4}{5x} = \dfrac{20(2)}{20(3x)} + \dfrac{15(3)}{15(4x)} - \dfrac{12(4)}{12(5x)} =$

$\dfrac{40}{60x} + \dfrac{45}{60x} - \dfrac{48}{60x} = \dfrac{37}{60x}$

13. $\dfrac{2}{x-1} - \dfrac{3}{x-2} = \dfrac{2(x-2)}{(x-1)(x-2)} - \dfrac{3(x-1)}{(x-2)(x-1)}$

$= \dfrac{2x-4-3x+3}{(x-1)(x-2)}$

$= \dfrac{-x-1}{(x-1)(x-2)}$

14. $\dfrac{2}{3a-3b} + \dfrac{3}{2a+2b} = \dfrac{2}{3(a-b)} + \dfrac{3}{2(a+b)}$

$= \dfrac{2(a+b)2}{2(a+b)3(a-b)} + \dfrac{3(a-b)3}{3(a-b)2(a+b)}$

$= \dfrac{4a+4b+9a-9b}{6(a+b)(a-b)}$

$= \dfrac{13a-5b}{6(a+b)(a-b)}$

15. $1 + \dfrac{2}{m} = \dfrac{m}{m}(1) + \dfrac{2}{m} = \dfrac{2+m}{m}$

16. $12 - \dfrac{4x-3y}{x-y} = 12\dfrac{x-y}{x-y} - \dfrac{4x-3y}{x-y}$

$= \dfrac{12x-12y}{x-y} - \dfrac{4x-3y}{x-y}$

$= \dfrac{8x-9y}{x-y}$

17. $\dfrac{y-1}{3y+15} - \dfrac{y+3}{5y+25} = \dfrac{y-1}{3(y+5)} - \dfrac{y+3}{5(y+5)}$

$= \dfrac{5(y-1)}{5(3)(y+5)} - \dfrac{3(y+3)}{3(5)(y+5)}$

$= \dfrac{5y-5-3y-9}{15(y+5)}$

$= \dfrac{2y-14}{15y(y+5)}$

18. $\dfrac{x}{x^2-2x+1} - \dfrac{1}{x-1} = \dfrac{x}{(x-1)(x-1)} - \dfrac{1}{x-1}$

$= \dfrac{x}{(x-1)(x-1)} - \dfrac{(x-1)1}{(x-1)(x-1)}$

$= \dfrac{x-x+1}{(x-1)(x-1)}$

$= \dfrac{1}{x^2-2x+1}$

19. $\dfrac{x+3}{x^2-4x} + \dfrac{x-2}{x-4} = \dfrac{x+3}{x(x-4)} + \dfrac{x-2}{x-4}$

$\qquad = \dfrac{x+3}{x(x-4)} + \dfrac{x(x-2)}{x(x-4)}$

$\qquad = \dfrac{x+3+x^2-2x}{x(x-4)}$

$\qquad = \dfrac{x^2-x+3}{x(x-4)}$

20. $\dfrac{6ab}{a^2-b^2} - \dfrac{a+b}{ab} = \dfrac{6ab}{(a+b)(a-b)} - \dfrac{a+b}{ab}$

$\qquad = \dfrac{ab(6ab)}{ab(a+b)(a-b)} - \dfrac{(a+b)(a-b)(a+b)}{(a+b)(a-b)ab}$

$\qquad = \dfrac{6a^2b^2 - a^3 + ab^2 - a^2b - b^3}{ab(a+b)(a-b)}$

$\qquad = \dfrac{-a^3 + 6a^2b^2 + ab^2 - a^2b - b^3}{ab(a+b)(a-b)}$

21. $\dfrac{x+3}{x^2-6x+9} - \dfrac{8x-24}{9-x^2}$

$\qquad = \dfrac{x+3}{(x-3)(x-3)} - \dfrac{8x-24}{(3-x)(3+x)}$

$\qquad = \dfrac{x+3}{(x-3)(x-3)} - \dfrac{8x-24}{-(x-3)(x+3)}$

$\qquad = \dfrac{(x+3)(x+3)}{(x+3)(x-3)(x-3)} + \dfrac{(8x-24)(x-3)}{(x-3)(x+3)(x-3)}$

$\qquad = \dfrac{x^2+6x+9+8x^2-48x+72}{(x+3)(x-3)(x-3)}$

$\qquad = \dfrac{9x^2-42x+81}{(x-3)^2(x+3)}$

22. $\dfrac{x^2+2}{x^2-x-2} + \dfrac{1}{x+1} - \dfrac{x}{x-2}$

$\qquad = \dfrac{x^2+2}{(x+1)(x-2)} + \dfrac{1}{x+1} - \dfrac{x}{x-2}$

$\qquad = \dfrac{x^2+2}{(x+1)(x-2)} + \dfrac{x-2}{(x+1)(x-2)} -$

$\qquad \dfrac{x(x+1)}{(x-2)(x+1)} = \dfrac{x^2+2+x-2-x^2-x}{(x+1)(x-2)} = 0$

23. $\dfrac{1}{x+y} + \dfrac{1}{x-y} - \dfrac{1}{x^2-y^2}$

$\qquad = \dfrac{1}{x+y} + \dfrac{1}{x-y} - \dfrac{1}{(x+y)(x-y)}$

$\qquad = \dfrac{x-y}{(x+y)(x-y)} + \dfrac{x+y}{(x-y)(x+y)} - \dfrac{1}{(x+y)(x-y)}$

$\qquad = \dfrac{2x-1}{(x+y)(x-y)}$

24a. $V_N = V_A - V_W$ (headwind)

$\qquad V_A = 500$ mi/h

$\qquad V_N = V_A + V_W$ (tailwind)

$\qquad t_{WE} = \dfrac{d}{V_A + V_W} = \dfrac{d}{500+100} = \dfrac{d}{600}$ h

$\qquad t_{EW} = \dfrac{d}{V_A - V_W} = \dfrac{d}{400}$ h

24b. $t_T = \dfrac{d}{600} + \dfrac{d}{400}$

24c. $t_T \dfrac{2 \cdot 300}{2 \cdot 600} + \dfrac{360 \cdot 3}{400 \cdot 3} = \dfrac{720 + 1080}{1200} = \dfrac{1800}{1200}$

$\qquad t_T = 1.5$ hrs

25a. $\text{A} \rightarrow 16$
$\qquad \text{B} \rightarrow 12$
$\qquad \text{C} \rightarrow h$

$\qquad t = (h-8); \dfrac{(h-8)}{16}, \dfrac{(h-8)}{12}, \dfrac{(h-8)}{h}$

25b. $\dfrac{(h-8)}{16} + \dfrac{(h-8)}{12} + \dfrac{(h-8)}{h}$

$\qquad = \dfrac{3h(h-8)}{3h(16)} + \dfrac{4h(h-8)}{4h(12)} + \dfrac{48(h-8)}{48h}$

$\qquad = \dfrac{3h^2-24h+4h^2-32h+48h-384}{48h}$

$\qquad = \dfrac{7h^2-8h-384}{48h}$

25c. $2 = h - 8$
$\qquad 10 = h$

$\qquad \dfrac{7(10)^2 - 8(10) - 384}{48(10)} = \dfrac{700 - 80 - 384}{480}$

$\qquad = \dfrac{236}{480} = \dfrac{59}{120}$

26. $n + \dfrac{1}{n}$, the LCD is n, therefore $\dfrac{n^2}{n} + \dfrac{1}{n} = \dfrac{n^2+1}{n}$

27. $\left(\dfrac{m^2+4m-5}{2m^2+m-3} \cdot \dfrac{m-1}{m+5} \right) - \dfrac{2}{m+2}$

$\qquad \dfrac{(m+5)(m-1)(m-1)}{(2m+3)(m-1)(m+5)} - \dfrac{2}{m+2}$

$\qquad = \dfrac{m-1}{2m+3} - \dfrac{2}{m+2}$

$\qquad = \dfrac{(m-1)(m+2) - 2(2m+3)}{(2m+3)(m+2)}$

$\qquad = \dfrac{m^2+m-2-4m-6}{(2m+3)(m+2)}$

$\qquad = \dfrac{m^2-3m-8}{(2m+3)(m+2)}$

28. $\left(4 + \dfrac{1}{n+4} \right)\left(\dfrac{n+4}{n-3} \right)$

$\qquad = \left(\dfrac{4(n+4)}{n+4} + \dfrac{1}{n+4} \right)\left(\dfrac{n+4}{n-3} \right)$

$\qquad = \left(\dfrac{4n+17}{n+4} \right)\left(\dfrac{n+4}{n-3} \right)$

$\qquad = \dfrac{4n+17}{n-3}$

29a. $\dfrac{1}{h} = \dfrac{\frac{1}{a} + \frac{1}{b}}{2}, h = \dfrac{2}{\frac{1}{a} + \frac{1}{b}} = \dfrac{2}{\frac{b}{ba} + \frac{a}{ba}} = \dfrac{2}{\frac{b+a}{ba}} = \dfrac{2ba}{b+a}$

$\qquad h = \dfrac{2(5)(20)}{20+5} = \dfrac{200}{25} = 8$

29b. $h = \dfrac{2(2)(6)}{2 + 6} = \dfrac{24}{8} = 3$

29c. $h = \dfrac{2(5)15}{5 + 15} = \dfrac{150}{20} = 7.5$

29d. $\dfrac{2ba}{b + a}$ this expression found in part a.

30a. $x_1, x_2 = \dfrac{-b \pm \sqrt{b^2 - 4ac}}{2a}$

30b. $x_1 + x_2 = \dfrac{-b \pm \sqrt{b^2 - 4ac}}{2a} + \dfrac{-b - \sqrt{b^2 - 4ac}}{2a}$

$= \dfrac{-2b}{2a} = -\dfrac{b}{a}$

30c. $x_1 - x_2 = \dfrac{-b + \sqrt{b^2 - 4ac}}{2a} - \dfrac{-b - \sqrt{b^2 - 4ac}}{2a}$

$x_1 - x_2 = \dfrac{2\sqrt{b^2 - 4ac}}{2a} = \dfrac{\sqrt{b^2 - 4ac}}{a}$

31. $\dfrac{x}{x^2 - 1} + \dfrac{A}{x^2 - 1} = \dfrac{1}{x - 1}$

$\dfrac{x}{(x - 1)(x + 1)} + \dfrac{A}{(x - 1)(x + 1)} = \dfrac{1(x + 1)}{(x - 1)(x + 1)}$

$\dfrac{x + A}{(x - 1)(x + 1)} = \dfrac{x + 1}{(x - 1)(x + 1)}$

$A = 1$

32. $\dfrac{1}{x + 2} + \dfrac{A}{x - 3} = \dfrac{4x + 3}{(x + 2)(x - 3)}$

$\dfrac{(x - 3)1}{(x - 3)(x + 2)} + \dfrac{A(x + 2)}{(x - 3)(x + 2)} = \dfrac{4x + 3}{(x + 2)(x - 3)}$

$x + 3 + A(x + 2) = 4x + 3$

$\quad A(x + 2) = 3x + 6$

$\quad A(x + 2) = 3(x + 2)$

$\quad A = 3$

MIXED REVIEW

33. $12 - 9 \div 3 = 12 - 3 = 9$

34. $7 + 8 \cdot 2 = 7 + 16 = 23$

35. $64 \div 8 \div 2 = 8 \div 2 = 4$

36. $5 \cdot 2^3 = 5 \cdot 8 = 40$

37. B; $x = -\dfrac{b}{2a} = -\dfrac{6}{2(1)} = -3$

38. $\dfrac{x}{x - 3} - \dfrac{3}{x - 3} = \dfrac{x - 3}{x - 3} = 1$

39. $\dfrac{2}{x^2 - 1} + \dfrac{2x}{x + 1} = \dfrac{2}{(x - 1)(x + 1)} + \dfrac{2x}{x + 1}$

$= \dfrac{2}{(x - 1)(x + 1)} + \dfrac{2x(x - 1)}{(x + 1)(x - 1)}$

$= \dfrac{2 + 2x^2 - 2x}{(x + 1)(x - 1)}$

$= \dfrac{2x^2 - 2x + 2}{x^2 - 1}$

40. $\dfrac{-5x + 5}{x^2 - x - 6} + \dfrac{3}{x + 2} - \dfrac{x - 1}{x - 3}$

$= \dfrac{-5x + 5}{(x + 2)(x - 3)} + \dfrac{3(x - 3)}{(x + 2)(x - 3)} + \dfrac{(-x + 1)(x + 2)}{(x + 2)(x - 3)}$

$= \dfrac{-5x + 5 + 3x - 9 + -x^2 - x + 2}{(x + 2)(x - 3)}$

$= \dfrac{-x^2 - 3x - 2}{(x + 2)(x - 3)}$

$= \dfrac{-1(x^2 + 3x + 2)}{(x + 2)(x - 3)} = \dfrac{-1(x + 2)(x + 1)}{(x + 2)(x - 3)}$

$= \dfrac{-x - 1}{x - 3}$

ALGEBRAWORKS

1a. Move farther from the subject.

1b. increase the f-stop or close the aperature.

1c. increase the shutter speed.

2. Answers will vary.

3. $1 = 1$ second $\qquad 15 = \dfrac{1}{15}$ second

$2 = \dfrac{1}{2}$ second $\qquad 30 = \dfrac{1}{30}$ second

$4 = \dfrac{1}{2}\left(\dfrac{1}{2}\right) = \dfrac{1}{4}$ second $\quad 60 = \dfrac{1}{60}$ second

$8 = \dfrac{1}{2}\left(\dfrac{1}{4}\right) = \dfrac{1}{8}$ second $\quad 125 = \dfrac{1}{125}$ second

$\qquad\qquad\qquad\qquad\qquad 250 = \dfrac{1}{250}$ second

$\qquad\qquad\qquad\qquad\qquad 500 = \dfrac{1}{500}$ second

$\qquad\qquad\qquad\qquad 1000 = \dfrac{1}{1000}$ second

4. 1000

Lesson 11.6, page 543–547

EXPLORE

1. $\dfrac{\dfrac{1}{16}}{\dfrac{3}{3} \cdot \dfrac{3}{4} + \dfrac{1}{12}} = \dfrac{\dfrac{1}{16}}{\dfrac{10}{12}} = \dfrac{1}{16} \cdot \dfrac{12}{10} = \dfrac{12}{160} = \dfrac{3}{40}$

2. $\dfrac{48\left(\dfrac{1}{16}\right)}{48\left(\dfrac{3}{4} + \dfrac{1}{12}\right)} = \dfrac{3}{36 + 4} = \dfrac{3}{40}$

LCD $= 48$; the answers are the same.

3a. $\dfrac{\dfrac{3}{3} \cdot \dfrac{1}{2} + \dfrac{1}{3} \cdot \dfrac{2}{2}}{\dfrac{3}{3} \cdot \dfrac{1}{4} + \dfrac{1}{6} \cdot \dfrac{1}{2}} = \dfrac{\dfrac{5}{6}}{\dfrac{5}{12}} = \dfrac{5}{6} \cdot \dfrac{12}{5} = 2$

$\dfrac{12\left(\dfrac{1}{2} + \dfrac{1}{3}\right)}{12\left(\dfrac{1}{4} + \dfrac{1}{6}\right)} = \dfrac{6 + 4}{3 + 2} = \dfrac{10}{5} = 2$

3b. $\dfrac{\dfrac{4}{4}\cdot\dfrac{2}{5}+\dfrac{3}{4}\cdot\dfrac{5}{5}}{\dfrac{2}{2}\cdot\dfrac{3}{10}+\dfrac{7}{20}}=\dfrac{\dfrac{8+15}{20}}{\dfrac{6+7}{20}}=\dfrac{\dfrac{23}{20}}{\dfrac{13}{20}}=\dfrac{23}{20}\cdot\dfrac{20}{13}=\dfrac{23}{13}$

4. Answers will vary. Students may find the second method is most efficient because you need to reduce the fraction later and if all fractions use the LCD, then it divides out easily.

TRY THESE

1. $\dfrac{\dfrac{2}{3}\cdot\dfrac{5}{5}}{\dfrac{8}{15}}=\dfrac{10}{15}\cdot\dfrac{15}{8}=\dfrac{10}{8}=\dfrac{5}{4}$

2. $\dfrac{1+\dfrac{5}{6}}{2-\dfrac{1}{4}}\cdot\dfrac{12}{12}=\dfrac{12+10}{24-3}=\dfrac{22}{21}$

3. $\dfrac{3+\dfrac{1}{y}}{2-\dfrac{2}{y}}\cdot\dfrac{y}{y}=\dfrac{3y+1}{2y-2}=\dfrac{3y+1}{2y-2}$

4. $\dfrac{\dfrac{2}{x}}{\dfrac{8}{x^2}}\cdot\dfrac{x^2}{x^2}=\dfrac{2x}{8}=\dfrac{x}{4}$

5. $\dfrac{\dfrac{1}{a}}{1-\dfrac{1}{a}}\cdot\dfrac{a}{a}=\dfrac{1}{a-1}$

6. $\dfrac{5-\dfrac{2}{m-1}}{3-\dfrac{1}{m-1}}\cdot\dfrac{m-1}{m-1}=\dfrac{5(m-1)-2}{3(m-1)-1}=\dfrac{5m-7}{3m-4}$

7. $\dfrac{1}{\dfrac{1}{a}+\dfrac{1}{b}}\cdot\dfrac{ab}{ab}=\dfrac{ab}{b+a}$

8. $\dfrac{\dfrac{2x-2}{x^2+2x}}{\dfrac{4x-4}{3x+6}}=\dfrac{\dfrac{2x-2}{x(x+2)}}{\dfrac{4x-4}{3(x+2)}}\cdot\dfrac{3x(x+2)}{3x(x+2)}=\dfrac{3(2x-2)}{x(4x-4)}=$

$\dfrac{6(x-1)}{4x(x-1)}=\dfrac{3}{2x}$

9. $\dfrac{\dfrac{x^2-1}{3x+3}}{\dfrac{x-1}{x+1}}=\dfrac{\dfrac{(x+1)(x-1)}{3(x+1)}}{\dfrac{x-1}{x+1}}\cdot\dfrac{3(x+1)}{3(x+1)}=$

$\dfrac{(x+1)(x-1)}{3(x-1)}=\dfrac{x+1}{3}$

10a. Deluxe: $\dfrac{200}{x}+\dfrac{300}{y}$

Cameraworks: $\dfrac{240}{x}+\dfrac{320}{y}$

10b. $\dfrac{\text{Deluxe}}{\text{Camera W.}}=\dfrac{\dfrac{200}{x}+\dfrac{300}{y}}{\dfrac{240}{x}+\dfrac{320}{y}}\cdot\dfrac{xy}{xy}=\dfrac{200y+300x}{240y+320x}$

$=\dfrac{10y+15x}{12y+16x}$

11. $\dfrac{\dfrac{1}{x}-1}{\dfrac{2}{x}+1}$

Method 1: add the rational numbers in the denominator & the divide into numerator.

Method 2: Find LCD of all rational numbers & then reduce.

Method 2 would be easiest

Method 1: $\dfrac{\dfrac{1}{x}-1}{\dfrac{2+x}{x}}=\left(\dfrac{1}{x}-1\right)\left(\dfrac{x}{2+x}\right)=\dfrac{1-x}{2+x}$

Method 2: $\dfrac{\dfrac{1}{x}-\dfrac{x}{x}}{\dfrac{2}{x}+\dfrac{x}{x}}\cdot\dfrac{x}{x}=\dfrac{1-x}{2+x}$

Method 2 was easiest because it only took one step.

1. $\dfrac{\dfrac{3}{4}}{\dfrac{3}{16}}\cdot\dfrac{16}{16}=\dfrac{12}{3}=4$

2. $\dfrac{1+\dfrac{5}{6}}{1-\dfrac{3}{8}}\cdot\dfrac{24}{24}=\dfrac{24+20}{24-9}=\dfrac{44}{15}$

3. $\dfrac{\dfrac{5}{4}-\dfrac{7}{8}}{\dfrac{1}{16}+\dfrac{1}{12}}\cdot\dfrac{16}{16}=\dfrac{20-14}{1+8}=\dfrac{6}{9}=\dfrac{2}{3}$

4. $\dfrac{\dfrac{6}{7}}{\dfrac{a}{a}}\cdot\dfrac{a}{a}=\dfrac{6}{7}$

5. $\dfrac{\dfrac{5}{x}+5}{\dfrac{3}{2x}+2}\cdot\dfrac{2x}{2x}=\dfrac{10+10x}{3+4x}$

6. $\dfrac{\dfrac{9}{c-d}}{\dfrac{4}{c-d}}\cdot\dfrac{c-d}{c-d}=\dfrac{9}{4}$

7. $\dfrac{2-\dfrac{1}{x+y}}{3-\dfrac{2}{x+y}}\cdot\dfrac{x+y}{x+y}=\dfrac{2x+2y-1}{3x+3y-2}$

8. $\dfrac{\dfrac{x^2-y^2}{xy}}{\dfrac{x-y}{y}}\cdot\dfrac{xy}{xy}=\dfrac{(x-y)(x+y)}{x(x-y)}=\dfrac{x+y}{x}$

9. $\dfrac{\dfrac{1}{xy}-\dfrac{1}{y^2}}{\dfrac{1}{x^2y}-\dfrac{1}{xy^2}}\cdot\dfrac{x^2y^2}{x^2y^2}=\dfrac{xy-x^2}{y-x}=\dfrac{x(y-x)}{y-x}=x$

10. $\dfrac{\dfrac{1}{y+2}-3}{2+\dfrac{2}{y+2}}\cdot\dfrac{y+2}{y+2}=\dfrac{1-3y-6}{2y+4+2}=\dfrac{-3y-5}{2y+6}$

11. $\dfrac{\dfrac{1}{x-4}+5}{\dfrac{2}{x-4}+3}\cdot\dfrac{x-4}{x-4}=\dfrac{1+5x-20}{2+3x-12}=\dfrac{5x-19}{3x-10}$

12.
$$\frac{\dfrac{2}{n-3}+\dfrac{3}{n-2}}{\dfrac{2}{n-2}+\dfrac{3}{n-3}}\cdot\frac{(n-3)(n-2)}{(n-3)(n-2)}=\frac{2n-4+3n-9}{2n-6+3n-6}$$
$$=\frac{5n-13}{5n-12}$$

13.
$$\frac{\dfrac{1}{x+5}+\dfrac{1}{x-3}}{\dfrac{2x^2-3x-5}{(x+5)(x-3)}}\cdot\frac{(x+5)(x-3)}{(x+5)(x-3)}=$$
$$\frac{x-3+x+5}{2x^2-3x-5}=\frac{2x+2}{2x^2-3x-5}=\frac{2(x+1)}{(2x-5)(x+1)}$$
$$=\frac{2}{2x-5}$$

14.
$$\frac{x+\dfrac{xy}{y-x}}{\dfrac{y^2}{x^2-y^2}+1}\cdot\frac{(x-y)(x+y)}{(x-y)(x+y)}=\frac{x(x^2-y^2)-xy(x+y)}{y^2+x^2-y^2}$$
$$=\frac{x^3-xy^2-x^2y-xy^2}{x^2}=\frac{x^2-2y^2-xy}{x}$$

15.
$$\frac{\dfrac{n^2-n-6}{n^2-5n-14}}{\dfrac{n^2+6n+5}{n^2-6n-7}}=\frac{\dfrac{(n-3)(n+2)}{(n-7)(n+2)}}{\dfrac{(n+5)(n+1)}{(n-7)(n+1)}}\cdot$$
$$\frac{(n-7)(n+2)(n+1)}{(n-7)(n+2)(n+1)}=\frac{(n-3)(n+2)(n+1)}{(n+5)(n+1)(n+2)}=$$
$$\frac{n-3}{n+5}$$

16.
$$\frac{\dfrac{t+6}{t}-\dfrac{1}{t+2}}{\dfrac{(t+3)(t+1)}{t(t+1)}}\cdot\frac{t(t+2)}{t(t+2)}=\frac{(t+6)(t+2)-t}{(t+3)(t+2)}$$
$$=\frac{t^2+8t+12-t}{(t+3)(t+2)}=\frac{(t^2+7t+12)}{(t+3)(t+2)}=$$
$$\frac{(t+3)(t+4)}{(t+3)(t+2)}=\frac{t+4}{t+2}$$

17.
$$\frac{\dfrac{5x}{(x-4)(x-2)}}{\dfrac{2}{x-4}+\dfrac{3}{x-2}}\cdot\frac{(x-4)(x-2)}{(x-4)(x-2)}=$$
$$\frac{5x}{2x-4+3x-12}=\frac{5x}{5x-16}$$

18.
$$\frac{\dfrac{1}{y+3}+\dfrac{2y}{(y+3)^3}}{\dfrac{5}{(y+3)^2}+\dfrac{4}{y+3}}\cdot\frac{(y+3)^3}{(y+3)}=\frac{(y+3)^2+2y}{5(y+3)+4(y+3)^2}$$
$$=\frac{y^2+6y+9+2y}{5y+15+4y^2+24y+36}=\frac{y^2+8y+9}{4y^2+29y+51}$$

19a. $E=Wt$

$$t=\frac{E}{W}$$

$$W_1\colon t_1=\frac{100}{n^2-4}$$

$$W_2\colon t_2=\frac{40}{n+2}$$

$$W_3\colon t_3=\frac{50}{n+2}$$

$$W_4\colon t_4=\frac{25}{n-2}$$

19b.
$$\frac{\dfrac{100}{(n+2)(n-2)}+\dfrac{40}{n+2}}{\dfrac{50}{n+2}+\dfrac{25}{n-2}}\cdot\frac{(n+2)(n-2)}{(n+2)(n-2)}=$$
$$\frac{100+40n-80}{50n-100+25n+50}=\frac{40n+20}{75n-50}=\frac{8n+4}{15n-10}$$

EXTEND

20. Find the LCD of both expressions and multiply by the LCD to simplify. Then exchange the numerator & denominator.

21. $a=\dfrac{b+1}{b-1},\ b=\dfrac{1}{1-c}$

$$a=\frac{\dfrac{1}{1-c}+1}{\dfrac{1}{1-c}-1}\cdot\frac{1-c}{1-c}=\frac{1+1-c}{1-1+c}=\frac{2-c}{c}$$

22.
$$\left(\frac{\dfrac{1}{a}-\dfrac{a-x}{a^2+x^2}}{\dfrac{1}{x}-\dfrac{x-a}{x^2+a^2}}+\frac{\dfrac{1}{a}-\dfrac{a+x}{a^2+x^2}}{\dfrac{1}{x}-\dfrac{x+a}{x^2+a^2}}\right)\cdot\frac{ax(a^2+x^2)}{ax(a^2+x^2)}=$$
$$\frac{x(a^2+x^2)-ax(a-x)}{a(a^2+x^2)-ax(x-a)}+\frac{x(a^2+x^2)-ax(a+x)}{a(a^2+x^2)-ax(a+x)}=$$
$$\frac{a^2+x^3-a^2x+ax^2}{a^3+ax^2-a^2x+a^2x}+\frac{a^2x+x^3-a^2x-ax^2}{a^3+ax^2-a^2x-ax^2}$$
$$\frac{x^3+ax^2}{a^3+a^2x}+\frac{x^3-ax^2}{a^3-a^2x}$$
$$\frac{x^2(x+a)}{a^2(x+a)}+\frac{x^2(x-a)}{a^2(a-x)}=\frac{x^2}{a^2}-\frac{x^2}{a^2}=0$$

23.
$$\frac{\dfrac{x}{2}\left(\dfrac{1}{x-y}\right)\dfrac{1}{(x+y)}\left(\dfrac{x^2-y^2}{x^2y+xy^2}\right)}{\dfrac{1}{x+y}}=\frac{\dfrac{x}{2}\left(\dfrac{1}{x+y}\right)\dfrac{(x-y)(x+y)}{xy(x+y)}}{\dfrac{1}{x+y}}$$
$$=\frac{\dfrac{1}{2}\dfrac{1}{y(x+y)}}{\dfrac{1}{x+y}}=\frac{1}{2y(x+y)}\cdot\frac{x+y}{1}=\frac{1}{2y}$$

24a.
$$R_T=\frac{1}{\dfrac{1}{R_1}+\dfrac{1}{R_2}+\dfrac{1}{R_3}}=\frac{1}{\dfrac{R_2R_3+R_1R_3+R_1R_2}{R_1R_2R_3}}=$$
$$\frac{R_1R_2R_3}{R_2R_3+R_1R_3+R_1R_2}$$

24b. $R_T=\dfrac{1\cdot2\cdot6}{2\cdot6+1\cdot6+1\cdot2}=\dfrac{12}{20}=\dfrac{3}{5}$

25. 1, 1.5, 1.4, 1.4167, 1.4138, 1.414

26. the number alternately gets larger then smaller, seems to approach 1.414.

27. It seems to approach the value of $\sqrt{2}$

MIXED REVIEW

28. $\begin{bmatrix} -3 & 5 \\ 5 & -8 \end{bmatrix}$

29. does not exist

30. $\begin{bmatrix} \frac{3}{11} & -\frac{5}{11} \\ \frac{1}{11} & \frac{2}{11} \end{bmatrix}$

31. $\begin{bmatrix} 0 & 1 \\ -1 & 0 \end{bmatrix}$

32. $x(x + 2) \geq 24$
$x^2 + 2x - 24 \geq 0$
$(x + 6)(x - 4) \geq 0$
$x \leq -6$
or $x \geq 4$

33. $x^2 + 4x + 3 \geq 0$
$(x + 3)(x + 1)$
$x \leq -3$
or $x \geq -1$

34. $x^2 - 25 > 0$
$(x - 5)(x + 5) > 0$
$x > 5$ or $x < -5$

35. $x^2 - 8x + 16 \geq 0$
$(x - 4)(x - 4) \geq 0$
$x \geq x$ or $x \leq 4$, so all real numbers

36. C; $\dfrac{\frac{2x + 4}{x^2 - 9}}{\frac{x + 2}{x + 3}} = \dfrac{2(x + 2)}{(x - 3)(x + 3)} \cdot \dfrac{x + 3}{(x + 2)} = \dfrac{2}{x - 3}$

Lesson 11.7, pages 548–553

EXPLORE

1. $y = 2x + 1$
$y = x + 4$
Graph both equations, the x-value of the point(s) of intersection are the solutions.
$x = 3$

2. Plot $y = 3$ and $y = \dfrac{x}{x + 2}$.
The x-value of the point of intersection is the solution.

3. $y = \dfrac{x}{x + 2}$
$y = 3$
therefore, $x = -3$

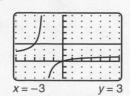
$x = -3 \qquad y = 3$

4. $x \neq 4$ because it is a point of "discontinuity" (asymptote)

5. The curves approach the asymptote.

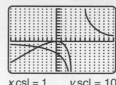
x csl = 1 $\qquad$ y scl = 10

TRY THESE

1. $\dfrac{x}{3} + \dfrac{x}{8} = 11$
$\dfrac{8x + 3x}{24} = 11$
$11x = 264$
$x = 24$

Check: $\dfrac{24}{3} + \dfrac{24}{8} = 11$
$8 + 3 = 11 ✓$

2. $\dfrac{x + 2}{5} + \dfrac{x - 1}{5} = \dfrac{11}{5}$
$2x + 1 = 11$
$2x - 10$
$x = 5$

Check: $\dfrac{5 + 2}{5} + \dfrac{5 - 1}{5} = \dfrac{11}{5}$
$7 + 4 = 11 ✓$

3. $\dfrac{m + 4}{6} = \dfrac{m - 3}{5}$
$5m + 20 = 6m - 18$
$38 = m$

Check: $\dfrac{38 + 4}{6} = \dfrac{38 - 3}{5}$
$7 = 7 ✓$

4. $3 - \dfrac{x - 1}{4} = \dfrac{x}{9}$
$36\left(3 - \dfrac{x - 1}{4}\right) = 36\left(\dfrac{x}{9}\right)$
$108 - 9x + 9 = 4x$
$117 = 13x$
$x = 9$

Check: $3 - \left(\dfrac{9 - 1}{4}\right) = \dfrac{9}{9}$
$3 - \dfrac{8}{4} = 1$
$3 - 2 = 1 ✓$

5. $\dfrac{5}{x} + x = -6$
$x\left(\dfrac{5}{x} + x\right) = -6x$
$5 + x^2 = -6x$
$x^2 + 6x + 5 = 0$
$(x + 5)(x + 1) = 0$
$x = -5, -1$

Check: $\dfrac{-5}{5} + (-5) = -6$
$-1 - 5 = -6 ✓$
$\dfrac{-5}{1} - 1 = -6$
$-5 - 1 = -6 ✓$

6. $\dfrac{18}{2x} - \dfrac{2x}{2x} = \dfrac{4x}{2x}$
$18 - 2x = 4x$
$18 = 6x$
$x = 3$

Check: $\dfrac{18}{2(3)} - 1 = 2$
$\dfrac{18}{6} - 1 = 2$
$3 - 1 = 2 ✓$

7. $\dfrac{1 - x}{1 + x} = 4$
$\dfrac{1 - x}{1 + x} = \dfrac{4(1 + x)}{1 + x}$
$1 - x = 4 + 4x$
$-3 = 5x; x = -\dfrac{3}{5}$

Check: $\dfrac{1 + \frac{3}{5}}{1 - \frac{3}{5}} = 4 = \dfrac{1.6}{0.4}$
$4 = 4 ✓$

8. $2 + \dfrac{3}{y-1} = \dfrac{5}{2y-2}$ Check:

$2(2)(y-1) + 3(2) = 5$ $2 + \dfrac{3}{-0.25} = \dfrac{5}{2(0.75)-2}$

$\qquad 4y - 4 + 6 = 5$ $\qquad 2 - 12 = -10$

$\qquad\qquad 4y = 3$ $\qquad\quad -10 = -10 \checkmark$

$\qquad\qquad\quad y = \dfrac{3}{4}$

9. $\dfrac{x-2}{x-4} = \dfrac{2}{x-4}$ Check: there is no solution

$x \neq 4,\ x - 2 = 2$

$x = 4,\text{ but } x \neq 4$

10a. $x = $ fast worker (minutes per oven)

$x + 3 = $ slow worker

At x minutes per oven, the faster worker assembles $\dfrac{60}{x}$ ovens per hour.

$\dfrac{60}{x} + \dfrac{60}{x+3} = 9$

10b. $60(x+3) + 60x = 9(x+3)x$

$60x + 180 + 60x - 9x^2 - 27x = 0$

$-9x^2 - 93x + 180 = 0$

$(3x + 5)(3x - 36) = 0$

$x = -\dfrac{5}{3},\ 12;$ reject the negative solution since it has no meaning in this problem,

$x = 12$ minutes

11a. $y = \dfrac{2x^3 - x^2 + 5}{x^2 + 2}$

$\dfrac{2x^3 - x^2 + 5}{x^2 + 2} = 5$

11b. $x = 3.25$ months

12a. solve graphically—too difficult to solve algebraically

12b. solve algebraically, easier

12c. solve algebraically, easier

Practice

1. $\dfrac{y}{2} + \dfrac{y}{4} = 12$ Check: $\dfrac{16}{2} + \dfrac{16}{4} = 12$

$2y + y = 48$ $\qquad\quad 8 + 4 = 12$

$\qquad 3y = 48$ $\qquad\qquad 12 = 12 \checkmark$

$\qquad\quad y = 16$

2. $\qquad 3x + \dfrac{3}{4} = \dfrac{5}{6}$ Check: $3\left(\dfrac{1}{36}\right) + \dfrac{3}{4} = \dfrac{5}{6}$

$12(3x) + 12\left(\dfrac{3}{4}\right) = \dfrac{5}{6}(12)$ $\qquad\quad \dfrac{1}{12} + \dfrac{3}{4} = \dfrac{5}{6}$

$\qquad\qquad 36x + 9 = 10$ $\qquad\quad \dfrac{1}{12} + \dfrac{9}{12} = \dfrac{5}{6}$

$\qquad\qquad\quad 36x = 1$ $\qquad\qquad\quad \dfrac{10}{12} = \dfrac{10}{12} \checkmark$

$\qquad\qquad\qquad x = \dfrac{1}{36}$

3. $\dfrac{1}{x} + \dfrac{2}{x} = 1$ Check: $\dfrac{1}{3} + \dfrac{2}{3} = 1$

$\dfrac{1}{x} + \dfrac{2}{x} = \dfrac{x}{x}$ $\qquad\quad \dfrac{3}{3} = 1 \checkmark$

$1 + 2 = x$

$\qquad x = 3$

4. $2x + 3 = \dfrac{x-1}{4}$ Check: $2\left(-\dfrac{13}{7}\right) + 3 = \dfrac{-\frac{13}{7} - 1}{4}$

$\dfrac{4(2x+3)}{4} = \dfrac{x-1}{4}$ $\qquad -\dfrac{26}{7} + 3 = \dfrac{-\frac{20}{7}}{4}$

$8x + 12 = x - 1$ $\quad -\dfrac{26}{7} + \dfrac{21}{7} = -\dfrac{5}{7} = -\dfrac{5}{7} \checkmark$

$\qquad 7x = -13$

$\qquad\quad x = -\dfrac{13}{7}$

5. $\dfrac{3}{n+2} + \dfrac{15}{n+2} = 2$ Check:

$\dfrac{3}{n+2} + \dfrac{15}{n+2} = \dfrac{2(n+2)}{n+2}$ $\dfrac{3}{7+2} + \dfrac{15}{7+2} = 2$

$\qquad\qquad 18 = 2n + 4$ $\qquad\quad \dfrac{3}{9} + \dfrac{15}{9} = 2$

$\qquad\qquad 2n = 14$ $\qquad\qquad\quad \dfrac{18}{9} = 2$

$\qquad\qquad\quad n = 7$ $\qquad\qquad\qquad 2 = 2 \checkmark$

6. $\dfrac{c+1}{c-3} = \dfrac{4}{c-3}$

$c + 1 = 4$

$c = 3,$ however, $c \neq 3,$ so there is no solution

7. $\dfrac{2}{x} = \dfrac{3}{x-4}$ Check: $\dfrac{2}{-8} = \dfrac{3}{-8-4}$

$\dfrac{2(x-4)}{x(x-4)} = \dfrac{3x}{x(x-4)}$ $\qquad -0.25 = \dfrac{3}{-12}$

$\qquad 2x - 8 = 3x$ $\qquad -0.25 = -0.25 \checkmark$

$\qquad\quad -8 = x$

8. $\dfrac{5}{2x+5} = \dfrac{4}{x+5}$

$\dfrac{5(x+5)}{(x+5)(2x+5)} = \dfrac{4(2x+5)}{(x+5)(2x+5)}$

$\qquad 5x + 25 = 8x + 120$

$\qquad\qquad 5 = 3x$

$\qquad\qquad x = \dfrac{5}{3}$

Check: $\dfrac{5}{2\left(\frac{5}{3}\right) + 5} = \dfrac{4}{\frac{5}{3} + 5}$

$\qquad\quad \dfrac{5}{\frac{25}{3}} = \dfrac{4}{\frac{20}{3}}$

$\qquad\quad \dfrac{15}{25} = \dfrac{12}{20}$

$\qquad\qquad \dfrac{3}{5} = \dfrac{3}{5} \checkmark$

9. $\dfrac{3x+12}{3x+23} = \dfrac{3}{4}$ Check: $\dfrac{3(7)+12}{3(7)+23} = \dfrac{3}{4}$

$\dfrac{4(3x+12)}{4(3x+23)} = \dfrac{3(3x+23)}{4(3x+23)}$ $\qquad\quad \dfrac{33}{44} = \dfrac{3}{4}$

$\quad 12x + 48 = 9x + 69$ $\qquad\quad \dfrac{3}{4} = \dfrac{3}{4} \checkmark$

$\qquad\quad 3x = 21$

$\qquad\qquad x = 7$

10.
$$\frac{1}{x+3} = \frac{2}{x} - \frac{3}{4x}$$
$$\frac{4x}{4x(x+3)} = \frac{2(x+3)4}{4x(x+3)} - \frac{3(x+3)}{4x(x+3)}$$
$$4x = 8x + 24 - 3x - 9$$
$$-x = 15$$
$$x = -15$$

Check: $\dfrac{1}{-15+3} = \dfrac{2}{-15} - \dfrac{3}{4(-15)}$
$$-\frac{1}{12} = -\frac{2}{15} + \frac{3}{60}$$
$$-\frac{1}{12} = -\frac{8}{60} + \frac{3}{60}$$
$$-\frac{1}{12} = -\frac{5}{60}$$
$$-\frac{1}{12} = -\frac{1}{12} \checkmark$$

11.
$$\frac{1}{x-5} + \frac{1}{x+5} = \frac{6}{(x-5)(x+5)}$$
$$\frac{x+5}{(x+5)(x-5)} + \frac{(x-5)}{(x-5)(x+5)} = \frac{6}{(x-5)(x+5)}$$
$$x + 5 + x - 5 = 6$$
$$2x = 6$$
$$x = 3$$

Check: $\dfrac{1}{3-5} + \dfrac{1}{3+5} = \dfrac{6}{(-2)(8)}$
$$\frac{1}{-2} + \frac{1}{8} = \frac{6}{-16}$$
$$-0.5 + 0.125 = -0.375$$
$$-0.375 = -0.375 \checkmark$$

12.
$$\frac{5}{x} + \frac{7}{x^2} = 2 \qquad \text{Check: } \frac{5}{\frac{7}{2}} + \frac{7}{\left(\frac{7}{2}\right)^2} = 2$$
$$\frac{5x}{x^2} + \frac{7}{x^2} = \frac{2x^2}{x^2} \qquad \frac{5 \cdot 2}{7} + \frac{7 \cdot 4}{49} = 2$$
$$5x + 7 = 2x^2 \qquad \frac{10}{7} + \frac{28}{49} = 2$$
$$2x^2 - 5x - 7 = 0 \qquad \frac{70}{49} + \frac{28}{49} = 2$$
$$(2x - 7)(x + 1) = 0 \qquad \frac{98}{49} = 2$$
$$2x - 7 = 0 \text{ or } x + 1 = 0 \qquad 2 = 2$$
$$x = \frac{7}{2} \text{ or } \qquad x = -1 \qquad \frac{5}{-1} + \frac{7}{(-1)^2} = 2$$
$$-5 + 7 = 2$$
$$2 = 2 \checkmark$$

13.
$$\frac{6}{x-1} - \frac{4}{x-2} = \frac{2}{x+1}$$
$$\frac{6(x-2)(x+1)}{(x-1)(x-2)(x+1)} - \frac{4(x-1)(x+1)}{(x-2)(x+1)(x-1)} =$$
$$\frac{2(x-1)(x-2)}{(x-2)(x-1)(x+1)}$$
$$6(x^2 - x - 2) - 4(x^2 - 1) = 2(x^2 - 3x + 2)$$
$$6x^2 - 6x - 12 - 4x^2 + 4 = 2x^2 - 6x + 4$$
$$2x^2 - 6x - 8 = 2x^2 - 6x + 4$$
$$-8 = 4 \text{ no solution}$$

14.
$$\frac{6x^2 + 5x - 11}{3x + 2} = 2x - 5$$
$$\frac{6x^2 + 5x - 11}{3x + 2} = \frac{(2x - 5)(3x + 2)}{3x + 2}$$
$$6x^2 + 5x - 11 = 6x^2 - 11x - 10$$
$$16x = 1$$
$$x = \frac{1}{16}$$

Check: $\dfrac{6\left(\frac{1}{16}\right)^2 + 5\left(\frac{1}{16}\right) - 11}{3\left(\frac{1}{16}\right) + 2} = 2\left(\frac{1}{16}\right) - 5$
$$\frac{\frac{6}{256} + \frac{5}{16} - 11}{\frac{3}{16} + 2} = \frac{2}{16} - 5$$
$$\frac{\frac{6}{256} + \frac{5 \cdot 16}{16 \cdot 16} - 11\frac{256}{256}}{\frac{3}{16} + \frac{32}{16}} = \frac{2}{16} - 5\frac{16}{16} \checkmark$$
$$\frac{\frac{6}{256} + \frac{80}{256} - \frac{2816}{256}}{\frac{35}{16}} = -\frac{78}{16}$$
$$-\frac{2730}{256} \cdot \frac{16}{35} = -\frac{78}{16}$$
$$-\frac{78}{16} = -\frac{78}{16} \checkmark$$

PRACTICE

15. $\dfrac{x}{x^2 - 1} = \dfrac{1}{2x - 2} - \dfrac{2}{x + 1}; x \neq 1, 0, -1$
$$\frac{x}{(x-1)(x+1)} = \frac{1}{2(x-1)} - \frac{2}{(x+1)}$$
$$\frac{2x}{2(x-1)(x+1)} = \frac{x+1}{2(x-1)(x+1)} -$$
$$\frac{2(x-1)2}{2(x+1)(x-1)}$$
$$2x = x + 1 - 4x + 4$$
$$2x = -3x + 5$$
$$5x = 5$$
$$x = 1, \text{ since } x \neq 1, \text{ there is not solution}$$

16a. Let $x + 3$ represent the velocity of the canoe traveling downstream.

Let $x - 3$ represent the velocity of the canoe traveling upstream. Use the formula $t = \dfrac{d}{r}$
$$\frac{10}{x + 3} = \frac{4}{x - 3}$$

16b. Find x:
$$\frac{10(x - 3)}{(x + 3)(x - 3)} = \frac{4(x + 3)}{(x + 3)(x - 3)}$$
$$10x - 30 = 4x + 12$$
$$6x = 42$$
$$x = 7 \text{ mi/h in still water}$$

17. Let V_R represent the velocity of Ray.
Let V_A represent the velocity of Aru.
Let t_R represent the time of Ray.
Let t_A represent the time of Aru.

$$\text{velocity} = \frac{\text{distance}}{\text{time}}$$

$$V_R = \frac{500}{t_R}$$

$$V_A = \frac{500}{t_A}$$

$$t_A = t_R - 4$$

$$V_A = 2.48\, V_R$$

$$\frac{500}{t_A} = 2.48\,\frac{500}{t_R}$$

$$\frac{500}{t_R - 4} = 2.48\,\frac{500}{t_R}$$

$$\frac{t_R\,500}{t_R\,(t_R - 4)} = 2.48\,\frac{500\,(t_R - 4)}{t_R\,(t_R - 4)}$$

$$500t_R = 1240(t_R - 4)$$

$$500t_R = 1240t_R - 4960$$

$$4960 = 740t_R$$

$$6.7 \approx t_R$$

$$V_R \approx \frac{500}{6.7} \approx 75 \text{ mi/h}$$

$$V_A \approx 2.48(75) \approx 186 \text{ mi/h}$$

18a. $\dfrac{x}{10} + \dfrac{x}{15} \to \dfrac{3x}{30} + \dfrac{2x}{30} = \dfrac{30}{30},\; 5x = 30$

$$x = 6 \text{ h}$$

Check: $\dfrac{6}{10} + \dfrac{6}{15} = 0.6 + 0.4 = 1\;\checkmark$

18b. $\dfrac{x}{10} + \dfrac{x}{15} + \dfrac{x}{24} = 1$

$$\frac{360x}{3600} + \frac{240x}{3600} + \frac{150x}{3600} = \frac{3600}{3600}$$

$$750x = 3600$$

$$x = 4.8 \text{ hours}$$

Check: $\dfrac{4.8}{10} + \dfrac{4.8}{15} + \dfrac{4.8}{24} = 0.48 + 0.32 + 0.2 = 1\;\checkmark$

19.
$$x + \frac{1}{x} = 4$$

$$\frac{x^2}{x} + \frac{1}{x} = \frac{4x}{x}$$

$$x^2 + 1 = 4x$$

$$x^2 - 4x + 1 = 0$$

$$x = 2 \pm \frac{\sqrt{12}}{2}$$

Check: $\left(2 + \dfrac{\sqrt{12}}{2}\right) + \dfrac{1}{\left(2 + \frac{\sqrt{12}}{2}\right)} \stackrel{?}{=} 4$

$$\frac{4 + \sqrt{12}}{2} + \frac{2}{4 + \sqrt{12}} \stackrel{?}{=} 4$$

$$\frac{4 + \sqrt{12}}{2} + \frac{2(4 - \sqrt{12})}{(4 + \sqrt{12})(4 - \sqrt{12})} \stackrel{?}{=} 4$$

$$\frac{8 + 2\sqrt{12}}{4} + \frac{8 - 2\sqrt{12}}{4} \stackrel{?}{=} 4$$

$$\frac{16}{4} \stackrel{?}{=} 4$$

$$4 = 4 \;\checkmark$$

$$\left(2 - \frac{\sqrt{12}}{2}\right) + \frac{1}{\left(2 - \frac{\sqrt{12}}{2}\right)} \stackrel{?}{=} 4$$

$$\frac{4 - \sqrt{12}}{2} + \frac{2}{4 - \sqrt{12}} \stackrel{?}{=} 4$$

$$\frac{4 - \sqrt{12}}{2} + \frac{2(4 + \sqrt{12})}{(4 - \sqrt{12})(4 + \sqrt{12})} \stackrel{?}{=} 4$$

$$\frac{8 - \sqrt{12}}{4} + \frac{8 + 2\sqrt{12}}{4} \stackrel{?}{=} 4$$

$$\frac{16}{4} \stackrel{?}{=} 4$$

$$4 \stackrel{?}{=} 4 \;\checkmark$$

20a. Plot
$$y_1 = \frac{x^3 + x^2 - 5x + 8}{x^2 + x + 1} \text{ and } y_2 = 4$$

$x = 0.39 \qquad y = 4$

the x-values of the intersections of the graphs reveals
$x = 4.74 \approx$ May 22 and $x = 0.42 \approx$ January 13

20b. A new vein was discovered February 16 where the graph slopes upward; $x \approx 1.58$.

21. $\dfrac{1}{x - 3} + \dfrac{1}{x + 3} = \dfrac{2x}{(x - 3)(x + 3)},\, x \neq 3, -3$

$$x + 3 + x - 3 = 2x$$

$$2x = 2x$$

The solution is all real numbers except $x = 3$ and $x = -3$.

EXTEND

22. $\dfrac{r_1}{L - d} = \dfrac{r_2}{d}$

$$\frac{r_1 d}{(L - d)d} = \frac{r_2(L - d)}{d(L - d)}$$

$$r_1 d = r_2 L - r_2 d$$

$$(r_1 + r_2)d = r_2 L$$

$$d = \frac{r_2 L}{r_1 + r_2}$$

23. $I = \dfrac{E}{R_L + r}$

$$\frac{I(R_L + r)}{R_L + r} = \frac{E}{R_L + r}$$

$$I(R_L + r) = E$$

$$IR_L + Ir = E$$

$$r = \frac{E - IR_L}{I}$$

24. $\left(P + \dfrac{a}{v_2}\right)(v_1 - b) = RT$

$P + \dfrac{a}{V_2} = \dfrac{RT}{V_1 - b}$

$P = \dfrac{RT}{V_1 - b} - \dfrac{a}{V_2}$

25. $\dfrac{4(x + 1)}{10} = \dfrac{12}{x}$

$\qquad 4x^2 + 4x = 120$

$\quad 4x^2 + 4x - 120 = 0$

$\quad (4x + 24)(x - 5) = 0$

$4x + 24 = 0 \quad \text{or} \quad x - 5 = 0$

$4x = -24 \quad \text{or} \quad x = 5$

$x = -6$

Reject $x = -6$ since negative lengths make no sense.

$\overline{EF} = 5$

$\overline{AB} = 24$

26a. Let x be the number of gallons in the tank. The tank can be emptied at $\dfrac{x}{48}$ gal/min. The tank can be filled at $\dfrac{x}{36}$ gal/min. The combination is $\dfrac{x}{36} - \dfrac{x}{48}$ or $\dfrac{x}{144}$ gal/min

26b. If the outlet flow is greater than inflow the tank will never fill.

27. $\dfrac{m - x}{m - n} = 2 + \dfrac{p_1 - x}{n - p}$

$(m - x)(n - p) = 2(m - n)(n - p) + (p - x)(m - n)$

$mn - xn - mp + xp = 2mn - 2n^2 - 2mp + 2np + pm - xm - np + xn$

$xm - xn - xn + xp = 2mn - mn - 2n^2 - 2mp + mp + 2np + mp - np$

$x(m - 2n + p) = mn - 2n^2 - mp + np + mp$

$x = \dfrac{-2n^2 + mn + np}{m - 2n + p}$

$x = \dfrac{n(-2n + m + p)}{m - 2n + p}$

$x = n$

28. $\dfrac{x - a}{x - b} = \dfrac{b - x}{a - x}$

$\dfrac{x - a}{x - b} = \dfrac{-(x - b)}{-(x - a)}$

$\dfrac{(x - a)(x - a)}{(x - b)(x - a)} = \dfrac{(x - b)(x - b)}{(x - b)(x - a)}$

$x^2 - 2ax + a^2 = x^2 - 2bx + b^2$

$-2ax + 2bx = b^2 - a^2$

$2x(b - a) = b^2 - a^2$

$x = \dfrac{b^2 - a^2}{2(b - a)}$

$\quad = \dfrac{(b - a)(b + a)}{2(b - a)}$

$\quad = \dfrac{b + a}{2}$

29. $|x| < 2$

$x < 2 \text{ or } x > -2$

$-2 < x < 2$

30. $4x + 2 \geq 6 \qquad \text{or} \quad 4x + 2 \leq -6$

$\quad 4x \geq 4 \qquad \text{or} \qquad 4x \leq -8$

$\quad x \geq 1 \qquad \text{or} \qquad x \leq -2$

31. $|2x - 7| > 10$

$2x - 7 > 10 \quad \text{or} \quad 2x - 7 < -10$

$\quad 2x > 17 \quad \text{or} \qquad 2x < -3$

$\quad x > \dfrac{17}{2} \quad \text{or} \qquad x < -\dfrac{3}{2}$

32. $|3x + 5| \geq 0$

$3x + 5 \geq 0 \qquad \text{or} \quad 3x + 5 \leq 0$

$\quad 3x \geq -5 \qquad \text{or} \qquad 3x \leq -5$

$\quad x \geq -\dfrac{5}{3} \quad \text{or} \qquad x \leq -\dfrac{5}{3}$

all real numbers

33. $i^2 = \sqrt{-1} \cdot \sqrt{-1} = -1$

34. $i^3 = \sqrt{-1} \cdot \sqrt{-1} \cdot \sqrt{-1} = -1\sqrt{-1} = -\sqrt{-1} = i$

35. $4i^{13} = ri^{12} = i^1 = 4i$

36. $\dfrac{2}{i} - \dfrac{2}{i} \cdot \dfrac{i}{i} = \dfrac{2i}{-1} = -2i$

37. A; $\quad 64 = 32^x$

$2^6 = 2^{5x}$

$6 = 5x$

$x = 1.2$

38. $\dfrac{c}{c + 2} = \dfrac{7}{9}$ $\qquad$ Check:

$\dfrac{9c}{9(c + 2)} = \dfrac{7(c + 2)}{9(c + 2)} \qquad \dfrac{7}{7 + 2} = \dfrac{7}{9}$

$9c = 7c + 14 \qquad\qquad \dfrac{7}{9} = \dfrac{7}{9} \checkmark$

$2c = 14$

$c = 7$

39. $x + \dfrac{3}{x} = 4$ $\qquad$ Check:

$\dfrac{x^2}{x} + \dfrac{3}{x} = \dfrac{4x}{x} \qquad\qquad 1 + \dfrac{3}{1} = 4$

$x^2 + 3 = 4x \qquad\qquad\quad 4 = 4 \checkmark$

$x^2 - 4x + 3 = 0 \qquad\quad 3 + \dfrac{3}{3} = 4$

$(x - 3)(x - 1) = 0 \qquad 3 + 1 = 4$

$x = 1, 3 \qquad\qquad\qquad 4 = 4 \checkmark$

40. $\dfrac{5}{x} + \dfrac{3}{x} = 1$ $\qquad$ Check:

$5 + 3 = x \qquad\qquad\qquad \dfrac{5}{8} + \dfrac{3}{8} = 1$

$8 = x \qquad\qquad\qquad\qquad \dfrac{8}{8} = 1$

$\qquad\qquad\qquad\qquad\qquad 1 = 1 \checkmark$

41. $2 - \dfrac{1}{a} = \dfrac{1}{a^2}$ $\qquad$ Check:

$\dfrac{2a^2}{a^2} - \dfrac{a}{a^2} = \dfrac{1}{a^2} \qquad\qquad 2 - \dfrac{1}{1} = \dfrac{1}{1}$

$2a^2 - a = 1 \qquad\qquad\qquad 2 - 1 = 1$

$2a^2 - a - 1 = 0 \qquad\qquad\quad 1 = 1 \checkmark$

$(2a + 1)(a - 1) = 0$

$a = 1, -\dfrac{1}{2}$

$2 + \dfrac{1}{\frac{1}{2}} = \dfrac{1}{\frac{1}{4}}$

$2 + 2 = 4$

$4 = 4 \checkmark$

Lesson 11.7, pages 554–557

EXPLORE THE PROBLEM

1. $\dfrac{1}{n}$

2. $\dfrac{1}{n}; \dfrac{2}{n}; \dfrac{n}{n}$

3. $\dfrac{1}{n} \cdot \dfrac{1}{n} = \dfrac{1}{n^2}; \dfrac{1}{n} \cdot \dfrac{2}{n} = \dfrac{2}{n^2}; \dfrac{1}{n} \cdot \dfrac{3}{n} = \dfrac{3}{n^2}; \dfrac{n}{n} \cdot \dfrac{1}{n} = \dfrac{n}{n^2}$

4. The sum of the areas of all the rectangles will be close to the actual area under the curve. It will be slightly more since the rectangles are all slightly greater than the curve height.

5. $A = \dfrac{n + 1}{2n} = \dfrac{5 + 1}{2(5)} = \dfrac{6}{10} = 0.6$

   $A = \dfrac{10 + 1}{2(10)} = 0.55, A = \dfrac{100 + 1}{2(100)} = 0.505$

   As n gets larger, A approaches 0.5.

6. This value approaches $\dfrac{1}{2}$ as n gets very large.

   $A = \dfrac{1}{2}bh, A = \dfrac{1}{2}(1)(1) = 5$

7.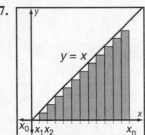

   This approximation is less than the actual area. As n gets very large the area approaches 0.5, $A = \dfrac{n + 1}{2n}$

INVESTIGATE FURTHER

8. $x_0 = 0; x_1 = 0.5; x_2 = 1.0; x_3 = 1.5; x_4 = 2.0$

9. $y_1 = (0.5)^2 = 0.25$

   $y_2 = (1)^2 = 1$

   $y_3 = (1.5)^2 = 2.25$

   $y_4 = (2.0)^2 = 4$

10. $A_T = (0.5)(0.25) + (0.5)(1) +$

    $(0.5)(2.25) + (0.5)(4.0)$

    $A_T = 0.125 + 0.5 + 1.125 + 2 = 3.75$

11. This approximation is larger than the actual area. It could be improved by increasing the number of rectangles.

12. There are three rectangles because it uses the functional value at the endpoint of each of the n intervals.

    $A = (x_2 - x_1)0.25 + (x_3 - x_2)1 + (x_4 - x_3)\dfrac{9}{4}$

    $= 0.5(0.25) - 0.5(1) + (0.5)\left(\dfrac{9}{4}\right)$

    $A = .125 + 0.5 + 1.125 = 1.75$

13. $1.75 < AA < 3.75$

APPLY THE STRATEGY

14a. $\dfrac{2}{8} = 0.25$: each interval

   $y = (0.25)^2 = 0.0625$

   $y = (0.5)^2 = 0.25$

   $y = (0.75)^2 = 0.5625$

   $y = (1)^2 = 1$

   $y = (1.25)^2 = 1.5625$

   $y = (1.5)^2 = 2.25$

   $y = (1.75)^2 = 3.0625$

   $y = (2.0)^2 = 4$

   $A = 0.25(0.0625) + 0.25(0.25) + 0.25(0.5625) +$
   $0.25(1) + 0.25(1.5625) + 0.25(2.25) +$
   $0.25(3.0625) + 0.25(4) = 3.1875$

14b. $\dfrac{8}{3} < 3.1875 < 3.75$

15a. $y = 4x - x^2$

   $y_1 = 4(0.5) - (0.5)^2 = 1.75$

   $y_2 = 4(1.0) - (1.0)^2 = 3$

   $y_3 = 4(1.5) - (1.5)^2 = 3.75$

   $y_4 = 4(2.0) - (2.0)^2 = 4$

   $A = (1.75)(0.5) + (3)(0.5) + (3.75)(0.5) + (4)(0.5)$

   $A = 0.875 + 4.5 + 1.875 + 2 = 6.25$

   $y_1 = 1.75$

   $y_2 = 3$

   $y_3 = 3.75$

   $A = 0.5(1.75) + 0.5(3) + 0.5(3.75) = 4.25$

15b. $4.25 < AA < 6.25$

PROBLEM SOLVING STRATEGIES

1. Rosa took 27 steps, but the ratio of baby steps to mother steps is $\dfrac{5}{2}$.

   $\dfrac{27}{x} = \dfrac{5}{2}$

   $5x = 54$

   $x = 10.8$

   Rosa had a headstart of 10.8 mother steps. Remember $d = rt$. Let d_m = distance of Mom and d_R = distance of Rosa. Then $d_m = \dfrac{\frac{1}{2}}{\frac{1}{5}}x = \dfrac{5}{2}x$ and $d_R = \dfrac{\frac{1}{5}}{\frac{1}{8}}x = \dfrac{8}{5}x$.

$$d_m = d_r + 10.8$$
$$\frac{5}{2}x = \frac{8}{5}x + 10.8$$
$$25x = 16x + 108$$
$$9x = 108$$
$$x = 12$$
$$d_m = \frac{5}{2}x = \frac{5}{2}(12) = 30 \text{ mother steps}$$

2a.

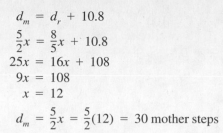

The combined distance pr represents the length of the entire lake.

2b. Three times the combined distance. The elapsed time represents three times the time of the first meeting.

2c. At the time of the first meeting Explorer traveled 500 yd. Since the time of the second meeting is 3 times as great, Explorer has traveled $500 \cdot 3 = 1500$ yd at the time of the second meeting.
At the second meeting Explorer has traveled the width of the lake plus 300 yd.
The width of the lake is $1500 - 300 = 1200$ yd.

2d. At the first meeting Explorer has traveled 500 yd and Voyager has traveled $1200 - 500 = 700$ yd. The ratio is $\frac{500}{700}$ or $\frac{5}{7}$. Since they both traveled the same time, this must also be the ratio of their speeds.

3a. 27 1×1 cubes + 8 2×2 cubes + 1 3×3 cube = 36

3c. 64 1×1 cubes + 27 2×2 cubes + 8 3×3 cubes + 1 4×4 cube = 100

3c. The general formula is $n^3 + (n - 1)^3 + (n - 2)^3 + \cdots + 1$.
If $n = 10$, $10^3 + 9^3 + 8^3 + 7^3 + 6^3 + 5^3 + 4^3 + 3^3 + 2^3 + 1 = 3025$

3d. Use trial and error in the general formula to find side = 20 units.

Chapter Review pages 558–559

1. d   **2.** a   **3.** e   **4.** c   **5.** b

6. All x, except $x = 4$

7. $x^2 - 25 = (x + 5)(x - 5)$ all x except $x = 5$ and $x = -5$

8. all real x

9. $\dfrac{x^2 - 64}{x - 8} = \dfrac{(x + 8)(x - 8)}{x - 8} = x + 8$
Since $x \neq 8$, the point of discontinuity is $(8, 16)$.

10. horizontal asymptote is $y = \dfrac{4}{1} = 4$
vertical asymptote is at $x = 1$

11. $y = \dfrac{k}{x}$ $4 = \dfrac{k}{9}$
$k = 36$ then $y = \dfrac{36}{12} = 3$

12. $y = kxz$ $100 = k(2.5)(10)$
$k = 4$
$y = 4(14)(7) = 392$

13. $\dfrac{5x^2y}{3z^2} \cdot \dfrac{6z^2}{10xy^2} = \dfrac{30x^2yz^2}{30xy^2z^2} = \dfrac{x}{y}$

14. $\dfrac{a^2 - b^2}{a^2 + 2ab + b^2} \cdot \dfrac{5a + 5b}{2a}$
$\dfrac{(a - b)(a + b)}{(a + b)(a + b)} \cdot \dfrac{5(a + b)}{2a} = \dfrac{5}{2}\left(\dfrac{a - b}{a}\right)$

15. $\dfrac{8e}{f} \div \dfrac{16e^2}{15f^2}$
$\dfrac{8e}{f} \cdot \dfrac{15f^2}{16e^2} = \dfrac{120ef^2}{16e^2f} = \dfrac{15f}{2e}$

16. $\dfrac{n + 3}{5} + \dfrac{9n + 2}{5} = \dfrac{10n + 5}{5} = 2n + 1$

17. $\dfrac{x + 3}{x - 3} - \dfrac{x - 3}{x + 3} = \dfrac{(x + 3)(x + 3)}{(x - 3)(x + 3)} - \dfrac{(x - 3)(x - 3)}{(x + 3)(x - 3)}$
$= \dfrac{x^2 + 6x + 9 - x^2 + 6x - 9}{x^2 - 9} = \dfrac{12x}{x^2 - 9}$

18. $\dfrac{2}{y^2 - 7y} + \dfrac{3}{y - 7} - \dfrac{3}{y} = \dfrac{2}{y(y - 7)} + \dfrac{3y}{(y - 7)y} - \dfrac{3(y - 7)}{y(y - 7)} = \dfrac{2 + 3y - 3y + 21}{y(y - 7)} = \dfrac{23}{y(y - 7)}$

19. $\dfrac{1 + \frac{7}{10}}{2 - \frac{3}{5}} = \dfrac{\frac{10}{10} + \frac{7}{10}}{\frac{20}{10} - \frac{6}{10}} = \dfrac{\frac{17}{10}}{\frac{14}{10}} = \dfrac{17}{10} \cdot \dfrac{10}{14} = \dfrac{17}{14}$

20. $\dfrac{5}{\frac{1}{x} - \frac{1}{y}} = \dfrac{\frac{5xy}{xy}}{\frac{y - x}{xy}} = \dfrac{5xy}{xy} \cdot \dfrac{xy}{y - x} = \dfrac{5xy}{y - x}$

21. $\dfrac{\frac{3x - 3}{x + 1}}{\frac{x^2 - 1}{(x + 1)(x + 1)}} = \dfrac{\frac{3(x - 1)}{x + 1}}{\frac{(x - 1)(x + 1)}{(x + 1)(x + 1)}} =$
$\dfrac{3(x - 1)}{x + 1} \cdot \dfrac{x + 1}{x - 1} = 3$

22. $\dfrac{n}{3} + \dfrac{n}{2} = 10$ Check:
$\dfrac{2n}{6} + \dfrac{3n}{6} = \dfrac{60}{6}$ $\dfrac{12}{3} + \dfrac{12}{2} = 10$
$5n = 60$ $4 + 6 = 10$
$n = 12$ $10 = 10 \checkmark$

23. $4 - \dfrac{x+1}{3} = \dfrac{x}{8}$ Check:

$96 - 8x + 8 = 3x$ $4 - \dfrac{8+1}{3} = \dfrac{8}{8}$

$88 = 11x$ $4 - 3 = 1$

$x = 8$ $1 = 1$ ✔

24. $\dfrac{x+1}{x-2} = \dfrac{3}{x-2}$

$x + 1 = 3$

$x = 2$

no solution, since $x \neq 2$

25. $y = -x^2 + 3x$ $0 \leq x \leq 2$

if $n = 4$,   then $x_1 = \dfrac{2}{4} = 0.5$

$x_2 = 1.0$

$x_3 = 1.5$

$x_4 = 2.0$

right hand: $y_1 = 1.25$

$y_2 = 2$

$y_3 = 2.25$

$y_4 = 2$

$A = 0.5(1.25) + 0.5(2.0) + 0.5(2.25) + 0.5(2)$

$= 3.75$

left hand:

$A = 0.5(1.25) + 0.5(2.0) + 0.5(2.25) = 2.75$

$2.75 < AA < 3.75$

Chapter Assessment, pages 560–561

CHAPTER TEST

1. All x, except $x = 0$

2. All x, except $x = -3$

3. $(x^2 - 5x + 4) = 0$

$(x - 4)(x - 1) = 0$

$x \neq 1, 4$

4. the domain is all x

5. $\dfrac{x^2 - 4x - 12}{x - 6} = \dfrac{(x-6)(x+2)}{x-6} = x + 2; x \neq 6$

discontinuity at $(6, 8)$

6. vertical asymptote at $x = -1$

horizontal asymptote at $y = 0$

7. $x^2 + x - 20 = (x + 5)(x - 4)$

vertical asymptotes at $x = -5$, $x = 4$

horizontal asymptote at $y = 0$

8. horizontal asymptote at $y = 8$

vertical asymptote at $x = -6$

9. $3x - 6 = 3(x - 2)$

vertical asymptote at $x = 2$

10. $y = \dfrac{k}{x}, 5 = \dfrac{k}{20}, k = 100$

$200 = \dfrac{100}{x}, x = 0.5$

11. $x = kyz, 128 = k(3.2)(5)$

$k = 8$

$x = 8(24)\left(\dfrac{1}{3}\right) = 64$

12. $y = \dfrac{k}{x^2}$ $0.25 = \dfrac{k}{(12)^2}, k = 36$

$y = \dfrac{36}{(3)^2} = 4$

13. $I = \dfrac{k}{d^2}$ $5 = \dfrac{k}{(24)^2}, k = 2880$

$I = \dfrac{2880}{(6)^2} = 80 \dfrac{\text{units}}{\text{ft}^2}$

14. To add two rational expressions, multiply all terms by the LCD. Then add.

15. Plot $y_1 = \dfrac{x+3}{x}$ and $y_2 = 2$. Find the point of intersection at $(3, 2)$. The answer is the x-value of the point of intersection, $x = 3$.

16. $\dfrac{m}{4} \cdot \dfrac{6m}{4} = \dfrac{6m^2}{16} = \dfrac{3m^2}{8}$

17. $\dfrac{10}{x+2} \cdot \dfrac{2(x+2)}{5} = 4$

18. $\dfrac{x^2 + 2x + 1}{x - 2} \cdot \dfrac{x^2 - 4}{x^2 + 3x + 2} =$

$\dfrac{(x+1)(x+1)}{x-2} \cdot \dfrac{(x-2)(x+2)}{(x+2)(x+1)} = x + 1$

19. $\dfrac{2k^2 - 2k}{k^2 - 25} \cdot \dfrac{2k^2 - 9k - 5}{k^3} =$

$\dfrac{2k(k-1)}{(k-5)(k+5)} \cdot \dfrac{(2k+1)(k-5)}{k^3} = \dfrac{2(k-1)(2k+1)}{k^2(k+5)}$

20. $\dfrac{y}{3} - \dfrac{6 - 2y}{3} = \dfrac{y - 6 + 2y}{3} = \dfrac{3y - 6}{3} =$

$\dfrac{3(y-2)}{3} = y - 2$

21. $\dfrac{5}{x} + \dfrac{2}{3x} + \dfrac{1}{3} = \dfrac{5(3)}{x(3)} + \dfrac{2}{3x} + \dfrac{x}{3x} = \dfrac{15 + 2 + x}{3x} =$

$\dfrac{x + 17}{3x}$

22. $\dfrac{3}{x-2} + \dfrac{8}{x+1} = \dfrac{3(x+1)}{(x-2)(x+1)} +$

$\dfrac{8(x-2)}{(x-2)(x+1)} = \dfrac{3x + 3 + 8x - 16}{(x+1)(x-2)} =$

$\dfrac{11x - 13}{(x-2)(x+1)}$

23. $\dfrac{1}{x + y} + \dfrac{1}{x - y} - \dfrac{1}{x^2 - y^2} =$

$\dfrac{x - y}{(x^2 - y^2)} + \dfrac{(x + y)}{(x^2 - y^2)} - \dfrac{1}{(x^2 - y^2)} =$

$\dfrac{x - y + x + y - 1}{(x^2 - y^2)} = \dfrac{2x - 1}{(x^2 - y^2)}$

24. $V_1 = \dfrac{300}{(x^2 - 4)}, \ V_2 = \dfrac{325}{x + 2}$

$\dfrac{300}{x^2 - 4} \div \dfrac{325}{x + 2} = \dfrac{300}{(x - 2)(x + 2)} \cdot \dfrac{x + 2}{325} =$

$\dfrac{12}{13(x - 2)}$

25. $\dfrac{\frac{1}{x} + \frac{1}{y}}{\frac{2}{x} - \frac{2}{y}} = \dfrac{\frac{y}{xy} + \frac{x}{xy}}{\frac{2y}{xy} - \frac{2x}{xy}} = \dfrac{\frac{x + y}{xy}}{\frac{2y - 2x}{xy}} =$

$\dfrac{x + y}{xy} \cdot \dfrac{xy}{2y - 2x} = \dfrac{x + y}{2y - 2x}$

26. $\dfrac{\frac{x}{x + 1} + \frac{1}{x}}{\frac{x}{x^2 + x}} = \dfrac{\frac{x \cdot x}{x(x + 1)} + \frac{x + 1}{x(x + 1)}}{\frac{x}{x^2 + x}} = \dfrac{\frac{x^2 + x + 1}{x^2 + x}}{\frac{x}{x^2 + x}} =$

$\dfrac{x^2 + x + 1}{x^2 + x} \cdot \dfrac{x^2 + x}{x} = \dfrac{x^2 + x + 1}{x}$

27. $\dfrac{12}{x} = \dfrac{x + 2}{4}$

$48 = x(x + 2)$

$48 = x^2 + 2x$

$0 = x^2 + 2x - 48$

$0 = (x + 8)(x - 6)$

$\begin{aligned} x + 8 &= 0 & x - 6 &= 0 \\ x &= -8 & x &= 6 \end{aligned}$

28. $\dfrac{6}{x - 4} + \dfrac{4}{x - 2} = \dfrac{9}{x - 4}$

$\dfrac{6(x - 2)}{(x - 4)(x - 2)} + \dfrac{4(x - 4)}{(x - 2)(x - 4)} = \dfrac{9(x - 2)}{(x - 4)(x - 2)}$

$6(x - 2) + 4(x - 4) = 9(x - 2)$

$6x - 12 + 4x - 16 = 9x - 18$

$x = 10$

29. B; $x \neq 2.$ $\dfrac{4}{x - 2} = \dfrac{x^2}{x - 2}$

$4 = x^2$

$\pm 2 = x$

PERFORMANCE ASSESSMENT

SPORTS STORY

The graph shows exponential decay
    horizontal asymptote is $y = 0$
    vertical asymptote is $x = 0$
    As x gets larger, 4 gets smaller
    as y gets larger, x gets smaller

EASY SIMPLIFICATION?

The method does not work in general $\left(\dfrac{13}{34} \neq \dfrac{1}{4}\right)$. Students should point out that only common *factors* can be divided from both the numerator and denominator of a rational expression. Digit cancellation" works only rarely and then by coincidence.

RATIONAL THOUGHTS

Among the similarities students may list and describe are the definitions of rational numbers and rational expressions, and methods of adding, subtracting, multiplying, and dividing the two.

Cumulative Review, page 562

1. No, $i + -i = 0$

2. A rational function is a ratio of two polynomials the domain is the set of values of x for which the denominator is not zero. To find the domain, set the denominator equal to zero, and solve. These values of x are excluded from the domain.

3. $A^{-1} = \begin{bmatrix} -4 & 5 \\ 5 & -6 \end{bmatrix}$

4. $3x^2 - 8x - 3 \geq 0$

$(3x + 1)(x - 3) \geq 0$

$\begin{aligned} 3x + 1 &\geq 0 & \text{or} & & x - 3 &\geq 0 \\ x &\leq -\tfrac{1}{3} & \text{or} & & x &\geq 3 \end{aligned}$

5. $f = kAV^2$

$3 = k(2)(6)^2$

$k = 0.042$

$f = 0.042(5)(20)^2$

$f = 83.3 \text{ lb}$

6. $x^2 + 8 + k = 0$

$(x + 4)(x + 4) = 0$

$k - 16$

7. Let $x =$ number of boys in original group.

Then $\dfrac{80}{x} =$ amount each boy chips in.

$\dfrac{80}{x + 1} = \dfrac{80}{x} - 4$

$80x = 80(x + 1) - 4x(x + 1)$

$80x = 80x + 80 - 4x^2 - 4x$

$80x = 76x + 80 - 4x^2$

$4x^2 + 4x - 80 = 0$

$4(x^2 + x - 20) = 0$

$4(x + 5)(x - 4) = 0$

$\begin{aligned} x + 5 &= 0 & x - 4 &= 0 \\ x &= -5 & x &= 4 \end{aligned}$

reject $x = -5$

$x = 4$ boys

8. $v = l \cdot w \cdot h$
$v = 5 \times 6 \times 7 = 210 \text{ in}^3$
$(5 + c)(6 + c))(7 + c) = 420$
$c^2 + 11c + 30)(c + 7) = 420$
$c^3 + 11c^2 + 30c + 7c^2 + 77c + 210 - 420 = 0$
$c^3 + 18c^2 + 107c - 210 = 0$
Solve graphically: $c = 1.53 \text{ in.}$
Check: $(6.53)(7.53)(8.53) = 420$
$420 = 420 \checkmark$

9. D; $\dfrac{3}{3x - 1} - \dfrac{1}{x} = \dfrac{3x}{(3x - 1)x} - \dfrac{1(3x - 1)}{x(3x - 1)}$

$= \dfrac{3x - 3x + 1}{x(3x - 1)}$

$= \dfrac{1}{(3x - 1)x}$

10.

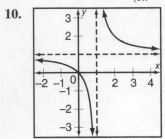

11. $v = 24{,}000(1 - 0.15)^t$
$10{,}000 = 24{,}000(0.85)^t$

$\dfrac{5}{12} = 0.85^t$

$\log \dfrac{5}{12} = \log 0.85^t$

$\log \dfrac{5}{12} = t \log 0.85$

$t = \dfrac{\log \dfrac{5}{12}}{\log 0.85}$

$\approx 5 \text{ years}$

12. Total number of students in class $=$
$3 + 6 + 8 + 5 + 3 = 25.$ Number of students
receiving B or A $= 3 + 6 = 9.$ Probability of random
student receiving a B or A $= \dfrac{9}{25}$.

13. $\dfrac{1 + x}{2} = \dfrac{1 + x + x^2}{3}$
$3(1 + x) = 2(1 + x + x^2)$
$3 + 3x = 2 + 2x + 2x^2$
$0 = 2x^2 - x - 1$
$0 = (2x + 1)(x - 1)$
$2x + 1 = 0 \qquad x - 1 = 0$
$2x = -1 \qquad x = 1$
$x = -\dfrac{1}{2}$

14. $3x + 2y = 7 \qquad\qquad kx - y = 5$
$y = \dfrac{7 - 3x}{2} \qquad kx - \left(\dfrac{7 - 3x}{2}\right) = 5$
$2kx - 7 + 3x = 10$
$2kx + 3x = 17$
$x(2k + 3) = 17$
$x = \dfrac{17}{2k + 3}$

15. $\dfrac{1 - \dfrac{1}{x}}{1 - \dfrac{1}{x^2}} = \dfrac{\dfrac{x^2}{x^2} - \dfrac{x}{x^2}}{\dfrac{x^2}{x^2} - \dfrac{1}{x^2}} = \dfrac{x^2 - x}{x^2 - 1} = \dfrac{x(x - 1)}{(x - 1)(x + 1)} =$

$\dfrac{x}{x + 1}$

16. D; $\log_2 5 = x$
$2^x = 5$
$\log 2^x = \log 5$
$x \log 2 = \log 5$
$x = \dfrac{\log 5}{\log 2}$
$x \approx 2.32$
$4^{2.32} = 25$

17. $x^4 - x^2 - 2 = 0$
let $m = x^2$
$m^2 - m - 2 = 0$
$(m + 1)(m - 2) = 0$
$m = -1 \qquad m = 2$
$x^2 = -1 \qquad x^2 = 2$
$x = \pm i \qquad x = \pm\sqrt{2}$ 2 real solutions, 2 imaginary

18. $\dfrac{x^3 - 64}{x^2 - 16} \cdot \dfrac{x^2 + 8x + 16}{x^2 + 4x + 16} = \dfrac{(x^3 - 64)}{(x - 4)(x + 4)} \cdot$

$\dfrac{(x + 4)(x + 4)}{(x^2 + 4x + 16)} = \dfrac{(x^3 - 64)(x + 4)}{(x^2 + 4x + 16)(x - 4)} =$

$\dfrac{(x - 4)(x^2 + 4x + 16)(x + 4)}{(x^2 + 4x + 16)(x - 4)} = x + 4$

Standardized Test, page 563

1. D

2. D; Graph the function

3. E; Use the Location Theorem. If $F(0)$ and $F(1)$ have
opposite signs then there is at least one real solution
between 0 and 1.
$F(0) = 0^3 + 2(0)^2 + 0 + k = k$
$F(1) = 1^3 + 2(1)^2 + 1 + k = k + 4$
So k and $k + 4$ must have opposite sign
So $-4 < k < 0$.

4. A; I is true because a rational function can be
expressed as the quotient of two polynomials.
II is true because the denominator cannot be 0.
III is false. The domain is all real numbers except
$x = -1$.

5. A; $2a + b = 11 \quad\rightarrow\quad 2a + b = 11$
$\underline{a + 2b = 10} \quad\rightarrow\quad \underline{-2a - 4b = -20}$
$\qquad\qquad\qquad\qquad\qquad\qquad -3b = -9$
$\qquad\qquad\qquad\qquad\qquad\qquad\quad b = 3$
$\qquad\qquad\qquad\qquad\qquad a + 2(3) = 10$
$\qquad\qquad\qquad\qquad\qquad\qquad\quad a = 10 - 6$
$\qquad\qquad\qquad\qquad\qquad\qquad\quad a = 4$

6. C; $\dfrac{(1 - x^{-2})}{1 + x^{-1}} = \dfrac{1 - \frac{1}{x^2}}{1 + \frac{1}{x}} = \dfrac{\frac{x^2 - 1}{x^2}}{\frac{x^2 + x}{x^2}} =$

$\dfrac{x^2 - 1}{x^2} \cdot \dfrac{x^2}{x^2 + x} = \dfrac{(x - 1)(x + 1)}{x(x + 1)} = \dfrac{x - 1}{x}$

7. E; $F(-1) = -13$, $F(0) = -10$, $F(1) = -21$, $F(2) = -28$, $F(3) = -13$, $F(4) = 42$. By the Location Theorem there is at least one real solution between 3 and 4.

8. B; $\log x + \log y = 2$
$\qquad\qquad \log y = 2 - \log x$
Substitute into the second equation:
$\log x^3 - \log y = 4$
$\log x^3 - (2 - \log x) = 4$
$\log x^3 - 2 + \log x = 4$
$3 \log x + \log x = 6$
$\qquad\quad 4 \log x = 6$
$\qquad\qquad \log x = \dfrac{6}{4}$
$\qquad\qquad \log x = \dfrac{3}{2}$

Solve for y: $\quad \log y = 2 - \log x$
$\qquad\qquad\qquad\quad = 2 - \dfrac{3}{2}$
$\qquad\qquad\qquad\quad = \dfrac{1}{2}$

$\log \dfrac{x}{y} = \log x - \log y = \dfrac{3}{2} - \dfrac{1}{2} = 1$

9. E; $y = \dfrac{k}{x^2}$ then $y = \dfrac{k}{(2x)^2} = \dfrac{k}{4x^2} = \dfrac{y}{4}$

10. D; The axis of symmetry is at $x = -\dfrac{b}{2a}$. Therefore the answer must be C, D or E. Since the graph is a solid line the answer must be D.

11. D; $\dfrac{9x^2yz^6}{27xy^2z^3} = \dfrac{xz^3}{3y}$

12. E; $\dfrac{2x}{x + 5} + \dfrac{1}{x - 5} = \dfrac{10}{x^2 - 25}$; $x \neq 5, -5$

$\dfrac{2x(x - 5) + x + 5}{(x + 5)(x - 5)} = \dfrac{10}{x^2 - 25}$

$2x^2 - 10x + x + 5 = 10$

$\qquad 2x^2 - 9x - 5 = 0$

$\qquad (2x + 1)(x - 5) = 0$

$2x + 1 = 0 \qquad\quad$ or $x - 5 = 0$

$\quad x = -\dfrac{1}{2} \qquad\qquad\qquad x = 5$

reject $x = 5$

Chapter 12 Conic Sections

Data Activity, pages 564–565

1. $\$1151 \div 312 \approx \$3.7$ million;
$\$14,229 \div 1581 \approx \$9$ million

2. $10,307 - 7,847 \approx 2000$

3. range $= 2.58 - 2.20 = 0.38$ min; median
$= \dfrac{2.33 + 2.38}{2} = 2.355$ min

4. 102

5. Let x-axis represent the year and y-axis represent the number of customers; coefficient of correlation ≈ 0.94.

6. When $x = 100$, $y \approx 37,677,369$.

7. Answers will vary; The trend shows a decrease in the amount of a montly bill and this would not continue.

Lesson 12.1, pages 567–571

EXPLORE

1.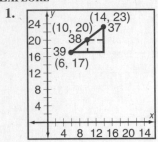

2. The triangle is a right triangle with legs 6 miles and 8 miles. The hypotenuse measures $\sqrt{6^2 + 8^2} = 10$ miles.

3. $(10, 20)$

4. $\sqrt{3^2 + 4^2} = 5$; By the Pythagorean theorem, Tower 38 is 5 miles from each tower.

5. Answers will vary. The x-coordinates decrease by 4. The y-coordinates decrease by 3. The x- and y-coordinates of Tower 38 are the means of the x- and y-coordinates of the other towers.

TRY THESE

1. $d = \sqrt{(14 - 2)^2 + (12 - 7)^2} = 13$;
mdpt $= \left(\dfrac{2 + 14}{2}, \dfrac{7 + 12}{2}\right) = (8, 9.5)$

2. $d = \sqrt{(11 - -9)^2 + (-6 - 15)^2} = 29$;
mdpt $= \left(\dfrac{-9 + 11}{2}, \dfrac{15 + -6}{2}\right) = (1, 4.5)$

3. $d = \sqrt{(9 - 11)^2 + (-3 - 3)^2} = \sqrt{40} = 2\sqrt{10}$
mdpt $= \left(\dfrac{11 + 9}{2}, \dfrac{3 + -3}{2}\right) = (10, 0)$

4. $d = \sqrt{(-5 - 0)^2 + (0 - 5)^2} = \sqrt{50} = 5\sqrt{2}$
mdpt $= \left(\dfrac{0 + -5}{2}, \dfrac{5 + 0}{2}\right) = \left(-\dfrac{5}{2}, \dfrac{5}{2}\right)$

5. $(-4, 7)$; The distance from the origin to $(0, 8)$ is 8 units and the distance to $(-4, 7)$ is $\sqrt{4^2 + 7^2}$, or $\sqrt{65}$, which is more than 8 units.

6. Let $d = 17$; $17 = \sqrt{(0-15)^2 + (y_2 - 6)^2}$
$289 = (-15)^2 + y^2 - 12y + 36$;
$y^2 - 12y - 28 = 0$; $(y - 14)(y + 2) = 0$;
$y = 14$ or -2; The points are $(0, 14)$ or $(0, -2)$.

7. $d = \sqrt{(32 - 20)^2 + (28 - 18)^2} \approx 15.6$ mi

8. $\overline{AB} = \sqrt{(7 - -1)^2 + (1 - 1)^2} = 8$;
$\overline{BC} = \sqrt{(7 - 7)^2 + (7 - 1)^2} = 6$;
$\overline{AC} = \sqrt{(7 - -1)^2 + (7 - 1)^2} = 10$;
$P = 8 + 6 + 10 = 24$ units

9. A number squared is always positive so $(x_1 - x_2)^2$ yields the same result as $(x_2 - x_1)^2$.

PRACTICE

1. $d = \sqrt{(3 - 9)^2 + (-3 - 5)^2} = 10$;
mdpt $= \left(\dfrac{9 + 3}{2}, \dfrac{5 + -3}{2}\right) = (6, 1)$

2. $d = \sqrt{(2 - -8)^2 + (-12 - 12)^2} = 26$;
mdpt $= \left(\dfrac{-8 + 2}{2}, \dfrac{12 + -12}{2}\right) = (-3, 0)$

3. $d = \sqrt{(-1 - 6)^2 + (3 - 2)^2} = \sqrt{50} = 5\sqrt{2}$;
mdpt $= \left(\dfrac{6 + -1}{2}, \dfrac{2 + 3}{2}\right) = (2.5, 2.5)$

4. $d = \sqrt{(-3 - 11)^2 + (0 - -4)^2} = \sqrt{80} = 4\sqrt{5}$;
mdpt $= \left(\dfrac{-11 + -3}{2}, \dfrac{-4 + 0}{2}\right) = (-7, -2)$

5. $d = \sqrt{(-11.8 - 15.2)^2 + (7 - 7)^2} = 27$
mdpt $= \left(\dfrac{15.2 + -11.8}{2}, \dfrac{7 + 7}{2}\right) = (1.7, 7)$

6. $d = \sqrt{(7 - -6)^2 + (6 - 2)^2} = \sqrt{185}$;
mdpt $= \left(\dfrac{-6 + 7}{2}, \dfrac{2 + 6}{2}\right) = (0.5, 4)$

7. $d = \sqrt{(2.3 - 2.3)^2 + (6.5 - -5.1)^2} = 11.6$;
mdpt $= \left(\dfrac{2.3 + 2.3}{2}, \dfrac{-5.1 + 6.5}{2}\right) = (2.3, 0.7)$

8. $d = \sqrt{\left(1 - \dfrac{1}{2}\right)^2 + \left(-\dfrac{1}{2} - 2\right)^2} = \sqrt{\dfrac{13}{2}} = \dfrac{1}{2}\sqrt{26}$;
mdpt $= \left(\dfrac{0.5 + 1}{2}, \dfrac{-0.5}{2}\right) = (0.75, 0.75)$

9. $d = \sqrt{(4\sqrt{6} - 2\sqrt{6})^2 + (7\sqrt{5} - 3\sqrt{5})^2} = \sqrt{104}$;
mdpt $= \left(\dfrac{2\sqrt{6} + 4\sqrt{6}}{2}, \dfrac{3\sqrt{5} + 7\sqrt{5}}{2}\right) = (3\sqrt{6}, 5\sqrt{5})$

10. The points are on the same vertical line so you can take $|y_2 - y_1|$.

11. Let $x = 3$; $3 = \dfrac{x_1 + 7}{2}$, so $x = -1$;
Let $y - 5$; $5 = \dfrac{y_1 + 11}{2}$; so $y = -1$;
The other endpoint is $(-1, -1)$.

12. Let $d = 29$; $29 = \sqrt{(-3 - 17)^2 + (k - -6)^2}$;
$841 = 400 + k^2 + 12k + 36$; $k^2 + 12k - 405 = 0$;
$(k - 15)(k + 27) = 0$; $k = 15$ or -27.

13. Let $d = 15$; $15 = \sqrt{(n - -11)^2 + (0 - 9)^2}$;
$225 = n^2 + 22n + 121 + 81$; $n^2 + 22n - 23 = 0$;
$(n + 23)(n - 1) = 0$; $n = -23$ or 1.

14. $\overline{AB} = \sqrt{(-2 - 5)^2 + (10 - 6)^2} = \sqrt{65}$
$\overline{BC} = \sqrt{(2 - -2)^2 + (4 - 10)^2} = \sqrt{52} = 2\sqrt{13}$
$\overline{AC} = \sqrt{(2 - 5)^2 + (4 - 6)^2} = \sqrt{13}$; It is a right triangle because $(\sqrt{52})^2 + (\sqrt{13})^2 = (\sqrt{65})^2$.

15. $\overline{AB} = \sqrt{(8 - 1)^2 + (5 - 3)^2} = \sqrt{53}$
$\overline{BC} = \sqrt{(3 - 8)^2 + (-4 - 5)^2} = \sqrt{106}$
$\overline{AC} = \sqrt{(3 - 1)^2 + (-4 - 3)^2} = \sqrt{53}$; It is isosceles because two sides have equal lengths.

16. $\overline{AB} = \sqrt{(45 - 0)^2 + (28 - 0)^2} = 53$
$\overline{BC} = \sqrt{(36 - 28)^2 + (60 - 45)^2} = 17$
$\overline{CD} = \sqrt{(47 - 36)^2 + (0 - 60)^2} = 61$
$\overline{AD} = \sqrt{(47 - 0)^2 + (0 - 0)^2} = 47$;
$P = 53 + 17 + 61 + 47 = 178$ units

17. $d = \sqrt{(12 - 3)^2 + (-5 - 6)^2} \approx 14°$

18. $70 \times 14 = 980$ mi

EXTEND

19. Find the hypotenuse of a right triangle. The first leg is
$(55)(0.6) = 33$; the second leg is $(62)(1.4) = 86.8$; $(33)^2 + (86.8)^2 = x^2$; $x \approx 92.9$ mi

20. mpdt of $\overline{AB} = \left(\frac{0 + 4}{2}, \frac{0 + 6}{2}\right) = (2, 3)$;
mdpt of $\overline{BC} = \left(\frac{4 + 6}{2}, \frac{6 + 2}{2}\right) = (5, 4)$;
$\overline{AC}$ has a slope of $\frac{1}{3}$ as does the segment joining the midpoints of $\overline{AB}$ and $\overline{BC}$ so they are parallel.
$\overline{AC} = \sqrt{(6 - 0)^2 + (2 - 0)^2} = \sqrt{40} = 2\sqrt{10}$;
the distance of the midpoints is
$d = \sqrt{(5 - 2)^2 + (4 - 3)^2} = \sqrt{10}$; which is half the length of $\overline{AC}$.

21. $d = \sqrt{(-6 - -3)^2 + (17 - 5)^2 + (-9 - -13)^2}$
$= 13$

22. $d = \sqrt{(10 - -2)^2 + (15 - -5)^2 + (-1 - 8)^2}$
$= 25$

THINK CRITICALLY

23. Point P is not necessarily the midpoint of the line segment $\overline{MN}$, since P may not lie on the line segment. For example, let M be $(0, 0)$, N be $(1, 12)$ and P be $(6, 8)$. Then P is 10 units from M and 10 units from N, but it is not the midpoint.

24. mdpt $= \left(\frac{a + 0}{2}, \frac{b + 0}{2}\right) = \left(\frac{a}{2}, \frac{b}{2}\right)$; $\overline{OM} = \overline{MB} = \overline{MA}$
because they all have the same distance of $\frac{\sqrt{a^2 + b^2}}{2}$.

25. $2^5 = x$; $x = 32$ **26.** $x^3 = 125$; $x = 5$

27. $49^{3/2} = x$; $x = (\sqrt{49})^3 = 343$

28. D; The midpoint of the diameter is
$\left(\frac{-12 + -2}{2}, \frac{7 + -17}{2}\right) = (-7, -5)$, and that would be the center of the circle.

Lesson 12.2, pages 572–574

THINK BACK

1. A line parallel to the given lines and midway between them.

2. One point that is the midpoint of the segment.

3. A line that is the perpendicular bisector of the segment.

EXPLORE

4. (see picture in text)

5. Place the plane parallel to the base of the cone.

6. Place the plane at an angle to the base so it is neither parallel or perpendicular to it.

7. The plane should be parallel to the side of the cone.

8. Place the plane perpendicular to the bases, but do not go through the point of intersection of the cones.

9. Shine the light perpendicular to the surface and you will create a circle; turning the light from the perpendicular will first create an ellipse, and then a parabola. The flat surface acts as a plane slicing the cone, which is the light beam.

MAKE CONNECTIONS

10. (point on paper)

11. The points will form a circle with center C and radius 5 cm.

12. It will always be a circle.

13. The distance from P to F_1 is 4 units and from P to F_2 is 5 units, so sum is 9 units.

14. Answers will vary.

15. The points graphed should form an ellipse when connected.

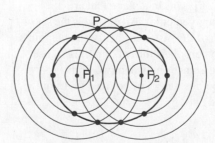

16. The ellipse has two lines of symmetry: one that contains F_1 and F_2 and one that is the perpendicular bisector of segment F_1F_2.

17. Answer will vary.   **18.** Answers will vary.

19. The graph should be a hyperbola.

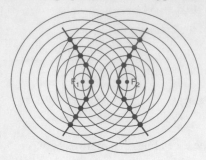

20. The hyperbola has the same lines of symmetry as the ellipse. (see number 16)

21. (see picture in text)   **22.** Answers will vary.

23. The graph is a parabola.

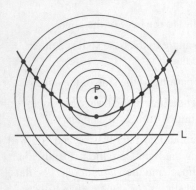

24. Its vertex is midway between P and L; the line of symmetry passes through P and the vertex.

SUMMARIZE

25. Answers will vary.

26.

circle   ellipse   parabola   hyperbola

A circle is the set of points in a plane that are equidistant from a given point. An ellipse is the set of points whose sum of the distances between two points is a constant. A hyperbola is the set of points whose difference of the distances between two points is a constant. A parabola is the set of points equidistant from a point and a line.

27.

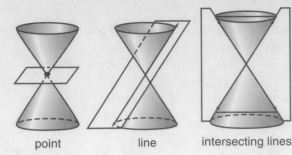

point   line   intersecting lines

EXPLORE

1.

2.

x	y	$x^2 + y^2$	$\sqrt{x^2 + y^2}$
6	8	100	10
6	−8	100	10
10	0	100	10
−10	0	100	10
−3.5	≈ 9.4	100	10
−3.5	≈ −9.4	100	10

3. If each point on the circle is connected with (0, 0) a right triangle is formed with legs whose measures are the x- and y-coordinates of the point and with a hypotenuse of 10. By the Pythagorean theorem, $x^2 + y^2 = 10^2 = 100$, giving the constant value of 100 in Column 3. $\sqrt{x^2 + y^2} = \sqrt{100} = 10$, giving the constant value of 10 in Column 4.

4. $x^2 + y^2 = 100$
$x^2 + (5.4)^2 = 100$
$\qquad x^2 = 70.84$
$\qquad\quad x \approx 8.4 \text{ or } -8.4$

5. $(-35)^2 + y^2 = 37^2; y = 12 \text{ or } -12$

TRY THESE

1. $(x - 0)^2 + (y - 0)^2 = 5^2; x^2 + y^2 = 25$

2. $(x - -2)^2 + (y - 3)^2 = 9^2;$
$(x + 2)^2 + (y - 3)^2 = 81$

3. $(x - 3.7)^2 + (y + 6.5)^2 = \pi^2$

4. $(x - 15)^2 + y^2 = 3$

5. $r = \sqrt{100} = 10;$
$C = (0, 0)$

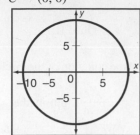

6. $r = \sqrt{4} = 2;$
$C = (0, -3)$

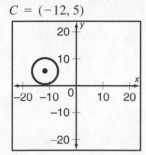

9. $C = (-1, 0); r = 5$

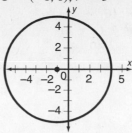

10. $C = (2, -3); r = \sqrt{35}$

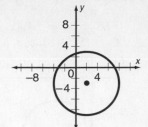

7. $r = \sqrt{20} = 2\sqrt{5};$
$C = (-12, 5)$

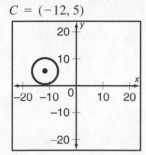

8. $(x^2 + 8x + 16) +$
$(y^2 - 2y + 1) =$
$-15 + 16 + 1;$
$(x + 4)^2 +$
$(y - 1)^2 = 2;$
$r = \sqrt{2};$
$C = (-4, 1)$

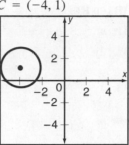

11. $(x^2 + 6x + 9)^2 + (y^2 - 4y + 4)^2 = 15 + 9 + 4;$
$(x + 3)^2 + (y - 2)^2 = 28;$
$C = (-3, 2); r = \sqrt{28} = 2\sqrt{7}$

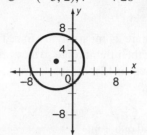

12. $(x^2 - 6x + 9)^2 + (y^2 + 6y + 9)^2 =$
$-2 + 9 + 9; (x - 3)^2 + (x + 3)^2 = 16;$
$C = (3, -3); r = 4$

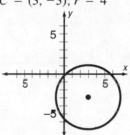

9. $x^2 + y^2 = (3960 + 22,300)^2; x^2 + y^2 = (26, 260)^2$

10. $x^2 + y^2 = 3^2; y = \pm\sqrt{9 - x^2}$

11. $C = (1.5, -1); r = 2.5;$
$(x - 1.5)^2 + (y + 1)^2 = (2.5)^2;$
$y + 1 = \sqrt{6.25 - (x - 1.5)^2};$
$y = -1 \pm \sqrt{6.25 - (x - 1.5)^2}$

12. All three graphs are circles. The first two graphs both have the same radius but different centers. The last two graphs have the same centers but different radii.

PRACTICE

1. $x^2 + y^2 = 196$ **2.** $(x - 6)^2 + (y - 7)^2 = 1$

3. $(x + 9)^2 + (y + 3)^2 = 49$

4. $x^2 + (y + 4)^2 = 7.84$

5. $(x - 4.3)^2 + (y + 7.7)^2 = 9\pi^2$

6. $\left(x - \frac{2}{3}\right)^2 + \left(y - \frac{1}{2}\right)^2 = \frac{4}{25}$

7. $C = (0, 0); r = 3$

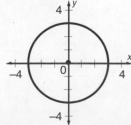

8. $C = (0, 3); r = 2$

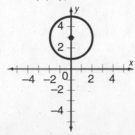

13. $x^2 + (y^2 + 14y + 49)^2 = -48 + 49;$
$x^2 + (y + 7)^2 = 1; C = (0, -7); r = 1$

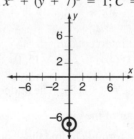

14. $x^2 + y^2 - 10x + 16y = -65$;
$(x^2 - 10x + 25) + (y^2 + 16y + 64) =$
$-65 + 25 + 64; (x - 5)^2 + (y + 8)^2 = 24$;
$C = (5, -8); r = \sqrt{24} = 2\sqrt{6}$

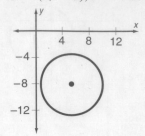

15. $C = (0, 0); r = 12; x^2 + y^2 = 144$

16. If $r = 12; d = 24$; They are 24 mi apart on a line. They make a semi-circle along their flight route so they are one-half the circumference, or 12π mi apart.

17. $240t = 24\pi, t = 0.1\pi$ h

18. $(-9)^2 + y^2 = 12^2; y^2 = 63; y = \sqrt{63} = 3\sqrt{7}$; ($y$ is positive in the second quadrant)

19. $x^2 + y^2 = 6.76; x^2 + y^2 = 60.84; x^2 + y^2 = 169$; $x^2 + y^2 = 331.24; x^2 + y^2 = 547.56$; (Take one-half of each diameter and square it; center is always $(0, 0)$.

20. Answers will vary.

EXTEND

21. $\sqrt{(0 - 5)^2 + (0 + 12)^2} = r; r^2 = 169$
$C = (0, 0); x^2 + y^2 = 169$

22. If the area is $36\pi, r = 6; (x + 3)^2 + (v - 4)^2 = 36$

23. $r = 5; \left(x - \dfrac{1}{21}\right)^2 + \left(y + \dfrac{3}{2}\right)^2 = 25$

24. $C = (3 - 7) r = \sqrt{15} (x - 3)^2 + (y + 7)^2 = 15$

25a. $(x + 2)^2 + (y - 5)^2 = 40$

25b. If $x = 0; 4 + (y - 5)^2 = 40; y = \pm\sqrt{36} + 5 = 11, -1;$ if $y = 0; (x + 2)^2 + 25 = 40; x = \pm\sqrt{15} - 2$; The intercepts are $(0, 11); (0, 1); (-2 + \sqrt{15}, 0); (-2 - \sqrt{15}, 0)$

26a. mdpt $= \left(\dfrac{-3 + 1}{2}, \dfrac{4 + 2}{2}\right) = (-1, 3);$ The midpoint of the diameter is the center of the circle.

26b. $r = \sqrt{5}$; Use the distance formula to find the distance from the midpoint to either of the two endpoints given.

26c. $(x + 1)^2 + (y - 3)^2 = 5$

27. $C = (1, 2)$ from midpoint formula; $r = \sqrt{34}$ from distance formula; $(x - 1)^2 + (y - 2)^2 = 34$

THINK CRITICALLY

28. $r = 2$ (distance from center to x-axis); $(x + 4)^2 + (y - 2)^2 = 4$

29. $(+, +)$ − Quadrant I; $(-, +)$ − Quad II; $(-, -)$ − Quad III; $(+, -)$ − Quad VI

30. The point (b, c) must satisfy the equation of the circle, $a^2 = x^2 + y^2$. Therefore, $a^2 = b^2 + c^2$; slope of $\overline{AB} = \dfrac{c}{b + a}$; slope of $\overline{BC} = \dfrac{c}{b - a}$ Multiply the slopes. slope of $\overline{AB}$ · slope of $\overline{BC} = \dfrac{c}{b + a} \cdot \dfrac{c}{b - a}$
$= \dfrac{c^2}{b^2 - a^2} = \dfrac{a^2 - b^2}{b^2 - a^2} = -1$

Since the product of the slopes of $\overline{AB}$ and $\overline{BC}$ is -1, the segments are perpendicular. Therefore, $\angle ABC$ is a right angle.

MIXED REVIEW

31. $m = 3; b = -5$

32. $2y = 4x + 5; y = 2x + \dfrac{5}{2}; m = 2; b = \dfrac{5}{2}$

33. $4x - 8y = 4y - 2; 12y = 4x + 2; y = \dfrac{1}{3}x + \dfrac{1}{6};$
$m = \dfrac{1}{3}; b = \dfrac{1}{6}$

34. $y = 5x; x = \dfrac{y}{5}; f^{-1}(x) = \dfrac{x}{5}$

35. $y = x + 4; x = y - 4; f^{-1}(x) = x - 4$

36. $y = \dfrac{1}{x + 1}; xy + y = 1; x = \dfrac{1 - y}{y}; f^{-1}(x) = \dfrac{1 - x}{x}$

37. B; $(x^2 - 10x + 25) + y^2 = 24 + 25; (x - 5)^2 + y^2 = 49; r = 7$

Lesson 12.4, pages 581–586

EXPLORE

1. The sum of the distances from the tacks to any point on the ellipse remains constant (the length of the string).

2. Major axis $= 2a =$ sum of distances from tacks $=$ length of string

3. foci

TRY THESE

1. 8; from -5 to 3   **2.** 2; from 0 to 2

3. $(-5, 1); (3, 1); (-1, 0); (-1, 2)$

4. $(-1, 1)$

5. If $2a = 8, a = 4;$ if $2b = 2, b = 1; c^2 = 4^2 - 1^2, c = \sqrt{15};$ from center, $(-1 + \sqrt{15}, 1); (-1 - \sqrt{15}, 1)$

6. $\frac{(x+1)^2}{16} + \frac{(y-1)^2}{1} = 1$

7. $a^2 = 100, a = 10; 2^2 = 20$ major x-axis
$b^2 = 36; b = 6; 2b = 12 =$ minor y-axis;
$c^2 = 100 - 36, c = 8 =$ distance of foci from center;
$C = (0, 0);$ foci $= (8, 0)$ and $(-8, 0);$
vertices $= (10, 0), (-10, 0), (0, 6), (0, -6)$

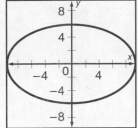

8. $b^2 = 9, b = 3, 2b = 6 =$ minor x-axis;
$a^2 = 25, a = 5, 2a = 10 =$ major y-axis;
$c^2 = 25 - 9; c = 4 =$ distance of foci from center;
$C = (0, 0);$ foci $= (0, 4)$ and $(0, -4);$
vertices $= (0, 5), (0, -5), (3, 0), (-3, 0)$

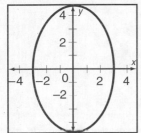

9. $\frac{x^2}{4} + \frac{y^2}{1} = 9; \frac{x^2}{36} + \frac{y^2}{9} = 1; C = (0, 0);$

$a^2 = 36, a = 6, 2a = 12 =$ major x-axis;

$b^2 = 9, b = 3, 2b = 6 =$ minor y-axis;

$c^2 = 36 - 9, c = \sqrt{27} = 3\sqrt{3} =$ distance of foci from center; foci $= (-3\sqrt{3}, 0)$ and $(-\sqrt{3}, 0);$
vertices $= (6, 0), (-6, 0), (0, 3), (0, -3)$

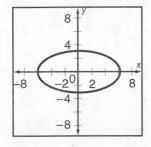

10. $a^2 = 169, a = 13, 2a = 26 =$ major x-axis;
$b^2 = 25, b = 5, 2b = 10 =$ minor y-axis;
$c^2 = 169 - 25, c = 12 =$ distance of foci from center; $C = (-2, 3);$ foci $= (10, 3)$ and $(-14, 3);$
vertices $= (11, 3), (-15, 3), (-2, 8), (-2, -2)$

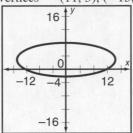

11. $9(x^2 + 6x + 9) + 16(y^2 - 2y + 1)$

$= 47 + 81 + 16; \frac{(x+3)^2}{16} + \frac{(y-1)^2}{9} = 1,$

$a^2 = 16, a = 4, 2a = 8 =$ major x-axis;
$b^2 = 9, b = 3, 2b = 6 =$ minor y-axis;
$c^2 = 16 - 9 = \sqrt{7} =$ distance of foci from center;
$C = (-3, 1);$ foci $= (-3 + \sqrt{7}, 1)$ and
$(-3 - \sqrt{7}, 1);$ vertices $= (1, 1), (-7, 1),$
$(-3, 4), (-3, -2).$

12a. The shape is an ellipse with a major axis of 30 miles. If $2a = 30, a = 15$ and $c = 8; 8^2 = 15^2 - b^2,$
$b = \sqrt{161}, 2b = 2\sqrt{161} =$ length of minor axis;

12b. $\frac{x^2}{225} + \frac{y^2}{161} = 1$ (equation of ellipse)

13. Answers will vary. The ellipse will get thinner. If the major axis and minor axis become the same length, the shape becomes a circle. The point where the foci coincide would be the center of the circle.

PRACTICE

1. $C = (0, 0);$ major horizontal axis is $a = 5$ and the

minor y-axis is $b = 2; \frac{x^2}{25} + \frac{y^2}{4} = 1$ $(a > b)$

2. $C = (0, 0);$ major vertical axis is $a = 8$ and the minor

x-axis is $b = 6; \frac{x^2}{36} + \frac{y^2}{64} = 1$

3. $C = (2, -3);$ major horizontal axis is $a = 3$ and the
minor y-axis is $b = 1; \frac{(x-2)^2}{9} + \frac{(y+3)^2}{1} = 1$

4. $a^2 = 25, a = 5, 2a = 10 = $ major x-axis;
$b^2 = 16; b = 4, 2b = 8 = $ minor y-axis;
$c^2 = 25 - 16, c = 3 = $ distance of foci from center;
$C = (0, 0);$ foci $= (3, 0)$ and $(-3, 0);$
vertices $= (5, 0), (-5, 0), (0, 4), (0, -4)$

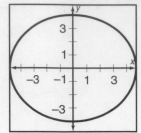

5. $a^2 = 36, a = 6, 2a = 12 = $ major x-axis;
$b^2 = 20, b = 2\sqrt{5}, 2b = 4\sqrt{5} = $ minor y-axis;
$c^2 = 36 - 20, c = 4 = $ distance of foci from center;
$C = (0, 0);$ foci $= (4, 0)$ and $(-4, 0);$
vertices $= (6, 0), (-6, 0), (0, 2\sqrt{5}), (0, -2\sqrt{5})$

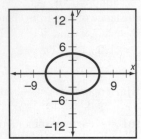

6. $\frac{x^2}{4} + \frac{y^2}{8} = 1; b^2 = 4, b = 2, 2b = 4 = $ minor x-axis;

$a^2 = 8, a = 2\sqrt{2}, 2a = 4\sqrt{2} = $ major y-axis;
$c^2 = 8 - 4, c = 2 = $ distance of foci from center;
$C = (0, 0);$ foci $= (0, 2)$ and $(0, -2);$
vertices $= (2, 0), (-2, 0), (0, 2\sqrt{2}), (0, -2\sqrt{2})$

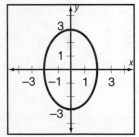

7. $\frac{x^2}{25} + \frac{y^2}{169} = 1; b^2 = 25, b = 5, 2b = 10 = $
minor x-axis; $a^2 = 169, a = 13, 2a = 26 = $ major
y-axis; $c^2 = 169 - 25, c = 12 = $ distance of foci
from center; $C = (0, 0);$ foci $= (0, 12), (0, -12);$
vertices $= (5, 0), (-5, 0); (0, 13), (0, -13)$

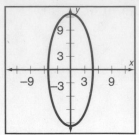

8. $a^2 = 9, a = 3, 2a = 6 = $ major x-axis;
$b^2 = 3, b = \sqrt{3}, 2b = 2\sqrt{3} = $ minor y-axis;
$c^2 = 9 - 3, c = \sqrt{6} = $ distance of foci from center;
$C = (-5, 2);$ foci $= (-5 + \sqrt{6}, 2)$ and
$(-5 - \sqrt{6}, 2);$ vertices $= (-2, 2), (-8, 2),$
$(-5, 2 + \sqrt{3}), (-5, 2 - \sqrt{3})$

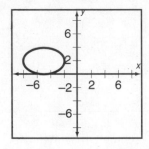

9. $9(x^2 + 8x + 16) + 25(y^2 - 6y + 9)$
  $= -144 + 144 + 225;$
$9(x + 4)^2 + 25(y - 3)^2 = 225;$

$\frac{(x + 4)^2}{25} + \frac{(y - 3)^2}{9} = 1;$
$a^2 = 25, a = 5, 2a = 10 = $ major x-axis;
$b^2 = 9, b = 3, 2b = 6 = $ minor y-axis;
$c^2 = 25 - 9, c = 4 = $ distance of foci from center;
$C = (-4, 3);$ foci $= (0, 3)$ and $(-8, 3);$
vertices $= (1, 3), (-9, 3), (-4, 6), (-4, 0)$

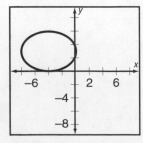

10a. $2a = 96, a = 48, a^2 = 2304;$

$2b = 46, b = 23, b^2 = 529; \dfrac{x^2}{2304} + \dfrac{y^2}{529} = 1$

10b. $c^2 = 2304 - 529, c \approx 42.13 =$ distance of foci from center of $(0, 0)$; foci $= (42.13, 0)$ and $(-42.13, 0)$

11. Answers will vary. From the numerators, find the center (h, k). From the denominators, find a and b. If the ellipse is horizontal, the vertices are $(h - a, k)$, $(h + a, k), (h, k + b),$ and $(h, k - b)$. If the ellipse is vertical, the vertices are $(h - b, k), (h + b k), (h, k + a),$ and $(h, k - a)$.

12. $2a = 20, a = 10, a^2 = 100; c = 6, c^2 = 36;$

$36 = 100 - b^2, b^2 = 64; \dfrac{x^2}{100} + \dfrac{y^2}{64} = 1$

13. $C = (-1, -6); 2a = |-6 - 4| = 10,$
$a = 5, a^2 = 25; 2b = |-8 - -4|$
$= 4, b = 2, b^2 = 4;$
$\dfrac{(x + 1)^2}{25} + \dfrac{(y + 6)^2}{4} = 1$

EXTEND

14. $a^2 = 25, a = 5; c^2 = 25 - 16, c = 3; e = \dfrac{3}{5} = 0.60$

15. $a^2 = 169, a = 13; c^2 = 169 - 144, c = 5;$
$e = \dfrac{5}{13} \approx 0.38$

16. $a^2 = 289, a = 17; c^2 = 289 - 64, c = 15;$
$e = \dfrac{15}{17} \approx 0.88$

17. $a^2 = b^2$ in a circle so $c^2 = 0, c = 0; e = \dfrac{0}{a} = 0;$ A circle has no "flatness."

18. The greatest eccentricity occurs in the theoretical case where $b = 0$, that is, when the minor axis shrinks to zero and the ellipse flattens to a line. Then $c = a$ and $e = 1$. The least eccentricity occurs when $a = b$, that is, when the ellipse is a circle. Then $c = 0$ and $e = 0$.

19. closest distance + farthest distance = length of major axis; $2a = 7.3$ billion mi; $a = 3.65$ billion mi; $a + c = 4.55$ billion mi, $c = 0.9$ billion mi;
$e = \dfrac{0.9}{3.65} \approx 0.25$

20. $(0.9)^2 = (3.65)^2 - b^2; b = 3.54;$
$a = 3.65; \dfrac{x^2}{3.65^2} + \dfrac{y^2}{3.54^2} = 1$

THINK CRITICALLY

21. $\dfrac{x^2}{25} + \dfrac{y^2}{25} = 1$, where $a = b = 5 =$ length of both axis.

MIXED REVIEW

22. $4\underline{|}\quad\begin{array}{cccc} 1 & -1 & -12 \\ & 4 & 12 \\ \hline 1 & 3 & 0 \end{array}$;

Quotient is $x + 3$.

23. $-2\underline{|}\quad\begin{array}{cccc} 1 & -7 & -13 & 10 \\ & -2 & 18 & -10 \\ \hline 1 & -9 & 5 & 0 \end{array}$;

Quotient is $x^2 - 9x + 5$.

24. $3\underline{|}\quad\begin{array}{ccccc} 3 & 0 & -25 & 0 & -18 \\ & 9 & 27 & 6 & 18 \\ \hline 3 & 9 & 2 & 6 & 0 \end{array}$;

Quotient is $3x^3 + 9x^2 + 2x + 6$.

25. $1\underline{|}\quad\begin{array}{cccc} 1 & 0 & 0 & -1 \\ & 1 & 1 & 1 \\ \hline 1 & 1 & 1 & 0 \end{array}$;

Quotient is $x^2 + x + 1$.

26. $9(x^2 + 6x + 9) + (y^2 - 10y + 25) =$
$-97 + 81 + 25; 9(x + 3)^2 + (y - 5)^2 = 9;$

$\dfrac{(x + 3)^2}{1} + \dfrac{(y - 5)^2}{9} = 1; b^2 = 1, b = 1; 2b = 2;$

$a^2 = 9, a = 3, 2a = 6;$ The answer is C; 6 and 2.

ALGEBRAWORKS

1. Answers will vary, but the number of people in the hall would change many factors such as temperature and absorption or reflection of the sound waves.

2. Answers will vary.

3. The sounds heard near the foci will be loud, but sounds heard anywhere else will not be near as loud.

4a. $d_A = \sqrt{(x - 6)^2 + (y - 0)^2};$
$d_B = \sqrt{(x - 12)^2 + (y - 0)^2}$

4b. $\dfrac{2}{(d_B)^2} = \dfrac{1}{(d_A)^2}; \dfrac{2}{(x - 12)^2 + y^2} = \dfrac{1}{(x - 6)^2 + y^2}$

4c. $\dfrac{2}{x^2 - 24x + 144 + y^2} = \dfrac{1}{x^2 - 12x + 36 + y^2}$;

$2(x^2 - 12x + 36 + y^2) = x^2 - 24x + 144 + y^2$
$2x^2 - 24x + 72 + 2y^2 = x^2 - 24x + 144 + y^2$
$x^2 + y^2 = 72;$ This equatioin indicates a circle with a radius of $6\sqrt{2}$. The volume for both speakers will be the same at any point on the circle described.

EXPLORE

1. use graphing calculator; hyperbola

2. The lines do not intersect the two pieces of the hyperbola.

3. (3, 4.47), (3, 6); (5, 9.17); (5, 10); (10, 19.60), (10, 20); (100, 199.96), (100, 200); (1000, 2000.00); As x gets larger, y_1 and y_3 are almost the same function.

TRY THESE

1. $5^2 = 4^2 + b^2$, $b^2 = 9$; the transverse axis is horizontal, so the equation is $\dfrac{x^2}{16} - \dfrac{y^2}{9} = 1$.

2. $C = (2, 3)$; $a = 3$, $a^2 = 9$; $b = 2$; $b^2 = 4$; $\dfrac{(x - 2)^2}{9} - \dfrac{(y - 3)^2}{4} = 1$

3. This is a rectangular hyperbola with $k < 0$; $k = -3$, so the equation is $xy = -3^2$ or $xy = -9$.

4. $2a = 12$, $a = 6$, $a^2 = 36$; $(2\sqrt{10})^2 = 36 + b^2$, $b^2 = 4$; $\dfrac{y^2}{36} - \dfrac{x^2}{4} = 1$.

5. The transverse axis is horizontal; $a = 15$. The vertices are $(15, 0)$ and $(-15, 0)$. $c^2 = 225 + 64$, $c = 17$. The foci are $(17, 0)$ and $(-17, 0)$. The asymptotes are $y = \pm\dfrac{8}{15}x$.

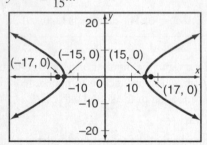

6. The transverse axis is vertical; $a = 3$. The center is $(-1, 3)$. The vertices are $(-1, 6)$ and $(-1, 0)$. $c^2 = 9 + 1$, $c = \sqrt{10}$. The foci are $(-1, 3 + \sqrt{10})$ and $(-1, 3, -\sqrt{10})$.

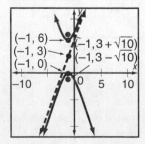

7. This is a rectangular hyperbola where $k > 0$. The branches lie in the I and III quadrants and its vertices are $(2, 2)$ and $(2, -2)$.

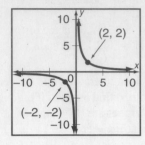

8. $(y^2 + 8y + 16) - 2(x^2 - 2x + 1) = -6 + 16 - 2$; $(y + 4)^2 - 2(x - 1)^2 = 8$; $\dfrac{(y + 4)^2}{8} - \dfrac{(x - 1)^2}{4} = 1$

$C = (1, -4)$; $a^2 = 8$, $a = 2\sqrt{2}$, so the vertices are $(1, -4 + 2\sqrt{2})$ and $(1, -4 - 2\sqrt{2})$. $C^2 = 8 + 4$, $C = 2\sqrt{3}$, so the foci are $(1, -4 + 2\sqrt{3})$ and $(1, -4 - 2\sqrt{3})$. The slopes of the asymptotes are $\pm\dfrac{a}{b} = \pm\dfrac{2\sqrt{2}}{2} = \pm\sqrt{2}$.

9. $-15(y - 3)^2 = 120 - 8x^2$; $(y - 3)^2 = \dfrac{120 - 8x^2}{-15}$;

$y - 3 = \pm\sqrt{\left(\dfrac{-120 + 8x^2}{15}\right)}$,

$y_2 = 3 - \sqrt{\left(\dfrac{-120 + 8x^2}{15}\right)}$

$y = 3 \pm \sqrt{\left(\dfrac{-120 + 8x^2}{15}\right)}$.

$y_1 = 3 + \sqrt{\left(\dfrac{-120 + 8x^2}{15}\right)}$,

$y_2 = 3 - \sqrt{\left(\dfrac{-120 + 8x^2}{15}\right)}$

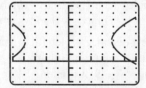

10. The transverse axis is the x-axis; $C = (0, 0)$; $a = 4$, $a^2 = 16$; $c = 6$, $c^2 = 36$; $36 = 16 + b^2$, $b = 2\sqrt{5}$, $b^2 = 20$; $\dfrac{x^2}{16} - \dfrac{y^2}{20} = 1$

11. Answers may vary. The slope of the asymptotes equal $\pm\dfrac{b}{a}$. Once a and b are found, the foci can be found by using the equation $c^2 = a^2 + b^2$.

1. This is a rectangular hyperbola where
$k = 6^2 = 36$; $xy = 36$ $(k > 0)$

2. $a = 2\sqrt{3}$, $a^2 = 12$; $c = \sqrt{21}$, $c^2 = 21$;
$21 = 12 + b^2$,
$b^2 = 9$; $\dfrac{x^2}{12} - \dfrac{y^2}{9} = 1$

3. $a = 3$, $a^2 = 9$; $b = 1$, $b^2 = 1$; $C = (3, 4)$;

$$\dfrac{(x - 3)^2}{9} - \dfrac{(y - 4)^2}{1} = 1$$

4. $a = 2\sqrt{5}$, $a^2 = 20$; $c = 3\sqrt{5}$, $c^2 = 45$;

$45 = 20 + b^2$, $b^2 = 25$; $C = (0, 0)$; $\dfrac{y^2}{20} - \dfrac{x^2}{25} = 1$

5. This is a rectangular hyperbola where
$k = -(8^2) = -64$; $xy = -64$ (k is negative because
the graph is in the II and IV Quad.)

6. $a = 8$; $a^2 = 64$; $b = 6$; $b^2 = 36$; center = $(2, -2)$:
So, $\dfrac{(y + 2)^2}{64} - \dfrac{(x - 2)^2}{36} = 1$

7. $C = (0, 0)$; $2a = 36$, $a = 18$; $a^2 = 324$;
$2b = 20$, $b = 10$, $b^2 = 100$; vertical, so

$$\dfrac{y^2}{324} - \dfrac{x^2}{100} = 1$$

8. $C = (5, -3)$; $c = 13 - 5 = 8$, $c^2 = 64$;
$2a = 10$, $a = 5$, $a^2 = 25$; $64 = 25 + b^2$, $b^2 = 39$;

horizontal, so $\dfrac{(x - 5)^2}{25} - \dfrac{(y + 3)^2}{39} = 1$

9.
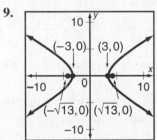
The transverse axis is horizontal; $a = 3$. The vertices are $(3, 0)$ and $(-3, 0)$. $c^2 = 9 + 4$; $c = \sqrt{13}$. The foci are $(\sqrt{13}, 0)$ and $(-\sqrt{13}, 0)$. The asymptotes are $y = \pm \dfrac{4}{9}x$. $C = (0, 0)$

10.
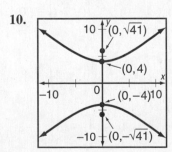
$\dfrac{y^2}{16} - \dfrac{x^2}{25} = 1$;

The transverse axis is vertical, $a = 4$. The vertices are $(0, 4)$ and $(0, -4)$. $c^2 = 16 + 25$, $c = \sqrt{41}$. The foci are $(0, \sqrt{41})$ and $(0, -\sqrt{41})$. $C = (0, 0)$

11.
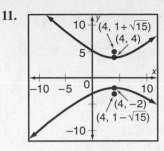
The transvers axis is vertical. $C = (4, 1)$; $a = \sqrt{6}$. The vertices are $(4, 1 + \sqrt{6})$ and $(4, 1 - \sqrt{6})$. $c^2 = 6 + 9$; $c = \sqrt{15}$. The foci are $(4, 1 + \sqrt{15})$ and $(4, 1 - \sqrt{15})$.

12.
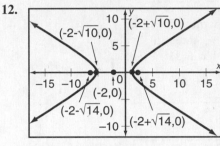
The transverse axis is horizontal. $C = (-2, 0)$; $a = \sqrt{10}$. The vertices are $(-2 + \sqrt{10}, 0)$ and $(-2 + \sqrt{10}, 0)$ and

$c^2 = 10 + 4$, $c = \sqrt{14}$. The foci are $(-2 + \sqrt{14}, 0)$ and $(2 - \sqrt{14}, 0)$.

13.
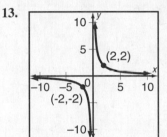
This is a rectangular hyperbola where $k = 4$. Since $k > 0$, the branches are in the first and third quadrants. $a^2 = 4$, $a = 2$; The vertices are $(2, 2)$ and $(-2, -2)$.

14.
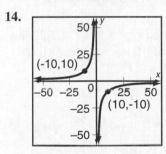
$xy = -100$; rectangular hyperbola, $k = -100$. The graph lies in the second and fourth quadrants. $a^2 = 100$, $a = 10$; The vertices are $(10, -10)$ and $(-10, 10)$.

15. $25(x^2 - 4x + 4) - 9(y^2 + 8y + 16) =$
$269 + 100 - 144$; $25(x - 2)^2 - 9(y - 4)^2 = 225$;
$\dfrac{(x - 2)^2}{9} - \dfrac{(y + 4)^2}{25} = 1$. $C = (2, -4)$; $a = 3$, so
the vertices are $(5, -4)$ and $(-1, -4)$.
$c^2 = 9 + 25$, $c = \sqrt{34}$; The foci are $(2 + \sqrt{34}, -4)$
and $(2 - \sqrt{34}, -4)$. The slope of the asymptotes are
$\pm \dfrac{b}{a} = \pm \dfrac{5}{3}$ (horizontal graph).

16. $(y^2 + 6y + 9) - 3(x^2 - 2x + 1) = 18 + 9 - 3;$

$(y + 3)^2 - 3(x - 1)^2 = 24; \dfrac{(y + 3)^2}{24} - \cdot \dfrac{(x - 1)^2}{8}$

$= 1. C = (1, -3); a^2 = 24, a = 2\sqrt{6},$
so the vertices are $(1, -3 + 2\sqrt{6})$ and $(1, -3 - 2\sqrt{6})$.
$c^2 = 24 + 8, c = 4\sqrt{2};$ The foci are $(1, -3 + 4\sqrt{2})$
and $(1, -3 - 4\sqrt{2})$. The slope is
$\pm \dfrac{a}{b} = \pm \dfrac{2\sqrt{6}}{2\sqrt{2}} = \pm\sqrt{3}.$

17. Multiply each term by the LCD,
$80.5 (y - 3)^2 - 4(x + 5)^2 = 80;$
$5(y - 3)^2 = 4(x + 5)^2 + 80;$

$(y - 3)^2 = \dfrac{4}{5}(x + 5)^2 + 16;$

$y = 3 \pm \sqrt{\dfrac{4}{5}(x + 5)^2 + 16};$

$y_1 = 3 \pm \sqrt{\dfrac{4}{5}(x + 5)^2 + 16}$

$y_2 = 3 \pm \sqrt{\dfrac{4}{5}(x + 5)^2 + 16}$

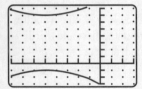

18. $\dfrac{V_A}{P_A} \dfrac{V_8}{P_B}; \dfrac{10}{14.7} = \dfrac{x}{58.7}; 14.7x = 587; x \approx 39.9 \text{ cm}^3$

19. Answers may vary but when $k > 0, xy > 0.$ This only happens in Quadrants I and III. If $k < 0, xy < 0$ and this occurs in Quadrants II and IV.

EXTEND

20. $\dfrac{x^2}{4} + \dfrac{y^2}{4} = 1; a = b,$ so graph is a circle.

21. $\dfrac{x^2}{6} + \dfrac{y^2}{6} = 1;$ hyperbola, because of minus sign.

22. $\dfrac{x^2}{16} + \dfrac{y^2}{9} = 1;$ ellipse.

23. $2c = 8800, c = 4400; 2a = 5 \cdot 1100 = 5500,$
$a = 2750; (4400)^2 = (2750)^2 + b^2, b \approx 3435;$
$\dfrac{x^2}{2750^2} - \dfrac{y^2}{3435^2} = 1.$

24. $\dfrac{x^2}{9} - \dfrac{y^2}{9} = 1; a = b = 3;$ The slopes of the
asymptotes are $\pm \dfrac{b}{a} = \pm \dfrac{3}{3} = \pm 1.$ The products
of the slope is $-1,$ so the asymptotes are perpendicular to each other.

25. Answers will vary, but anytime $a = b$ the asymptotes will be perpendicular.

THINK CRITICALLY

26. No, because you do not know if the slope represents
$\pm \dfrac{b}{a}$ or $\pm \dfrac{a}{b}.$

27. $m = \pm \dfrac{3}{4}; a = 4, b = 3; C = (7 - 4, -2)$
$= (3, -2). \dfrac{(x - 3)^2}{16} - \dfrac{(y + 2)^2}{9} = 1$

28. Answers will vary, but using a graphing calculator or picking points and sketching will show both graphs with a center of $(0, 0)$ and a horizontal transverse axis. However, because $a > b,$ the vertices of the second graph will be closer to the origin and its branches will open wider.

MIXED REVIEW

29. $\dfrac{1}{8} = 0.125$

30. $\dfrac{4}{8} = 0.5$

31. $\dfrac{2}{8} = 0.25$

32. $\dfrac{3}{8} = 0.375$

33. $f(g(x)) = 2(3x + 1) - 3 = 6x + 2 - 3 = 6x - 1;$
B is the correct answer.

Lesson 12.6, pages 592–597

EXPLORE

1. The graph is a parabola.

2. It is the same graph but it goes down instead of up. It is reflected over the x-axis.

3. The graphs are the same but one opens left and the other opens right. They are reflected over the y-axis.

TRY THESE

1. $(x - 0)^2 = -20(y - 0); (x - 0)^2 = 4(-5)(y - 0)$

2. $(x - 0)^2 = 4\left(\dfrac{1}{4}\right)(y - 0)$

3. $x^2 - 12x + 60 = 8y; (x^2 - 12x + 36) + 24 = 8y;$
$(x - 6)^2 = 8y - 24; (x - 6)^2 = 8(y - 3);$
$(x - 6)^2 = 4(2)(y - 3)$

4. $(x^2 + 4x + 10) = -2y; (x^2 + 4x + 4) + 6 = -2y;$
$(x + 2)^2 = -2y - 6; (x + 2)^2 = -2(y + 3);$

$(x - (-2))^2 = 4\left(-\dfrac{1}{2}\right)(y - (-3))$

5.

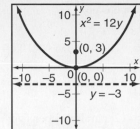

$V = (0, 0)$; $4p = 12$, $p = 3$, focus $= (0, 3)$; The directrix is the line $y = -3$.

6.

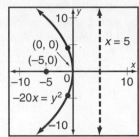

$V = (0, 0)$; $4p = -20$, $p = -5$, focus $= (-5, 0)$; The directrix is the line $x = 5$.

7.

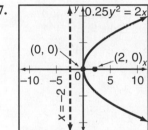

$y^2 = 8x$; $V = (0, 0)$; $4p = 8$, $p = 2$, focus $= (2, 0)$; The directrix is the line $x = -2$.

8.

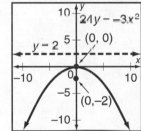

$x^2 = -8y$; $V = (0, 0)$; $4p = -8$, $p = -2$, focus $= (0, -2)$; The directrix is the line $y = 2$.

9. $p = -5$, $4p = -20$; parabola opens down; $x^2 = -20y$

10. $p = 7$, $4p = 28$; parabola opens right; $y^2 = 28x$.

11. $p = 9$, $4p = 36$; parabola opens up; $x^2 = 36y$

12. $p = -3$; $4p = -12$; parabola opens left; $(y + 5)^2 = -12(x - 4)$

13. $(x^2 + 4x + 4) + 16 = 8y$; $(x + 2)^2 = 8y - 16$; $(x + 2)^2 = 8(y - 2)$. The vertex is $(-2, 2)$. $4p = 8$, $p = 2$, focus $= (-2, 4)$; The directrix is the line $y = 0$.

14. $p = 3$, $4p = 12$; $x^2 = 12y$

15. Answers will vary. If x is squared and $p > 0$, the parabola opens up. If $p < 0$, it opens down. If y is squared and $p > 0$, the parabola opens right. If $p < 0$, it opens left.

1.

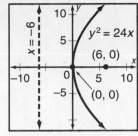

$V = (0, 0)$; $4p = 24$, $p = 6$, focus $= (6, 0)$; The directrix is the line $x = -6$.

2.

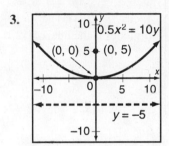

$V = (0, 0)$; $4p = 32$, $p = 8$, focus $= (0, 8)$; The directrix is the line $y = -8$.

3.

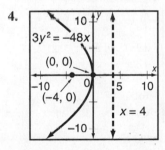

$x^2 = 20y$; $V = (0, 0)$; $4p = 20$, $p = 5$, focus $= (0, 5)$; The directrix is the line $y = -8$.

4.

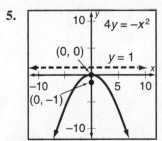

$y^2 = -16x$; $V = (0, 0)$; $4p = -16$, $p = -4$, focus $= (-4, 0)$; The directrix is the line $x = 4$.

5.

$x^2 = -4y$; $V = (0, 0)$; $4p = -4$, $p = -1$, focus $= (0, -1)$; The directrix is the line $y = 1$.

6. 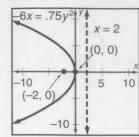 $y^2 = -8x$; $V = (0, 0)$;
$4p = -8, p = -2$,
focus $= (-2, 0)$; The
directrix is the line $x = 2$.

7. 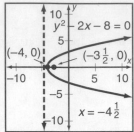 $y^2 = 2x + 8$,
$y^2 = 2(x + 4)$;
$V = (-4, 0)$; $4p$
$= 2, p = \frac{1}{2}$,
focus $= (-3.5, 0)$; The
directrix is the line
$x = -4.5$

8. 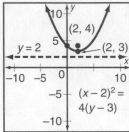 $V = (2, 3)$; $4p = 4, p = 1$,
focus $= (2, 4)$; The
directrix is the line $y = 2$.

9. 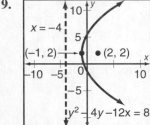 $y^2 - 4y = 12x + 8$;
$(y^2 - 4y + 4) =$
$12x + 8 + 4, (y - 2)^2 =$
$12x + 12, (y - 2)^2 =$
$12(x + 1)$; $V = (-1, 2)$;
$4p = 12, p = 3$,
focus $= (2, 2)$; The
directrix is the line $x = -4$.

10. 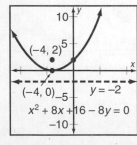 $x^2 + 8x + 16 =$
$8y, (x + 4)^2 = 8y$;
$V = (-4, 0)$; $4p = 8$,
$p = 2$, focus $= (-4, 2)$;
The directrix is the line
$y = -2$.

11. $p = 3, 4p = 12$; parabola opens right; $y^2 = 12x$

12. $p = -10$; $4p = -40$; parabola opens down;
$x^2 = -40y$

13. $p = -6, 4p = -24$; parabola opens left; $y^2 = -24x$

14. $p = 5, 4p = 20$; parabola opens up; $x^2 = 20y$

15. $p = 8, 4p = 32$; parabola opens up;
$(x - 1)^2 = 32(y + 1)$

16. $V = (-4, 6)$; $p = 1, 4p = 4$; parabola opens right;
$(y - 6)^2 = 4(x + 4)$

17. $V = (8, 2)$; $p = -2, 4p = -8$; parabola opens down;
$(x - 8)^2 = -8(y - 2)$.

18. $V = (-2, 0)$; $p = 3, 4p = 12$; parabola opens up;
$(x + 2)^2 = 12y$

19. You would first put the equation in the form
$(x - h)^2 = 4p(y - k)$. The vertex is (h, k), the focus is
$(h, k + p)$, and the directrix is the line $y = k - p$.

20. $x^2 + 6x = -4y - 5$,
$(x^2 + 6x + 9) = -4y - 5 + 9$,
$(x + 3)^2 = -4y + 4$; $(x + 3)^2 = -4(y - 1)$;
$V = (-3, 1)$; $4p = -4, p = -1$, focus $= (-3, 0)$;
The directrix is the line $y = 2$.

21. $y^2 - 6y + 9 = 4x - 17 + 9$,
$(y - 3)^2 = 4(x - 2)$; $V = (2, 3)$; $4p = 4, p = 1$,
focus $= (3, 3)$; The directrix is the line $x = 1$

22. $2(y^2 + 2y) = 2x - 2$; $2(y^2 + 2y + 1) =$
$2x - 2 + 2$; $2(y + 1)^2 = 2x$;
$(y + 1)^2 = \frac{2x}{2}$; $(y + 1)^2 = x$;
$V = (0, -1)$; $4p = 1, p = \frac{1}{4}$,

focus $= \left(\frac{1}{4}, -1\right)$; The directrix is the line $x = -\frac{1}{4}$.

23. $x^2 + 4x = y - 1$; $(x^2 + 4x + 4) = y - 1 + 4$;
$(x + 2)^2 = (y + 3)$; $V = (-2, -3)$; $4p = 1, p = \frac{1}{4}$,

focus $= \left(-2, -\frac{11}{4}\right)$; The directrix is the line $y = -\frac{13}{4}$.

24. $x^2 = 31.25y$; $4p = 31.25, p = 7.8125$ ft $=$ distance
of the oil pipe above the mirror.

EXTEND

25. The parabola opens up or down so the axis of symmetry
is vertical, $x = h$.

26. The parabola opens left or right so the axis of symmetry
is horizontal, $y = k$.

27. $(x - 1)^2 = y + 3$; The parabola opens up so the axis
of symmetry is vertical, $x = 1$.

28. $(y + 5)^2 = x + 2$; The parabola opens right so the
axis is horizontal, $y = -5$.

29. $x^2 - 8x + 16 = y + 16$; $(x - 4)^2 = y + 16$; The
parabola opens up so the axis is vertical, $x = 4$.

30. $y^2 + 4y = -x - 1$; $y^2 + 4y + 4 = -x - 1 + 4$;
$(y + 2)^2 = -(x - 3)$; The parabola opens left so the
axis is vertical, $y = -2$.

31a.

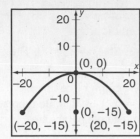

The equation is
$$x^2 = -\frac{80}{3}y$$

31b. If width $= 16$ ft; $x = \frac{16}{2} = 8$

$$(8)^2 = -\frac{80}{3}y$$

$$y = -2, 4; 2.4 \text{ ft}$$

32a. The parabola gets wider as the absolute value of p increases.

32b. Yes, the results should be the same.

THINK CRITICALLY

33. $b(20) = 2(9-3)^2 - 4; 20b = 72 - 4;$
$b = \frac{68}{20} = \frac{17}{5}.$

34a.

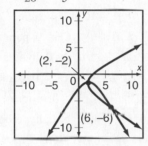

34b. In the horizontal parabola, $(y + 2)^2 = 4p(x - 2)$.
$(-6 + 2)^2 = 4p(6 - 2); 16 = 16p; p = 1; 4p = 4;$
$(y + 2)^2 = 4(x - 2)$. In the vertical parabola,
$(x - 2)^2 = 4p(y + 2)$. $(6 - 2)^2 = 4p(-6 + 2)$,
$16 = -16p, p = -1, 4p = -4; (x - 2)^2$
$= -4(y + 2)$.

35. $\overline{FM} = \overline{MQ}$ (definition of a paraola); $\overline{FD} = \overline{MQ}$
(opposite sides of a rectangle are equal);
$\overline{FV} + \overline{DV} = \overline{MQ}$; By substitution,
$\overline{FM} = \overline{FV} + \overline{DV}; \overline{FV} = \overline{DV} = p$, so
$\overline{FM} = p + p = 2p$. $\overline{FN}$ also equals $2p$, so
$\overline{MN} = 2p + 2p = 4p$.

MIXED REVIEW

36. $\begin{bmatrix} 1 & -1 \\ 3 & 0 \end{bmatrix}\begin{bmatrix} 2 & 3 \\ -1 & 4 \end{bmatrix} =$

$\begin{bmatrix} (1)(2) + (-1)(-1) & (1)(3) + (-1)(4) \\ (3)(2) + (0)(-1) & (3)(3) + (0)(4) \end{bmatrix} =$

$\begin{bmatrix} 3 & -1 \\ 6 & 9 \end{bmatrix}$

37. $\begin{bmatrix} 2 & 3 \\ -1 & 4 \end{bmatrix}\begin{bmatrix} 1 & -1 \\ 3 & 0 \end{bmatrix} =$

$\begin{bmatrix} (2)(1) + (3)(3) & (2)(-1) + (3)(0) \\ (-1)(1) + (4)(3) & (-1)(-1) + (4)(0) \end{bmatrix} =$

$\begin{bmatrix} 11 & -2 \\ 11 & 1 \end{bmatrix}$

38. $\begin{bmatrix} 2 & 3 \\ -1 & 4 \end{bmatrix}$ **39.** $\begin{bmatrix} 2 & 3 \\ -1 & 4 \end{bmatrix}$

40. sum $= -9$, product $= 20$.

41. sum $= 5$, product $= -13$.

42. $x^2 + 3x + \frac{2}{3} = 0$; sum $= -3$, product $= \frac{2}{3}$.

43. $x^2 - \frac{1}{5}x - \frac{3}{5} = 0$; sum $= \frac{1}{5}$, product $= -\frac{3}{5}$.

44. $x = 7^2 = 49$

45. no solution

46. $2x + 3 = 5^2$; $2x = 22$; $x = 11$

47. $(n - 5)^2 = n + 7, n^2 - 10n + 25 = n + 7,$
$n^2 - 11n + 18 = 0, (n - 9)(n - 2) = 0, n = 9;$
(2 is an extraneous root)

48. $x^2 - 6x = 4y - 29, x^2 - 6x + 9 = 4y - 29 + 9,$
$(x - 3)^2 = 4(y - 5); V = (3, 5)$. A is the correct
answer.

ALGEBRAWORKS

1.

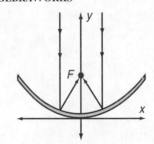

2.

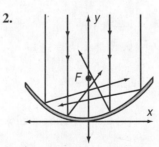

3.

4.

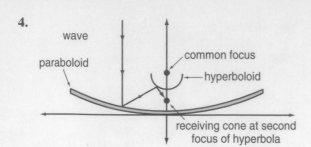

wave
paraboloid
common focus
hyperboloid
receiving cone at second
focus of hyperbola

Lesson 12.7, pages 598–602

EXPLORE

1. Answers will vary, but A must always equal C.

2. Answers will vary, but A and C must have the same sign. (Their values are not equal)

3. Answers will vary, but A and C must have different signs.

4. Answers will vary, but either $A = 0$ or $C = 0$. (Both cannot equal zero, though)

5. If $B = 0$ and: $A = C$, it is a circle; A and C have the same sign, it is an ellipse; A and C have opposite signs, it is a hyperbola; $A = 0$ or $C = 0$, it is a parabola. If B does not equal zero, you cannot use this strategy.

TRY THESE

1. $5x^2 + 6xy + 4y^2 + (-3x) + (-2y) + 1 = 0$;
   $A = 5, B = 6, C = 4, D = -3, E = -2, F = 1$.

2. $x^2 + 0xy + 0y^2 + (-7x) + 0y + 12 = 0$;
   $A = 1, B = 0, C = 0, D = -7, E = 0, F = 12$.

3. $4(x - 1)^2 + 9(y + 2)^2 = 36$;
   $4(x^2 - 2x + 1) + 9(y^2 + 4y + 4) = 36$;
   $4x^2 - 8x + 4 + 9y^2 + 36y + 36 = 36$;
   $4x^2 + 0xy + 9y^2 + (-8x) + 36y + 4 = 0$;
   $A = 4, B = 0, C = 9, D = -8, E = 36, F = 4$.

4. $y = -3x^2 - 6xy + 15 + 2x^2 - 2xy + 2$;
   $1x^2 + 8xy + 0y^2 + 0x + y + (-17) = 0$
   $A = 1, B = 8, C = 0, D = 0, E = 1, F = -17$

5. $A = 25, B = 0, C = 16, B^2 - 4AC = -1600$; $-1600 < 0$; The equation represents an ellipse.

6. $A = 3, B = 0, C = 0$; $B^2 - 4AC = 0$; The equation represents a parabola.

7. $A = 1, B = 0, C = -1$; $B^2 - 4AC = 4$; The equation represents a hyperbola.

8. $A = 1, B = 0, C = 1$; $B^2 - 4AC = -4$; The equation represents an ellipse. Since $A = C$, it is a circle.

9. $A = 6, B = 4, C = 6$; $B^2 - 4AC = -128$; The equation is an ellipse.

10. $A = 1, B = -5, C = -2$; $B^2 - 4AC = 33$; The equation represents a hyperbola.

11. $A = -2, B = -3, C = 3$; $B^2 - 4AC = 33$; The equation represents a hyperbola.

12. $A = -3, B = 6, C = -3$; $B^2 - 4AC = 0$; The equation is a parabola.

13. 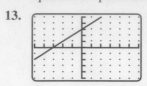 The graph is the line $y = x + 3$. It could represent a degenerate parabola.

14. Identify A, B, and C by putting the equation in general form. Find the value of the discriminant, $B^2 - 4AC$. If it equals zero, you have a parabola. If it is less than zero, you have an ellipse and you should check to see if $A = C$ and $B = 0$. If it does, you have a circle. If the discriminant is greater than zero, the conic is a hyperbola.

PRACTICE

1. $A = 2, B = 0, C = 0$; $B^2 - 4AC = 0$; parabola

2. $A = 1, B = 0, C = 8$; $B^2 - 4AC < 0$ ellipse

3. $A = 4, B = 0, C = -8$; $B^2 - 4AC > 0$; hyperbola

4. $A = 1, B = 0, C = 1$; $B^2 - 4AC < 0$; circle

5. $A = -2, B = 8, C = -8$; $B^2 - 4AC = 0$; parabola

6. $A = 1, B = 1, C = -3$; $B^2 - 4AC > 0$; hyperbola

7. $A = 25, B = -10, C = 0$; $B^2 - 4AC = 0$; parabola

8. $A = 9, B = 2, C = 1$; $B^2, - 4AC < 0$; ellipse

9. $A = 0, B = 12, C = 0$; $B^2 - 4AC > 0$; hyperbola

10. $A = 1, B = -1, C = 1$; $B^2 - 4AC < 0$; ellipse

11. $A = \dfrac{1}{225}, B = 0, C = -\dfrac{1}{64}$; $B^2 - 4AC > 0$; hyperbola

12. $A = 1, B = 0, C = 1$; $B^2 - 4AC < 0$; where $A = C$, so it is a circle.

13. $A = 15, B = 0, C = 15$; $B^2 - 4AC < 0$; ellipse where $A = C$, so it is a circle.

14. $A = 3, B = 0, C = 0$; $B^2 - 4AC = 0$; parabola

15. $A = 3, B = -4, C = 0$; $B^2 - 4AC > 0$; hyperbola

16. $A = 2, B = 5, C = 5$; $B^2 - 4AC < 0$; ellipse

17. $A = 1, B = 0, C = 0$; $B^2 - 4AC = 0$; parabola

18. $A = \dfrac{1}{10}, B = 0, C = \dfrac{1}{18}$; $B^2 - 4AC < 0$; ellipse

19. 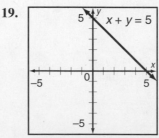 The graph is a line with a slope of -1 and a y-intercept of 5, a degenerate parabola.

a degenerate parabola which is a line

20.

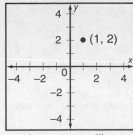

a degenerate ellipse
which is the point (1, 2)

The graph is the point (1, 2), which is a degenerate ellipse.

21.

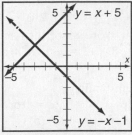

a degenerate hyperbola
which is two lines

The graph is two intersecting lines, $y = x + 5$ and $y = -x - 1$, a degenerate hyperbola.

22. If it is a parabola, $B^2 - 4AC = 0$. $A = -5$, $C = -2$, and B is the erased coefficient.
$B^2 - 4(-5)(-2) = 0$, $B^2 - 40 = 0$,
$B^2 = 40$, $B = \pm 2\sqrt{10}$.

23. Answers will vary, but it is easiest to choose positive values for A and C and solve for a B value using the idea of the value of the discriminant being greater than, less than, or equal to zero. Values for D, E, and F can be chosen randomly.

EXTEND

24. It is a line so it is a degenerate parabola.

25. It is the point (0, 0) so it is a degenerate ellipse.

26. It is two intersecting lines, $y = 2x$ and $y = -2x$, so it is a degenerate hyperbola.

27. $B^2 - 4AC = 0$, $B^2 - 4(4)(9) = 0$,
$B^2 - 144 = 0$, $B = \pm 12$

28. $B^2 - 4AC > 0$, $B^2 - 4(3)(2) > 0$, $B^2 - 24 > 0$,
$B^2 > 24$; $B > 2\sqrt{6}$ or $B < -2\sqrt{6}$

29. If it were an ellipse, $B^2 - 4AC < 0$, $B^2 - 4(-3)(5) = B^2 + 60$. Any value of b squared would be positive and an answer less than zero would never ocur. This equation could not be an ellipse.

THINK CRITICALLY

30a. $4y^2 + (4xy + y) + (x^2 - 2x - 9) = 0$;
$4y^2 + (4x + 1)y + (x^2 - 2x - 9) = 0$

30b. $a = 4$, $b = 4x + 1$, $c = x^2 - 2x - 9$;
$$x = \frac{-(4x + 1) \pm \sqrt{(4x + 1)^2 - 4(4)(x^2 - 2x - 9)}}{2(4)};$$
$$y_1 = \frac{-4x - 1 + \sqrt{40x + 145}}{8};$$
$$y_2 = \frac{-4x - 1 - \sqrt{40x + 145}}{8}$$

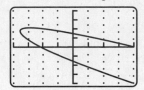

31. $Cy^2 + (Bx + E)y + (Ax^2 + Dx + F) = 0$; Using the quadratic formula and letting $x = 0$, the y-intercepts would be $\dfrac{-E \pm \sqrt{E^2 - 4CF}}{2C}$.

32. $\dfrac{-10 \pm \sqrt{10^2 - 4(2)(8)}}{2(2)} = -1$ or -4

MIXED REVIEW

33. no, because $1 + 1 = 2$ **34.** yes

35. no, because $\pm 1 \div 0$ is undefined. **36.** $5^{\frac{1}{2}}$

37. $x^{\frac{3}{6}} = x^{\frac{1}{2}}$ **38.** $2^{\frac{1}{3}}(n^5)^{\frac{1}{3}} = 2\frac{1}{3}n^{\frac{5}{3}}$

39. $(h^3)^{\frac{1}{5}}(k^{15}) = h^{\frac{3}{5}}k^3$ **40.** $2^x = 16$, $x = 4$

41. $3^{2x} = 3^3$; $2x = 3$; $x = \dfrac{3}{2}$ **42.** $10^{-2} = x$, $x = 0.01$

43. $x^{\frac{5}{3}} = 32$; $\sqrt[5]{32} = 2$; $2^3 = 8$; $x = 8$

44. $A = 1$, $B = 2$, $C = 1$; $B^2 - 4AC = 0$; parabola; The correct answer is D.

Lesson 12.8, pages 603–609

EXPLORE

1. 0, 1, or 2 points. Configurations will vary.

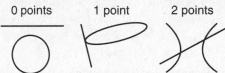

0 points   1 point   2 points

2. 0, 1, 2, 3, or 4 points. Configurations will vary.

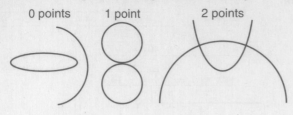

0 points   1 point   2 points

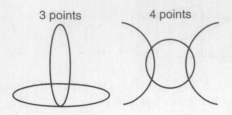

3 points   4 points

TRY THESE

1. A line and an ellipse; By substitution,
$4x^2 + (2x - 1)^2 = 25$; $4x^2 + 4x^2 - 4x + 1 = 25$;
$8x^2 - 4x - 24 = 0$; $4(2x^2 - x - 6) = 0$;
$4(2x + 3)(x - 2) = 0$; $x = 2$ or -1.5; solving for
y, $y = 2(2) - 1 = 3$ or
$y = 2(-1.5) - 1 = -4$; $(2, 3)$ and $(-1.5, -4)$

2. A hyperbola and a circle; $x = \dfrac{15}{y}$; by substitution,

$\left(\dfrac{15}{y}\right)^2 + y^2 = 34$, $\dfrac{225}{y^2} + y^2 = 34$,

$225 + y^4 = 34y^2$, $y^4 - 34y^2 + 225 = 0$,
$(9y^2 - 9)(y^2 - 25) = 0$, $y = \pm 3$ or ± 5; solving
for y in $xy = 15$, $(5, 3)$, $(-5, -3)$, $(3, 5)$, $(-3, -5)$

3. An ellipse and a hyperbola; multiplying the first
equation by 4 and adding the result to the other
equation, $13x^2 = 0$ and $x = 0$; solving for
y, $4y^2 = 16$, $y = \pm 2$; $(0, 2)$ and $(0, -2)$

4. A line and a circle; By substitution where
$x = 5 - y$, $(5 - y)^2 + y^2 = 17$,
$25 - 10y + y^2 + y^2 = 17$, $2y^2 - 10y + 8 = 0$,
$2(y^2 - 5y + 4) = 0$, $2(y - 4)(y - 1) = 0$, $y = 4$
or 1; solving for x, $x + 4 = 5$, $x = 1$ or
$x + 1 = 5$, $x = 4$; $(1, 4)$ and $(4, 1)$

5. A hyperbola and a line; By substitution where
$y = 2x - 3$, $(2x - 3)^2 - x^2 = 9$,
$4x^2 - 12x + 9 - x^2 = 9$, $3x^2 - 12x = 0$,
$x(3x - 12) = 0$, $x = 0$ or 4; solving for
y, $y = 2(0) - 3 = -3$ or $y = 2(4) - 3 = 5$; $(0, -3)$
and $(4, 5)$

6. An ellipse and a hyperbola; multiplying the first
equation by -1 and adding the result to the othe
equation, $-14x^2 = -14$, $x^2 = 1$, $x = \pm 1$; solving
for y, $y^2 - 9(\pm 1)^2 = 16$, $y^2 = 25$, $y = \pm 5$; $(1, 5)$,
$(1, -5)$, $(-1, 5)$, $(-1, -5)$.

7. Let $x =$ width, then the length $= \dfrac{42 - 2x}{2} = 21 - x$;
In the right triangle, $x^2 + (21 - x)^2 = 15^2$,
$x^2 + 441 - 42x + x^2 = 225$, $2x^2 - 42x + 216$
$= 0$, $2(x - 9)(x - 12) = 0$; length and width are 12
yd and 9 yd; The area is 108 yd^2.

8. Answers will vary, but graphing does not always yield
exact solutions.

PRACTICE

1. A line and a circle; by substitution,
$x^2 + (x + 2)^2 = 100$, $x^2 + x^2 + 4x + 4 = 100$,
$2x^2 + 4x - 96 = 0$, $2(x^2 + 2x - 48) = 0$,
$2(x - 6)(x + 8) = 0$, $x = 6$ or -8; solving for
y $y = 6 + 2 = 8$ or $y = -8 + 2 = -6$; $(6, 8)$ and
$(-8, -6)$

2. A line and a parabola; by substitution,
$5x = x^2$, $x^2 - 5x = 0$, $x(x - 5) = 0$, $x = 0$ or 5;
solving for y, $y = (5)(0) = 0$ or $y = (5)(5) =$
25; $(0, 0)$ and $(5, 25)$

3. A hyperbola and a line; by substitution where
$x = \dfrac{54}{y}$, $3\left(\dfrac{54}{y}\right) = 2y$, $162 = 2y^2$, $2(y^2 - 81) = 0$,
$y = \pm 9$; solving for x, $x(9) = 54$ and $x = 6$ or
$x(-9) = 54$ and $x = -6$; $(6, 9)$ and $(-6, -9)$

4. Two parabolas; multiplying the first equation by -2 and
adding the equations, $-11y = -33$ and $y = 3$; solving
for x, $x^2 + 3(3) = 18$, $x = \pm 3$; $(3, 3)$ and $(-3, 3)$.

5. A parabola and a line; multiplying the second equation
by -1 and adding the equations, $4x^2 - 2x = 0$,
$2x(2x - 1) = 0$, $x = 0$ or $\dfrac{1}{2}$; solving for y, $y = -3$

or $-\dfrac{10}{3}$; $\left(\dfrac{1}{2}, -3\right)$ and $\left(0, -\dfrac{10}{3}\right)$

6. An ellipse and a hyperbola; multiplying the first
equation by 3 and the second equation by 5 and adding,
$29x^2 = 725$, $x^2 = 25$, $x = \pm 5$; solving for
y, $y = \pm 2$; $(5, 2)$, $(5, -2)$, $(-5, 2)$, $(-5, -2)$

7. In the first equation, $A = 1$, $B = -1$, and
$C = 1$; $B^2 - 4AC < 0$ so it is an ellipse. The second
equation is a line. By substitution where
$x = 8 - y$, $(8 - y)^2 - (8 - y)y + y^2 = 43$,
$3(y^2 - 8y + 7) = 0$, $3(y - 7)(y - 1) = 0$, $y = 7$
or 1; solving for x, $x = 1$ or 7, $(1, 7)$ and $(7, 1)$

8. A line and a hyperbola $(B^2 - 4AC > 0)$; By
substitution, $x^2 - 2x(-3x + 1) - 5 = 0$,
$7x^2 - 2x - 5 = 0$, $(7x + 5)(x - 1) = 0$, $x = 1$ or

$-\dfrac{5}{7}$; solving for y, $y = -2$ or $\dfrac{22}{7}$; $(1, -2)$ and $\left(-\dfrac{5}{7}, \dfrac{22}{7}\right)$

9. A line and an ellipse; By substitution where
$y = 2 - (2.5)x$, $x^2 + 4(2 - (2.5)x)^2 = 16$
$26x^2 - 40x = 0$, $2x(13x - 20) = 0$, $x = 0$ or $\frac{20}{13}$;

solving for y, $y = 2$ or $-\frac{24}{13}$; $(0, 2)$ and $\left(\frac{20}{13}, -\frac{24}{13}\right)$

10. A line and an ellipse; writing the second equation in standard form,

$9x^2 + 25y^2 - 54x - 200y + 256 = 0$;
By substitution where $x = \frac{44}{3} - \frac{5}{3}y$,
$50y^2 - 550y + 1400 = 0$,
$10(5y^2 - 55y + 140) = 0$, $10(5y - 35)(y - 4) = 0$, $y = 7$ or 4; solving for x, $x = 3$ or 8; $(3, 7)$ and $(8, 4)$

11. Two hyperbolas ($B^2 - 4AC > 0$ in both equations);

substituting where, $y = -\frac{18}{x} - \frac{1}{2}$,
$x^2 + 23x - 108 = 0$, $(x + 27)(x - 4) = 0$, $x = -27$ or 4; solving for y, $y = \frac{1}{6}$ or -5; $\left(-27, \frac{1}{6}\right)$ and $(4, -5)$

12. Two hyperbolas ($B^2 - 4AC > 0$ in both equations);

By substitution where $y = \frac{1}{24x}$, $24x^2 - 11x + 1 = 0$, $(8x - 1)(3x - 1) = 0$, $x = \frac{1}{8}$ or $\frac{1}{3}$; solving for y, $y = \frac{1}{3}$
$\frac{1}{8}$; $\left(\frac{1}{8}, \frac{1}{3}\right)$ and $\left(\frac{1}{3}, \frac{1}{8}\right)$

13. Let the length $= x$ and the width $= \frac{360}{x}$;

$x^2 + \left(\frac{360}{x}\right)^2 = 41^2$; $x^4 - 1681x^2 + 129{,}600 = 0$, $(x^2 - 81)(x^2 - 1600) = 0$; $x = 9$ or $40 =$ the length or the width; 9 cm and 40 cm

14. Using the point (x, y) for the epicenter, apply the distance formula to two of the three points given. Solve the system to get the answer $(45, 22)$.

15. By substitution where $x = \frac{320}{y}$, $64y^4 - 1280y^2 - 512{,}000 = 0$, $64(y^4 - 20y^2 - 8000) = 0$, $64(y^2 - 100)(y^2 + 80) = 0$, $y = \pm10$ or $\pm4\sqrt{5}$;

solving for x, $x = \pm32$ or $\pm16\sqrt{5}$; looking at the map, the answer is $(-32°, -10°)$ or $(32°$ W, $10°$ S$)$

16. Yes, only if both equations represent the same graph.

EXTEND

17. By substitution, $2^x = 2^{2x} - 12$, $2^{2x} - 2^x - 12 = 0$, $(2^x + 3)(2^x - 4) = 0$, $2^x \neq -3$, $2^x = 4$, $x = 2$; solving for y, $y = 4$; $(2, 4)$

18. By substitution, $4^x = 4^{2x+1} - 3$ and the only solution for x is 0; solving for y, $y = 1$; $(0, 1)$

19. The first equation is a circle with $C = (5, 0)$ and $r = 4$. The second equation is a circle with $C = (-5, 0)$ and $r = 4$. These two circles do not intersect so the system has no solution.

20. Write two equations in two variables, $x^2 + y^2 = 1145$ and $x^2 - y^2 = 423$. Adding the two equations, $2x^2 = 1568$, $x = \pm28$; solving for y, $y = \pm19$; The answer is 28 and 19 or -28 and -19.

21. Write two equations in two variables, $x + y = 24$ and $3xy = 357$: By substitution where $x = 24 - y$, $3y(24 - y) = 357$, $3y^2 - 72y + 357 = 0$, $3(y^2 - 24y + 119) = 0$, $3(y - 7)(y - 17) = 0$, $y = 7$ or $17 = x =$ two numbers

22a. The equation of the circle is $x^2 + y^2 = 2500$. If $x = 110 - 2y$, by substitution $y = 48$ or 40; Solving for x, $x = 14$ or 30; The plane will be picked up between the coordinates $(14, 48)$ and $(30, 40)$.

22b. Use the distance formula on the two points in part a and $d = 8\sqrt{5}$. Since $r \times t = d$, $240t = 8\sqrt{5}$ so $t = \frac{\sqrt{5}}{30} \approx 0.07$ hours or 4.5 min.

23. Answers will vary.

THINK CRITICALLY

24a. Treat the equation of the hyperbola and an equation of the asymptotes as a system and solve. There should be no solution to the system.

24b. By substitution, $\frac{x^2}{a^2} - \frac{\left(\frac{b}{a}x\right)^2}{b^2} = 1$, $\frac{x^2}{a^2} - \frac{b^2x^2}{a^2b^2} = 1$,

$\frac{x^2}{a^2} - \frac{x^2}{a^2} = 1$, $0 = 1$; Since $0 \neq 1$, there is no solution and the graphs do not intersect.

25. $A = lw \qquad P = 2l + 2w$

$w = \frac{A}{l} \qquad l = \frac{1}{2}P - w \qquad$ Solve for w and l

$w = \dfrac{A}{\frac{1}{2}P - w} \qquad\qquad$ Substitute for l

$\frac{1}{2}Pw - w^2 = A \qquad\qquad$ Simplify

$w^2 - \frac{1}{2}Pw + A = 0$

$w = \dfrac{P \pm \sqrt{P^2 - 16A}}{4} \qquad$ quadratic formula

$l = \frac{1}{2}P - \frac{A}{l} \qquad\qquad$ Substitute for w

$l^2 - \frac{1}{2}Pl + A = 0 \qquad\qquad$ Simplify

$l = \dfrac{P \pm \sqrt{P^2 - 16A}}{4} \qquad$ quadratic formula

$A = lw = \left(\dfrac{P + \sqrt{P^2 - 16A}}{4}\right)$

$$\left(\frac{P - \sqrt{P^2 - 16A}}{4}\right) =$$

$$\frac{P^2 - (P^2 - 16A)}{16} = A$$

$$A = lw = \left(\frac{P - \sqrt{P^2 - 16A}}{4}\right)$$

$$\frac{P + \sqrt{P^2 - 16A}}{4} = \frac{P^2 - (P^2 - 16A)}{16} = A$$

So, when $l = \dfrac{P + \sqrt{P^2 - 16A}}{4}$ $w = \dfrac{P - \sqrt{P^2 - 16A}}{4}$

and when $l = \dfrac{P - \sqrt{P^2 - 16A}}{4}$ $w = \dfrac{P + \sqrt{P^2 - 16A}}{4}$.

MIXED REVIEW

26. mean $= 112 \div 7 = 16$; median $= 14$; mode $= 23$

27. mean $= 1849 \div 4 = 462.25$;
median $= \dfrac{450 + 481}{2} = 465.5$; There is no mode.

28. mean $= -15.4 \div 5 = -3.08$; median $= -5.6$;
There is no mode.

29. mean $= \dfrac{7}{3} \div 4 = \dfrac{7}{12}$; median $= \left(\dfrac{1}{2} + \dfrac{2}{3}\right) \div 2 = \dfrac{7}{12}$;
There
is no mode.

30. $y = mx + b; y = 3x - 2$

31. $y = mx + b; 4 = 2(-3) + b, b = 10; y = 2x + 10$

32. $m = -\dfrac{A}{B} = \dfrac{-1}{-2} = \dfrac{1}{2}; y = mx + b; 1 = \dfrac{1}{2}(4) + b,$
$b = -1; y = \dfrac{1}{2}x - 1$

33. $m = 3$ (opposite reciprocal);
$y = mx + b; 1 = 3(1) + b, b = -2; y = 3x - 2$

34. A; $x = \dfrac{ky}{z}; 8 = \dfrac{k(4)}{-1};$
$k = -2; x = \dfrac{-2y}{z}; x = \dfrac{-2(-2)}{4}; x = 1$

35. By substitution, $x(-x^2) = 1, -x^3 = 1, x = -1$;
solving for $y, y = -1; (-1, -1)$

36. By substitution, $3x = x^2 + 2, x^2 - 3x + 2 = 0,$
$(x - 2)(x - 1) = 0, x = 2$ or 1; solving for $y, y = 4$
or $1; (2, 4)$ and $(1, 1)$

37. Multiplying the first equation by -16 and adding the
equations, $9x^2 = 0, x = 0$; solving for
$y, y = \pm 5; (0, 5)$ and $(0, -5)$

ALGEBRAWORKS

1.

2.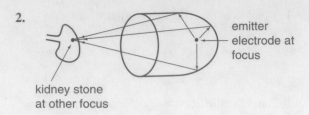

emitter
electrode at
focus

kidney stone
at other focus

3a.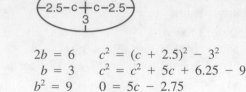

$2b = 6$ $c^2 = (c + 2.5)^2 - 3^2$
$b = 3$ $c^2 = c^2 + 5c + 6.25 - 9$
$b^2 = 9$ $0 = 5c - 2.75$
$5c = 2.75$
$c = 0.55$

$a = c + 2.5$

$a = 0.55 + 2.5$ $\dfrac{x^2}{(3.05)^2} + \dfrac{y^2}{9} = 1$

$a = 3.05$

3b. $c = 0.55$ $2(0.55) = 1.1$ cm

Lesson 12.9, pages 610–613

EXPLORE THE PROBLEM

1. $2a = 43.4 + 28.6 = 72$ million mi $=$ length of
major axis; $\dfrac{1}{2}(72) = 36$ million mi $=$ length of
semi-major axis

2. The center of the ellipse is $\left(\dfrac{43.3 + 28.6}{2}, 0\right) = (36, 0)$;
$c = 43.4 - 36$ or $c = 36 - 28.6; c = 7.4$ million mi

3. $(36)^2 = b^2 + (7.4)^2, b \approx 35.2$ million mi

4. $\dfrac{x^2}{(36)^2} + \dfrac{y^2}{(35.2)^2} = 1$

5. The length of time it takes Mercury to move from
M_1 to M_2 equals the length of time it takes the planet to
move from M_3 to M_4. Therefore, by Kepler's Second
Law, the shaded region with vertifes at M_1, M_2, and the
Sun must have the same area as the shaded region with
vertices at M_3, M_4, and the Sun. Mercury is closer to
the Sun at M_3 and M_4 so it must move along a greater
length of the circumference from M_3 to M_4 than it does
from M_1 to M_2 in order for the radius vector to sweep
over an equal area in an equal time. Students may
reason that since the radius vector at M_3 is about one-
third the radius vector at M_1, the angle about the sum
should be three times greater to sweep out equal areas.

6. At perihelion. As the figure for Exercise 5 shows, the planet must move fastest when it is closest to the Sun because the radius vector is shortest there. A planet increases its speed as it approaches the Sun, allowing the short radius vector to sweep over the same areas that longer radius vectors swept over when the planet was farther away but moving more slowly.

7. Since the length of the semi-major axis is 36 million mi, (from #1), and 1 $A.U.$ = 93 million mi, the length of the semi-major axis is $\frac{36}{93} \approx 0.387\ A.U.$

8a. $(0.387)^3 = x^2$; $x \approx 0.241$ y

8b. 0.241 y ≈ 88 d; $(0.241)(365)$

9a. $C = 3\pi(36 + 35.2) - \pi\sqrt{(141.6)(143.2)}$;
$C \approx 223.7$ million mi or 223,700,000 mi

9b. 88 d $= 88\ (24)\ (60)\ (60)$ s $= 7,603,200$ s
rate $= 223,700,000$ mi per 7,603,200 s, or about 29.4 mi/s

10. Answers will vary, but the probability of a collision is very small.

INVESTIGATE FURTHER

11. Only comets with elliptical orbits will be periodic. Since 40% of comets are elliptical, $12(0.4) = 4.8 \approx 5$ comets that are likely to be periodic in a given year.

12. Parabolic (one squared term); $4p = 32$, $p = 8$ million mi $=$ perihelion distance

13. hyperbolic; $16y^2 - 9x^2 + 800y + 4600 = 0$
$-9x^2 + 0xy + 16y^2 + 0x + 800y + 4600 = 0$
$B^2 - 4AC = 0^2 - 4(-9)(16) - 576$
$576 > 0$
The Perihelion for a hyperbola is on half the focal length minus one half the transverse axis length
$16y^2 + 800y - 9x^2 = -4600$
$16(y^2 + 50y + 625) - 9x^2 = 5400$

$\dfrac{(y + 25)^2}{337.5} - \dfrac{x^2}{600} = 1$

$a^2 = 337.5$ $b^2 = 600$
$a^2 + b^2 = c^2$
$937.5 = c^2$
$c - a = \sqrt{937.5} - \sqrt{337.5} \approx 12.2$ million mi

14. $9x^2 + 25y^2 = 562,500$
$\dfrac{x^2}{250^2} + \dfrac{y^2}{150^2} = 1$; ellipse
The Perihelion for an ellipse is one half the major axis minus one half the focal length

$c^2 = a^2 - b^2$
$c^2 = 250^2 - 150^2$
$c = 200$
$a - c = 250 - 200 = 50$ million mi

APPLY THE STRATEGY

	Name	Type	Location of Focus	Semi-major axis (millions of miles)	Orbital period (years)
15.	Mars	Planet	**Sun**	141.8	**1.88**
16.	Halley	Comet	**Sun**	**1,669.4**	76.05
17.	Crommelin	Comet	**Sun**	**851.0**	27.68
18.	Tuttle	Comet	**Sun**	530.1	**13.61**
19.	Ceres	Asteroid	**Sun**	257.0	**4.59**
20.	Taurids	Meteors	**Sun**	**204.5**	3.26

21. Apogee distance $= 560 + 3963 = 4523$
Perigee distance $= 146 + 3963 = 4109$

22. Semi-major axis $= \frac{1}{2}$(apogee distance $+$ perigee distance)

$$a = \tfrac{1}{2}(4523 + 4109)$$

$$a = 4316$$

REVIEW PROBLEM SOLVING STRATEGIES

1. The freight car (F) and the locomotive (F) can be switched in five moves:
L pulls F to the right
L pushes F into the tunnel
L return to the main track and moves left
L hooks up with F and pulls it back to the main track
L pushes F to the right and returns to original position of F
Plan a method for recording the moves. The simpler problem in Part b should give students the idea to hook up the locomotive and passenger cars and work with the combination LP. The steps are: Hook up LP
Switch LP and F using the steps in Part b
Push F to the right
LP moves left and is pushed into the tunnel
Unhook L and L returns to the main track and moves right
L picks up P in the tunnel at right and moves it to the main track
L pushes P to the left and the switch is complete

2. 13; On a circular track, the runners in front of Lucas are the same as the runners behind him, or $n - 1$ the total number (there are n runners including Lucas); so,
$\frac{1}{3}(n - 1) + \frac{3}{4}(n - 1) = n$ which gives $n = 13$.

3. Since the two trains are moving at 60 miles per hour in opposite directions, the relative velocity of one train to the other is $60 + 60 = 120$ miles per hour.

$$\frac{120 \text{ mi}}{1 \text{ hr}} \times \frac{5280 \text{ ft}}{1 \text{ mi}} \times \frac{1 \text{hr}}{3600 \text{ sec}} = 176 \text{ feet per second}$$

$$\frac{176 \text{ ft}}{1 \text{ sec}} \times 6 \text{ sec} = 1056 \text{ ft.}$$

Empire special is 1056 feet long.

Chapter Review, pages 614–615

1. The definition of an ellipse; C.

2. The definition of a hyperbola; D.

3. The definition of a circle; A.

4. The definition of a parabola; B.

5. $d = \sqrt{(9 - 5)^2 + (-1 - 2)^2} = 5$;

mdpt $= \left(\dfrac{5 + 9}{2}, \dfrac{2 + -1}{2}\right) = \left(7, \dfrac{1}{2}\right)$

6. $d = \sqrt{(-4 - -12)^2 + (-8 - 7)^2} = 17$;

mdpt $= \left(\dfrac{-12 + -4}{2}, \dfrac{7 + -8}{2}\right) = \left(-8, -\dfrac{1}{2}\right)$

7. $d = \sqrt{(-3 - 3)^2 + (-1 - 3)^2} = \sqrt{52} = 2\sqrt{13}$;

mdpt $= \left(\dfrac{3 + -3}{2}, \dfrac{3 + -1}{2}\right) = (0, 1)$

8. $(x - 3)^2 + (y - -7)^2 = (\sqrt{5})^2$;
$(x - 3)^2 + (y + 7)^2 = 5$

9. $(x^2 - 2x + 1) + (y^2 + 6y + 9) = 39 + 1 + 9$;
$(x - 1)^2 + (y + 3)^2 = 49$; $C = (1, -3)$; $r = 7$

10. $9(x^2 - 10x + 25) + 16(y^2 - 6y + 9) = -225 + 225 + 144$;
$9(x - 5)^2 + 16(y - 3)^2 = 144$;
$\dfrac{(x - 5)^2}{16} + \dfrac{(y - 3)^2}{9} = 1$; $C = (5, 3)$; $a = 4, b = 3$,
vertices are $(9, 3), (1, 3), (5, 6)$ and $(5, 0)$;
$c^2 = 16 - 9, c = \sqrt{7}$, foci are $(5 + \sqrt{7}, 3)$ and $(5 - \sqrt{7}, 3)$; horizontal ellipse;

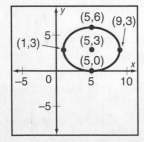

11. $36(y^2 - 10y + 25) - 64(x^2 + 4x + 4) = 1668 + 900 - 256$; $36(y - 5)^2 - 64(x + 2)^2 = 2304$; $\dfrac{(y - 5)^2}{64} - \dfrac{(x + 2)^2}{36} = 1$; $C = (-2, 5)$; $a = 8$,
$b = 6$ so vertices are $(-2, 13)$ and $(-2, -3)$; vertical hyperbola; $c^2 = 64 + 36, c = 10$, foci are $(-2, 15)$

and $(-2, -5)$; slope of asymptotes $= \pm\dfrac{a}{b} = \pm\dfrac{8}{6} = \pm\dfrac{4}{3}$

12. $a = 5$ and slope of asymptotes $= \pm\dfrac{2}{5}$ so $b = 2$ and the hyperbola is horizontal $\left(m = \pm\dfrac{b}{a}\right)$; $\dfrac{x^2}{25} - \dfrac{y^2}{4} = 1$

13. $V = (0, 0)$; $4p = 48, p = 12$, focus $= (12, 0)$; The equation of the directrix is $x = -12$. The parabola opens right.

14. $(x^2 + 10x + 25) = 12y - 61 + 25$;
$(x + 5)^2 = 12(y - 3)$; $V = (-5, 3)$; $4p = 12, p = 3$,
focus $= (-5, 6)$; The equation of the directrix is $y = 0$. The parabola opens up.

15. $A = 3, B = -5, C = 2$; $B^2 - 4AC > 0$, so the equation represents a hyperbola.

16. $A = 2, B = 4, C = 2$; $B^2 - 4AC = 0$, so the equation represents a parabola.

17. By substitution where $y = 1 - x$,
$x^2 + (1 - x)^2 = 25, x^2 + 1 - 2x + x^2 = 25$,
$2x^2 - 2x - 24 = 0, 2(x^2 - x - 12) = 0$,
$2(x - 4)(x + 3) = 0, x = 4$ or -3; solving for
$y, y = -3$ or 4; $(4, -3)$ and $(-3, 4)$

18. By substitution where $y = 2x - 1$,
$4x^2 + (2x - 1)^2 = 25, 4x^2 + 4x^2 - 4x + 1 = 25$,
$8x^2 - 4x - 24 = 0, 4(2x^2 - x - 6) = 0$,
$4(2x + 3)(x - 2) = 0, x = 2 - 6$ or
$-\dfrac{3}{2}$; solving for $y, y = 3$ or -4; $(2, 3)$ and $\left(-\dfrac{3}{2}, -4\right)$.

19. Multiplying the first equation by -2 and adding the equations, $-11y = -33, y = 3$; solving for
$x, x = \pm3$; $(3, 3)$ and $(-3, 3)$

20. $(1.52)^3 = x^2, x \approx 1.9$ yrs

Chapter Test, page 616

1. $d = \sqrt{(-10 - 2)^2 + (9 - 4)^2} = 13$

2. mdpt $= \left(\dfrac{2 + -10}{2}, \dfrac{4 + 9}{2}\right) = \left(-4, \dfrac{13}{2}\right)$

3. Using opposite vertices $(2, 5)$ and $(4, 11)$,
$d = \sqrt{(4 - 2)^2 + (11 - 5)^2} = \sqrt{40} = 2\sqrt{10} =$
length of a diagonal; Diagonals of a rectangle are equal and they bisect each other, so they intersect at the
midpoint which is $\left(\dfrac{2 + 4}{2}, \dfrac{5 + 11}{2}\right)$ or $(3, 8)$.

4. $x^2 + (y^2 - 8y + 16) = 84 + 16$;
$x^2 + (y - 4)^2 = 100$; $C = (0, 4)$; $r = 10$

5. $(x + 4)^2 + y^2 = (6.2)^2$; $(x + 4)^2 + y^2 = 38.44$

6. $9(x^2 + 8x + 16) + 25(y^2 - 2y + 1) = 56 + 144 + 25$; $9(x + 4)^2 + 25(y - 1)^2 = 225$
$\dfrac{(x + 4)^2}{25} + \dfrac{(y - 1)^2}{9} = 1$; $C = (-4, 1)$; $a = 5$,

$b = 3, c^2 = a^2 - b^2, c = 4$, foci are $(-8, 1), (0, 1)$; vertices are $(1, 1), (-9, 1), (-4, 4), (-4, -2)$; $2a = 10 = $ length of major axis; $2b = 6 = $ length of minor axis; horizontal ellipse

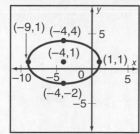

7. This is a horizontal ellipse where $2a = 11 - 3 = 8$, $a = 4, a^2 = 16; 2b = 2, b = 1, b^2 = 1; C = (7, 4)$;
$$\frac{(x - 7)^2}{16} + \frac{(y - 4)^2}{1} = 1$$

8. $2a = 28{,}000, a = 14{,}000; 2b = 20{,}000, b = 10{,}000$;
$$\frac{x^2}{(14{,}000)^2} + \frac{y^2}{(10{,}000)^2} = 1; \text{ horizontal ellipse}$$

9. $9(x^2 - 6x + 9) - 16(y^2 + 2y + 1) =$
$79 + 81 - 16; 9(x - 3)^2 - 16(y + 1)^2 = 144$;
$$\frac{(x - 3)^2}{16} - \frac{(y + 1)^2}{9} = 1; C = (3, -1); a^2 = 16,$$
$a = 4$, vertices are $(7, -1), (-1, -1); c^2 = 16 + 9$,
$c = 5$, foci are $(8, -1)$ and $(-2, -1); m = \pm\frac{b}{a} = \pm\frac{3}{4}$; horizontal

10. $m = \pm\frac{b}{u} = \pm\frac{2}{5}$ so $b - 2$ and $a = 5$;
$$\frac{(x - 2)^2}{25} - \frac{(y - 3)^2}{4} - 1$$

11.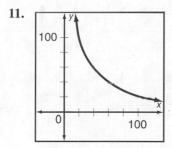

12. $V = (0, 0); 4p = 16, p = 4$, focus $= (0, 4)$; The equation of the directrix is $y = -4$; parabola opens up

13. $V = (0, 0); 4p = 1, p = \frac{1}{4}$, focus $= \left(\frac{1}{4}, 0\right)$; The equation of the directrix is $x = -\frac{1}{4}$; parabola opens rights

14. $(x^2 - 8x + 16) = 8y + 16; (x - 4)^2 = 8(y + 2)$; $V = (4, -2); 4p = 8, p = 2$, focus $= (4, 0)$; The equation of the directrix is $y = -4$; parabola opens up

15. $p = 7, 4p = 28; x^2 = 28y$; parabola opens up

16. $B^2 - 4AC < 0$, and $A = B$, so it is a circle, D.

17. By substitution where $y = 14 - x$,
$x^2 + (14 - x)^2 = 106$;
$x^2 + 196 - 28x + x^2 = 106; 2x^2 - 28x + 90 = 0$,
$2(x^2 - 14x + 45) = 0, 2(x - 9)(x - 5) = 0, x = 9$ or 5; solving for $y, y = 5$ or 9; (9, 5) and (5, 9)

18. By substitution where $y = \frac{2}{3}x - \frac{10}{3}$,
$$4x^2 - x\left(\frac{2}{3}x - \frac{10}{3}\right) = 0; 4x^2 - \frac{2}{3}x^2 + \frac{10}{3}x = 0,$$
$12x^2 - 2x^2 + 10x = 0, 10x^2 + 10x = 0$,
$10x(x + 1) = 0, x = 0$ or -1; solving for
$y, y = -\frac{10}{3}$ or $-4; \left(0, -\frac{10}{3}\right)$ and $(-1, -4)$

19. let $w = \frac{1200}{l}; (w + 1.5)(l - 3) = 1260$;
$$\left(\frac{1200}{l} + 1.5\right)(l - 3) = 1260;$$
$1.5l^2 - 64.5l - 3600 = 0$;
$15(l^2 - 43l - 2400) = 0; 15(l - 75)(l + 32) = 0$,
$l = 75$ yd; $w = \frac{1200}{75} = 16$ yd

20. Answers will vary. Students should mention that each conic section can be formed by intersecting a plane and a double-napped cone (two cones placed apex-to-apex). If the plane is parallel to the bases of the cone, the cross-section is a circle. By continually changing the angle of intersection, ellipses, parabolas, and finally, with the plane perpendicular to the bases of the cone, hyperbolas are formed.

21. $\frac{(5.2)^3}{(0.39)^3} = \frac{(11.87)^2}{x^2}, x \approx 0.24$ y

Cumulative Review, page 618

1. The asymptotes are horizontal $y = 1$ and vertical at $x = -2$; The rational function that has these asymptotes is C.

2. All three of these statements are true; A.

3. First take the common log of both sides: $\log 5^x = \log 17$. Bring the exponent in front and divide: $x \log 5 = \log 17, x \approx 1.8$

4. $(2i)(5) = 10i$; D

5. $3\left(x^2 + \frac{4}{3}x + \frac{16}{36}\right) + 3\left(y^2 - \frac{10}{3}y + \frac{100}{36}\right) =$
$7 + \frac{16}{12} + \frac{100}{12}; 3\left(x^2 + \frac{4}{6}\right)^2 + 3\left(y^2 - \frac{10}{6}\right)^2 = \frac{200}{12}$;
$\left(x^2 + \frac{2}{3}\right)^2 + \left(y^2 - \frac{5}{3}\right)^2 = \frac{50}{9}; C = \left(-\frac{2}{3}, \frac{5}{3}\right)$;
$r = \sqrt{\frac{50}{9}} = \frac{5\sqrt{2}}{3} = \frac{5}{3}\sqrt{2}$

6. $x^4 = 1; x = (\pm 1, \pm i)$

7. 1 is a rational factor;

$$\begin{array}{c|ccc} 1 & 6 & -5 & -2 \end{array} \; ; x^2 + 7x + 2 = 0,$$

$$\begin{array}{c|ccc} \downarrow & 1 & 7 & 2 \\ \hline & 1 & 7 & 2 & 0 \end{array} \quad x = \dfrac{-7 \pm \sqrt{41}}{2}$$

(using quadratric formula); $\left(1, \dfrac{-7 \pm \sqrt{41}}{2}\right)$

8. $x = \left(\dfrac{1 \pm 3i}{2}\right)$ (using quadratic formula)

9. 2 is a rational root;

$$\begin{array}{c|cccc} 2 & 1 & 1 & -16 & 20 \\ \downarrow & & 2 & 6 & -20 \\ \hline & 1 & 3 & -10 & 0 \end{array}$$

$x^2 + 3x - 10 = 0, (x + 5)(x - 2) = 0, x = 2, -5;$

$x = (5, 2)$

10. $(3^2)^{2x} = 3^{3x+1}; 4x = 3x + 1, x = 1$

11. Multiplying each term by the LCD of $(x + 1)(x - 1)$,
$x(x + 1) + 2 = 8(x - 1); x^2 + x + 2 = 8x - 8,$
$x^2 - 7x + 10 = 0, (x - 5)(x - 2) = 0, x = 5, 2;$
$x = \{5, 2\}$

12. Adding the equations, $3y = 12, y = 4$; solving for
$x, x = 2; \{(2, 4)\}$

13. By substitution where $x = 6 + y$,
$(6 + y)^2 + y^2 = 26; 36 + 12y + y^2 = 26,$
$2y^2 + 12y + 10 = 0, 2(y^2 + 6y + 5) = 0,$
$2(y + 5)(y + 1) = 0, y = 5$ or -1; solving for
$x, x = 1$ or $5; (1, -5)$ and $(5, -1)$

14. Let $x = $ rate per hour of job B; $x - 3 = $ rate per hour
of job A; $\dfrac{1200}{x} = \dfrac{1080}{x - 3} - 20; x = \$12 = $ hourly rate
of job $B; x - 3 = \$9 = $ hourly rate of job A

15. Set up two equations in two variables: $2x + 2y = 150$
and $xy = 1000$. Solve the system by substituting
$x = \dfrac{1000}{y}; 2\left(\dfrac{1000}{y}\right) + 2y = 150,$
$2000 + 2y^2 = 150y, 2(y^2 - 75y + 1000) = 0,$
$y \approx 57.7$ft (using quadratic formula); solving for
$x, x \approx 17.3$ ft

16. Set up three equations in three variables:

$x + y + z = 20,000$
$0.04x = 0.06z$

$0.04x + 0.05y + 0.06z = 980$; Solve the system and
$x = \$6,000, y = \$10,000$, and $z = \$4,000$.

17. $y = \dfrac{5}{x}$; inversely

18. $y = \dfrac{-50}{x}$; inversely

19. $y = 4x$; directly

20. $y = -\dfrac{x}{10z}$; jointly

Standardized Test, page 619

1. D; There is not enough information.

2. C; $\dfrac{x^2}{16} + \dfrac{y^2}{9} = 1; a^2 = 4, 2a = 8 = $ major axis;

$\dfrac{y^2}{16} - \dfrac{x^2}{9} = 1; a^2 = 4, 2a = 8 = $ transverse axis;

The measures are equal.

3. A; $x = \dfrac{2 \pm \sqrt{12}}{2} = \dfrac{2 \pm 2\sqrt{3}}{2} = 1 \pm \sqrt{3}$; The sum is
$1 + \sqrt{3} + 1 - \sqrt{3} = 2$; The product is
$(1 + \sqrt{3})(1 - \sqrt{3}) = 1 - 3 = -2; 2 > -2.$

4. C; The bounds are equal.

5. B; $i^{18} = -1; i^{12} = 1$

6. A; By graphing, the first column has a larger area.

7. C; $m = -\dfrac{3}{2}, b = -\dfrac{3}{2}$; They are equal.

8. B; $\overline{AC} = \sqrt{(4 - -4)^2 + (7 - 6)^2} = \sqrt{65}$;
$\overline{AB} = \sqrt{(3 - -4)^2 + (1 - 6)^2} = \sqrt{74}$;
$\sqrt{74} > \sqrt{65}.$

9. B; By multiplying the second equation by -2 and
adding, $y = -1$; solving for $x, x = 4$; the y-value is
$-1; -1 > -3$

10. B; $V = (-4, -3)$ and the y-value is $-3; 4p = 8, p = 2,$
focus is $(-4, -1)$ and the y-value is $-1; -1 > -3$

11. C; $2^x = 32, x = 5; x^4 = 625, x = 5$; They are equal.

12. A; $(x^2 - 8x + 16) + (y^2 + 2y + 1) = 19 + 17$
$(x - 4)^2 + (y + 1)^2 = 36$
$C = (4 - 1)$
$4 > -1$

Data Activity, pages 620-621

1. $C = \pi d$, $\pi \approx 3.14$

   ring size: $6\frac{1}{2} \rightarrow$ diameter $= 0.666$

   therefore, $C = \pi \cdot 0.666 = 2.092$

2. Sizes increase by 0.016 in.

Size	Diameter
10	0.778
$10\frac{1}{2}$	0.794
11	0.810
$11\frac{1}{2}$	0.826
12	0.842
$12\frac{1}{2}$	0.858
13	0.874
$13\frac{1}{2}$	0.890

3. $m = \dfrac{y_1 - y_2}{x_1 - x_2} = \dfrac{3 - 2\frac{1}{2}}{0.554 - 0.538} = 31.25$

4. $y = mx + b$, where x represents diameter and y represents size.

   $3 = (31.25)(0.554) + b$

   $b = -14.3125$

   $y = 31.25x - 14.3125$ or $S = 31.25d - 14.3125$

5. Answers will vary; but several different ring sizes should be listed and a conclusion be stated.

Lesson 13.1, pages 623–625

THINK BACK

1. $f(x) = 2x + 2$

   $f(0) = 2 \cdot 0 + 2 = 2$

   $f(1) = 2 \cdot 1 + 2 = 4$

   $f(2) = 2 \cdot 2 + 2 = 6$

   $f(3) = 2 \cdot 3 + 2 = 8$

   $f(4) = 2 \cdot 4 + 2 = 10$

   $f(5) = 2 \cdot 5 + 2 = 12$

2.

x	0	1	2	3	4	5
$f(x)$	2	4	6	8	10	12

3. $g(x) = 2^{x-1}$

   $g(1) = 2^{1-1} = 1$

   $g(2) = 2^{2-1} = 2$

   $g(3) = 2^{3-1} = 4$

   $g(4) = 2^{4-1} = 8$

   $g(5) = 2^{5-1} = 16$

4.

x	1	2	3	4	5
$g(x)$	1	2	4	8	16

5. Substitute the appropriate values for x in the function.

6. Look for patterns that can be continued. For example: In problem 2, for the first function, as x increases by 1, $f(x)$ increases by 2. In problem 4, as x increases by 1, $g(x)$ is doubled.

EXPLORE

7. Answers will vary; some students may notice that each figure has two more line segments than the one preceding it, other students may notice that the numbers of line segments in each figure are consecutive even numbers, starting with 2.

8. 10; Add 2 to 8, the number of line segments in the last figure.

9. Answers will vary; some students may notice that each figure has twice the number of circles that are in the preceding figure, other students may notice that the number of circles in each figure increases sequentially as a power of 2, starting with 2^0.

10. 32; $2^{6-1} = 32$; Multiply the number of circles in the last figure by 2.

11. $2^{8-1} = 128$; $2^{10-1} = 512$

12. The values of $f(x)$ are the same as the number of line segments.

13. The values of $g(x)$ are the same as the number of circles.

MAKE CONNECTIONS

14. $a_1 = 1$; $a_2 = 2$; $a_5 = 5$

15. Each term increases by 1; $a_n = a_{n-1} + 1$

16. $5 + 1 = 6$

17. $a_n = a_{n-1} + 7$; each term is 7 greater than the previous term.

18. $a_5 = 30$; $a_6 = 30 + 7 = 37$; $a_7 = 37 + 7 = 44$

19. $a_n = 3a_{n-1}$; each term is obtained by multiplying the previous term by 3.

20. $a_5 = 81$; $a_6 = 3 \cdot 81 = 243$; $a_7 = 3 \cdot 243 = 729$; $a_8 = 3 \cdot 729 = 2187$

SUMMARIZE

21. $a_n = a_{n-1} + 2$

   $a_n = 2$; $a_2 = 2 + 2 = 4$

   $\qquad\quad a_3 = 4 + 2 = 6$

   $\qquad\quad a_4 = 6 + 2 = 8$

   $\qquad\quad a_5 = 8 + 2 = 10$

   $\qquad\quad a_6 = 10 + 2 = 12$

   $\qquad\quad a_7 = 12 + 2 = 14$

   $\qquad\quad a_8 = 14 + 2 = 16$

   $\qquad\quad a_9 = 16 + 2 = 18$

   $\qquad\quad a_{10} = 18 + 2 = 20$

22. $a_n = 2a_{n-1}$

$a_1 = 1; a_2 = 2 \cdot 1 = 2; \quad a_3 = 2 \cdot 2 = 4;$
$\qquad a_4 = 2 \cdot 4 = 8; \quad a_5 = 2 \cdot 8 = 16;$
$\qquad a_6 = 2 \cdot 16 = 32; \quad a_7 = 2 \cdot 32 = 64;$
$\qquad a_8 = 2 \cdot 64 = 128; a_9 = 2 \cdot 128 = 256;$
$\qquad a_{10} = 2 \cdot 256 = 512$

23a. $a_1 = 1980;$
$a_2 = 1980 + 4 = 1984$
$a_3 = 1984 + 4 = 1988$
$a_4 = 1988 + 4 = 1992$
$a_5 = 1992 + 4 = 1996$

23b. $a_n = a_{n-1} + 4$

23c. No; all answers must be even, in fact, all leap years must be divisible by 4.; or continue sequence with $a_6 = 2000; a_7 = 2004; a_8 = 2008$ and 2007 was not included in sequence.

24. $a_n = 2a_{n-1} + 3$; multiply previous term by 2 and then add 3.

25. $a_5 = 77$
$a_6 = 2 \cdot 77 + 3 = 157$
$a_7 = 2 \cdot 157 + 3 = 317$
$a_8 = 2 \cdot 317 + 3 = 637$

26. Answers will vary; some students may notice that each term in the sequence is the square of its position in the sequence, others may notice that the differences between terms are consecutive odd numbers starting with 3.

$a_n = n^2; a_6 = 6^2 = 36; a_n = 7^2 = 49; a_8 = 8^2 = 64$

27. $a_1 = 5$
$a_2 = 2 \cdot 5 - 3 = 7$
$a_3 = 2 \cdot 7 - 3 = 11$
$a_4 = 2 \cdot 11 - 3 = 19$
$a_5 = 2 \cdot 19 - 3 = 35$

28. $a_1 = 5$

$a_2 = \frac{1}{2}(5) = \frac{5}{2}$

$a_3 = \frac{1}{2}\left(\frac{5}{2}\right) = \frac{5}{4}$

$a_4 = \frac{1}{2}\left(\frac{5}{4}\right) = \frac{5}{8}$

$a_5 = \frac{1}{2}\left(\frac{5}{8}\right) = \frac{5}{16}$

29. Answers will vary, but may include the following:
- each term is created by adding the same constant to previous term;
- each term is created by multiplying the previous term by same constant;
- each term is created by a combination of multiplying and adding constant to previous term
- each term is created by indexing the exponent.

30. No; there are many possible patterns using the three terms, for example: squares 1, 2, 4, 8, 16 or increasing

index being added 1, 2, 4, 7, 11. More information is needed to decide which pattern is used.

31. Each term beyond the second is found by adding the previous two terms; 34, 55, 89, 144, 233, 377.

Lesson 13.2, pages 626–632

EXPLORE

1.

Generation Number	0	1	2	3	4	5	6
Number of Bacteria	1	2	4	8	16	32	64

2. The number of bacteria in one generation is 2 times the number of bacteria in the previous generation.

3. $a_n = 2a_{n-1}$

4. The number of bacteria in a generation is 2 raised to the exponent equivalent to the generation number.

5. $a_n = 2^n$

6. Answers will vary; to find the number of bacteria in the seventh generation, use $a_n = 2a_{n-1}$. Since a_6 is known, you are able to double a_6 to find a_7. To find the bacteria in the 15th generation use $a_n = 2^n$. This formula gives the answer directly and does not require listing the previous terms. Number of bacteria in $a_7 = 128$; $a_{15} = 32{,}768$.

7. $2^{19} = 524{,}288 \qquad 2^{20} = 1{,}048{,}576$
20 generations will produce more than one million descendants.

8. $20 \text{ generations} \times 30 \frac{\text{minutes}}{\text{generations}} \times \frac{1 \text{ hour}}{60 \text{ minutes}} = 10 \text{ hours}$

TRY THESE

1. explicit **2.** explicit **3.** recursive

4. recursive **5.** explicit **6.** recursive

7. $a_n = n - 1; 0, 1, 2, 3, 4$

8. $a_n = \frac{n}{n+3}; \frac{1}{4}, \frac{2}{5}, \frac{3}{6}, \frac{4}{7}, \frac{5}{8}$

9. $a_n = (-2)^n; -2, 4, -8, 16, -32$

10. $a_1 = 7;$
$a_n = a_{n-1} - 4; 7, 3, -1, -5, -9$

11. $a_1 = 2;$
$a_n = 7 - 2a_{n-1}; 2, 3, 1, 5, -3$

12. $a_1 = 4$
$a_n = 1 + \frac{1}{a_{n-1}}; 4, 1\frac{1}{4}, 1\frac{4}{5}, 1\frac{5}{9}, 1\frac{9}{14}$

13a. \$5.65, \$5.90, \$6.15, \$6.40, \$6.65, \$6.90

13b. $a_1 = 5.65; a_n = a_{n-1} + 0.25$

14. $a_n = 5n - 6; a_8 = 5 \cdot 8 - 6 = 34$

15. $a_n = 5n^3; a_{11} = 5 \cdot 11^3 = 6655$

16. $a_n = \frac{3n + 7}{2n - 5}; a_{14} = \frac{3 \cdot 14 + 7}{2 \cdot 14 - 5} = \frac{49}{23}$

17. Answers and examples will vary, but should include the idea that an infinite sequence has no last term and that a finite sequence has a last term.

18. $a_n = 4n$ or $a_n = a_{n-1} + 4$ **19.** $a_n = \dfrac{n+1}{n+2}$

20. $a_n = 4^{n-1}$ or $a_n = 4a_{n-1}$

PRACTICE

1. $a_n = 4n + 5$;
$a_1 = 4 \cdot 1 + 5 = 9; a_2 = 4 \cdot 2 + 5 = 13$;
$a_3 = 4 \cdot 3 + 5 = 17; a_4 = 4 \cdot 4 + 5 = 21$;
$a_5 = 4 \cdot 5 + 5 = 25$

2. $a_n = \dfrac{n+3}{n}$;
$a_1 = \dfrac{1+3}{1} = 4; a_2 = \dfrac{2+3}{2} = \dfrac{5}{2}; a_3 = \dfrac{3+3}{3} = 2$;
$a_4 = \dfrac{4+3}{4} = \dfrac{7}{4}; a_5 = \dfrac{5+3}{5} = \dfrac{8}{5}$

3. $a_n = (-1)^n$;
$a_1 = (-1)^1 = -1; a_2 = (-1)^2 = +1$;
$a_3 = (-1)^3 = -1; a_4 = (-1)^4 = +1$;
$a_5 = (-1)^5 = -1$

4. $a_1 = 9; a_n = 2 + 10a_{n-1}$
$a_1 = 9; a_2 = 2 + 10 \cdot 9 = 92$;
$a_3 = 2 + 10 \cdot 92 = 922$;
$a_4 = 2 + 10 \cdot 922 = 9222$;
$a_5 = 2 + 10 \cdot 9222 = 92{,}222$

5. $a_1 = 2; a_n = (a_{n-1})^2$
$a_1 = 2; a_2 = 2^2 = 4; a_3 = 4^2 = 16; a_4 = 16^2 = 256$;
$a_5 = 256^2 = 65{,}536$

6. $a_1 = 4; a_n = n + \dfrac{a_{n-1}}{2}$
$a_1 = 4; a_2 = 2 + \dfrac{4}{2} = 4; a_3 = 3 + \dfrac{4}{2} = 5$;
$a_4 = 4 + \dfrac{5}{2} = 6\dfrac{1}{2}; a_5 = 5 + \dfrac{\frac{13}{2}}{2} = 8\dfrac{1}{4}$

7. $a_n = 9n + 2$
$a_8 = 9 \cdot 8 + 2 = 74$

8. $a_n = (3n - 4)(2n + 5)$;
$a_7 = (3 \cdot 7 - 4)(2 \cdot 7 + 5) = 323$

9. $a_n = 3n^2(4n - 5)$;
$a_5 = 3 \cdot 5^2(4 \cdot 5 - 5) = 1125$

10. $a_n = 3n$;
$a_1 = 3 \cdot 1 = 3\text{ ft}; a_2 = 3 \cdot 2 = 6\text{ ft}$;
$a_3 = 3 \cdot 3 = 9\text{ ft}; a_4 = 3 \cdot 4 = 12\text{ ft}$;
$a_5 = 3 \cdot 5 = 15\text{ ft}; a_{10} = 3 \cdot 10 = 30\text{ ft}$

11. $a_n = 2n - 1$

12. $a_n = -(3n - 2)$ or $a_n = -3n + 2$ **13.** $a_n = \dfrac{1}{3^n}$

14. $a_1 = 1$ $a_n = -7a_{n-1}$

15. $a_1 = 1; a_n = a_{n-1} + n$ **16.** $a_1 = 4; a_n = \dfrac{1}{4}a_{n-1}$

17. Yes; the recursive formula is $a_1 = 3, a_n = 3 \cdot a_{n-1}$; the explicit formula is $a_n = 3^n$.

18a. \$4500; \$4050; \$3645; \$3280.50; \$2952.45

18b. $a_1 = 4500$
$a_n = a_{n-1} - 0.1a_{n-1}$ or $a_n = 0.9a_{n-1}$

EXTEND

19. $a_n = n^2 - n + 41; a_1 = 41; a_2 = 43; a_3 = 47$;
$a_4 = 53; a_5 = 61$; all are odd and are prime numbers.
$a_{41} = 1681$, which is $(41)^2$; the prime number pattern does not continue starting at $n = 41$ in sequence of $a_n = n^2 - n + 41.20$.

20. $a_n = (n - 1)\sqrt{x^n}$ **21.** $a_n = \dfrac{x^n}{2^{n-1}}$

22. $a_n = (-1)^n x^{2n-1}$

23. $a_1 = 0.4; a_n = 0.1[3(2^{n-2}) + 4]$
$a_1 = 0.4\text{AU}; a_2 = 0.7\text{AU}; a_3 = 1\text{AU}; a_4 = 1.6\text{AU}$
$a_1 = 37{,}200{,}000\text{ mi}; a_2 = 65{,}100{,}000\text{ mi}$
$a_3 = 93{,}000{,}000\text{ mi}; a_4 = 148{,}800{,}000\text{ mi}$

24a. $\dfrac{1}{4}, \dfrac{3}{16}, \dfrac{9}{64}, \dfrac{27}{256}, \dfrac{81}{1024}$ **24b.** $a_n = \dfrac{3^{n-1}}{4^n}$

THINK CRITICALLY

25a. 5, 2.236, 1.495, 1.223, 1.106, 1.052, 1.025, 1.013, 1.006, 1.003; the terms are decreasing and approaching 1.

25b. Each term decreases and approaches 1.

25c. $a_n = x^{\frac{1}{n}}$ when $x = 5$ so $a_1 = 5^1, a_2 = 5^{\frac{1}{2}}, a_3 = 5^{\frac{1}{3}}$, . . . Therefore, as n increases, the exponent approaches 0, then $5^0 = 1$.

26. a linear function   **27.** an exponential function

28a. $a_1 = a_2 = 1, a_n = a_{n-1} + a_{n-2}$ where $n \geq 3$.
1, 1, 2, 3, 5, 8, 13, 21, 34, 55, 89, 144, 233, 377, 610, 987, 1597

28b. $1 + 1 + 2 + 3 + 5 = 12$
12, the sum is 1 less than seventh term.

28c. $1 + 1 + 2 + 3 + 5 + 8 + 13 + 21 = 54$
54, the sum is 1 less than tenth term.

28d. The sum of the first n terms is 1 less than the $(n + 2)$ term.

28e. The sum is 1 less than the seventeenth term;
$1597 - 1 = 1596$;
Check: $1 + 1 + 2 + 3 + 5 + 8 + 13 + 21 + 21 + 34 + 55 + 89 + 144 + 233 + 377 + 610 = 1596$

MIXED REVIEW

29. $-2, 1, 3$

30. sum: $(2n - 5) + (3n + 2) = 5n - 3$
difference: $(2n - 5) - (3n + 2) = -n - 7$
product: $(2n - 5)(3n + 2) = 6n^2 - 11n - 10$

31. sum: $(a - 6b) + (2c + d) = a - 6b + 2c + d$
difference: $(a - 6b) - (2c + d) =$
$a - 6b - 2c - d$
product: $(a - 6b)(2c + d) =$
$2ac + a - 12bc - 6bd$

32. C

PROJECT CONNECTION

1. 81 cm; 108 cm; 135 cm   **2.** 27 cm; 36 cm; 45 cm

3.

Length of Segment	Number of Segments	Perimeter of		
		Triangle	Square	Pentagon
27 cm	1	81 cm	108 cm	135 cm
9 cm	3	27 cm	36 cm	45 cm
3 cm	5	9 cm	12 cm	15 cm
1 cm	7	3 cm	4 cm	5 cm

4. Length of Segment is a geometric sequence; since multiply by a ratio of $\frac{1}{3}$; $a_n = 3^{4-n}$ or $a_n = \frac{1}{3}a_{n-1}$

Number of Segments is an arithmetic sequence; since increase by adding 2; $a_n = 2(n - 1) + 1$ or $a_n = a_{n-1} + 2$

Perimeter of Triangle is a geometric sequence; since multiply by a ratio of $\frac{1}{3}$; $a_n = 3 \cdot 3^{4-n}$ or $a_n = \frac{1}{3}a_{n-1}$

Perimeter of Square is a geometric sequence; since multiply by a ratio of $\frac{1}{3}$; $a_n = 4 \cdot 3^{4-n}$ or $a_n = \frac{1}{3}a_{n-1}$

Perimeter of Pentagon is a geometric sequence; since multiply by a ratio of $\frac{1}{3}$; $a_n = 5 \cdot 3^{4-n}$ or $a_n = \frac{1}{3}a_{n-1}$

ALGEBRAWORKS

1. 6   **2.** 12   **3.** 1, 6, 12, 18, 24

4. $a_n = 6(n - 1)$, where $n > 1$

5. $a_{n+1} = a_n + 6$, where $a_1 = 6$

6a. Counting 1×1 squares, there are 9 squares.
Counting 2×2 squares, there are 4 squares.
Counting 3×3 squares, there is 1 square.

6b. 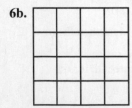 Counting (1×1) squares, there are 16 squares;
Counting (2×2) squares, there are 9 squares;
Counting (3×3) squares, there are 4 squares;
Counting (4×4) squares, there is 1 square.

6c. In a 7×7 grid: there would be 49 (1×1) squares; 36 (2×2) squares; 25 (3×3) squares; 16 (4×4) squares; 9 (5×5) squares; 4 (6×6) squares; and 1 (7×7) square.

7. Answers will vary. Possible answer: number of dots in n complete "windmill" designs is $12n$ generating the sequence 12, 24, 36, ...

Lesson 13.3, pages 633–638

EXPLORE

1. Plan 1: $a_n = a_{n-1} + 3715$, $a_0 = 17{,}000$

Plan 2: $a_n = \frac{6}{5}a_{n-1}$, $a_0 = 17{,}000$

2. Salary Plan 1:

Year	0	1	2	3	4
Salary	17,000	20,715	24,430	28,145	31,860

Salary Plan 2:

Year	0	1	2	3	4
Salary	17,000	20,400	24,480	29,376	35,251

3. Salary Plan 1:   Salary Plan 2:

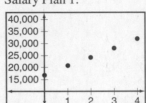 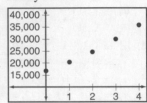

4. A linear equation

5. A nonlinear (exponential) equation.

6. The best salary over a 3 year period is Salary Plan 1. The best salary over a 8 year period is Salary Plan 2. Reason: Plan 1, you would make $265 more in 3 years. Plan 2, you would make $66,841 more in 8 years.

TRY THESE

1. arithmetic; $d = 5$ **2.** geometric; $r = \frac{3}{2}$

3. geometric; $r = 5$ **4.** arithmetic; $d = 150$

5. arithmetic; $d = \frac{1}{2}$ **6.** geometric; $r = \frac{1}{2}$

7. $d = 2$; $a_n = a_1 + (n - 1)d$
$a_{25} = 20 + (25 - 1)2 = 68$ seats in the 25th row.

8. $a_n = a_1 + (n - 1)d$
$a_{12} = 5 + (12 - 1)(-3) = -28$

9. $a_{17} = -14 + (17 - 1)(4) = 50$

10. $a_1 = 4$
$a_7 = 76 = 4 + (7 - 1)d$; $d = 12$
$a_{30} = 4 + (30 - 1) \cdot 12 = 352$

11. $a_4 = -8$, $r = -\frac{1}{2}$

$a_4 = -8 = a_1\left(-\frac{1}{2}\right)^{4-1}$

$a_1 = 64$

$a_{10} = 64\left(-\frac{1}{2}\right)^{10-1} = -\frac{1}{8}$

12. No. There will be a ratio between every successive pair of terms in every type of sequence, but not necessarily the same ratio.

Example: $1, 3, 5, 7, \ldots$

$\dfrac{a_2}{a_1} = \dfrac{3}{1} = 3 \neq \dfrac{a_3}{a_2} = \dfrac{5}{3}$

13. Arithmetic:

$a_1 = 4$

$a_4 = 500 = 4 + (4 - 1)d$

$496 = 3d$

$165\dfrac{1}{3} = d$

$a_2 = 4 + (2 - 1)165\dfrac{1}{3} = 169\dfrac{1}{3}$

$a_3 = 4 + (3 - 1)165\dfrac{1}{3} = 334\dfrac{2}{3}$

Geometric:

$a_1 = 4$

$a_4 = 500 - 4 \cdot r^{4-1}; \$125 = r^3; r = 5$

$a_2 = 4 \cdot 5^{2-1} = 20$

$a_3 = 4 \cdot 5^{3-1} = 100$

PRACTICE

1. $d = 3$; next term is 14   **2.** $d = -4$; next term is -7

3. $d = 2.15$; next term is 12.02

4. $r = 4$; next term is 1024   **5.** $r = -\dfrac{1}{3}$; next term is $\dfrac{1}{9}$

6. $r = 0.75$; next term is 1645.3125

7. Arithmetic sequence: 7, 14, 21, 28, 35
Geometric sequence: 7, 14, 28, 56, 112

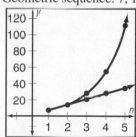

The arithmetic sequence is a linear equation. The geometric sequence is non-linear equation. In fact, it is an exponential equation.

8. $a_1 = 2; d = 4; a_{10} = 2 + (10 - 1)4 = 38$

9. $a_1 = 6; d = -2; a_{12} = 6 + (12 - 1)(-2) = -16$

10. $a_1 = 11; d = 1.4; a_4 = 11 + (4 - 1)(1.4) = 15.2$

11. $a_6 = 17; 17 = a_1 + (6 - 1)d; a_1 = 17 - 5d$

$a_{12} = 29; 29 = a_1 + (12 - 1)d; a_1 = 29 - 11d$

$17 - 5d = 29 - 11d; d = 2; a_1 = 17 - 5(2) = 7$

$a_{30} = 7 + (30 - 1)2 = 65$

12. $a_2 = \dfrac{5}{6}; \dfrac{5}{6} = a_1 + (2 - 1)d; \dfrac{5}{6} = a_1 + d;$

$a_1 = \dfrac{5}{6} - d$

$a_6 = \dfrac{13}{6}; \dfrac{13}{6} = a_1 + (6 - 1)d; \dfrac{13}{6} = a_1 + 5d;$

$a_1 = \dfrac{13}{6} - 5d$

$\dfrac{5}{6} - d = \dfrac{13}{6} - 5d; d = \dfrac{8}{24} = \dfrac{1}{3}; a_1 = \dfrac{5}{6} - \dfrac{1}{3} = \dfrac{1}{2}$

$a_{10} = \dfrac{1}{2} + (10 - 1)\dfrac{1}{3} = \dfrac{7}{2}$

13. $a_1 = 7.5; a_9 = -8.5; -8.5 = 7.5 + (9 - 1)d;$
$d = -2$

$a_6 = 7.5 + (6 - 1)(-2) = -2.5$

14. $r = \dfrac{-6}{4} = \dfrac{-3}{2}; a_n = a_1 r^{n-1};$

$a_5 = 4 \cdot \left(\dfrac{-3}{2}\right)^{5-1} = 20.25$

15. $a_4 = (-6)\left(\dfrac{1}{2}\right)^{(4-1)} = \dfrac{-3}{4}$ or -0.75

16. $a_{10} = 81\left(\dfrac{1}{2}\right)^{10-1} = \dfrac{81}{512}$ or 0.158

17. $a_5 = 48 = a_1 r^{(5-1)}; a_8 = -384 = a_1 r^{(8-1)}$

$\dfrac{48}{r^4} = \dfrac{-384}{r^7}; \dfrac{r^7}{r^4} = \dfrac{-384}{48}; r^3 = \dfrac{-384}{48} = -8; r = -2$

$a_5 = 48 = a_1(-2)^{5-1}; a_1 = \dfrac{48}{(-2)^4} = 3$

$a_{10} = 3 \cdot (-2)^{10-1} = -1536$

18. $a_3 = \dfrac{1}{9} = a_1\left(\dfrac{1}{3}\right)^{3-1}; \dfrac{1}{9} = a_1\dfrac{1}{9}; a_1 = 1$

19. $a_5 = 300 \cdot r^{(5-1)}; \dfrac{0.03}{300} = r^4; r = 0.1$

$a_n = a_1 r^{n-1}; a_n = 300 \cdot (0.1)^{n-1}$

20. $a_0 = 0.01; a_1 = 0.02; a_2 - 0.04$

$a_2 = (0.02)(2)^{n-1}$

$a_{15} = (0.02)2^{15-1} = 327.68$ in.

21. $a_n = 16 + (n - 1)32$

$a_{10} = 16 + (10 - 1)32 = 304$ ft.

22. $a_1 = 2; a_7 = 10; 10 = 2 + (7 - 1)d; d = \dfrac{4}{3}$

$2, \dfrac{10}{3}, \dfrac{14}{3}, 6, \dfrac{22}{3}, \dfrac{26}{3}, 10$

23. $a_5 = 512 = 2r^{5-1}; 256 = r^4; r = 4$ or $r = -4$
2, 8, 32, 128, 512 or 2, -8, 32, -128, 512

EXTEND

24. b; Answer is a multiple of 7 with remainder of 4.

25. c; Answer is even and a multiple of 9.

26. Use the formula for the general term, substitute in the known values for a_n, a_1, and d. Then solve for n. Examples will vary.

27. $a_n = 35 + (n - 1)(-4); (n - 1)(-4) = -32; n = 9$
$a_n = 35 - 32 = 3$
3 blocks in the 9th row

28. $a_2 = a_1 + (2 - 1)d = (4p - 3q);$
$a_1 = (4p - 3q) - 1d$
$a_4 = a_1 + (4 - 1)d = (10p + q);$
$a_1 = (10p + q) - 3d$
$(4p - 3q) - 1d = (10p + q) - 3d; d = (3p + 2q)$
$a_1 = (4p - 3q) - 1(3p + 2q) = p - 5q$

29. $a_n = (100\%)\left(\frac{1}{2}\right)^{(n-1)}$; 11th stroke is the 12th in the

sequence; $a_{12} = 100\left(\frac{1}{2}\right)^{12-1} = \frac{1}{2048}$ or $\frac{25}{512}\%$ or

approximately 0.05%

THINK CRITICALLY

30. Using arithmetic sequence, a series with three elements:
$a_1 = a, a_2,$ and $a_3 = b$. Using the general form gives,

$a_3 = a_1 + (3 - 1)d; d = \frac{b - a}{2}$

So, $a_2 = a + (2 - 1)\left(\frac{b - a}{2}\right)$

$a_2 = \frac{2a}{2} + \frac{(b - a)}{2} = \frac{a + b}{2}$

31. Using geometric sequence, a series with three elements:

$a_1 = a, a_2,$ and $a_3 = b$. Using the general geometric

form gives, $a_n = a_1 r^{(n-1)}; a_3 = ar^{(3-1)}; b = ar^2;$

$r = \pm\sqrt{\frac{b}{a}}$

So, $a_2 = a\sqrt{\frac{b}{a}} = \sqrt{\frac{a^2}{1^2} \cdot \frac{b}{a}} = \sqrt{ab}$, or

$a_2 = a\left(-\sqrt{\frac{b}{a}}\right)^{(2-1)} = -\sqrt{\frac{a^2}{1^2} \cdot \frac{b}{a}} = -\sqrt{ab}$

32. $a_n = a_1 r^{(n-1)}$; in 5 years would be the 5 in the series.
$a_6 = 1.1(1.06)^{6-1}$
   $= 1.472$ or about 1.5 billion KwH
Possible response: By using graphing calculator guess
and check by changing powers of $1.1 \times (1.06)^n$ when
exponent is entered as 10, result is 1.9699 which gives
$(n - 1) = 10$ and $n = 11$ and when exponent is
entered as 11, result is 2.088 which gives $(n - 1) = 11$
and $n = 12$. So the electrical usage will double in
approximately 12 years.

33. $a_n = a_1 r^{(n-1)}$; increasing rate will be $(1 + r)$, so
$a_n = a_1(1 + r)^{(n-1)}$; let $a_1 = 100$ and $a_6 = 200$;
$200 = 100(1 + r)^5; 2 = (1 + r)^5; \sqrt[5]{2} = \sqrt[5]{(1 + r)^5};$
$114.9 = 1 + r; r = 14.9\%$ increase

MIXED REVIEW

34. Equation is symmetric about y-axis.

35. Equation is symmetric about both axes and origin.

36. Not symmetric to axis or origin.

37. Equation is symmetric about origin.

38. $\sqrt[2]{19} = 19^{\frac{1}{2}}$ **39.** $\sqrt[4]{c^3} = c^{\frac{3}{4}}$ **40.** $\left(\sqrt[6]{11}\right)^5 = 11^{\frac{5}{6}}$

41. $\sqrt[3]{(n - 1)^7} = (n - 1)^{\frac{7}{3}}$

42. A; Given an arithmetic sequence where $a_1 = -30$ and
$d = 6$
$a_n = a_1 + (n - 1)d;$
$a_{18} = -30 + (18 - 1)6 = 72$

Lesson 13.4, pages 639–645

EXPLORE

1. Performance, each student needs to have two shapes.

2. dimension: (3×4); area 12 squares.
Sum $= \frac{1}{2}$ (area of rectangle) or Sum $= \frac{1}{2}(3 \times 4)$

3. dimension: (4×5); area 20 squares
Sum $= \frac{1}{2}$ (area of rectangle) or Sum $= \frac{1}{2}(4 \times 5)$

4. Sum $= \frac{1}{2}$ (area of rectangle) or Sum $= \frac{1}{2}(6 \times 7)$

5. There would be n rows.
There would be $(n + 1)$ columns.

6. dimension: $n \times (n + 1)$

7. Area $= n(n + 1) = n^2 + n$; Sum $= \frac{1}{2}$ (area),
Sum $= \frac{1}{2}n(n + 1)$ or $\frac{n^2 + n}{2}$

8. Sum $= \frac{1}{2}[n(n + 1)]$, where $a_1 = 1$ and $a_n = n$ so,
Sum $= \frac{n}{2}(n + 1)$ or $\frac{n}{2}(a_n + a_1)$

TRY THESE

1. $a_1 = 2; d = 8$; use $S_n = \frac{n}{2}(2a_1 + (n - 1)d)$
$S_{10} = \frac{10}{2}(2 \cdot 2 + (10 - 1)8) = 380$

2. $a_1 = 5; d = 3$
$S_{20} = \frac{20}{2}(2 \cdot 5 + (20 - 1)3) = 670$

3. $a_1 = -7; d = 3$
$S_{26} = \frac{26}{2}(2(-7) + (26 - 1)3) = 793$

4. $a_1 = 6; a_6 = 21$; use $S_n = \frac{n}{2}(a_1 + a_n)$
$S_6 = \frac{6}{2}(6 + 21) = 81$

5. $a_1 = 2; d = 5; n = 20$
$S_{20} = \frac{20}{2}(2 \cdot 2 + (20 - 1)5) = 990$

6. $a_1 = 1; a_{100} = 100$
$S_{100} = \frac{100}{2}(1 + 100) = 5050$

7. When you are given number of terms, the first term, and
the last term, use $S_n = \frac{n}{2}(a_1 + a_n)$.
When you are given or can determine the first term, the
common difference, and the number of terms, use
$S_n = \frac{n}{2}(2a_1 + (n - 1)d)$

8. $a_1 = 3; r = 5;$ use $S_n = \dfrac{a_1(1 - r^n)}{(1 - r)}$

$S_7 = \dfrac{3(1 - 5^7)}{(1 - 5)} = 58{,}593$

9. $a_1 = 2; r = -\dfrac{1}{2}$

$S_{10} = \dfrac{2\left(1 - \left(-\frac{1}{2}\right)^{10}\right)}{\left(1 - \left(-\frac{1}{2}\right)\right)} = \dfrac{341}{256}$ or about 1.332

10. $a_1 = 1; r = 0.3; S_5 = \dfrac{1(1 - 0.3^5)}{(1 - 0.3)} = 1.4251$

11. $a_1 = 1; r = 2; S_{20} = \dfrac{1(1 - 2^{20})}{(1 - 2)} = 1{,}048{,}575$

12. $S_{14} = \left(\dfrac{1}{64}\right)\dfrac{(1 - (-2)^{14})}{(1 - (-2))} = -\dfrac{5461}{64}$ or about -85.328

13. $S_5 = \dfrac{-2\left(1 - \left(-\frac{1}{2}\right)^5\right)}{\left(1 - \left(-\frac{1}{2}\right)\right)} = -\dfrac{11}{8}$ or -1.375

14.

Multiply
S_n by r

$\quad S_n r = \qquad\qquad a_1 r + a_1 r^2 + a_1 r^3 + \cdots + a_1 r^n$

Subtract S_n

$\quad -S_n = -a_1 - a_1 r - a_1 r^2 - a_1 r^3 - \ldots - a_1 r^{n-1}$

difference of $\quad S_n r - Sn = -a_1$

$\qquad\qquad S_n(r - 1) = -a_1 + a_1 r^n$

$\qquad\qquad S_n(r - 1) = a_1(-1 + r^n)$

$\qquad\qquad\qquad S_n = \dfrac{a_1(r^n - 1)}{(r - 1)}$

---

15. $a_1 = 24; n = 15; a_{15} = 10;$ use $S_n = \dfrac{n}{2}(a_1 + a_n)$

$S_n = \dfrac{15}{2}(24 + 10) - 255$ logs

16. $a_1 = 0.01; r = 2; n = 30;$ use $S_n = \dfrac{a_1(1 - r^n)}{(1 - r)}$

$S_{30} = \dfrac{0.01(1 - 2^{30})}{(1 - 2)} = \$10{,}737{,}418.23$

17. $\displaystyle\sum_{n=1}^{7} (2n - 1); S_7 = \dfrac{7}{2}(1 + 13) = 49$

18. $\displaystyle\sum_{n=1}^{19} -(3n - 1)\ \ ; S_{19} = \dfrac{19}{2}(-2 - 56) = -551$

19. $\displaystyle\sum_{n=1}^{6} 2^n; S_6 = \dfrac{2(1 - 2^6)}{1 - 2} = \dfrac{2(-63)}{-1} = 126$

20. $\displaystyle\sum_{n=1}^{9} 4^{n-1}; S_9 = \dfrac{1(1 - 4^9)}{1 - 4} = 87{,}381$

PRACTICE

1. $a_1 = 1; n = 13; d = 2;$ use $S_n = \dfrac{n}{2}(2a_1 + (n - 1)d)$

$S_{13} = \dfrac{13}{2}(2 \cdot 1 + (13 - 1)2) = 169$

2. $a_1 = 2; n = 16; d = 2$

$S_{16} = \dfrac{16}{2}(2 \cdot 2 + (16 - 1)2) = 272$

3. $a_1 = 2; n = 8; d = -3$

$S_8 = \dfrac{8}{2}(2 \cdot 2 + (8 - 1)(-3)) = -68$

4. $a_1 = -3; n = 20; d = 4$

$S_{20} = \dfrac{20}{2}(2(-3) + (20 - 1)4) = 700$

5. $a_1 = 15; n = 10; d = 1.5$

$S_{10} = \dfrac{10}{2}(2 \cdot 15 + (10 - 1)1.5) = 217.5$

6. $a_1 = 2\dfrac{2}{3}; n = 27\ d = \dfrac{5}{6}$

$S_{27} = \dfrac{27}{2}\left(2 \cdot \left(2\dfrac{2}{3}\right) + (27 - 1)\dfrac{5}{6}\right) = 364\dfrac{1}{2}$

7. $a_1 = 3; a_{13} = 51;$ use $S_n = \dfrac{n}{2}(a_1 + a_n)$

$S_{13} = \dfrac{13}{2}(3 + 51) = 351$

8. $a_1 = 0.8; n = 9; d = -0.2$

$S_9 = \dfrac{9}{2}(2 \cdot (0.8) + (9 - 1)(-0.2)) = 0$

9. $a_1 = 7; a_{15} = 63$

$S_{15} = \dfrac{15}{2}(7 + 63) = 525$

10. $a_1 = 9; n = 5; r = \dfrac{1}{3};$ use $S_n = \dfrac{a_1(1 - r^n)}{(1 - r)}$

$S_5 = \dfrac{9\left(1 - \left(\frac{1}{3}\right)^5\right)}{\left(1 - \frac{1}{3}\right)} = \dfrac{121}{9}$ or about 13.4

11. $a_1 = 1; n = 9; r = \dfrac{1}{2}$

$S_9 = \dfrac{1\left(1 - \left(\frac{1}{2}\right)^9\right)}{\left(1 - \frac{1}{2}\right)} = \dfrac{511}{256}$ or about 2.0

12. $a_1 = 16; n = 7; r = -\dfrac{1}{2}$

$S_7 = \dfrac{16\left(1 - \left(-\frac{1}{2}\right)^7\right)}{\left(1 - \left(-\frac{1}{2}\right)\right)} = \dfrac{43}{4}$ or about 10.8

13. $a_1 = 2; n = 10; r = \sqrt{2}$

$S_{10} = \dfrac{2(1 - (\sqrt{2})^{10})}{(1 - \sqrt{2})} = $ approximately 149.7

or by using two series.
one being $2 + 4 + 8 + 16 + 32 = 62$ and a second
being $2\sqrt{2} + 4\sqrt{2} + 8\sqrt{2} + 16\sqrt{2} + 32\sqrt{2} = 62\sqrt{2}$; an exact answer of $62 + 62\sqrt{2}$

14. $a_1 = 4; n = 8; r = \dfrac{1}{10}$

$S_8 = \dfrac{4\left(1 - \frac{1}{10}^8\right)}{\left(1 - \frac{1}{10}\right)} = 4.4444444$ or about 4.4

15. $a_1 = 4; n = 12; r = 1.25$

$S_{12} = \dfrac{4(1 - (1.25)^{12})}{(1 - 1.25)} \approx 216.8$

16. $a_1 = 1; n = 6; r = -\dfrac{2}{3}$

$S_6 = \dfrac{1\left(1 - \left(-\frac{2}{3}\right)^6\right)}{\left(1 - \left(-\frac{2}{3}\right)\right)} = \dfrac{133}{243}$ or about 0.5

17. $a_1 = 3, a_2 = 6, a_3 = 12; r = 2; n = 8$
$S_8 = \dfrac{3(1 - 2^8)}{(1 - 2)} = 765$

18. $a_1 = 2, r = -3$
$S_7 = \dfrac{2(1 - (-3)^7)}{(1 - (-3))} = 1094$

19. $a_1 r^{n-1} = a_n; 36r^{5-1} = 2.25; r = \pm 0.5$

$S_5 = \dfrac{36(1 - (0.5)^5)}{(1 - (0.5))} = \dfrac{279}{4}$ or 69.75 or

$S_5 = \dfrac{36(1 - (-0.5)^5)}{(1 - (-0.5))} = \dfrac{99}{4}$ or 24.75

20. $a_7 = 128; r = 2; a_1 r^{n-1} = a_n; a_1 2^{7-1} = 128$

$a_1 = 2; S_7 = \dfrac{2(1 - 2^7)}{(1 - 2)} = 254$

21. $a_1 = \$32,000 \; r = 1.05$

$S_6 = \dfrac{32,000(1 - 1.05^6)}{(1 - 1.05)} = \$217,661.21$

22. $a_1 = 72; d = 2; n = 50$

$S_{50} = \dfrac{50}{2}(2 \cdot 72 + (50 - 1)2) = 6050$ seats

23. $\displaystyle\sum_{n=1}^{5}(4n - 3) = \dfrac{5}{2}(1 + 17) = 45$

24. $\displaystyle\sum_{n=1}^{10}(-2n + 22) = \dfrac{10}{2}(20 + 2) = 110$

25. $\displaystyle\sum_{n=1}^{21}(3n + 2) = \dfrac{21}{2}(5 + 65) = 735$

26. $\displaystyle\sum_{n=1}^{8}3^{(n-1)} = \dfrac{1(1 - 3^8)}{1 - 3} = 3280$

27. $\displaystyle\sum_{n=1}^{9}[5(2^n)] = \dfrac{10(1 - 2^9)}{1 - 2} = 5110$

28. $\displaystyle\sum_{n=1}^{6}16\left(-\dfrac{1}{2}\right)^{(n-1)} = \dfrac{16(1 - \left(-\frac{1}{2}\right)^6)}{\left(1 - \left(-\frac{1}{2}\right)\right)} = 10.5$

29. $a_1 = \$4.00 \quad d = 0.25$
$4 + 3.75 + 3.50 + 3.25 + 3 + 2.75 + 2.50 + 2.25 + 2 = \$27.00$
Megan bought 9 books.

30. $a_1 = \$2,000 \quad d = \$500$

$2,000 + 2,500 + 3,000 + \cdots = \$92,000$
16 days

31. A sequence is an ordered set of numbers that are related mathematically. A series is the sum of the term of a sequence.

EXTEND

32. $a_1 = 4x - y; d = -3x + 4y; n = 10$

$S_{10} = \dfrac{10}{2}(2(4x - y) + (10 - 1)(-3x + 4y)) = -95x + 170y$

33. $a_1 = 1; d = 2$

$a_n = \dfrac{n}{2}(2 + (n - 1)2) = n^2$

34. $a_1 = 2; r = a^2; n = 12$

$S_{12} = \dfrac{2(1 - (a^2)^{12})}{1 - a^2} = \dfrac{2 - 2a^{24}}{1 - a^2}$

35. $d = 5; S_{348} = 3534$

$3534 = \dfrac{348}{2}(2a_1 + (348 - 1)5); a_1 = -\dfrac{24,863}{29}$ or

about -857.345

36. When you have number of terms, the first term, and the common ratio, you will use $S_n = \dfrac{a_1(1 - r^n)}{1 - r}$. When you have first term, last term, and common ratio, you will use $S_n = \dfrac{a_1 - ra_n}{1 - r}$

37. $a_1 + (a_1 + d) + (a_1 + 2d) = 30$
$$3a_1 + 3d = 30$$
$$a_1 + d = 10$$
$$d = 10 - a_1$$

$$a_1(a_1 + d)(a_1 + 2d) = 360$$
$$10a_1(a_1 + 2(10 - a_1)) = 360$$
$$10a_1(a_1 + 20 - 2a_1) = 360$$
$$10a_1(20 - a_1) = 360$$
$$200a_1 - 10a_1{}^2 = 360$$
$$0 = 10a_1{}^2 - 200a_1 + 360$$
$$0 = a_1{}^2 - 20a_1 + 36$$
$$0 = (a_1 - 2)(a_1 - 18)$$
$$2 = a_1 \text{ or } 18 = a_1$$
$d = 10 - 2 = 8 \text{ or } d = 10 - 18 = -8$
So, the terms are 2, 10, 18.

38. $a_1 + a_2 + a_3 = 52; a_3 = 2.25(a_1 + a_2); a_3 = 36$
$a_1 + a_2 = 16$; since $a_n = a_1 r^n$, $16 = a_1 + a_1 r$ and
$36 = a_1 r^2$; so $\dfrac{36}{r^2} = a_1$; Substituting,

$$16 = \frac{36}{r^2} + \frac{36r}{r^2}$$
$$16r^2 = 36 + 36r$$
$$4r^2 = 9 + 9r$$
$$4r^2 - 9r - 9 = 0$$
$$r = \frac{-(-9) \pm \sqrt{81 - 4(4)(-9)}}{2(4)}$$
$$r = \frac{9 \pm \sqrt{225}}{8}$$
$$r = \frac{9 + 15}{8} = \frac{24}{8} = 3$$
$$a_1 = \frac{36}{3^2} = 4$$

For $a_1 = 4$ and $r = 3$, the terms are:
4, 12, 36
also,
$$r = \frac{9 - 15}{8} = \frac{-6}{8} = -\frac{3}{4}$$
$$a_1 = \frac{36}{\left(-\frac{3}{4}\right)^2} = 64$$

For $a_1 = 64$ and $r = -\dfrac{3}{4}$ the terms are:
64, −48, 36

39. Let the three terms be $a_1, a_1 + d, a_1 + 2d$.
$$a_1 + (a_1 + d) + (a_1 + 2d) = 42$$
$$3a_1 + 3d = 42$$
$$a_1 + d = 14$$
$$a_1 = 14 - d$$
$$(a_1 - 4)r = (a_1 + d) - 2$$
$$(14 - d - 4)r = (14 - d + d) - 2$$
$$(10 - d)r = 12$$
$$r = \frac{12}{10 - d}$$

$$(a_1 - 4)r^2 = (a_1 + 2d) + 2$$
Substitute $14 - d$ for a_1.
$$(14 - d - 4)r^2 = (14 - d + 2d) + 2$$
$$(10 - d)r^2 = 16 + d$$
Substitute $\dfrac{12}{10 - d}$ for r.
$$(10 - d)\left(\frac{12}{10 - d}\right)^2 = 16 + d$$
$$\frac{144}{10 - d} = 16 + d$$
$$144 = (10 - d)(16 + d)$$
$$144 = 160 - 6d - d^2$$
$$d^2 + 6d - 16 = 0$$
$$(d + 8)(d - 2) = 0$$
$$d = -8 \quad \text{or} \quad d = 2$$

$a_1 = 14 - (-8) = 22$ or	$a_1 = 14 - 2 = 12$
$a_1 + d = 22 - 8 = 14$	$a_1 + d = 12 + 2 = 14$
$a_1 + 2d = 22 - 16 = 6$	$a_1 + 2d = 12 + 4 = 16$

The terms are 22, 14, 6 or 12, 14, 16.

MIXED REVIEW

40. $x - 8$

41. $x^2 - 2x - 1 + \dfrac{-8}{x - 2}$

42. $x^2 + 6$

43. $x^3 + x + 1$

44. C; $\displaystyle\sum_{n=1}^{6} 2^n = 2 + 4 + 8 + 16 + 32 + 64 = 126$

PROJECT CONNECTION

1–4. The perimeter of each new figure is $\dfrac{4}{3}$ the perimeter of the previous figure. The number of sides in each new figure is $3 \cdot (4)^{n-1}$, and the length of each side is $\dfrac{l}{3^{n-1}}$, where l is the length of the side of the original equilateral triangle.

5. No, the perimeters are unbounded.

ALGEBRAWORKS

1. Spiral A is arithmetic; Spiral B is geometric.

2. For Spiral A: $a_1 = 180$ units; $d = -3; n = 30$
$a_n = a_1 + (n - 1)d; a_{30} = 180 + (30 - 1)(-3) = 93$ units
For Spiral B: $a_1 = 180$ units; $r = 0.9; n = 30$
$a_n = a_1 r^{n-1}; a_{30} = 180 \cdot 0.9^{30-1} \approx 8.5$ units

3. For Spiral A:
$S_{30} = \dfrac{30}{2}(2 \cdot 180 + (30 - 1)(-3)) = 4095$ units
For Spiral B:
$S_{30} = \dfrac{180(1 - 0.9^{30})}{1 - 0.9} \approx 1723.7$ units

4. $A = \pi r^2$
$\pi \times 60^2 - \pi \times 20^2 = 10{,}053$ mm^2

5. $A = l \times w$
$10{,}053 = l \times 0.0015$
$l = 6{,}702{,}000$ mm

6. $6{,}702{,}000$ mm $\times \dfrac{1 \text{ in.}}{25.4 \text{ mm}} \times \dfrac{1 \text{ ft}}{12 \text{ in}} \times \dfrac{1 \text{ mile}}{5{,}280 \text{ ft}} \approx 4.2$ mi

Lesson 13.5, pages 646–652

EXPLORE

1. $128; \frac{1}{2}$ 2. $64; \frac{1}{4}$ 3. $32; \frac{1}{8}; 16; \frac{1}{16}; 8; \frac{1}{32}; 4; \frac{1}{64}$

4. $128 + 64 + 32 + 16 + 8 + 4; \frac{1}{2} + \frac{1}{4} + \frac{1}{8} +$
$\frac{1}{16} + \frac{1}{32} + \frac{1}{64}$; the series are geometric because each
term is obtained by multiplying the preceding term by
the common ratio $\frac{1}{2}$.

5. $252; 0.984375$

6. Answers will vary. Students should recognize that they
will never shade the entire square no matter how many
times they repeat the process.

TRY THESE

1. converges, $\left|\frac{1}{3}\right| < 1$

2. converges, $\left|-\frac{1}{5}\right| < 1$

3. diverges, $|1.5| > 1$ 4. $\dfrac{18}{1 - \frac{1}{3}} = 27$

5. $\dfrac{8}{1 - \left(-\frac{1}{2}\right)} = 5\frac{1}{3}$ 6. $\dfrac{1}{1 - \frac{1}{4}} = \frac{4}{3}$

7. diverges, $|-2| > 1$

8. $\displaystyle\sum_{n=1}^{\infty} 6(0.4)^{n-1} = 6 + 2.4 + 0.96 + \ldots = \dfrac{6}{1 - 0.4} = 10$

9. $\dfrac{10}{1 - (-0.1)} = \dfrac{100}{11} = 9.09$

10. $\dfrac{7}{11}$ 11. $\dfrac{29}{9}$ 12. $\dfrac{1}{11}$ 13. $\dfrac{428}{333}$

14. $\dfrac{40}{1 - 0.8} = 200 \text{ cm}$

15. Answers will vary. An infinite geometric series
converges if the common ratio has an absolute value
less than 1. The sum equals first term divided by
$(1 - r)$, where r is the common ratio.

PRACTICE

1. converges, $\left|-\frac{1}{2}\right| < 1$ 2. diverges, $|2| > 1$

3. converges, $\left|\frac{1}{10}\right| < 1$ 4. converges, $\left|\frac{1}{100}\right| < 1$

5. converges, $\left|\frac{2}{3}\right| < 1$ 6. diverges, $|-2| > 1$

7. $\dfrac{4}{1 - \frac{1}{2}} = 8$ 8. $\dfrac{1}{1 - \left(-\frac{1}{3}\right)} = \frac{3}{4}$

9. diverges since $|r| > 1$, that is $\left|\frac{5}{3}\right| > 1$.

10. $\dfrac{12}{1 - \left(-\frac{1}{3}\right)} = 9$ 11. $\dfrac{2}{1 - \frac{1}{4}} = \frac{8}{3}$

12. $\dfrac{10}{1 - \frac{1}{5}} = \dfrac{25}{3}$ 13. $|r| = \left|-\frac{4}{3}\right| > 1$,
diverges

14. $\dfrac{-5}{1 - \frac{3}{5}} = -\dfrac{25}{2}$ or -12.5 15. $\dfrac{108}{1\left(-\frac{3}{4}\right)} = \dfrac{432}{7}$

16. $|r| = |1| = 1$, diverges   17. $\dfrac{27}{1 - \left(-\frac{1}{3}\right)} = \dfrac{81}{4}$

18. $|r| = |-1| = 1$, diverges   19. $\dfrac{1}{1 - \frac{1}{2}} = 2$

20. $\dfrac{3}{1 - 0.9} = 30$ 21. $\dfrac{100}{1 - \frac{2}{5}} = \dfrac{500}{3}$

22. $\dfrac{\frac{1}{2}}{1 - \frac{7}{8}} = 4$ 23. $\dfrac{31}{33}$ 24. $\dfrac{59}{9}$ 25. $\dfrac{25}{33}$ 26. $\dfrac{41}{333}$

27. $\dfrac{73}{111}$ 28. $\dfrac{26}{37}$

29. $n = 0.2\overline{8}$
$$\begin{array}{r} 100n = 28.888\ldots \\ -\ 10n = -2.888\ldots \\ \hline 90n = 26 \end{array}$$
$n = \dfrac{26}{90} = \dfrac{13}{45}$

30. $n = 0.2\overline{740}$
$$\begin{array}{r} 10000n = 2740.\overline{740} \\ -\ 10n = -2.\overline{740} \\ \hline 9990n = 2738 \end{array}$$
$n = \dfrac{2738}{9990} = \dfrac{37}{135}$

31. $\dfrac{3,000,000}{1 - \frac{3}{50}} \approx 3{,}191{,}489$ people

32. $\dfrac{80}{1 - \frac{3}{4}} = 320$ revolutions

33. $\dfrac{60}{1 - \frac{3}{5}} = 150$ ft down motions.

$150 - 60 = 90$ ft up motion
$150 + 90 = 240$ ft total vertical motion

34. Answer will vary. Students should note that successive
terms in both types of series differ by a common ratio. A
finite series terminates and has a sum. An infinite series
does not terminate and may or may not have a sum,
depending on the common ratio.

EXTEND

35. $\dfrac{\frac{1}{6}}{1 - \frac{1}{6}} = \dfrac{1}{5}$

36. $\dfrac{\sqrt{2}}{1 - \frac{\sqrt{2}}{2}} = 2\sqrt{2} + 2$ or about 4.828

37. $\dfrac{\frac{\sqrt{3}}{\sqrt{3}+1}}{1 - \left(\frac{\sqrt{3}+1}{\sqrt{3}+3}\right)} = \dfrac{3}{2}$

320 Solutions Manual   **Algebra 2: An Integrated Approach**

39. $\frac{41}{99}$ **40.** $\frac{67}{99}$ **41.** $\frac{89}{99}$ **42.** $\frac{4}{99}$

43. A two digit repeating decimal, write the two digits over 99.

44. A three digit repeating decimal, write the three digits over 999.

45. $\frac{a}{1 - r}$, for $|r| < 1$

46. If $n = 20$, $S_{20} = 0.9999990463$, which is within 0.0000009 of 1, satisfying the condition.

MIXED REVIEW

47. 16   **48.** 81   **49.** $\frac{1}{16}$ **50.** 4   **51.** $\frac{2\sqrt{3}}{3}$

52. $\frac{\sqrt{2}}{2}$ **53.** $-2 + \sqrt{7}$ **54.** $\frac{2\sqrt[3]{6}}{3}$

55. B; $\dfrac{9}{1 - \left(-\frac{2}{3}\right)} = \dfrac{27}{5}$

56. $x^2 - 7x + 12 = 0$
$(x - 3)(x - 4) = 0$
$x = 3, 4$

57. $6x^2 - x - 1 = 0$
$(2x - 1)(3x + 1) = 0$
$x = -\dfrac{1}{3}, \dfrac{1}{2}$

58. $18x^2 + 3x - 10 = 0$
$(6x + 5)(3x - 2) = 0$
$x = \dfrac{5}{6}, \dfrac{2}{3}$

59. $3x^2 - 5x - 3 = 0$

$x = \dfrac{-b \pm \sqrt{b^2 - 4ac}}{2a}$

$\dfrac{5 + \sqrt{25 - 4 \cdot 3(-3)}}{6} \approx 2.135$

$\dfrac{5 - \sqrt{25 - 4 \cdot 3(-3)}}{6} \approx -0.468$

ALGEBRAWORKS

1. Colosseum: $\dfrac{x^2}{310^2} + \dfrac{y^2}{255^2} = 1$

Arena floor: $\dfrac{x^2}{145^2} + \dfrac{y^2}{90^2} = 1$

2a. The floor is a circle. The intersection of a plane with a cone is a circle.

2b. $x^2 + y^2 = (1.5)^2$

3. $y = 71 + \sqrt{576 - x^2}$

4. $(x - 320)^2 = 4(-40)(y - 320)$

5.

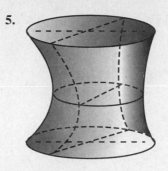

Lesson 13.6, pages 653–657

EXPLORE

1. Answers will vary. Among features students may point out are that each row begins and ends with 1, each row and the entire triangle are symmetric about a line drawn vertically through the center of the triangle, diagonals that begin with a 1 in the second row contain the natural numbers in order, and each number is the sum of the two numbers diagonally above it.

2. 1, 6, 15, 20, 15, 6, 1
Begin and end with 1, and add pairs of adjacent numbers.

3. 1, 2, 4, 8, 16, 32. The sums are powers of 2, starting with the zero power.
$2^{10 - 1} = 2^9 = 512$

TRY THESE

1. 18   **2.** a^{17} **3.** 6   **4.** 10, 7   **5.** $\dfrac{9!}{3!6!} = 84$

6. $(a + b)^4 = a^4 + 4a^3b^1 + 6a^2b^2 + 4a^1b^3 + b^4$

7. $(d - 2)^6 = d^6 - 12d^5 + 60d^4 - 160d^3 + 240d^2 - 192d^1 + 64$

8. $(3p + q)^5 =$
$243p^5 + 405p^4q^1 + 270p^3q^2 + 90p^2q^3 + 15p^1q^4 + q^5$

9. $10x^2y^3$ **10.** $15m^7n^4$ **11.** $4032c^5d^4$ **12.** $1760a^3b^9$

13. $(x + 4)^3 = (x^3 + 12x^2 + 48x + 64)$ in.^{3}

14. Answer will vary. Students should mention that each entry in triangle is found by adding the two adjacent numbers in the row above, that every row begins and ends with one, and that the coefficients of the expansion of $(x + y)^{10}$ will be in the 11th row of the pascal triangle.

PRACTICE

1. $(a + b)^6 = a^6 + 6a^5b^1 + 15a^4b^2 + 20a^3b^3 + 15a^2b^4 + 6a^1b^5 + b^6$

2. $(2a + x)^5 = (2a)^5 + 5(2a)^4x^1 + 10(2a)^3x^2 + 10(2a)^2x^3 + 5(2a)^1x^4 + x^5$
$32a^5 + 80a^4x + 80a^3x^2 + 40a^2x^3 + 10ax^4 + x^5$

3. $(c - 5d)^4 = c^4 + 4(c)^3(-5d)^1 + 6(c)^2(-5d)^2 + 4(c)^1(-5d)^3 + (-5d)^4$
$c^4 - 20c^3d^1 + 150c^2d^2 - 500c^1d^3 + 625d^4$

4. $(1 - x)^8 = 1 - 8x^1 + 28x^2 - 56x^3 + 70x^4 - 56x^5 + 28x^6 - 8x^7 + x^8$

5. $(2 - 3b)^4 = (2)^4 + 4(2)^3(-3b)^1 + 6(2)^2(-3b)^2 + 4(2)^1(-3b)^3 + (-3b)^4$
   $16 - 96b^1 + 216b^2 - 216b^3 + 81b^4$

6. $(a - b^2)^7 = a^7 + 7a^6(-b^2)^1 + 21a^5(-b^2)^2 + 35a^4(-b^2)^3 + 35a^3(-b^2)^4 + 21a^2(-b^2)^5 + 7a^1(-b^2)^6 + (-b^2)^7$
   $a^7 - 7a^6b^2 + 21a^5b^4 - 35a^4b^6 + 35a^3b^8 - 21a^2b^{10} + 7a^1b^{12} - b^{14}$

7. $(b^2 - ac)^3 = (b^2)^3 + 3(b^2)^2(-ac) + 3(b^2)(-ac)^2 + (-ac)^3$
   $b^6 - 3a^1b^4c^1 + 3a^2b^2c^2 - a^3c^3$

8. $(3a^3 - 2b^3)^4 = 1(3a^3)^4 + 4(3a^3)^3(-2b^3) + 6(3a^3)^2(-2b^3)^2 + 4(3a^3)(-2b^3)^3 + (-2b^3)^4$
   $= 81a^{12} - 216a^9b^3 + 216a^6b^6 - 96a^3b^9 + 16b^{12}$

9. $(30 + 1)^3 + 30^3 + 3(30)^2(1)^1 + 3(30)^1(1)^2 + 1$
   $27000 + 2700 + 90 + 1 = 29,791$

10. $(100 - 2)^4 = 100^4 + 4(100)^3(-2)^1 + 6(100)^2(-2)^2 + 4(100)^1(-2)^3 + (-2)^4$
    $100,000,000 - 8,000,000 + 240,000 - 3,200 + 16$
    $92,000,000 + 236,800 + 16 = 92,236,816$

11. $\left(a - \dfrac{3}{b}\right)^4 = a^4 + 4a^3\left(-\dfrac{3}{b}\right)^1 + 6a^2\left(-\dfrac{3}{b}\right)^2 + 4a^1\left(-\dfrac{3}{b}\right)^3 + \left(-\dfrac{3}{b}\right)^4$
    $a^4 - \dfrac{12a^3}{b} + \dfrac{54a^2}{b^2} - \dfrac{108a}{b^3} + \dfrac{81}{b^4}$

12. $\left(n + \dfrac{1}{2}\right)^5 = n^5 + 5n^4\left(\dfrac{1}{2}\right)^1 + 10n^3\left(\dfrac{1}{2}\right)^2 + 10n^2\left(\dfrac{1}{2}\right)^3 + 5n^1\left(\dfrac{1}{2}\right)^4 + \left(\dfrac{1}{2}\right)^5$
    $= n^5 + \dfrac{5}{2}n^4 + \dfrac{5}{2}n^3 + \dfrac{5}{4}n^2 + \dfrac{5}{16}n + \dfrac{1}{32}$

13. $\dfrac{7!}{(7 - 3)!3!}a^4b^3 = 35a^4b^3$

14. $\dfrac{8!}{(8 - 4)!4!}a^4(-b)^4 = 70a^4b^4$

15. $\dfrac{9!}{(9 - 6)!6!}(3)^3(-x)^6 = 2268x^6$

16. $\dfrac{6!}{(6 - 3)!3!}x^3(-3y)^3 = -540x^3y^3$

17. $\dfrac{7!}{(7 - 5)!5!}(2a)^2(b)^5 = 84a^2b^5$

18. $\dfrac{4!}{(4 - 1)!1!}(3a^2)^3(-2b)^1 = -216a^6b$

19. $\dfrac{14!}{(14 - 11)!11!}(n)^3(-2)^{11} = -745,472n^3$

20. $\dfrac{5!}{(5 - 5)!0!}(x^2)^0(-2y)^5 = -32y^5$

21. $\dfrac{6!}{(6 - 2)!2!}(3a)^4(-b)^2 = 1215a^4b^2$

22. $\dfrac{5!}{(5 - 3)!3!}(3x)^2(-2y)^3 = -720x^2y^3$

23. $\dfrac{6!}{(6 - 4)!4!}(2m^2)^2(-3n^2)^4 = 4860m^4n^8$

24. $\dfrac{6!}{(6 - 3)!3!}\left(\dfrac{a}{b}\right)^3\left(\dfrac{b}{a}\right)^3 = 20$

25. First four terms:
    $(1 + 0.02)^{10} = 1^{10} + 10(1)^9(0.02) + 45(1)^8(0.02)^2 + 120(1)^7(0.02)^3$

    $1 + 0.2 + 0.018 + 0.00096 = 1.21896$
    $1.02^{10} = 1.21899442$
    remaining terms 0.00003442

26. $(0.97)^{12} = (1 - 0.03)^{12}$

27. First, determine the power of the binomial expansion; add the exponents, $4 + 7 = 11$.
    In the next term, the exponent on a decreases by 1 to 3 and the exponent on b increases by 1 to 8. The coefficient is given by $\dfrac{n!}{(\text{exponent of } a)!(\text{exponent of } b)!}$. The next term is $165a^3b^8$.

EXTEND

28. $\dfrac{20!}{18!} = 20 \cdot 19 = 380$

29. $\dfrac{45!}{42!} = 45 \cdot 44 \cdot 43 = 85,140$

30. $\dfrac{25!32!}{24!33!} = \dfrac{25}{33}$

31. $36 \cdot 35! = 36!$

32. $\dfrac{n!}{(n - 1)!} = n$

33. $(p + 5)!(p + 6) = (p + 6)!$

34. $\dfrac{(x + 1)!}{(x - 1)!} = x(x + 1)$

35. $\dfrac{(x - y)!}{(x - y - 1)!} = (x - y)$

36. Rewrite $1.015^{30} = (1 + 0.015)^{30}$
    Finding the first two terms; $1 + 30(0.015) = 1.45$. All of the remaining terms would be positive, their sum is > 0, so the sum of the whole series would be greater than 1.45.

37. $\dfrac{9!}{(9 - 4)!4!}(c)^4(2d)^5 = 4032c^4d^5$

38. $\dfrac{8!}{(8 - 4)!4!}x^4\left(-\dfrac{1}{x}\right)^4 = 70$

39. $(1 - i)^5 = 1^5 - 5i + 10i^2 - 10i^3 + 5i^4 - i^5 = -4 + 4i$

THINK CRITICALLY

40. Yes; if $n = 6$ then $(6 - 4)! = 2!$ and
    $(6 - 4)(6 - 5)(6 - 6)! = 2 \cdot 1 \cdot 0 = 2!$
    or if $n = 7$ then $(7 - 4)! = 3!$ and
    $(7 - 4)(7 - 5)(7 - 6)! = 3 \cdot 2 \cdot 1 = 3!$

41. $1^{\frac{1}{2}} + \dfrac{1}{2}1^{\frac{1}{2} - 1}b^1 + \dfrac{\frac{1}{2}\left(\frac{1}{2} - 1\right)}{1 \cdot 2}1^{\frac{1}{2} - 2}b^2 +$

    $\dfrac{\frac{1}{2}\left(\frac{1}{2} - 1\right)\left(\frac{1}{2} - 2\right)}{1 \cdot 2 \cdot 3}1^{\frac{1}{2} - 3}b^3 =$

    $1 + \dfrac{1}{2}b^1 - \dfrac{1}{8}b^2 + \dfrac{1}{16}b^3$

42. $\sqrt{6} = 2\left(1 + \frac{1}{2}\right)^{\frac{1}{2}}$

$= 2\left[1 + \frac{1}{2}\left(\frac{1}{2}\right) - \frac{1}{8}\left(\frac{1}{2}\right)^2 + \frac{1}{16}\left(\frac{1}{2}\right)^3\right]$

$= 2\left[1 + \frac{1}{4} - \frac{1}{32} + \frac{1}{128}\right] \approx 2.45$

MIXED REVIEW

43. $(f + g)(x) = (2x + 1) + (x^2 + x) = x^2 + 3x + 1$

44. $(f \cdot g)(x) = (2x + 1) \cdot (x^2 + x) = 2x^3 + 3x^2 + x$

45. $f(g(x)) = 2(x^2 + x) + 1 = 2x^2 + 2x + 1$

46. $g(f(x)) = (2x + 1)^2 + (2x + 1) = 4x^2 + 6x + 2$

47. 4 **48.** 144 **49.** $\frac{2}{3}$ **50.** 864

51. $9x^2 + 4y^2 + 54x - 8y + 49 = 0$
$9(x^2 + 6x) + 4(y^2 - 2y) = -49$
$9(x^2 + 6x + 9) + 4(y^2 - 2y + 1) = 36$
$9(x + 3)^2 + 4(y - 1)^2 = 36$
$\frac{(x + 3)^2}{4} + \frac{(y - 1)^2}{9} = 1$
Center $(-3, 1)$

52. B; $\frac{9!}{(9 - 6)!6!}x^3y^6 = 84x^3y^6$

53. $(1, 2, 3)$ **54.** $(2, 4, 1)$ **55.** $(0, -6, 1)$

PROJECT CONNECTION

1–3, 5, 6. Answers will vary.

4. The number of square tiles form a geometric sequence $3, 9, 27, 81, \ldots$ where $r = 3$, the next value is 243.

Lesson 13.7, pages 658–661

EXPLORE THE PROBLEM

1. $1 + \frac{1}{2} + \frac{1}{3} + \ldots + \frac{1}{n}$

2. To demonstrate that a series is not arithmetic,

$a_2 - a_1 = \frac{1}{2} - 1 = -\frac{1}{2}$

$a_3 - a_2 = \frac{1}{3} - \frac{1}{2} = -\frac{1}{6}$.

To demonstrate that the series is not geometric,

$\frac{a_2}{a_1} = \frac{\frac{1}{2}}{1} = \frac{1}{2}$ but $\frac{a_3}{a_2} = \frac{\frac{1}{3}}{\frac{1}{2}} = \frac{2}{3}$

3.

n	1	2	3	4	5	6	7	8	9	10
$\sum\limits_{n=1}^{n} \frac{1}{n}$	1	1.5	1.83	2.08	2.28	2.45	2.59	2.72	2.83	2.93

4. Answers will vary. Students may point out that differences between and ratios of successive terms decrease as n increases. Also, that the total does not exceed 4 grams in the first ten years but that it is impossible to decide from the values whether the total will exceed 4 grams at some future time.

5.

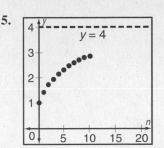

Answers will vary but should include the fact that the data continue to be inclusive.

6. Beginning after the second term, the number of terms in each successive group is doubled.

7a. $\frac{1}{4} + \frac{1}{4} = \frac{1}{2}; \frac{1}{3} > \frac{1}{4}$; so $\frac{1}{3} + \frac{1}{4} > \frac{1}{2}$
(addition property)

b. $\frac{1}{8} + \frac{1}{8} + \frac{1}{8} + \frac{1}{8} = \frac{1}{2}; \frac{1}{5}, \frac{1}{6}$, and $\frac{1}{7}$ each exceed $\frac{1}{8}$; so $\frac{1}{5} + \frac{1}{6} + \frac{1}{7} + \frac{1}{8} > \frac{1}{2}$ (addition property)

8. Each group is greater than or equal to $\frac{1}{2}$. Since the grouping process can be continued, the sum is greater than the sum of an infinite number of $\frac{1}{2}$s.

9. It is a geometric series with $r = 1$.

10. For above $1 + \frac{1}{2} + \frac{1}{3} + \ldots > \frac{1}{2} + \frac{1}{2} + \frac{1}{2} + \ldots$ since Question 9 diverges, Question 1 diverges and will eventually exceed any number, in particular 4.

11. The partial sums exceed 4 beginning at about $n = 30$. Therefore, Synergy will not meet council's intent in 30 years.

INVESTIGATING FURTHER

12. $\frac{n + 1}{2^{n+1}}$ **13.** $\frac{\frac{n + 1}{2^{n+1}}}{\frac{n}{2^n}} = \frac{n + 1}{2n}$ **14.** $\frac{1 + \frac{1}{n}}{2}$

15. As n becomes very large, the fraction $\frac{1}{n}$ is negligible (approaches a value of 0) and the expression $\frac{1 + \frac{1}{n}}{2}$ is approximately equal to $\frac{1}{2}$.

16. $\frac{\frac{n + 1 + 1}{n + 1}}{\frac{n + 1}{n}} = \frac{n^2 + 2n}{n^2 + 2n + 1}$; divide each term by n,

$\frac{n + 2}{n + 2 + \frac{1}{n}}$; diverges

17. $\frac{\frac{n + 1}{10^{n+1}}}{\frac{n}{10^n}} = \frac{n + 1}{10n}$; divide each term by n, $\frac{1 + \frac{1}{n}}{10}$;

converges

18. $\dfrac{\frac{1}{2(n+1)}}{\frac{1}{2n}} = \dfrac{n}{n+1}$; divide each term by n, $\dfrac{1}{1+\frac{1}{n}}$;

converges

19. $\dfrac{\frac{2^{n+1}}{(n+1)^2}}{\frac{2^n}{n^2}} = \dfrac{2n^2}{n^2+2n+1}$; divide each term by n,

$\dfrac{2n}{n+2+\frac{1}{n}}$; diverges

20. $\dfrac{\frac{1}{(n+1)!}}{\frac{1}{n!}} = \dfrac{1}{n+1}$; converges.

21a. An infinite arithmetic series increase or decrease without limit, and it diverges.

21b. An infinite geometric series converges if $|r| < 1$. Otherwise, it diverges.

21c. Answers will vary. Students should review methods from lesson: make a table, construct a graph to see if a limit is approached, group terms, and check to see if ratio of consecutive terms is less than one.

APPLY THE STRATEGY

22. $\dfrac{1}{3} + \dfrac{2}{9} + \dfrac{3}{27} + \ldots + \dfrac{n}{3^n} + \dfrac{n+1}{3^{n+1}}$

Using the ratio test: $\dfrac{\frac{n+1}{3^{n+1}}}{\frac{n}{3^n}} = \dfrac{n+1}{3n} = \dfrac{1+\frac{1}{n}}{3}$

For very large n, the ratio equals $\dfrac{1}{3}$.

Since $\dfrac{1}{3} < 1$, the series converges.

Students may explain with a table or a graph.

23. An infinite geometric series converges when $|r| < 1$. A reciprocal series would have a ratio $\left|\dfrac{1}{r}\right| \geq 1$, therefore this series would diverge.

24. No, because the series diverges. The students' explanations may vary.

25a.

n	1	2	3	4	5	6	7	8
S_n	1	2	2.5	2.667	2.708	2.717	2.718	2.718

25b.

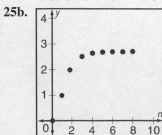

25c. ≈ 2.718 **25d.** e

1. Two possible arrangements are:

1, 8, 15	$d = 7$	1, 8, 15	$d = 7$
4, 9, 14	$d = 5$	2, 7, 12	$d = 5$
2, 6, 10	$d = 4$	6, 10, 14	$d = 4$
3, 5, 7	$d = 2$	9, 11, 13	$d = 2$
11, 12, 13	$d = 1$	3, 4, 5	$d = 1$

2. Olivia; students should show that the only supposition that does not lead to a contradiction with the other information is "Ned left before Olivia, then Olivia went." For example, if Marsha left before Ned, then Ned went, but if Marsha left before Ned, then Marsha left before Paul, so Marsha went (contradiction). If Olivia left before Paul, then Paul went, but if Olivia left before Paul, then Marsha left before Paul, so Marsha went (contradiction). If Marsha left before Paul, then Marsha went, but if Marsha left before Paul, then Marsha may have left before or after Ned; If Marsha left before Ned, then Ned went (contradiction); If Marsha left before Paul, but after Ned and Ned left before Olivia, there is no contradiction and Olivia went.

3.

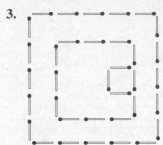

Chapter Review, pages 662–663

1. b   **2.** e   **3.** c   **4.** a   **5.** d

6. explicit   **7.** recursive

8. $a_n = n^2 - 1$; 0, 3, 8, 15, 24

9. $a_1 = 3$, $a_n = a_{n-1} - 1$; 3, 2, 1, 0, −1

10. $a_n = 2^{n-1}$ **11.** $a_n = 1 + 2n$ or $-(2n - 1)$

12. $a_1 = 4$; $a_n = a_{n-1} + 4$

13. $a_1 = 1$; $a_n = a_{n-1} + a_{n-2}$

14. $a_n = a_1 + (n-1)d$; $a_1 = 3$, $d = 4$
$a_{15} = 3 + (15 - 1)4 = 59$

15. $a_1 = 70$, $d = -3$
$a_{20} = 70 + (20 - 1)(-3) = 13$

16. $a_1 = 5$, $a_{18} = 56$
$56 = 5 + (18 - 1)d$; $d = 3$

17. $a_n = a_1 r^{n-1}$; $a_1 = -64$, $r = -\dfrac{1}{2}$
$a_6 = (-64)\left(-\dfrac{1}{2}\right)^{6-1} = 2$

18. $a_1 = \frac{1}{27}, r = 3$

$a_8 = \frac{1}{27} \cdot (3)^{8-1} = 81$

19. $a_1 = 1, r = 2, a_n = 1024$
$a_n = a_1 r^{n-1}$
$1024 = 1 \cdot 2^{n-1}$
$2^{10} = 2^{n-1}$
$10 = n - 1$
$11 = n$

20. Arithmetic; $a_1 = 1, d = 5$
$S_{30} = \frac{30}{2}[2(1) + (30 - 1)5] = 2205$

21. Arithmetic; $a_1 = 102, d = -2$
$S_{17} = \frac{17}{2}[2 \cdot 102 + (17 - 1)(-2)] = 1462$

22. Arithmetic; $a_1 = 1, d = 1$
$S_{200} - \frac{200}{2}[2 \cdot 1 + (200 - 1)1] = 20,100$

23. Geometric; $a_1 = 32, r = -\frac{1}{2}$

$S_{10} = \frac{32\left(1 - \left(-\frac{1}{2}\right)^{10}\right)}{\left(1 - \left(-\frac{1}{2}\right)\right)} = 21.3125$

24. Geometric; $a_1 - 1, r - 3$

$S_7 = \frac{1(1 - 3^7)}{(1 - 3)} = 1093$

25. Geometric; $a_1 = 1280, r = \frac{1}{4}$

$S_6 = \frac{1280\left(1 - \left(\frac{1}{4}\right)^6\right)}{\left(1 - \left(\frac{1}{4}\right)\right)} = 1706.25$

26. $\sum_{i=1}^{7}(3i + 1)$ **27.** $\sum_{i=1}^{6}(2i - 1)$

28. converges, since $r = \frac{1}{2}$ **29.** diverges, since $r = 2$

30. converges, since $r = -\frac{2}{3}$

31. $a_1 = 64, r = \frac{1}{4}$ **32.** $a_1 = 12, r = -\frac{1}{2}$

$S = \frac{64}{1 - \frac{1}{4}} = 85\frac{1}{3}$ $S = \frac{12}{1 - \left(-\frac{1}{2}\right)} = 8$

33. $a_1 = 1.5, r = 0.3$

$S = \frac{1.5}{1 - 0.3} \approx 2.14$

34. $0.454545 \ldots = \frac{45}{99} = \frac{5}{11}$

35. $(a + b)^4 = a^4 + 4a^3b^1 + 6a^2b^2 + 4a^1b^3 + b^4$

36. $\frac{7!}{5!2!}m^5n^2 = 21m^5n^2$ **37.** $\frac{8!}{3!5!}a^3(-b)^5 = -56a^3b^5$

38. $\frac{5!}{1!4!}(2x^2)^1(-y^3)^4 = 10x^2y^{12}$

Chapter Assessment, pages 664–665

CHAPTER TEST

1. explicit   **2.** explicit   **3.** 19, 18, 17, 16, 15

4. 2, 4, 0, 16, 144   **5.** $a_n = 5n - 2$

6. $a_1 = 1, a_n = 2a_{n-1} + 1$

7. $a_1 = 5, d = 4$
$a_{30} = 5 + (30 - 1)4 = 121$

8. $a_1 = -98, d = 7$
$a_{21} = -98 + (21 - 1)7 = 42$

9. $a_1 = 243, r = -\frac{1}{3}$

$a_7 = 243 \cdot \left(-\frac{1}{3}\right)^{7-1} = \frac{1}{3}$

10. $a_1 = 52{,}000, r = 0.1$
$a_7 = 52{,}000 \cdot (0.1)^6 = 0.052$

11. 12, _, _, _, 22; $a_5 = a_1 + (n - 1)d$
$22 = 12 + (5 - 1)d$;
$10 = 4d; d = 2.5$
12, 14.5, 17, 19.5, 22

12. 4, _, _, 500; $a_4 = a_1 r^{4-1}$
$500 = 4r^3$
$125 = r^3; r = 5$
4, 20, 100, 500

13. $a_1 = 2, d = 6$

$S_{40} = \frac{40}{2}[2 \cdot 2 + (40 - 1)6] = 4760$

14. $a_1 = 209, d = -7$

$S_{25} = \frac{25}{2}[2 \cdot 209 + (25 - 1)(-7)] = 3125$

15. $a_1 = 1200, r = 0.1$

$S_8 = \frac{1200(1 - 0.1^8)}{(1 - 0.1)} = 1333.333$

16. $a_1 = 1, r = 2$

$S_8 = \frac{1(1 - 2^8)}{(1 - 2)} = 255$

17. $\sum_{i=1}^{5} 3i + 9$

18a. 1, 3, 5, 7, . . . 41
$a_n = a_1 + (n - 1)d$
$41 = 1 + (n - 1)2$
$20 = n - 1; n = 21$ rows

18b. $S_{21} = \frac{21}{2}(2 \cdot 1 + (21 - 1)2) = 441$ cartons

19. Answers will vary. Students should state that an arithmetic sequence is a sequence in which each term differs from the previous term by a common difference. An arithmetic series is the sum of the terms of an arithmetic sequence. A geometric sequence, on the other hand, is a sequence in which each term and previous term form a common ratio. A geometric series is the sum of the terms of a geometric sequence.

20. $r = -\frac{1}{2}$ and $\left|-\frac{1}{2}\right| < 1$; converges

21. $r = 10$ and 10 is not less than one; diverges

22. $a_1 = 1, r = \left(-\frac{1}{3}\right)$ **23.** $a_1 = 1.6, r = 0.2$

$S = \dfrac{1}{1 - \left(-\frac{1}{3}\right)} = \dfrac{3}{4}$ $S = \dfrac{1.6}{(1 - 0.2)} = 2$

24. $0.2222\ldots = \dfrac{2}{9}$

25. $100 + 2\displaystyle\sum_{i=1}^{\infty} 0.56^i (100) =$

$100 + 2\left(\dfrac{56}{(1 - 0.56)}\right) \approx 354.55n.$

26. $(m - n)^5 =$

$m^5 - 5m^4n + 10m^3n^2 - 10m^2n^3 + 5mn^4 - n^5$

27. A; $\dfrac{6!}{4!2!}(3x)^4(-y)^2 = 1215x^4y^2$

Cumulative Review, page 666

1. $1, 3, 9, \ldots; a_1 = 1, r = 3$
$a_{10} = a_1r^{10-1}; a_{10} = 1 \cdot (3)^9 = 3^9$ or 19,683

2. $a_n = a_1r^{n-1}; n = 2001 - 1995 = 6,$
$a_1 = 19,995, r = (1 - 0.13)$
$a_6 = 19,995(1 - 0.13)^{6-1} \approx \$9966.$

3. $\dfrac{i}{\left(\frac{2i^2}{3i^3}\right)} = \left(\dfrac{i}{1}\right)\left(\dfrac{3i^3}{2i^2}\right) = \dfrac{3}{-2} = 3x; x = -\dfrac{1}{2}$

4. $\displaystyle\sum_{n=1}^{\infty}\left(\dfrac{1}{4}\right)^n; a_1 = \dfrac{1}{4}; r = \dfrac{1}{4}$

$S = \dfrac{\frac{1}{4}}{\left(1 - \frac{1}{4}\right)} = \dfrac{1}{3}$

5.
$$x^2 + 4x - 5 + \dfrac{-10}{2x - 1};$$
$2x - 1\overline{)2x^3 + 7x^2 - 14x - 5}$
$\underline{-2x^3 + x^2}$
$8x^2 - 14x$
$\underline{-8x^2 + 4x}$
$-10x - 5$
$\underline{10x - 5}$
-10
$(2x - 1)$ is not a factor

6. $y = \dfrac{x - 1}{x^2 - x - 6} = \dfrac{(x - 1)}{(x - 3)(x + 2)}; x \neq -2, 3$

7. $a_1 = 1, a_n = (a_{n-1})^2 + 3$ for $n \geq 2$

8. $A = l \cdot w; 1 = (\log_2 8) \cdot w = 3w; w = \dfrac{1}{3}$ or

$\dfrac{1}{3} = \log_8 2;$ C

---

9. For an ellipse, $a^2 = b^2 + c^2$.
Since a focus is at $(4, 0)$, $c = 4$ and the major axis is horizontal, on the x-axis. Since the minor axis equals 8, $b = 4$.
$a^2 = 4^2 + 4^2$
$a^2 = 32$

For this ellipse, $\dfrac{x^2}{a^2} + \dfrac{y^2}{b^2} = 1$

$\dfrac{x^2}{32} + \dfrac{y^2}{16} = 1$

$x^2 + 2y^2 = 32$

10. $a_{80} = -56; d = -2; a_n = a_1 + (n - 1)d$
$-56 = a_1 + (80 - 1)(-2); a_1 = 102$

11. $1100(1.06)^4 + 2000(1.06)^3 + 2200(1.06)^2 +$
$2700(1.06)^1 + 3000$
$= \$12,104.67$

12. $1800\text{ lb} \times 15\text{ ft} \times \dfrac{1}{3\text{ in.}} \times \dfrac{1}{(6\text{ in.})^2} = x\text{ lb} \times 10\text{ ft}$

$\times \dfrac{1}{4\text{ in.}} \times \dfrac{1}{(2\text{ in.})^2}$

$x = 400\text{ lb.}$

13. $\overline{OM} = \overline{MB} = \overline{MA} = \sqrt{a^2 + b^2};$
The median to the hypotenuse of a right triangle is equal to half the length of the hypotenuse.

Standardized Test, page 667

1. C; $\log_2 x = 0.25 \log_2 y$ **2.** C, $a_1 = 7; r = 2$
$x = \sqrt[4]{y}; x^4 = y$

3. D; $f(x) = x + 1; f(f(f(x))) = (((x + 1) + 1) + 1) = x + 3$

4. B; $x =$ one side of triangle
$x + 6 =$ second side of triangle

Area of Triangle $= 10\text{ ft}^2 = \dfrac{1}{2}x \cdot (x + 6); x \approx 2.4$
Hypotenuse $= \sqrt{2.4^2 + (2.4 + 6)^2} \approx 8.7\text{ ft}$

5. C; $\dfrac{1}{x} + \dfrac{1}{y} = 1; \dfrac{y + x}{xy} = 1; x + y = xy; \dfrac{1}{x + y} = \dfrac{1}{xy}$

6. C   **7.** A, $\dfrac{5!}{3!2!}a^3(4b)^2 = 160a^3b^2$

8. D; $\left(1 + \dfrac{1}{n}\right) \div \left(\dfrac{n + 1}{n^2}\right); \dfrac{(n + 1)}{n} \cdot \dfrac{n^2}{(n + 1)} = n$

9. D   **10.** B

11. B; $a_1 = 10; a_2 = \dfrac{10}{2} = 5; a_3 = 3(5) + 1 = 16;$

$a_4 = \dfrac{16}{2} = 8; a_5 = \dfrac{8}{2} = 4; a_6 = \dfrac{4}{2} = 2;$

$a_7 = \dfrac{2}{2} = 1; a_8 = 3(1) + 1 = 4; a_9 = \dfrac{4}{2} = 2;$

$a_{10} = \dfrac{2}{2} = 1$

12. B   **13.** A   **14.** C

Data Activity, page 669

1. x number of fires in 1993

   $x(1 - 0.052) = 2,054,500$; about $2,167,194$ fires

2. $\dfrac{3425}{4275} \times 100 \approx 80\%$ civilian fire deaths occurred in homes.

3. 1 year $= 525,600$ minutes

   $\dfrac{4275 \text{ deaths}}{\text{year}} = \dfrac{4275 \text{ deaths}}{525,600 \text{ minutes}} \approx \dfrac{1 \text{ death}}{123 \text{ min}}$

   $\dfrac{27,250 \text{ injury}}{\text{year}} = \dfrac{27,250}{525,600 \text{ min}} \approx \dfrac{1 \text{ injury}}{19 \text{ minutes}}$

4. Property Damage $\$8.151$ billion
   Building -6.867 billion
   Vehicle -0.156 billion
   Other Categories $\$1.128$ billion

5. $\$6.867 \times 0.211 \approx \$1.449$ billion lost

6. Answers will vary. Students should include information found in table of "Fires in the United States Selected Data, 1994."

7. Answers will vary. Insurance companies may be another source of information.

Lesson 14.1, pages 671–678

EXPLORE

1.

| 1st place | 2nd place | 3rd place | finishing orders |

2.
PABC	PBAC	PCAB	ABPC	BAPC	CAPB
PACB	PBCA	PCBA	ACPB	BCPA	CBPA
APBC	BPAC	CPAB	ABCP	BACP	CABP
APCB	BPCA	CPBA	ACBP	BCAP	CBAP

3. There are 6 ways to order 3 runners. There are 24 ways to order 4 runners.

4. Result for four $= 4 \times$ result for three

5. Result for five $= 5 \times$ result for four $= 5(24) = 120$

TRY THESE

1. $4 \cdot 3 \cdot 2 = 24$ 2. $\dfrac{6!}{1!} = 720; \dfrac{6 \cdot 5 \cdot 4 \cdot 3 \cdot 2}{1} = 720$

3. $\dfrac{9!}{3!} = 60,480; \dfrac{9 \cdot 8 \cdot 7 \cdot 6 \cdot 5 \cdot 4 \cdot 3!}{3!} = 60,480$

4. $\dfrac{12!}{2!4!} = 9,979,200$

   $\dfrac{12 \cdot 11 \cdot 10 \cdot 9 \cdot 8 \cdot 7 \cdot 6 \cdot 5 \cdot 4!}{4! \cdot 2} = 9,979,200$

5. $\dfrac{n!}{(n-1)!} = \dfrac{n \cdot (n-1)!}{(n-1)!} = n$

6. b. You need a sequence of n factors, each being one less than the preceding factor. Choice a has only two factors. Choice c is equivalent to $\dfrac{1}{n+1}$.

7. $_6P_6 = 6! = 720$;

   $\dfrac{6!}{(6-6)!} = \dfrac{6!}{0!} = \dfrac{6 \cdot 5 \cdot 4 \cdot 3 \cdot 2}{1} = 720$

8. $_6P_1 = \dfrac{6!}{(6-1)!} = \dfrac{6!}{5!} = 6$ 9. $_{10}P_3 = \dfrac{10!}{7!} = 720$

10. $_nP_{n-1} = \dfrac{n!}{(n-(n-1))!} = \dfrac{n!}{(n-n+1)!} = n!$

11. $8! = 40,320$ possible positions

12. $7! = 5040$ possible schedules

13. $_4P_2 = \dfrac{4!}{(4-2)!} = \dfrac{4!}{2!} = 12$ possible numbers

14. $_8P_3 = \dfrac{8!}{(8-3)!} = \dfrac{8!}{5!} = 336$ possible 3-letter arrangements

15a. $5! = 120$ 5-digit zip codes without reused digits

15b. $5^5 = 3125$ 5-digit zip codes with reused digits

16. 8 letters, 2 'E's, 2 'N's, 2 'T's

   $\dfrac{8!}{2!2!2!} = 5040$ different arrangements

17. 12 letters, 2 'O's, 2 'N's, 3 'T's, 2 'I's

   $\dfrac{12!}{2!2!3!2!} = 9,979,200$ different arrangements

18. 8 Flags; $P = (n-1)!$ for circular permutations

   $(8-1)! = 5040$ possible arrangements

PRACTICE

1. $4 \cdot 3 = 12$ possible ways

2. $3 \cdot 6 \cdot 4 = 72$ possible lunch choices

3. $5 \cdot 5 = 25$ possible ways

4. $12 \cdot 11 = 132$ possible ways

5. $5 \cdot 4 \cdot 3 = 60$ possible ways

6. $8 \cdot 7 \cdot 6 \cdot 5 = 1680$ possible ways

7. $\dfrac{9!}{3!2!} = 30,240; \dfrac{9 \cdot 8 \cdot 7 \cdot 6 \cdot 5 \cdot 4 \cdot 3!}{3! \cdot 2} = 30,240$

8. $\dfrac{11!}{2!5!} = 166,320$ 9. $\dfrac{(8-2)!}{6!} = \dfrac{6!}{6!} = 1$

10. $\dfrac{(r+2)!}{r!} = \dfrac{(r+2)(r+1)(r)!}{r!} = (r+2)(r+1)$

11. No, $(6-2)! = 4! = 24$ while $6! - 2! = 720 - 2 = 718$.

12. $_8P_8 = 8! = 40,320$ 13. $_8P_1 = \dfrac{8!}{(8-1)!} = 8$

14. $_{12}P_4 = \dfrac{12!}{(12-4)!} = \dfrac{12!}{8!} = 11,880$

15. $_{(n+1)}P_{(n-1)} = \dfrac{(n+1)!}{(n+1-(n-1))!} = \dfrac{(n+1)!}{2!}$

or $\dfrac{(n+1)!}{2}$

16. $5! = 120$ **17.** $(5-1)! = 24$

18. $_6P_4 = \dfrac{6!}{(6-4)!} = 360$ **19.** $_6P_4 = \dfrac{6!}{(6-4)!} = 360$

20. $3! = 6$ **21.** $6! = 720$ **22.** $\dfrac{6!}{2!3!} = 60$

23. $\dfrac{10!}{3!2!4!} = 12,600$ **24.** $_4P_4 = 4! = 24$

25. $_3P_2 = \dfrac{3!}{(3-2)!} = 6$ **26.** $_3P_3 = 3! = 6$

27. $_5P_3 = \dfrac{5!}{(5-3)!} = 60$

28. $6!4! = 17,280$ possible arrangements

29. $\dfrac{14!}{5!3!6!} = 168,168$ possible arrangements

30. $(11-1)! = 3,628,800$ possible arrangements

EXTEND

31. $1 \cdot {}_5P_5 = 1 \cdot 5! = 120$ possible arrangements

32. $_3P_3 \cdot {}_3P_3 = 3! \cdot 3! = 36$ possible arrangements

33. $1 \cdot 1 \cdot {}_5P_5 = 1 \cdot 1 \cdot 5! = 120$ possible arrangements

34. $1 \cdot 1 \cdot {}_3P_1 \cdot {}_4P_4 = 1 \cdot 1 \cdot 3 \cdot 4! = 72$
possible arrangements

35. $1 \cdot 1 \cdot {}_2P_1 \cdot {}_4P_4 = 1 \cdot 1 \cdot 2 \cdot 4! = 48$
possible arrangements

36. $4 \cdot {}_5P_2 = 4 \cdot 5 \cdot 4 = 80$ possible arrangements

37. $2 \cdot {}_5P_2 = 2 \cdot 5 \cdot 4 = 40$ possible arrangements

38. $2 \cdot {}_5P_2 = 2 \cdot 5 \cdot 4 = 40$ possible arrangements

39. $_3P_1 \cdot {}_3P_1 \cdot {}_2P_1 = 3 \cdot 3 \cdot 2 = 18$ possible numbers

40. $3 \cdot {}_6P_6 = 3 \cdot 6! = 2160$ possible arrangements

41. $1 \cdot {}_6P_6 = 1 \cdot 6! = 720$ possible arrangements

42. $4 \cdot {}_6P_6 = 4 \cdot 6! = 2880$ possible arrangements

43. $_4P_2 \cdot {}_5P_5 = 4 \cdot 3 \cdot 5! = 1440$ possible arrangements

44. $_3P_3 \cdot {}_4P_4 = 3!4! = 144$ possible arrangements

45. $2 \cdot {}_3P_3 \cdot {}_3P_3 = 2 \cdot 3! \cdot 3! = 72$ possible firing orders

46. $(n-1)!n! = 3! \cdot 4! = 144$ possible arrangements

THINK CRITICALLY

47. $x = 5;$ $_xP_2 = 20$

$\dfrac{x!}{(x-2)!} = 20$

$\dfrac{x(x-1)(x-2)!}{(x-2)!} = 20$

$x(x-1) = 20$

$x^2 - x - 20 = 0$

$(x-5)(x+4) = 0$

$x = 5$ or -4; reject -4

48. $n = k; {}_nP_r = k({}_{n-1}P_{r-1})$

$\dfrac{n!}{(n-r)!} = k\left(\dfrac{(n-1)!}{((n-1)-(r-1))!}\right)$

$\dfrac{n!}{(n-r)!} = k\left(\dfrac{(n-1)!}{(n-r)!}\right)$

$n! = k(n-1)!$

$n(n-1)! = k(n-1)!$

$n = k$

49. No; leads to a quadratic equation whose roots are irrational. x must be a positive integer.

50. $n = 8; {}_nP_3 = \dfrac{2}{5}({}_{n-1}P_4)$

$\dfrac{n!}{(n-3)!} = \dfrac{2}{5}\left(\dfrac{(n-1)!}{(n-1-4)!}\right)$

$\dfrac{n(n-1)(n-2)(n-3)!}{(n-3)!} =$

$\dfrac{2}{5} \cdot \dfrac{(n-1)(n-2)(n-3)(n-4)(n-5)!}{(n-5)!}$

$n(n-1)(n-2) = \dfrac{2}{5}(n-1)(n-2)(n-3)(n-4)$

$5n = 2(n^2 - 7n + 12)$

$2n^2 - 19n + 24 = 0$

$(n-8)(2n-3) = 0$

$n = 8$ or $\dfrac{3}{2}$; reject $\dfrac{3}{2}$

51. $_nP_r - {}_nP_{r-1} = (n-r) \cdot {}_nP_{r-1}$

$\dfrac{n!}{(n-r)!} - \dfrac{n!}{(n-(r-1))!} = (n-r)\dfrac{n!}{(n-r+1)!}$

$\dfrac{(n-r+1)}{(n-r+1)} \cdot \dfrac{n!}{(n-r)!} - \dfrac{n!}{(n-r+1)!} = \dfrac{n!(n-r)}{(n-r+1)!}$

$\dfrac{n!(n-r+1) - n!}{(n-r+1)!} =$

$\dfrac{n![(n-r+1) - 1]}{(n-r+1)!} =$

$\dfrac{n!(n-r)}{(n-r+1)!}$

MIXED REVIEW

52. $\dfrac{3}{6}$ or $\dfrac{1}{2}$ or 0.5 or 50%   **53.** $\dfrac{2}{6}$ or $\dfrac{1}{3}$ or about 33%

54. C; $|2x-1| < 7$

$-7 < 2x - 1 < 7$

$-6 < 2x < 8$

$-3 < x < 4$

ALGEBRAWORKS

1. $9^5 = 59,049$ possible keys.

2. $9^6 - 9^3 = 530,712$ more pin permutations possible

3. Then the key would have low spots of all the same height, therefore a straight piece of metal wire could possibly get the pins to line up and the lock could easily be picked.

4. The second arrangement places the low spots in a straight-line diagonal position, and this makes picking the lock easier.

EXPLORE/WORKING TOGETHER

1. $_4P_3 = \dfrac{4!}{(4-3)!} = 24$

   24 ways; Assume the group members are A, B, C, and D. Let P = President. Let V = Vice President. Let T = Treasurer.
   Then there are 24 different arrangements for PVT.

PVT	PVT	PVT	PVT
■ ABC	BAC	CAB	DAB
■ ABD	BAD	CAD	DAC
ACB	BCA	CBA	DBA
■ ACD	■ BCD	CBD	DBC
ADB	BDA	CDA	DCA
ADC	BDC	CDB	DCB

2. 4 ways; There are 4 different ways to select a committee of 3 from a group of 4. The 4 possible ways are marked (■) in the list for Explore 1.

3a. No. In the first list, the order of the members is important; for the second list, order or position was not important. To get the second list from the first list, eliminate repetitions. For example, $ABC = ACB = BAC = BCA = CAB = CBA$.

3b. The student should state that the possible number of outcomes in the first list is $_nP_r$ or $\dfrac{4!}{(4-3)!} = 24$.
   While the second situation is the first list divided by $_rP_r$ or $\dfrac{_nP_r}{_rP_r}$ or $\dfrac{4!}{(4-3)!\,3!} - 4$.
   Therefore, the general form would be $\dfrac{_nP_r}{_rP_r}$ or $\dfrac{n!}{(npr)!(r)!}$.

4a. $_5P_3 = \dfrac{5!}{(5-3)!} = 60$

4b. $\dfrac{5!}{3!(5-3)!} - 10$

4c. Assume the 5th group member is E. There are 60 different arrangements for PVT.

PVT	PVT	PVT	PVT	PVT
■ ABC	BAC	CAB	DAB	EAB
■ ABD	BAD	CAD	DAC	EAC
■ ABE	BAE	CAE	DAE	EAD
ACB	BCA	CBA	DBA	EBA
■ ACD	■ BCD	CBD	DBC	EBC
■ ACE	■ BCE	CBE	DBE	EBD
ADB	BDA	CDA	DCA	ECA
ADC	BDC	CDB	DCB	ECB
■ ADE	■ BDE	■ CDE	DCE	ECD
AEB	BEA	CEA	DEA	EDA
AEC	BEC	CEB	DEB	EDB
AED	BED	CED	DEC	EDC

There are only 10 different ways to select a committee of 3 from the group of 5. See the items marked ■ in the preceding list.

TRY THESE

1. $_7C_3 = \dfrac{7!}{3!4!} = 35$

2. $\dbinom{6}{5} = \dfrac{6!}{5!1!} = 6$

3. $_{20}C_{20} = \dfrac{20!}{20!1!} = 1$

4. $\dbinom{20}{1} = \dfrac{20!}{1!19!} = 20$

5. $_{20}C_0 = \dfrac{20!}{0!\,20!} = 1$

6. Gus; In the formula $_nC_r = \dfrac{n!}{r!(n-r)!}$, replace r by n
   $_nC_n = \dfrac{n!}{n!(n-n)!} = \dfrac{n!}{n!0!} = \dfrac{n!}{n!\cdot 1} = 1$

7. $_7C_2 = \dfrac{7!}{2!5!} = 21$ lines

8. $_{17}C_4 = \dfrac{17!}{4!10!} = 2380$ possible displays

9. $_9C_4 = \dfrac{9!}{4!5!} = 126$ possible committees

10. $_8C_3 = \dfrac{8!}{3!5!} = 56$ possible committees with Marie-Jean Buyle as a member

11. $_8C_4 = \dfrac{8!}{4!4!} = 70$ possible committees without Sanford Crosley as a member.

12. $_{52}C_{13}$ 13. $_4C_2 \cdot _{48}C_{11}$ 14. $_{13}C_5 \cdot _{39}C_8$

15. $_{39}C_{13}$ 16. $_{50}C_{11}$

17. $_9C_3 + _9C_4 + _9C_5 + _9C_6 + _9C_7 + _9C_8 + _9C_9 = 466$

18. $_9C_1 + _9C_2 + _9C_3 + _9C_4 + _9C_5 = 381$

PRACTICE

1. $_8C_6 = \dfrac{8!}{6!2!} = 28$

2. $_{10}C_5 = \dfrac{10!}{5!5!} = 252$

3. $\dbinom{12}{4} = \dfrac{12!}{4!8!} = 495$

4. $_{100}C_{100} = 1$

5. $\dbinom{1000}{1} = 1000$

6. $_7C_2 = \dfrac{7!}{2!5!} = 21$ and $_7C_5 = \dfrac{7!}{5!2!} = 21$, making $_7C_2 = _7C_5$; When you select 2 elements from a set of 7, you are simultaneously creating two subsets, one with 2 elements and another with 5.

   In general: $_nC_r = _nC_{n-r}$
   $_nC_r = \dfrac{n!}{r!(n-r)!} \overset{?}{=} \dfrac{n!}{(n-r)!(n-(n-r))!} =$
   $\dfrac{n!}{r!(n-r)!} = _nC_{(n-r)}\ _nC_r = _nC_{(n-r)}$

7. $_{22}C_4 = \dfrac{22!}{4!8!} = 7315$ ways

8. $_5C_2 = \dfrac{5!}{2!3!} = 10$ lines possible

9. $_{20}C_6 = \dfrac{20!}{6!14!} = 38{,}760$ possible safety patrols

10. $_{12}C_5 = \dfrac{12!}{5!7!} = 792$ possible question combinations

11. $_{12}C_4 = \dfrac{12!}{4!8!} = 495$

12. $_{11}C_3 = \dfrac{11!}{3!8!} = 165$

13. $_{10}C_4 = \dfrac{10!}{4!6!} = 210$

14. $_9C_3 = \dfrac{9!}{3!6!} = 84$

15. $_{12}C_6 = \dfrac{12!}{6!6!} = 924$ **16.** $_{11}C_5 = \dfrac{11!}{5!6!} = 462$

17. $_{11}C_6 = \dfrac{11!}{6!5!} = 462$ **18.** $_{10}C_5 = \dfrac{10!}{5!5!} = 252$

19. $_5C_3 \cdot {}_7C_3 = \dfrac{5!}{3!2!} \cdot \dfrac{7!}{3!4!} = 350$

20. $_5C_3 \cdot {}_6C_2 = \dfrac{5!}{3!2!} \cdot \dfrac{6!}{2!4!} = 150$

21. $_4C_3 \cdot {}_6C_2 = \dfrac{4!}{3!1!} \cdot \dfrac{6!}{2!4!} = 60$ **22.** $_{39}C_{13} = \dfrac{39!}{13!26!}$

23. $_4C_2 \cdot {}_{48}C_{11} = \dfrac{4!}{2!2!} \cdot \dfrac{48}{11!32!}$ **24.** $_{26}C_{13} = \dfrac{26!}{13!13!}$

25. $_{13}C_4 \cdot {}_{13}C_3 \cdot {}_{26}C_6 = \dfrac{13!}{4!9!} \cdot \dfrac{13!}{3!10!} \cdot \dfrac{26!}{6!20!}$

26. $_{10}C_7 + {}_{10}C_8 + {}_{10}C_9 + {}_{10}C_{10} = 176$

27. $_{10}C_0 + {}_{10}C_1 + {}_{10}C_2 + {}_{10}C_3 + {}_{10}C_4 = 386$

29. $_{10}C_0 + {}_{10}C_1 + {}_{10}C_2 + {}_{10}C_3 + {}_{10}C_4 + {}_{10}C_5 = 638$

EXTEND

30. $(_4C_1(_{13}C_5)) \cdot (_3C_1(_{13}C_8)) = \dfrac{4!}{1!3!} \cdot \dfrac{13!}{5!8!} \cdot \dfrac{4!}{1!3!} \cdot \dfrac{13!}{8!5!}$
    $= 19,876,428$

31. $_4C_2(_{13}C_5 \cdot {}_{13}C_5) \cdot {}_2C_1(_{13}C_3) =$
$\left(\dfrac{4!}{2!2!} \cdot \dfrac{13!}{5!8!} \cdot \dfrac{13!}{5!8!} \right) \cdot \left(\dfrac{2!}{1!1!} \cdot \dfrac{13!}{3!10!} \right) = 5,684,658,408$

32. $_nC_2 = \dfrac{n!}{2!(n-2)!} = 45$

$\dfrac{n(n-1)(n-2)!}{2!(n-2)!} = 45$

$n(n-1) = 90$

$n^2 - n - 90 = 0$

$(n-10)(n+9) = 0$

$n = 10$ or -9; reject -9, since n must be a positive integer.

33. No; the solution leads to a quadratic equation whose solutions are irrational. n must be a positive integer.

34. $14; \ _nC_2 = \dfrac{n!}{2!(n-2)!} = 91$

$n(n-1) = 182$

$n^2 - n - 182 = 0$

$(n-14)(n+13) = 0$

$n = 14$ or -13; reject -13, n must be positive integer.

35. $30; \ _nC_2 = \dfrac{n!}{2!(n-2)!} = 435$

$n(n-1) = 870$

$n^2 - n - 870 = 0$

$(n-30)(n+29) = 0$

$n = 30$ or -29; reject -29, n must be positive integer.

---

36. $(a+b)^0$ $\dbinom{0}{0}$ 1

     $(a+b)^1$ $\dbinom{1}{0} \ \dbinom{1}{1}$ $1a + 1b$

     $(a+b)^2$ $\dbinom{2}{0} \ \dbinom{2}{1} \ \dbinom{2}{2}$ $1a^2 + 2ab + 1b^2$

     $(a+b)^3$ $\dbinom{3}{0} \ \dbinom{3}{1} \ \dbinom{3}{2} \ \dbinom{3}{3}$ $1a^3 + 3a^2b + 3ab^2 + 1b^3$

     $(a+b)^4$ $\dbinom{4}{0} \ \dbinom{4}{1} \ \dbinom{4}{2} \ \dbinom{4}{3} \ \dbinom{4}{4}$ $1a^4 + 4a^3b + 6a^2b^2 + 4ab^3 + 1b^4$

37. $(a+b)^8 = a^8 + 8a^7b + 28a^6b^2 + 56a^5b^3 + 70a^4b^4 + 56a^3b^5 + 28a^2b^6 + 8ab^7 + b^8$

38a. $\dbinom{8}{3} = 56 = \dbinom{8}{3}$

38b. These are symmetrically placed values in Pascal's Triangle and are therefore equal in value.

38c. $\dbinom{n}{r} = \dbinom{n}{n-r}$

39a. $\dbinom{8}{3} = 56; \ \dbinom{7}{3} = 35; \ \dbinom{7}{2} = 21$

39b. $\dbinom{8}{3} = \dbinom{7}{3} + \dbinom{7}{2}; \ 56 = 35 + 21$

39c. From their position in Pascal's Triangle, you add those two terms in Row 7 to get the terms in Row 8.

39d. $\dbinom{n}{r} = \dbinom{n-1}{r} + \dbinom{n-1}{r-1}$

THINK CRITICALLY

40. $1.19 in 3 quarters, 4 dimes, and 4 pennies

41. $_nC_{(n-1)} = \dfrac{n!}{(n-1)!(n-(n-1))!} = \dfrac{n(n-1)!}{(n-1)!2!} = n$

42. $_nC_{(n-2)} = \dfrac{n!}{(n-2)!(n-(n-2))!} =$
$\dfrac{n(n-1)(n-2)!}{(n-2)!2!} = \dfrac{n(n-1)}{2}$

43. $_{n+3}C_{n+2} \ \dfrac{(n+3)!}{(n+2)!((n+3)-(n+2))!} =$
$\dfrac{(n+3)(n+2)!}{(n+2)!1!} = n+3$

44. $\binom{5}{3} = \frac{5!}{3!2!}$

$\binom{5}{2} - \frac{5}{2!3!}$; so $\binom{5}{3} = \binom{5}{2}$.

In general,

$\binom{n}{r} = \frac{n!}{r!(n-r)!}$

$\binom{n}{n-r} = \frac{n!}{(n-r)!(n-(n-r))!} = \frac{n!}{(n-r)!r!}$

So, $\binom{n}{r} = \binom{n}{n-r}$.

45. $n = 4, r = 2;$

$_nC_r = \frac{n!}{r!(n-r)!} = 6$

$_nP_r = \frac{n!}{(n-r)!} = 12$

$\frac{r!}{r!}\frac{n!}{(n-r)!} = 12$

$r!\frac{n!}{r!(n-r)!} = 12$

$\qquad r! \cdot 6 = 12$ Substitute from $_nC_r$ equation.

$\qquad\quad r! = 2$

$\qquad\quad r = 2;$

$_nC_2 = \frac{n!}{2!(n-2)!} = \frac{n(n-1)(n-2)!}{2!(n-2)!} = 6;$

$\qquad \frac{n(n-1)}{2} = 6$

$\qquad n^2 - n = 12$

$\qquad n^2 - n - 12 = 0$

$\qquad (n-4)(n+3) = 0$

$\qquad\qquad n = 4 \text{ or } -3$

reject -3, since n must be positive integer

46. $_nC_r + {}_nC_{r-1} = {}_{n+1}C_r$

$\frac{n!}{r!(n-r)!} + \frac{n!}{(r-1)!(n-r+1)!} \overset{?}{=} \frac{(n+1)!}{r!(n+1-r)!}$

$n!\left(\frac{1}{r!(n-r)!} + \frac{1}{(r-1)(n-r+1)!}\right) \overset{?}{=}$

$n!\left(\frac{1}{r!(n-r)!} \cdot \frac{(n-r+1)}{(n-r+1)} + \right.$

$\left.\frac{1}{(r-1)!(n-r+1)!} \cdot \frac{r}{r}\right) \overset{?}{=}$

$n!\left(\frac{n-r+1}{r!(n-r+1)!} + \frac{r}{r!(n-r+1)!}\right) \overset{?}{=}$

$\frac{n!(n+1)}{r!(n+1-r)!} = {}_{n+1}C_r$

47. $_nC_r = {}_{n-1}C_r + {}_{n-1}C_{r-1}$

$\frac{n!}{r!(n-r)!} = \frac{(n-1)!}{r!(n-1-r)!} +$

$\qquad \frac{(n-1)!}{(r-1)!(n-1-r+1)!}$

$= (n-1)!\left(\frac{1}{r!(n-r-1)!} + \frac{1}{(r-1)!(n-r)!}\right)$

$= (n-1)!\left(\frac{1}{r!(n-r-1)!} \cdot \frac{(n-r)}{(n-r)} + \right.$

$\frac{(r)}{(r)} \cdot \frac{1}{(r-1)!(n-r)!}\right)$

$= (n-1)!\left(\frac{n-r}{r!(n-r)!} + \frac{r}{r!(n-r)!}\right)$

$= \frac{(n-1)!\,n}{r!(n-r)!}$

$_nC_r = \frac{n!}{r!(n-r)!}$

MIXED REVIEW

48. $\frac{3}{6} = \frac{1}{2}$ or 50% or 0.5 **49.** $\frac{26}{52} = \frac{1}{2}$ or 50% or 0.5

50. A; A. 2, 6, 6, 10; mean $\frac{24}{4} = 6$

$\qquad\qquad$ median $\frac{6+6}{2} = 6$

$\qquad\qquad$ mode 6

$\qquad$ B. 2, 2, 4, 10, 12; mean $\frac{30}{5} = 6$

$\qquad\qquad$ median 4

$\qquad\qquad$ mode 2

$\qquad$ C. 2, 2, 2, 4, 10; mean $\frac{20}{5} = 4$

$\qquad\qquad$ median 2

$\qquad\qquad$ mode 2

$\qquad$ D. 2, 2, 6, 10, 20; mean $\frac{40}{5} = 8$

$\qquad\qquad$ median 6

$\qquad\qquad$ mode 2

51. D $\qquad\qquad\qquad$ **52.** yes

53. No. Suppose Bob sits between Alice and Chris. Then Bob sits next to Alice and Bob sits next to Chris, but Alice does not sit next to Chris.

54. yes

55. $-2\begin{bmatrix} -4 & 7 \\ 3 & 5 \\ 0 & 1 \end{bmatrix} = \begin{bmatrix} -2(-4) & -2(7) \\ -2(3) & -2(5) \\ -2(0) & -2(1) \end{bmatrix} = \begin{bmatrix} 8 & -14 \\ -6 & -10 \\ 0 & -2 \end{bmatrix}$

56. $3\begin{bmatrix} 4 & -2 \\ -2 & 1 \end{bmatrix} + \begin{bmatrix} 7 & 9 \\ 3 & -5 \end{bmatrix} = \begin{bmatrix} 12 & -6 \\ -6 & 3 \end{bmatrix} +$

$\begin{bmatrix} 7 & 9 \\ 3 & -5 \end{bmatrix} = \begin{bmatrix} 19 & 3 \\ -3 & -2 \end{bmatrix}$

Lesson 14.3, pages 686–694

EXPLORE

1a. A1 A2 A3 A4 A5 B1 B2 B3 B4 B5 C1 C2
C3 C4 C5 D1 D2 D3 D4 D5 E1 E2 E3 E4 E5

1b. A1 A2 A3 A4 A5 E1 E2 E3 E4 E5

1c. 25; 10 $\qquad\qquad$ **1d.** $\frac{10}{25}$ or $\frac{2}{5}$ or 0.4 or 40%

2. B1 B2 B3 B4 B5 C1 C2 C3 C4

C5 D1 D2 D3 D4 D5; $\frac{15}{25}$ or $\frac{3}{5}$

3a. Parts of the total possible. Complementary events.

3b. The sum is 1.

4. A1 A2 A3 A4 A5

5a. $P = \frac{1}{5}$; Out of 5 possible only one is even and less than 4.

5b. A, even: A2 A4 or A, less than 4: A1, A2, A3
So A, even or less than 4: A1, A2, A3, A4
$P = \frac{4}{5}$

TRY THESE

1a. Yes; no other outcomes are possible and mutually exclusive.

1b. No; there are other outcomes possible and mutually exclusive.

1c. Yes; no other outcomes are possible and mutually exclusive.

2. $P(\text{not red}) = \frac{9}{12} = \frac{3}{4}; \frac{5}{12} + \frac{4}{12} = \frac{9}{12}$

3. $P(\text{not black}) = \frac{12}{12} = 1$

4. $P(\text{red or white}) = \frac{7}{12}; \frac{3}{12} + \frac{4}{12} = \frac{7}{12}$

5. Odds (white) 4:8 or 1:2   **6.** Odds (not blue) 7:5

7. $P(\text{even and greater than 5}) = \frac{2}{8} = \frac{1}{4}$

8. $P(\text{odd and less than 5}) = \frac{2}{8} = \frac{1}{4}$

9. $P(\text{less than 2 and even}) = \frac{0}{8} = 0$

10. $P(\text{less than 5 or even}) = \frac{6}{8} = \frac{3}{4}$

11. $P(\text{odd or is 4}) = \frac{5}{8}$ **12.** $P(\text{less than 3 or is 7}) = \frac{3}{8}$

13. $P(\text{New York Democrat}) = \frac{{}_1C_1 \cdot {}_{10}C_4}{{}_{10}C_5} = 50\%$
Choose the New York Democrat, plus 4 other Democrats. Divide by the total number of ways to choose 5 Democrats, Republican selections do not matter here.

14. $P(\text{Kim}) = \frac{1}{2}$; $P(\text{Kim or Jamal}) = \frac{3}{4}$; and P (Kim and Jamal) $= \frac{1}{3}$
$P(\text{Kim or Jamal}) = P(\text{Kim}) + P(\text{Jamal}) - P$ (Kim and Jamal)
$\frac{3}{4} = \frac{1}{2} + P(J) - \frac{1}{3}$

PRACTICE

1. Arnie; Complementary since no other outcomes are possible; mutually exclusive since they cannot occur together.

2. $P(\text{not a 7}) = \frac{48}{52} = \frac{12}{13}$

3. $P(\text{King or queen}) = \frac{8}{52} = \frac{2}{13}$

4. $P(\text{not a jack or an ace}) = \frac{44}{52} = \frac{11}{13}$

5. $P(\text{red 10}) = \frac{2}{52} = \frac{1}{26}$

6. $P(\text{Black or a 9}) = \frac{(26 + 2)}{52} = \frac{7}{13}$

7. 26 red cards to 26 black cards; 26:26 or 1:1

8. 4 threes and 4 queens to 44 other cards; 8:44 or 2:11

9. 2 black eights to 50 other cards; 2:50 or 1:25

10. not (six or jack or queen) means $52 - 12 = 40$
odds would be 40:12 or 10:3

11. not (red or seven) means $52 - (26 + 2) = 24$
odds would be 24:28 or 6:7

12. $P(\text{ends in Y and begins with J}) = \frac{2}{12}$ or $\frac{1}{6}$

13. $P(\text{ends in Y or R}) = \frac{8}{12}$ or $\frac{2}{3}$

14. $P(\text{contains R and at least 2 vowels}) = \frac{7}{12}$

15. $P(\text{begins with a vowel or ends with R}) = \frac{6}{12}$ or $\frac{1}{2}$

16. $P(\text{fewer than 10 letters or begins with ak}) = \frac{12}{12}$ or 1

17. $P(\text{begin with P}) = \frac{{}_4P_4}{{}_5P_5} = \frac{24}{120} = \frac{1}{5}$

18. $P(\text{begin and ends with vowels})$
$= \frac{2 \cdot {}_3P_3 \cdot 1}{{}_5P_5} = \frac{12}{120}$ or $\frac{1}{10}$

19. $P(\text{begins with P and ends with a vowel})$
$= \frac{1 \cdot {}_3P_3 \cdot 2}{{}_5P_5} = \frac{12}{120} = \frac{1}{10}$

20. $P(\text{begins with } P \text{ or ends with a vowel}) = P(\text{begins with P}) + P(\text{ends with vowel}) - P(\text{begins with P and ends with vowel})$
$\frac{1}{5} + \frac{4}{10} - \frac{1}{10} = \frac{5}{10}$ or $\frac{1}{2}$

21. $P(\text{all liberal arts}) = \frac{{}_7P_5}{{}_{16}C_5} = \frac{21}{4368} = \frac{1}{208}$

22. $P(\text{3 liberal arts and 2 science})$
$= \frac{{}_7C_3 \cdot {}_9C_2}{{}_{16}C_5} = \frac{35 \cdot 36}{4368} = \frac{15}{52}$

23. $P(\text{all science}) = \frac{{}_9C_5}{{}_{16}C_5} = \frac{126}{4368} = \frac{3}{104}$

24. $P(\text{4 science and 1 liberal arts})$
$= \frac{{}_9C_4 \cdot {}_7C_1}{{}_{16}C_5} = \frac{126 \cdot 7}{4368} = \frac{21}{104}$

25. $P(\text{not hamburger}) = \frac{26 \cdot 19}{26} = \frac{7}{26}$

26. $P(\text{not hamburger or pizza})$
$= \frac{26}{26} - \left(\frac{19}{26} + \frac{17}{26} - \frac{13}{26}\right) = \frac{3}{26}$

27. $P(\text{rain}) = \frac{12}{30}$ or $\frac{2}{5}$ **28.** $P(\text{high temp}) = \frac{15}{30}$ or $\frac{1}{2}$

29. $P(\text{rain or high temp}) = \frac{12}{30} + \frac{15}{30} - \frac{8}{30} = \frac{19}{30}$

30. $P(\text{water retention or hair loss}) = P(\text{water retention}) + P(\text{hair loss}) - P(\text{both})$
$35\% = 25\% + P(\text{hair loss}) - 5\%$
$P(\text{hair loss}) = 15\%$

EXTEND

31. $P(\text{female}) = \frac{14}{25}$ **32.** $P(\text{senior}) = \frac{13}{25}$

33. $P(\text{male and junior}) = \frac{3}{25}$

34. $P(\text{female or senior}) = P(\text{female}) + P(\text{senor}) - P(\text{female and senior})$
$= \frac{14}{25} + \frac{13}{25} - \frac{5}{25} = \frac{22}{25}$

35. $P(\text{malc or junior}) = P(\text{male}) + P(\text{junior}) - P(\text{male and junior})$
$= \frac{11}{25} + \frac{12}{25} - \frac{3}{25} = \frac{20}{25}$ or $\frac{4}{5}$

36. $P(\text{romance}) = \frac{18}{30}$ or $\frac{3}{5}$

37. $P(\text{longevity or career}) = \frac{6}{30}$ or $\frac{1}{5}$

38. x is number of washcloths; $x + 5$ is number of towels.
$P(\text{washcloth}) = \frac{2}{3} = \frac{x + 5}{2x + 5}; 4x + 10 = 3x + 15$
$x = 5$ washcloths and
$x + 5 = 10$ towels

39. $P(\text{wool or green}) = P(\text{wool}) + P(\text{green}) - P(\text{wool and green})$
$= \frac{12}{24} + \frac{8}{24} - \frac{4}{24} = \frac{16}{24}$ or $\frac{2}{3}$

40. $P(\text{cotton or not red}) = P(\text{cotton}) + P(\text{not red}) - P(\text{cotton and not red})$
$= \frac{12}{24} + \frac{16}{24} - \frac{8}{24} = \frac{20}{24}$ or $\frac{5}{6}$

THINK CRITICALLY

41. $P(\text{female or grade 10}) = \frac{3}{5} = P(\text{female}) + P(\text{grade 10}) - P(\text{female and grade 10})$
$P(\text{female}) = \frac{x + 13}{x + 32}$
$P(\text{grade 10}) = \frac{10}{x + 32}$
$P(\text{female and grade 10}) = \frac{5}{x + 32}$
$\frac{3}{5} = \frac{x + 13}{x + 32} + \frac{10}{x + 32} + \frac{5}{x + 32} = \frac{x + 18}{x + 32}$
$3x + 96 = 5x + 90$
$2x = 6$
$x = 3$

42. $P(\text{male or not grade 12})$
$= P(\text{male}) + P(\text{not grade 12}) - P(\text{male and not grade 12})$
$= \frac{19}{35} + \frac{25}{35} - \frac{13}{35} = \frac{31}{35}$

43. Students taking Italian: $28 = 11 + 4 + 5 + x$
Students taking Italian and French $(x): x = 8$
Students taking French: $25 = 8 + 4 + 8 + y$
Students taking French and Spanish $(y): y = 5$
Students taking Spanish: $23 = z + 5 + 4 + 5; z = 9$
Number of Students taking a language $= 50$

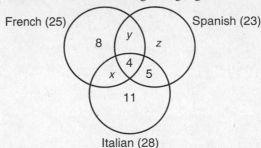

French (25)   Spanish (23)

Italian (28)

$P(\text{taking three languages}) = \frac{4}{50} = \frac{2}{25}$

44–47. To determine number of each carton: A, B, C, D, and E.
$A + B + C + D + E = 360; A = 2B + 80;$
$B = C - 40; C = D + E; D = E; E = x$
$D = x; C = 2x; B = 2x - 40; A = 4x$
so, $(4x) + (2x - 40) + (2x) + (x) + (x) = 360;$
$\qquad\qquad\qquad\qquad\qquad 10x - 40 = 360$
$\qquad\qquad\qquad\qquad\qquad\qquad 10x = 400$
$\qquad\qquad\qquad\qquad\qquad\qquad\quad x = 40$

$A = 160; B = 40; C = 80; D = 40; E = 40$

44. $P(A) = \frac{160}{360}$ or $\frac{4}{9}$ **45.** $P(\text{not } C) = \frac{280}{360}$ or $\frac{7}{9}$

46. $P(B \text{ or } D) = \frac{80}{360}$ or $\frac{2}{9}$ **47.** $P(C \text{ or } A) = \frac{240}{360}$ or $\frac{1}{3}$

48. $m = \dfrac{\begin{vmatrix} 2 & -5 \\ 1 & -7 \end{vmatrix}}{\begin{vmatrix} 2 & -5 \\ 3 & -7 \end{vmatrix}} = \dfrac{-9}{1} = -9$

$n = \dfrac{\begin{vmatrix} 2 & 2 \\ 3 & 1 \end{vmatrix}}{\begin{vmatrix} 2 & -5 \\ 3 & -7 \end{vmatrix}} = \dfrac{-4}{1} = -4$

49. $a = \dfrac{\begin{vmatrix} 16 & -2 & 1 \\ 2 & -1 & 5 \\ 8 & 0 & -3 \end{vmatrix}}{\begin{vmatrix} 6 & -2 & 1 \\ 2 & -1 & 5 \\ 2 & 0 & -3 \end{vmatrix}} = \dfrac{-36}{-12} = 3$

$b = \dfrac{\begin{vmatrix} 6 & 16 & 1 \\ 2 & 2 & 5 \\ 2 & 8 & -3 \end{vmatrix}}{\begin{vmatrix} 6 & -2 & 1 \\ 2 & -1 & 5 \\ 2 & 0 & -3 \end{vmatrix}} = \dfrac{-8}{-12} = \dfrac{2}{3}$

$c = \dfrac{\begin{vmatrix} 6 & -2 & 16 \\ 2 & -1 & 2 \\ 2 & 0 & 8 \end{vmatrix}}{\begin{vmatrix} 6 & -2 & 1 \\ 2 & -1 & 5 \\ 2 & 0 & -3 \end{vmatrix}} = \dfrac{8}{-12} = -\dfrac{2}{3}$

50. C.

A. $2^{-1} + 3^{-1}$ does not equal 5^{-1}

B. $5^{\circ} - 3^{\circ}$ does not equal 2°

D. $\dfrac{1}{20}$ does not equal $\dfrac{2}{1^{\circ}}$

AlgebraWorks

1. $\dfrac{28}{23{,}456} \times 100 \approx 0.1194\%$

2. 1.194 per thousand

3. $23{,}456 - 28 = 23{,}428$ policy holders

4. $23{,}428 \times 0.00133 \approx 31$ claims for 31 year old males; to the nearest integer.

5. $\$100{,}000 \times 31 = \$3{,}100{,}000$ payed out

6. $\$3{,}100{,}000 \div 23{,}428 \approx \$132.32$ annual premium

7. $\dfrac{7.37}{1000} = 0.737\%$

Lesson 14.4, pages 695–702

Explore

1a. (1, 1), (1, 2), (1, 3), (1, 4), (1, 5), (1, 6),
(2, 1), (2, 2), (2, 3), (2, 4), (2, 5), (2, 6)
(3, 1), (3, 2), (3, 3), (3, 4), (3, 5), (3, 6)
(4, 1), (4, 2), (4, 3), (4, 4), (4, 5), (4, 6)
(5, 1), (5, 2), (5, 3), (5, 4), (5, 5), (5, 6)
(6, 1), (6, 2), (6, 3), (6, 4), (6, 5), (6, 6)

1b. 36

2a. 6 **2b.** $P(6 \text{ on first cube}) = \dfrac{6}{36} \text{ or } \dfrac{1}{6}$

3a. 6 **3b.** $P(6 \text{ on second cube}) = \dfrac{6}{36} \text{ or } \dfrac{1}{6}$

4a. 1 **4b.** $P(4 \text{ on both cubes}) = \dfrac{1}{36}$

5. $\dfrac{1}{6} \cdot \dfrac{1}{6} = \dfrac{1}{36}$

Try These

1. Independent, if she returns the first marble before second draw; Dependent, if she does not return the first marble.

2a. $P(2 \text{ black cards, with replacement}) = \dfrac{26}{52} \cdot \dfrac{26}{52} = \dfrac{1}{4}$

2b. $P(2 \text{ black cards, without replacement})$
$= \dfrac{26}{52} \cdot \dfrac{25}{51} = \dfrac{25}{102}$

3a. $P(\text{first card is Red and second card Black})$
$= \dfrac{26}{52} \cdot \dfrac{26}{52} = \dfrac{1}{4}$

3b. $P(\text{first Red and second Black, without replacement})$
$= \dfrac{26}{52} \cdot \dfrac{26}{51} = \dfrac{13}{51}$

4a. $P(\text{two red queens}) = \dfrac{2}{52} \cdot \dfrac{2}{52} = \dfrac{1}{676}$

4b. $P(\text{two red queens, without replacement})$
$= \dfrac{2}{52} \cdot \dfrac{1}{51} = \dfrac{1}{1326}$

5a. $P(\text{two orange}) = \dfrac{3}{5} \cdot \dfrac{3}{5} = \dfrac{9}{25}$

5b. $P(\text{two orange, without replacement})$
$= \dfrac{3}{5} \cdot \dfrac{2}{4} = \dfrac{3}{10}$

6a. $P(\text{not a 3}) = \dfrac{4}{5} \cdot \dfrac{4}{5} = \dfrac{16}{25}$

6b. $P(\text{not a 3, without replacement}) = \dfrac{4}{5} \cdot \dfrac{3}{4} = \dfrac{3}{5}$

7a. $P(\text{green}) = \dfrac{25}{25} \cdot \dfrac{0}{25} = 0$

7b. $P(\text{one green, without replacement})$
$= \dfrac{25}{25} \cdot \dfrac{0}{24} = 0$

8. Let Bean $= B$; Carrots $= C$; Radishes $= R$
Marigolds $= M$; Petunias $= P$.
Sample Space: $(B, M) \ (C, M) \ (R, M)$
$(B, P) \ (C, P) \ (R, P)$

9. $P(\text{not Beans}) = \dfrac{4}{6} \text{ or } \dfrac{2}{3}$ 10. $P(\text{Petunia}) = \dfrac{3}{6} \text{ or } \dfrac{1}{2}$

11. $P(A) = \dfrac{1}{2}; P(A \text{ and } B) = \dfrac{1}{8}, P(B|A) = \dfrac{\frac{1}{8}}{\frac{1}{2}} = \dfrac{1}{2}$

12. $P(A) = \dfrac{7}{8}; P(A \text{ and } B) = \dfrac{4}{8}; P(B|A) = \dfrac{\frac{4}{8}}{\frac{7}{8}} = \dfrac{4}{7}$

13. $P(\text{son on end}) = \dfrac{4}{6} = \dfrac{2}{3}$

$P(\text{son on end and father in middle}) = \dfrac{2}{6} \text{ or } \dfrac{1}{3}$

$P(\text{father in middle given son on end}) = \dfrac{\frac{1}{3}}{\frac{2}{3}} = \dfrac{1}{2}$

14. $P(5 \text{ on both cubes}) = \frac{1}{6} \cdot \frac{1}{6} = \frac{1}{36}$

15. $P(2 \text{ even numbers}) = \frac{3}{6} \cdot \frac{3}{6} = \frac{9}{36}$ or $\frac{1}{4}$

16. $P(4 \text{ or lower on both cubes}) = \frac{4}{6} \cdot \frac{4}{6} = \frac{16}{36}$ or $\frac{4}{9}$

17. $P(\text{even on first and odd on second}) = \frac{3}{6} \cdot \frac{3}{6} = \frac{9}{36}$ or $\frac{1}{4}$

18. $P(\text{odd on first and less than 3 on second})$

$= \frac{3}{6} \cdot \frac{2}{6} = \frac{6}{36}$ or $\frac{1}{6}$

PRACTICE

1. Mutually exclusive events cannot occur at the same time. Independent events can occur at the same time but have no effect on the probability of each other.

2. $P(\text{both vowels}) = \frac{3}{7} \cdot \frac{2}{6} = \frac{1}{7}$

3. $P(\text{both constants}) = \frac{4}{7} \cdot \frac{3}{6} = \frac{2}{7}$

4. $P(\text{first vowel and second constant}) = \frac{3}{7} \cdot \frac{4}{6} = \frac{2}{7}$

5. $P(\text{begin with P}) = \frac{24}{120} \cdot \frac{23}{119} = \frac{23}{595}$

6. $P(\text{both begin with constants}) = \frac{96}{120} \cdot \frac{95}{119} = \frac{76}{119}$

7. $P(\text{first begins with P and second begins with M})$

$= \frac{24}{120} \cdot \frac{24}{119} = \frac{24}{595}$

8. $P(\text{three beef}) = \frac{{}_7C_3}{{}_{24}C_3} = \frac{35}{2024}$

9. $P(\text{three liver}) = \frac{{}_6C_3}{{}_{24}C_3} = \frac{20}{2024}$ or $\frac{5}{506}$

10. $P(2 \text{ veal and 1 beef}) = \frac{{}_5C_2 \cdot {}_7C_1}{{}_{24}C_3} = \frac{70}{2024}$ or $\frac{35}{1012}$

11. $P(1 \text{ tuna and 2 liver}) = \frac{{}_6C_1 \cdot {}_6C_2}{{}_{24}C_3} = \frac{90}{2024}$ or $\frac{45}{1012}$

12. $P(1 \text{ beef, 1 tuna, 1 veal})$

$= \frac{{}_7C_1 \cdot {}_6C_1 \cdot {}_5C_1}{{}_{24}C_3} = \frac{210}{2024}$ or $\frac{105}{1012}$

13. $P(2 \text{ humanities}) = \frac{{}_8C_2}{{}_{13}C_2} = \frac{28}{78}$ or $\frac{14}{39}$

14. $P(1 \text{ humanity and 1 science}) = \frac{{}_8C_1 \cdot {}_5C_1}{{}_{13}C_2} = \frac{40}{78}$ or $\frac{20}{39}$

15. $P(2 \text{ red}) = \frac{{}_2C_2}{{}_5C_2} = \frac{1}{10}$

16. $P(2 \text{ red or 2 white})$

$= \frac{{}_3C_2}{{}_5C_2} + \frac{{}_2C_2}{{}_5C_2} = \frac{3}{10} + \frac{1}{10} = \frac{4}{10}$ or $\frac{2}{5}$

17. $P(\text{first red and second white}) = \frac{{}_3C_1 \cdot {}_2C_1}{{}_5C_2} = \frac{6}{10}$ or $\frac{3}{5}$

18. $P(\text{one head}) = \frac{3}{4}$; $P(\text{one head and one tail}) = \frac{1}{2}$

$P(B|A) = \dfrac{\frac{1}{2}}{\frac{3}{4}} = \frac{2}{3}$

19. $P(\text{no heads}) = \frac{1}{4}$; $P(\text{no heads and no tails}) = \frac{0}{4}$

$P(B|A) = \dfrac{\frac{0}{4}}{\frac{1}{4}} = 0$

20. $P(\text{flosses regularly and has healthy gums}) = P(\text{flosses regularly}) \cdot P(\text{has healthy gums after flossing}) = 0.3 \cdot 0.6 = 0.18$

$P(\text{does not floss regularly and has healthy gums} = P(\text{does not floss regularly}) \cdot P(\text{has healthy gums after not flossing}) = 0.7 \cdot 0.2 = 0.14$

$P(\text{flosses regularly}) = \frac{0.18}{0.18 + 0.14} = 0.5625 = \frac{9}{16}$

21. B, B   1, 1   2, 2   3, 3   4, 4   5, 5   6, 6
B, 1   1, 2   2, 3   3, 4   4, 5   5, 6
B, 2   1, 3   2, 4   3, 5   4, 6
B, 3   1, 4   2, 5   3, 6
B, 4   1, 5   2, 6
B, 5   1, 6
B, 6
28 titles

22. $P(5 \text{ doubles}) = \frac{{}_7C_5}{{}_{28}C_5} = \frac{21}{98{,}280} = \frac{1}{4680}$

or about 2.14×10^{-4}

23. $P(\text{blank or } 6) = \frac{13}{28}$

24. $P(2 \text{ dominoes, both have a } 3) = \frac{{}_7C_2}{{}_{28}C_2} = \frac{21}{378}$ or $\frac{1}{18}$

25. $P(\text{sum at most is } 3) = \frac{6}{28}$ or $\frac{3}{14}$

EXTEND

26. $P(\text{high and tested positive}) = P(\text{high}) \cdot P(\text{tested positive given high}) = 0.18 \cdot 0.85 = 0.153$

$P(\text{normal and tested positive}) = P(\text{normal}) \cdot P(\text{tested positive given normal}) = 0.82 \cdot 0.1 = 0.082$

$P(\text{high}) = \frac{0.153}{0.153 + 0.082} \approx 0.65$

27. $P(2 \text{ candies with same color})$

$= \frac{{}_3C_2 + {}_7C_2}{{}_{10}C_2} = \frac{24}{45} = \frac{8}{15}$

28. $P(2 \text{ candies with different colors})$

$= \frac{{}_3C_1 \cdot {}_7C_1}{{}_{10}C_2} = \frac{21}{45} = \frac{7}{15}$

29. $P(5 \text{ candies with at least 3 silver})$

$= \frac{{}_7C_3 \cdot {}_3C_2 + {}_7C_4 \cdot {}_3C_1 + {}_7C_5 \cdot {}_3C_0}{{}_{10}C_5} = \frac{231}{252}$ or $\frac{11}{12}$

30. $P(\text{all 3 correct}) = P(A) \cdot P(B) \cdot P(C) = 0.144$

31. $P(\text{at least 2 correct}) = P(\text{all 3 correct}) + P(A) \cdot P(B) \cdot P(\text{not C}) + P(A) \cdot P(\text{not B}) \cdot P(C) + P(\text{not A}) \cdot P(B) \cdot P(C) = 0.144 + 0.336 + 0.096 + 0.036 = 0.612$

32. $P(\text{at least 1 correct}) = 1 - P(\text{none correct})$
$= 1 - 0.056 = 0.944$

THINK CRITICALLY

33. $P(B|A) = \dfrac{2}{6} = \dfrac{1}{3}$

34. $P(B|A) = \dfrac{1}{4}$

35. $P(B|A) = \dfrac{4}{11}$

36. Area (20-point circle) $= \pi(2)^2 = 4\pi$
Area (15-point circle)
$= \pi(4)^2 - 4\pi = 16\pi - 4\pi = 12\pi$
Area (10-point circle)
$= \pi(6)^2 - 16\pi = 36\pi - 16\pi = 20\pi$
Area (5-point circle)
$= \pi(8)^2 - 36\pi = 64\pi - 36\pi = 28\pi$
$P(25 \text{ points})$
$= P(20, 5) + P(15, 10) + P(10, 15) + P(5, 20)$
$= P(20) \cdot P(5) + P(15) \cdot P(10) +$
$P(10) \cdot P(15) + P(5) \cdot P(20)$
$= \dfrac{4\pi}{64\pi} \cdot \dfrac{28\pi}{64\pi} + \dfrac{20\pi}{64\pi} \cdot \dfrac{12\pi}{64\pi} + \dfrac{12\pi}{64\pi} \cdot \dfrac{20\pi}{64\pi} + \dfrac{28\pi}{64\pi} \cdot \dfrac{4\pi}{64\pi}$
$= \dfrac{11}{64} \approx 0.17 = 17\%$

MIXED REVIEW

37. $x^2 - 6x - 7 = 0$
$x^2 - 6x \qquad = 7$
$x^2 - 6x + (-3)^2 \overset{?}{=} 7 + (-3)^2$
$(x - 3)^2 \qquad = 16$
$x - 3 \qquad = \pm 4$
$x = 7 \text{ or } -1$

38. $x^2 + 6x - 4 = 0$
$x^2 + 6x \qquad = 4$
$x^2 + 6x + (3)^2 = 4 + (3)^2$
$(x + 3)^2 \qquad = 13$
$x + 3 \qquad = \pm\sqrt{13}$
$x = -3 + \sqrt{13} \text{ or } -3 - \sqrt{13}$

39. D

40. $x^2 + 9x + 8 = 0$
$(x + 1)(x + 8)$
$x = -1 \text{ or } x = -8$
$x < -8 \text{ or } x > -1$

$x + 1$	$-$	$-$	$+$
$x + 8$	$-$	$+$	$+$
$(x+1)(x+8)$	$+$	$-$	$+$

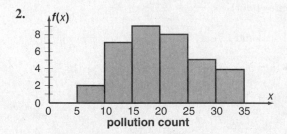

41. $x^2 + 4x < 0$
$x(x + 4) < 0$
$x < 0$

42. $2(27)^0 + (27)^{\frac{2}{3}} = 2(1) + (\sqrt[3]{27})^2 = 2 + 9 = 11$

43. $(8)^{-1}(3(8)^{\frac{1}{3}} + (8)^0) = \dfrac{1}{8}(3(2) + 1) = \dfrac{7}{8}$

44. $2x^2 - 50$
$2(x + 5)(x - 5)$

45. $ax^2 + 7ax + 12a$
$a(x + 3)(x + 4)$

46. $g^2 - g - 12h - 3gh$
$(g - 4)(g - 3h)$

47. B; $x^3 - x^2 - 2x = x(x^2 - x - 2) =$
$x(x + 1)(x - 2) = 0; x = 0, -1, 2$

Lesson 14.5, pages 703–709

EXPLORE/WORKING TOGETHER

1–3. Answers will vary.

4. Answers will vary; tallest bars are in the middle, the bars at the ends are shortest.

5. The tallest bars.

TRY THESE

1.

Pollution Count	0 – 4.9	5 – 9.9	10 – 14.9	15 – 19.9	20 – 24.9	25 – 29.9	30 – 34.9
Frequency	0	2	7	9	8	5	4

2.

3. $15 - 19.\overline{9}$

4. $P(5 < x < 15) = \dfrac{\text{Area of (Bar 2 + Bar 3)}}{\text{Total Area}}$; note Bar 2 appears as the first bar, since first bar has zero area.
$= \dfrac{5 \cdot 2 + 5 \cdot 7}{175} = \dfrac{9}{35}$

5. $P(x > 20) = \dfrac{\text{Area of (Bar 5 + Bar 6 + Bar 7)}}{\text{Total Area}}$
$= \dfrac{5 \cdot 8 + 5 \cdot 5 + 5 \cdot 4}{175} = \dfrac{17}{35}$

6. Sally; since graph A intersects the x-axis, it cannot represent a normal curve, which is asymptotic to the x-axis.

7.

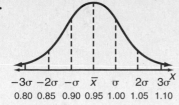

$-3\sigma \ -2\sigma \ -\sigma \ \ \bar{x} \ \ \sigma \ \ 2\sigma \ \ 3\sigma$
0.80 0.85 0.90 0.95 1.00 1.05 1.10

8. $P(0.90 < x < 1.00) = 68\%$

9. $P(x < 0.90) = 13.5 + 2.4 = 15.9$
$0.50 \times 15.9 = 7.95$
So, about 8 cartons.

10a. 13.5%   **10b.** 2.4%

11.

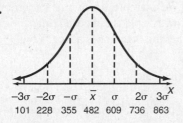

$-3\sigma \ -2\sigma \ -\sigma \ \ \bar{x} \ \ \sigma \ \ 2\sigma \ \ 3\sigma$
101 228 355 482 609 736 863

12. $P(482 < x < 736) = 34 + 13.5 = 47.5\%$

13. 84th percentile

14. Yes, the 67th percentile is equivalent to a score of 545.5; Joel's score was higher.

15. 0.3413, about 34%   **16.** 0.3413, about 34%

17. 0.1587, about 15.9%

PRACTICE

1.

Time	2.5–2.9	3.0–3.4	3.5–3.9	4.0–4.4	4.5–4.9	5.0–5.4	5.5–5.9
Frequency	1	5	8	12	7	4	3

2.

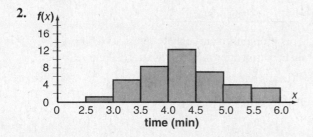

3. 4.0–4.4

4. $P(3.0 < x < 4.5) = \dfrac{\text{Area of (Bar 2 + Bar 3 + Bar 4)}}{\text{Total Area}}$

$= \dfrac{0.5 \cdot 4 + 0.5 \cdot 8 + 0.5 \cdot 12}{20} = \dfrac{5}{8}$

5. $P(x < 4.0) = \dfrac{\text{Area of (Bar 1 + Bar 2 + Bar 3)}}{\text{Total Area}}$

$= \dfrac{0.5 \cdot 1 + 0.5 \cdot 5 + 0.5 \cdot 8}{20} = \dfrac{7}{20}$

6.

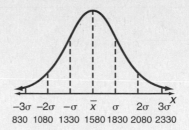

$-3\sigma \ -2\sigma \ -\sigma \ \ \bar{x} \ \ \sigma \ \ 2\sigma \ \ 3\sigma$
830 1080 1330 1580 1830 2080 2330

7. $34 + 13.5 = 47.5\%$

8. $13.5 + 2.4 = 15.9\%$
500 bulbs $\times$ 15.9% $\approx$ 79 bulbs.

9a. 13.5%   **9b.** 15.9%   **9c.** 90.65%

10. C   **11.** A   **12.** D   **13.** A   **14.** C

15. No; 70th percentile means that Seymour's score was equal to or better than 70% of the students who took the test; we do not know what his actual score was.

16.

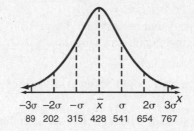

$-3\sigma \ -2\sigma \ -\sigma \ \ \bar{x} \ \ \sigma \ \ 2\sigma \ \ 3\sigma$
89 202 315 428 541 654 767

17. $34 + 34 = 68\%$ **18.** 16th percentile

19. Yes. The 90th percentile is equivalent to a score under 710.5; Cesar's score was higher.

20. 0.3413, about 34%

21. 0.4641, about 46%

22. 0.1359, about 13.6%

23. 0.1587, about 15.9%

24. $y = 0.16e^{-0.08(x-67)^2}$

25. $Y = (2.5\sqrt{2\pi})^{-1}e^{\wedge}(-0.5((x - 67)/2.5)^2)$

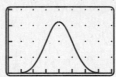

26. 30.6%; The area under the curve that lies in the interval $67 \le x \le 68$ is approximately a trapezoid with width 1 unit and parallel sides of lengths 0.147 and 0.159. The area of this trapezoid is 0.153, meaning that about 15.3% of the men were between 67 and 68 inches tall. Using symmetry, 15.3% of the men were between 66 and 67 inches tall.

27. $\dfrac{1}{\overline{x}} = \dfrac{15 + 32 + 57 + 44}{4} = 37$

28. $\dfrac{\sum\limits_{L=1}^{n}(x_1 - \overline{x})}{n} = \dfrac{|15 - 37| + |32 - 37| + |57 - 37| + |44 - 37|}{4} = 13.5$

29. $\dfrac{\sum\limits_{L=1}^{n}(x_1 - \overline{x})^2}{n} = \dfrac{(15 - 37)^2 + (32 - 37)^2 + (57 - 37)^2 + (44 - 37)^2}{4} = 239.5$

30. $\sqrt{\dfrac{\sum\limits_{L=1}^{n}(x_1 - \overline{x})^2}{n}} = \sqrt{239.5} \approx 15.5$

MIXED REVIEW

31. C

Lesson 14.6, pages 710–716

EXPLORE/WORKING TOGETHER

1–2. Answers will vary, but results should begin to resemble a normal distribution.

3. 3 should be the most likely.

4. The heights of the bars should be lower as you move from 3.

5. The symmetries should be better and better resemble a normal distribution.

6. The results should be closer to a normal distribution.

TRY THESE

1. Yes
- two outcomes: Heads or Tails
- fixed number of trials: 8
- trials are independent
- same probability for each trial: $P(\text{Heads}) = \dfrac{1}{2}$, $P(\text{tails}) = \dfrac{1}{2}$

2. Yes
- two outcomes: correct or incorrect
- fixed number of trials: 10 questions
- trials are independent
- same probability for each trial: $P(\text{correct}) = \dfrac{1}{5}$, $P(\text{incorrect}) = \dfrac{4}{5}$

3. No, unless each ball selected is replaced before the next selection is made. Without replacement, the trials are not independent and the probability of each trial is not the same.

4. Answers will vary. Example: A binomial experiment is rolling number cubes 10 times and keep track of whether the outcome is 3 or 6.
$P(3 \text{ or } 6) = \dfrac{2}{6}$ or $\dfrac{1}{3}$, $P(\text{not 3 or not 6}) = \dfrac{4}{6}$ or $\dfrac{2}{3}$.

5. Answers will vary. An example, a binomial experiment is to draw 10 cards with replacement each time and keep track of whether the outcome is a Ace, King, Queen, or Jack.

$P(\text{Ace, King, Queen, or Jack}) = \dfrac{4}{13}$,

$P(\text{not (Ace, King, Queen, or Jack)}) = \dfrac{9}{13}$

6. $_3C_2\left(\dfrac{3}{5}\right)^2\left(\dfrac{2}{5}\right)^1 = \dfrac{54}{125}$ or 0.432

7.

Number of Odd Outcomes	0	1	2	3
Probability	$\dfrac{8}{125}$	$\dfrac{36}{125}$	$\dfrac{54}{125}$	$\dfrac{27}{125}$

8.

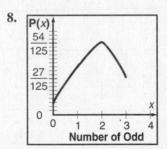

9. This probability curve that represents a binomial distribution approximates the normal curve.

10. $P(\text{heads exactly 3 times}) = {}_8C_3\left(\dfrac{1}{2}\right)^3\left(\dfrac{1}{2}\right)^5 \approx 21.9\%$

11. $P(\text{tails at least 6 times})$

$= {}_8C_6\left(\dfrac{1}{2}\right)^6\left(\dfrac{1}{2}\right)^2 + {}_8C_7\left(\dfrac{1}{2}\right)^7\left(\dfrac{1}{2}\right)^1 + {}_8C_8\left(\dfrac{1}{2}\right)^8\left(\dfrac{1}{2}\right)^0 \approx 14.5\%$

12. $P(\text{heads no more than 4 times})$

$= {}_8C_0\left(\dfrac{1}{2}\right)^0\left(\dfrac{1}{2}\right)^8 + {}_8C_1\left(\dfrac{1}{2}\right)^1\left(\dfrac{1}{2}\right)^7 + {}_8C_2\left(\dfrac{1}{2}\right)^2\left(\dfrac{1}{2}\right)^6 +$

$\quad {}_8C_3\left(\dfrac{1}{2}\right)^3\left(\dfrac{1}{2}\right)^5 + {}_8C_4\left(\dfrac{1}{2}\right)^4\left(\dfrac{1}{2}\right)^4 \approx 63.7\%$

13. $P(\text{exactly 6 hits}) = {}_9C_6\left(\dfrac{3}{10}\right)^6\left(\dfrac{7}{10}\right)^3 \approx 2.1\%$

14. $P(\text{no more than 3 hits})$

$$= {}_9C_0\left(\frac{3}{10}\right)^0\left(\frac{7}{10}\right)^9 + {}_9C_1\left(\frac{3}{10}\right)^1\left(\frac{7}{10}\right)^8 + {}_9C_2\left(\frac{3}{10}\right)^2\left(\frac{7}{10}\right)^7 +$$

$${}_9C_3\left(\frac{3}{10}\right)^3\left(\frac{7}{10}\right)^6 \approx 73.0\%$$

15. $P(\text{at least 7 hits})$

$$= {}_9C_7\left(\frac{3}{10}\right)^7\left(\frac{7}{10}\right)^2 + {}_9C_8\left(\frac{3}{10}\right)^8\left(\frac{7}{10}\right)^1 + {}_9C_9\left(\frac{3}{10}\right)^9\left(\frac{7}{10}\right)^0$$
$$\approx 0.4\%$$

PRACTICE

1. Yes
   - two outcomes: makes the basket or not
   - fixed number of trials: 7
   - trials are independent
   - same probability for each trial: $P(\text{basket}) = \frac{3}{8}$,
     $P(\text{no basket}) = \frac{5}{8}$

2. No, unless Sally replaces each card she selected before she makes the next selection, the trials are not independent and the probability of each trial is not the same.

3. Yes
   - two outcomes: red or white
   - fixed number of trials: 8
   - trials are independent
   - same probability for each trial: $P(\text{red}) = \frac{15}{28}$
     $P(\text{white}) = \frac{13}{28}$

4. Answers will vary. One binomial experiment is to take 10 spins on the board and keep track of whether the outcomes is 5 or more. $P(\text{5 or more}) = \frac{3}{7}$
   $P(\text{of less than 5}) = \frac{4}{7}$

5. Answers will vary. One binomial experiment is to take 6 spins on the board and keep track of whether the outcomes is blue or not blue. $P(\text{blue}) = \frac{1}{3}$
   $P(\text{not blue}) = \frac{2}{3}$

6. $P(\text{the number 4 on exactly 2 cubes})$
   $$= {}_3C_2\left(\frac{1}{6}\right)^2\left(\frac{5}{6}\right)^1 = \frac{5}{72}$$

7. $P(\text{number 6 on exactly 2 cubes})$
   $$= {}_4C_2\left(\frac{1}{6}\right)^2\left(\frac{5}{6}\right)^2 = \frac{25}{216}$$

8.

Number of Heads	0	1	2	3	4	5	6	7	8
Probability	$\frac{1}{256}$	$\frac{8}{256}$	$\frac{28}{256}$	$\frac{56}{256}$	$\frac{70}{256}$	$\frac{56}{256}$	$\frac{28}{256}$	$\frac{8}{256}$	$\frac{1}{256}$

9.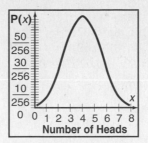

10. Use the line of Pascal's Triangle that has values as the numerators of the probability fractions. This line of Pascal's Triangle corresponds to the binomial $(a + b)^3$. The denominators of the probability fractions are all 2^3.

$$P(\text{1}^{st}\text{ compartment}) = \frac{1}{2^3}$$

$$P(\text{2}^{nd}\text{ compartment}) = \frac{3}{2^3}$$

$$P(\text{3}^{rd}\text{ compartment}) = \frac{3}{2^3}$$

$$P(\text{4}^{th}\text{ compartment}) = \frac{1}{2^3}$$

11. $\frac{1}{2^7}, \frac{7}{2^7}, \frac{21}{2^7}, \frac{35}{2^7}, \frac{35}{2^7}, \frac{21}{2^7}, \frac{7}{2^7}, \frac{1}{2^7}$

12. $P(\text{exactly 1 win in 3 games}) = {}_3C_1\left(\frac{2}{3}\right)^1\left(\frac{1}{3}\right)^2 - \frac{2}{9}$ or about 0.222

13. $P(\text{at least 1 win in 3 games}) =$
   $${}_3C_1\left(\frac{2}{3}\right)^1\left(\frac{1}{3}\right)^2 + {}_3C_2\left(\frac{2}{3}\right)^2\left(\frac{1}{3}\right)^1 + {}_3C_3\left(\frac{2}{3}\right)^3 = \frac{26}{27}$$
   or about 0.963

14. $P(\text{no wins in 5 games}) = {}_5C_0\left(\frac{2}{3}\right)^0\left(\frac{1}{3}\right)^5 = \frac{1}{243}$ or about 0.004

15. $P(\text{exactly 3 wins in 5 games}) = {}_5C_3\left(\frac{2}{3}\right)^3\left(\frac{1}{3}\right)^2 = \frac{80}{243}$ or about 0.329

16. $P(\text{no more than 3 wins in 5 games}) =$

$${}_5C_0\left(\frac{2}{3}\right)^0\left(\frac{1}{3}\right)^5 + {}_5C_1\left(\frac{2}{3}\right)^1\left(\frac{1}{3}\right)^4 + {}_5C_2\left(\frac{2}{3}\right)^2\left(\frac{1}{3}\right)^3 + {}_5C_3\left(\frac{2}{3}\right)^3\left(\frac{1}{3}\right)^2$$

$$= \frac{131}{243} \approx 0.539$$

17. $P(\text{exactly 1 with gene}) = {}_4C_1(0.27)^1(0.73)^4 \approx 0.420$

18. $P(\text{exactly 3 with gene}) = {}_4C_3(0.27)^3(0.73)^1 \approx 0.057$

19. $P(\text{no more than 1 with gene}) =$
   ${}_4C_0(0.27)^0(0.73)^4 + {}_4C_1(0.27)^1(0.73)^3 \approx 0.704$

20. $P(\text{zero}) = {}_4C_0(0.27)^0(0.73)^4 \approx 0.284$

21. $P(\text{exactly 5 foul shots in 10 shots}) =$
   ${}_{10}C_5(0.38)^5(0.62)^5 \approx 0.183$

22. P(at least 5 foul shots in 10 shots) $=$
$_{10}C_5(0.38)^5(0.62)^5 + {}_{10}C_6(0.38)^6(0.62)^4 +$
$_{10}C_7(0.38)^7(0.62)^3 + {}_{10}C_8(0.38)^8(0.62)^2 +$
$_{10}C_9(0.38)^9(0.6)^1 + {}_{10}C_{10}(0.38)^0(0.62)^{10} \approx 0.318$

23. P(exactly 12 passengers show up) $=$

$_{16}C_{12}\left(\frac{4}{5}\right)^{12}\left(\frac{1}{5}\right)^4 \approx 20\%$ or 0.200

24. P("overbooked") $=$

$_{16}C_{13}\left(\frac{4}{5}\right)^{13}\left(\frac{1}{5}\right)^3 + {}_{16}C_{14}\left(\frac{4}{5}\right)^{14}\left(\frac{1}{5}\right)^2 + {}_{16}C_{15}\left(\frac{4}{5}\right)^{15}\left(\frac{1}{5}\right) +$

$_{16}C_{16}\left(\frac{4}{5}\right)^{16}\left(\frac{1}{5}\right)^0 \approx 0.598$ or 59.8%

25. P(aligned, begin aligned) $= 0.89$ or 89%

26. P(aligned, begin unaligned) $= 0.88$ or 88%

27. P(unaligned, begin aligned) $= 0.11$ or 11%

28. P(unaligned, begin unaligned) $= 0.12$ or 12%

29.
$$\begin{array}{c} \text{next stage} \\ \begin{array}{cc} D & R \end{array} \end{array}$$

$$\begin{array}{c} \text{beginning state} \end{array} \begin{array}{c} D \\ R \end{array} \begin{bmatrix} \frac{5}{6} & \frac{1}{6} \\ \frac{1}{3} & \frac{2}{3} \end{bmatrix}$$

Multiply to find second stage

$$\begin{bmatrix} \frac{5}{6} & \frac{1}{6} \\ \frac{1}{3} & \frac{2}{3} \end{bmatrix}\begin{bmatrix} \frac{5}{6} & \frac{1}{6} \\ \frac{1}{3} & \frac{2}{3} \end{bmatrix} = \begin{bmatrix} 0.75 & 0.25 \\ 0.5 & 0.5 \end{bmatrix}$$

Probability of a Republican mayor after two elections is 0.25 or $\frac{1}{4}$.

30. $\begin{bmatrix} 0.75 & 0.25 \\ 0.5 & 0.5 \end{bmatrix}\begin{bmatrix} \frac{5}{6} & \frac{1}{6} \\ \frac{1}{3} & \frac{2}{3} \end{bmatrix} = \begin{bmatrix} 0.708\overline{3} & 0.291\overline{6} \\ 0.58\overline{3} & 0.41\overline{6} \end{bmatrix}$

Probability of a Democratic mayor after three elections is $0.58\overline{3}$ or $\frac{7}{12}$.

THINK CRITICALLY

31. P(3 green) $= {}_3C_3(0.6)^3(0.4)^0 = 0.216$
P(1 red) $= {}_3C_2(0.6)^2(0.4)^1 = 0.432$
P(2 red) $= {}_3C_1(0.6)^1(0.4)^1 = 0.288$
P(3 red) $= {}_3C_0(0.6)^0(0.4)^3 = 0.064$

32. Their sum should be 1.
$0.216 + 0.432 + 0.288 + 0.064 = 1$

33. more likely to be stopped at one or fewer lights:
$P(> 1) = 0.288 + 0.064 = 0.352$ or 35.2%
$P(\leq 1) = 0.216 + 0.432 = 0.648$ or 64.8%

34. P(series ending in exactly 5 games) $= \frac{1}{4}$

$4 \cdot P(\text{National})^4 \cdot P(\text{American})^1 =$

$4 \cdot \left(\frac{1}{2}\right)^4\left(\frac{1}{2}\right)^1 = 0.125$

$4 \cdot P(\text{American})^4 \cdot P(\text{National})^1 =$

$4 \cdot \left(\frac{1}{2}\right)^4\left(\frac{1}{2}\right) = 0.125$

$0.125 + 0.125 = 0.25$ or $\frac{1}{4}$

35. Probability that National League wins in:

(4 games) $= 1 \cdot \left(\frac{3}{5}\right)^4\left(\frac{2}{5}\right)^0 = 0.1296$ or 13% or $\frac{81}{625}$

(5 games) $= 4 \cdot \left(\frac{3}{5}\right)^4\left(\frac{2}{5}\right)^1 = \frac{648}{3125}$ or ≈ 0.207 or 20.7%

(6 games) $= 10 \cdot \left(\frac{3}{5}\right)^4\left(\frac{2}{5}\right)^2 = \frac{648}{3125}$ or ≈ 0.207 or 20.7%

(7 games) $= 20 \cdot \left(\frac{3}{5}\right)^4\left(\frac{2}{5}\right)^3 \approx 0.166$ or 16.6%

Probability that American League wins in:

(4 games) $= 1 \cdot \left(\frac{2}{5}\right)^4\left(\frac{3}{5}\right)^0 = \frac{16}{625}$ or 0.0256 or 2.56%

(5 games) $= 4 \cdot \left(\frac{2}{5}\right)^4\left(\frac{3}{5}\right)^1 = \frac{192}{3125}$ or ≈ 0.061 or 6.1%

(6 games) $= 10 \cdot \left(\frac{2}{5}\right)^4\left(\frac{3}{5}\right)^2 = \frac{288}{3125}$ or ≈ 0.092 or 9.2%

(7 games) $= 20\left(\frac{2}{5}\right)^4\left(\frac{3}{5}\right)^3 \approx 0.111$ or 11.1%

Probability that World Series lasts:
(4 games) $= 13\% + 2.56\% = 15.56\%$
(5 games) $= 20.7\% + 6.1\% = 26.8\%$
(6 games) $= 20.7\% + 9.2\% = 29.9\%$
(7 games) $= 16.6\% + 11.1\% = 27.7\%$
The probability that the National League will win in 5 games is $\frac{648}{3127}$ or 0.207 or 20.7%. The most probable length of the World Series is 6 games.

MIXED REVIEW

36. $2\sqrt{108} = 2\sqrt{2^2 \cdot 3^3} = 12\sqrt{3}$

37. $\sqrt{72a^6b} = 6a^3b^4\sqrt{2b}$

28. $5\sqrt[3]{16} = 10\sqrt[3]{2}$

39. C

40. $\log_5 x = 2; 5^2 = x; x = 25$

41. $\log_9 x = \frac{3}{2}; 9^{\frac{3}{2}} = x; x = 27$

42. $\log_x 81 = 4; x^4 = 81; x = 3$

1. $50.00

2. $150.00

3. The probability of the company having to make that benefit payment is much greater, since additional areas are covered.

4. Answers will vary.

Lesson 14.7, pages 717–723

EXPLORE/WORKING TOGETHER
1–4. Performance and table will vary.

5. Answers will vary.
   card: sideways more likely;
   thumbtack: up more likely;
   cup: sideways more likely;
   card: red or black equally likely

6. $P(\text{red}) = \frac{1}{2}$ 7. Answers will vary.

8. The result should be closer.

9. No; you will come closer to theoretical probability as the number of trials increases, but highly unlikely that a group of 50 trials would have the same outcomes.

TRY THESE
1. Use a number cube by letting the outcomes 1, 3, 5 represent "the drug works" and the outcomes 2, 4 represent "the drug does not work." Disregard the outcome 6. One trial is to roll one cube twice and record the results.

2. Use a standard deck of playing cards by letting a heart represent a graduating senior who is not going to college and getting a card of any of the other suits represent a senior who is college bound. One trial is to pick four cards from a full deck. Record the trial and return cards to deck.

3. Toss 6 coins at once, letting heads represent black and tails represent white. Record the trial.

4. From a standard deck of playing cards remove one suit, say spades. When drawing a card from the 39-card deck, let getting a heart represent a hit and getting either a diamond or club represent not hit. Draw 5 cards, with replacement. Record results.

5. From a table of random numbers choose 31 digits, accepting only the digits 1–7 as choices. Let the occurrence of a 1 or a 2 represent catching a cold and the occurrence of 3, 4, 5, or 7 represent not catching a cold. Record the results.

6. $60\% = \frac{3}{5}$. To establish a ratio of $\frac{3}{5}$, consider only the five outcomes 1, 2, 3, 4, 5 (disregard an outcome of 6). Let the outcomes 1, 2, 3 represent rain and the outcomes 4, 5 represent no rain. To represent the 7-day period, roll

7 cubes as one trial. Count as a success a trial in which 1, 2, or 3 occurs on exactly three of the seven cubes.

7. Answers will vary.

8. Since 60% is equivalent to $\frac{6}{10}$, choose 6 of the 10 digits to represent rain, say 1–6. Then 7, 8, 9, 0 represent no rain. From a table, randomly mark off a block of 7-digit numbers. Count as a success a 7-digit number in which 1–6 appear as exactly three of the digits.

9. Answers will vary.

10. $P(\text{rain}) = {}_7C_3\left(\frac{3}{5}\right)^3\left(\frac{2}{5}\right)^4 \approx 0.194$

PRACTICE
1. Maury; with a device with more than 2 outcomes, define only 2 outcomes, on a number cube: define even or odd as the outcomes.

2. Standard deck of cards:
   heart = does not break
   other suit = does break
   Draw 4 cards, with replacements each time.
   Record 25 trials. Answers will vary.
   $P(\text{break}) = {}_4C_1\left(\frac{3}{4}\right)^1\left(\frac{1}{4}\right)^3 \approx 0.047$

3. Roll 5 number cubes; outcomes of 1 = not returning, outcomes of 2, 3, 4, or 5 = return, disregard outcome of 6.
   Record 25 trials. Answers will vary.
   $P(\text{exactly 4 return}) = {}_5C_4\left(\frac{4}{5}\right)^4\left(\frac{1}{5}\right)^1 \approx 0.41$

4. Roll 20 number cubes; outcomes of 1–5 = cured, outcomes of 6 = not cured.
   Record 25 trials, answers will vary.
   $P(\text{exactly 14 cures}) = {}_{20}C_{14}\left(\frac{5}{6}\right)^{14}\left(\frac{1}{6}\right)^6 \approx 0.065$

5. $P(\text{exactly 10, declared major})$
   $= {}_{20}C_{10}\left(\frac{2}{5}\right)^{10}\left(\frac{3}{5}\right)^{10} \approx 0.12$

6. $P(\text{exactly 2 defectives}) = {}_{10}C_2\left(\frac{1}{10}\right)^2\left(\frac{9}{10}\right)^8 \approx 0.19$

7. $P(\text{complete 20}) = {}_{30}C_{20}\left(\frac{3}{5}\right)^{20}\left(\frac{2}{5}\right)^{10} \approx 0.12$

8. The outcomes 2 and 12 appear with about the same frequency. So do rolls of 3 and 11, 4 and 10, 5 and 9, 6 and 8. Outcome 7 has the highest frequency. By dividing each frequency by total number of trials (32, 476), the probabilities are similar to Explore of Lesson 14.4.

9. Answers will vary. Students should find probabilities about the same.

10. Use 2 number cubes. Associate each of the 36 outcomes with a math card. Roll the 2 cubes 100 times. Record the results in a 36-cell matrix, corresponding to the number of elements in the sample space. To show at least 34 different math cards, the matrix must have no more than 2 empty cells. Repeat the simulation 20 times.

11. Answers will vary. Students may compare results with the theoretical probability of about 0.6.

12. Toss 10 coins; heads = correct, tails = incorrect; success = exactly 7 heads on the 10 coins.

13. $P(\text{exactly 7 correct answers}) = {}_{10}C_7\left(\frac{1}{2}\right)^7\left(\frac{1}{2}\right)^3 \approx 0.117$

14. Roll a number cube 96 times. Let an outcome of 1 = no show, and outcomes 2–5 = will show. Record whether the outcome 1 occurs fewer than 11 times. Repeat the simulation 20 times and record the fraction of simulations in which there are fewer than 11 occurrences of an outcome of 1.

15. ${}_{96}C_0\left(\frac{5}{6}\right)^{96} + {}_{96}C_1\left(\frac{5}{6}\right)^{95}\left(\frac{1}{6}\right)^1 + \ldots + {}_{96}C_{10}\left(\frac{5}{6}\right)^{86}\left(\frac{1}{6}\right)^{10}$

16. Answers will vary, but should approach π

EXTEND

17. 28.27 square units

Run a computer program to randomly choose R points in the square $ABCD$, test these points in the inequality $x^2 + y^2 < 9$ that defines the area enclosed by the circle, and count the number of points, N, that satisfy the inequality.

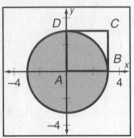

Area of circle $= 36 \cdot \dfrac{N}{R}$

18. 32/3 square units

Run a computer program to randomly choose R points in the rectangle $ABCD$, test these points in the inequality $y < -x^2 + 4$ that defines the area enclosed by the parabola and the x-axis, and count the number of points, N, that satisfy the inequality.

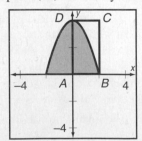

Area enclosed by parabola and x-axis $= 16 \cdot \dfrac{N}{R}$

19. for i: 1 . . . 100

20. put "# of 1's is", count 1
put "# of 3's is", count 3
put "# of 5's is", count 5

21. rand int (coin, 0, 1)   **22.** rand int (card, 1, 13)

MIXED REVIEW

23. $5\sqrt{27} - \sqrt{108} - \sqrt{75}$
$= 15\sqrt{3} - 6\sqrt{3} - 5\sqrt{3} = 4\sqrt{3}$

24. $\sqrt[3]{192} - \sqrt[3]{0.375}$
$= 4\sqrt[3]{3} - 0.5\sqrt[3]{3} = 3.5\sqrt[3]{3}$

25. B
$\sqrt{-4} \cdot \sqrt{-25}$
$= 2i \cdot 5i = -10$

Lesson 14.8, pages 724–727

1. Whether or not a family has children in need of SAT preparation.

2. To satisfy a ratio of 30% or 3:10, you want to survey 3 families with high-school age children for every 7 families without high-school age children. So, for the stratified sample of 20 families, 6 should have high-school age children (30% or 20) and 14 should not (70% of 20).

3. Answers will vary. One method is to print out two lists of all the families in the neighborhood, one list containing those with high-school age children and the other list containing those without. Assign numbers to the families on each list. Using 6 numbers generated randomly by a computer, choose 6 families from the first list; similarly, choose 14 families from the second list.

4. Limiting the sampling to those families with high-school age children will show whether there is a strong enough suppose among the actual target group to generate the needed energies and money. Perhaps this should not be an effort expected from the general community.

INVESTIGATE FURTHER

5. When a small sample is taken from a population, it is very likely that the members of the sample will come from the center of the population distribution. Most students who are to take the SAT will want some special preparation.

6a.

Sample of Size 3		Sample Mean	
75	85	80	80
75	85	97	$85\frac{2}{3}$
75	80	97	84
85	80	97	$87\frac{1}{3}$

6b. $\dfrac{80 + 85\frac{2}{3} + 84 + 87\frac{1}{3}}{4} = 84.25$
Mean of sample means.

$\dfrac{75 + 85 + 80 + 97}{4} = 84.25$
Mean of population of 4

The mean of the distribution that consists of the sample means is equal to the mean of the population.

6c. Any sample taken from a population must have a mean that is smaller than the largest data point of the population and larger than the smallest data point of the population. In this example, each sample mean is smaller than 97 and larger than 75. Thus, the sample means differ from the mean of the population by less than the original data and the standard deviation of the set of sample means must be smaller than the standard deviation of the population.

6d. σ(set of sample means) ≈ 2.72
σ(population of 4) ≈ 8.17

APPLY THE STRATEGY

7. Poll the whole population. The data can be found by a show of hands in each senior homeroom.

8. Sample a taste but be sure the dough is well mixed.

9. Too small a sample; people walking by building may have business in building and may not be representative.

10. A local telephone directory limits the sample to a geographic area; many women do not list their names in a directory; most women with full time jobs are not home on a weekday morning.

11. Not random sample; if, knowing that every 20th appliance is to be inspected, workers take more care with these particular appliances.

12. Assign a number from 1 through 300 to the address of each household. Use a computer, calculator, or table to generate 25 random number from 1 through 300.

13. Be sure the candies are well mixed; do not look as you draw each of the 25 sample candies from the carton; reach into different areas of the carton.

14. Use 2 number cubes, assign each student an ordered pair of rolls.

15. Use a standard deck of cards, assigning each week to a card.

16. Stratified sample; the college would want the library to be used by all levels of the student population.

17. 145 males of whom: 80 are undergraduates,
50 are graduates, and
15 are PhD
105 females of whom: 70 are undergraduates
25 are graduates
10 are PhD

18. $4.4 + 4.2 + 3 \cdot 4.1 + 6 \cdot 4 + 8 \cdot 3.9 + 5 \cdot 3.8 +$
$4 \cdot 3.7 + 3.6 + 3.5 = 117; \frac{117}{30} = 3.9$ oz mean
for sample

19. If the selection was random, then the mean weight of the sample should be close to the mean of the entire population.

20. $S_x = 0.1838$, if this data is a sample
$\sigma_x = 0.1807$, if this data is the population

21a.

Sample of Size 2	Mean
{12, 15}	13.5
{12, 16}	14.0
{12, 17}	14.5
{15, 16}	15.5
{15, 17}	16.0
{16, 17}	16.5

21b. sample mean
$= \frac{13.5 + 14 + 14.5 + 15.5 + 16.0 + 16.5}{6} = 15$
sample deviation: $\sigma_x \approx 1.0801$

21c. for the data set:
mean, $\frac{12 + 15 + 16 + 17}{4} = 15$;
Standard deviation, $\sigma_x = 1.8708$

21d. The mean of the population is the same as the mean of the samples. The standard deviation of the samples is smaller than the standard deviation of the population.

REVIEW PROBLEM SOLVING STRATEGIES

1a. $P(\text{win}) = \frac{1}{2}$

1b. $P(\text{win}) = \frac{1}{2}$

1c. $P(\text{win}) = \frac{1}{3}$

1d. Put 1 green disk in one box and 9 green disks and 10 red disks in the other box, then
$P(\text{green}) = \frac{1}{2} \cdot 1 + \frac{1}{2} \cdot \frac{9}{19} = \frac{14}{19}$

2. The minimum number of weighings is three as follows:

1) Using just the balance, divide the 180 ounces between the two pans. Each will hold 90 ounces.

2) Again, using just the balances, divide one 90 ounces batch into two 45-ounce batches.

3) Remove 5 ounces from one 45-ounce batch by balancing against the two weights. Use this 40-ounce for one batch and combine all the rest of the sugar for a second batch of 140 ounces.

3. The baskets cannot be exactly equal since the total worth of $236 is not divisible by 3.
One solution is—
Basket 1: 1 soap, 4 lotion, 1 perfume—worth $78
Basket 2: 2 soap, 2 lotion, 2 perfume—worth $78
Basket 3: 1 soap, 3 lotion, 2 perfume—worth $80
Another solution is—
Basket 1: 1 soap, 4 lotion, 1 perfume—worth $78
Basket 2: 1 soap, 4 lotion, 1 perfume—worth $78
Basket 3: 2 soap, 1 lotion, 3 perfume—worth $80

Chapter Review, pages 728–729

1. d **2.** a **3.** b **4.** e **5.** c

6. $_6P_6 = \dfrac{6!}{(6-6)!} = 720$

7. $_6C_2 = \dfrac{6!}{2!4!} = 15$

8. $P(\text{odd}) = \dfrac{5}{10}$ or $\dfrac{1}{2}$

$P(\text{greater than 6}) = \dfrac{4}{10}$ or $\dfrac{2}{5}$

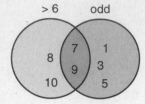

$P(\text{odd and greater than 6}) = \dfrac{2}{10}$ or $\dfrac{1}{5}$

$P(\text{odd or greater than 6}) = \dfrac{5}{10} + \dfrac{4}{10} - \dfrac{2}{10} = \dfrac{7}{10}$

9. $P(\text{both black}) = \dfrac{1}{2} \cdot \dfrac{1}{2} = \dfrac{1}{4}$

10. $P(\text{both black}) = \dfrac{1}{2} \cdot \dfrac{25}{51} = \dfrac{25}{102}$

11. $P(\text{Black}) = \dfrac{5}{8}$

$P(1^{\text{st}}\text{ bag and Black}) = \dfrac{1}{4}$

$P(1^{\text{st}}\text{ bag given Black}) = \dfrac{\frac{1}{4}}{\frac{5}{8}} = \dfrac{2}{5}$

12. $2\sigma = 68\%$

13. third $\sigma = 13.5\%$ **14.** $_4C_2\left(\dfrac{1}{6}\right)^2\left(\dfrac{5}{6}\right)^2 = \dfrac{25}{216} \approx 11.6\%$

15. $_4C_2\left(\dfrac{1}{6}\right)^2\left(\dfrac{5}{6}\right)^2 + {_4C_3}\left(\dfrac{1}{6}\right)^3\left(\dfrac{5}{6}\right)^1 + {_4C_1}\left(\dfrac{1}{6}\right)^4\left(\dfrac{5}{6}\right)^0$

$= \dfrac{19}{144}$ or about 13.2%

16. $P(\text{detection}) = 0.9$
$P(\text{no detection}) = 0.1$
$P(3\text{ failure}) = 0.1 \cdot 0.1 \cdot 0.1 = 0.001$

17. One model is to use a standard deck of cards. Let getting a spade represent "does not cure" and getting a card of any other suit represent "does cure." One trial is to pick one card from the full deck.

18. One model is to use a standard deck of cards. Let getting a spade represent "is a freshman" and getting a card of any other suit represent "is not a freshman." One trial is to pick one card from the full deck. The probability is 0.144.

19. $50 \times 60\% = 30$
$50 \times 30\% = 15$
$50 \times 10\% = 5$

20.

Sample of Size 2	Mean
{7, 10}	8.5
{7, 13}	10.0
{7, 18}	12.5
{10, 13}	11.5
{10, 18}	14.0
{13, 18}	15.5

For the set of sample means,
the mean is 12
the standard deviation is about 2.345

Chapter Assessment, page 730

1. $_5C_2 \cdot {_{21}C_4} = 59{,}850$ **2.** $\dfrac{6!}{3!} = 120$

3a. $_7P_7 = 5040$ **3b.** $(7-1)! = 720$

4. $2 \cdot {_7P_7} \cdot {_7P_7} = 50{,}803{,}200$ **5.** D

6. If the first of the two events has no effect on the second, the events are independent. For example, if the first of the two cards drawn from the standard deck of cards is replaced before the second card is drawn, the sample space for the second card is the same as that of the first, and the drawings of the two cards are independent events. If, however, the first card is not replaced before the second card is drawn, the sample space for the second card is different from that of the first, and the drawings of the two cards are dependent events. Events that cannot occur at the same time are mutually exclusive. For example, a single card drawn from a deck cannot be both red and black.

7. $P(2\text{ red}) = \dfrac{1}{3} \cdot \dfrac{1}{3} = \dfrac{1}{9}$

8. $P(2\text{ are not red}) = \dfrac{2}{3} \cdot \dfrac{2}{3} = \dfrac{4}{9}$

9. $P(\text{red, then Blue}) = \dfrac{1}{3} \cdot \dfrac{2}{5} = \dfrac{2}{15}$

10. $P(\text{red, followed by red without replacement})$

$= \dfrac{1}{3} \cdot \dfrac{9}{29} = \dfrac{3}{29}$

11. $P(4\text{ on 3}^{\text{rd}}\text{ roll}) = \dfrac{1}{6}$

12. 6 ways to arrange 2 boys and 2 girls in a line and 3 ways to arrange with one girl on an end and the girls are separated.

GBGB	GGBB
GBBG	BBGG
BGBG	BGGB

$P(\text{girls separated}) = \dfrac{3}{6} = \dfrac{1}{2}$

13. D; $P(\text{exact 1 male, in family of 6 children})$

$= {_6C_1} \cdot \left(\dfrac{1}{2}\right)^5\left(\dfrac{1}{2}\right)^2 = \dfrac{6}{64} = \dfrac{3}{32}$

14. $P(\text{exactly 1 wrong}) = {_5C_1} \cdot \left(\dfrac{1}{2}\right)^4\left(\dfrac{1}{2}\right)^1 = \dfrac{5}{32}$

15. $P(\text{exactly } 1, 2\text{A's}) = 3\left(\frac{1}{4}\right)^2\left(\frac{3}{4}\right) = \frac{9}{64}$

16. $_3C_0\left(\frac{2}{3}\right)^0\left(\frac{1}{3}\right)^3 + {}_3C_1\left(\frac{2}{3}\right)^1\left(\frac{1}{3}\right)^2 = \frac{7}{27}$ or about 0.259

17. $P(\text{Region A exactly twice}) = {}_3C_2\left(\frac{1}{6}\right)^2\left(\frac{5}{6}\right)^1 = \frac{5}{72}$
    or about 0.0694

18. $P(\text{Region D at least twice})$

$= {}_3C_2\left(\frac{1}{3}\right)^2\left(\frac{2}{3}\right)^1 + {}_3C_3\left(\frac{1}{3}\right)^3\left(\frac{2}{3}\right)^0 = \frac{7}{27}$ or about 0.259

19. C

Cumulative Review, page 732

1. $(7 - 1)! = 720$

2. $f(-x) = x^4 + 2x^3 + x^2 - 2$

$$\underline{1\rfloor} \quad 1 \quad 2 \quad 1 \quad 0 \quad -2$$
$$\quad\quad\quad\quad 1 \quad 3 \quad 4 \quad 4$$
$$\overline{\quad\quad 1 \quad 3 \quad 4 \quad 4 \quad 2}$$

Lower bound is -1.

3. B

4. Answers will vary. On way to establish a ratio of 40% is to let an outcome of 1 or 2 represent *rain* and an outcome of 3, 4, or 5 represent *no rain*. Disregard an outcome of 6. Roll 5 cubes at a time to represent the 5 days.

5. Since directrix is horizontal and below the focus, parabola is in the form of $(x - h)^2 = 4p(y - k)$.

$h = 4; k = 1; p = \dfrac{1 - (-3)}{2} = 2$

$(x - 4)^2 = 4(2)(y + 1)$

$x^2 - 8x + 16 = 8y + 8$

$8y = x^2 - 8x + 8$

$y = \dfrac{1}{8}x^2 - x + 1$

6.

y-value	50		30		16		8		6		10		20
1st order diff.		−20		14		8		2		1		10	
2nd order diff.			6		6		6		6		6		
3rd order diff.				0		0		0		0			

$2a = 6$
$a = 3$

$3a + b = 4$
$3(3) + b = 4$
$b = -5$

$a + b + c = 6$
$3 + (-5) + c = 6$
$c = 8$

$y = 3x^2 - 5x + 8$
$3(20)^2 - 5(20) + 8 = 1108$

7. $68\% \cdot 25 = 17$

8. $y = -x^2$

$\dfrac{y + 1}{x + 1} = \dfrac{1}{2}$

$\dfrac{-x^2 + 1}{x + 1} = \dfrac{1}{2}$

$-2x^2 + 2 = x + 1$

$0 = 2x^2 + x - 1$

$0 = (2x - 1)(x + 1)$

$2x - 1 = 0$ or $x + 1 = 0$

$2x = 1$

$x = \dfrac{1}{2}$ or $\quad x = -1$

Since $x = -1$ makes the denominator of $\dfrac{y + 1}{x + 1}$ equal to 0, it is not a solution.

$y = -x^2; y = -\left(\dfrac{1}{2}\right)^2; y = -\dfrac{1}{4}$

$x = \dfrac{1}{2}, y = -\dfrac{1}{4}$

9. Let $x =$ time for plane
$66 + x =$ time for train
$r =$ rate for train

$r(66 + x) = 12rx$
$x = 6$ h
$66 + x = 72$ h

$\dfrac{3240 \text{ mi}}{72 \text{ h}} = 45$ mph for train

10. $\dfrac{4}{11} \cdot \dfrac{3}{10} + \dfrac{5}{11} \cdot \dfrac{4}{11} + \dfrac{2}{11} \cdot \dfrac{1}{10} = \dfrac{34}{110}$ or $\dfrac{17}{55}$

11. $\displaystyle\sum_{n=1}^{5}[(n + 1)^3 - n^3] = (2^3 - 1^3) + (3^3 - 2^3) +$

$(4^3 - 3^3) + (5^3 - 4^3) + (6^3 - 5^3)$

$= 7 + 19 + 37 + 61 + 91 = 215$

12. D; $\dfrac{x^2}{x - 3} + \dfrac{9}{3 - x} = \dfrac{x^2}{x - 3} - \dfrac{9}{x - 3} = \dfrac{x^2 - 9}{x - 3} = $

$\dfrac{(x + 3)(x - 3)}{x - 3} = x + 3$

13.

Sample of Size 3			Sample Mean
145	149	168	154
145	149	154	149.3
145	168	154	155.7
149	168	154	157

Mean of sample means $= 154$

14. $a_1 = 6$

$a_2 = \dfrac{2}{3}a_1 = \dfrac{2}{3}(6) = 4$

$a_3 = \dfrac{2}{3}a_2 = \dfrac{2}{3}(4) = \dfrac{8}{3}$

$a_4 = \dfrac{2}{3}a_3 = \dfrac{2}{3}\left(\dfrac{8}{3}\right) = \dfrac{16}{9}$

$a_5 = \dfrac{2}{3}a_4 = \dfrac{2}{3}\left(\dfrac{16}{9}\right) = \dfrac{32}{27}$

15. A quadrilateral is a parallelogram of any one of the following conditions is satisfied.
The two diagonals have the same midpoint.
Both pairs of opposite sides are parallel (use slope).
Both pairs of opposite side are equal lengths.
One pair of opposite sides are parallel and equal.

16.
$$3 | \quad 1 \quad -1 \quad -15$$
$$ \quad \quad \quad 3 \quad \quad 6$$
$$\overline{ \quad 1 \quad \quad 2 \quad \quad -9} \quad R = -9$$

17.
$$-1 | \quad 1 \quad \quad 0 \quad \quad 0 \quad -1$$
$$ \quad \quad \quad -1 \quad \quad 1 \quad -1$$
$$\overline{ \quad 1 \quad -1 \quad \quad 1 \quad -2} \quad R = -2$$

Standardized Test, page 733

1. -1

2. 2; -2 and 0

3. $\dfrac{8!}{2!5!} = \dfrac{40,320}{2 \cdot 120} = 168$

4. $_8P_3 + {}_4P_0 = 8 \cdot 7 \cdot 6 + 1 = 337$

5. $_5C_2 + {}_{10}C_3 = 10 + 120 = 130$

6. $3x^2 - 4x - 4 = 0$
$$x^2 - \frac{4}{3}x - \frac{4}{3} = 0$$
$$s_1 + s_2 = \frac{4}{3}$$
$$s_1 s_2 = -\frac{4}{3}$$
$$\text{sum} = \frac{4}{3} - \frac{4}{3} = 0$$
or
$$3x^2 - 4x - 4 = 0$$
$$(3x + 2)(x - 2) = 0$$
$$3x + 2 = 0 \text{ or } x - 2 = 0$$
$$3x = -2$$
$$x = -\frac{2}{3} \text{ or } x = 2$$
$$\text{Sum} = -\frac{2}{3} + 2 = \frac{4}{3}$$
$$\text{Product} = -\frac{2}{3} \cdot 2 = -\frac{4}{3}$$
$$\frac{4}{3} + \left(-\frac{4}{3}\right) = 0$$

7. $16x^2 + y^2 = 64$
$$\frac{x^2}{4} + \frac{y^2}{64} = 1$$
$$a^2 = 64$$
$$a = 8$$
Major axis $= 2a = 2(8) = 16$

8. $(x - 5)^2 + y^2 = 25$
$$y - x = 5$$
$$y = x + 5$$
$$(x - 5)^2 + (x + 5)^2 = 25$$
$$x^2 - 10x + 25 + x^2 + 10x + 25 = 25$$
$$2x^2 + 50 = 25$$

$$2x^2 = -25$$
$$x^2 = -\frac{25}{2}$$
Since no real number squared can be negative, the system has 0 solutions.

9. $\dfrac{2x}{x + 5} + \dfrac{1}{x - 5} = \dfrac{10}{x^2 - 25}$

$$\frac{2x}{x + 5} \cdot \frac{x - 5}{x - 5} + \frac{1}{x - 5} \cdot \frac{x + 5}{x + 5} = \frac{10}{x^2 - 25}$$
$$\frac{2x^2 - 10x}{x^2 - 25} + \frac{x + 5}{x^2 - 25} = \frac{10}{x^2 - 25}$$
$$2x^2 - 10x + x + 5 = 10$$
$$2x^2 - 9x + 5 = 10$$
$$2x^2 - 9x - 5 = 0$$
$$(2x + 1)(x - 5) = 0$$
$$2x + 1 = 0 \text{ or } x - 5 = 0$$
$$2x = -1$$
$$x = -\frac{1}{2} \text{ or } x = 5$$

Since $x = 5$ makes the denominator of $\dfrac{1}{x - 5}$ equal to 0, it is not a solution. The only solution is $-\dfrac{1}{2}$.

10. The silver dollar in combination with any of the other coins are the only possibilities.
$$\frac{1}{5} \cdot \frac{4}{4} + \frac{4}{5} \cdot \frac{1}{4} = \frac{1}{5} + \frac{1}{5} = \frac{2}{5}$$

11. $\dfrac{4!}{3!1!}(3)^1(1)^3 = 12$

12. $y = I(0.9996)^t = 54(0.9996)^{50} \approx 52.9$

13. $16^0 + 16^{-\frac{1}{2}} = 1 + \dfrac{1}{\sqrt{16}} = 1 + \dfrac{1}{4} = \dfrac{5}{4}$

14. $1.95^x = 54$
$$\log_{1.95} 1.95^x = \log_{1.95} 54$$
$$x = \frac{\log 54}{\log 1.95}$$
$$x = 5.97$$

15.

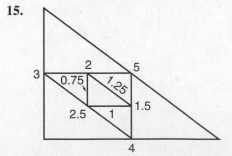

$$3 + 4 + 5 = 12$$
$$1.5 + 2 + 2.5 = 6$$
$$0.75 + 1 + 1.25 = 3$$
Geometric series; $a_1 = 12$; $r = \dfrac{1}{2}$

$$\sum_{n=1}^{\infty} 12\left(\frac{1}{2}\right)^{n-1} = \frac{a_1}{1 - r} = \frac{12}{1 - \dfrac{1}{2}} = 12 \cdot 2 = 24$$

16. $_5C_3 \cdot \left(\frac{1}{3}\right)^3 \left(\frac{2}{3}\right)^2 = \frac{40}{243} \approx 0.165$

17.
$$9x^2 - 4y^2 - 144 = 0$$
$$9x^2 - 4y^2 = 144$$
$$\frac{x^2}{16} - \frac{y^2}{36} = 1$$
$$a^2 = 16$$
$$a = 4$$

Vertices: $(4, 0), (-4, 0)$

18.

```
3│  2   -11    14    -2    12    9
        6    -15    -3   -15   -9
3│  2    -5    -1    -5    -3    0
        6      3     6     3
1│  2    1      2     1     0
2
       -1      0    -1
    2   0      2     0
```

$x^2 + 2$ has no real solutions

The only rational solutions are 3, 3, and $-\frac{1}{2}$.

The largest is 3.

19. $\frac{3 + 3i}{2 - 4i} = \frac{3 + 3i}{2 - 4i} \cdot \frac{2 + 4i}{2 + 4i} = \frac{6 + 12i + 6i + 12i^2}{4 + 8i - 8i - 16i^2} =$

$\frac{6 + 18i - 12}{4 + 16} = \frac{-6 + 18i}{20} = -\frac{3}{10} + \frac{9}{10}i$

20. Let x = number of leather suits

Let y = number of suede suits

$c = 80x + 160y$

Restraints: $20 \leq x \leq 90$
$$y \leq 100$$
$$51 \leq x + y \leq 120$$
$$y \leq \frac{1}{2}x$$

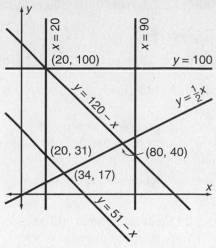

$C(20, 100) = 80(20) + 160(100) = \$17,600$
$C(80, 40) = 80(80) + 160(40) = \$12,800$
$C(34, 17) = 80(34) + 160(17) = \$5440$
$C(20, 31) = 80(20) + 160(31) = \$6560$
Minimum investment is \$5440.

21. By similar triangles,
$$\frac{x}{6} = \frac{6}{x + 4}$$
$$x^2 + 4x = 36$$
$$x^2 + 4x - 36 = 0$$
$$x = \frac{-4 \pm \sqrt{4^2 - 4(-36)}}{2}$$
$$x = \frac{-4 \pm 12.6}{2} = \frac{8.6}{2} = 4.3, \text{ negative solution is not a length}$$

So hypotenuse is $4.3 + 4.3 + 4 = 12.6$ cm

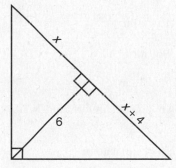

Chapter 15 Trigonometric Functions and Graphs

Data Activity, pages 734–735

1. Range is the difference between the greatest and least data value.
$193,796,104 - 37,170,223 = 156,625,881$

2. Answers will vary.
$117,000,000 + 46,000,000 + 45,000,000 + 43,000,000 + 37,000,000 = 290,000,000$ tons to the nearest 10 million

3. 9 out of 20 ports are in Louisiana or Texas.
$\dfrac{9}{20} = \dfrac{45}{100} = 45\%$

4. Four of the top 20 are in Texas. Since 20% of 25 is $(0.2)(25) = 5$, there must be 1 more Texas port in the top 25. Therefore, 5 are in the extended list.

5. $\dfrac{24,202,970}{54,320,932} \approx 0.4455 = 44.6\%$ to the nearest tenth

Lesson 15.1, pages 737–742

EXPLORE

1a. $m\angle EON = 90°$ **1b.** $m\angle EOW = 180°$

1c. $m\angle EOS = 270°$

2. $\dfrac{1}{4}, \dfrac{1}{2}, \dfrac{3}{4}$ **3.** $C = 2\pi r$

4. $\dfrac{1}{4}C = \dfrac{1}{4}(2\pi r) = \dfrac{\pi}{2}r$

$\dfrac{1}{2}C = \dfrac{1}{2}(2\pi r) = \pi r$

$\dfrac{3}{4}C = \dfrac{3}{4}(2\pi r) = \dfrac{3}{2}\pi r$

5. $\dfrac{1}{4}C = \dfrac{\pi}{2}(3960) = 1980\pi$ mi

$\dfrac{1}{2}C = \pi(3960) = 3960\pi$ mi

$\dfrac{3}{4}C = \dfrac{3}{2}\pi(3960) = 5940\pi$ mi

6. Multiply the fraction by the circumference C which equals $2\pi r$.

7. The arc from Bangor to the equator is $\dfrac{45}{360} = \dfrac{1}{8}$ of the whole circle. Therefore, its length is
$\dfrac{1}{8}(2\pi r) = \dfrac{1}{8}(2\pi \cdot 3960) \approx 3110$ mi.

The arc from Austin to the equator is $\dfrac{30}{360} = \dfrac{1}{12}$ of the whole circle. Therefore, its length is
$\dfrac{1}{12}(2\pi r) = \dfrac{1}{12}(2\pi \cdot 3960) \approx 2073$ mi.

TRY THESE

1. $180° < \theta < 270°$

2. Quadrant III **3.** Quadrant I **4.** Quadrant II

5. $30° = 30 \cdot \dfrac{\pi}{180} = \dfrac{\pi}{6}$ radians

6. $315° = 315 \cdot \dfrac{\pi}{180} = \dfrac{7\pi}{4}$ radians

7. $200° = 200 \cdot \dfrac{\pi}{180} = \dfrac{10\pi}{9}$ radians

8. $\dfrac{\pi}{3}$ radians $= \dfrac{\pi}{3} \cdot \dfrac{180}{\pi} = 60°$

9. $\dfrac{5\pi}{4}$ radians $= \dfrac{5\pi}{4} \cdot \dfrac{180}{\pi} = 225°$

10. 0.7π radians $= 0.7\pi\left(\dfrac{180}{\pi}\right) = 126°$

11. $s = \theta r$
$s = \dfrac{5\pi}{12}(13) \approx 17.0$ in. to the nearest tenth.

12. $30°$ is $\dfrac{30}{360}$ or $\dfrac{1}{12}$ of a complete rotation. So, the length of the arc is $\dfrac{1}{12}$ the circumference of the circle, $C = 2\pi r$

Therefore, $\dfrac{1}{12}(2\pi r) = 6$

$\dfrac{\pi}{6}r = 6$

$r = 6\left(\dfrac{6}{\pi}\right)$

$r \approx 11.5$ m

13. Answers will vary. Since radian measure is defined as the ratio of arc length to radius, $\theta = \dfrac{s}{r}$, arc length is easily found using the formula $s = \theta r$.

PRACTICE

1. $270° < \theta < 360°$

2. Quadrant II **3.** Quadrant III **4.** Quadrant IV

5. $90° = 90 \cdot \dfrac{\pi}{180} = \dfrac{\pi}{2}$ radians

6. $150° = 150 \cdot \dfrac{\pi}{180} = \dfrac{5\pi}{6}$ radians

7. $330° = 330 \cdot \dfrac{\pi}{180} = \dfrac{11\pi}{6}$ radians

8. $60° = 60 \cdot \dfrac{\pi}{180} = \dfrac{\pi}{3}$ radians

9. $225° = 225 \cdot \dfrac{\pi}{180} = \dfrac{5\pi}{4}$ radians

10. $72° = 72 \cdot \dfrac{\pi}{180} = \dfrac{2\pi}{5}$ radians

11. $340° = 340 \cdot \dfrac{\pi}{180} = \dfrac{17\pi}{9}$ radians

12. $6° = 6 \cdot \dfrac{\pi}{180} = \dfrac{\pi}{30}$ radians

13. $54° = 54 \cdot \dfrac{\pi}{180} = \dfrac{3\pi}{10}$ radians

14. $255° = 255 \cdot \dfrac{\pi}{180} = \dfrac{17\pi}{12}$ radians

15. $81° = 81 \cdot \dfrac{\pi}{180} = \dfrac{9\pi}{20}$ radians

16. $2° = 2 \cdot \dfrac{\pi}{180} = \dfrac{\pi}{90}$ radians

17. 2π radians $= 2\pi \cdot \dfrac{180}{\pi} = 360°$

18. $\dfrac{2\pi}{3}$ radians $= \dfrac{2\pi}{3} \cdot \dfrac{180}{\pi} = 120°$

19. $\dfrac{3\pi}{2}$ radians $= \dfrac{3\pi}{2} \cdot \dfrac{180}{\pi} = 270°$

20. $\dfrac{3\pi}{5}$ radians $= \dfrac{3\pi}{5} \cdot \dfrac{180}{\pi} = 108°$

21. $\dfrac{\pi}{4}$ radians $= \dfrac{\pi}{4} \cdot \dfrac{180}{\pi} = 45°$

22. $\dfrac{7\pi}{6}$ radians $= \dfrac{7\pi}{6} \cdot \dfrac{180}{\pi} = 210°$

23. $\dfrac{11\pi}{8}$ radians $= \dfrac{11\pi}{8} \cdot \dfrac{180}{\pi} - 247.5°$

24. $\dfrac{5\pi}{9}$ radians $= \dfrac{5\pi}{9} \cdot \dfrac{180}{\pi} = 100°$

25. $\dfrac{16\pi}{9}$ radians $= \dfrac{16\pi}{9} \cdot \dfrac{180}{\pi} = 320°$

26. $\dfrac{\pi}{90}$ radians $= \dfrac{\pi}{90} \cdot \dfrac{180}{\pi} = 2°$

27. 0.95π radians $= 0.95\pi\left(\dfrac{180}{\pi}\right) = 171°$

28. 1.66π radians $= 1.66\pi\left(\dfrac{180}{\pi}\right) = 298.8°$

29. $90 - 48 = 42°$ **30.** $\dfrac{\pi}{2} - \dfrac{\pi}{8} = \dfrac{3\pi}{8}$

31. $\dfrac{\pi}{2} - \dfrac{4\pi}{9} = \dfrac{9\pi}{18} - \dfrac{8\pi}{18} = \dfrac{\pi}{18}$

32. $180 - 35.72 = 144.28°$ **33.** $\pi - \dfrac{5\pi}{12} = \dfrac{7\pi}{12}$

34. $\pi - 0.87\pi = 0.13\pi$

35. The numbers on a clock divide the circle into 12 equal parts. So, when the minute hand turns from 6 at 2:30 to 11 at 2:55 it has turned through an arc that is $\dfrac{5}{12}$ of the circle. Therefore, the radian measure of the angle through which it turns is $\dfrac{5}{12}$ of 2π or $\dfrac{5\pi}{6}$. To find how far it travels use $s = \theta r$. So,
$$s = \dfrac{5\pi}{6} \cdot 6 = 5\pi \text{ in.}$$

36. The sum of the angles of a triangle is $180°$ or π radians. Let the measures of the angles of the triangle be represented by $3x$, $4x$, and $5x$.
$$3x + 4x + 5x = \pi$$
$$12x = \pi$$
$$x = \dfrac{\pi}{12}$$
So, the measures are $3x = \dfrac{3\pi}{12} = \dfrac{\pi}{4}$, $4x = \dfrac{4\pi}{12} = \dfrac{\pi}{3}$, and $5x = \dfrac{5\pi}{12}$.

37a. Let θ represent the angle the Earth moves in 1 day.
$$1 \text{ day} = \dfrac{1}{365} \text{ of a year}$$
Therefore, $\theta = \dfrac{1}{365}(2\pi) \approx 0.017$ radians

37b. $s = \theta r$
$s = (0.017)(9.3 \times 10^7) = 1.581 \times 10^6 \approx$ 1,600,000 mi

38a. $s = \theta r$
$r = 3960$ mi
$\theta = 62.7 - 62.5 = 0.2°$
Change degrees to radians: $\quad 0.2\left(\dfrac{\pi}{180}\right)$
So, $\quad s = (0.2)\left(\dfrac{\pi}{180}\right)(3960) \approx 13.8$ mi

38b. $s = \theta r$
$20 = \theta(3960)$
$\theta = \dfrac{20}{3960} = \dfrac{1}{196}°$
$\theta = \dfrac{1}{196} \cdot \dfrac{180}{\pi} \approx 0.3$
Passenger liner at latitude 62.5°N and coast guard ship at latitude $62.5 - 0.3 = 62.2°$N

39. $3° = 3 \cdot \dfrac{\pi}{180} = \dfrac{\pi}{60}$ radians
$\dfrac{\pi}{60}$ radians $\neq 3$ radians

EXTEND

40. $35°15' = \left(35\dfrac{15}{60}\right) = \left(35\dfrac{1}{4}\right) = 35.25°$

41. $80°24' = \left(80\dfrac{24}{60}\right) = \left(80\dfrac{2}{5}\right) = 80.4°$

42. $114°18'48'' = \left(114 + \dfrac{18}{60} + \dfrac{48}{3600}\right) =$
$\left(114 + \dfrac{3}{10} + \dfrac{1}{75}\right) \approx 114.31°$
Note, since $1' = \dfrac{1}{60}°$ and $1'' = \dfrac{1}{60}'$, then $1'' = \dfrac{1}{3600}°$.

43. $48°46'12'' = \left(48 + \dfrac{46}{60} + \dfrac{12}{3600}\right) =$
$\left(48 + \dfrac{23}{30} + \dfrac{1}{300}\right) = 48.77°$
$48.77°\left(\dfrac{\pi}{180}\right) \approx 0.271\pi$ radians

44. $1.49\pi \cdot \dfrac{180}{\pi} = 268.2°$

$268.2° = \left(268 + \dfrac{2}{10}\right) = \left(268\dfrac{12}{60}\right) = 268°12'$

45. $C = \pi d = \pi(30) = 30\pi$ in.
Each revolution a tire travels 30π in.
$1 \text{ mi} = 5280 \text{ ft} = 5280(12)\text{in.} = 63,360$ in.
Therefore, the tire makes $\dfrac{63,360}{30\pi} = 672.27$ revolutions.

46. If it is assumed that the intercepted arc is approximately the moon's diameter then
$$s = r\theta$$
$$2160 = r(0.00872)$$
$$r = \dfrac{2160}{0.00872} \approx 247,700 \text{ mi}$$

THINK CRITICALLY

47. Equations may vary but should be some form of the equation $\dfrac{r}{\pi} = \dfrac{d}{180}$.

48. Call the center of the Earth C. The sum of the measure of the angles of quadrilateral ACBS is 360°. So,
$\angle C = 360 - 16 - 90 - 90 = 164°$.
Therefore the length of arc AB is
$$L = \theta r = (164)\left(\dfrac{\pi}{180}\right)(3960) \approx 11,335 \text{ mi}$$

MIXED REVIEW

49. mean: $\dfrac{28 + 19 + 33 + 40 + 19}{5} = 27.8$
median: $19, 19, 28, 33, 40 \quad$ median $= 28$
mode: 19

50. mean: $\dfrac{1.6 + 1.8 - 1.9 - 0.5 + 0.2}{5} = 0.24$
median: $-1.9, -0.5, 0.2, 1.6, 1.9 \quad$ median $= 0.2$
mode: none

51. mean: $4\dfrac{1}{2} + 3\dfrac{3}{4} + 6\dfrac{1}{4} + 2\dfrac{1}{2}4 = \dfrac{17}{4} = 4\dfrac{1}{4}$
median: $2\dfrac{1}{2}, 3\dfrac{3}{4}, 4\dfrac{1}{2}, 6\dfrac{1}{4} \quad$ median $= \dfrac{3\frac{3}{4} + 4\frac{1}{2}}{2} = 4\dfrac{1}{8}$
mode: none

52. $f(x) = 5x$
$x = 5f^{-1}(x)$
$f^{-1}(x) = \dfrac{x}{5}$

53. $f(x) = x - 6$
$x = f^{-1}(x) - 6$
$f^{-1}(x) = x + 6$

54. $f(x) = -\dfrac{x}{3}$
$x = -\dfrac{f^{-1}(x)}{3}$
$3x = -f^{-1}(x)$

55. $f(x) = \dfrac{3x + 2}{4}$
$x = \dfrac{3f^{-1}(x) + 2}{4}$
$4x = 3f^{-1}(x) + 2$
$4x - 2 = 3f^{-1}(x)$
$f^{-1}(x) = \dfrac{4x - 2}{3}$

56. $d = \sqrt{(x_2 - x_1)^2 + (y_2 - y_1)^2} =$
$\sqrt{(-3 - 9)^2 + (4 - (-1))^2} =$
$\sqrt{(-12)^2 + 5^2} = \sqrt{144 + 25} = \sqrt{169} = 13$

57. $x = \dfrac{-3 + 9}{2} = \dfrac{6}{2} = 3$
$y = \dfrac{4 + (-1)}{2} = \dfrac{3}{2} = 1.5$
midpoint is (3, 1.5)

58. B; $240\left(\dfrac{\pi}{180}\right) = \dfrac{4\pi}{3}$

PROJECT CONNECTION

1–2, 4. Answers will vary.

3. The latitude is the complimentary angle of that read from the quadrant.

Lesson 15.2, pages 743–745

THINK BACK

1. The vertex is O.　　　　**2.** The initial side is $\overline{OA}$.

3. The terminal side is $\overline{OB}$.　　**4.** Quadrant IV

5. $360 - 30 = 330°$　　　**6.** $330° = 330\left(\dfrac{\pi}{180}\right) = \dfrac{11\pi}{6}$

EXPLORE

7. $m\angle AOB = 70 + 360 = 430°$

8. $m\angle DOC = -180 + (-25) = -205°$

9. III; $600 - 360 = 240°$　**10.** II; $\dfrac{11\pi}{4} - 2\pi = \dfrac{3\pi}{4}$

11. I; $400 - 360 = 40°$　　**12.** IV　　**13.** II

14. IV; $-\dfrac{13\pi}{6} = -2\pi + \left(-\dfrac{\pi}{6}\right)$

15. I; $4.2\pi - 4\pi = 0.2\pi$

16. III; $-1200° = 3(-360) + (-120)$

17. yes; $40 + 360 = 400°$

18. yes; $-90 + 360 = 270°$

19. yes; $\dfrac{3\pi}{2} + 2\pi = \dfrac{3\pi}{2} + \dfrac{4\pi}{2} = \dfrac{7\pi}{2}$

20. no; $250 + n(360) \neq 950°$

21. yes; $-1.6\pi + 3(-2\pi) = -7.6\pi$

22. no; $\dfrac{7\pi}{6} + n(2\pi) \neq -\dfrac{7\pi}{6} = -210°$

23. $-234 + 360 = 126°$

24. $638 + (-360) = 278°$

25. $1101 + (-3)(360) = 1101 - 1080 = 21°$

26. $-977 + 3(360) = -977 + 1080 = 103°$

27. $2351 + (-6)(360) = 2351 - 2160 = 191°$

28. $-3601 + (11)(360) = -3601 + 3960 = 359°$

29. $y = 60 + 360 = 420°$
$z = 60 + (-360) = -300°$

30. $30 + 360 = 420°$
$30 + 2(360) = 30 + 720 = 750°$
$30 + 3(360) = 30 + 1080 = 1110°$
$30 + 4(360) = 30 + 1440 = 1470°$
$30 + 5(360) = 30 + 1800 = 1830°$

31. Each measure exceeds the preceeding one by 360°.

32. $\frac{\pi}{2} + 2\pi = \frac{\pi}{2} + \frac{4\pi}{2} = \frac{5\pi}{2}$

$\frac{\pi}{2} + 2(2\pi) = \frac{\pi}{2} + 4\pi = \frac{\pi}{2} + \frac{8\pi}{2} = \frac{9\pi}{2}$

$\frac{\pi}{2} + 3(2\pi) = \frac{\pi}{2} + 6\pi = \frac{\pi}{2} + \frac{12\pi}{2} = \frac{13\pi}{2}$

$\frac{\pi}{2} + 4(2\pi) = \frac{\pi}{2} + 8\pi = \frac{\pi}{2} + \frac{16\pi}{2} = \frac{17\pi}{2}$

$\frac{\pi}{2} + 5(2\pi) = \frac{\pi}{2} + 10\pi = \frac{\pi}{2} + \frac{20\pi}{2} = \frac{21\pi}{2}$

33. Each measure exceeds the preceeding one by 2π.

34. For degree measures, add and subtract multiples of 360° to and from the measure of the given angle. For radian measures, add and subtract multiples of 2π to and from the measure of the given angle.

35. $100,000 \div 360 = 277.8$
Therefore let $n = 278$ in the expression $50 + n(360)$ and obtain $50 + (278)(360) =$
$50 + 100,080 = 100,130°$
Let $n = 279$ and obtain another possible answer of
$50 + (279)(360) = 50 + 100,440 = 100,490°$

36. $500 \div 2 = 250$
Therefore, let $n = 250$ in the expression $\frac{\pi}{4} + n(2\pi)$ and obtain $\frac{\pi}{4} + (250)(2\pi) = \frac{\pi}{4} + 500\pi =$
$\frac{\pi}{4} + \frac{2000\pi}{4} = \frac{2001\pi}{4}$

37. Answers will vary. Students should mention that coterminal angles are angles in standard position with coinciding terminal sides. They might say that coterminal angles may have measures greater than 360° (2π) or less than 0°. Angles with measures of 400° $\left(\frac{20\pi}{9}\right)$ and $-320°$ $\left(-\frac{16\pi}{9}\right)$ are both coterminal with a 40° $\left(\frac{2\pi}{9}\right)$ angle.

38a. $\frac{3\pi}{8} + (-2\pi) = \frac{3\pi}{8} - \frac{16\pi}{8} = -\frac{13\pi}{8}$

38b. $\frac{17\pi}{16} + (-2\pi) = \frac{17\pi}{16} + \frac{-32\pi}{16} = -\frac{15\pi}{16}$

38c. $n - 2\pi$

39a. Sketches will vary. Two typical examples are shown.

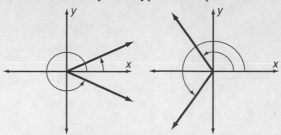

39b. Answers will vary. Students should see that each terminal side is a reflection of the other across the x-axis.

40a. 1 year later, January 1, 1996

40b. Because Saturn takes 29.5 years to complete one orbit of the sun, it will revolve $\frac{1}{29.5}$ of one 360° orbit, or about 12.2°, during the 1 year that it takes Earth to complete one orbit. Therefore, on January 1, 1997, the Earth will be back where it started, while Saturn will be about 12.2° beyond its January 1, 1996 position.

40c. They will not be realigned when $t = 1$. During the following year ($1 < t < 2$), Earth will catch up to Saturn.

40d. Earth moves through an angle of rotation of $360t$ and Saturn moves through an angle of rotation of
$\frac{1}{29.5} \cdot 360t \approx 12.2t$

40e. For $1 < t < 2$ there is a value of t such that the expression $360(t - 1) = 360t - 360$ represents the time at which Earth will catch up to Saturn.

40f. $\frac{360t}{29.5} = 360t - 360$
$\frac{720t}{59} = 360t - 360$
$720t = 59(360t - 360)$
$720t = 21,240t - 21,240$
$21,240 = 20,520t$
$t = \frac{21,240}{20,520}$
$t \approx 1.035$
$0.035(365) \approx 13$ days. So, the date would be January 13, 1997.

Lesson 15.3, pages 746–753

1. $\overline{AB} = 2, \overline{AD} = 1, \overline{BD} = \sqrt{3}$

2. In a 30–60–90 triangle the length of the hypotenuse is twice the length of the shorter leg and the length of the longer side is $\sqrt{3}$ times the length of the shorter side.

3. $\overline{AB} = 1, \overline{BC} = 1, \overline{AC} = \sqrt{2}$

4. In a 45–45–90 triangle the legs are equal in length and the hypotenuse is equal to $\sqrt{2}$ times the length of a leg.

1.

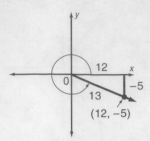

$r = \sqrt{12^2 + (-5)^2} = 13 \qquad x = 12, y = -5$

$\sin: \dfrac{y}{r} = \dfrac{-5}{13} = -\dfrac{5}{13} \qquad \csc: \dfrac{r}{y} = \dfrac{13}{-5} = -\dfrac{13}{5}$

$\cos: \dfrac{x}{r} = \dfrac{12}{13} \qquad\qquad \sec: \dfrac{r}{x} = \dfrac{13}{12}$

$\tan: \dfrac{y}{x} = \dfrac{-5}{12} = -\dfrac{5}{12} \qquad \cot: \dfrac{x}{y} = \dfrac{12}{-5} = -\dfrac{12}{5}$

2. $\tan 30 = \dfrac{\sqrt{3}}{2}$

3. $\csc(-45) = -\sqrt{2}$

4. $\sec \dfrac{4\pi}{3} = \sec 240°$

reference angle $= 240 - 180 = 60°$

$\sec 240 = -2$

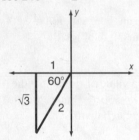

5. $\sin(-1080)$

reference angle is 0 since $3(360) = 1080°$.

$\sin(-1080) = 0$

6. 0.2924 **7.** 0.9998 **8.** 4.8459 **9.** 1.5557

10. positive, terminal side in Quadrant III

11. positive, terminal side in Quadrant I

12. negative, terminal side in Quadrant II

13. negative, terminal side in Quadrant IV

14. $L = 6087 - 30 \cos 2\theta = 6087 - 30 \cos 2(22.5) =$

$6087 - 30 \cos 45 = 6087 - 30\left(\dfrac{\sqrt{2}}{2}\right) =$

$(6087 - 15\sqrt{2})$ ft

15. Since the measure of the angle is between 270° and 360°, its terminal side lies in Quadrant IV. A reference triangle can be sketched in Quadrant IV by drawing a perpendicular to the *x*-axis from the point $(\sqrt{3} \ -1)$ on the terminal side of the angle. The hypotenuse of the reference triangle measures 2. From the definition of sine, $\sin 330° = \dfrac{y}{r} = -\dfrac{1}{2}$.

1. $r = \sqrt{(15)^2 + (-8)^2} = 17 \qquad x = 15, y = -8$

$\sin: \dfrac{y}{r} = \dfrac{-8}{17} = -\dfrac{8}{17} \qquad \csc: \dfrac{r}{y} = \dfrac{17}{-8} = -\dfrac{17}{8}$

$\cos: \dfrac{x}{r} = \dfrac{15}{17} \qquad\qquad \sec: \dfrac{r}{x} = \dfrac{17}{15}$

$\tan: \dfrac{y}{x} = \dfrac{-8}{15} = -\dfrac{8}{15} \qquad \cot: \dfrac{x}{y} = \dfrac{15}{-8} = -\dfrac{15}{8}$

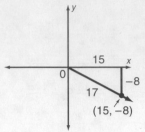

2. $r = \sqrt{(-4)^2 + 3^2} = 5 \qquad x = -4, y = 3$

$\sin: \dfrac{y}{r} = \dfrac{3}{5} \qquad\qquad\qquad \csc: \dfrac{r}{y} = \dfrac{5}{3}$

$\cos: \dfrac{x}{r} = \dfrac{-4}{5} = -\dfrac{4}{5} \qquad \sec: \dfrac{r}{x} = \dfrac{5}{-4} = -\dfrac{5}{4}$

$\tan: \dfrac{y}{x} = \dfrac{3}{-4} = -\dfrac{3}{4} \qquad \cot: \dfrac{x}{y} = \dfrac{-4}{3} = -\dfrac{4}{3}$

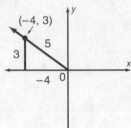

3. $r = \sqrt{1^2 + 2^2} = \sqrt{5} \qquad x = 1, y = 2$

$\sin: \dfrac{y}{r} = \dfrac{2}{\sqrt{5}} = \dfrac{2\sqrt{5}}{5} \qquad \csc: \dfrac{r}{t} = \dfrac{\sqrt{5}}{2}$

$\cos: \dfrac{x}{r} = \dfrac{1}{\sqrt{5}} = \dfrac{\sqrt{5}}{5} \qquad \sec: \dfrac{r}{x} = \dfrac{\sqrt{5}}{1} = \sqrt{5}$

$\tan: \dfrac{y}{x} = \dfrac{2}{1} = 2 \qquad\qquad \cot: \dfrac{x}{y} = \dfrac{1}{2}$

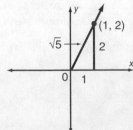

4. $r = \sqrt{4^2 + 0^2} = 4$ $\qquad$ $x = 4, y = 0$

$\sin: \dfrac{y}{r} = \dfrac{0}{4} = 0$ $\qquad$ $\csc: \dfrac{r}{y} = \dfrac{4}{0}$ undefined

$\cos: \dfrac{x}{r} = \dfrac{4}{4} = 1$ $\qquad$ $\sec: \dfrac{r}{x} = \dfrac{4}{4} = 1$

$\tan: \dfrac{y}{x} = \dfrac{0}{4} = 0$ $\qquad$ $\cot: \dfrac{x}{y} = \dfrac{4}{0}$ undefined

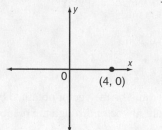

5. $\sin 225 = -\dfrac{\sqrt{2}}{2}$;

reference angle $= 225 - 180 = 45°$

6. $\cos 150 = -\dfrac{\sqrt{3}}{2}$;

reference angle $= 180 - 150 = 30°$

7. $\cot \dfrac{7\pi}{5} = \sqrt{3}$;

reference angle $= \dfrac{7}{6}\pi - \pi = \dfrac{\pi}{6} = 30°$

8. $\sin(-30) = -\dfrac{1}{2}$; reference angle $= 30°$

9. $\tan(-60) = -\sqrt{3}$; reference angle $= 60°$

10. $\sec(405) = \sqrt{2}$;

reference angle $= 405 - 360 = 45°$

11. $\csc(-480) = -\dfrac{2\sqrt{3}}{3}$;

reference angle $= 480 - 360 = 120°$;

$180 - 120 = 60°$

12. $\sin \dfrac{13\pi}{6} = \dfrac{1}{2}$;

reference angle $= \dfrac{13\pi}{6} - 2\pi = \dfrac{\pi}{6} = 30°$

13. $\tan 300 = -\sqrt{3}$; reference

angle $= 360 - 300 = 60°$

14. $\sin 180 = 0$

15. $\cos 0 = 1$

16. $\sec(-630)$: undefined; reference angle is $90° = \dfrac{630}{7}$

17. 0.9877

18. -1.0045

19. -0.4960

20. -1.2790

21. $n = \dfrac{\sin i}{\sin r} = \dfrac{\sin 42.7}{\sin 27.1} \approx 1.49$

22. $h = \dfrac{(v \sin \theta)^2}{64} = \dfrac{(150 \sin 0.18\pi)^2}{64} \approx 100.9$ ft

23. Answers will vary. Students should point out that for any point on the terminal side of an angle in quadrant III, the x-coordinate is negative and the y-coordinate is negative. The value of r is always positive. Calling the angle θ $\cos \theta = \dfrac{x}{r} = \dfrac{\text{negative}}{\text{positive}} = $ negative.

EXTEND

24. Since $\sin \theta = \dfrac{2}{5}$, use a reference triangle with leg opposite θ of length 2 and hypotenuse of length 5. Use the Pythagorean theorem to find the length x of the leg adjacent to θ.

$x^2 + 2^2 = 5^2$
$x^2 + 4 = 25$
$x^2 = 21$
$x = \sqrt{21}$

$\cos \theta = \dfrac{\text{adjacent side}}{\text{hypotenuse}} = \dfrac{\sqrt{21}}{5}$

$\tan \theta = \dfrac{\text{opposite side}}{\text{adjacent side}} = \dfrac{2}{\sqrt{21}} = \dfrac{2\sqrt{21}}{21}$

25. Since θ is in Quadrant IV and $\cot \theta = -\dfrac{5}{12}$, use a reference triangle with leg opposite θ of length 12 and leg adjacent to θ of length 5. Use the Pythagorean theorem to find the length x of the hypotenuse.

$x^2 = 5^2 + 12^2 = 25 + 144 = 169$
$x = \sqrt{169} = 13$

$\sec \theta = \dfrac{\text{hypotenuse}}{\text{adjacent side}} = \dfrac{13}{5}$

$\csc \theta = \dfrac{\text{hypotenuse}}{\text{opposite side}} = -\dfrac{13}{12}$

Note in Quadrant IV sec is positive and csc is negative.

26. $\tan \theta = \sqrt{3}$; reference angle $60°$

$\theta = 60°, 240°$

27. $\sec \theta = -\sqrt{2}$; reference angle $45°$

$\theta = 135°, 225°$

28. $\sin x = -1$; x is a quadrant angle.

$x = 270°$

29. $\sin \theta = \dfrac{y}{r}$, $\cos \theta = \dfrac{x}{r}$ so $\dfrac{\sin \theta}{\cos \theta} = \dfrac{\frac{y}{r}}{\frac{x}{r}}$

$= \dfrac{y}{r} \cdot \dfrac{r}{x} = \dfrac{y}{x} = \tan \theta$

THINK CRITICALLY

30. $\sin \theta = \dfrac{\text{opposite}}{\text{hypotenuse}}$. The hypotenuse is the longest side of a right triangle except at $\theta = 90°$ when, theoretically, the side opposite θ is congruent to the hypotenuse. The point $(0, 1)$ lies on the terminal side of this angle, $r = 1$, and $\sin \theta = \dfrac{1}{1} = 1$.

Therefore, 1 is the maximum value of $\sin \theta$.

31a. positive; $\cot \theta = \dfrac{1}{\tan \theta}$

31b. Since $\tan \theta$ is positive and $\tan \theta = \dfrac{\sin \theta}{\cos \theta}$, $\sin \theta$ and $\cos \theta$ must have the same signs. This happens in Quadrants I and III. So $\cos \theta$ is positive if θ is in Quadrant I and is negative if θ is in Quadrant III.

32. $\overline{CL} \parallel \overline{OE}$, so m$\angle CLO$ = m$\angle EOL$ because the angles are alternate interior angles of two parallel lines intersected by a transversal. Therefore, m$\angle CLO$ = θ. In $\triangle CLO$, r is adjacent to $\angle CLO$ and R is the hypotenuse. Therefore,

cos $\angle CLO$ = cos θ = $\dfrac{\text{adjacent}}{\text{hypotenuse}}$ = $\dfrac{r}{R}$. Multiplying by R gives $R \cos \theta = r$.

MIXED REVIEW

33. Area of the rectangular scoreboard is
$A_1 = lw = (15)(24) = 360 \text{ ft}^2$.
Area of circular bull's eye is
$A_2 = \pi r^2 = \pi(3)^2 = 9\pi \text{ ft}^2$.
The fractional part of the scoreboard that is covered by the bull's eye is $\dfrac{9\pi}{360} = \dfrac{\pi}{40}$. Therefore, the probability of the ball landing in the bull's eye is $\dfrac{\pi}{40}$.

34. $y = -4x + 7$ is in slope-intercept form $y = mx + b$. So, slope $m = -4$.

35. Put equation in slope-intercept form.
$5x - 10y - 4 = 0$
$\qquad 10y = 5x - 4 \quad$ add 10 to each side
$\qquad y = \dfrac{1}{2}x - \dfrac{2}{5} \quad$ divide both sides by 10

So, slope $m = \dfrac{1}{2}$.

36. $y = 8$ is the equation of a horizontal line. So, slope $m = 0$.

37. $x = -3$ is a vertical line so slope is undefined.

38. $m = \dfrac{y_2 - y_1}{x_2 - x_1} = \dfrac{4 - (-6)}{4 - 0} = \dfrac{10}{4} = \dfrac{5}{2}$

39. Put equation in slope-intercept form.
$x + y = 9$
$\qquad y = -x + 9 \quad$ subtract x from both sides

So, slope $m = -1$. Since parallel lines have the same slope, any line parallel to this line has slope -1.

40. A; $\sin \theta = \dfrac{\text{opposite side}}{\text{hypotenuse}} = \dfrac{2}{5}$
Use Pythagorean theorem to find adjacent side.
$x^2 + 2^2 = 5^2$
$\qquad x^2 = 5^2 - 2^2 = 25 - 4 = 21$
$\qquad x = \sqrt{21}$
So, $\cos \theta = -\dfrac{\sqrt{21}}{5}$. In Quadrant II cosine is negative.

PROJECT CONNECTION

1–2, 4. Answers will vary.

3. The height of the reference point h, is determined by the equation $h = (b \tan E) + h'$, where E is the angle of elevation, b is the distance from the base of the reference point, and h' is the height of the center of the quadrant above the ground.

ALGEBRAWORKS

1. Extend the vector representing the plane bearing and obtain another 45° angle. Then subtract to show that the plane bearing vector and wind vector form a right angle. $315° - 45° - 180° = 90°$

2. Let x represent magnitude of ground speed vector.
$x^2 = 100^2 + 600^2 = 10{,}000 + 360{,}000 = 370{,}000$
$\quad x = \sqrt{370{,}000} \approx 608 \text{ mi/h}$

3. $\tan \theta = \dfrac{100}{600} = \dfrac{1}{6} = 0.1666$
$\quad \theta = \tan^{-1} 0.1666$
$\quad \theta \approx 9.5$

4. The resultant vector would be the vector obtained by reflecting the old resultant vector through the plane's bearing vector.

5. $0°, 270°, 315°, 135°$

6.

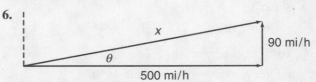

Let x represent the course vector.
$x^2 = 500^2 + 90^2 = 258100$
$\quad x \approx 508$ So, the ground speed is 508 mi/h.
$\tan \theta = \dfrac{90}{500}$
$\quad \theta \approx 9.2°$ Therefore, the plane's course is $90 - 9.2 = 79.8°$

7. $\theta = 90 - 72.3 = 17.7°$
Let y represent wind speed.
$\dfrac{y}{500} = \tan 17.7$
$y = 500(\tan 17.7) \approx 159.6 \text{ mi/h}$
Let z represent ground speed.
$\dfrac{z}{500} = \sec 17.7°$
$z = 500(\sec 17.7) \approx 524.8 \text{ mi/h}$
or use Pythagorean theorem to find z
$z^2 = (159.6)^2 + (500)^2 = 275{,}472.16$
$z = \sqrt{275{,}472.16}$
$z \approx 524.8 \text{ mi/h}$

Lesson 15.4, pages 754–757

EXPLORE THE PROBLEM

1. $\overline{LR}$ lies opposite $\angle S$; $\overline{LS}$ lies adjacent to $\angle S$

2. tangent: $\dfrac{\text{opposite side}}{\text{adjacent side}}$ **3.** $\tan 10.3 = \dfrac{0.807 \text{ mi}}{d}$

4. $(\tan 10.3) d = 0.807 \text{ mi}; d = \dfrac{0.807}{\tan 10.3} \approx 4.4 \text{ mi}$

INVESTIGATE FURTHER

5. $\cos \theta = \dfrac{x}{5}$ **6.** $\sin \theta = \dfrac{5}{x}$

7. $\cos \theta = \dfrac{5}{x}$ **8.** $\tan \theta = \dfrac{5}{x}$

9a. They are complementary.

9b. Subtract the measure of the known angle from 90°.

9c. $m\angle B = 90 - m\angle A = 90 - 27.9 = 62.1°$

10a. Use the Pythagorean theorem.

10b. $\overline{BC}^2 + \overline{AC}^2 = \overline{AB}^2;$
$\overline{BC}^2 = \overline{AB}^2 - \overline{AC}^2 = (6.5)^2 - (5.6)^2 = 10.89;$
$\overline{BC} = \sqrt{10.89} \approx 3.3$

11. $\sin A = \dfrac{3}{4} = 0.75$

12. $m\angle A = 48.6°$
$m\angle B = 90 - m\angle A = 90 - 48.6 = 41.4°$

13a. $h = 322 \sin 26.7°$

13b. $h \approx 322(0.44932) \approx 144.7$ ft

APPLY THE STRATEGY

14. Let x represent the width of the lot.
$\tan 47.3 = \dfrac{x}{162.25 \text{ ft}};$
$x = \tan 47.3(162.25 \text{ ft}) \approx 175.83$ ft

15. Let h represent the height of the plane.
$\tan 21.9 = \dfrac{h}{8.6 \text{ mi}}; h = \tan 21.9(8.6 \text{ mi}) \approx 3.5$ mi

16. Let $\theta =$ angle of elevation of sun.
$\tan \theta = \dfrac{230}{298.4}; \theta = \tan^{-1}\dfrac{230}{398.4} \approx 30°$

17. Let $n = \overline{AC}$ and call eagle's nest E. By the Pythagorean theorem $\overline{BC} = \sqrt{300^2 + n^2}$.
In $\triangle DCE$, $\tan 30 = \dfrac{h}{\overline{BC}} = \dfrac{h}{\sqrt{300^2 + n^2}},$
$h = \tan 30\sqrt{300^2 + n^2}$. In $\triangle ACE$, $\tan 60 = \dfrac{h}{n}$;
$h = n \tan 60$. Substitute to get
$n \tan 60 = \tan 30 \sqrt{300^2 + n^2}$
$\sqrt{3}n = \dfrac{\sqrt{3}}{3}\sqrt{300^2 + n^2}$
$n^2 = \dfrac{300^2 + n^2}{9}$
$8n^2 = 300^2$
$n = \dfrac{300}{\sqrt{8}}$
$n = 75\sqrt{2}$
$h = n \tan 60 = 75\sqrt{2}(\sqrt{3}) = 75\sqrt{6} \approx 183.7$ ft

18. Answers will vary. Students should indicate that you need to know the location of the right angle and either the lengths of two sides or the length of one side and the measure of one acute angle.

REVIEW PROBLEM SOLVING STRATEGIES

1a. yes; Vertical angles are equal so the angles of both triangles at vertex S are equal and both triangles have a right angle. Therefore, since the measures of two angles are equal, the triangle are similar.

1b. 2 to 1: line segment $\overline{AD}$ is 4 units long and line segment $\overline{BC}$ is 2 units long. Corresponding sides of similar triangles are proportional.

1c. from Al $6\frac{2}{3}$ blocks; from Bob, $3\frac{1}{3}$ blocks; from Charlie, $2\frac{2}{3}$ blocks; from Dan, $5\frac{1}{3}$ blocks.

2. Answers will vary. Students may use guess and check. One possible arrangement is shown. 7 was not used.

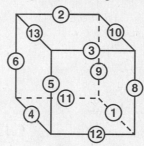

3. Paul started with $25.50. Possible strategy to use, work problem backwards.

Lesson 15.5, pages 758–764

EXPLORE

1. The triangle cannot be constructed because it violates the triangle inequality. The combined lengths of the two sides of 3 and 4 are less than the length of the third side, 9.

2–3. A unique triangle can be constructed.

4. As the figure below shows, this is an ambiguous case. Two triangles satisfying the description can be constructed. $\triangle ABC$ and $\triangle ABD$. Therefore, a unique triangle cannot be constructed.

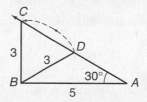

5. As infinite number of triangles with this description can be constructed, all of them similar.

Try These

1. $m\angle C = 180 - (m\angle A + m\angle B) =$
$180 - (36 + 48) = 96°$

$$\frac{\sin B}{b} = \frac{\sin A}{a} \qquad \frac{\sin C}{c} = \frac{\sin A}{a}$$

$$\frac{\sin 48}{b} = \frac{\sin 36}{12} \qquad \frac{\sin 96}{c} = \frac{\sin 366}{12}$$

$(\sin 36)b = 12 \sin 48 \quad (\sin 36)c = 12 \sin 96$

$$b = \frac{12 \sin 48}{\sin 36} \qquad c = \frac{12 \sin 96}{\sin 36}$$

$$b \approx 15.2 \qquad c \approx 20.3$$

2. $a^2 = b^2 + c^2 - 2bc \cos A;$
$(6.6)^2 = (5.4)^2 + (5.6)^2 - 2(5.4)(5.6) \cos A;$
$60.48 \cos A = 16.96; \angle A \approx 73.7°$
$b^2 = a^2 + c^2 - 2ac \cos B;$
$(5.4)^2 = (6.6)^2 + (5.6)^2 - 2(6.6)(5.6) \cos B;$
$73.92 \cos B = 45.76; \angle B \approx 51.8°$
$m\angle C = 180 - m\angle A - m\angle B =$
$180 - 73.7 - 51.8 = 54.5°$

3. Let K represent area of $\triangle ABC$
$K = \frac{1}{2}ab \sin C = \frac{1}{2}(3)(4) \sin 115 = 5.4 \text{ cm}^2$

4. Use Heron's formula to find area of triangle.
$a = 9 \text{ ft}, b = 10 \text{ ft}, c = 13 \text{ ft},$
$s = \frac{1}{2}(9 + 10 + 13) = 16$
Area $K = \sqrt{s(s - a)(s - b)(s - c)} =$
$\sqrt{16(16 - 9)(16 - 10)(16 - 13)} \approx 44.9 \text{ ft}^2$
Since the cost is \$8.50 per ft^2, the total cost is
$(8.50)(44.9) = \$381.65$

5. Let x represent the length of the pond and use the law of cosines.
$x^2 = 60^2 + 40^2 - 2(60)(40) \cos 60 = 2800;$
$x = \sqrt{2800} \approx 52.9 \text{ yd}$

6. $\angle A = \angle B = 45°, \angle C = 90°$
$a = b = 2, c = 2\sqrt{2}$
$\frac{\sin 45}{2} \approx 0.354 \approx \frac{\sin 90}{2\sqrt{2}}$

7. Answers will vary. If the triangle is a right triangle, or if a side and the altitude to that side are known, $a = \frac{1}{2}bh$ will probably be best because both b and h will be known. If three sides are known, Heron's formula will be best because the semi-perimeter can easily be calculated and the square root evaluated. A sine formula will be best if two sides and an included angle are known or can be found.

Practice

1. $\angle A = 180 - \angle B - \angle C =$
$180 - 125.2 - 30 = 24.8°$
$\frac{\sin A}{a} = \frac{\sin B}{b}; \frac{\sin 24.8}{a} = \frac{\sin 125.2}{8};$
$a(\sin 125.2) = 8 \sin 24.8; a = \frac{8 \sin 24.8}{\sin 125.2} \approx 4.1$
$\frac{\sin C}{c} = \frac{\sin B}{b}; \frac{\sin 30}{c} = \frac{\sin 125.2}{8};$
$c(\sin 125.2) = 8 \sin 30; c = \frac{8 \sin 30}{\sin 125.2} \approx 4.9$

2. $a^2 = b^2 + c^2 - 2bc \cos A =$
$12^2 + 10^2 - 2(12)(10) \cos 60$
$a = \sqrt{12^2 + 10^2 - 2(12)(10) \cos 60} \approx 11.1$
$\frac{\sin B}{b} = \frac{\sin A}{a}; \frac{\sin B}{12} = \frac{\sin 60}{11.1}; \sin B = \frac{12 \sin 60}{11.1}$
$\angle B = \sin^{-1}\left(\frac{12 \sin 60}{11.1}\right) \approx 69°$
$\angle C = 180 - \angle A - \angle B = 180 - 60 - 69 = 51°$

3. $c^2 = a^2 + b^2 - 2ab \cos C;$
$(20)^2 = (14)^2 + (12.5)^2 - 2(14)(12.5) \cos C;$
$\cos C = \frac{(20)^2 - (14)^2 - (12.5)^2}{-2(14)(12.5)} \approx -0.1364;$
$\angle C \approx 97.8°$
$\frac{\sin A}{a} = \frac{\sin C}{c}; \frac{\sin A}{14} = \frac{\sin 97.8}{20};$
$\sin A = \frac{14 \sin 97.8}{20} \approx 0.6935; \angle A \approx 43.9°$
$\angle B = 180 - \angle A - \angle C =$
$180 - 43.9 - 97.8 = 38.3°$

4. $\angle C = 180 - \angle A - \angle B =$
$180 - 30.7 - 31.3 = 118°$
$\frac{\sin A}{a} = \frac{\sin C}{c}; \frac{\sin 30.7}{a} = \frac{\sin 118}{80};$
$a \sin 118 = 80 \sin 30.7; a = \frac{80 \sin 30.7}{\sin 118} \approx 46.3$
$\frac{\sin B}{b} = \frac{\sin C}{c}; \frac{\sin 31.3}{b} = \frac{\sin 118}{80};$
$b \sin 118 = 80 \sin 31.3; b = \frac{80 \sin 31.3}{\sin 118} \approx 47.1$

5. $b^2 = a^2 + c^2 - 2ac \cos B =$
$8^2 + 9^2 - 2(8)(9) \cos 74.3 \approx 106.034$
$b \approx 10.3$
$\frac{\sin A}{a} = \frac{\sin B}{b}; \frac{\sin A}{8} = \frac{\sin 74.3}{10.3}; \sin A = \frac{8 \sin 74.3}{10.3}$
$\angle A \approx 48.4°$
$\angle C = 180 - \angle B - \angle A =$
$180 - 74.3 - 48.4 = 57.3°$

6. $\angle C = 180 - \angle A - \angle B = 180 - 30 - 120 = 30°$
If two angles of a triangle are congruent then the sides opposite them are congruent.. So, $c = 12$
$\frac{\sin B}{b} = \frac{\sin A}{a}; \frac{\sin 120}{b} = \frac{\sin 30}{12}; b \sin 30 = 12 \sin 120;$
$b = \frac{12 \sin 120}{\sin 30} \approx 20.8$

7. This is an ambiguous case. There are two possibilities.
Possibility 1: $\angle C$ is acute.
$$\frac{\sin C}{c} = \frac{\sin A}{a}; \frac{\sin C}{8} = \frac{\sin 30}{6}; \sin C = \frac{8 \sin 30}{6};$$
$\angle C \approx 41.8°$

$\angle B = 180 - \angle A - \angle C =$
$180 - 30 - 41.8 = 108.2°$
$$\frac{\sin B}{b} = \frac{\sin A}{a}; \frac{\sin 108.2}{b} = \frac{\sin 30}{6};$$
$b \sin 30 = 6 \sin 108.2; b = \dfrac{6 \sin 108.2}{\sin 30} \approx 11.4$

Possibility 2: $\angle C$ is obtuse.
$\angle C = 180 - 41.8 = 138.2°$
$\angle B = 180 - \angle A - \angle C = 180 - 30 - 138.2 = 11.8$
$$\frac{\sin B}{b} = \frac{\sin A}{a}; \frac{\sin 11.8}{b} = \frac{\sin 30}{6}; b \sin 30 = 6 \sin 11.8;$$
$b = \dfrac{6 \sin 11.8}{\sin 30} \approx 2.5$

8. $c^2 = a^2 + b^2 = 2ab \cos C;$
$17^2 = 8^2 + 15^2 - 2(8)(15) \cos C;$
$\cos C = \dfrac{17^2 - 8^2 - 15^2}{-2(8)(15)} = 0; \angle C = 90°$

$$\frac{\sin A}{a} = \frac{\sin C}{c}; \frac{\sin A}{8} = \frac{\sin 90}{17}; 17 \sin A = 8 \sin 90;$$
$\sin A = \dfrac{8 \sin 90}{17} = \dfrac{8}{17}; \angle A = 28.1°$
$\angle B = 90 - 28.1 = 61.9°$

9. Area $= \dfrac{1}{2} bc \sin A = \dfrac{1}{2}(12)(15) \sin 42 \approx 60.2$

10. Area $= \dfrac{1}{2} ac \sin B = \dfrac{1}{2}(19.6)(55.5) \sin 104.0 \approx 527.7$

11. Let x represent the distance between the lighthouses.
Apply the law of cosines.
$x^2 = (62.7)^2 + (115.9)^2 - 2(62.7)(115.9) \cos 125.3$
$x^2 = 25,762.60181$
$\ \ x \approx 160.5$ mi

12. First change angle measures to degrees.
$\angle FAB = \dfrac{2\pi}{15} \cdot \dfrac{180}{\pi} = 24°, \angle FBA = \dfrac{4\pi}{9} \cdot \dfrac{180}{\pi} = 80°$
$\angle AFB = 180 - 24 - 80 = 76°$
Apply law of sines.
Let x represent the distance from A to F and y represent the distance from B to F.
$$\frac{\sin B}{x} = \frac{\sin E}{6}; \frac{\sin 80}{x} = \frac{\sin 76}{6};$$
$x \sin 76 = 6 \sin 80; x = \dfrac{6 \sin 80}{\sin 76} \approx 6.1$ mi
$$\frac{\sin A}{y} = \frac{\sin F}{6}; \frac{\sin 24}{y} = \frac{\sin 76}{6}; y \sin 76 = 6 \sin 24;$$
$y = \dfrac{6 \sin 24}{\sin 76} \approx 2.5$ mi

13. Apply Heron's formula to find the area of 1 bracket.
$a = 11.5$ cm. $b = 9.6$ cm, $c = 4.3$ cm,
$s = \dfrac{1}{2}(11.5 + 9.6 + 4.3) = 12.7$
area $= \sqrt{s(s - a)(s - b)(s - c)}$
$\quad = \sqrt{12.7(12.7 - 11.5)(12.7 - 9.6)(12.7 - 4.3)}$
$\quad = 19.9$ cm^2
cost $= 19.9(1000)(.0023) = \$45.82$

14. In the drawing $\overline{AB}$ represents the cliff and $\overline{CD}$ represents the light house. $\overline{AH}$ is horizontal.

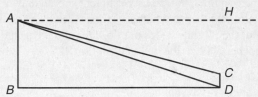

$\angle HAC = 14.9°$
$\angle HAD = 18.3°$
$\overline{CD} = 50$ ft
$\angle CAD = 18.3 - 14.9 = 3.4°$
$\angle DAB = 90 - 18.3 = 71.7°$
$\angle DAB$ and $\angle CDA$ are alternate interior angles of parallel lines. So, $\angle CDA = 71.7°$.
$\angle ACD = 180 - 3.4 - 71.7 = 104.9°$
Use the law of sines to find $\overline{AD}$.
$\dfrac{\sin 104.9}{\overline{AD}} = \sin 3.450; \overline{AD} \sin 3.4 = 50 \sin 104.9;$
$\overline{AD} = \dfrac{50 \sin 104.9}{\sin 3.4} \approx 814.7321$
$\angle BDA$ and $\angle DAH$ are alternate interior angles of parallel lines. So, $\angle BDA = 18.3°$
In $\triangle ABD$, $\sin \angle BDA = \dfrac{\overline{AB}}{\overline{AD}}$; $\sin 18.3 = \dfrac{\overline{AB}}{814.7321}$;
$\overline{AB} = 814.7321 \sin 18.3 \approx 255.8$ ft

15. The drawing shows the two possible positions for the plane, either at P or Q.

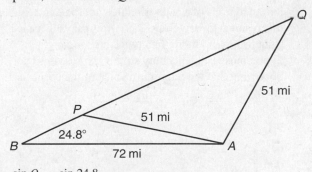

$\dfrac{\sin Q}{72} = \dfrac{\sin 24.8}{51}$
$\sin Q = \dfrac{72 \sin 24.8}{51} = 0.5922; \angle Q \approx 36.3°$
and $\angle BAQ = 180 - 24.8 - 36.3 = 118.9°$
So, $\dfrac{\sin 118.9}{\overline{BQ}} = \dfrac{\sin 24.8}{51}; \overline{BQ} = \dfrac{51 \sin 118.9}{\sin 24.8} \approx 106$ mi

Or, instead of $\triangle BQA$ the situation is really $\triangle BPA$ and
$\angle BPA = 180 - 36.3 = 143.7°$ and
$\angle BAP = 180 - 24.8 - 143.7 = 11.5°$

So, $\dfrac{\sin 11.5}{\overline{BP}} = \dfrac{\sin 143.7}{72}$; $\overline{BP} = \dfrac{72 \sin 11.5}{\sin 143.7}$;

$\overline{BP} \approx 24$ mi

16. Answers will vary. Students should mention that the law of sines will be needed first if two angles and any side or two sides and an angle opposite one of them are given. In the latter case, the student should draw a diagram to decide whether there are zero, one, or two solutions. If three sides or two sides and the included angle are given, the law of cosines will be needed first.

EXTEND

17. Triangle 1 is a right triangle.

area $= \dfrac{1}{2}(6)(8) = 24$

perimeter $= 6 + 8 + 10 = 24$

18. Triangle 2 is a right triangle.

area $= \dfrac{1}{2}(5)(12) = 30$

perimeter $= 5 + 12 + 13 = 30$

19. area $= \sqrt{s(s-a)(s-b)(s-c)} =$

$\sqrt{30(30-6)(30-25)(30-29)} = 60$

perimeter $= 6 + 25 + 29 = 60$

20. area $= \sqrt{21(21-7)(21-15)(21-20)} = 42$

perimeter $= 7 + 15 + 20 = 42$

21. area $= \sqrt{18(18-9)(18-10)(18-17)} = 36$

perimeter $= 9 + 10 + 17 = 36$

22. The sides have integer lengths and the areas and perimeters have the same numerical values.

THINK CRITICALLY

23. Answers will vary. If A is the angle opposite the side of 1, then $\cos A = 2.575$. Since the cosine compares the length of a leg of a right triangle with the hypotenuse, this result is clearly impossible, for it says that a leg of a right triangle is more than twice the length of the hypotenuse. Students may simply express this by observing that the maximum value of the cosine is 1.

24. $\dfrac{\sin 60}{10} = \dfrac{\sin C}{6\sqrt{3}}$; $\sin C = \dfrac{6\sqrt{3} \sin 60}{10} = 0.9$;

$\angle C \approx 64.2°$

or $\angle C = 115.8°$.

therefore, 2 solutions

25. $\dfrac{\sin 60}{5\sqrt{3}} = \dfrac{\sin B}{12}$; $\sin B = \dfrac{12 \sin 60}{5\sqrt{3}} \approx 1.2$

This is an impossibility, $\sin B \not> 1$, therefore no solutions.

26. Total Payment $= 151.57(12)(4) = \$7275.36$

Cost $= 7275.36 - 6000 = \$1275.36$

27. $x^2 - 5x + 8 = 0$, $a = 1$, $b = -5$, $c = 8$

sum $= -\dfrac{b}{a} = -\left(\dfrac{-5}{1}\right) = 5$

product $= \dfrac{c}{a} = \dfrac{8}{1} = 8$

28. $5x^2 - 9x - 3 = 0$, $a = 5$, $b = -9$, $c = -3$

sum $= -\dfrac{b}{a} = -\left(\dfrac{-9}{5}\right) = \dfrac{9}{5}$

product $= \dfrac{c}{a} = -\dfrac{3}{5}$

29. $-2x + 6 = 3x - 4x^2$

$4x^2 - 5x + 6 = 0$, $a = 4$, $b = -5$, $c = 6$

sum $= -\dfrac{b}{a} = -\left(\dfrac{-5}{4}\right) = \dfrac{5}{4}$

product $= \dfrac{c}{a} = \dfrac{6}{4} = \dfrac{3}{2}$

30. $y = x^2 - 6x + 8$

$y = x^2 - 6x + 9 - 1$

$y = (x - 3)^2 - 1$

vertex at $(3, -1)$

y-intercept at $(0, 8)$

$x^2 - 6x + 8 = 0$

$(x - 4)(x - 2) = 0$

$x = 4, x = 2$

x-intercepts at $(2, 0)$ and $(4, 0)$

31. C; $a^2 = b^2 + c^2 - 2bc \cos A =$

$3^2 + 2^2 - 2(3)(2) \cos 30 =$

$9 + 4 - 12\left(\dfrac{\sqrt{3}}{2}\right) = 13 - 6\sqrt{3}$

$a = \sqrt{13 - 6\sqrt{3}}$

ALGEBRAWORKS

1a. arc: $\theta = 58° = 58 \cdot \dfrac{\pi}{180} = \dfrac{29\pi}{90}$

$s = \theta r = \dfrac{29\pi}{90} \cdot 3960 \approx 4008.7$ mi

straight: Let x represent $\overline{PW}$

$x^2 = 3960^2 + 3960^2 - 2(3960)(3960) \cos 58°$

$x^2 = 14{,}743{,}236.13$

$x = 3839.7$ mi

1b. $\dfrac{3839.7}{57} = 67.3$ days

$70.3 - 67.3 = 3.0$ days

2. Let x represent the distance $\overline{SL}$

$x^2 = 50^2 + 35^2 - 2(50)(35) \cos 23 \approx 503.2330$

$x \approx 22.4$ mi

3a. $\cos d = (\sin a \sin b) + (\cos a \cos b \cos L)$
  $\cos d = (\sin 38 \sin 21) + (\cos 38 \cos 21 \cos (158.122))$
  $\cos d \approx 0.8158$
  $d \approx 35.3°$

3b. $s = \theta r = 35.3\left(\dfrac{\pi}{180}\right)(3960) \approx 2439.8 \text{ mi}$

Lesson 15.6, pages 765–769

THINK BACK

1a. $\sin 0 = 0$
  $\cos 0 = 1$

1b. $\sin 90 = 1$
  $\cos 90 = 0$

1c. $\sin \pi = 0$
  $\cos \pi = -1$

1d. $\sin \dfrac{3\pi}{2} = -1$
  $\cos \dfrac{3\pi}{2} = 0$

2. For each element of the domain, there is exactly one element of the range. That is, for each angle measure there is exactly one value for the sine and one value for the cosine.

3. The domain is the set of angles of rotation, usually measured in degrees or radians. The range is the sine or cosine of these angles. Although it has not been formally established yet, students should see that the maximum and minimum values of the range are 1 and -1, because the sine and cosine functions express the ratios of the lengths of legs of right triangles to the lengths of hypotenuses, and legs can never exceed hypotenuses in length.

4. Angles will vary. 390°, 750°, 1110°. No matter which angles are listed, the sine of each should be $\frac{1}{2}$ and the cosine of each should be $\frac{\sqrt{3}}{2}$.

5. No. For any member of the range of either function, there are an infinite number of members of the domain, all of them coterminal. An example is the range member $\frac{1}{2}$, which, as noted above, has 30°, 390°, 750°, and 1110° (plus all angles $[30 + 360n]°$, where n is an integer) as domain members. However, this does not violate the definition of a function.

EXPLORE / WORKING TOGETHER

6a.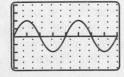

Answers will vary. Students should mention that the sections of the graph from 0°–360° and from 360°–720° are congruent. Each point in the first section has a match 360° to the right.

6b. max: 1, min: -1

6c. 360°. The pattern should repeat because angles between 360° and 720° are coterminal with angles from 0° to 360° and therefore should have the same sines.

6d. The y-intercept is (0, 0). There are x-intercepts at (0, 0) and every 180° to the right and left.

6e. The graph confirms the answers.

7. Answers will vary. Students should mention that the cosine graph looks like the sine graph shifted 90° to the left. Both graphs have maxima and minima of 1 and -1 and horizontal repeating patterns 360° in width. Because of the horizontal shift of the cosine graph, the y-intercept is (0, 1). There are x-intercepts at (90, 0) and every 180° to the left and right.

8. The general shape, the width of the repeating pattern, and the intercepts do not change. However, the maxima and minima change to $\pm A$. For negative A, the graph is reflected across the x-axis.

9. The effects are the same.

10a.

10b. The coefficient of x appears to stretch or shrink the graph horizontally. The coefficient gives the number of complete sine curves that appear in a 2π interval.

11a.

11b. The constant appears to translate the graph up (for a positive constant) or down (for a negative constant) by an amount equal to the constant.

12a.

12b. The constants appear to translate the graph left (for a positive constant) or right (for a negative constant) by an amount equal to the constant.

13. Students should obtain the same general results. The graph of a cosine function is increased in maxima, decreased in minima, shrunk, expanded, and translated up, down, left, and right in accordance with the same changes in the algebraic expression of the function that produce analogous results for the sine function.

14. 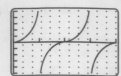 Answers will vary. Students should note that the tangent graph is markedly different from those of the sine and cosine. Like the sine, the tangent has x-intercepts at 0, π, and 2π. Like both the sine and cosine, the tangent appears to repeat itself, though at π rather than 2π intervals. Unlike the sine and cosine, the tangent has no maximum or minimum values and the graph is not continuous.

15. The graph appears to approach vertical asymptotes at π intervals. Explanations for this will vary. At $\frac{\pi}{2}$, $\frac{3\pi}{2}$, and so on, the terminal side of an angle of rotation coincides with either the positive or negative y-axis. Therefore, the point $(0, -1)$ or $(0, 1)$ will lie on the terminal side. For these angles, $\tan x = \frac{-1}{0} =$ or $\tan x = \frac{1}{0}$, both of which are undefined. Using the methods of Chapter 12, students can use tables of values to show that $\tan x$ grows arbitrarily large as x approaches $\frac{\pi}{2}$ from the left, and arbitrarily small as x approaches $\frac{\pi}{2}$ from the right. The same will be true at $\frac{3\pi}{2}$ and all other angles coterminal with $\frac{\pi}{2}$ and $\frac{3\pi}{2}$.

16. 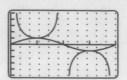 The graph is steeper and straighter than the graph of $y = \tan x$.

17. The 2π graph of $y = \tan x$ has been compressed into π. Two such compressed graphs appear in the 2π interval shown.

18. The graph of $y = \tan x$ has been translated $\frac{\pi}{4}$ to the right.

19. The graph of $y = \tan x$ has been translated up 2 units.

20. Graph cosecant: $y = \frac{1}{\sin x}$; secant: $y = \frac{1}{\cos x}$; cotangent: $y = \frac{1}{\tan x}$

21. Asymptotes will appear wherever the reciprocal function equals zero. For the cosecant, this will be at $0°$, $180°$, $360°$, and so on. For the secant, asymptotes will occur at $90°$, $270°$, $450°$, and so on. For the cotangent, they will occur at $0°$, $180°$, $360°$, and so on.

22. Answers will vary. Because of the reciprocal relationship of the two functions, one increases as the other decreases, and vice-versa. Both are positive together and negative together. They intersect only at $y = 1$ and $y = -1$ because 1 and -1 are the only numbers that are their own reciprocals.

23. Answers will vary. Because of the reciprocal relationship of the two functions, one increases as the other decreases, and vice-versa. Both are positive together and negative together. They intersect only at $y = 1$ and $y = -1$ because 1 and -1 are the only numbers that are their own reciprocals.

24. Answers will vary. The tangent function increases continually except at its "crossover" asymptote points. Because the cotangent is the reciprocal of the tangent, it decreases continually except at its "crossover" asymptote points. They intersect only at $y = 1$ and $y = -1$ because 1 and -1 are the only numbers that are their own reciprocals.

MAKE CONNECTIONS

25. Leave the x-intercepts the same but change the maxima and minima to $\pm N$.

26. Leave the maxima and minima unchanged but compress or expand the graph so that there are N complete curves in $360°$.

27. Translate the curve left (for positive N) or right (for negative N) by $N°$.

28. Translate the curve up (for positive N) or down (for negative N) by N units.

29. d   **30.** b   **31.** a   **32.** c   **33.** a   **34.** c

35. d   **36.** b   **37.** c   **38.** b   **39.** a   **40.** d

SUMMARIZE

41. Make $y = \cos x$ steeper so that its maximum is 5 and its minimum is -5. Compress the new graph so that four complete curves (with the new maximum of 5 and minimum of -5) are squeezed into a $360°$ interval. Translate the new graph $30°$ to the left and up 2 units.

42. Equations may vary: $y = 3 \sin x$, $y = 3 \sin (x + 2\pi)$ or $y = 3 \sin (x - 2\pi)$

$$y = 3 \cos \left(x - \frac{\pi}{2}\right) \text{ or } y = 3 \cos \left(x + \frac{3\pi}{2}\right).$$

43a.

x	0°	45°	90°	135°	180°	225°	270°	315°	360°
sin x	0	0.71	1	0.71	0	−0.71	−1	−0.71	0
cos x	1	0.71	0	−0.71	−1	−0.71	0	0.71	1
sin x + cos x	1	1.42	1	0	−1	−1.42	−1	0	1

43b.

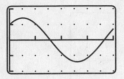

43c.

x	0°	45°	90°	135°	180°	225°	270°	315°	360°
sin $2x$	0	1	0	−1	0	1	0	−1	0
cos $2x$	1	0	−1	0	1	0	−1	0	1
sin $2x$ + cos $2x$	1	1	−1	−1	1	1	−1	−1	1

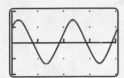

44. $y = \sin x$ is odd.
$y = \cos x$ is even.
$y = \tan x$ is odd.

Lesson 15.7, pages 770–777

EXPLORE

1.

time (sec)	0	1	2	3	4	5	6	7
height (ft)	0	14.1	20	14.1	0	−14.1	−20	−14.1

time (sec)	8	9	10	11	12	13	14	15
height (ft)	0	14.1	20	14.1	0	−14.1	−20	−14.1

time (sec)	16	17	18	19	20	21	22	23	24
height (ft)	0	14.1	20	14.1	0	−14.1	−20	−14.1	0

Calculations are simplified after $t = 8$ because each angle is coterminal with an angle whose sine has already been calculated. Therefore, the values of y cycle repeatedly through an 8-value pattern.

2.

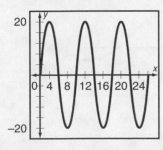

3. Descriptions will vary. Students should recognize that the graph is a sine curve with a maximum of 20 ft, a minimum of −20 ft, and a horizontal pattern that repeats every 360° or 8 seconds.

TRY THESE

1. amplitude: 1, period: $\frac{2\pi}{3}$ or $\frac{2\pi}{3} \cdot \frac{180}{\pi} = 120°$

2. amplitude: 3, period: 2π or 360°

3. amplitude: $|-2| = 2$, period: $\frac{2\pi}{\pi} = 2$

4. amplitude: $\frac{1}{4}$, period: $\frac{2\pi}{\frac{1}{2}} = 4\pi$ or 720°

5. down 3 units   **6.** up 5 units   **7.** right 200°

8. left $\frac{\pi}{2}$ and down 6 units

9.

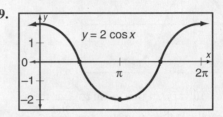

10.

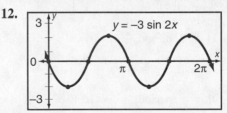

11.

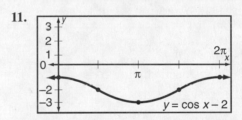

12.

13a. $y = 4 \sin\left(\pi t - \frac{\pi}{2}\right)$

at $t = 0$, $y = 4 \sin\left(0 - \frac{\pi}{2}\right) =$

$4 \sin\left(-\frac{\pi}{2}\right) = 4(-1) = -4$

13b. 4 in.

13c. $4 \sin\left(\pi t - \frac{\pi}{2}\right) = 0$; $\sin\left(\pi t - \frac{\pi}{2}\right) = 0$;

$\pi t - \frac{\pi}{2} = 0$; $\pi t = \frac{\pi}{2}$; $t = \frac{1}{2} = 0.5$ sec

13d. $\frac{2\pi}{\pi} = 2$ radians

13e.

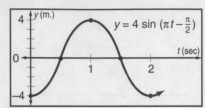

14. Explanations will vary. In $y = a \sin bx$, a is the amplitude, 5 and b is the value such that $\frac{360}{b} = 180°$, the period. This gives $b = 2$ and an equation, before translation, of $y = 5 \sin 2x$. In $y = a \sin (bx + c) + d$, d is the downward translation, -3 and c is the value such that $\frac{c}{b} = 90°$, the translation to the left. Since $b = 2$, this gives $c = 180°$. Therefore, the equation is $y = 5 \sin (2x + 180) - 3$.

PRACTICE

1. amplitude: 1, period: 2π

2. amplitude: $|-2| = 2$, period: 2π

3. amplitude: 1, period: $\frac{2\pi}{2} = \pi$

4. amplitude: 1.5, period: $\frac{2\pi}{4} = \frac{\pi}{2}$

5. amplitude: $|-3| = 3$, period: $\frac{2\pi}{2\pi} = 1$

6. amplitude: 0.4, period: $\frac{2\pi}{0.5} = 4\pi$

7. amplitude: $|-9| = 9$, period: $\frac{2\pi}{\frac{1}{5}} = 10\pi$

8. amplitude: 4, period: $\frac{2\pi}{|-45|} = \frac{2\pi}{45}$

9. down 2 units   10. up 6 units   11. left π units

12. right 15°   13. factor 2 to get $2(x + 75)$; left 75°

14. left $\frac{\pi}{2}$ and up 3 units

15. factor 3 to get $3(x - 30)$; right 30° and down 4 units

16.

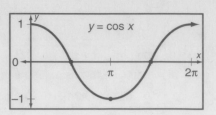

17.

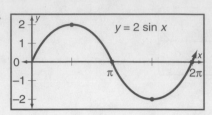

18.

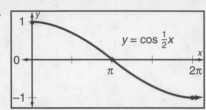

19.

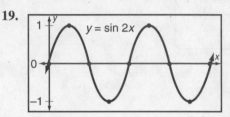

20.

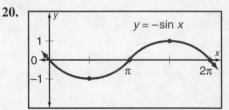

21.

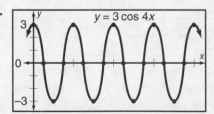

22.

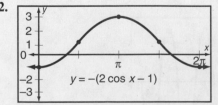

23.

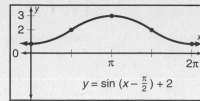

$y = \sin\left(x - \frac{\pi}{2}\right) + 2$

24a. 1996: $n = 8000 + 5000 \sin \pi(0) =$
$8000 + 5000(0) = 8000$
1997: $n = 8000 + 5000 \sin \pi(1) =$
$8000 + 5000(0) = 8000$
1998: $n = 8000 + 5000 \sin \pi(2) =$
$8000 + 5000(0) = 8000$

24b. amplitude of $5000 \sin \pi t$ is 5000 maximum
population $8000 + 5000 = 13,000$ first reached
when $\sin \pi t = 1$ at $t = 0.5$ July 1, 1996

24c. minimum population reached when $\sin \pi t = -1$ at
$t = \frac{3}{2} = 1.5$
minimum population: 3000 on July 1, 1997

24d. period: $\frac{2\pi}{\pi} = 2$ years

24e.

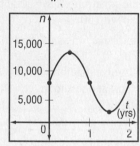

25a. the second note was loudest; its amplitude is greater

25b. period: $\frac{2\pi}{55\pi} = \frac{2}{55}$ sec

frequency: $\frac{55}{2} = 27\frac{1}{2}$ vibrations/sec

26. Answers will vary. Translate the curve into "standard
position," with the beginning of a single cycle of a sine
curve at $(0, 0)$. From the amount of up or down
translation determine d, reversing the rules given in
"Translations of a Sine or Cosine Graph." Determine a
from the height of the highest point of the curve above
the horizontal axis. Determine b by dividing $360°$ or 2π
by the period. Determine $|c|$ by multiplying b by the
amount you translated the graph to move it into
standard position. c will be positive if you moved it
right, negative if you moved it left. Examples will vary.

EXTEND

27. maximum reached when $\sin 2x = 1$
This happens when $2x = \frac{\pi}{2}$; $x = \frac{\pi}{4}$.

It happens again when $2x = \frac{\pi}{2} + 2\pi$;

$x = \frac{\pi}{4} + \pi = \frac{5\pi}{4}$. maximum value

$y = -4 + 3 \sin 2\left(\frac{\pi}{4}\right) = -4 + 3 \sin \frac{\pi}{2} =$
$-4 + 3 = -1$

28. minimum reached when $\sin 2x = -1$
This happens when $2x = \frac{3\pi}{2}$; $x = \frac{3\pi}{4}$.

It happens again when $2x = \frac{3\pi}{2} + 2\pi$;

$x = \frac{3\pi}{4} + \pi = \frac{7\pi}{4}$. minimum value

$y = -4 + 3 \sin 2\left(\frac{3\pi}{4}\right) = -4 + 3(-1) = -7$

29. $2 \cos x + \sqrt{3} = 0$; $2 \cos x = -\sqrt{3}$; $\cos x = -\frac{\sqrt{3}}{2}$;

$x = \frac{5\pi}{6}, \frac{7\pi}{6}$

30. $-2 \sin x = 1$; $\sin x = -\frac{1}{2}$; $x = \frac{7\pi}{6}, \frac{11\pi}{6}$

31. $x \approx 1.29, 4.34$

32. $(\sec x)^2 - 2 = 0$; $(\sec x)^2 = 2$; $\sec x = \pm \sqrt{2}$
$x = \frac{\pi}{4}, \frac{3\pi}{4}, \frac{5\pi}{4}, \frac{7\pi}{4}$

THINK CRITICALLY

33. $a = 3$, period: $\frac{2\pi}{b} = \pi$; $b = 2$
$y = 3 \sin 2x$

34. $a = 1$, period: $90 = 90 \cdot \frac{\pi}{180} = \frac{\pi}{2}$; $\frac{2\pi}{b} = \frac{\pi}{2}$; $b = 4$,
translate: 6 down
$y = \sin 4x = 6$

35. $a = 2$, period: $\frac{2\pi}{b} = \frac{3\pi}{2}$; $b = \frac{4}{3}$, translate: 4 up
$y = 2 \sin \frac{4}{3}x + 4$

36. $a = 4$, period: $45 = 45\left(\frac{\pi}{180}\right) = \frac{\pi}{4}$; $\frac{2\pi}{b} = \frac{\pi}{4}$; $b = 8$,

translate: 3 down, $30°$ left or $\frac{\pi}{6}$ left

$y = 4 \sin\left(8x + \frac{4\pi}{3}\right) - 3$ or
$y = 4 \sin(8x + 240°) - 3$

37. $a = 1$, period: $\frac{2\pi}{b} = \frac{5\pi}{2}$; $b = \frac{4}{5}$, translate: 1 up,

$225°$ right or $\frac{5\pi}{4}$ right

$y = \sin\left(\frac{4}{5}x - \pi\right) + 1$ or $y = \sin\left(\frac{4}{5}x - 180°\right) + 1$

38a. Sept. 15: $m = 8.5$

$T = 57.3 + 24.8 \sin\left(\frac{\pi}{6}(8.5) + 4\right) \approx 77.8°F$

38b. Maximum achieved when $\sin\left(\frac{\pi}{6}m + 4\right) = 1$;

$$\left(\frac{\pi}{6}m + 4\right) = \frac{\pi}{2};$$

$$\frac{\pi}{6}m = \frac{\pi}{2} - 4; m = \frac{\frac{\pi}{2} - 4}{\frac{\pi}{6}} \approx -4.639.$$

period: $p = \frac{2\pi}{\frac{\pi}{6}} = 12$

So, the first positive m-value that produces a maximum is $m = -4.639 + 12 = 7.361$. Maximum value is

$$T = 57.3 + 24.8 \sin\left(\frac{\pi}{6}(7.361) + 4\right) \approx 82.1°F$$

and it occurs approximately August 11.

38c. Minimum achieved when $\sin\left(\frac{\pi}{6}m + 4\right) = -1$;

$$\left(\frac{\pi}{6}m + 4\right) = \frac{3\pi}{2}; \frac{\pi}{6}m = \frac{3\pi}{2} - 4;$$

$$m = \frac{\frac{3\pi}{2} - 4}{\frac{\pi}{6}} \approx 1.316$$

Minimum value is

$$T = 57.3 + 24.8 \sin\left(\frac{\pi}{6}(1.361) + 4\right) \approx 32.5°F$$

and it occurs approximately February 10.

MIXED REVIEW

39. $A = \frac{1}{2}h(b_1 + b_2)$

$$\frac{1}{2}h(b_1 + b_2) = A$$

$$h(b_1 + b_2) = 2A$$

$$b_1 + b_2 = \frac{2A}{h}$$

$$b_2 = \frac{2A}{h} - b_1$$

40. $A = P + Prt$

$$P + Prt = A$$

$$P(1 + rt) = A$$

$$P = \frac{A}{1 + rt}$$

41. $F = \frac{mv^2}{r}$

$$\frac{mv^2}{r} = F$$

$$mv^2 = Fr$$

$$m = \frac{Fr}{v^2}$$

42. $(x^2 - 81) = (x + 9)(x - 9)$

43. $(x^2 + 6x + 9) = (x + 3)(x + 3) = (x + 3)^2$

44. $(6x^2 + 7x - 3) = (2x + 3)(3x - 1)$

45. $3x^2 + 9x - 30 = 3(x^2 + 3x - 10) = 3(x + 5)(x - 2)$

46.
$$x^2 + 4x + y^2 - 6y = 3$$
$$x^2 + 4x + 4 + y^2 - 6y + 9 = 16$$
$$(x + 2)^2 + (y - 3)^2 = 16$$
center: $(-2, 3)$, radius: 4

47. $A; p = \frac{2\pi}{b} = \frac{2\pi}{3} = \frac{2\pi}{3} \cdot \frac{180}{\pi} = 120°$

48. $L(x, y)$ equidistant from $M(-2, 1)$ and $N(4, 3)$ means that $\overline{LM} = \overline{LN}$ or

$$\sqrt{(x - (-2))^2 + (y - 1)^2} = \sqrt{(x - 4)^2 + (y - 3)^2}$$
$$(x + 2)^2 + (y - 1)^2 = (x - 4)^2 + (y - 3)^2$$
$$x^2 + 4x + 4 + y^2 - 2y + 1 =$$
$$x^2 - 8x + 16 + y^2 - 6y + 9$$
$$4x - 2y + 5 = -8x - 6y + 25$$
$$4y = -12x + 20$$
$$y = -3x + 5$$

Lesson 15.8, pages 778–781

EXPLORE

1. $\sin x = 0$ at $x = 30°, 150°$

2. Answers will vary. They can be any number of the form $30 + 360n$ or $150 + 360n$ where n is any integer.

3. Suppose that x_1 is a solution to the equation. Then, since coterminal angles have the same sines, every angle coterminal with x_1 will also be a solution to the equation. Therefore, the equation does not have a unique solution. (This was shown in Questions 1–2: $x = 30°$, for example, is a solution to the equation $\sin x = 0.5$, but so is every angle coterminal with $30°$.)

4. Sets will vary, but all should have 0.5 as the first term of each ordered pair. $\{(0.5, 30), (0.5, 150), (0.5, 390), (0.5, 510)\}$ Students should recognize that the set is not a function since each first element does not have a unique second element paired with it.

TRY THESE

1. $a = \text{Tan}^{-1}b$ or $a = \text{Arctan } b$

2. $\angle A = \text{Sin}^{-1}\frac{3}{6} = \text{Sin}^{-1}\frac{1}{2} = \frac{\pi}{6}$

3. $\angle A = \text{Cos}^{-1}\frac{4}{4\sqrt{2}} = \text{Cos}^{-1}\frac{\sqrt{2}}{2} = \frac{\pi}{4}$

4. By the converse of the Pythagorean theorem $\triangle ABC$ is a right triangle with right angle A. So, $\angle A = 90° = \frac{\pi}{2}$.

5. $\frac{\pi}{4}$ 6. $\frac{\pi}{4}$ 7. $\frac{\pi}{3}$ 8. $-\frac{\pi}{2}$

9. Sketch a right triangle with hypotenuse 13 and one leg 5 and call the angle whose sine is $\frac{5}{13}$ θ. Use Pythagorean theorem to find other leg; $x^2 = 13^2 - 5^2 = 144; x = 12$. So, $\tan \theta = \frac{5}{12}$.

10. The cosine of the angle whose cosine is 0.3 is 0.3.

11. The angle that has a sine equal to the sin 32 is 32°.

12. Sketch a right triangle with hypotenuse 5 and one leg 3 and call the angle whose sine is $\frac{3}{5}$ θ. Use the Pythagorean theorem to find other leg. So, $\cos \theta = \frac{4}{5}$

13.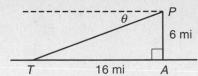

Let P represent plane flying 6 mi above point A, T represent the airport which is 16 mi from A, and θ represent angle of depression. First find $\angle TPA = \text{Tan}^{-1} \frac{16}{6} \approx 69.4°$. $\angle TPA$ and θ are complementary. So, $\theta = 90 - 69.4 = 20.6°$

14. Answers will vary. $y = \text{Arcsin } x$ is the inverse of the function $y = \text{Sine } x$ for $-1 \le x \le 1$ and $-\frac{\pi}{2} \le y \le \frac{\pi}{2}$. Arcsin x means "the angle whose sine is x." If $x = \sin y$, then Arcsin $x = y$.

PRACTICE

1. $\frac{\pi}{3}$; 30-60-90 triangle   **2.** $\frac{\pi}{2}$; 45-45-90 triangle

3. $\frac{5\pi}{6}$; 30-60-90 triangle   **4.** $\frac{\pi}{2}$; quadrant angle

5. $\frac{3\pi}{4}$; 45-45-90 triangle

6. $\frac{5}{6}$ **7.** $23.6°$ **8.** $\frac{1}{2}$; 30-60-90 triangle

9. $\text{Sin}^{-1}\left(\cos\frac{\pi}{2}\right) = \text{Sin}^{-1}(0) = 0$

10. $\text{Arcsin}\left(\cos\frac{\pi}{6}\right) = \text{Arcsin}\left(\frac{\sqrt{3}}{2}\right) = \frac{\pi}{3}$

11. $\sin\left(2 \text{ Sin}^{-1}\frac{1}{2}\right) = \sin(2 \cdot 30) = \sin 60 = \frac{\sqrt{3}}{2}$

12. $\text{Arccos}\left(\tan\frac{3\pi}{4}\right) = \text{Arccos}(-1) = \pi$

13. Let $\angle A$ lie opposite the 3-unit side and $\angle B$ opposite the 4-unit side. $\angle A = \text{Arcsin}\frac{3}{5} \approx 36.9°$, $\angle B = \text{Arcsin}\frac{4}{5} \approx 53.1°$

14. Let $\angle A$ lie opposite the 5-unit side and $\angle B$ opposite the 12-unit side.
$\angle A = \text{Arctan}\frac{5}{12} \approx 22.6°$
$\angle B = \text{Arctan}\frac{12}{5} \approx 67.4°$

15. Let $\angle A$ lie opposite the 8-unit side and $\angle B$ opposite the 15-unit side. $\angle A = \text{Cos}^{-1}\frac{15}{17} \approx 28.1°$,
$\angle B = \text{Cos}^{-1}\frac{8}{17} = 61.9°$

16. Let $\angle A$ lie opposite the 2.8-unit side and $\angle B$ opposite the 4.5-unit side.
$\angle A = \text{Sin}^{-1}\frac{2.8}{5.3} = 31.9°$, $\angle B = \text{Tan}^{-1}\frac{4.5}{2.8} = 58.1°$

17. Let θ represent the angle the ladder makes with the ground.
$\theta = \text{Arcsin}\frac{13\frac{5}{12}}{18} \approx 48.2°$

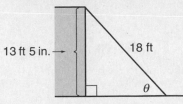

18. Let θ represent the maximum angle.
$\theta = \text{Tan}^{-1}\frac{1}{1.266} \approx 38.3°$

19. Explanations will vary. For a given function, any interval may be chosen as long as there is one and only one value of the function for each value of x in the interval. This is true for the sine and tangent functions in the interval $-\frac{\pi}{2} \le x \le \frac{\pi}{2}$. (In other texts, different intervals may be used.) The cosine, however, is not a function in this interval (e.g., $\cos -\frac{\pi}{2} = \cos\frac{\pi}{2}$), so a different interval is used to restrict cosine values.

EXTEND

20. $\sin\left(\text{Arcsin } 1 - \text{Arccos}\frac{1}{2}\right) = \sin\left(\frac{\pi}{2} - \frac{\pi}{3}\right) = \sin\left(\frac{\pi}{6}\right) = \frac{1}{2}$

21. $\tan(\text{Cos}^{-1} 0 + \text{Tan}^{-1} 1) = \tan\left(\frac{\pi}{2} + \frac{\pi}{4}\right) = \tan\frac{3\pi}{4} = -1$

22. $\cos(\text{Sin}^{-1} 0 - \text{Arccos } 1) = \cos(0 - 0) = \cos 0 = 1$

23. $\cos\left(\text{Cos}^{-1}\left(-\frac{\sqrt{2}}{2}\right) - \frac{\pi}{2}\right) = \cos\left(\frac{3\pi}{4} - \frac{\pi}{2}\right) = \cos\left(\frac{\pi}{4}\right) = \frac{\sqrt{2}}{2}$

THINK CRITICALLY

24. $\frac{(v\sin\theta)^2}{64} = h$; $\frac{(40\sin\theta)^2}{64} = 13.5$;
$(40\sin\theta)^2 = (64)(13.5)$; $40\sin\theta = \sqrt{864}$;
$\sin\theta = \frac{\sqrt{864}}{40}$; $\theta = \text{Arcsin}\frac{\sqrt{864}}{40} \approx 47.3°$

25. $5\sin^2\theta + 3\sin\theta - 1 = 0$; Use the quadratic formula $x = \frac{-b \pm \sqrt{b^2 - 4ac}}{2a}$.
$\sin\theta = \frac{-3 \pm \sqrt{3^2 - 4(5)(-1)}}{2(5)} = \frac{-3 \pm \sqrt{29}}{10}$
$\theta = \text{Sin}^{-1}\left(\frac{-3 + \sqrt{29}}{10}\right), \text{Sin}^{-1}\left(\frac{-3 - \sqrt{29}}{10}\right)$
$\theta \approx 0.2408, -0.9946$

26. $2\tan^2 x + 9\tan x + 3 = 0$

$\tan x = \dfrac{-9 \pm \sqrt{9^2 - 4(2)(3)}}{2(2)} = \dfrac{-9 \pm \sqrt{57}}{4}$

$x = \text{Tan}^{-1}\left(\dfrac{-9 + \sqrt{57}}{4}\right), \text{Tan}^{-1}\left(\dfrac{-9 - \sqrt{57}}{4}\right)$

$x = -0.3478, -1.3346$

But $-1.3336 < -\dfrac{\pi}{2}$ and is not in the domain of x so it is excluded. So, the solution is $x = -0.3478$

Mixed Review

27. distributive property

28. additive identity property

29. commutative property of addition

30. $\sqrt{x} - 5 = 2; \sqrt{x} = 7; x = 35$
Check: $\sqrt{35} - 5 = 7 - 5 = 2$ ✓

31. $\sqrt{3x - 5} = 4; 3x - 5 = 16; 3x = 21; x = 7$
Check: $\sqrt{3 \cdot 7 - 5} = \sqrt{21 - 5} = \sqrt{16} = 4$ ✓

32. $1 + \sqrt{2x + 7} = \sqrt{x + 15};$
$(1 + \sqrt{2x + 7})^2 = (\sqrt{x + 15})^2;$
$1 + 2\sqrt{2x + 7} + 2x + 7 = x + 15;$
$2\sqrt{2x + 7} = -x + 7; (2\sqrt{2 + 7})^2 = (-x + 7)^2;$
$4(2x + 7) = x^2 - 14x + 49;$
$8x + 28 = x^2 - 14x + 49; 0 = x^2 - 22x + 21;$
$(x - 21)(x - 1) = 0; x - 21 = 0 \text{ or } x - 1 = 0;$
$x = 21 \text{ or } x = 1$
Check: $1 + \sqrt{2(21) + 7} \overset{?}{=} \sqrt{21 + 15}$
$\qquad\quad 1 + \sqrt{42 + 7} \overset{?}{=} \sqrt{36}$
$\qquad\qquad\qquad\quad 8 \neq 6$
$x = 21$ does not work.
$1 + \sqrt{2(1) + 7} \overset{?}{=} \sqrt{1 + 15}$
$\quad 1 + \sqrt{2 + 7} \overset{?}{=} \sqrt{16}$
$\qquad\quad 1 + \sqrt{9} \overset{?}{=} 4$
$\qquad\qquad 1 + 3 \overset{?}{=} 4$
$\qquad\qquad\qquad 4 = 4$ ✓
$x = 1$ is the solution.

33. B; $\cos\left(\text{Arcsin}\left(\dfrac{5}{13}\right)\right)$

Let θ be the angle whose sine is $\dfrac{5}{13}$ and draw a right triangle with one angle θ, hypotenuse 13 and leg opposite θ 5. By the Pythagorean theorem the other leg is 12. So, $\cos\left(\text{Arcsin}\left(\dfrac{5}{13}\right)\right) = \cos\theta = \dfrac{12}{13}$.

Chapter Review, pages 782–783

1. e   **2.** d   **3.** a   **4.** c   **5.** b

6. $90° = 90\left(\dfrac{\pi}{180}\right) = \dfrac{\pi}{2}$ **7.** $225° = 225\left(\dfrac{\pi}{180}\right) = \dfrac{5\pi}{4}$

8. $-150° = -150\left(\dfrac{\pi}{180}\right) = -\dfrac{5\pi}{6}$

9. $900° = 900\left(\dfrac{\pi}{180}\right) = 5\pi$

10. $\dfrac{3\pi}{4} = \dfrac{3\pi}{4}\left(\dfrac{180}{\pi}\right) = 135°$

11. $0.7\pi = 0.7\pi\left(\dfrac{180}{\pi}\right) = 126°$

12. $\dfrac{2\pi}{3} = \dfrac{2\pi}{3}\left(\dfrac{180}{\pi}\right) = 120°$

13. $-\dfrac{8\pi}{5} = -\dfrac{8\pi}{5}\left(\dfrac{180}{\pi}\right) = -288°$

14. $500 - 360 = 140°$ **15.** $-13 + 360 = 347°$

16. $1035 - 360(2) = 315°$

17. $-1199 + 360(4) = 241°$

18. 0.5; 30-60-90 triangle   **19.** -1; quadrant angle

20. $-\sqrt{3}$; 30-60-90 triangle   **21.** $\sqrt{2}$; 45-45-90 triangle

22. 1.0355   **23.** 0.4695   **24.** -2.3048 **25.** 0.3090

26.
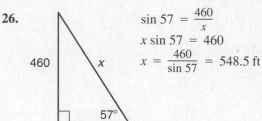

$\sin 57 = \dfrac{460}{x}$

$x \sin 57 = 460$

$x = \dfrac{460}{\sin 57} = 548.5 \text{ ft}$

27. $\angle C = 180 - 85.9 - 30 = 64.1°$
$\dfrac{\sin A}{a} = \dfrac{\sin B}{b}; \dfrac{\sin 85.9}{a} = \dfrac{\sin 30}{10};$
$a = \dfrac{10 \sin 85.9}{\sin 30} \approx 19.9$
$\dfrac{\sin C}{c} = \dfrac{\sin B}{b}; \dfrac{\sin 64.1}{c} = \dfrac{\sin 30}{10};$
$c = \dfrac{10 \sin 64.1}{\sin 30} \approx 18.0$

28. $a^2 = b^2 + c^2 - 2bc \cos A;$
$14^2 = 6.5^2 + 11.1^2 - 2(6.5)(11.1) \cos A;$
$\cos A = \dfrac{14^2 - 6.5^2 - 11.1^2}{-2(6.5)(11.1)} = \dfrac{30.54}{-144.3};$
$A = \text{Cos}^{-1} \dfrac{30.54}{-144.3}; \angle A = 102.2°$
$\dfrac{\sin B}{b} = \dfrac{\sin A}{a}; \dfrac{\sin B}{6.5} = \dfrac{\sin 102.2}{14};$
$\sin B = \dfrac{6.5 \sin 102.2}{14} \approx 0.4538$
$\angle B = \text{Sin}^{-1} 0.4538 \approx 27.0°$
$\angle C = 180 - \angle A - \angle B =$
$180 - 102.2 - 27.0 = 50.8°$

29. amplitude: 3, period: 2π

30. amplitude: $|-2| = 2$, period: 2π

31. amplitude: 1, period: $\dfrac{2\pi}{4} = \dfrac{\pi}{2}$

32. amplitude: 5, period: $\dfrac{2\pi}{\frac{1}{3}} = 6\pi$

33.

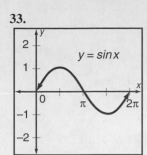

34.

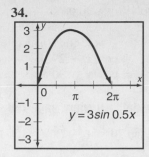

35.

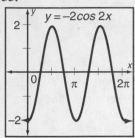

36.

$y = 0.5 \cos \frac{1}{4} x$

37. $\frac{\pi}{6}$; 30-60-90 triangle   **38.** $\frac{3\pi}{4}$; 45-45-90 triangle

39. $\tan \left(\text{Arcsin} \frac{2}{3} \right) = \tan (41.8103) \approx 0.8944$

Chapter Assessment, pages 784–785

CHAPTER TEST

1. $270° = 270 \left(\frac{\pi}{180} \right) = \frac{3\pi}{2}$

2. $-60° = -60 \left(\frac{\pi}{180} \right) = -\frac{\pi}{3}$ or $\frac{5\pi}{3}$

3. $\frac{\pi}{4} = \frac{\pi}{4} \left(\frac{180}{\pi} \right) = 45°$ **4.** $\frac{11\pi}{6} = \frac{11\pi}{6} \left(\frac{180}{\pi} \right) = 330°$

5. $779 - 360(2) = 779 - 720 = 59°$

6. $-644 + 360(2) = -644 + 720 = 76°$

7. $\cos 45 = \frac{\sqrt{2}}{2}$ **8.** $\sin 90 = 1$ **9.** $\tan \frac{5\pi}{6} = -\frac{\sqrt{3}}{3}$

10. $\csc \left(-\frac{2\pi}{3} \right) = -\frac{2\sqrt{3}}{3}$

11. $\tan 30° = \frac{x}{80}$
$x = 80 \tan 30$
$x \approx 46.2$ ft

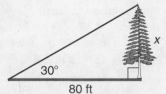

12. $\sin 56 = \frac{4.2}{x}$

$x = \frac{4.2}{\sin 56}$
$x \approx 5.1$ m

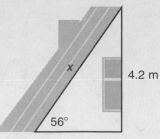

13. $n = \frac{\sin 45}{\sin 30} = \frac{\frac{\sqrt{2}}{2}}{\frac{1}{2}} = \sqrt{2} \approx 1.414$

14. $a^2 = b^2 + c^2 - 2bc \cos A$;
$4^2 = 3^2 + 4.5^2 - 2(3)(4.5) \cos A$;
$\cos A = \frac{4^2 - 3^2 - 4.5^2}{-2(3)(4.5)} \approx 0.4907$

$\angle A \approx 60.6°$

$\frac{\sin B}{b} = \frac{\sin A}{a}; \frac{\sin B}{3} = \frac{\sin 60.6}{4};$

$\sin B = \frac{3 \sin 60.6°}{4} \approx 0.6534$

$\angle B \approx 40.8°$
$\angle C = 180 - \angle A - \angle B =$
$180 - 60.6 - 40.8 = 78.6°$

15. $\frac{\sin A}{a} = \frac{\sin C}{c}; \frac{\sin A}{32} = \frac{\sin 98}{45};$

$\sin A = \frac{32 \sin 98}{45} \approx 0.7042$

$\angle A \approx 44.8°$
$\angle B = 180 - \angle A - \angle C =$
$180 - 44.8 - 98 = 37.2°$

$\frac{\sin B}{b} = \frac{\sin C}{c}; \frac{\sin 37.2}{b} = \frac{\sin 98}{45};$

$b = \frac{45 \sin 37.2}{\sin 98} \approx 27.5$

16. area $= \frac{1}{2} ac \sin B = \frac{1}{2}(7.8)(9.6) \sin 77.1 \approx 36.5$

17. area $= \frac{1}{2} bc \sin A = \frac{1}{2}(11.6)(15.1) \sin 58.2 \approx 74.4$

18. amplitude: 1, period: 2π

19. amplitude: 3, period: $\frac{2\pi}{2} = \pi$

20. amplitude: 0.7, period: $\frac{2\pi}{\frac{1}{2}} = 4\pi$

21. amplitude: $|-5| = 5$, period: $\frac{2\pi}{3}$.

22. translate up 3 units

23. translate to the right $\frac{\pi}{4}$ radians

24. translate to the left $\frac{\pi}{6}$ radians and down 2 units.

25.

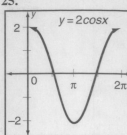

26.

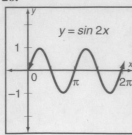

27.

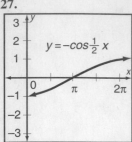

28.

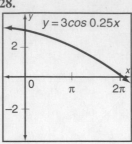

29. $\frac{\pi}{4}$ or 45°   **30.** $\frac{\pi}{3}$ or 60°   **31.** $-\frac{\pi}{3}$ or −60°

32. Use a 3-4-5 right triangle. $\text{Arcsin}\,\frac{4}{5} = \theta$ where θ is opposite the side of length 4. So,

$$\cos\left(\text{Arcsin}\,\frac{4}{5}\right) = \cos\theta = \frac{3}{5}$$

33. 0.39π

34. No. Explanations will vary. Students should mention that because the sine function is periodic, functions values of a given angle are identical for all angles coterminal with the given angle. For example, sin 30° = sin 390° = sin 750° = . . . = 0.5. As a result, each member of the domain of the inverse sine function (e.g., 0.5) will be paired with an infinite number of range members (30°, 390°, 750°, . .). Therefore, the inverse of the sine function is not a function.

35. $\frac{2\pi}{5}$ radians is $\frac{1}{5}$ of 2π radians which is $\frac{1}{5}$ of a revolution. One revolution of the minute hand is 1 h = 60 min. So $\frac{1}{5}(60\text{ min}) = 12$ min and the time is 3:12; A.

36. Let θ represent the angle.

$$\cos\theta = \frac{6}{14} = \frac{3}{7}$$

$$\theta = \text{Cos}^{-1}\frac{3}{7} = 64.6°$$

PERFORMANCE ASSESSMENT

CHOOSE YOUR METHOD

Students choice for best value for the area will vary according to student's accuracy. Estimating the number of squares is probably more difficult then estimating lengths of sides.

CURVY EQUATIONS

$y = \sin x, y = \sin(x + 360), y = \sin(x + 720), \dots,$
$y = (x + 360n)$ (n an integer); $y = \cos(x - 90),$
$y = \cos(x + 270), y = \cos(x + 630), \dots,$
$y = \cos(x - 90 + 360n)$ (n an integer)

Cumulative Review, page 786

1. $\tan 70 = \frac{\overline{PC}}{150}; \overline{PC} = 150\tan 70 \approx 412.12$

$\sin 50 = \frac{\overline{PC}}{\overline{AP}} = \frac{412.12}{\overline{AP}}, \overline{AP} = \frac{412.12}{\sin 50} \approx 538$

2. $\text{Arccos}\,\frac{1}{2} = \frac{\pi}{3}$ or 60°

3. Let θ represent the angle whose sine is $\frac{8}{17}$. Draw a right triangle with one angle θ, hypotenuse 17, and leg opposite θ, 8. Use Pythagorean theorem to find other leg x. $x^2 = 17^2 - 8^2 = 225; x = 15$
So, $\tan\left(\text{Arcsin}\,\frac{8}{17}\right) = \tan\theta = \frac{8}{15}$.

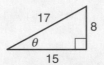

4.

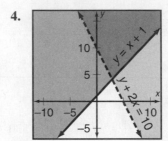

5. Let $x = \overline{KT}$ and since $\overline{KT} + \overline{KS} = 18, 18 - x = \overline{KS}$. $\overline{KR} = \overline{KT} = x$ and $\overline{RS} = 12$ so, use the Pythagorean theorem to write the equation: $(18 - x)^2 = 12^2 + x^2$; $324 - 36x + x^2 = 144 + x^2; 36x = 180; x = 5$ and $18 - x = 13$; So, $\overline{KT} = 5$ km and $\overline{KS} = 13$ km.

6. Let x represent the rate of the stream. The boat travels 9 mi upstream at the rate $8 - x$. So, since $d = rt, 9 = (8 - x)t$ where t is the time. The boat travels 15 mi downstream at the rate $8 + x$. So, $15 = (8 + x)t$. Solve these two equations for t.
$t = \frac{9}{8 - x}$ and $t = \frac{15}{8 + x}$
Since the times are the same substitute and get
$\frac{9}{8 - x} = \frac{15}{8 + x}; 9(8 + x) = 15(8 - x);$
$72 + 9x = 120 - 15x; 24x = 48; x = 2$
Therefore, the rate of the stream is 2 mi/h.

7. $4x^2 - 9y^2 + 36y = 72$
$4x^2 - 9(y^2 - 4y + 4) = 72 - 36$
$4x^2 - 9(y - 2)^2 = 36$
$\dfrac{x^2}{9} - \dfrac{(y - 2)^2}{4} = 1$
$a = 3, b = 2$
Slopes for the asymptotes are $\pm\dfrac{b}{a} = \pm\dfrac{2}{3}$.

8. Two triangles ($\triangle ABC$ and $\triangle AB\,'C$), each containing acute $\angle A$, are determined when ab and $a > h$ (the altitude from C).

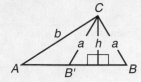

9. By the Pythagorean theorem $c^2 = a^2 + b^2$. Take the log of both sides and apply the properties of logarithms; $\log c^2 = \log(a^2 + b^2); 2\log c = \log(a^2 + b^2)$. Therefore, D.

10. $\dfrac{5\pi}{12} = \dfrac{5\pi}{12}\left(\dfrac{180}{\pi}\right) = 75°$

11. $y = 3x^2 - 5x + 8, a = 3, b = -5, c = 8$
axis of symmetry is $x = -\dfrac{b}{2a}; x = -\dfrac{-5}{2(3)}; x = \dfrac{5}{6}$

12. $d = 9 - 19\sin\dfrac{2\pi}{15}t = 9 - 19\sin\dfrac{2\pi}{15}(2) \approx -5.12$ m

13. Complete the squares.
$x^2 - 6x + 9 + y^2 + 4y + 4 = 51 + 9 + 4$
$(x - 3)^2 + (y + 2)^2 = 64$
Circle has center at $(3, -2)$

14. $x = 1 + \sqrt{x + 5}; x - 1 = \sqrt{x + 5};$
$(x - 1)^2 = (\sqrt{x + 5})^2; x^2 - 2x + 1 = x + 5;$
$x^2 - 3x - 4 = 0; (x - 4)(x + 1) = 0$
$x = 4, -1$
Check:
$4 = 1 + \sqrt{4 + 5} = 1 + \sqrt{9} = 1 + 3 = 4$ ✓
$-1 \neq 1 + \sqrt{1 + 5} = 1 + \sqrt{6}$
The solution is 4.

15. Substitute, $\dfrac{\frac{1}{3} + \frac{1}{7}}{1 - \frac{1}{3}\left(\frac{1}{7}\right)}$; multiply numerator and

denominator by the LCD of $\frac{1}{3}$ and $\frac{1}{7}$,

$\left(\dfrac{\frac{1}{3} + \frac{1}{7}}{1 - \frac{1}{3}\left(\frac{1}{7}\right)}\right)\left(\dfrac{21}{21}\right) = \dfrac{7 + 3}{21 - 1} = \dfrac{10}{20} = \dfrac{1}{2}$

Standardized Test, page 787

1. B; $s = r\theta$
$5 = 5\theta$
$1 = \theta$

2. C; $\dfrac{\log_4 64}{\log_4 8} = \dfrac{\log_4 8^2}{\log_4 8} = \dfrac{2\log_4 8}{\log_4 8} = 2$

3. E; $\dfrac{x}{x - 2} - \dfrac{2}{x + 4} = \dfrac{12}{x^2 + 2x - 8}$
$x(x + 4) - 2(x - 2) = 12$
$x^2 + 4x - 2x + 4 = 12$
$x^2 + 2x - 8 = 0$
$(x + 4)(x - 2) = 0$
$x = -4$ or $x = 2$
But $x \neq 2$ and $x \neq -4$ since both are excluded from domain

4. A; $\sin P = \dfrac{p}{r}$; $\sin Q = \dfrac{q}{s}$; $p > q$; $\dfrac{p}{r} > \dfrac{q}{r}$; so $\sin P > \sin Q$.
$\cos P = \dfrac{q}{r}$; $\cos Q = \dfrac{p}{r}$ $p > q$, so $\cos P < \cos Q$.
$\sin P = \dfrac{p}{r}$ and $\cos Q = \dfrac{p}{r}$, so $\sin P = \cos Q$

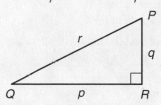

5. D; $(x - 1)(x - (1 + i))(x - (1 - i)) =$
$(x - 1)(x^2 - (1 + i)x - (1 - i)x +$
$(1 + i)(1 - i)) =$
$(x - 1)(x^2 - x - ix - x + ix + 1 - i^2) =$
$(x - 1)(x^2 - 2x + 2) =$
$x^3 - 2x^2 + 2x - x^2 + 2x - 2 =$
$x^3 - 3x^2 + 4x - 2$

6. C; $\left(\dfrac{1}{10}\right)\left(\dfrac{1}{10}\right)\left(\dfrac{1}{10}\right) = \dfrac{1}{1000}$

7. B; amplitude $= \dfrac{1}{2}$; period $= \dfrac{2\pi}{\pi} = 2; y = \dfrac{1}{2}\sin 2x$
$p(-1) = -13$

8. E; $P(-1) = -13; P(0) = -10; P(1) = -21;$
$P(2) = -28, P(3) = -13, P(4) = 42;$ Change of sign between $P(3)$ and $P(4)$

9. B; From Pascal's triangle
1   8   28   56   70   56   28   8   1

10. E; The graphs do not intersect.
$y = -\dfrac{12}{x}$
$y = \sqrt{\dfrac{225 - 25x^2}{9}}$
$y = -\sqrt{\dfrac{225 - 25x^2}{9}}$

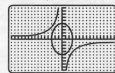

Chapter 16 Trigonometric Identities and Equations

Data Activity, pages 788–789

1. List the 11 entries in order. Median is the 6th one. 3650 m/s (brick). Range for this group is
$5200 - 332 = 4868$ m/s.

2. $4110 - 3650 = 460$ m how much farther in 1s
$0.25(460) = 115$ m how much farther in 0.25s

3. copper: 3810 m/s; $\dfrac{10,000}{3810} = 2.6$ s

4. linear; $\dfrac{344 - 332}{20 - 0} = \dfrac{362 - 344}{50 - 20} = \dfrac{3}{5}$

5. $\dfrac{\text{Change in } S}{\text{Change in } T} = \dfrac{344 - 332}{20 - 0} = \dfrac{3}{5} = 0.6$ m/s for each 1°C
$65 - 50 = 15°$, increase in temperature
$15(0.6) = 9$ m/s, increase in speed
$362 + 9 = 371$ m/s, speed at 65°C

6. The choice of scale used on the axis will affect the appearance of the graph.

Lesson 16.1, pages 791–797

EXPLORE

1. $\sin \theta \csc \theta = \dfrac{y}{r} \cdot \dfrac{r}{y} = 1$

2. $\cos \theta \sec \theta = \dfrac{x}{r} \cdot \dfrac{r}{x} = 1$

3. $\tan \theta \cot \theta = \dfrac{y}{x} \cdot \dfrac{x}{y} = 1$

4. $\dfrac{\sin \theta}{\cos \theta} = \dfrac{\frac{y}{r}}{\frac{x}{r}} = \dfrac{y}{x}$
5. $\dfrac{\cos \theta}{\sin \theta} = \dfrac{\frac{x}{r}}{\frac{y}{r}} = \dfrac{x}{y}$

6. $\dfrac{\sec \theta}{\csc \theta} = \dfrac{\frac{r}{x}}{\frac{r}{y}} \cdot \left(\dfrac{xy}{xy}\right) = \dfrac{ry}{rx} = \dfrac{y}{x}$

7. $\csc \theta = \dfrac{1}{\sin \theta}$; $\sec \theta = \dfrac{1}{\cos \theta}$; $\cot \theta = \dfrac{1}{\tan \theta}$

8. $\tan \theta$

9. $\cot \theta$

10.

θ	$\sin \theta$	$\cos \theta$	$\tan \theta$
36°	0.5878	0.8090	0.7265
−36°	−0.5878	0.8090	−0.7265
164°	0.2756	−0.9613	−0.2867
−164°	−0.2756	−0.9613	0.2867
285°	−0.9659	0.2588	−3.7321
−285°	0.9659	0.2588	3.7321

11. $\sin (-\theta) = -\sin \theta$; $\cos (-\theta) = \cos \theta$;
$\tan (-\theta) = -\tan \theta$

12.

θ	$\sin^2 \theta$	$\cos^2 \theta$	$\cos^2 \theta + \sin^2 \theta$
36°	0.3455	0.6545	1
−36°	0.3455	0.6545	1
164°	0.0760	0.9240	1
−164°	0.0760	0.9240	1
285°	0.9330	0.0670	1
−285°	0.9330	0.0670	1

$\cos^2 \theta + \sin^2 \theta = 1$

TRY THESE

1. $\csc^2 \theta = 1 + \cot \theta$; $\csc^2 \theta - \cot^2 \theta = 1$

2. 1

3. $\dfrac{\sin \theta}{\sin (-\theta)} = \dfrac{\sin \theta}{-\sin \theta} = -1$

4. $\dfrac{\cos (-\theta)}{\cos \theta} = \dfrac{\cos \theta}{\cos \theta} = 1$

5. Answers will vary. Cos and sec are even and they have similar rules in that for each one $f(-x) = f(x)$. Sin, tan, csc, and cot are odd and they all have similar rules in that $f(-x) = -f(x)$.

6. $\cot \theta \sin \theta = \dfrac{\cos \theta}{\sin \theta} \cdot \sin \theta = \cos \theta$

7. $\dfrac{\cos \theta}{\cot \theta} = \dfrac{\cos \theta}{\frac{\cos \theta}{\sin \theta}} = \cos \theta \cdot \left(\dfrac{\frac{\sin \theta}{\cos \theta}}{\frac{\sin \theta}{\cos \theta}}\right) = \sin \theta$

8. $\dfrac{\tan \theta}{\cos \theta} = \dfrac{\frac{\sin \theta}{\cos \theta}}{\cos \theta} = \dfrac{\sin \theta}{\cos^2 \theta} = \dfrac{\sin \theta}{1 - \sin^2 \theta}$

9. $(\sec^2 \theta - 1) \cos^2 \theta = \tan^2 \theta \cos^2 \theta = \dfrac{\sin^2 \theta}{\cos^2 \theta} \cdot \cos^2 \theta = \sin^2 \theta$

10. $(1 - \cos \alpha)(1 + \cos \alpha) = 1 - \cos^2 \alpha = \sin^2 \alpha$

11. $\sin (-\beta) \cot (-\beta) = (-\sin \beta)(-\cot \beta) = \sin \beta \left(\dfrac{\cos \beta}{\sin \beta}\right) = \cos \beta$

12. $\cot (90° - \theta) \cos (-\theta) = \tan \theta(\cos \theta) = \dfrac{\sin \theta}{\cos \theta} \cdot \cos \theta = \sin \theta$

13. $(1 - \sec \beta)(1 + \sec \beta) = 1 - \sec^2 \beta = -\tan^2 \beta$

14. 15 ft Let θ represent the angle of elevation and x represent the distance between house and garage.

θ x

14a. $\sin^2 \theta + \cos^2 \theta = 1$; $(0.2222)^2 + \cos^2 \theta = 1$;
$\cos^2 \theta = 1 - (0.2222)^2$; $\cos^2 \theta = 0.9506$;
$\cos \theta \approx 0.9750$

14b. $\cos \theta = \dfrac{x}{15}$; $x = 15 \cos \theta = 15(0.975) \approx 14.63$ ft

14c. $\theta = \text{Sin}^{-1} 0.2222 \approx 12.8°$

15. $\dfrac{\cos \alpha \tan \alpha}{\sin \alpha} = \left(\dfrac{\cos \alpha}{\sin \alpha}\right) \cdot \tan \alpha = \cot \alpha \tan \alpha = 1$

16. $\csc \beta (1 - \cos^2 \beta) = \dfrac{1}{\sin \beta} \cdot \sin^2 \beta = \sin \beta$

17. $\cos \alpha \csc \alpha = \cos \alpha \cdot \dfrac{1}{\sin \alpha} = \dfrac{\cos \alpha}{\sin \alpha} = \cot \alpha = \dfrac{1}{\tan \alpha}$

18. $\dfrac{\cos \beta - \sin \beta - \cos^3 \beta}{\cos \beta} = \dfrac{\cos \beta - \cos^3 \beta - \sin \beta}{\cos \beta} =$

$\dfrac{\cos \beta (1 - \cos^2 \beta) - \sin \beta}{\cos \beta} = \dfrac{\cos \beta \sin^2 \beta - \sin \beta}{\cos \beta} =$

$\dfrac{\cos \beta \sin^2 \beta}{\cos \beta} - \dfrac{\sin \beta}{\cos \beta} = \sin^2 \beta - \tan \beta$

PRACTICE

1. $-\cos^2 \theta - \sin^2 \theta = -(\cos^2 \theta + \sin^2 \theta) = -1$

2. $\sec(-\theta) \cos(-\theta) = \sec \theta \cos \theta =$

$\dfrac{1}{\cos \theta} \cdot \cos \theta = 1$

3. $\dfrac{\cot(-\theta)}{\cot \theta} = \dfrac{-\cot \theta}{\cot \theta} = -1$

4. $\dfrac{\cos \theta}{\sin \theta \cot \theta} \cdot \left(\dfrac{\dfrac{1}{\sin \theta}}{\dfrac{1}{\sin \theta}}\right) = \dfrac{\cot \theta}{\cot \theta} = 1$

5. Answers will vary. Students may suggest starting with the side of the identity that appears more complicated and trying to simplify it, applying various identities that will simplify expressions, and using various algebraic techniques to simplify expressions.

6. $\sec \theta \tan \theta = \dfrac{1}{\cos \theta} \cdot \dfrac{\sin \theta}{\cos \theta} = \dfrac{\sin \theta}{\cos^2 \theta} = \dfrac{\sin \theta}{1 - \sin^2 \theta}$

7. $\dfrac{\cos \theta}{\sec \theta} = \dfrac{\cos \theta}{\dfrac{1}{\cos \theta}} = \cos^2 \theta$

8. $\dfrac{\sin \theta}{\cot \theta} = \dfrac{\sin \theta}{\dfrac{\cos \theta}{\sin \theta}} = \dfrac{\sin^2 \theta}{\cos \theta} = \dfrac{1 - \cos^2 \theta}{\cos \theta}$

9. $(\csc^2 \theta - 1)(\sin^2 \theta) = \csc^2 \theta \sin^2 \theta - \sin^2 \theta =$
$1 - \sin^2 \theta = \cos^2 \theta$

10. $\dfrac{\cot(-\alpha)}{\csc \alpha} = \dfrac{\dfrac{-\cos \alpha}{\sin \alpha}}{\dfrac{1}{\sin \alpha}} = -\cos \alpha$

11. $\cos(-\beta) \tan(\beta) = \cos \beta(-\tan \beta) =$

$-\cos \beta \left(\dfrac{\sin \beta}{\cos \beta}\right) = -\sin \beta$

12. $\tan(90° - \theta) \sin(-\theta) = \cot \theta(-\sin \theta) =$

$-\dfrac{\cos \theta}{\sin \theta} \cdot \sin \theta = -\cos \theta$

13. $(1 - \csc \beta)(1 + \csc \beta) = 1 - \csc^2 \beta = -\cot^2 \beta$

14a.

415 ft   89.6 ft   Let θ represent the angle of elevation.

$\csc \theta = \dfrac{415}{89.6} \approx 4.6317$

14b. $\cot^2 \theta = \csc^2 \theta - 1 = (4.6317)^2 - 1 = 20.4526;$
$\cot \theta \approx 4.5225$

14c. $\theta = \text{Csc}^{-1} 4.6317 \approx 12.5°$

15. $\sec^2 \alpha - 3 = 1 + \tan^2 \alpha - 3 = \tan^2 \alpha - 2$

16. $\csc^2 \beta + 6 = 1 + \cot^2 \beta + 6 = \cot^2 \beta + 7$

17. $\dfrac{1}{\sin^2 \theta} + \dfrac{1}{\cos^2 \theta} = \dfrac{\cos^2 \theta}{\sin^2 \theta \cos^2 \theta} + \dfrac{\sin^2 \theta}{\sin^2 \theta \cos^2 \theta} =$
$\dfrac{\cos^2 \theta + \sin^2 \theta}{\sin^2 \theta \cos^2 \theta} = \dfrac{1}{\sin^2 \theta \cos^2 \theta}$

18. $\cos^2 \alpha = 1 - \sin^2 \alpha = 1 - \dfrac{1}{\csc^2 \alpha}$

19. $\dfrac{\sin^2 \theta - 4 \sin \theta + 4}{\sin^2 \theta - 4} = \dfrac{(\sin \theta - 2)(\sin \theta - 2)}{(\sin \theta + 2)(\sin \theta - 2)} =$
$\dfrac{(\sin \theta - 2)}{(\sin \theta + 2)}$

20. $\dfrac{\cos^2 \beta - 10 \cos \beta + 25}{\cos^2 \beta - 25} =$
$\dfrac{(\cos \beta - 5)(\cos \beta - 5)}{(\cos \beta + 5)(\cos \beta - 5)} = \dfrac{\cos \beta - 5}{\cos \beta + 5}$

21. $\dfrac{1}{\cot^2 \alpha} - \dfrac{1}{\csc^2 \alpha} = \dfrac{\csc^2 \alpha}{\cot^2 \alpha \csc^2 \alpha} - \dfrac{\cot^2 \alpha}{\cot^2 \alpha \csc^2 \alpha} =$
$\dfrac{\csc^2 \alpha - \cot^2 \alpha}{\cot^2 \alpha \csc^2 \alpha} = \dfrac{1}{\cot^2 \alpha \csc^2 \alpha} =$

$\dfrac{1}{\dfrac{\cos^2 \alpha}{\sin^2 \alpha} \cdot \dfrac{1}{\sin^2 \alpha}} = \dfrac{1}{\dfrac{\cos^2 \alpha}{\sin^4 \alpha}} = \dfrac{\sin^4 \alpha}{\cos^2 \alpha}$

22. $\cot \beta + \tan \beta = \dfrac{\cos \beta}{\sin \beta} + \dfrac{\sin \beta}{\cos \beta} =$
$\dfrac{\cos^2 \beta}{\sin \beta \cos \beta} + \dfrac{\sin^2 \beta}{\sin \beta \cos \beta} = \dfrac{\cos^2 \beta + \sin^2 \beta}{\sin \beta \cos \beta} =$
$\dfrac{1}{\sin \beta \cos \beta} = \dfrac{1}{\sin \beta} \cdot \dfrac{1}{\cos \beta} = \csc \beta \sec \beta$

23. $(\tan \theta + 1)^2 = \tan^2 \theta + 2 \tan \theta + 1 =$
$\sec^2 \theta + 2 \tan \theta$

24.

$\dfrac{1 - \cos^2 \alpha}{\cos \alpha - 1}$	$\dfrac{-\sec \alpha - 1}{\sec \alpha}$
$\dfrac{(1 + \cos \alpha)(1 - \cos \alpha)}{\cos \alpha - 1}$	$\dfrac{-\sec \alpha}{\sec \alpha} - \dfrac{1}{\sec \alpha}$
$\dfrac{-1(1 + \cos \alpha)}{-1 - \cos \alpha}$	$-1 - \cos \alpha$
$-1 - \cos \alpha$	$-1 - \cos \alpha$

25.

$\dfrac{\sec \beta \sin \beta}{\csc \beta \cos \beta}$	$\sec^2 \beta - 1$
$\dfrac{\dfrac{1}{\cos \beta} \cdot \sin \beta}{\dfrac{1}{\sin \beta} \cdot \cos \beta}$	$\tan^2 \beta$
$\dfrac{\tan \beta}{\cot \beta}$	
$\dfrac{\tan \beta}{\dfrac{1}{\tan \beta}}$	
$\tan^2 \beta$	

26. $(\cot \theta - \csc \theta)^2 =$

$\cot^2 \theta - 2 \cot \theta \csc \theta + \csc^2 \theta =$

$\dfrac{\cos^2 \theta}{\sin^2 \theta} - \left(\dfrac{2 \cos \theta}{\sin \theta} \cdot \dfrac{1}{\sin \theta} \right) + \dfrac{1}{\sin^2 \theta} =$

$\dfrac{\cos^2 \theta - 2 \cos \theta + 1}{\sin^2 \theta} = \dfrac{(1 - \cos \theta)^2}{1 - \cos^2 \theta} =$

$\dfrac{(1 - \cos \theta)(1 - \cos \theta)}{(1 + \cos \theta)(1 - \cos \theta)} = \dfrac{1 - \cos \theta}{1 + \cos \theta}$

27. $\dfrac{\sin \alpha}{\csc \alpha + \cot \alpha} = \dfrac{\sin \alpha}{\dfrac{1}{\sin \alpha} + \dfrac{\cos \alpha}{\sin \alpha}} = \dfrac{\sin \alpha}{\dfrac{1 + \cos \alpha}{\sin \alpha}} =$

$\dfrac{\sin^2 \alpha}{1 - \cos \alpha} = \dfrac{1 - \cos^2 \alpha}{1 + \cos \alpha} =$

$\dfrac{(1 + \cos \alpha)(1 - \cos \alpha)}{1 + \cos \alpha} = 1 - \cos \alpha$

28. $\dfrac{\cos \theta}{\sec \theta + \tan \theta} = \dfrac{\cos \theta}{\dfrac{1}{\cos \theta} + \dfrac{\sin \theta}{\cos \theta}} = \dfrac{\cos \theta}{\dfrac{1 + \sin \theta}{\cos \theta}} =$

$\dfrac{\cos^2 \theta}{1 + \sin \theta} = \dfrac{1 - \sin^2 \theta}{1 + \sin^2 \theta} = \dfrac{(1 + \sin \theta)(1 - \sin \theta)}{1 + \sin \theta} =$

$1 - \sin \theta$

29a. $\tan \theta_p = \dfrac{n_2}{n_1}$

$\dfrac{1}{\cot \theta_p} = \dfrac{n_2}{n_1}$

$\cot \theta_p = \dfrac{n_1}{n_2}$

29b. $\tan \theta_p = \dfrac{n_2}{n_1} = \dfrac{2.42}{1.0}$; $\Theta_p = \text{Tan}^{-1} 2.42 \approx 67.5°$

EXTEND

30. No   **31.** Yes   **32.** Yes   **33.** No

Answers will vary for 34–37. Possible examples are given.

34. $\theta = 240°$; $\sqrt{\cos^2 \theta} \overset{?}{=} \cos \theta$

$\sqrt{\cos^2 240} \overset{?}{=} \cos 240$

$0.5 \neq -0.5$

35. $\theta = 300°$; $\sqrt{\sin^2 \theta} \overset{?}{=} \sin \theta$

$\sqrt{\sin^2 300} \overset{?}{=} \sin 300$

$\dfrac{\sqrt{3}}{2} \neq -\dfrac{\sqrt{3}}{2}$

36. $\theta = 150°$; $\sqrt{1 + \tan^2 \theta} \overset{?}{=} \sec \theta$

$\sqrt{1 + \tan^2 150} \overset{?}{=} \sec 150$

$\dfrac{2\sqrt{3}}{3} \neq -\dfrac{2\sqrt{3}}{3}$

37. $\theta = 330°$; $\sqrt{1 + \cot^2 \theta} \overset{?}{=} \csc \theta$

$\sqrt{1 + \cot^2 330} \overset{?}{=} \csc 330$

$2 \neq -2$

THINK CRITICALLY

38. $\sin(-\beta) = -\sin \beta$ So, $\beta = -1, 0, 1$

39. Answers will vary. Here is a possible answer.

$\sin \theta = \cos \left(\dfrac{\pi + 0.01}{2} - \theta \right)$

40. Answers will vary. Examples include $\sin \alpha = n, n > 1$; $\csc \alpha = n, -1 < n < 1$; $\sec \alpha = n, -1 < n < 1$.

41.

$\sec^2 \theta - \sin^2 \theta$	$\cos^2 \theta + \tan^2 \theta$
$\dfrac{1}{\cos^2 \theta} - \sin^2 \theta$	$\cos^2 \theta + \dfrac{\sin^2 \theta}{\cos^2 \theta}$
$\dfrac{1 - \sin^2 \theta \cos^2 \theta}{\cos^2 \theta}$	$\dfrac{\cos^4 \theta + \sin^2 \theta}{\cos^2 \theta}$
$\dfrac{1 - (1 - \cos^2 \theta)(\cos^2 \theta)}{\cos^2 \theta}$	$\dfrac{\cos^4 \theta + 1 - \cos^2}{\cos^2 \theta}$
$\dfrac{1 - [\cos^2 \theta - \cos^4 \theta]}{\cos^2 \theta}$	$\dfrac{\cos^4 \theta - \cos^2 \theta + 1}{\cos^2 \theta}$
$\dfrac{\cos^4 \theta - \cos^2 \theta + 1}{\cos^2 \theta}$	$\dfrac{\cos^4 \theta - \cos^2 \theta + 1}{\cos^2 \theta}$

42. $4 \tan^2 \beta - 1 = 4(\tan^2 \beta + 1) - 1 - 4 =$
$4 \sec^2 \beta - 5$

MIXED REVIEW

43. $\log_5 x = 4$; $x = 5^4 = 625$

44. $\log_x 81 = 4$; $x^4 = 81 = 3^4$; $x = 3$

45. D; $\sqrt{(-36)(-100)} = \sqrt{(36)(100)} = \pm 6(10) = \pm 60$

46. $_{52}C_8 = \dfrac{52!}{8!(52 - 8)!} = \dfrac{52!}{8!44!} =$

$\dfrac{52 \cdot 51 \cdot 50 \cdot 49 \cdot 48 \cdot 47 \cdot 46 \cdot 45}{8 \cdot 7 \cdot 6 \cdot 5 \cdot 4 \cdot 3 \cdot 2} = 752{,}538{,}150$

47. $\dfrac{1}{1 - \cos \theta} + \dfrac{1}{1 + \cos \theta} =$

$\dfrac{1 + \cos \theta}{(1 - \cos \theta)(1 + \cos \theta)} + \dfrac{1 - \cos \theta}{(1 - \cos \theta)(1 + \cos \theta)} =$

$\dfrac{2}{1 - \cos^2 \theta} = \dfrac{2}{\sin^2 \theta} = 2 \csc^2 \theta$

48. $\dfrac{\sin \beta}{\tan \beta} = \dfrac{\sin \beta}{\dfrac{\sin \beta}{\cos \beta}} = \dfrac{\cos \beta \sin \beta}{\sin \beta} = \cos \beta$

Lesson 16.2, pages 798–803

EXPLORE

1. $y_1 = 3 \sin x, y_2 = \sqrt{3} + \sin x$

2.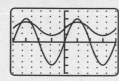

3. $-300°, -240°, 60°, 120°$ **4.** $420°, 480°, 780°, 840°$

5. They differ by 360° or are coterminal.

6. They differ by 360° or are coterminal.

7. $-600°, -660°$

TRY THESE

1. 2   2. 2   3. 4   4. 1

5. Answers will vary. A trigonometric identity is true for all values for which the functions are defined and an equation may only be true for certain values of the variable.

6. $2 \sin x + \sqrt{3} = 0; 2 \sin x = -\sqrt{3}; \sin x = -\dfrac{\sqrt{3}}{2};$
   $x = 240°, 300°$

7. $\sqrt{3} \csc x - 2 = 0; \sqrt{3} \csc x = 2; \csc x = \dfrac{2\sqrt{3}}{3};$
   $x = 60°, 120°$

8. $2 \sin^2 x = 3 \cos x; 2(1 - \cos^2 x) = 3 \cos x;$
   $2 - 2 \cos^2 x = 3 \cos x; 2 \cos^2 x + 3 \cos x - 2 = 0;$
   $(2 \cos x - 1)(\cos x + 2) = 0; \cos x = -2$
   (impossible) or $\cos x = \dfrac{1}{2}, x = 60°, 300°$

9. $3 \sec^2 x - 7 \sec x + 2 = 0;$
   $(3 \sec x - 1)(\sec x - 2) = 0; \sec x = \dfrac{1}{3}$ (impossible)
   or $\sec x = 2; x = 60°, 300°$

10. $\sqrt{2} \sin x \cos x - \sin x = 0; \sin x (\sqrt{2} \cos x - 1) = 0;$
    $\sin x = 0$ or $\cos x = \dfrac{\sqrt{2}}{2}; x = 0, \dfrac{\pi}{4}, \pi, \dfrac{7\pi}{4}$

11. $\tan^2 x - 4 \sec x + 5 = 0;$
    $\sec^2 x - 1 - 4 \sec x + 5 = 0;$
    $\sec^2 x - 4 \sec x + 4 = 0;$
    $(\sec x - 2)^2 = 0; \sec x = 2; x = \dfrac{\pi}{3}, \dfrac{5\pi}{3}$

12. $6 \cos^2 x + 7 \sin x = 3;$
    $6(1 - \sin^2 x) + 7 \sin x - 3 = 0;$
    $6 - 6 \sin^2 x + 7 \sin x - 3 = 0;$
    $6 \sin^2 x - 7 \sin x - 3 = 0;$
    $(2 \sin x - 3)(3 \sin x + 1) = 0; \sin x = \dfrac{3}{2}$
    (impossible) or $\sin x = -\dfrac{1}{3};$
    $x = 3.48, 5.94$
    A better way to solve this equation may be to use a graphing utility.

13. $16 \sec^2 x - 42 \tan x = 11;$
    $16(1 + \tan^2 x) - 42 \tan x = 11;$
    $16 + 16 \tan^2 x - 42 \tan x - 11 = 0;$
    $16 \tan^2 x - 42 \tan x + 5 = 0;$
    $(8 \tan x - 1)(2 \tan x - 5) = 0; \tan x = \dfrac{1}{8}$ or
    $\tan x = \dfrac{5}{2}; x = 0.12, 1.19, 3.27, 4.33$
    This equation may also be solved using a graphing utility.

14. $x = 1.94$

15. $\tan x = \sqrt{3}; x = \dfrac{\pi}{3} + n\pi, n$ is any integer

16. $\cos x = \cot x; \cos x = \dfrac{\cos x}{\sin x}; \sin x = 1$
    $x = \dfrac{\pi}{2} + 2n\pi, n$ is any integer

17. $20 \sin^2 x - 13 \sin x + 2 = 0;$
    $(5 \sin x - 2)(4 \sin x - 1) = 0; \sin x = \dfrac{2}{5}$ or
    $\sin x = \dfrac{1}{4}; x = 0.41 + 2n\pi,$
    $0.25 + 2n\pi, n$ is any integer.

18. $10 \cos^2 x - 23 \cos x - 5 = 0;$
    $(5 \cos x + 1)(2 \cos x - 5) = 0; \cos x = -\dfrac{1}{5}$ or
    $\cos x = \dfrac{5}{2}$ (impossible); $x = 1.77 + 2n\pi,$
    $4.51 + 2n\pi, n$ is any integer

PRACTICE

1. 2 solutions   2. 4 solutions

3. 2 solutions   4. 1 solution

5. Answers will vary. Strategies include factoring, using identities so that each factor is written in terms of only one trigonometric function, using the quadratic formula, using a graphing utility, etc.

6. $2 \cos x - \sqrt{3} = 0; \cos x = \dfrac{\sqrt{3}}{2}; x = 30°, 330°$
   Check:
   $2 \cos 30 - \sqrt{3} = 2\left(\dfrac{\sqrt{3}}{2}\right) - \sqrt{3} = \sqrt{3} - \sqrt{3} = 0$ ✓
   $2 \cos 330 - \sqrt{3} = 2\left(\dfrac{\sqrt{3}}{2}\right) - \sqrt{3} = \sqrt{3} - \sqrt{3} = 0$ ✓

7. $\sqrt{2} \csc x = 2; \csc x = \dfrac{2}{\sqrt{2}}; x = 45°, 135°$
   Check: $\sqrt{2} \csc 45 = \sqrt{2}\left(\dfrac{2}{\sqrt{2}}\right) = 2$ ✓
   $\sqrt{2} \csc 135 = \sqrt{2}\left(\dfrac{2}{\sqrt{2}}\right) = 2$ ✓

8. $-2 \cos^2 x = 3 \sin x; -2(1 - \sin^2 x) = 3 \sin x;$
   $-2 + 2 \sin^2 x = 3 \sin x; 2 \sin^2 x - 3 \sin x - 2 = 0;$
   $(\sin x - 2)(2 \sin x + 1) = 0; \sin x = 2$ (impossible)
   or $\sin x = -\dfrac{1}{2}; x = 210°, 330°$

Check: $-2\cos^2 210 \overset{?}{=} 3\sin 210$

$$-2\left(-\frac{\sqrt{3}}{2}\right)^2 \overset{?}{=} 3\left(-\frac{1}{2}\right)$$

$$-\frac{3}{2} = -\frac{3}{2} \checkmark$$

$$-2\cos^2 330 = 3\sin 330$$

$$-2\left(\frac{\sqrt{3}}{2}\right)^2 \overset{?}{=} 3\left(-\frac{1}{2}\right)$$

$$-\frac{3}{2} = -\frac{3}{2} \checkmark$$

9. $4\csc^2 x + 5\csc x = 6$; $4\csc^2 x + 5\csc x - 6 = 0$;
$(4\csc x - 3)(\csc x + 2) = 0$;

$\csc x = \frac{3}{4}$ (impossible) or $\csc x = -2$

$x = 210°, 330°$

Check: $4\csc^2 210 + 5\csc 210 =$
$$4(-2)^2 + 5(-2) = 16 - 10 = 6 \checkmark$$

$$4\csc^2 330 + 5\csc 330 =$$
$$4(-2)^2 + 5(-2) = 16 - 10 = 6 \checkmark$$

10. $3\tan^2 x - 1 = 0$; $\tan^2 x = \frac{1}{3}$; $\tan x = \pm\frac{\sqrt{3}}{3}$;
$x = \frac{\pi}{6}, \frac{5\pi}{6}, \frac{7\pi}{6}, \frac{11\pi}{6}$

11. $4\sin^2 x - 1 = 0$; $\sin^2 x = \frac{1}{4}$; $\sin x = \pm\frac{1}{2}$;
$x = \frac{\pi}{6}, \frac{5\pi}{6}, \frac{7\pi}{6}, \frac{11\pi}{6}$

12. $15\sin^2 x - 17\cos x - 11 = 0$;
$15(1 - \cos^2 x) - 17\cos x - 11 = 0$;
$15 - 15\cos^2 x - 17\cos x - 11 = 0$;
$15\cos^2 x + 17\cos x - 4 = 0$;
$(5\cos x - 1)(3\cos x + 4) = 0$;
$\cos x = \frac{1}{5}$ or $\cos x = -\frac{4}{3}$ (impossible); $x = 1.37, 4.91$

13. $18\cos^2 x + 27\sin x - 22 = 0$;
$18(1 - \sin^2 x) + 27\sin x - 22 = 0$;
$18 - 18\sin^2 x + 27\sin x - 22 = 0$;
$18\sin^2 x - 27\sin x + 4 = 0$;
$(6\sin x - 1)(3\sin x - 4) = 0$; $\sin x = \frac{4}{3}$
(impossible)

or $\sin x = \frac{1}{6}$; $x = 0.17, 2.97$

14. $3\cot x\cos x + \cos x = 0$; $\cos x(3\cot x + 1) = 0$;
$\cos x = 0$ or $\cot x = -\frac{1}{3}$; $x = \frac{\pi}{2}, \frac{3\pi}{2}, 1.89, 5.03$

15. $5\tan x\sin x + \sin x = 0$; $\sin x(5\tan x + 1) = 0$;
$\sin x = 0$ or $\tan x = -\frac{1}{5}$; $x = 0, 2.94, \pi, 6.09$

16. $A = \frac{1}{2}s^2\sin\alpha$; $55 = \frac{1}{2}(12)^2\sin\alpha$; $\sin\alpha = \frac{(55)^2}{(12)^2}$
$\sin\alpha = 0.7639$; $\alpha = 0.87$

17. $9\sin^2 x - 6\sin x + 1 = 0$; $(3\sin x - 1)^2 = 0$;
$\sin x = \frac{1}{3}$; $x = 0.34 + 2n\pi, 2.80 + 2n\pi$

Check: $9\sin^2(0.34 + 2n\pi) -$
$$6(\sin(0.34 + 2n\pi)) + 1 =$$
$$9(0.347)^2 - 6(0.347) + 1 = 0 \checkmark$$
$$9\sin^2(2.80 + 2n\pi) - 6(\sin 2.80) + 1 = 0 \checkmark$$

18. $4\sin x - 3 = 0$; $\sin x = \frac{3}{4}$;
$x = 0.85 + 2n\pi, 2.29 + 2n\pi$
Check: $4(\sin 0.85 + 2n\pi) - 3 =$
$$4(0.75) - 3 = 0 \checkmark$$
$$4(\sin 2.29 + 2n\pi) - 3 =$$
$$4(0.75) - 3 = 0 \checkmark$$

19. $4\cos^2 x - 4\cos x + 1 = 0$; $(2\cos x - 1)^2 = 0$;
$\cos x = \frac{1}{2}$; $x = \frac{\pi}{3} + 2n\pi, \frac{5\pi}{3} + 2n\pi$
Check:
$$4\left(\cos^2\frac{\pi}{3} + 2n\pi\right) - 4\cos\left(\frac{\pi}{3} + 2n\pi\right) + 1 =$$
$$4\left(\frac{1}{2}\right)^2 - 4\left(\frac{1}{2}\right) + 1 = 0 \checkmark$$
$$4\left(\cos^2\frac{5\pi}{3} + 2n\pi\right) - 4\cos\left(\frac{5\pi}{3} + 2n\pi\right) =$$
$$= 4\left(\frac{1}{2}\right)^2 - 4\left(\frac{1}{2}\right) + 1 = 0 \checkmark$$

20. $5\sec x + 9 = 0$; $\sec x = -\frac{9}{5}$;
$x = 2.16 + 2n\pi, 4.12 + 2n\pi$
Check: $5\sec(2.16 + 2n\pi) + 9 =$
$$5(-1.80) + 9 = 0 \checkmark$$
$$5\sec(4.12 + 2n\pi) + 9 =$$
$$5(-1.80) + 9 = 0 \checkmark$$

21. $x = 3.50, 5.93$ 22. $x = 0.56, 2.47, 3.70, 5.61$

23. $x = 0.21, 4.28$ 24. $x = 1.20, 2.14, 4.34, 5.28$

25. $W = Pd\cos\theta$; $\cos\theta = \frac{W}{Pd} = \frac{27}{1.9(20)} = 0.711$;
$\theta \approx 45°$

EXTEND

26. $1 - \tan x = \sec x$; Square both sides and obtain
$1 - 2\tan x + \tan^2 x = \sec^2 x$;
$1 - 2\tan x + \tan^2 x = \tan^2 x + 1$; $-2\tan x = 0$;
$\tan x = 0$; $x = 0°$
Check: $1 - \tan 0 \overset{?}{=} \sec 0$
$$1 - 0 \overset{?}{=} 1$$
$$1 = 1 \checkmark$$

27. $1 - \cot x = \csc x$; Square both sides and obtain
$1 - 2\cot x + \cot^2 x = \csc^2 x$;
$1 - 2\cot x + \cot^2 x = 1 + \cot^2 x$; $-2\cot x = 0$;
$\cot x = 0$; $x = 90°$
Check: $1 - \cot 90 \overset{?}{=} \csc 90$
$$1 - 0 \overset{?}{=} 1$$
$$1 = 1 \checkmark$$

28. $\sin x \tan x = \sin x$; $\sin x \tan x - \sin x = 0$;
$\sin x(\tan x - 1) = 0$; $\sin x = 0$ or $\tan x = 1$; $x = 0°$,
$45°, 180°, 225°$
Check: $\sin 0 \tan 0 \overset{?}{=} \sin 0$
$$0 \cdot 0 \overset{?}{=} 0$$
$$0 = 0 \checkmark$$
$$\sin 45 \tan 45 \overset{?}{=} \sin 45$$
$$\frac{\sqrt{2}}{2} \cdot 1 \overset{?}{=} \frac{\sqrt{2}}{2}$$
$$\frac{\sqrt{2}}{2} = \frac{\sqrt{2}}{2} \checkmark$$
$$\sin 180 \tan 180 \overset{?}{=} \sin 180$$
$$0(0) \overset{?}{=} 0$$
$$0 = 0 \checkmark$$
$$\sin 225 \tan 225 \overset{?}{=} \sin 225$$
$$-\frac{\sqrt{2}}{2}(1) \overset{?}{=} -\frac{\sqrt{2}}{2}$$
$$-\frac{\sqrt{2}}{2} = -\frac{\sqrt{2}}{2} \checkmark$$

29. $\cos x \cot x = \cos x$; $\cos x \cot x - \cos x = 0$;
$\cos x(\cot x - 1) = 0$, $\cos x = 0$ or $\cot x = 1$;
$x = 45°, 90°, 225°, 270°$
Check: $\cos 45 \cot 45 \overset{?}{=} \cos 45$
$$\frac{\sqrt{2}}{2}(1) \overset{?}{=} \frac{\sqrt{2}}{2}$$
$$\frac{\sqrt{2}}{2} = \frac{\sqrt{2}}{2} \checkmark$$
$$\cos 90 \cot 90 \overset{?}{=} \cos 90$$
$$0 \cdot 0 \overset{?}{=} 0$$
$$0 = 0 \checkmark$$
$$\cos 225 \cot 225 \overset{?}{=} \cos 225$$
$$-\frac{\sqrt{2}}{2}(1) \overset{?}{=} -\frac{\sqrt{2}}{2}$$
$$-\frac{\sqrt{2}}{2} = -\frac{\sqrt{2}}{2} \checkmark$$
$$\cos 270 \cot 270 \overset{?}{=} \cos 270$$
$$0 \cdot 0 \overset{?}{=} 0$$
$$0 = 0 \checkmark$$

30. $8 \cos \frac{\pi}{2}(t - 1) = 2$; $\cos \frac{\pi}{2}(t - 1) = \frac{1}{4}$;

$\frac{\pi}{2}(t - 1) \approx \pm 1.32$; $t - 1 \approx 0.839$ or
$$t - 1 \approx -0.839;$$

$t = 1.839$ or $t = 0.16$. The first time $y = 2$ is at
$t = 0.16$ s. This exercise can be worked using a
graphing utility.

31. $8 \cos \frac{\pi}{2}(t - 1) = 4$; $\cos \frac{\pi}{2}(t - 1) = \frac{1}{2}$;

$\frac{\pi}{2}(t - 1) = \pm 1.047$; $t = 1.67$ or $t = 0.33$. The

first time is $t = 0.33$ s. This exercise can be worked
using a graphing utility.

THINK CRITICALLY

32. If you solve them on the interval $0° \le x \le 360°$, you
will obtain the same answer twice because $0°$ and $360°$
are coterminal.

33. $(\cos x + \sin x)^2 + (\cos x - \sin x)^2 =$
$\cos^2 x + 2 \cos x \sin x + \sin^2 x + \cos^2 x -$
$2 \cos x \sin x + \sin^2 x =$
$(\cos^2 x + \sin^2 x) + (2 \cos x \sin x - 2 \cos x \sin x) +$
$(\cos^2 x + \sin^2 x) = 2$

34. Use a graphing utility to graph $y = 2 \sin x$ and $y = 1$
on the same set of axes. These graphs intersect at
$2 \sin x = 1$; $\sin x = \frac{1}{2}$; $x = 30°, 150°$. Using these
intersection values and the graphs determine that
$2 \sin x \le 1$ for $0° \le x \le 30°$ or $150° \le x < 360°$

35. Use a graphing utility to graph $y = 2 \cos x$ and
$y = \sqrt{3}$ on the same set of axes. These graphs intersect
at $2 \cos x = \sqrt{3}$; $\cos x = \frac{\sqrt{3}}{2}$; $x = 30°, 330°$. Using
these intersection values and the graphs determine that
$2 \cos x \le \sqrt{3}$ for $30° \le x \le 330°$

36. Use a graphing utility to graph $y = |\cos x|$ and
$y = \frac{\sqrt{3}}{2}$ on the same set of axes. These graphs intersect
at $\cos x = \frac{\sqrt{3}}{2}$ or $\cos x = -\frac{\sqrt{3}}{2}$. So, $x = 30°, 150°$,
$210°, 330°$. Using these values and the graphs
determine that $|\cos x| = \frac{\sqrt{3}}{2}$ when $30° \le x \le 150°$ or
$210° \le x \le 330°$

37. Use a graphing utility to graph $y = |\sin x|$ and $y = \frac{1}{2}$
on the same set of axes. These graphs intersect at
$\sin x = \frac{1}{2}$ or $\sin x = -\frac{1}{2}$. So, $x = 30°, 150°, 210°$,
$330°$. Using the values and the graphs determine that
$|\sin x| \ge \frac{1}{2}$ when $30° \le x \le 150°$ or
$210° \le x \le 330°$.

MIXED REVIEW

38.
$$\begin{bmatrix} 2(5) + (-1)(-4) + 4(2) & 2(6) + (-1)(3) + 4(0) \\ 0(5) + (-2)(-4) + (-3)(2) & 0(6) + (-2)(3) + (-3)(0) \\ 1(5) + (1)(-4) + (1)(2) & 1(6) + 1(3) + (1)(0) \end{bmatrix} =$$
$$\begin{bmatrix} 22 & 9 \\ 2 & -6 \\ 3 & 9 \end{bmatrix}$$

39. $\begin{bmatrix} 8(2) & 8(5) \\ 6(2) & 6(5) \\ 3(2) & 3(5) \end{bmatrix} = \begin{bmatrix} 16 & 40 \\ 12 & 30 \\ 6 & 15 \end{bmatrix}$

40. $e^{8 + (-2)} = e^6$

41. $(27)^{\frac{1}{3}} \cdot e^{18\left(\frac{1}{3}\right)} = 3e^6$

42. C; $_{12}C_2 = \dfrac{12!}{2!(12 - 2)!} = \dfrac{12!}{2!10!} = \dfrac{12 \cdot 11}{2} = 66$

43. $x = 0.12, 3.02$

44. $x = 1.93, 4.35$

45. $(3 + 5) + (-4 + (-1))i = 8 + (-5)i = 8 - 5i$

46. Since $23 \div 4$ has a remainder $3i^{23} = i^4 = 1$. Therefore, $i^{23} - 1 = 1 - i$

47. $\dfrac{8}{2i} = \dfrac{4}{i}\left(\dfrac{i}{i}\right) = \dfrac{4i}{-1} = -4i$

Lesson 16.3, pages 804–811

EXPLORE

1. $\alpha - \beta$

2. $AB = \sqrt{1^2 + 1^2 - 2(1)(1)\cos(\alpha - \beta)} = \sqrt{2 - 2\cos(\alpha - \beta)}$

3. $AB = \sqrt{(\cos\alpha - \cos\beta)^2 + (\sin\alpha - \sin\beta)^2}$
$= \sqrt{\cos^2\alpha - 2\cos\alpha\cos\beta + \cos^2\beta + \sin^2\alpha - 2\sin\alpha\sin\beta + \sin^2\beta}$
$= \sqrt{2 - 2\cos\alpha\cos\beta - 2\sin\alpha\sin\beta}$

4. $\sqrt{2 - 2\cos(\alpha - \beta)} = \sqrt{2 - 2\cos\alpha\cos\beta - 2\sin\alpha\sin\beta}$
$2 - 2\cos(\alpha - \beta) = 2 - 2\cos\alpha\cos\beta - 2\sin\alpha\sin\beta$
$\cos(\alpha - \beta) = \cos\alpha\cos\beta - \sin\alpha\sin\beta$

TRY THESE

1. $\sin 65\cos 55 - \cos 65\sin 55 =$
$\sin(65 - 55) = \sin 10°$

2. $\cos 99\cos 22 + \sin 99\sin 22 =$
$\cos(99 - 22) = \cos 77°$

3. $\dfrac{\tan\frac{5\pi}{7} + \tan\frac{\pi}{7}}{1 - \tan\frac{5\pi}{7}\tan\frac{\pi}{7}} = \tan\left(\frac{5\pi}{7} + \frac{\pi}{7}\right) = \tan\frac{6\pi}{7}$

4. $\dfrac{\tan\frac{4\pi}{3} - \tan\frac{\pi}{3}}{1 + \tan\frac{4\pi}{3}\tan\frac{\pi}{3}} = \tan\left(\frac{4\pi}{3} - \frac{\pi}{3}\right) = \tan\pi$

5. $\sin 75 = \sin(45 + 30) = \sin 45\cos 30 +$
$\cos 45\sin 30 = \dfrac{\sqrt{2}}{2} \cdot \dfrac{\sqrt{3}}{2} + \dfrac{\sqrt{2}}{2} \cdot \dfrac{1}{2} =$
$\dfrac{\sqrt{6}}{4} + \dfrac{\sqrt{2}}{4} = \dfrac{\sqrt{6} + \sqrt{2}}{4}$

6. $\cos 105 = \cos(45 + 60) = \cos 45\cos 60 -$
$\sin 45\sin 60 = \dfrac{\sqrt{2}}{2} \cdot \dfrac{1}{2} - \dfrac{\sqrt{2}}{2} \cdot \dfrac{\sqrt{3}}{2} =$
$\dfrac{\sqrt{2}}{4} - \dfrac{\sqrt{6}}{4} = \dfrac{\sqrt{2} - \sqrt{6}}{4}$

7. $\tan 195 = \tan(150 + 45) = \dfrac{\tan 150 + \tan 45}{1 - \tan 150\tan 45} =$

$\dfrac{-\frac{\sqrt{3}}{3} + 1}{1 - \left(-\frac{\sqrt{3}}{3}\right)(1)} = \dfrac{1 - \frac{\sqrt{3}}{3}}{1 + \frac{\sqrt{3}}{3}} = \dfrac{3 - \sqrt{3}}{3 + \sqrt{3}}\left(\dfrac{3 - \sqrt{3}}{3 - \sqrt{3}}\right) =$

$\dfrac{9 - 6\sqrt{3} + 3}{9 - 3} = \dfrac{12 - 6\sqrt{3}}{6} = 2 - \sqrt{3}$

8. $\sin\dfrac{11\pi}{12} = \sin\left(\dfrac{2\pi}{3} + \dfrac{\pi}{4}\right) = \sin\dfrac{2\pi}{3}\cos\dfrac{\pi}{4} +$
$\cos\dfrac{2\pi}{3}\sin\dfrac{\pi}{4} = \dfrac{\sqrt{3}}{2}\left(\dfrac{\sqrt{2}}{2}\right) + \left(-\dfrac{1}{2}\right)\left(\dfrac{\sqrt{2}}{2}\right) = \dfrac{\sqrt{6}}{4} -$
$\dfrac{\sqrt{2}}{4} = \dfrac{\sqrt{6} - \sqrt{2}}{4}$

9. $\cos\dfrac{19\pi}{12} = \cos\left(\dfrac{3\pi}{12} + \dfrac{16\pi}{12}\right) = \cos\left(\dfrac{\pi}{4} + \dfrac{4\pi}{3}\right) =$
$\cos\dfrac{\pi}{4}\cos\dfrac{4\pi}{3} - \sin\dfrac{\pi}{4}\sin\dfrac{4\pi}{3} = \dfrac{\sqrt{2}}{2}\left(-\dfrac{1}{2}\right) -$
$\dfrac{\sqrt{2}}{2}\left(-\dfrac{\sqrt{3}}{2}\right) = -\dfrac{\sqrt{2}}{4} + \dfrac{\sqrt{6}}{4} = \dfrac{\sqrt{6} - \sqrt{2}}{4}.$

10. $\tan\dfrac{9\pi}{4} = \tan\left(2\pi + \dfrac{\pi}{4}\right) = \dfrac{\tan 2\pi + \dfrac{\pi}{4}}{1 - \tan 2\pi\tan\dfrac{\pi}{4}} =$
$\dfrac{0 + 1}{1 - 0 \cdot 1} = 1$

11. No. $\cos(\alpha + \beta) = \cos\alpha\cos\beta - \sin\alpha\sin\beta$, which is not equal to $\cos\alpha + \cos\beta$. For example, $\cos(0° + 90°) = \cos 0°\cos 90° - \sin 0°\cos 90° = 0$, which is not equal to $\cos 0° + \cos 90°$, which is 1.

For 12–14, if $\cos \alpha = \frac{4}{5}$ and $\sin \beta = -\frac{5}{13}$ then

$\sin \alpha = -\sqrt{1 - \left(\frac{4}{5}\right)^2} = -\frac{3}{5}$ (negative because

$270° < \alpha < 360°$) $\cos \beta = -\sqrt{1 - \left(-\frac{5}{13}\right)^2} = -\frac{12}{13}$

(negative because $180° < \beta < 270°$), $\tan \alpha = \frac{\sin \alpha}{\cos \alpha} =$

$\dfrac{-\frac{3}{5}}{\frac{4}{5}} = -\frac{3}{4}$, and $\tan \beta = \frac{\sin \beta}{\cos \beta} = \dfrac{-\frac{5}{13}}{-\frac{12}{13}} = \frac{5}{12}$.

12. $\cos \alpha \cos \beta + \sin \alpha \sin \beta = \frac{4}{5}\left(-\frac{12}{13}\right) + \left(-\frac{3}{5}\right)\left(-\frac{5}{13}\right) =$

$-\frac{33}{65}$

13. $\sin \alpha \cos \beta - \cos \alpha \sin \beta = \left(-\frac{3}{5}\right)\left(-\frac{12}{13}\right) -$

$\frac{4}{5}\left(-\frac{5}{13}\right) = \frac{56}{65}$

14. $\dfrac{\tan \alpha + \tan \beta}{1 - \tan \alpha \tan \beta} = \dfrac{-\frac{3}{4} + \frac{5}{12}}{1 - \left(-\frac{3}{4}\right)\left(\frac{5}{12}\right)} \cdot \left(\frac{48}{48}\right) =$

$\frac{-36 + 20}{48 - (-15)} = \frac{-16}{63} = -\frac{16}{63}$

15.

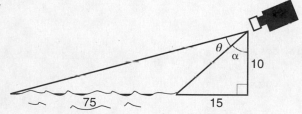

Let $\alpha =$ the measure of the angle opposite the 15 ft
side. $\tan \alpha = \frac{15}{10} = \frac{3}{2}$

$\tan (\theta + \alpha) = \dfrac{\tan \theta + \tan \alpha}{1 - \tan \theta \tan \alpha} = \dfrac{\tan \theta + \frac{3}{2}}{1 - \frac{3}{2} \tan \theta} =$

$\frac{90}{10} = 9$

$\tan \theta + \frac{3}{2} = 9\left(1 - \frac{3}{2} \tan \theta\right)$

$\tan \theta + \frac{3}{2} = 9 - \frac{27}{2} \tan \theta$

$\frac{29}{2} \tan \theta = \frac{15}{2}$

$\tan \theta = \frac{15}{29}$

$\theta = 0.48$

16. $\sin\left(x + \frac{7\pi}{3}\right) + \sin\left(x - \frac{7\pi}{3}\right) = 1$

$\sin x \cos \frac{7\pi}{3} + \cos x \sin \frac{7\pi}{3} + \sin x \cos \frac{7\pi}{3} -$

$\cos x \sin \frac{7\pi}{3} = 1$

$2 \sin x \cos \frac{7\pi}{3} = 1$

$\sin x = 1$

$x = \frac{\pi}{2}$

17. $\cos\left(x + \frac{\pi}{6}\right) = \cos\left(x - \frac{\pi}{6}\right) + 1$

$\cos x \cos \frac{\pi}{6} - \sin x \sin \frac{\pi}{6} = \cos x \cos \frac{\pi}{6} +$

$\sin x \sin \frac{\pi}{6} + 1$

$2 \sin x \sin \frac{\pi}{6} = -1$

$\sin x = -1$

$x = \frac{3\pi}{2}$

18. $\cos (180 + \alpha) = \cos 180 \cos \alpha - \sin 180 \sin \alpha = -\cos \alpha - 0 = -\cos \alpha$

19. $-\sin \beta = \sin (-\beta) = \cos (-\beta - 90) = \cos (360 - \beta - 90) = \cos (270 - \beta)$

PRACTICE

1. $\sin \frac{7\pi}{9} \cos \frac{5\pi}{9} - \cos \frac{7\pi}{9} \sin \frac{5\pi}{9} = \sin\left(\frac{7\pi}{9} - \frac{5\pi}{9}\right)$

$= \sin\left(\frac{2\pi}{9}\right)$

2. $\cos \frac{6\pi}{5} \cos \frac{2\pi}{5} - \sin \frac{6\pi}{5} \sin \frac{2\pi}{5} = \sin\left(\frac{6\pi}{5} + \frac{2\pi}{5}\right)$

$= \sin\left(\frac{8\pi}{5}\right)$

3. $\dfrac{\tan 62 + \tan 75}{1 - \tan 62 \tan 75} = \tan (62 + 75) = \tan 137°$

4. $\dfrac{\tan 111 - \tan 87}{1 + \tan 111 \tan 87} = \tan (111 - 87) = \tan 24°$

5. $\sin 345 = \sin (300 + 45)$

$= \sin 300 \cos 45 + \cos 300 \sin 45$

$$= -\frac{\sqrt{3}}{2} \cdot \frac{\sqrt{2}}{2} + \frac{1}{2} \cdot \frac{\sqrt{2}}{2}$$

$$= \frac{-\sqrt{6} + \sqrt{2}}{4}$$

6. $\tan 255 = \tan (300 - 45) = \dfrac{\tan 300 - \tan 45}{1 + \tan 300 \tan 45} =$

$\dfrac{-\sqrt{3} - 1}{1 + (-\sqrt{3})(1)} = \dfrac{-\sqrt{3} - 1}{1 - \sqrt{3}} \cdot \dfrac{1 + \sqrt{3}}{1 + \sqrt{3}} = 2 + \sqrt{3}$

7. $\sin (-75) = \sin (-45 + -30) =$

$\sin (-45) \cos (-30) + \cos (-45) \sin (-30)$

$= \left(-\dfrac{\sqrt{2}}{2}\right) \cdot \left(\dfrac{\sqrt{3}}{2}\right) + \left(\dfrac{\sqrt{2}}{2}\right) \cdot \left(-\dfrac{1}{2}\right) = \dfrac{-\sqrt{6} - \sqrt{2}}{4}$

8. $\cos (-285) = \cos (-225 - 60) =$

$\cos (-225) \cos 60 + \sin (-225) \sin 60$

$= \left(-\dfrac{\sqrt{2}}{2}\right) \cdot \left(\dfrac{1}{2}\right) + \left(\dfrac{\sqrt{2}}{2}\right)\left(\dfrac{\sqrt{3}}{2}\right) = \left(\dfrac{-\sqrt{2} + \sqrt{6}}{4}\right)$

9. $\cos \dfrac{7\pi}{12} = \cos \left(\dfrac{3\pi}{12} + \dfrac{4\pi}{12}\right) = \cos \left(\dfrac{\pi}{4} + \dfrac{\pi}{3}\right) =$

$\cos \dfrac{\pi}{4} \cos \dfrac{\pi}{3} - \sin \dfrac{\pi}{4} \sin \dfrac{\pi}{3} = \left(\dfrac{\sqrt{2}}{2}\right) \cdot \left(\dfrac{1}{2}\right) -$

$\left(\dfrac{\sqrt{2}}{2}\right) \cdot \left(\dfrac{\sqrt{3}}{2}\right) = \dfrac{\sqrt{2} - \sqrt{6}}{4}$

10. $\tan \dfrac{\pi}{12} = \tan \left(\dfrac{4\pi}{12} - \dfrac{3\pi}{12}\right) = \tan \left(\dfrac{\pi}{3} - \dfrac{\pi}{4}\right) =$

$\dfrac{\tan \dfrac{\pi}{3} - \tan \dfrac{\pi}{4}}{1 + \tan \dfrac{\pi}{3} \tan \dfrac{\pi}{4}} = \dfrac{\sqrt{3} - 1}{1 + (\sqrt{3})(1)} = \dfrac{\sqrt{3} - 1}{1 + \sqrt{3}} =$

$\dfrac{\sqrt{3} - 1}{1 + \sqrt{3}} \cdot \dfrac{1 - \sqrt{3}}{1 - \sqrt{3}} = 2 - \sqrt{3}$

11. Disagree; $\sin 2\alpha = \sin (\alpha + \alpha) =$
   $= \sin \alpha \cos \alpha + \cos \alpha \sin \alpha = 2 \sin \alpha \cos \alpha$;
   So, $2 \sin \alpha \cos \alpha \neq 2 \sin \alpha$.
   However, some students may say agree, pointing out
   that it is true for $\alpha = n\pi$, where n is an integer.

For 12–14, $\sin^2 \alpha + \left(-\dfrac{\sqrt{2}}{2}\right)^2 = 1$; $\sin \alpha = -\dfrac{\sqrt{2}}{2}$ (the value

is negative because $180° < \alpha < 270°$) and

$\cos 2\beta + \left(\dfrac{15}{17}\right)^2 = 1$; $\cos \beta = -\dfrac{8}{17}$ (the value is negative
because $90° < B < 180°$).

12. $\cos (\alpha - \beta) = \cos \alpha \cos \beta + \sin \alpha \sin \beta =$

$\left(-\dfrac{\sqrt{2}}{2}\right) \cdot \left(-\dfrac{8}{17}\right) + \left(-\dfrac{\sqrt{2}}{2}\right) \cdot \left(\dfrac{15}{17}\right) = -\dfrac{7\sqrt{2}}{34}$

13. $\cos (\alpha + \beta) = \cos \alpha \cos \beta - \sin \alpha \sin \beta =$

$\left(-\dfrac{\sqrt{2}}{2}\right) \cdot \left(-\dfrac{8}{17}\right) - \left(-\dfrac{\sqrt{2}}{2}\right) \cdot \left(\dfrac{15}{17}\right) = \dfrac{23\sqrt{2}}{34}$

14. $\sin (\alpha - \beta) = \sin \alpha \cos \beta - \cos \alpha \sin \beta =$

$\left(-\dfrac{\sqrt{2}}{2}\right) \cdot \left(-\dfrac{8}{17}\right) - \left(-\dfrac{\sqrt{2}}{2}\right) \cdot \left(\dfrac{15}{17}\right) = \dfrac{23\sqrt{2}}{34}$

15.

$\text{Tan}^{-1} x = \dfrac{8}{7}$ $\qquad \sqrt{8^2 + 7^2} = 10.6$

$x = 0.85$ $\qquad\qquad c = 10.6$

$2\pi - 0.85 = y$

$\qquad\qquad\qquad d = \sqrt{35^2 + c^2 - 2(35)(c)\cos y}$

$y = 5.43$ $\qquad\qquad d = 42.7$

$\qquad\qquad\qquad \dfrac{42.7}{\sin y} = \dfrac{35}{\sin \beta}; \beta = 0.66$

16. $\sin x \cos \dfrac{\pi}{4} + \cos x \sin \dfrac{\pi}{4} =$

$\sin x \cos \dfrac{\pi}{4} - \cos x \sin \dfrac{\pi}{4} - 1$

$2 \cos x \sin \dfrac{\pi}{4} = -1$

$2(\cos x)\left(\dfrac{\sqrt{2}}{2}\right) = -1$

$\cos x = -\dfrac{\sqrt{2}}{2}$

$x = \dfrac{3\pi}{4}, \dfrac{5\pi}{4}$

17. $\cos x \cos \dfrac{\pi}{4} - \sin x \sin \dfrac{\pi}{4} =$

$1 - \left(\cos x \cos \dfrac{\pi}{4} + \sin x \sin \dfrac{\pi}{4}\right)$

$2 \cos x \cos \dfrac{\pi}{4} = 1$

$\sqrt{2} \cos x = 1$

$\cos x = \dfrac{\sqrt{2}}{2}$

$x = \dfrac{\pi}{4}, \dfrac{7\pi}{4}$

18. $\cos x \cos \dfrac{\pi}{8} - \sin x \sin \dfrac{\pi}{8} =$

$\cos x \cos \dfrac{\pi}{8} + \sin x \sin \dfrac{\pi}{8} =$

$-2 \sin x \sin \dfrac{\pi}{8} = 0$

$\sin x = 0$

$x = 0, \pi$

19. $\sin x \cos \dfrac{\pi}{7} + \cos x \sin \dfrac{\pi}{7} =$

$\sin x \cos \dfrac{\pi}{7} - \cos x \sin \dfrac{\pi}{7} - \dfrac{1}{2}$

$2 \cos x \sin \dfrac{\pi}{7} = -\dfrac{1}{2}$

$\cos x = -\dfrac{1}{4 \sin \dfrac{\pi}{7}} = -0.5762$

$x = 2.18, 4.10$

20.
$$\dfrac{\tan x + \tan \dfrac{3\pi}{4}}{1 - \tan x \tan \dfrac{3\pi}{4}} = \cos x \cos \dfrac{3\pi}{4} + \sin x \sin \dfrac{3\pi}{4}$$

$$\dfrac{\tan x - 1}{\tan x + 1} = -\dfrac{\sqrt{2}}{2} \cos x + \dfrac{\sqrt{2}}{2} \sin x$$

$$\dfrac{\dfrac{\sin x}{\cos x} - 1}{\dfrac{\sin x}{\cos x} + 1} = -\dfrac{\sqrt{2}}{2} \cos x + \dfrac{\sqrt{2}}{2} \sin x$$

$$\dfrac{\sin x - \cos x}{\sin x + \cos x} = \dfrac{\sqrt{2}}{2} (\sin x - \cos x)$$

$$\dfrac{\sin^2 x - 2 \sin x \cos x + \cos^2 x}{\sin^2 x + 2 \sin x \cos x + \cos^2 x} =$$

$$\dfrac{1}{2}(\sin^2 x - 2 \sin x \cos x + \cos^2 x)$$

$$\dfrac{1 - 2 \sin x \cos x}{1 + 2 \sin x \cos x} = \dfrac{1}{2}(1 - 2 \sin x \cos x)$$

$$1 - 2 \sin x \cos x = \dfrac{1}{2}(1 - 4 \sin^2 x \cos^2 x)$$

$$1 - 2 \sin x \cos x = \dfrac{1}{2} - 2 \sin^2 x \cos^2 x$$

$$2 \sin^2 x \cos^2 x - 2 \sin x \cos x + \dfrac{1}{2} = 0$$

$$4 \sin^2 x \cos^2 x - 4 \sin x \cos x + 1 = 0$$
$$(2 \sin x \cos x - 1)^2 = 0$$

$$2 \sin x \cos x = 1$$

$$\sin x \cos x = \dfrac{1}{2}$$

$$\sqrt{1 - \cos^2 x} \, \cos x = \dfrac{1}{2}$$

$$(1 - \cos^2 x)\cos^2 x = \dfrac{1}{4}$$

$$\cos^2 x - \cos^4 x = \dfrac{1}{4}$$

$$4 \cos^4 x - 4 \cos^2 x + 1 = 0$$
$$(2 \cos^2 x - 1)^2 = 0$$

$$\cos^2 x = \dfrac{1}{2}$$

$$\cos x = \pm \dfrac{\sqrt{2}}{2}$$

$$x = \dfrac{\pi}{4}, \dfrac{3\pi}{4}, \dfrac{5\pi}{4}, \dfrac{7\pi}{4}$$

$\dfrac{3\pi}{4}$ and $\dfrac{7\pi}{4}$ are extraneous solutions. Therefore,

$$x = \dfrac{\pi}{4}, \dfrac{5\pi}{4}$$

21. $\tan \left(x - \dfrac{\pi}{4}\right) = \sin \left(x + \dfrac{\pi}{4}\right)$

Solve using a graphing utility by finding the

intersection of $y = \tan \left(x = \dfrac{\pi}{4}\right)$ and $y = \sin \left(x + \dfrac{\pi}{4}\right)$

$x = 1.45, 3.26$

22. $\sin (\pi + \theta) = \sin \pi \cos \theta + \cos \pi \sin \theta = $
$(0) (\cos \theta) + (-1) (\sin \theta) = -\sin \theta$

23. $\cos \left(\dfrac{3\pi}{4} + \alpha\right) = \cos \dfrac{3\pi}{4} \cos \alpha - \sin \dfrac{3\pi}{4} \sin \alpha = $

$(0) (\cos \alpha) + (-1) \sin \alpha = -\sin \alpha$

24. $\cos \left(\dfrac{3\pi}{4} + \beta\right) = \cos \dfrac{3\pi}{4} \cos \beta - \sin \dfrac{3\pi}{4} \sin \beta = $

$\left(-\dfrac{\sqrt{2}}{2}\right)(\cos \beta) - \left(\dfrac{\sqrt{2}}{2}\right) (\sin \beta) = $

$-\dfrac{\sqrt{2}}{2} (\cos \beta + \sin \beta)$

25. $\sin \left(\dfrac{\pi}{4} + \theta\right) = \sin \dfrac{\pi}{4} \cos \theta + \cos \dfrac{\pi}{4} \sin \theta = $

$\left(\dfrac{\sqrt{2}}{2}\right) (\cos \theta) + \left(\dfrac{\sqrt{2}}{2}\right) (\sin \theta) = \dfrac{\sqrt{2}}{2} (\cos \theta + \sin \theta)$

26. $\tan \left(\dfrac{\pi}{4} - \alpha\right) = \dfrac{\tan \dfrac{\pi}{4} - \tan \alpha}{1 + \tan \dfrac{\pi}{4} \tan \alpha} = $

$\dfrac{1 - \tan \alpha}{1 + (1)(\tan \alpha)} = \dfrac{1 - \tan \alpha}{1 + \tan \alpha}$

27. $\tan \left(\dfrac{\pi}{4} + \beta\right) = \dfrac{\tan \dfrac{\pi}{4} + \tan \beta}{1 - \tan \dfrac{\pi}{4} \tan \beta} = $

$\dfrac{1 + \tan \beta}{1 - (1) (\tan \beta)} = \dfrac{1 + \tan \beta}{1 - \tan \beta}$

28. $\tan (\alpha + \beta) = \dfrac{\sin (\alpha + \beta)}{\cos (\alpha + \beta)}$

$= \dfrac{\sin \alpha \cos \beta + \cos \alpha \sin \beta}{\cos \alpha \cos \beta - \sin \alpha \sin \beta}$

$= \dfrac{\dfrac{\sin \alpha \cos \beta}{\cos \alpha \cos \beta} + \dfrac{\cos \alpha \sin \beta}{\cos \alpha \cos \beta}}{\dfrac{\cos \alpha \cos \beta}{\cos \alpha \cos \beta} - \dfrac{\sin \alpha \sin \beta}{\cos \alpha \cos \beta}}$

$= \dfrac{\dfrac{\sin \alpha}{\cos \alpha} + \dfrac{\sin \beta}{\cos \beta}}{1 - \left(\dfrac{\sin \alpha}{\cos \alpha}\right)\left(\dfrac{\sin \beta}{\cos \beta}\right)}$

$= \dfrac{\tan \alpha + \tan \beta}{1 - \tan \alpha \tan \beta}$

29. $\tan (\alpha - \beta) = \tan (\alpha + (-\beta)) = $

$\dfrac{\tan \alpha + \tan (-\beta)}{1 - \tan \alpha \tan (-\beta)} = \dfrac{\tan \alpha - \tan \beta}{1 + \tan \alpha \tan \beta}$

30.

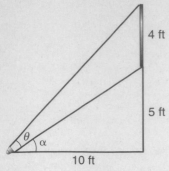

4 ft

5 ft

θ α

10 ft

Let α = measure of the angle opposite the 5 ft side.

$$\tan \alpha = \frac{5}{10} = \frac{1}{2}$$

$$\tan (\theta + \alpha) = \frac{\tan \theta + \tan \alpha}{1 - \tan \theta \tan \alpha} =$$

$$\frac{\tan \theta + \frac{1}{2}}{1 - (\tan \theta)\left(\frac{1}{2}\right)} = \frac{2 \tan \theta + 1}{2 - \tan \theta} = \frac{9}{10}$$

$$10(2 \tan \theta + 1) = 9(2 - \tan \theta)$$

$$20 \tan \theta + 10 = 18 - 9 \tan \theta$$

$$29 \tan \theta = 8$$

$$\tan \theta = \frac{8}{29}$$

$$\theta = 0.27$$

EXTEND

31. $\sin \alpha \cos \beta + \cos \alpha \sin \beta = \sin (\alpha + \beta)$

$\underline{+ \ [\sin \alpha \cos \beta - \cos \alpha \sin \beta = \sin (\alpha - \beta)]}$

$2 \sin \alpha \cos \beta = \sin (\alpha + \beta) - \sin (\alpha - \beta)$

32. $\sin \alpha \cos \beta + \cos \alpha \sin \beta = \sin (\alpha + \beta)$

$\underline{- \ [\sin \alpha \cos \beta - \cos \alpha \sin \beta = \sin (\alpha - \beta)]}$

$2 \cos \alpha \sin \beta = \sin (\alpha + \beta) + \sin (\alpha - \beta)$

33. $2 \sin 60 \cos 12 = \sin (60 + 12) + \sin (60 - 12)$

$= \sin 72° + \sin 48°$

34. $2 \sin 42 \sin 14 = \cos (42 - 14) - \cos (42 + 14)$

$= \cos 28° - \cos 56°$

35. $2 \cos 72 \cos 15 = \cos (72 - 15) + \cos (72 + 15)$

$= \cos 57° + \cos 87°$

36. $2 \cos 100 \sin 16 = \sin (100 + 16) - \sin (100 - 16)$

$= \sin 116° - \sin 84°$

37. $\dfrac{\sin 5\theta + \sin 3\theta}{\cos 5\theta + \cos 3\theta} = \dfrac{2 \cos 4\theta \sin \theta}{2 \cos 4\theta \cos \theta} = \dfrac{\sin \theta}{\cos \theta}$

$= \tan \theta$

38. $\dfrac{\cos 3\theta + \cos \theta}{\sin 3\theta - \sin \theta} = \dfrac{2 \cos 2\theta \cos \theta}{2 \cos 2\theta \sin \theta} = \dfrac{\cos \theta}{\sin \theta}$

$= \cot \theta$

THINK CRITICALLY

39. $\dfrac{1 - \tan \alpha \tan \beta}{\sec \alpha \sec \beta} = \dfrac{1 - \tan \alpha \tan \beta}{\dfrac{1}{\cos \alpha \cos \beta}}$

$(1 - \tan \alpha \tan \beta) (\cos \alpha \cos \beta) =$

$\cos \alpha \cos \beta - (\tan \alpha \tan \beta \cos \alpha \cos \beta) =$

$\cos \alpha \cos \beta - \sin \alpha \sin \beta = \cos (\alpha + \beta)$

40. $= \dfrac{\cot \alpha + \cot \beta}{\csc \alpha \csc \beta} = \dfrac{\cot \alpha + \cot \beta}{\dfrac{1}{\sin \alpha \sin \beta}}$

$(\cot \alpha + \cot \beta) (\sin \alpha \sin \beta) =$

$\cot \alpha \sin \alpha \sin \beta + \cot \beta \sin \alpha \sin \beta =$

$\cos \alpha \sin \beta + \cos \beta \sin \alpha =$

$\sin \alpha \cos \beta + \cos \alpha \sin \beta = \sin (\alpha + \beta)$

41. $\cot (\alpha + \beta) = \dfrac{1}{\tan (\alpha - \beta)}$

$\dfrac{1}{\dfrac{\tan \alpha - \tan \beta}{1 + \tan \alpha \tan \beta}} = \dfrac{1 + \tan \alpha \tan \beta}{\tan \alpha - \tan \beta} =$

$\dfrac{1 + \tan \alpha \tan \beta}{\tan \alpha - \tan \beta} \left(\dfrac{\cot \alpha \cot \beta}{\cot \alpha \cot \beta}\right) = \dfrac{\cot \alpha \cot \beta + 1}{\cot \beta - \cot \alpha}$

$= \dfrac{-1 - \cot \alpha \cot \beta}{\cot \alpha - \cot \beta}$

42. $\tan (\alpha + \beta) = \dfrac{\tan \alpha + \tan \beta}{1 - \tan \alpha \tan \beta} =$

$\dfrac{\tan \alpha + \tan \beta}{1 - \tan \alpha \tan \beta} \left(\dfrac{\cot \alpha \cot \beta}{\cot \alpha \cot \beta}\right) =$

$\dfrac{\cot \beta + \cot \alpha}{\cot \alpha \cot \beta - 1} = \dfrac{\cot \alpha + \cot \beta}{\cot \alpha \cot \beta - 1}$

MIXED REVIEW

43. C; $m \cdot n = -1$

$m(-4) = -1$

$m = \dfrac{1}{4}$

44. $x^2 + 8x - 5 = 0, x^2 + 8x + 16 = 5 + 16;$

$(x + 4)^2 = 21; x + 4 = \pm \sqrt{21}; x = -4 \pm \sqrt{21}$

45. $2x^2 - 7x + 1 = 0; 2\left(x^2 - \dfrac{7}{2}x\right) = -1;$

$2\left(x^2 - \dfrac{7}{2}x + \dfrac{49}{16}\right) = -1 + \dfrac{49}{8}; 2\left(x - \dfrac{7}{4}\right)^2 = \dfrac{41}{8};$

$\left(x - \dfrac{7}{4}\right)^2 = \dfrac{41}{16}; x - \dfrac{7}{4} = \dfrac{\pm\sqrt{41}}{4};$

$x = \dfrac{7}{4} \pm \dfrac{\sqrt{41}}{4} = \dfrac{7 \pm \sqrt{41}}{4}$

46. $(2x + 3)^6 = (2x)^6 + \dfrac{6}{1}(2x)^5(3)^1 + \dfrac{6 \cdot 5}{1 \cdot 2}(2x)^4(3)^2 +$

$\dfrac{6 \cdot 5 \cdot 4}{1 \cdot 2 \cdot 3}(2x)^3(3)^3 + \dfrac{6 \cdot 5 \cdot 4 \cdot 3}{1 \cdot 2 \cdot 3 \cdot 4}(2x)^2(3)^4 +$

$\dfrac{6 \cdot 5 \cdot 4 \cdot 3 \cdot 2}{1 \cdot 2 \cdot 3 \cdot 4 \cdot 5}(2x)^1(3)^5 + (3)^6 =$

$64x^6 + 576x^5 + 2160x^4 + 4320x^3 + 4860x^2 +$

$2916x + 729$

47. $\tan (24 + 65) = \tan 89°$

48. $\tan (122 - 22) = \tan 100°$

1. The pitch of the sound increases as the length of the string between the pencils decreases.

2. The frequency of the sound waves increases as the length of the string between the pencils decreases.

3. A has a higher pitch because it has a greater frequency.

ALGEBRAWORKS

1. $y = \cos(2(256)\pi t) = \cos(512\pi t)$

2. $y = \cos(2(440)\pi t) = \cos(880\pi t)$

3. period $= \dfrac{2\pi}{512\pi} = 0.00391$s

   period $= \dfrac{2\pi}{880\pi} = 0.00227$s

4. $y = 2\cos\left(\dfrac{512\pi t + 392\pi t}{2}\right)\cos\left(\dfrac{512\pi t - 392\pi t}{2}\right)$

   $= 2\cos(452\pi t)\cos(60\pi t)$

5. $y = 2\cos\left(\dfrac{1024\pi t + 256\pi t}{2}\right)\cos\left(\dfrac{1024\pi t - 256\pi t}{2}\right)$

   $= 2\cos(640\pi t)\cos(384\pi t)$

6. let $x = \alpha + \beta$; $y = \alpha + \beta$; so $\alpha = \dfrac{x+y}{2}$; substitute in the identity $\cos(\alpha - \beta) + \cos(\alpha + \beta) = 2\cos\alpha\cos\beta$; $\cos y + \cos x = 2\cos\left(\dfrac{x+y}{2}\right)\cos\left(\dfrac{x-y}{2}\right)$

Lesson 16.4, pages 812–820

EXPLORE

1. α

2. $\sin 2\alpha = \sin(\alpha + \alpha) = \sin\alpha\cos\alpha + \cos\alpha\sin\alpha = 2\sin\alpha\cos\alpha$

3. $\cos 2\alpha = \cos(\alpha + \alpha) = \cos\alpha\cos\alpha - \sin\alpha\sin\alpha = \cos^2\alpha - \sin^2\alpha$

4. $\cos 2\alpha = (1 - \sin^2\alpha) - \sin^2\alpha = 1 - 2\sin^2\alpha$
   $\cos 2\alpha = \cos^2\alpha - (1 - \cos^2\alpha) = \cos^2\alpha - 1 + \cos^2\alpha = 2\cos^2\alpha - 1$

5. $\tan 2\alpha = \tan(\alpha + \alpha) = \dfrac{\tan\alpha + \tan\alpha}{1 - \tan\alpha\tan\alpha} = \dfrac{2\tan\alpha}{1 - \tan^2\alpha}$

TRY THESE

1. $\sin 75 = \sin\dfrac{150}{2} = \pm\sqrt{\dfrac{1 - \cos 150}{2}} =$

   $\pm\sqrt{\dfrac{1 + \frac{\sqrt{3}}{2}}{2}} = \pm\sqrt{\dfrac{2 + \sqrt{3}}{4}} =$

   $\pm\dfrac{\sqrt{2 + \sqrt{3}}}{2}$. Since 75 is in Quadrant I, sin 75 is positive. So $\sin 75° = \dfrac{\sqrt{2 + \sqrt{3}}}{2}$

2. $\cos 67.5 = \cos\dfrac{135}{2} = \pm\sqrt{\dfrac{1 + \cos 135}{2}} =$

   $\pm\sqrt{\dfrac{1 + \left(\frac{\sqrt{2}}{2}\right)}{2}} = \pm\sqrt{\dfrac{2 - \sqrt{2}}{4}} =$

   $\pm\dfrac{\sqrt{2 - \sqrt{2}}}{2}$. Since 67.5 is in Quadrant I, cos 67.5 is

   positive. So, $\cos 67.5° = \dfrac{\sqrt{2 - \sqrt{2}}}{2}$.

3. $\tan 22.5 = \tan\dfrac{45}{2} = \dfrac{\sin 45}{1 + \cos 45} = \dfrac{\frac{\sqrt{2}}{2}}{1 + \frac{\sqrt{2}}{2}} =$

   $\dfrac{\sqrt{2}}{2 + \sqrt{2}} = \dfrac{\sqrt{2}}{2 + \sqrt{2}}\left(\dfrac{2 - \sqrt{2}}{2 - \sqrt{2}}\right) = \dfrac{2\sqrt{2} - 2}{4 - 2} = \dfrac{2(\sqrt{2} - 1)}{2} = \sqrt{2} - 1$

4. $\sin\dfrac{7\pi}{12} = \sin\dfrac{\frac{7\pi}{6}}{2} = \pm\sqrt{\dfrac{1 + \cos\frac{7\pi}{6}}{2}} =$

   $\pm\sqrt{\dfrac{1 - \left(\frac{\sqrt{3}}{2}\right)}{2}} = \pm\sqrt{\dfrac{2 + \sqrt{3}}{4}} = \pm\dfrac{\sqrt{2 + \sqrt{3}}}{2}$

   Since $\dfrac{7\pi}{12}$ is in Quadrant II, $\sin\dfrac{7\pi}{12}$ is positive.

   So, $\sin\dfrac{7\pi}{12} = \dfrac{\sqrt{2 + \sqrt{3}}}{2}$

5. $\cos\dfrac{7\pi}{8} = \cos\dfrac{\frac{7\pi}{4}}{2} = \pm\sqrt{\dfrac{1 + \cos\frac{7\pi}{4}}{2}} =$

   $\pm\sqrt{\dfrac{1 - \frac{\sqrt{2}}{2}}{2}} = \pm\sqrt{\dfrac{2 + \sqrt{2}}{4}} = \pm\dfrac{\sqrt{2 + \sqrt{2}}}{2}$

   Since $\dfrac{7\pi}{8}$ is in Quadrant II, $\cos\dfrac{7\pi}{8}$ is negative. So,

   $\cos\dfrac{7\pi}{8} = -\dfrac{\sqrt{2 + \sqrt{2}}}{2}$

6. $\tan\dfrac{11\pi}{12} = \tan\dfrac{\frac{11\pi}{6}}{2} = \dfrac{1 - \cos\frac{11\pi}{6}}{\sin\frac{11\pi}{6}} = \dfrac{1 - \frac{\sqrt{3}}{2}}{-\frac{1}{2}} = -2 + \sqrt{3}$

7. Answers may vary. Find $\cos\theta$. Then use $+\sqrt{\dfrac{1 - \cos\theta}{2}}$ or $-\sqrt{\dfrac{1 - \cos\theta}{2}}$ depending on the quadrant in which $\dfrac{\theta}{2}$ lies.

8. $\cos \alpha = -\sqrt{1 - \left(\dfrac{7}{25}\right)^2} = -\sqrt{1 - \dfrac{49}{625}} =$

$-\sqrt{\dfrac{576}{625}} = -\dfrac{24}{25}$

$\sin 2\alpha = 2 \sin \alpha \cos \alpha = 2\left(\dfrac{7}{25}\right)\left(-\dfrac{24}{25}\right) = -\dfrac{336}{625}$

9. $\cos 2\alpha = 1 - 2 \sin^2 \alpha = 1 - 2\left(\dfrac{7}{25}\right)^2 =$

$1 - \dfrac{98}{625} = \dfrac{527}{625}$

10. $\tan \alpha = \dfrac{\sin \alpha}{\cos \alpha} = \dfrac{\dfrac{7}{25}}{-\dfrac{24}{25}} = -\dfrac{7}{24}$

$\tan 2\alpha = \dfrac{2 \tan \alpha}{1 - \tan^2 \alpha} = \dfrac{2\left(-\dfrac{7}{24}\right)}{1 - \left(-\dfrac{7}{24}\right)^2} = \dfrac{-\dfrac{7}{12}}{1 - \dfrac{49}{576}} =$

$\dfrac{-336}{576 - 49} = -\dfrac{336}{527}$

11. $\sin \dfrac{\alpha}{2} = \sqrt{\dfrac{1 - \cos \alpha}{2}} = \sqrt{\dfrac{1 - \left(-\dfrac{24}{25}\right)}{2}} = \sqrt{\dfrac{\dfrac{49}{25}}{2}} =$

$\sqrt{\dfrac{49}{50}} = \dfrac{7}{5\sqrt{2}} = \dfrac{7\sqrt{2}}{10}$

12. $\cos \dfrac{\alpha}{2} = \sqrt{\dfrac{1 + \cos \alpha}{2}} = \sqrt{\dfrac{1 + \left(-\dfrac{24}{25}\right)}{2}} = \sqrt{\dfrac{\dfrac{1}{25}}{2}} =$

$\sqrt{\dfrac{1}{50}} = \dfrac{1}{5\sqrt{2}} = \dfrac{\sqrt{2}}{10}$

13. $\tan \dfrac{\alpha}{2} = \dfrac{\sin \alpha}{1 + \cos \alpha} = \dfrac{\dfrac{7}{25}}{1 + \left(-\dfrac{24}{25}\right)} = \dfrac{7}{25 - 24} = 7$

14. $\tan \theta = \dfrac{h}{30}; \tan 2\theta = \dfrac{h}{14}; \tan 2\theta = \dfrac{2 \tan \theta}{1 - \tan^2 \theta}$

Substitute to get $\dfrac{h}{14} = \dfrac{2\left(\dfrac{h}{30}\right)}{1 - \left(\dfrac{h}{30}\right)^2}; \dfrac{h}{14} = \dfrac{\dfrac{h}{15}}{1 - \dfrac{h^2}{900}} =$

$\dfrac{60h}{900 - h^2}; 14(60h) = h(900 - h^2);$

$840h = 900h - h^3; h^3 - 60h = 0; h(h^2 - 60) = 0;$

$h = 0$ (extraneous), $h^2 = 60; h = \sqrt{60} \approx 8$ ft

15. $\cos 2x + 3 \cos x + 2 = 0;$

$2 \cos^2 x - 1 + 3 \cos x + 2 = 0$

$2 \cos^2 x + 3 \cos x + 1 = 0;$

$(2 \cos x + 1)(\cos x + 1) = 0$

$\cos x = -\dfrac{1}{2}$ or $\cos x = -1; x = \dfrac{2\pi}{3}, \pi, \dfrac{4\pi}{3}$

16. $\cos 2x + 3 \sin x - 2 = 0;$

$1 - 2 \sin^2 x - 3 \sin x - 2 = 0;$

$-2 \sin^2 x - 3 \sin x - 1 = 0;$

$2 \sin^2 x + 3 \sin x + 1 = 0;$

$(2 \sin x + 1)(\sin x + 1) = 0; \sin x = -\dfrac{1}{2}$ or

$\sin x = -1; x = \dfrac{7\pi}{6}, \dfrac{3\pi}{2}, \dfrac{11\pi}{6}$

17. $2 \sin x + \sin 2x = 0; 2 \sin x + 2 \sin x \cos x = 0$

$\sin x(1 + \cos x) = 0; \sin x = 0$ or $\cos x = -1;$

$x = 0, \pi$

18. $2 \cos x + 2 \sin 2x = 0; 2 \cos x + 2 \cos x \sin x = 0;$

$\cos x(1 + \sin x) = 0; \cos x = 0$ or $\sin x = -1;$

$x = \dfrac{\pi}{2}, \dfrac{3\pi}{2}$

19. $2 \sin^2 \alpha = 1 - \cos 2\alpha$

$\qquad 1 - (1 - 2 \sin^2 \alpha)$

$\qquad 1 - 1 + 2 \sin^2 \alpha$

$2 \sin^2 \alpha = 2 \sin^2 \alpha$

20. $\dfrac{(\sin \beta + \cos \beta)^2 - 1}{\sin^2 \beta - 2 \sin \beta \cos \beta + \cos^2 \beta - 1} \quad \dfrac{= \sin 2\beta}{2 \sin \beta \cos \beta}$

$\qquad \dfrac{\sin^2 \beta + \cos^2 \beta + 2 \sin \beta \cos \beta - 1}{}$

$\qquad\qquad \dfrac{1 + 2 \sin \beta \cos \beta - 1}{2 \sin \beta \cos \beta}$

21. $h = \dfrac{v^2 \sin^2 \Theta}{2g} = \dfrac{60^2 \sin^2 40}{2(32)} = 23$ ft

PRACTICE

1. $\sin \dfrac{210}{2} = \sqrt{\dfrac{1 - \cos 210}{2}} = \sqrt{\dfrac{1 - \left(-\dfrac{\sqrt{3}}{2}\right)}{2}} =$

$\dfrac{\sqrt{2 + \sqrt{3}}}{2}$

2. $\cos \dfrac{225}{2} = -\sqrt{\dfrac{1 + \cos 225}{2}} = -\sqrt{\dfrac{1 + \left(-\dfrac{\sqrt{2}}{2}\right)}{2}} =$

$-\dfrac{\sqrt{2 - \sqrt{2}}}{2}$

3. $\tan \dfrac{510}{2} = \dfrac{\sin 510}{1 + \cos 510} = \dfrac{\dfrac{1}{2}}{1 + \left(-\dfrac{\sqrt{3}}{2}\right)} = \dfrac{1}{2 - \sqrt{3}} =$

$\dfrac{2 + \sqrt{3}}{4 - 3} = 2 + \sqrt{3}$

4. $\sin \dfrac{\dfrac{5\pi}{4}}{2} = \sqrt{\dfrac{1 - \cos \dfrac{5\pi}{4}}{2}} = \sqrt{\dfrac{1 - \left(-\dfrac{\sqrt{2}}{2}\right)}{2}} = \dfrac{\sqrt{2 + \sqrt{2}}}{2}$

5. $\cos \dfrac{\dfrac{11\pi}{4}}{2} = -\sqrt{\dfrac{1 + \cos \dfrac{11\pi}{4}}{2}} = -\sqrt{\dfrac{1 + \left(-\dfrac{\sqrt{2}}{2}\right)}{2}} =$

$-\dfrac{\sqrt{2 - \sqrt{2}}}{2}$

6. $\tan\dfrac{\frac{7\pi}{6}}{2} = \dfrac{\sin\frac{7\pi}{6}}{1 + \cos\frac{7\pi}{6}} = \dfrac{-\frac{1}{2}}{1 + \left(-\frac{\sqrt{3}}{2}\right)} =$

$\dfrac{-1}{2 - \sqrt{3}} = \dfrac{-1(2 + \sqrt{3})}{4 - 3} = -2 - \sqrt{3}$

7. $\sin(2(30)) = \sin 60° = \dfrac{\sqrt{3}}{2}$ and $\sin(2(60)) =$

$\sin 120° = \dfrac{\sqrt{3}}{2}$; thus, 30° and 60° will produce the

same answer in the formula $x = \dfrac{v^2 \sin 2\theta}{g}$.

For Exercises 8–13, $\sin \alpha = -\dfrac{9}{41}$ and $270° < \alpha < 360°$.
So

$\cos\alpha = \sqrt{1 - \left(-\dfrac{9}{41}\right)^2} = \dfrac{40}{41}$ and $\tan\alpha = \dfrac{\sin\alpha}{\cos\alpha} = \dfrac{-\frac{9}{41}}{\frac{40}{41}}$

$= -\dfrac{9}{40}$

8. $\sin 2\alpha = 2\sin\alpha\cos\alpha = 2\left(-\dfrac{9}{41}\right)\left(\dfrac{40}{41}\right) = -\dfrac{720}{1681}$

9. $\cos 2\alpha = 2\cos^2\alpha - 1 - 2\left(\dfrac{40}{41}\right)^2 - 1 = \dfrac{1519}{1681}$

10. $\tan 2\alpha = \dfrac{2\tan\alpha}{1 - \tan^2\alpha} = \dfrac{2\left(-\frac{9}{40}\right)}{1 - \left(-\frac{9}{40}\right)^2} = -\dfrac{720}{1519}$

11. $\sin\dfrac{\alpha}{2} = \sqrt{\dfrac{1 - \cos\alpha}{2}} = \sqrt{\dfrac{1 - \frac{40}{41}}{2}} =$

$\sqrt{\dfrac{41 - 40}{82}} = \sqrt{\dfrac{1}{82}} = \dfrac{\sqrt{82}}{82}$

12. $\cos\dfrac{\alpha}{2} = -\sqrt{\dfrac{1 + \cos\alpha}{2}} = -\sqrt{\dfrac{1 + \frac{40}{41}}{2}} =$

$-\sqrt{\dfrac{41 + 40}{82}} = -\sqrt{\dfrac{81}{82}} = \dfrac{-9\sqrt{82}}{82}$

13. $\tan\dfrac{\alpha}{2} = \dfrac{\sin\alpha}{1 + \cos\alpha} = \dfrac{-\frac{9}{41}}{1 + \frac{40}{41}} =$

$\dfrac{-9}{41 + 40} = \dfrac{-9}{81} = -\dfrac{1}{9}$

For Exercises 14–19 $\sin\alpha = -\dfrac{12}{13}$ and $180° < \alpha < 270°$

So, $\cos\alpha = -\sqrt{1 - \left(-\dfrac{12}{13}\right)^2} = -\dfrac{5}{13}$ and $\tan\alpha = \dfrac{\sin\alpha}{\cos\alpha} =$

$\dfrac{-\frac{12}{13}}{-\frac{5}{13}} = \dfrac{12}{5}$.

14. $\sin 2\alpha = 2\sin\alpha\cos\alpha = 2\left(-\dfrac{12}{13}\right)\left(-\dfrac{5}{13}\right) = \dfrac{120}{169}$

15. $\cos 2\alpha = 2\cos^2\alpha - 1 = 2\left(-\dfrac{5}{13}\right)^2 - 1 = -\dfrac{119}{169}$

16. $\tan 2\alpha = \dfrac{2\tan\alpha}{1 - \tan^2\alpha} = \dfrac{2\left(\frac{12}{5}\right)}{1 - \left(\frac{12}{5}\right)^2} = -\dfrac{120}{119}$

17. $\sin\dfrac{\alpha}{2} = \sqrt{\dfrac{1 - \cos\alpha}{2}} = \sqrt{\dfrac{1 - \left(-\frac{5}{13}\right)}{2}} =$

$\sqrt{\dfrac{13 + 5}{26}} = \sqrt{\dfrac{9}{13}} = \dfrac{3\sqrt{13}}{13}$

18. $\cos\dfrac{\alpha}{2} = -\sqrt{\dfrac{1 + \cos\alpha}{2}} = -\sqrt{\dfrac{1 + \left(-\frac{5}{13}\right)}{2}} -$

$-\sqrt{\dfrac{13 - 5}{26}} = -\sqrt{\dfrac{4}{13}} = -\dfrac{2\sqrt{13}}{13}$

19. $\tan\dfrac{\alpha}{2} = \dfrac{1 - \cos\alpha}{\sin\alpha} = \dfrac{1 - \left(-\frac{5}{13}\right)}{-\frac{12}{13}} = \dfrac{13 + 5}{-12} = -\dfrac{3}{2}$

20. $t = \dfrac{2(65)\sin 41}{32} = 2.7$ s

21. $\cos 2x + 5\cos x + 3 = 0$;
$2\cos^2 x - 1 + 5\cos x + 3 = 0$;
$2\cos^2 x + 5\cos x + 2 = 0$;

$(2\cos x + 1)(\cos x + 2) = 0$; $\cos x = -\dfrac{1}{2}$ or

$\cos x = -2$ (impossible); $x = \dfrac{2\pi}{3}, \dfrac{4\pi}{3}$

22. $\cos 2x + 9\sin x + 4 = 0$;
$1 - 2\sin^2 x + 9\sin x + 4 = 0$;
$2\sin^2 x - 9\sin x - 5 = 0$;

$(2\sin x + 1)(\sin x - 5) = 0$; $\sin x = -\dfrac{1}{2}$ or

$\sin x = 5$ (impossible); $x = \dfrac{7\pi}{6}, \dfrac{11\pi}{6}$

23. $2\tan x - \tan 2x = 0$; $2\tan x - \dfrac{2\tan x}{1 - \tan^2 x} = 0$;
$2\tan x(1 - \tan^2 x) - 2\tan x = 0$;
$2\tan x - 2\tan^3 x - 2\tan x = 0$; $-2\tan^3 x = 0$;
$\tan x = 0$; $x = 0, \pi$

24. $\cos x - \cos 2x = 0$; $\cos x - (2\cos^2 x - 1) = 0$;
$\cos x - 2\cos^2 x + 1 = 0$;
$2\cos^2 x - \cos x - 1 = 0$;

$(2\cos x + 1)(\cos x - 1) = 0$; $\cos x = -\dfrac{1}{2}$ or

$\cos x = 1$; $x = 0, \dfrac{2\pi}{3}, \dfrac{4\pi}{3}$

25. $\cot \dfrac{\beta}{2} = \dfrac{1}{\tan \dfrac{\beta}{2}} = \dfrac{1}{\dfrac{\sin \beta}{1 + \cos \beta}} = \dfrac{1 + \cos \beta}{\sin \beta}$

26. $\cot \dfrac{\alpha}{2} = \dfrac{1}{\tan \dfrac{\alpha}{2}} = \dfrac{1}{\dfrac{1 - \cos \alpha}{\sin \alpha}} = \dfrac{\sin \alpha}{1 - \cos \alpha}$

27. $\tan^2 \dfrac{\theta}{2} = \left(\pm \sqrt{\dfrac{1 - \cos \theta}{1 + \cos \theta}} \right)^2 = \dfrac{1 - \cos \theta}{1 + \cos \theta}$

28. $\left(\cos \dfrac{\beta}{2} + 2 \sin \dfrac{\beta}{2} \right)^2 =$

$\cos^2 \dfrac{\beta}{2} + 2 \sin \dfrac{\beta}{2} \cos \dfrac{\beta}{2} + \sin^2 \dfrac{\beta}{2} =$

$\cos^2 \dfrac{\beta}{2} + \sin^2 \dfrac{\beta}{2} + \sin \beta = 1 + \sin \beta = \sin \beta + 1$

29. $\dfrac{1 + \cos 2\alpha}{\sin 2\alpha} = \dfrac{1 + 2\cos^2 \alpha - 1}{2 \sin \alpha \cos \alpha} = \dfrac{2 \cos^2 \alpha}{2 \sin \alpha \cos \alpha} =$

$\dfrac{\cos \alpha}{\sin \alpha} = \cot \alpha$

30. $\dfrac{\cot \theta - \tan \theta}{\cot \theta + \tan \theta} = \dfrac{\dfrac{\cos \theta}{\sin \theta} - \dfrac{\sin \theta}{\cos \theta}}{\dfrac{\cos \theta}{\sin \theta} + \dfrac{\sin \theta}{\cos \theta}} \cdot \dfrac{\sin \theta \cos \theta}{\sin \theta \cos \theta} =$

$\dfrac{\cos^2 \theta - \sin^2 \theta}{\cos^2 \theta + \sin^2 \theta} = \cos^2 \theta - \sin^2 \theta = \cos 2\theta$

31. From page 816,
$x = \dfrac{v^2 \sin 2\theta}{g} = \dfrac{60^2 \sin (2(39))}{32} = 110 \text{ ft}$

EXTEND

32. $\sin 3\theta = \sin (2\theta + \theta) = \sin 2\theta \cos \theta + \cos 2\theta \sin \theta =$
$2 \sin \theta \cos \theta \cos \theta + (1 - 2 \sin^2 \theta) \sin \theta =$
$2 \sin \theta \cos^2 \theta + \sin \theta - 2 \sin^3 \theta = 2 \sin \theta (1 - \sin^2 \theta) + \sin \theta - 2 \sin^3 \theta =$
$2 \sin \theta - 2 \sin^3 \theta + \sin \theta - 2 \sin^3 \theta = 3 \sin \theta - 4 \sin^3 \theta$
$\sin 3\theta = 3 \sin \theta - 4 \sin^3 \theta$

33. $\cos 3\theta = \cos (2\theta + \theta) = \cos 2\theta \cos \theta - \sin 2\theta \sin \theta =$
$(2 \cos^2 \theta - 1) \cos \theta - 2 \sin \theta \cos \theta \sin \theta =$
$2 \cos^3 \theta - \cos \theta - 2 \cos \theta \sin^2 \theta = 2 \cos^3 \theta - \cos \theta - 2 \cos \theta (1 - \cos^2 \theta) =$
$2 \cos^3 \theta - \cos \theta - 2 \cos \theta + 2 \cos^3 \theta = 4 \cos^3 \theta - 3 \cos \theta$
$\cos 3\theta = 4 \cos^3 \theta - 3 \cos \theta$

34. $\sin 3\theta = \sin \theta - 4 \sin^3 \theta = 3\left(\dfrac{4}{5}\right) - 4\left(\dfrac{4}{5}\right)^3 = \dfrac{44}{125}$

$\cos 3\theta = 4 \cos^3 \theta - 3 \cos \theta = 4\left(-\dfrac{3}{5}\right)^3 - 3\left(-\dfrac{3}{5}\right) = -\dfrac{117}{125}$

Note: Since $\sin \theta = \dfrac{4}{5}$ and $90° < \theta < 180°$,

$\cos \theta = -\sqrt{1 - \left(\dfrac{4}{5}\right)^2} = -\dfrac{3}{5}$.

35a. $A = \left(12 \sin \dfrac{38}{2}\right)\left(12 \cos \dfrac{38}{2}\right) \approx 44 \text{ in.}^2$

35b. $A = \left(4 \sin \dfrac{67}{2}\right)\left(4 \cos \dfrac{67}{2}\right) \approx 7 \text{ m}^2$

35c. maximum area $= 8 \text{ in.}^2$ at $\alpha = \dfrac{\pi}{2}$ or $90°$

36. $t = \dfrac{2v \sin \theta}{g}$; $2v \sin \theta = gt$; $\sin \theta = \dfrac{gt}{2v}$;

$\sin \theta = \dfrac{32(1)}{2(10)} = \dfrac{8}{9}$; $\theta \approx 63°$

37. greatest height achieved at $\theta = 90°$ or $\dfrac{\pi}{2}$

THINK CRITICALLY

38. $\cos 75 = \cos \dfrac{150}{2} = \sqrt{\dfrac{1 + \cos 150}{2}} = \sqrt{\dfrac{1 + \left(-\dfrac{\sqrt{3}}{2}\right)}{2}} =$

$\dfrac{\sqrt{2 - \sqrt{3}}}{2} \approx 0.2588$

$\cos 75 = \cos (120 - 45) =$
$\cos 120 \cos 45 + \sin 120 \sin 45 =$

$-\dfrac{1}{2}\left(\dfrac{\sqrt{2}}{2}\right) + \dfrac{\sqrt{3}}{2}\left(\dfrac{\sqrt{2}}{2}\right) = \dfrac{\sqrt{6} - \sqrt{2}}{4} \approx 0.2588$

39. Use a graphing utility. $\theta = 45°$

40. $\cot 2\alpha = \dfrac{1}{\tan 2\alpha} = \dfrac{1 - \tan^2 \alpha}{2\tan\alpha}$; since

$\tan 2\alpha = \dfrac{2\tan\alpha}{1 - \tan^2\alpha}$; $\cot 2\alpha = \dfrac{1 - \dfrac{1}{\cot^2\alpha}}{2\left(\dfrac{1}{\cot\alpha}\right)}$

$= \dfrac{\cot^2\alpha - 1}{2\cot\alpha}$; $\alpha \neq n\pi$

41. $\dfrac{1}{2}(\cot\theta - \tan\theta) = \dfrac{1}{2}\left(\dfrac{\cos\theta}{\sin\theta} - \dfrac{\sin\theta}{\cos\theta}\right) =$

$\dfrac{1}{2}\left(\dfrac{\cos^2\theta - \sin^2\theta}{\sin\theta\cos\theta}\right) = \dfrac{\cos^2\theta - \sin^2\theta}{2\sin\theta\cos\theta} = \dfrac{\cos 2\theta}{\sin 2\theta}$

MIXED REVIEW

42. $(f \circ g)(-1) = f(2(-1) + 4) = f(2) = -(2)^2 = -4$
$(g \circ f)(-1) = g(-(-1)^2) = g(-1) = 2(-1) + 4 = 2$

43. $(f \circ g)(-1) = f(-1 - 7) = f(-8) = 2(-8)^2 = 128$
$(g \circ f)(-1) = g(2(-1)^2) - g(2) = 2 - 7 = -5$

44. Graph $y = 5^x$ and $y = 13$ and find x where they intersect.
$x = 1.594$

45. Graph $y = 3^x$ and $y = 2$ and find x where they intersect.
$x = 0.631$

46. D; $(2 + 5i)(6 - 2i) = (12 + 10) + (30 - 4)i =$
$= 22 + 26i$

47. $S_n = \dfrac{a_1(1 - r^n)}{(1 - r)}$; $\dfrac{80(1 - 2^7)}{(1 - 2)} = 10{,}160$

48. If $\sin\alpha = \dfrac{3}{5}$ and $0° < \alpha < 90°$ then

$\cos\alpha = \sqrt{1 - \left(\dfrac{3}{5}\right)^2} = \dfrac{4}{5}$. So, $\sin 2\alpha =$

$2\cos\alpha\sin\alpha = 2\left(\dfrac{4}{5}\right)\left(\dfrac{3}{5}\right) = \dfrac{24}{25}$

PROJECT CONNECTION

1. Answers will vary.

2. The amplitude of the string and the sound level of the sound it produces increases with the force used to pluck it.

3a. amplitude: 2   **3b.** amplitude: 300

3c. amplitude: 20   **3d.** amplitude: 0.3
     d, a, c, b

ALGEBRAWORKS

1. $M = \dfrac{1200}{743} \approx 1.6$

2. Let s represent speed at Mach 5.
   $s = 5(743 \text{ mi/h}) = 3715 \text{ mi/h}$
   To achieve hypersonic region the plane would have to fly at a speed ≥ 3715 mi/h.

3. $\sin\dfrac{50}{2} = \dfrac{1}{M}$

$M = \dfrac{1}{\sin 25} \approx 2.4$

4. $\sin\dfrac{25}{2} = \dfrac{1}{M}$

$M = \dfrac{1}{\sin 12.5} \approx 4.6$

5. $\sin\dfrac{\theta}{2} = \dfrac{1}{4}$

$\dfrac{\theta}{2} \approx 14.48$

$\theta \approx 29°$

6. $\sin\dfrac{\theta}{2} = \dfrac{1}{2}$

$\dfrac{\theta}{2} = 30$

$\theta = 60°$

7. Speed of sound $= 743$ mi/h $= \dfrac{743}{3600}$ mi/s ≈ 0.2 mi/s.

Count the seconds divide by 5 (which is the same as multiplying by 0.2 and this equals the distance in miles).

Lesson 16.5, pages 821–827

EXPLORE

1. Answers will vary. Some possible examples are $(3, 390°)$ and $(3, -330°)$.

2. Infinitely many

3.

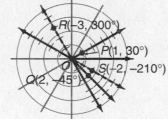

$P(3, 300°)$

4. Yes; $P(-3, 120°)$ **5.** Yes; $P(3, -60°)$

6. Yes; $P(-3, -240°)$

TRY THESE

1–4.

$R(-3, 300°)$
$P(1, 30°)$
$S(-2, -210°)$
$Q(2, -45°)$

5. $r = \sqrt{2^2 + (-2)^2} = \sqrt{8} = 2\sqrt{2}$;
$\theta = \text{Arctan}\left(\dfrac{-2}{2}\right) = 315°$; $P(2\sqrt{2}, 315°)$

6. $r = \sqrt{(-4)^2 + (4\sqrt{3})^2} = \sqrt{16 + 48} = 8$;
$\theta = \text{Arctan}\left(\dfrac{4\sqrt{3}}{-4}\right) + 180 =$
$\text{Arctan}(-\sqrt{3}) + 180 = 120°$; $Q(8, 120°)$

7. $r = \sqrt{9^2 + 40^2} = \sqrt{1681} = 41$;
$\theta = \text{Arctan}\left(\dfrac{40}{9}\right) \approx 77°$; $R(41, 77°)$

8. $r = \sqrt{(-7)^2 + 24^2} = \sqrt{49 + 576} = \sqrt{625} = 25$;
$\theta = \text{Arctan}\left(\frac{24}{-7}\right) + 180 \approx 106°$; $S(25, 106°)$

9. If Arctan $\theta = 0$ and $x > 0$, $y = 0$. If $y = 0$, the point is on the x-axis with a positive value for x. Therefore, $\theta = 0°$. If Arctan $\theta = 0$, and $x < 0$, $y = 0$. If $y = 0$, the point is on the x-axis with a negative value for x. Therefore, $\theta = 180°$; $U(3, 0°)$, $V(7, 180°)$.

10. $x = 5 \cos 45 = 5\left(\frac{\sqrt{2}}{2}\right) = \frac{5\sqrt{2}}{2}$

$y = 5 \sin 45 = 5\left(\frac{\sqrt{2}}{2}\right) = \frac{5\sqrt{2}}{2}$

$P\left(\frac{5\sqrt{2}}{2}, \frac{5\sqrt{2}}{2}\right)$

11. $x = (-6) \cos (-60) = (-6)\left(\frac{1}{2}\right) = -3$

$y = (-6) \sin (-60) = (-6)\left(-\frac{\sqrt{3}}{2}\right) = 3\sqrt{3}$

$Q(-3, 3\sqrt{3})$

12. $x = 3 \cos \left(-\frac{2\pi}{3}\right) = 3\left(-\frac{1}{2}\right) = -\frac{3}{2}$

$y = 3 \sin \left(-\frac{2\pi}{3}\right) = 3\left(-\frac{\sqrt{3}}{2}\right) = -\frac{3\sqrt{3}}{2}$

$R\left(-\frac{3}{2}, -\frac{3\sqrt{3}}{2}\right)$

13. $x = -2 \cos \left(-\frac{4\pi}{3}\right) = -2\left(-\frac{1}{2}\right) = 1$

$y = -2 \sin \left(-\frac{4\pi}{3}\right) = -2\left(\frac{\sqrt{3}}{2}\right) = -\sqrt{3}$

$S(1, -\sqrt{3})$

14. $r = \sqrt{(-4)^2 + 3^2} = \sqrt{16 + 9} = 5$

$\theta = \text{Arctan}\left(-\frac{3}{4}\right) + 180 \approx 143°$

$5 (\cos 143° + i \sin 143°)$

15. $r = \sqrt{5^2 + (-12)^2} = \sqrt{25 + 144} = 13$

$\theta = \text{Arctan}\left(-\frac{12}{5}\right) \approx 293°$

$13 (\cos 293° + i \sin 293°)$

16. $r = \sqrt{(-8)^2 + 0^2} = 8$

$\theta = \text{Arctan } 0 + 180 = 0 + 180 = 180°$

$8 (\cos 180° + i \sin 180°)$

17. $r = \sqrt{(10)^2 + 0^2} = 10$

$\theta = \text{Arctan } 0 = 0°$

$10 (\cos 0° + i \sin 0°)$

18. $r = \dfrac{3}{2 - \sin \theta}$; $r(2 - \sin \theta) = 3$; $2r - r \sin \theta = 3$;

$2\sqrt{x^2 + y^2} - y = 3$; $2\sqrt{x^2 + y^2} = y + 3$;

$4(x^2 + y^2) = y^2 + 6y + 9$;

$4x^2 + 4y^2 = y^2 + 6y + 9$;

$4x^2 + 3y^2 - 6y - 9 = 0$

19. $\frac{1}{2} + \frac{i\sqrt{3}}{2}$ **20.** $5\left(-\frac{\sqrt{2}}{2} + i \frac{\sqrt{2}}{2}\right) = -\frac{5\sqrt{2}}{2} + \frac{5i\sqrt{2}}{2}$

21. $6\left(-\frac{\sqrt{3}}{2} + i\left(\frac{1}{2}\right)\right) = -3\sqrt{3} + 3i$

22. $4\left(-\frac{\sqrt{3}}{2} + i\left(-\frac{1}{2}\right)\right) = -2\sqrt{3} - 2i$

23. heart-shaped

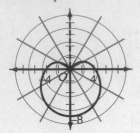

PRACTICE

1–4.

5. $r = \sqrt{5^2 + 5^2} = \sqrt{50} = 5\sqrt{2}$

$\theta = \text{Arctan } \frac{5}{5} = \text{Arctan } 1 = 45°$

$P(5\sqrt{2}, 45°)$

6. $r = \sqrt{9^2 + (-9)^2} = 9\sqrt{2}$

$\theta = \text{Arctan }\left(\frac{-9}{9}\right) = \text{Arctan } (-1) = 315°$

$Q(9\sqrt{2}, 315°)$

7. $r = \sqrt{8^2 + 0^2} = 8$

$\theta = \text{Arctan } 0 = 0°$

$R(8, 0°)$

8. $r = \sqrt{(-7)^2 + 0^2} = 7$

$\theta = \text{Arctan } 0 + 180 = 180°$

$S(7, 180°)$

9. $r = \sqrt{(-2)^2 + (2\sqrt{3})^2} = \sqrt{4 + 12} = 4$

$\theta = \text{Arctan }\left(\frac{2\sqrt{3}}{-2}\right) + 180 = \text{Arctan } (-\sqrt{3}) + 180$

$= -60 + 180 = 120°$; $T(4, 120°)$

10. $r = \sqrt{3^2 + (-3\sqrt{3})^2} = \sqrt{9 + 27} = 6$

$\theta = \text{Arctan }\left(\frac{-3\sqrt{3}}{3}\right) = \text{Arctan } (-\sqrt{3}) = 300°$

$U(6, 300°)$

11. $r = \sqrt{(-9)^2 + (-40)^2} = \sqrt{81 + 1600} =$

$\sqrt{1681} = 41$; $\theta = \text{Arctan }\left(\frac{-40}{-9}\right) + 180 \approx 257°$

$V(41, 257°)$

12. $r = \sqrt{(-12)^2 + (-5)^2} = \sqrt{144 + 25} = 13$

$\theta = \text{Arctan}\left(\frac{-5}{-12}\right) + 180 \approx 203°$

$W(13, 203°)$

13. Its modulus is $\sqrt{a^2 + 0} = \sqrt{a^2} = |a|$. Its argument is $0°$ if $a > 0$, $180°$ if $a < 0$ and any angle measure if $a = 0$.

14. $x = 2 \cos 135 = 2\left(-\frac{\sqrt{2}}{2}\right) = -\sqrt{2}$

$y = 2 \sin 135 = 2\left(\frac{\sqrt{2}}{2}\right) = \sqrt{2}$

$P(-\sqrt{2}, \sqrt{2})$

15. $x = -2 \cos(-120) = -2\left(-\frac{1}{2}\right) = 1$

$y = -2 \sin(-120) = -2\left(-\frac{\sqrt{3}}{2}\right) = \sqrt{3}$

$Q(1, \sqrt{3})$

16. $x = 4 \cos\left(-\frac{3\pi}{4}\right) = 4\left(-\frac{\sqrt{2}}{2}\right) = -2\sqrt{2}$

$y = 4 \sin\left(-\frac{3\pi}{4}\right) = 4\left(-\frac{\sqrt{2}}{2}\right) = -2\sqrt{2}$

$R(-2\sqrt{2}, -2\sqrt{2})$

17. $x = -5 \cos\frac{7\pi}{4} = -5\left(\frac{\sqrt{2}}{2}\right) = \frac{-5\sqrt{2}}{2}$

$y = -5 \sin\frac{7\pi}{4} = -5\left(\frac{-\sqrt{2}}{2}\right) = \frac{5\sqrt{2}}{2}$

$S\left(\frac{5\sqrt{2}}{2}, \frac{5\sqrt{2}}{2}\right)$

18. $r = \sqrt{(-8)^2 + (15)^2} = \sqrt{289} = 17$

$\theta = \text{Tan}^{-1}\left(\frac{15}{-8}\right) + 180 \approx 118°$

$17 (\cos 118° + i \sin 118°)$

19. $r = \sqrt{9^2 + (-40)^2} = \sqrt{1681} = 41$

$\theta = \text{Tan}^{-1}\left(\frac{-40}{9}\right) \approx 283°$

$41 (\cos 283° + i \sin 283°)$

20. $r = \sqrt{(-7)^2 + (24)^2} = \sqrt{625} = 25$

$\theta = \text{Tan}^{-1}\left(\frac{24}{-7}\right) + 180 \approx 106°$

$25 (\cos 106° + i \sin 106°)$

21. $r = \sqrt{6^2 + 8^2} = \sqrt{100} = 10$

$\theta = \text{Tan}^{-1}\left(\frac{8}{6}\right) \approx 53°$

$10 (\cos 53° + i \sin 53°)$

22. $r = |12| = 12$

$\theta = 0°$

$12 (\cos 0° + i \sin 0°)$

23. $r = |-15| = 15$

$\theta = 180°$

$15 (\cos 180° + i \sin 180°)$

24. $r = \sqrt{3^2} = 3$

$\theta = 90°$

$3 (\cos 90° + i \sin 90°)$

25. $r = \sqrt{(-4)^2} = 4$

$\theta = 270°$

$4(\cos 270° + i \sin 270°)$

26. $r = \sqrt{x^2 + y^2}; x = r \cos \theta; \cos \theta = \frac{x}{r}$

Substitute to obtain

$\sqrt{x^2 + y^2} = 2 + 2\left(\frac{x}{\sqrt{x^2 + y^2}}\right)$

$x^2 + y^2 = 2\sqrt{x^2 + y^2} + 2x$

27. $8\left(-\frac{\sqrt{2}}{2} + i\left(-\frac{\sqrt{2}}{2}\right)\right) = -4\sqrt{2} - 4i\sqrt{2}$

28. $5\left(\frac{\sqrt{3}}{2} + i\left(\frac{1}{2}\right)\right) = \frac{5\sqrt{3}}{2} + \frac{5}{2}i$

29. $2\left(\frac{1}{2} + i\left(-\frac{\sqrt{3}}{2}\right)\right) = 1 - i\sqrt{3}$

30. $3\left(\frac{\sqrt{2}}{2} + i\left(-\frac{\sqrt{2}}{2}\right)\right) = \frac{3\sqrt{2}}{2} - \frac{3i\sqrt{2}}{2}$

31. $6(1 + i(0)) = 6$ **32.** $7(-1 + i(0)) = -7$

33. $r = \sqrt{x^2 + y^2}; x = r \cos \theta; \cos \theta = \frac{x}{r}$

$\sqrt{x^2 + y^2} = \dfrac{225{,}000}{40 - 2\left(\dfrac{x}{\sqrt{x^2 + y^2}}\right)}$

$\sqrt{x^2 + y^2} = \dfrac{225{,}000\sqrt{x^2 + y^2}}{40\sqrt{x^2 + y^2} - 2x}$

$1 = \dfrac{225{,}000}{40\sqrt{x^2 + y^2} - 2x}$

$40\sqrt{x^2 + y^2} - 2x = 225{,}000$

EXTEND

34. $5(6) (\cos(80 + 20) + i \sin(80 + 20)) =$
$30 (\cos 100° + i \sin 100°)$

35. $8(9) (\cos(122 + 18) + i \sin(122 + 18)) =$
$72 (\cos 140° + i \sin 140°)$

36. $9(10) \left(\cos\left(\frac{\pi}{3} + \frac{\pi}{4}\right) + i \sin\left(\frac{\pi}{3} + \frac{\pi}{4}\right)\right) =$

$90 \left(\cos\frac{7\pi}{12} + i \sin\frac{7\pi}{12}\right)$

37. $7(7) \left(\cos\left(\frac{5\pi}{3} + \frac{3\pi}{4}\right) + i \sin\left(\frac{5\pi}{3} + \frac{3\pi}{4}\right)\right) =$

$49 \left(\cos\frac{29\pi}{12} + i \sin\frac{29\pi}{12}\right) =$

$49 \left(\cos\frac{5\pi}{12} + i \sin\frac{5\pi}{12}\right)$

38. $r_2 \neq 0$

39. $\frac{12}{3}$ (cos (65 − 25) + i sin (65 − 25))

4 (cos 40° + i sin 40°)

40. $\frac{15}{3}$ (cos (54 − 38) + i sin (54 − 38))

3 (cos 16° + i sin 16°)

41. $\frac{20}{4}\left(\cos\left(\frac{2\pi}{3} - \frac{\pi}{4}\right) + i\sin\left(\frac{2\pi}{3} - \frac{\pi}{4}\right)\right)$

$5\left(\cos\frac{5\pi}{12} + i\sin\frac{5\pi}{12}\right)$

42. $\frac{18}{9}\left(\cos\left(\frac{4\pi}{5} - \frac{\pi}{7}\right) + i\sin\left(\frac{4\pi}{5} - \frac{\pi}{7}\right)\right)$

$2\left(\cos\frac{23\pi}{35} + i\sin\frac{23\pi}{35}\right)$

43. 4 petals   **44.** 3 petals

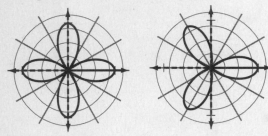

45. 8 petals

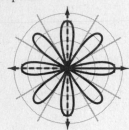

THINK CRITICALLY

46. If n is even, the rose has $2n$ petals. If n is odd, the rose has n petals.

47. the polar axis

48. $r_1(\cos\alpha + i\sin\alpha)r_2(\cos\beta + i\sin\beta) =$
$r_1r_2(\cos\alpha\cos\beta + \cos\alpha + i\sin\beta +$
$i\sin\alpha\cos\beta + i^2\sin\alpha\sin\beta) =$
$r_1r_2[(\cos\alpha\cos\beta - \sin\alpha\sin\beta) +$
$i(\sin\alpha\cos\beta + \cos\alpha\sin\beta)] =$
$r_1r_2[\cos(\alpha + \beta) + i\sin(\alpha + \beta)]$

49. The line $\theta = \frac{\pi}{2}$

50. $\alpha = \beta$ or α and β differ by a multiple of 2π.

51. The number of different arrangements for the 8 books is $8! = 40{,}320$. Therefore the probability of getting them in alphabetic order is $\frac{1}{40{,}320}$, since there is only one way for it to be in alphabetical order.

52. $\log_4 6f^2 = \log_4 6 + \log_4 f^2 = \log_4 6 + 2\log_4 f$

53. $\log_5 \frac{u}{v} = \log_5 u - \log_5 v$

54. B; $x^2 + 8x + 25 = 0$; $x = \dfrac{-8 \pm \sqrt{8^2 - 4(1)(25)}}{2(1)}$;

$x = \dfrac{-8 \pm \sqrt{-36}}{2} = \dfrac{-8 \pm 6i}{2} = -4 \pm 3i$

55. Axis of symmetry: $x = -\dfrac{b}{2a}$

$x = \dfrac{8}{2\left(\frac{1}{2}\right)}$

$x = 8$

56. Axis of symmetry: $x = -\dfrac{b}{2a}$

$x = \dfrac{-100}{2(2)}$

$x = -25$

57. $12\left(-\dfrac{\sqrt{3}}{2} + i\left(\dfrac{1}{2}\right)\right) = -6\sqrt{3} + 6i$

58. $15\left(-\dfrac{\sqrt{2}}{2} + i\left(\dfrac{\sqrt{2}}{2}\right)\right) = -\dfrac{15\sqrt{2}}{2} + \dfrac{15i\sqrt{2}}{2}$

Lesson 16.6, pages 828–831

1. $d = rt$; $32 = 16t$; $t = 2$h

2.

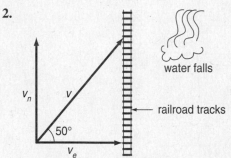

3. $\cos 50 = \dfrac{v_e}{v}$; $v_e = v\cos 50 \approx 10.3$ mi/h

4. $\sin 50 = \dfrac{v_n}{v}$; $v_n = v\sin 50 \approx 12.3$ mi/h

5. $d = vt$

$t = \dfrac{d}{v}$

$V_e = 16\cos 50$; $V_n = 16\sin 50$

Since the tracks are 32 mi due east of Middlebury, take the east component of the blimp's velocity and use it in the equation.

$t = \dfrac{32}{16\cos 50}$

$t \approx 3.1$h

6. $d = vt$

$d = 32(3.1)$

$d = 99.2$ mi   Answers will vary due to rounding.

INVESTIGATE FURTHER

7.
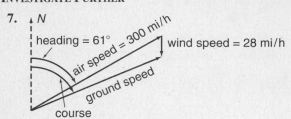

8. Let x represent the ground speed.

$x^2 = 300^2 + 28^2 - 2(300)(28) \cos 61 \approx 82{,}639.2$

$x \approx 287.5$ mi/h

9. $\dfrac{\sin \theta}{28} = \dfrac{\sin 61}{287.5}$; $\sin \theta = \dfrac{28 \sin 61}{287.5} \approx 0.085$

$\theta \approx 5°$

10. course $\approx 61 + 5 \approx 66°$

11. Answers may vary. Law of sines may be used if two sides and an angle opposite one of them is known or if two angles and a side opposite one of them is known. The law of cosines is used if two sides and an included angle are known.

APPLY THE STRATEGY

12. $V_x = r \cos \theta = 40 \cos 45 = 40\left(\dfrac{\sqrt{2}}{2}\right) = 20\sqrt{2}$ ft/s

$V_y = r \sin \theta = 40 \sin 45 = 40\left(\dfrac{\sqrt{2}}{2}\right) = 20\sqrt{2}$ ft/s

13.
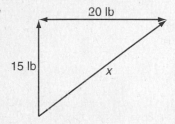

$x^2 = 15^2 + 20^2 - 225 + 400 = 625$

$x = \sqrt{625} = 25$ lb.

14.
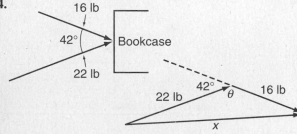

$\theta = 180 - 42 - 138°$

$x^2 = 16^2 + 22^2 - 2(16)(22) \cos 138$

$x^2 \approx 1263.2$

$x \approx 36$ lb

15.
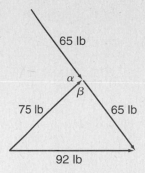

$92^2 = 75^2 + 65^2 - 2(75)(65) \cos \beta$;

$\cos \beta = \dfrac{92^2 - 75^2 - 65^2}{-2(75)(65)} = \dfrac{231}{1625}$

$\beta \approx 82°$

$\alpha \approx 180 - 82 = 98°$

16. 0°; the forces are applied in the same directions and the resultant is the sum of the forces.

17. 180°; 10 lb; the resultant is minimized if the forces are applied in opposite directions.

18. Since $||u|| > 0$ and $||v|| > 0$ then $u \cdot v = 0$ if and only if $\cos \theta = 0$, so $\theta = 90°$ and the vectors are perpendicular.

REVIEW PROBLEM SOLVING STRATEGIES

1. White. Assume that the fan in the first row is wearing a blue cap. Then the fan in the third row would have seen either two blue caps or one white cap and one blue cap. If he saw two blue caps, he would have known his own cap was white. But he did not know. If the fan in the second row had seen one blue and one white cap, he also would have known that the fan in the third row must have seen a blue cap and a white cap and therefore the fan in the second row would have known that his own cap was white. But he did not know. Therefore, the fan in the first row could not have been wearing a blue cap.

2. The time for the second half of the trip was as long as the whole trip would have taken him on foot, so no matter how fast the car was, he lost as much time as the car ride took. If his walking rate is r, the car's rate is $15r$; since the time t walking the entire distance d is $t = d/r$, then the time for half the trip in the car is

$\dfrac{1}{2}d$

$\dfrac{2}{15r} = \dfrac{d}{30r} = \dfrac{t}{30}$

3. Alice is 14, Betty is 20, and Cindy is 32.

CHAPTER REVIEW, PAGES 832–833

1. c   **2.** d   **3.** b   **4.** a

5. $\cos \theta \cot \theta + \sin \theta = \cos \theta \left(\dfrac{\cos \theta}{\sin \theta}\right) + \sin \theta =$

$\dfrac{\cos^2 \theta}{\sin \theta} + \dfrac{\sin^2 \theta}{\sin \theta} = \dfrac{1}{\sin \theta} = \csc \theta$

6.

$\dfrac{\csc \beta + 1}{-\csc \beta}$	$\dfrac{1 - \sin^2 \beta}{\sin \beta - 1}$
$\dfrac{\dfrac{1}{\sin \beta} + 1}{-\dfrac{1}{\sin \beta}}$	$\dfrac{(1 - \sin \beta)(1 + \sin \beta)}{\sin \beta - 1}$
$\dfrac{1 + \sin \beta}{\sin \beta}(-\sin \beta)$	$-1(1 + \sin \beta)$
$-1 - \sin \beta$	$-1 - \sin \beta$

7. $\csc^2 x - 4 = 0$; $\csc^2 x = 2$; $\csc x = \pm 2$;

$x = \dfrac{\pi}{6}, \dfrac{5\pi}{6}, \dfrac{7\pi}{6}, \dfrac{11\pi}{6}$

8. $4 \sec^2 x + 5 \sec x - 6 = 0$;
$(\sec x + 2)(4 \sec x - 3) = 0$

$\sec x = -2$ or $\sec x = \dfrac{3}{4}$ (impossible); $x = \dfrac{2\pi}{3}, \dfrac{4\pi}{3}$

9. $12 \sin^2 x - 9 \cos x - 7 = 0$;
$12(1 - \cos^2 x) - 9 \cos x - 7 = 0$;
$12 - 12 \cos^2 x - 9 \cos x - 7 = 0$;
$12 \cos^2 x + 9 \cos x - 5 = 0$

$\cos x = \dfrac{-9 \pm \sqrt{81 - 4(12)(-5)}}{2(12)} = \dfrac{-9 \pm \sqrt{321}}{24}$;

$x \approx 1.19, 5.09$

10. $9 \cos^2 x - 14 \sin x - 11 = 0$
$9(1 - \sin^2 x) - 14 \sin x - 11 = 0$
$9 - 9 \sin^2 x - 14 \sin x - 11 = 0$
$9 \sin^2 x + 14 \sin x + 2 = 0$

$\sin x = \dfrac{-14 \pm \sqrt{14^2 - 4(9)(2)}}{2(9)}$

$\sin x = \dfrac{-14 \pm \sqrt{124}}{28}$

$x = 3.30, 6.12$

In Exercises 11–13. if $\cos \alpha = \dfrac{8}{17}$, $\sin \beta = \dfrac{3}{5}$,

$270° < \alpha < 360°$ and $90° < \beta < 180°$ then

$\sin \alpha = -\sqrt{1 - \left(\dfrac{8}{17}\right)^2} = -\dfrac{15}{17}$,

$\cos \beta = -\sqrt{1 - \left(\dfrac{3}{5}\right)^2} = -\dfrac{4}{5}$, $\tan \alpha = \dfrac{-\dfrac{15}{17}}{\dfrac{8}{17}} = -\dfrac{15}{8}$, and

$\tan \beta = \dfrac{\dfrac{3}{5}}{-\dfrac{4}{5}} = -\dfrac{3}{4}$.

11. $\cos (\alpha + \beta) = \dfrac{8}{17}\left(-\dfrac{4}{5}\right) - \left(-\dfrac{15}{17}\right)\left(\dfrac{3}{5}\right) = \dfrac{13}{85}$

12. $\sin (\alpha - \beta) = \left(-\dfrac{15}{17}\right)\left(-\dfrac{4}{5}\right) - \dfrac{8}{17}\left(\dfrac{3}{5}\right) = \dfrac{36}{85}$

13. $\tan (\alpha - \beta) = \dfrac{\tan \alpha - \tan \beta}{1 + \tan \alpha \tan \beta} =$

$\dfrac{-\dfrac{15}{8} - \left(-\dfrac{3}{4}\right)}{1 + \left(-\dfrac{15}{8}\right)\left(-\dfrac{3}{4}\right)} = -\dfrac{36}{77}$

14. $\cos (345) = \cos (300 + 45) =$
$\cos 300 \cos 45 - \sin 300 \sin 45 =$

$\dfrac{1}{2}\left(\dfrac{\sqrt{2}}{2}\right) - \left(-\dfrac{\sqrt{3}}{2}\right)\left(\dfrac{\sqrt{2}}{2}\right) = \dfrac{\sqrt{2}}{4} + \dfrac{\sqrt{6}}{4} = \dfrac{\sqrt{2} + \sqrt{6}}{4}$

15. $\sin (-15) = \sin (30 - 45) = \sin 30 \cos 45 -$
$\cos 30 \sin 45 =$

$\dfrac{1}{2}\left(\dfrac{\sqrt{2}}{2}\right) - \dfrac{\sqrt{3}}{2}\left(\dfrac{\sqrt{2}}{2}\right) = \dfrac{\sqrt{2}}{4} - \dfrac{\sqrt{6}}{4} = \dfrac{\sqrt{2} - \sqrt{6}}{4}$

16. $\tan 195 = \tan (150 + 45) = \dfrac{\tan 150 + \tan 45}{1 - \tan 150 \tan 45} =$

$\dfrac{-\dfrac{\sqrt{3}}{3} + 1}{1 - \left(-\dfrac{\sqrt{3}}{3}\right)(1)} = \dfrac{-\sqrt{3} + 3}{3 + \sqrt{3}} = \dfrac{3 - \sqrt{3}}{3 + \sqrt{3}}$

17. $\sin \left(\dfrac{3\pi}{2} + \alpha\right) = \sin \dfrac{3\pi}{2} \cos \alpha +$

$\cos \dfrac{3\pi}{2} \sin \alpha = (-1) \cos \alpha + (0) \sin \alpha = -\cos \alpha$

18. $\tan \left(\dfrac{3\pi}{4} - \beta\right) = \dfrac{\tan \dfrac{3\pi}{4} - \tan \beta}{1 + \tan \dfrac{3\pi}{4} \tan \beta} = \dfrac{-1 - \tan \beta}{1 + (-1) \tan \beta}$

$= \dfrac{-1 - \tan \beta}{1 - \tan \beta} = \dfrac{\tan \beta + 1}{\tan \beta - 1}$

In Exercises 19–21. if $\cos \alpha = -\dfrac{40}{41}$ and

$180° < \alpha < 270°$ then $\sin \alpha = -\sqrt{1 - \left(-\dfrac{40}{41}\right)^2} = -\dfrac{9}{41}$

and $\tan \alpha = \dfrac{\sin \alpha}{\cos \alpha} = \dfrac{-\dfrac{9}{41}}{\dfrac{-40}{41}} = \dfrac{9}{40}$.

19. $\sin 2\alpha = 2 \sin \alpha \cos \alpha = 2\left(-\dfrac{9}{41}\right)\left(-\dfrac{40}{41}\right) = \dfrac{720}{1681}$

20. $\tan 2\alpha = \dfrac{2 \tan \alpha}{1 - \tan^2 \alpha} = \dfrac{2\left(\dfrac{9}{40}\right)}{1 - \left(\dfrac{9}{40}\right)^2} = \dfrac{720}{1519}$

21. $\cos \dfrac{\alpha}{2} = -\sqrt{\dfrac{1 + \cos \alpha}{2}} = -\sqrt{\dfrac{1 + \left(-\frac{40}{41}\right)}{2}} =$

$-\sqrt{\dfrac{41 - 40}{82}} = \sqrt{\dfrac{1}{82}} = \dfrac{\sqrt{82}}{82}$

22. $\cos \dfrac{315}{2} = -\sqrt{\dfrac{1 + \cos 315}{2}} = -\sqrt{\dfrac{1 + \frac{\sqrt{2}}{2}}{2}} =$

$-\sqrt{\dfrac{2 + \sqrt{2}}{4}} = -\dfrac{\sqrt{2 + \sqrt{2}}}{2}$

23. $\sin \dfrac{570}{2} = -\sqrt{\dfrac{1 - \cos 570}{2}} =$

$-\sqrt{\dfrac{1 - \left(-\frac{\sqrt{3}}{2}\right)}{2}} = \dfrac{-\sqrt{2 + \sqrt{3}}}{2}$

24. $\tan \dfrac{225}{2} = \dfrac{\sin 225}{1 + \cos 225} = \dfrac{-\frac{\sqrt{2}}{2}}{1 + \left(-\frac{\sqrt{2}}{2}\right)} =$

$= \dfrac{-\sqrt{2}}{2 - \sqrt{2}} = \dfrac{-\sqrt{2}(2 + \sqrt{2})}{4 - 2} =$

$\dfrac{-2\sqrt{2} - 2}{2} = -1 - \sqrt{2}$

25. $r = \sqrt{(-9)^2 + (9\sqrt{3})^2} = 18;$

$\theta = \text{Arctan} \left(\dfrac{9\sqrt{3}}{-9}\right) + 180 = 120°$

$P(18, 120°)$

26. $r = \sqrt{12^2 + (-12)^2} = 12\sqrt{2};$

$\theta = \text{Arctan} \left(\dfrac{-12}{12}\right) = 315°$

$Q(12\sqrt{2}, 315°)$

27. $x = r \cos \theta = 5 \cos 30 = \dfrac{5\sqrt{3}}{2}$

$y = r \sin \theta = 5 \sin 30 = \dfrac{5}{2}$

$R\left(\dfrac{5\sqrt{3}}{2}, \dfrac{5}{2}\right)$

28. $x = 8 \cos 135 = 8\left(-\dfrac{\sqrt{2}}{2}\right) = -4\sqrt{2}$

$y = 8 \sin 135 = 8\left(\dfrac{\sqrt{2}}{2}\right) = 4\sqrt{2}$

$S(-4\sqrt{2}, 4\sqrt{2})$

29. $8\left(-\dfrac{1}{2} + i\left(\dfrac{\sqrt{3}}{2}\right)\right) = -4 + 4i\sqrt{3}$

30. $4\left(-\dfrac{\sqrt{3}}{2} + i\left(\dfrac{1}{2}\right)\right) = -2\sqrt{3} + 2i$

31. $r = \sqrt{(-15)^2 + 8^2} = 17;$

$\theta = \text{Arctan} \left(\dfrac{8}{-15}\right) + 180 \approx 152°$

$17 (\cos 152° + i \sin 152°)$

32. $r = \sqrt{9^2 + 40^2} = 41; \theta = \text{Arctan} \dfrac{40}{9} \approx 77°$

$41 (\cos 77° + i \sin 77°)$

33. $r = |-5| = 5; \theta = 180°$

$5 (\cos 180° + i \sin 180°)$

34. $r = 12; \theta = 0°$

$12 (\cos 0° + i \sin 0°)$

35. $V_e = 18 \cos 42 \approx 13.4 \text{ mi/h}$

$V_n = 18 \sin 42 \approx 12.0 \text{ mi/h}$

36.

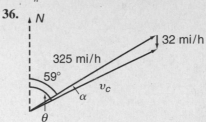

$V_c^2 = 325^2 + 32^2 - 2(325)(32) \cos 59$

$V_c = \sqrt{325^2 + 32^2 - 2(325)(32) \cos 59}$

$V_c = 309.7 \text{ mi/h}$

$\dfrac{\sin \alpha}{32} = \dfrac{\sin 59}{309.7}$

$\sin \alpha = \dfrac{32 \sin 59}{309.7}$

$\alpha \approx 5°$

$\theta = 59 + 5 = 64°$

Chapter Assessment, pages 834–835

CHAPTER TEST

1. C

2. Answers will vary. Techniques include using various identities to transform expressions and using algebraic methods to simplify and change expressions.

3. $\sin \alpha \sec \alpha + \cot \alpha = \sin \alpha\left(\dfrac{1}{\cos \alpha}\right) + \dfrac{\cos \alpha}{\sin \alpha}$

$= \dfrac{\sin \alpha}{\cos \alpha} + \dfrac{\cos \alpha}{\sin \alpha}$

$= \dfrac{\sin^2 \alpha + \cos^2 \alpha}{\cos \alpha \sin \alpha}$

$= \dfrac{1}{\cos \alpha \sin \alpha}$

$= \dfrac{1}{\cos \alpha} \cdot \dfrac{1}{\sin \alpha}$

$= \sec \alpha \csc \alpha$

4. $\dfrac{\sin \beta}{\csc \beta} + \dfrac{\cos \beta}{\sec \beta} = \dfrac{\sin \beta}{\frac{1}{\sin \beta}} + \dfrac{\cos \beta}{\frac{1}{\cos \beta}}$

$= \sin^2 \beta + \cos^2 \beta = 1$

$\sec^2 \beta - \tan^2 \beta = 1$

therefore, $\dfrac{\sin \beta}{\csc \beta} + \dfrac{\cos \beta}{\sec \beta} = \sec^2 \beta - \tan^2 \beta$

5. $\sin(270 + \theta) = \sin 270 \cos \theta + \cos 270 \sin \theta$
$\qquad = (-1)(\cos \theta) + (0)(\sin \theta)$
$\qquad = -\cos \theta$

6. $\sin 2x = 2 \cos x; 2 \sin x \cos x = 2 \cos x;$
$2 \sin x \cos x - 2 \cos x = 0; \cos x (\sin x - 1) = 0;$

$\cos x = 0$ or $\sin x = 1; x = \dfrac{\pi}{2}, \dfrac{3\pi}{2}$

7. $\cos 2x + \sin x = 1; 1 - 2 \sin^2 x + \sin x = 1,$
$2 \sin^2 x - \sin x = 0; \sin x(2 \sin x - 1) = 0;$

$\sin x = 0$ or $\sin x = \dfrac{1}{2}; x = 0, \dfrac{\pi}{6}, \dfrac{5\pi}{6}, \pi$

8. $\cos\left(x + \dfrac{\pi}{4}\right) = \sin\left(x - \dfrac{\pi}{4}\right)$

$\cos x \cos \dfrac{\pi}{4} - \sin x \sin \dfrac{\pi}{4} = \sin x \cos \dfrac{\pi}{4} - \cos x \sin \dfrac{\pi}{4}$

$\dfrac{\sqrt{2}}{2} \cos x - \dfrac{\sqrt{2}}{2} \sin x = \dfrac{\sqrt{2}}{2} \sin x - \dfrac{\sqrt{2}}{2} \cos x$

$\sqrt{2} \cos x - \sqrt{2} \sin x = 0$

$\cos x - \sin x = 0$

$\cos^2 x - 2 \cos x \sin x + \sin^2 x = 0 \qquad$ square both sides

$1 - 2 \cos x \sin x = 0$

$1 - \sin 2x = 0$

$\sin 2x = 1$

$2x = \dfrac{\pi}{2}, \dfrac{5\pi}{2}$

$x = \dfrac{\pi}{4}, \dfrac{5\pi}{4}$

9. $8 \sin^2 x - 2 \cos x - 5 = 0;$
$8(1 - \cos^2 x) - 2 \cos x - 5 = 0;$
$8 - 8 \cos^2 x - 2 \cos x - 5 = 0;$
$8 \cos^2 x + 2 \cos x - 3 = 0;$

$(4 \cos x + 3)(2 \cos x - 1) = 0; \cos x = -\dfrac{3}{4}$ or

$\cos x = \dfrac{1}{2}; x = 1.05, 2.42, 3.86, 5.24$

10. D; $\sin 75 = \sin(30 + 45) =$
$\sin 30 \cos 45 + \cos 30 \sin 45$

$= \dfrac{1}{2}\left(\dfrac{\sqrt{2}}{2}\right) + \dfrac{\sqrt{3}}{2}\left(\dfrac{\sqrt{2}}{2}\right) = \dfrac{\sqrt{2} + \sqrt{6}}{4}$ (B) and (C)

$\sin \dfrac{150}{2} = \sqrt{\dfrac{1 - \cos 150°}{2}} = \sqrt{\dfrac{1 - \left(-\dfrac{\sqrt{3}}{2}\right)}{2}} = \dfrac{\sqrt{2 + \sqrt{3}}}{2}$

(A)

D; 0.9659 is an approximation

11. $\sin 105 = \sin(60 + 45) =$
$\sin 60 \cos 45 + \cos 60 \sin 45 =$

$\dfrac{\sqrt{3}}{2}\left(\dfrac{\sqrt{2}}{2}\right) + \dfrac{1}{2}\left(\dfrac{\sqrt{2}}{2}\right) = \dfrac{\sqrt{6} + \sqrt{2}}{4}$

$\sin \dfrac{210}{2} = \sqrt{\dfrac{1 - \cos 210}{2}}\sqrt{\dfrac{1 - \left(-\dfrac{\sqrt{3}}{2}\right)}{2}} =$

$\dfrac{\sqrt{2 + \sqrt{3}}}{2}$

12. $\cos 285 = \cos(225 + 60) =$
$\cos 225 \cos 60 - \sin 225 \sin 60 =$

$-\dfrac{\sqrt{2}}{2}\left(\dfrac{1}{2}\right) - \left(-\dfrac{\sqrt{2}}{2}\right)\left(\dfrac{\sqrt{3}}{2}\right) =$

$-\dfrac{\sqrt{2}}{4} + \dfrac{\sqrt{6}}{4} = \dfrac{\sqrt{6} - \sqrt{2}}{4}$

$\cos \dfrac{570}{2} = \sqrt{\dfrac{1 + \cos 570}{2}} = \sqrt{\dfrac{1 + \left(-\dfrac{\sqrt{3}}{2}\right)}{2}} =$

$\dfrac{\sqrt{2 - \sqrt{3}}}{2}$

13. $\cos(-15) = \cos(45 - 60) =$
$\cos 45 \cos 60 + \sin 45 \sin 60 =$

$\dfrac{\sqrt{2}}{2}\left(\dfrac{1}{2}\right) + \dfrac{\sqrt{2}}{2}\left(\dfrac{\sqrt{3}}{2}\right) = \dfrac{\sqrt{2} + \sqrt{6}}{4}$

$\cos\left(\dfrac{-30}{2}\right) = \sqrt{\dfrac{1 + \cos(-30)}{2}} = \sqrt{\dfrac{1 + \dfrac{\sqrt{3}}{2}}{2}} =$

$= \dfrac{\sqrt{2 + \sqrt{3}}}{2}$

14. If $\sin \alpha = \dfrac{\sqrt{3}}{2}, \tan \beta = -1, 90° < \alpha < 180°$, and $90° < \beta < 180°$ then $\alpha = 120°$ and $\beta = 135°$. Therefore, $\cos(\alpha - \beta) = \cos(120 - 135) =$
$\cos 120 \cos 135 + \sin 120 \sin 135 =$
$-\dfrac{1}{2}\left(-\dfrac{\sqrt{2}}{2}\right) + \dfrac{\sqrt{3}}{2}\left(\dfrac{\sqrt{2}}{2}\right) = \dfrac{\sqrt{2}}{4} + \dfrac{\sqrt{6}}{4} = \dfrac{\sqrt{2} + \sqrt{6}}{4}$

15. If $\sin \alpha = \dfrac{40}{41}, \cos \beta = -\dfrac{5}{13}, 0° < \alpha < 90°$, and

$180° < \beta < 270°$ then $\cos \alpha = \sqrt{1 - \left(\dfrac{40}{41}\right)^2} = \dfrac{9}{41}$,

$\sin \beta = -\sqrt{1 + \left(-\dfrac{5}{13}\right)^2} = -\dfrac{12}{13}, \tan \alpha = \dfrac{\sin \alpha}{\cos \alpha} = \dfrac{40}{9}$,

and $\tan \beta = \dfrac{\sin \beta}{\cos \beta} = \dfrac{12}{5}$. Therefore,

$\tan(\alpha + \beta) = \dfrac{\tan \alpha + \tan \beta}{1 - \tan \alpha \tan \beta} = \dfrac{\dfrac{40}{9} + \dfrac{12}{5}}{1 - \dfrac{40}{9}\left(\dfrac{12}{5}\right)} =$

$\dfrac{200 + 108}{45 - 480} = \dfrac{308}{-435} = -\dfrac{308}{435}$

In Exercises 16–18. if $\cos \alpha = -\frac{8}{17}$ and $90° < \alpha < 180°$

then $\sin \alpha = \sqrt{1 - \left(-\frac{8}{17}\right)^2} = \frac{15}{17}$ and $\tan \alpha = \frac{\sin \alpha}{\cos \alpha} =$

$\dfrac{\frac{15}{17}}{-\frac{8}{17}} = -\frac{15}{8}$.

16. $\sin 2\alpha = 2 \sin \alpha \cos \alpha = 2\left(\frac{15}{17}\right)\left(-\frac{8}{17}\right) = -\frac{240}{289}$

17. $\cos \frac{\alpha}{2} = \sqrt{\frac{1 + \cos \alpha}{2}} = \sqrt{\frac{1 + \left(-\frac{8}{17}\right)}{2}} =$

$\sqrt{\frac{17 - 8}{34}} = \sqrt{\frac{9}{34}} = \frac{3\sqrt{34}}{34}$

18. $\tan 2\alpha = \frac{2 \tan \alpha}{1 - \tan^2 \alpha} = \frac{2\left(-\frac{15}{8}\right)}{1 - \left(-\frac{15}{8}\right)^2} = \frac{240}{161}$

19. From page 816, $x = \frac{v^2 \sin 2\theta}{g}$. Thus,

$x = \frac{(68)^2 \sin 2(27)}{32} \approx 117 \text{ ft}$

20. $r = \sqrt{(-3)^2 + (3\sqrt{3})^2} = \sqrt{9 + 27} = 6;$

$\theta = \text{Arctan}\left(\frac{3\sqrt{3}}{-3}\right) + 180 = -60 + 180 = 120°$

$(6, 120°)$

21. $r = \sqrt{(-5)^2 + 0^2} = 5; \theta = \text{Arctan } 0 = 180°$

$(5, 180°)$

22. $x = 6 \cos (-135) = 6\left(-\frac{\sqrt{2}}{2}\right) = -3\sqrt{2}$

$y = 6 \sin (-135) = 6\left(-\frac{\sqrt{2}}{2}\right) = -3\sqrt{2}$

$(-3\sqrt{2}, -3\sqrt{2})$

23. $x = -8 \cos 315 = -8\left(\frac{\sqrt{2}}{2}\right) = -4\sqrt{2}$

$y = -8 \sin 315 = -8\left(-\frac{\sqrt{2}}{2}\right) = 4\sqrt{2}$

$(-4\sqrt{2}, 4\sqrt{2})$

24. $r = \sqrt{(-5)^2 + 12^2} = 13;$

$\theta = \text{Arctan}\left(\frac{12}{-5}\right) + 180 \approx 113°$

$13 (\cos 113° + i \sin 113°)$

25. $r = \sqrt{4^2 + 3^2} = 5; \theta = \text{Arctan } \frac{3}{4} \approx 37°$

$5 (\cos 37° + i \sin 37°)$

26. $4\left(-\frac{1}{2} + i\left(-\frac{\sqrt{3}}{2}\right)\right) = -2 - 2i\sqrt{3}$

27. $5\left(-\frac{\sqrt{2}}{2} + i\left(-\frac{\sqrt{2}}{2}\right)\right) = -\frac{5\sqrt{2}}{2} - \frac{5i\sqrt{2}}{2}$

28. $\sqrt{x^2 + y^2} = \dfrac{210,000}{38 - \dfrac{2x}{\sqrt{x^2 + y^2}}};$

$38\sqrt{x^2 + y^2} - 2x = 210,000$

29.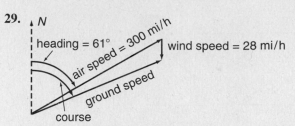

Let x represent the ground speed and Θ, the course of the plane.

$x^2 = 260^2 + 25^2 - 2(260)(25) \cos 55$

$x = \sqrt{260^2 + 25^2 - 2(260)(25) \cos 55}$

$x \approx 247 \text{ mi/h}$

$\dfrac{\sin \alpha}{25} = \dfrac{\sin 55}{247}$

$\sin \alpha = \dfrac{25 \sin 55}{247} \approx 0.083$

$\alpha \approx 5°$

$\theta = 55 + \alpha = 55 + 5 = 60°$

30.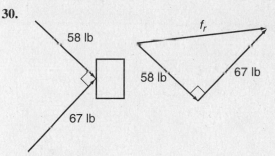

Let $f_r =$ the resultant force

$f_r^2 = 58^2 + 67^2$

$f_r = \sqrt{58^2 + 67^2}$

$f_r \approx 89 \text{ lb}$

EQUATION CREATION

Sample answers include:

For $x = 57°$: $\sin x = 0.8387, x = 123°$

$\cos x = 0.5446, x = 303°$

$\tan x = 1.5399, x = 237°$

For $x = 36°$: $\sin x = 0.5878, x = 144°$

$\cos x = 0.8090, x = 324°$

$\tan x = 0.7265, x = 216°$

EQUATION ART

A limacon of the form $y = a + b \sin \theta$ is symmetric with respect to a vertical line. A limacon of the form $y = a + b \cos \theta$ is symmetric with respect to a horizontal line. If the sign of b is changed the limacon is reflected through its axis of symmetry. If a is increased the inner loop gets smaller and if b is increased the outer loop gets larger. When $a = b$ the inner loop vanishes and the curve is called a cardioid because of its heart-like shape.

TABLE YOUR POOL

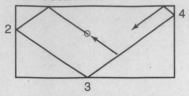

Answers will vary. Some factors that affect the reflections are the spin that may be placed on the ball, the levelness of the table, and the straightness of the rails. The speed of the ball will affect the number of reflections.

CUMULATIVE REVIEW, PAGE 836

1. If $\angle x$ is in Quadrant II and $\cos x = -\frac{4}{5}$ then

$$\sin x = \sqrt{1 - \left(-\frac{4}{5}\right)^2} = \frac{3}{5}.$$

Therefore, $\sin 2x = 2 \sin x \cos x = 2\left(\frac{3}{5}\right)\left(-\frac{4}{5}\right) = -\frac{24}{25}$

2. A; The logarithmic function is the inverse of the exponential function.

3. sum of the roots $= -\frac{b}{a}$

   product of the roots $= \frac{c}{a}$

   $-\frac{k}{6} = -\frac{3}{6}$
   $k = 3$

4. "Sine" is a function not a quantity. Since "sine" is not a quantity, it cannot be used as a multiplier and the Distributive Property has no meaning in connection with it. "Sine" is used in conjunction with an angle, designating the specific ration $\frac{y}{r}$, where y and r are the sides of a reference triangle for the angle in question. If, for example, $\alpha = 30°$ and $\beta = 45°$, then sin $(\alpha + \beta) = \sin (75°) = 0.9659$, which is not equivalent to $\sin 30° + \sin 45° = (0.5 + 0.7071) = 1.2071$. A similar circumstance exists with the log function because $\log (x + y) \neq \log x + \log y$

5. B

6. $\dfrac{(4.6 \times 10^3) \times (3 \times 10^{-2})}{2 \times 10^4} = \dfrac{(4.6)(3) \times 10}{2 \times 10^4} =$

   $= \dfrac{(4.6)(3)}{2} \times 10^{-3} =$

   $(2.3)(3) \times 10^{-3} = 6.9 \times 10^{-3} = 0.0069$

7. $|\sin \theta - 3| = 2$; $\sin \theta - 3 = 2$ or $\sin \theta - 3 = -2$; $\sin \theta = 5$ (impossible) or $\sin \theta = 1$; $\theta = 90°$

8. $2 \sec^2 \theta - 3 \tan \theta - 5 = 0$,
   $2(\tan^2 \theta + 1) - 3 \tan \theta - 5 = 0$;
   $2 \tan^2 \theta + 2 - 3 \tan \theta - 5 = 0$;
   $2 \tan^2 \theta - 3 \tan \theta - 3 = 0$;

   $$\tan \theta = \frac{3 \pm \sqrt{(-3)^2 - 4(2)(-3)}}{2(2)} \approx 2.186, -0.686$$

   $\theta = 65°, 146°, 245°, 326°$

9. Substitute into the standard form of the equation of a circle. $(x - h)^2 + (y - k)^2 = r^2$

   $(7, 1)$:
   $$(7 - h)^2 + (1 - k)^2 = r^2$$
   $$h^2 - 14h + 49 + k^2 - 2k + 1 = r^2$$
   $$h^2 + k^2 - 14h - 2k + 50 = r^2 \quad (1)$$

   $(6, 8)$:
   $$(6 - h)^2 + (8 - k)^2 = r^2$$
   $$h^2 - 12h + 36 + k^2 - 16k + 64 = r^2$$
   $$h^2 + k^2 - 12h - 16k + 100 = r^2 \quad (2)$$

   $(-1, 7)$:
   $$(-1 - h)^2 + (7 - k)^2 = r^2$$
   $$h^2 + 2h + 1 + k^2 - 14k + 49 = r^2$$
   $$h^2 + k^2 + 2h - 14k + 50 = r^2 \quad (3)$$

   Solve the system of equations for h, k, and r.
   Subtract (1) from (2).
   $$\begin{cases} h^2 + k^2 - 12h - 16k + 100 = r^2 \\ h^2 + k^2 - 14h - 2k + 50 = r^2 \end{cases}$$
   $$2h - 14k + 50 = 0$$
   $$h - 7k + 25 = 0 \quad (4)$$

   Subtract (1) from (3).
   $$\begin{cases} h^2 + k^2 + 2h - 14k + 50 = r^2 \\ h^2 + k^2 - 14h - 2k + 50 = r^2 \end{cases}$$
   $$16h - 12k = 0$$
   $$4h - 3k = 0 \quad (5)$$

   Solve equations (4) and (5) for h and k.
   $$\begin{cases} h - 7k + 25 = 0 \\ 4h - 3k = 0 \end{cases}$$

   $$\begin{cases} 4h - 28k + 100 = 0 \\ 4h - 3k = 0 \end{cases} \quad \text{multiply both sides by 4}$$
   $$-25k + 100 = 0 \quad \text{subtract}$$
   $$-25k = -100$$
   $$k = 4$$

   Substitute into (4) to solve for h.
   $$h - 7(4) + 25 = 0$$
   $$h - 28 + 25 = 0$$
   $$h = 3$$

   Substitute into (1) to solve for r^2.
   $$3^2 + 4^2 - 12(3) - 16(4) + 100 = r^2$$
   $$r^2 = 25$$

   So, the equation of the circle is
   $$(x - 3)^2 + (y - 4)^2 = 25$$

10. $6 + 3 + 3 + \frac{3}{2} + \frac{3}{2} + \frac{3}{4} + \frac{3}{4} = 16\frac{1}{2}$ ft

11. Let x = the number of minutes it takes for both scanners working simultaneously.

$\frac{1}{x}$ = the part of the job both scanners do in 1 min

$\frac{1}{45}$ = the part of the job the older scanner does in 1 min

$\frac{1}{30}$ = the part of the job the newer scanner does in 1 min

$\frac{1}{45} + \frac{1}{30} = \frac{1}{x}$

$30x + 45x = 45(30)$ multiply both sides by $(45(30)x)$

$75x = 1350$

$x = 18$ min

12. $d = 6 + 7 \cos \frac{\pi}{5}(5 - 2) = 6 + 7 \cos \frac{3\pi}{5} = 3.8$ ft

13.

$$
\begin{array}{r}
x^3 \; - \; x^2y \; + \; xy^2 - y^3 + \dfrac{2y^4}{x + y} \\
\hline
x + y \overline{)\; x^4 + 0x^3y + 0x^2y^2 + 0xy^3 + y^4} \\
\underline{x^4 + \; x^3y} \\
-x^3y + 0x^2y^2 \\
\underline{-x^3y - \; x^2y^2} \\
x^2y^2 + 0xy^3 \\
\underline{x^2y^2 + \; xy^3} \\
-xy^3 + y^4 \\
\underline{-xy^3 - y^4} \\
2y^4
\end{array}
$$

14. A

Standardized Test, page 837

1. C; I: $\tan \theta \cos \theta - \tan \theta = 0$; $\tan \theta (\cos \theta - 1) = 0$;
$\tan \theta = 0$ or $\cos \theta = 1$ (not possible for this domain)
$\theta = \pi$
II: $|2 \cos \theta - 3| = 5$; $2 \cos \theta - 3 = 5$ or
$2 \cos \theta - 3 = -5$; $2 \cos \theta = 8$ or $2 \cos \theta = -2$;
$\cos \theta = 4$ (impossible) or $\cos \theta = -1$; $\theta = \pi$

2. A; I: $r = \dfrac{\log 9}{\log 81} = \dfrac{\log 9}{\log 9^2} = \dfrac{\log 9}{2 \log 9} = \dfrac{1}{2}$

II: $d = 4 - 4.5 = -0.5 = -\dfrac{1}{2}$

3. D; 2 equations, 3 unknowns, not enough information to solve

4. A; $x + 1 = 0$ or $x + 2 = 0$; $x = -1$ or $x = -2$
I: If $x = -1$, $x(x + 1) = (-1)(-1 + 1) = 0$
If $x = -2$, $x(x + 1) = -2(-2 + 1) = 2$
II: If $x = -1$, $x(x + 2) = -1(-1 + 3) = -2$
If $x = -2$; $x(x + 2) = -2(-2 + 2) = 0$

5. C; I: $a + bi$: $r = \sqrt{a^2 + b^2}$
II: $a - bi$: $r = \sqrt{a^2 + (-b)^2} = \sqrt{a^2 + b^2}$

6. B; I: $a_{22} = 0.25$; $A = \begin{bmatrix} 2 & 0 & 0 \\ 0 & 4 & 0 \\ 0 & 0 & 5 \end{bmatrix}$; $A^{-1} = \begin{bmatrix} 0.5 & 0 & 0 \\ 0 & 0.25 & 0 \\ 0 & 0 & 0.2 \end{bmatrix}$

II: 1

7. C; I: $\dfrac{\pi}{2} - \dfrac{3\pi}{8} = \dfrac{4\pi}{8} - \dfrac{3\pi}{8} = \dfrac{\pi}{8}$

II: $\pi - \dfrac{7\pi}{8} = \dfrac{8\pi}{8} - \dfrac{7\pi}{8} = \dfrac{\pi}{8}$

8. B; I: $N = 50(5)\left(\dfrac{1}{10} + \dfrac{1}{25}\right) = 25 + 10 = 35$

II: $N = 50(3)\left(\dfrac{1}{5} + \dfrac{1}{10}\right) = 30 + 15 = 45$

9. C; I: period $= 2\pi$

II: period $= \dfrac{\pi}{\frac{1}{2}} = 2\pi$

10. B;

I.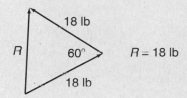

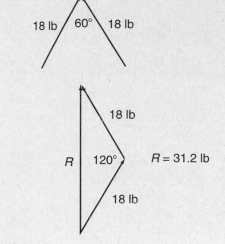

$R = 18$ lb

II.

$R = 31.2$ lb

11. C; I: $\log_8 8^5 = 5 \log_8 8 = 5$
II: $8^{\log_8 5}$ by definition this is 5

12. A; I: $\dfrac{3}{x}$; 1 vertical asymptote $x = 0$

II: $y = \dfrac{x - 2}{x^3 - x - 6}$; no vertical asymptotes

13. B; I:
$-5\rfloor$	1	1	k	$k + 20 = 0$
		-5	20	$k \quad = -20$
	1	-4	$k + 20$	

II:
$-4\rfloor$	1	1	k	$k + 12 = 0$
		-4	12	$k \quad = -12$
	1	-3	$k + 12$	